Hormonally Active Brain Peptides
Structure and Function

BIOCHEMICAL ENDOCRINOLOGY

Series Editor: Kenneth W. McKerns

STRUCTURE AND FUNCTION OF THE GONADOTROPINS
Edited by Kenneth W. McKerns

SYNTHESIS AND RELEASE OF ADENOHYPOPHYSEAL
HORMONES
Edited by Marian Jutisz and Kenneth W. McKerns

REPRODUCTIVE PROCESSES AND CONTRACEPTION
Edited by KennethW. McKerns

HORMONALLY ACTIVE BRAIN PEPTIDES: Structure and Function
Edited by Kenneth W. McKerns and Vladimir Pantić

Hormonally Active Brain Peptides

Structure and Function

Edited by

Kenneth W. McKerns

The International Foundation for Biochemical Endocrinology
Blue Hill Falls, Maine

and

Vladimir Pantić

Institute for Biological Research
Belgrade, Yugoslavia

SPRINGER SCIENCE+BUSINESS MEDIA, LLC

Library of Congress Cataloging in Publication Data

Main entry under title:

Hormonally active brain peptides.

(Biochemical endocrinology)
Sponsored by the International Foundation for Biochemical Endocrinology.
Includes bibliographical references and index.
1. Neuroendocrinology — Congresses. 2. Peptides hormones — Congresses. 3.
Endorphins — Congresses. 4. Brain chemistry — Congresses. I. McKerns, Kenneth
W. II. Pantić, Vladimir. III. International Foundation for Biochemical Endocrinology.
IV. Series. [DNLM: 1. Brain — Physiology — Congresses. 2. Hormones — Physiology —
Congresses. 3. Peptides — Physiology — Congresses. WL 300 H811 1980]
QP356.4.H68 1982   612'.8   82-9147
                                                                                    AACR2

ISBN 978-1-4615-9250-1   ISBN 978-1-4615-9248-8 (eBook)
DOI 10.1007/978-1-4615-9248-8

© 1982 Springer Science+Business Media New York
Originally published by Plenum Press, New York in 1982
Softcover reprint of the hardcover 1st edition 1982

Contributors

Mukund Aiyer, MRC Brain Metabolism Unit, University Department of Pharmacology, Edinburgh EH8 9JZ, Scotland

E. Arnauld, INSERM U 176, 33077 Bordeaux Cedex, France

Sándor Bajusz, Institute for Drug Research, Budapest, Hungary

Nicholas Barden, Department of Molecular Endocrinology, Le Centre Hospitalier de l'Université Laval, Quebec G1V 4G2, Canada

Michael Bienert, Institute of Drug Research, GDR Academy of Sciences, Berlin, GDR

Irwin M. Chaiken, Laboratory of Chemical Biology, National Institute of Arthritis, Metabolism and Digestive Diseases, National Institutes of Health, Bethesda, Maryland 20205

Sharon Chiappa, MRC Brain Metabolism Unit, University Department of Pharmacology, Edinburgh EH8 9JZ, Scotland

Maïthé Corbani, Laboratoire des Hormones Polypeptidiques, CNRS, 91190 Gif-sur-Yvette, France

Raymond Counis, Laboratoire des Hormones Polypeptidiques, CNRS, 91190 Gif-sur-Yvette, France

Mikhail S. Davidov, Regeneration Research Laboratory, Bulgarian Academy of Sciences, 1431 Sofia, Bulgaria

Sophia V. Drouva, Unité 159 de Neuroendocrinologie, Centre Paul Broca de l'INSERM, 75014 Paris, France

P. M. Dubois, Laboratoire d'Histologie-Embryologie, Faculté de Medecine Lyon-Suc, Oullins, France

B. Dufy, INSERM U 176, 33077 Bordeaux Cedex, France

L. Dufy-Barbe, INSERM U 176, 33077 Bordeaux Cedex, France

André Dupont, Department of Molecular Endocrinology, Le Centre Hospitalier de l'Université Laval, Quebec G1V 4G2, Canada

Philippa M. Edwards, International Institute of Cellular and Molecular Pathology, B-1200 Brussels, Belgium

Jacques Epelbaum, Unité 159 de Neuroendocrinologie, Centre Paul Broca de l'INSERM, 75014 Paris, France

Serge Fermandjian, Service de Biochimie, Département de Biologie, Centre d'Etudes Nucléaires de Saclay, F-91191 Gif-sur-Yvette Cedex, France

Goerge Fink, MRC Brain Metabolism Unit, University Department of Pharmacology, Edinburgh EH8 9JZ, Scotland

Ernst A. Fischer, Laboratory of Chemical Biology, National Institute of Arthritis, Metabolism and Digestive Diseases, National Institutes of Health, Bethesda, Maryland 20205

Pierre Fromageot, Service de Biochimie, Département de Biologie, Centre d'Etudes Nucléaires de Saclay, F-91191 Gif-sur-Yvette Cedex, France

Jürgen Gaues, Max-Planck-Institut für experimentelle Endokrinologie, D-3000 Hannover 61, FRG

K. M. Gautvik, Institute of Physiology, University of Oslo, Oslo 1, Norway

Linda C. Guidice, Laboratory of Chemical Biology, National Institute of Arthritis, Metabolism and Digestive Diseases, National Institutes of Health, Bethesda, Maryland 20205

Linda Görlich, Max-Planck-Institut für experimentelle Endokrinologie, D-3000 Hannover 61, FRG

László Gráf, Institute for Drug Research, and Organic Chemistry, Department of L Eötvös University, Budapest, Hungary

E. Huag, Institute of Physiology, University of Oslo, Oslo 1, Norway, and Hormone and Isotope Laboratory, Aker Hospital, Oslo 5, Norway

Simon Henderson, MRC Brain Metabolism Unit, University Department of Pharmacology, Edinburgh, EH8 9JZ, Scotland

Miklós Hollósi, Institute for Drug Research, and Organic Chemistry Department of L Eötvös University, Budapest, Hungary

Anikó Horváth, 1st Institute of Biochemistry, Semmelweis University Medical School, Budapest, Hungary

Christopher J. Hough, Laboratory of Chemical Biology, National Institute of Arthritis, Metabolism and Digestive Diseases, National Institutes of Health, Bethesda, Maryland 20205

Alun Hughes, Max-Planck-Institut für experimentelle Endokrinologie, D-3000 Hannover 61, FRG

Richard Ivell, Institut für Physiologische Chemie, Abteilung Zellbiochemie, Universität Hamburg, 2 Hamburg 20, FRG

J.-G. Iversen, Institute of Physiology, University of Oslo, Oslo 1, Norway

Murray Jameson, MRC Brain Metabolism Unit, University Department of Pharmacology, Edinburgh EH8 9JZ, Scotland

Peter W. Jungblut, Max-Planck-Institut für experimentelle Endokrinologie, D-3000 Hannover 61, FRG

Marian Jutisz, Laboratoire des Hormones Polypeptidiques, CNRS, 91190 Gif-sur-Yvette, France

Erhard Kallweit, Max-Planck-Institut für experimentelle Endokrinologie, D-3000 Hannover 61, FRG

Satya P. Kalra, Department of Obstetrics and Gynecology, University of Florida College of Medicine, Gainesville, Florida 32610

Holger Kalthoff, Institut für Physiologische Chemie, Abteilung Zellbiochimie, Universität Hamburg, 2 Hamburg 20, FRG

Dušan T. Kanazir, Department of Biochemistry and Molecular Biology, Faculty of Natural Sciences and Mathematics, University of Belgrade, Belgrade, Yugoslavia

Martin J. Kelly, University of Oregon Health Sciences Center, Portland, Oregon 97204

B. Kerdelhué, Laboratoire des Hormones Polypeptidiques, CNRS, 91190 Gif-sur-Yvette, France

György Kéri, 1st Institute of Biochemistry, Semmelweis University Medical School, Budapest, Hungary

Janet Kielhorn, Max-Planck-Institut für experimentelle Endokrinologie, D-3000 Hannover 61, FRG

Erhard Klauschenz, Institute of Drug Research, GDR Academy of Sciences, Berlin, GDR

Claude Kordon, Unité 159 de Neuroendocrinologie, Centre Paul Broca de l'INSERM, 75014 Paris, France

M. Kriz, Institute of Physiologie, University of Oslo, Oslo 1, Norway

.*M. Laburthe,* Unité INSERM U 55, Hôpital Saint-Antoine, 12000 Paris, France

Victor Levy-Perez, MRC Brain Metabolism Unit, University Department of Pharmacology, Edinburgh EH8 9JZ, Scotland

Karl Lintner, Service de Biochimie, Département de Biologie, Centre d'Etudes Nucléaires de Saclay, F-91191 Gif-sur-Yvette Cedex, France

Melvin Little, Max-Planck-Institut für experimentelle Endokrinologie, D-3000 Hannover 61, FRG

Sharon McCann, Max-Planck-Institut für experimentelle Endokrinologie, D-3000 Hannover 61, FRG

Kenneth W. McKerns, International Foundation for Biochemical Endocrinology, Blue Hill Falls, Maine 04615

Gábor B. Makara, Institute of Experimental Medicine, Hungarian Academy of Sciences, Budapest, Hungary

Itzhak Maschler, Max-Planck-Institut für experimentelle Endokrinologie, D-3000 Hannover 61, FRG

Heinrich H. D. Meyer, Max-Planck-Institut für experimentelle Endokrinologie, D-3000 Hannover 61, FRG

Imre Mezö, 1st Institute of Biochemistry, Semmelweis University Medical School, Budapest, Hungary

Josif R. Milin, Institute of Histology and Embryology, Medical Faculty, Novi Sad, Yugoslavia

Radivoy Miline, Institute of Histology and Embryology, Medical Faculty, Novi Sad, Yugoslavia

Karoly Nikolics, 1st Institute of Biochemistry, Semmelweis University Medical School, Budapest, Hungary

M. Palkovits, 1st Department of Anatomy, Semmelweis University Medical School, Budapest, Hungary

Vladimir R. Pantić, Institute for Biological Research 'Siniša Stanković,' Belgrade, Yugoslavia

Fritz Parl, Max-Planck-Institut für experimentelle Endokrinologie, D-3000

C. Pasqualini, Laboratoire des Hormones Polypeptidiques, CNRS, 91190 Gif-sur-Yvette, France

E. J. Peck, Jr., Department of Cell Biology, Baylor College of Medicine, Houston, Texas 77030

Anthony Pickering, MRC Brain Metabolism Unit, University Department of Pharmacology, Edinburgh EH8 9JZ, Scotland

François Piriou, Service de Biochimie, Département de Biologie, Centre d'Etudes Nucléaires de Saclay, F-91191 Gif-sur-Yvette Cedex, France

Panteley G. Popov, State Institue for Drugs Control, 1040 Sofia, Bulgaria

Geneviève Ribot, Laboratoire des Hormones Polypeptidiques, CNRS, 91190 Gif-sur-Yvette, France

Dietmar Richter, Institut für Physiologische Chemie, Abteilung Zellbiochimie, Universität Hamburg, 2 Hamburg 20, FRG

Gary C. Rosenfeld, Max-Planck-Institut für expermintelle Endokrinologie, D-3000 Hannover 61, FRG

G. Rosselin, Unité INSERM U. 55, Hôpital Saint-Antoine, 75571, Paris, Cedex 12 France

W. Rotsztejn, Unité INSERM U 55, Hôpital Saint-Antoine, 75571, Paris, Cedex 12 France

Guy C. Rousseau, International Institute of Cellular and Molecular Pathology, B-1200 Brussels, Belgium

M. R. Sairam, Reproduction Research Laboratory, Clinical Research Institute of Montreal, Montreal, Quebec H2W 1R7, Canada

O. Sand, Department of Physiology, Veterinary College of Norway, Oslo 1, Norway

Dipak Sarkar, MRC Brain Metabolism Unit, University Department of Pharmacology, Edinburgh EH8 9JZ, Scotland

Hartwig Schmale,   Institut für Physiologische Chemie, Abteilung Zellbiochemie, Universität Hamburg, 2 Hamburg 20, FRG

János Seprődi,   1st Institute of Biochemistry, Semmelweis University Medical School, Budapest, Hungary

Nancy Sherwood,   MRC Brain Metabolism Unit, University Department of Pharmacology, Edinburgh, EH8 9JZ, Scotland

Walter Sierralta,   Max-Planck-Institut für experimentelle Endokrinologie, D-3000 Hannover 61, FRG

Alison Speight,   MRC Brain Metabolism Unit, University Department of Pharmacology, Edinburgh EH8 9JZ, Scotland

Ervin Stark,   Institute of Experimental Medicine, Hungarian Academy of Sciences, Budapest, Hungary

Grant Stone,   Max-Planck-Institut für experimentelle Endokrinologie, D-3000 Hannover 61, FRG

Dimitar Strashimirov,   Department of Physiology, Medical Faculty, Pleven, Bulgaria

Pablo Szendro,   Max-Planck-Institut für experimentelle Endokrinologie, D-3000 Hannover 61, FRG

Balázs Szőke,   1st Institute of Biochemistry, Semmelweis University Medical School, Budapest, Hungary

István Teplán,   1st Institute of Biochemistry, Semmelweis University Medical School, Budapest, Hungary

Cecilia Terán,   Max-Planck-Institut für experimentelle Endokrinologie, D-3000 Hannover 61, FRG

Flavio Toma,   Service de Biochimie, Département de Biologie, Centre d'Etudes Nucléaires de Saclay, F-91191 Gif-sur-Yvette Cedex, France

Anne J. Truitt,   Max-Planck-Institut für experimentelle Endokrinologie, D-3000 Hannover 61, FRG

Kalina I. Vaptzarova,   Regeneration Research Laboratory, Bulgarian Academy of Sciences, 1431 Sofia, Bulgaria

J. D. Vincent,   INSERM U 176, 33077 Bordeaux Cedex, France

Rüdiger K. Wagner,   Max-Planck-Institut für experimentelle Endokrinologie, D-3000 Hannover 61, FRG

Alan Watts,   MRC Brain Metabolism Unit, University Department of Pharmacology, Edinburgh EH8 9JZ, Scotland

Preface

Near the end of September 1980, the contributors to *Hormonally Active Brain Peptides: Structure and Function* met to discuss their chapters for the monograph. This meeting was the eighth sponsored by the International Foundation for Biochemical Endocrinology and was held at the Hotel Plakir in Dubrovnik, Yugoslavia. Several months were allowed after the meeting for the contributors to revise their manuscripts and for editing.

Professor Dr. Vladimir Pantić and the Serbian Academy of Sciences and Arts were in charge of the local arrangements and social activities. The Foundation is grateful for the splendid job that was done and for the outstanding scientific, cultural, and social activities. I thank the Serbian Academy of Sciences and Arts and the Yugoslav Council of Academies for sponsoring my stay in Yugoslavia. I greatly enjoyed giving lectures, visiting laboratories, and discussing research projects in a number of centers in Yugoslavia.

A diverse group of topics was presented in Dubrovnik concerning certain brain peptides. The topics included "Structure, Function, and Conformation of Neuropeptides," "Distribution of Peptides in the Brain," "Regulation of the Release of Peptide Hormones," "The Active Center of Gonadotropins," "*In Vitro* Synthesis of Hypothalamic Hormones," "Studies on Gonadotropin-Releasing Hormone," "Studies on the Biosynthesis, Release, and Degradation of LH-RH," "Genesis and Properties of Pituitary ACTH, Prolactin, and Growth Hormone Producing Cells," "Precursors to Oligopeptide Hormones," and "Stress Effects on Peptide Hormone Synthesis." The moderators for the sessions were Kenneth McKerns, Vladimir Pantić, R. Miline, K. Nikolics, E. J. Peck, Jr., Asbjørn Aakvaag, G. Makora, Dorothy Villee, Claude Villee, Jr., Sophia Drouva, S. P. Kalra, K. Vaptzarova, Edward Herbert, and Dušan Kanazir.

The next monograph of the Foundation will deal with the topic, "Regulation of Gene Expression by Hormones." A meeting for late September

1982 is being planned based on a monograph to be entitled *Regulatory Mechanisms in Target Cell Responsiveness*. It will be held in Geilo, Norway. Professor Asbjørn Aakvaag and Dr. Vidar Hanssen will be in charge of local arrangements.

Blue Hill Falls, Maine   Kenneth W. McKerns

Contents

9   Stress, Corticoliberin (CRF), and Glucocorticoids in the Regulations of ACTH Release   157

Ervin Stark and Gábor B. Makara

10   The Role of Steroid Receptors in the Regulation and Integration of Steroid and Peptide Hormone Actions in Common Target Cells: Facts and Speculations   181

Dušan T. Kanazir

11   Effects of Estrogens on Receptor "Nucleotropy" and "Activation"   215

Walter Sierralta, Jürgen Gaues, Linda Görlich, Alun Hughes, Erhard Kallweit, Janet Kielhorn, Melvin Little, Itzhak Maschler, Sharon McCann, Fritz Parl, Gary C. Rosenfeld, Grant Stone, Pablo Szendro, Cecilia Terán, Anne J. Truitt, Rüdiger K. Wagner, and Peter W. Jungblut

Structure and Function Relationships among Synthetic Enkephalin Analogues

Sándor Bajusz

1. Introduction

The discovery of Met- and Leu-enkephalin, the two brain peptides, by Hughes, Kosterlitz, and their co-workers (Hughes *et al.*, 1975c) demonstrated first the existence of hypothetical endogenous substances that can interact with morphine (or opiate) receptors, the so-called *endorphins*. It was also disclosed that Met-enkephalin is identical to the fragment 61–65 of the pituitary hormone β-lipotropin, β-LPH-(61–65). This result set off a boom of research related to endorphins, or *opioid peptides* as named recently. Very soon these findings of Hughes *et al.* (1975c) were confirmed (e.g., Simantov and Synder, 1976), and also other opioid peptides derivable from β-LPH were found (Fig. 1). Namely, β-LPH-(61–91), isolated earlier as a contiguous fragment of melanotropins (Bradbury *et al.*, 1975; Li and Chung, 1976), was shown to possess morphinelike activity (Bradbury *et al.*, 1976a; Cox *et al.*, 1976) and was named β-endorphin (Li and Chung, 1976). The isolation and opioid activity of β-LPH-(61–76) (Guillemin *et al.*, 1976) and β-LPH-(61–77) (Lazarus *et al.*, 1976), called α- and γ-endorphin, respectively, were also described. Of these peptides the two enkephalins and β-endorphin are regarded as endogenously formed opioid peptides while α- and γ-endorphin as well as the other fragments isolated such as β-LPH-(61–79) (Gráf *et al.*, 1976) and β-LPH-(61–87) (Guillemin *et al.*, 1977) are considered degradation products of β-endorphin (e.g.,

---

Sándor Bajusz • Institute for Drug Research, Budapest, Hungary.

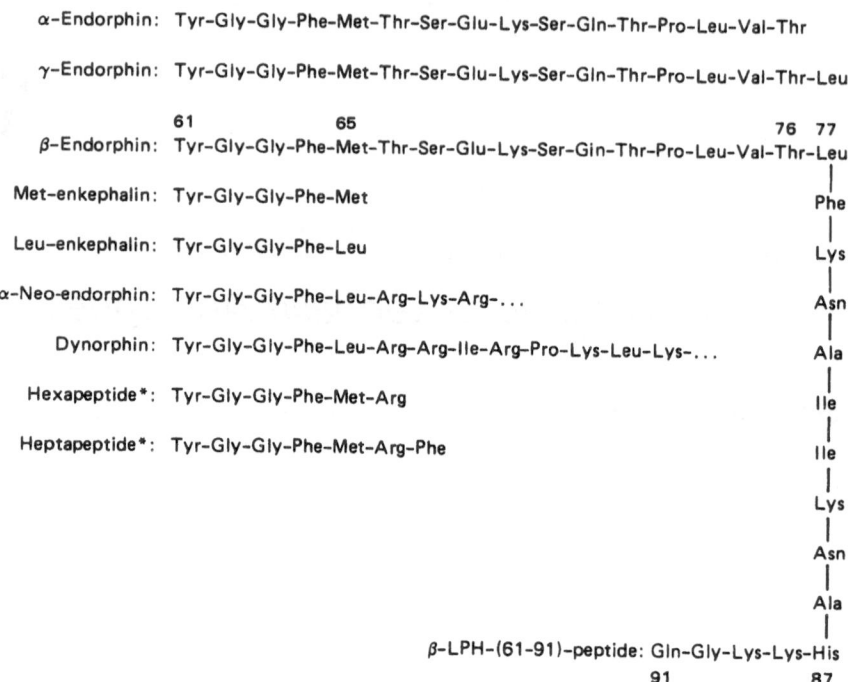

Figure 1. Amino acid sequence of natural peptides with morphinelike activity and of β-LPH-(61–91) peptide. (*) Fragment of pro-Met-enkephalin?

Gráf *et al.*, 1977; Beaumont and Hughes, 1979). The other four peptides in Fig. 1 were isolated in 1979. Neither of them is related to β-LPH. α-Neo-endorphin (Kangawa *et al.*, 1979) and dynorphin (Goldstein *et al.*, 1979) contain Leu-enkephalin while the hexapeptide (Huang *et al.*, 1979) and heptapeptide (Stern *et al.*, 1979) have Met-enkephalin as N terminus. The structure of the latter two peptides may indicate the existence of additional endogenous opioid peptide(s).

The epoch-making report of Hughes *et al.* (1975c) also launched an immense synthetic work, i.e., the synthesis of hundreds of analogues in order to obtain new compounds with improved opioid activity and to get insight into structure–activity relationships.

The intense and wide-ranging research on the chemistry and biology of opioid peptides has been discussed in different aspects by some recent reviews (e.g., Beddell *et al.*, 1977a,b; Frederickson, 1977; Miller and Cuatrecasas, 1978; Terenius, 1978; Beaumont and Hughes, 1979; Morgan, 1979; Morley, 1980). Structure–activity relationships of analogues derived by single substitution at each amino acid position of Met/Leu-enkephalin have been fully reviewed by Morley (1980). The present review concentrates primarily on enkephalin analogues derived by double (or multiple) changes, in particular on those reported recently, and attempts are made to delineate some structural features that seem to determine the enkephalinlike or morphinelike activity of enkephalin analogues.

The following examination of structure–activity relationships is based particulary on activities measured in the two most widely used opiate-sensitive isolated organ preparations—mouse vas deferens (MVD) (Hughes *et al.*, 1975a) and longitudinal muscle strip of guinea pig ileum (GPI) (Kosterlitz *et al.*, 1970)—and also a commonly used analgesic assay, i.e., the tail-flick test in rat or mouse introduced by D'Amour and Smith (1941). Additional data obtained in other *in vitro* models, e.g., the rat vas deferens preparation, or *in vivo* tests, e.g., the mouse hot plate test, have also been included in this review. And, of course, the results obtained in receptor binding assay have been taken into consideration.

2. Opiate Activity of Enkephalins and β-Endorphin

The early finding that "Met-enkephalin is twenty times more active than normorphine in the vas deferens and equipotent with normorphine in the guinea pig ileum" (Hughes *et al.*, 1975c) indicated a marked difference in the activity pattern of opioid peptide(s) and morphine congeners. Practically the same results were reported in a subsequent paper (Waterfield *et al.*, 1977), wherein activity of Leu-enkephalin, morphine, and β-endorphin were also presented. The *in vivo* studies revealed that enkephalins, even after central administration, showed extremely low and transient analgesic activity (Beluzzi *et al.*, 1976) whereas β-endorphin, given either centrally or intravenously, possessed higher analgesic potency than morphine (Gráf *et al.*, 1976: Tseng *et al.*, 1976; Loh *et al.*, 1976; Feldberg and Smyth, 1976). A comparison of activity of enkephalins, β-endorphin, and morphine on the rat vas deferens preparation (Lemaire *et al.*, 1978) has given further evidence of differences in their opioid character. Namely, β-endorphin is the most potent agonist, Leu-enkephalin is less active, and morphine and normorphine are inactive. These data are summarized in Table I. *In vitro* activities are expressed in $1/IC_{50}$ values calculated from data presented in original papers. Analgesic potencies are related to morphine as in the publications quoted. In addition, the MVD/GPI potency ratios calculated from IC_{50} values are also presented. Accordingly, the MVD/GPI ratio for Met- and Leu-enkephalin are 7.5 and 61.0 respectively, and for morphine and normophine, 0.14 and 0.16. β-Endorphin shows a third type of activity pattern, being about equipotent in these two *in vitro* models.

Table I.   Comparison of Activities of Opioid Peptides,
Morphine, and Normorphine

Substance	MVD^a	GPI^a	$\dfrac{MVD}{GPI}$	RVD^b	MTF[c] i.c.v.	MTF[c] i.v.
Met-enkephalin	78.1^d	10.4^d	7.5^d	1.6	0.02	0
Leu-enkephalin	128.2	2.1	61.0	0.3	0.02	0
β-Endorphin	17.9	13.9	1.28	7.7	34.7	4.2
Morphine	2.03	14.4	0.14	0	1	1
Normorphine	2.27	13.7	0.16	0	—	—

[a,b] In vitro activities are expressed in $10^{-6}/IC_{50}$ M values calculated from the data of Waterfield et al. (1977), Lemaire et al. (1978).
[c] In vivo activities for analgesia measured by the mouse tail-flick test (MTF) are related to morphine as published by Tseng et al. (1976), Loh et al. (1976).
[d] From recent data of the same laboratory (Kosterlitz et al. 1980), values of 68.0, 6.4, and 10.7, respectively, were calculated.

These observations can be explained, at least in part, by accepting the concept of Lord et al. (1977) that peptide and nonpeptide opioids interact with different subclasses of opiate receptors existing both in the brain and in peripheral tissues. Thus, MVD is thought to contain mainly δ receptors, which are particulary sensitive to enkephalins, while GPI contains a greater proportion of typical morphine-sensitive receptors, designated as μ receptors. Enkephalins, in particular Leu-enkephalin, have apparently high affinity for the δ receptors. β-Endorphin, the only endogenous opioid peptide showing analgesic activity, seems to have high affinity for both δ and μ receptors, the latter being considered to be of particular interest for antinociception (Law and Loh, 1978; Rónai and Berzétei, 1978; Kosterlitz, 1978).

Nevertheless, the most intriguing problem has remained unresolved: why has Met-enkephalin, with considerable affinity for μ receptors, only weak and transient analgesic activity? In July 1976, Pert and his colleagues (Pert et al., 1976a,b) proposed that this discrepancy between in vitro and in vivo potencies of enkephalins might be explained by rapid enzymatic degradation in the brain just as observed previously in the ileum (Hughes, 1975). This view was supported by the finding that β-endorphin showed considerable resistance to enzymatic degradation in brain homogenate. It was also reasoned that an aminopeptidase could be responsible for deactivation, mediated by cleavage of the Tyr-Gly bond. The same conclusion was drawn from studies on breakdown of enkephalin by rat and human plasma and rat brain homogenate (Hambrook et al., 1976). It has also been supposed that partial resistance of β-endorphin can be due to its tertiary structure wherein the labile Tyr-Gly bond is protected from aminopeptidases by intramolecular interactions (Hambrook *et al.*, 1976; Rónai *et al.*, 1977).

In our opinion (Bajusz *et al.*, 1978) the suggestion that the lack of significant *in vivo* activity of enkephalins is due to their rapid metabolism is untenable since all natural peptides undergo rapid biodegradation but no difficulty has so far been encountered in demonstration of their *in vivo* activity. Nevertheless, the hypothesis of Hughes *et al.* (1975c) that "enkephalins act as neurotransmitters or neuromodulators at synaptic junctions" might be correct, and consequently these peptides should be rapidly inactivated by specific membrane-bound enzyme existing in the vicinity of opiate receptors.

Let us mention here that Hughes *et al.* (1975b) designated the brain peptide(s) by the name "enkephalin" which only refers to the source of the peptide(s). The designation "opioid peptide" was recommended (Beaumont and Hughes, 1979) for all endogenous or synthetic peptides with morphine-like activity to avoid confusion created by the generic name "endorphin."

3. Design and Synthesis of Enkephalin Analogues with Improved Opioid Activity

It has long been known that the biological activity of natural peptides can be enhanced by suitable structural modifications. As for Met/Leu-enkephalin, it was generally accepted that the analogues had to be enzyme resistant.

The labile Tyr-Gly bond could be protected from aminopeptidase by two obvious alterations, namely, by methylation of the terminal H_2N, i.e., preparation of MeTyr analogues, and introduction of D-Ala in place of Gly^2. It is interesting to note that replacement of Gly^2 by D-Leu or D-Phe proved to be unfavorable (Coy *et al.*, 1976). To protect the peptides from carboxypeptidase, the COOH group was amidated ($CONH_2$), reduced to carbinol (CH_2OH), or the terminal Met/Leu was replaced by the D-isomer. In spite of their resistance to enzymes, these analogues failed to elicit analgesia upon systemic administration. In order to obtain intravenously, subcutaneously, or orally active analogues, multiple alterations had to be applied. For instance, substitution of Gly^2 by D-Ala, *N*-methylation of Phe^4, oxidation of Met^5 to sulfoxide and reduction of its COOH group to carbinol led to a "more stable analogue" of Met-enkephalin, i.e., [D-Ala^2,$MePhe^4$,Met(O)-ol^5] -enkephalin (Roemer *et al.*, 1977; Pless *et al.*, 1979). In the mouse tail-flick test, this compound when given intravenously and subcutaneously, respectively, showed analgesic activity 6.4 and 3.2

Table II. Effect on *in Vitro* Activity of Replacement of Xxx Residue by Yyy in Enkephalins*

Replacement of by		Met-enkephalin			Leu-enkephalin		
Xxx	Yyy	MVD	GPI	$\frac{MVD}{GPI}$	MVD	GPI	$\frac{MVD}{GPI}$
—	—	1.0	1.0	10.7[d]	1.0	1.0	51.6[d]
Tyr[1]	MeTyr	0.2	1.0	2[g]	0.62	7.17	3[a,f]
Gly[2]	D-Ala	4.97	5.61	9.5[d]	5.41	15.85	17.6[d]
Met[5]	D-Met	0.002	0.105	2[b,e]	—	—	—
Leu[5]	D-Leu	—	—	—	1.2	0.41	150[b,g]
COOH	CONH$_2$	0.53	1.3	5[c]			
Gly[2] Phe[4] Met[5]	D-Ala+ MePhe+ Met(O)-ol	0.94	2.13	0.47[d]			

*Potencies compared to those of Met- and Leu-enkephalin and the MVD/GPI ratios were taken or calculated from the literature as follows: [a]Beddell *et al.* (1977a); [b]Coy *et al.* (1976); [c]Frederickson *et al.* (1976); [d]Kosterlitz *et al.* (1980); [e]Ling *et al.* (1977); [f]Miller *et al.* (1978); [g]Morgan *et al.* (1977).

times higher than that of morphine. Upon intracerebroventricular (i.c.v.) administration it was about 1000 times stronger than morphine. In addition, it also showed a moderate per os activity.

In Table II relative potencies referred to Met- and Leu-enkephalin, respectively, show the effects of modifications on the *in vitro* activity. Accordingly, these alterations, with one exception (introduction of D-Leu[5]), produce a greater increase or a smaller decrease in the relative potency in the GPI than in the MVD. Consequently, the MVD/GPI ratio of the analogues is lower than that of the enkephalins. In this respect [D-Ala[2],MePhe[4],Met(O)-ol[5]] -enkephalin, the highly potent peptide analgesic, is more similar to morphine than to the parent compound Met-enkephalin (Kosterlitz *et al.*, 1980); thus, its significant morphinelike activity may also be explained by the altered activity pattern.

According to another strategy of designing new analogues (Bajusz *et al.*, 1976), the difference between the binding affinities of β-endorphin and Met-enkephalin to opiate receptors was considered. This property of the two peptides could also be paralleled with their *in vivo* potencies. Namely, the binding capacity of β-endorphin as determined by displacement of [^{3}H]naloxone is about 30-fold higher than that of Met-enkephalin (Bradbury *et al.*, 1976a,b). It was attempted to enhance the affinity of enkephalin to the receptors by replacing Gly[2] and/or Gly[3] by D-amino acid residue(s), which modification had been successfully applied in our previous study on thrombin inhibitors (Bajusz *et al.,* 1975). It was expected that the side chain of D-amino acid substituent—as a "guest side chain" of enkephalin—might find a complementary one in some portion of the receptors, i.e., an additional binding site might be formed. The *in vitro* activity of the parent molecule (i.e., Nle⁵-enkephalin) could be substantially increased by introducing D-Met or D-Nle, i.e., residues with straight side chain, into position 2. Hereupon the C-terminal residue, Nle, was replaced by Pro, which was expected to render resistance to carboxypeptidase owing to its amino acid character and, in addition, to form strong hydrophobic bonds as does Met (Nemethy and Scheraga, 1962). Finally, the terminal COOH group of the peptide was "neutralized" by amidation to let the fifth residue similar to its counterpart in the peptide chain of β-endorphin (Bajusz *et al.,* 1976, 1977). Biological evaluation revealed that Nle⁵-enkephalins showed no significant analgesic activity tested by the rat tail-flick method. However, the Pro⁵ congeners did induce analgesia even upon systemic administration. The most potent compound, [D-Met²,Pro⁵]-enkephalinamide, possessed 5.9 times higher activity than morphine on intravenous injection and was 78.5 times more potent than morphine when given i.c.v. (Székely *et al.,* 1977).

The *in vitro* activities (in $10^{-6}/IC_{50}$ M values) and the corresponding MVD/GPI ratio of the analogues are summarized in Table III. The data of Met-enkephalin, β-endorphin, and morphine are also presented for comparison. Apparently, D-amino acid substitution, in particular D Met² and D-Nle², can enhance potency of the parent compounds both in the MVD and in the GPI assays but to different extents as indicated by the alteration of the MVD/GPI ratios (Rónai *et al.,* 1979a,b). As for the Nle⁵-enkephalin

Table III. *In Vitro* Activity of Analogues of Type Tyr-Xxx-Gly-Phe-Yyy [a]

Assay:	MVD	GPI	MVD/GPI	MVD	GPI	MVD/GPI	MVD	GPI	MVD/GPI
Xxx Yyy:		Nle			Pro			Pro-NH₂	
Gly	48.5	2.74	17.7	0.155	0.027	5.7	0.107	0.123	0.87
D-Ala	1000.0	15.3	65.4	3.6	2.5	1.4	3.7	5.3	0.70
D-Nle	781.2	16.8	46.4	144.9	45.2	3.2	52.3	67.6	0.78
D-Met	1666.7	26.0	64.6	86.9	18.8	4.6	43.3	45.4	0.95
Reference substances	142.9	5.4	26.2	13.9	11.6	1.19	2.04	14.3	0.14
		Met-enkephalin			β-endorphin				

[a] Activities are expressed in $10^{-6}/IC_{50}$ M values calculated from the data of Bajusz *et al.* (1976, 1977), Székely *et al.* (1977), Rónai *et al.* (1977, 1979a,b).

analogues, the high value of MVD/GPI was further increased (from 17.7 to 65.4) by introducing D-amino acid residues into position 2 but in the Pro^5 congeners the low ratio was either further diminished (from 5.7 to 1.4) or, in the case of Pro-NH_2 terminus, hardly influenced (0.8 vs. 0.70–0.95). MVD/GPI ratios denote that the characteristic activity pattern of enkephalins has been markedly changed upon replacing Nle^5 (or Met/Leu^5) by Pro. In these peptides the free COOH group can save to a certain degree the original "enkephalinoid" character which is then entirely abolished by amidation, i.e., neutralization of the C terminus. As a consequence, Pro^5-enkephalins have low MVD/GPI ratios, e.g., the most potent analogue *in vivo,* [D-Met^2,Pro^5]-enkephalinamide, has a ratio of 0.95, and their biological profile is more similar to those of β-endorphin or morphine than to Met-enkephalin with potency ratios of 1.19, 0.14, and 26.2, respectively. The same has been observed with the analogue [D-Ala^2,$MePhe^4$,$Met(O)$-ol^5]-enkephalin, having a low potency ratio (0.47) and strong analgesic potency *in vivo.*

Of course, some deviations may occur in these values upon measuring them in different laboratories (and even in the same one but at different times) yet the tendency of alterations in the activity pattern can be characterized by the MVD/GPI ratios (e.g., Kosterlitz *et al.,* 1980).

Based on data presented in Tables II and III, one may conclude that the only way leading to peptide analgesics is to apply structural modifications until the original enkephalinlike pattern is transformed into a β-endorphinlike or rather into a morphinelike one.

Such modifications can certainly confer enzyme resistance to the peptides; this should, however, be regarded as a favorable contribution to the *in vivo* properties of peptides in general rather than the determinant of analgesic (or of any biological) activity.

It appears to be a more important factor that both highly potent analgesic analogues mentioned above, i.e., [D-Ala^2,$MePhe^4$,$Met(O)$-ol^5]-enkephalin and [D-Met^2,Pro^5]-enkephalinamide, are basic derivatives containing C-terminal CH_2OH or $CONH_2$ while Met/Leu-enkephalin are free peptides of zwitterionic character. Kosterlitz *et al.* (1980) have recently discussed the significance of the terminal COOH group in the maintenance of the enkephalinlike activity pattern including the relative potencies in the GPI and MVD assays and receptor binding assay using [^3H]-Leu-enkephalin and [^3H]naltrexone as ligands.

Hereafter we discuss the changes in activity pattern on alterations at the basic N and acidic C termini of enkephalins while considering the influence of different second and fifth residues, i.e., the effect of interdependence existing between these two residues as observed recently (Bajusz *et al.,* 1980b).

4. Modification at the N Terminus

4.1. Methylation

The *N*-methyl derivative of Met/Leu-enkephalin has already been mentioned among the "stabilized" analogues. For obtaining more information we present the activities of 12 methylated derivatives compared to the parent compounds with terminal H_2N group as well as the MVD/GPI ratios (Table IV).

Apparently the effect of this modification is markedly influenced by the residues at positions 2 and 5. Methylation of Leu-OH[5] and Leu-OMe[5]-enkephalin (I, II) decreased the potency in the MVD by 25–38% and caused a six- to sevenfold increase of the activity in the GPI. In the presence of D-Ala[2] (III, V), however, the loss of activity in the MVD was about 80%. Considering the results that dimethylation (IV) diminished the activity in the MVD by 90% and in the GPI by about 80%, it could be supposed that monomethylation did not increase the potency of D-Ala[2]-containing Leu-enkephalins either in the GPI. Comparing the activities of compounds V–VIII, it can be seen that the activity-reducing effect of methylation is not influenced by the configuration and the carboxyl function of Leu[5] (this

Table IV. *In Vitro* Activity of Analogues of Type MeTyr-Xxx-Gly-Phe-Yyy Compared to the Parent Compounds with Terminal Tyr Residue[a]

No.	Xxx	Yyy	MVD	GPI	MVD/GPI Tyr	MVD/GPI MeTyr	Reference
I	Gly	Leu-OH	0.62	7.20	37.25	2.75	c,g
II	Gly	Leu-OMe	0.75	5.86	8.68	1.09	c,g
III	D-Ala	Leu-OMe	0.22	—	3.28	—	c,e
IV	D-Ala	Leu-OMe[b]	0.10	0.18		1.86	c,g
V	D-Ala	Leu-OH	0.23	—	17.63	—	c,f
VI	D-Ala	desCOOH-Leu	0.25	0.59	0.64	0.27	f
VII	D-Ala	D-Leu-OH	0.19	0.51	78.38	28.57	c,d
VIII	D-Ala	D-Leu-OMe	0.25	0.51	1.44	0.69	c,d,g
IX	Gly	Met-OH	0.20	1.00	10.62	2.12	d,f
X	Gly	Met-NH_2	0.19	4.07	4.39	0.20	f,h
XI	D-Ala	D-Met-OH	0.66	0.44	30.97	46.65	c,d
XII	D-Ala	D-Met-NH_2	1.25	1.04	4.52	5.41	d
XIII	D-Ala	D-Met-OMe	0.26	1.36	10.48	2.03	c,d,g

[a] Relative potencies and the MVD/GPI ratios have been calculated from the corresponding $1/IC_{50}$ values.

[b] Values related to the Me_2Tyr derivative.

[c–h] Data are taken or calculated from the literature as follows: [c]Beddell *et al.* (1979a); [d]Beddell *et al.* (1977b); [e]Dutta *et al.* (1977); [f]Kosterlitz *et al.* (1980); [g]Miller *et al.* (1978); [h]Morgan *et al.* (1977).

conclusion may also be related to other residues with alkyl side chain). But methylation affects the potencies of analogues terminated by Met^5 or D-Met^5 in another way. Namely, D-Ala^2 substituent seems to moderate the change in activity in both the MVD and the GPI assays (cf. IX/XI and X/XII); in addition, the C-terminal functional group also influences the activity (cf. XI–XIII). In certain cases, methylation results in analogues having higher MVD/GPI ratios than the parent compounds (e.g., XI and XII).

It may be mentioned here that the receptor-binding potency of enkephalins was attributed to certain structural analogies with morphine. Namely, the phenol ring and the tertiary nitrogen of morphine alkaloids were compared to the side chain and H_2N group of Tyr (e.g., Bradbury et al., 1976a,c; Roques et al., 1976; Horn and Rodgers, 1977). But this speculation disregards the lack of stereochemical correspondence of α-carbon of the tyramine unit in enkephalins and morphine, respectively (Bajusz et al., 1976). Namely, the phenol ring and the tertiary amino group in the morphine molecule corresponds to the orientation of the side chain and the α-amino group of D-Tyr (R configuration) and not to that of the L isomer (S configuration) found in opioid peptides. Moreover, methylation of enkephalins should increase the similarity between the peptides and alkaloids. Yet only four of the nine methylated analogues examined showed higher potency in the GPI assay than their parent compounds. The others have unaltered or diminished activity. This fact seems to challenge the identification of the peptide H_2N group with the N-methyl group of the alkaloids. It is very probable that receptor binding of the "L" and "D" tyramine units found in enkephalins and morphine alkaloids, respectively, are different (cf. Maryanoff and Zelesko, 1978).

4.2. Acylation

Acetylation (Dutta et al., 1977) or carbamylation (Bradbury et al., 1976a) of enkephalin, i.e., modifications that remove the positive charge from the terminal nitrogen, result in inactive analogues. However, N-terminal extension of enkephalins with one or more amino acid residues results in peptides that retain certain activity. For instance, Gacel et al. (1979) found [Xxx^0,Met^5]-enkephalins (Xxx = Arg, Gly, Lys, Phe, or Tyr) to show about 20% of the potency of Met-enkephalin in both assays. According to Ling et al. (1978), Met-enkephalin analogues obtained by addition of Arg, Lys-Arg, or Asp-Lys-Arg, i.e., β-LPH fragments 60–65, 59–65, and 58–65, respectively, have potencies relative to Metenkephalin (= 100) of 90, 7, and 4.4, respectively, in the GPI assay. In contrast, attachment of β-alanine or its homologues, i.e., H_2N-$(CH_2)_n$-CO- (n = 3 and 5), to the N terminus of Met-enkephalin leads to inactive analogues (Gacel et al.,

1979). These findings clearly indicate that not only the presence of the free N terminus but also its distance from the Tyr-Gly bond in the analogues are of great importance for the opioid activity.

4.3. Amidination

Amidination of enkephalins leads to guanidino derivatives. Such compounds have recently been prepared (Bajusz *et al.*, 1980a). Based on the supposition that the peptide N terminus would combine with a carboxyl group on the receptors and considering that interaction between carboxylate and guanidinium ions could be stabilized by H-bridges in addition to the electrostatic bond, replacement of $H_3\overset{+}{N}$- by $(H_2N)_2\overset{+}{C}$-NH- was expected to improve the peptide-receptor binding. However, such alteration, i.e., replacing the terminal H-Tyr residue by $(H_2N)_2\overset{+}{C}$-Tyr moiety in enkephalins, results in a distal shift of the positive charge ($\sim$ 1.2 Å). A similar modification at the ϵ-amino group of Lys side chain in peptides, i.e., conversion of Lys into homoarginine (Har) residue, frequently inhibits the biological or biochemical reactions, e.g., enzyme–inhibitor or enzyme–substrate interactions. Thus, the Har-Xxx bonds are resistant to trypsin, indicating that the enzyme is unable to combine simultaneously with both the carbonyl group at the α-carbon and the positive charge of the basic residue at the side chain terminus when the distance between them is greater than within a Lys or Arg residue [the guanidino group of Arg, $(H_2N)_2\overset{+}{C}$-NH-, corresponds to the $H_3\overset{+}{N}$-CH$_2$- grouping in the side chain of Lys].

Results obtained with the guanidino derivative of two tetrapeptideamide and two pentapeptideamide analogues are presented in Tables V and VI.

In this series of experiments the potencies in the GPI preparation were assessed by using the so-called "single-dose" method (Kosterlitz and Watt, 1968) as well as the regular procedure of Kosterlitz *et al.* (1970), which can be termed the "multiple-dose" method. It is apparent (Table V) that amidination increased the *in vitro* potencies of pentapeptideamides and those of the tetrapeptideamides as well, and that the D-Met^2-containing peptides possessed somewhat higher activities than the D-Nle^2 analogues. Assessment of IC$_{50}$ values by the multiple-dose and single-dose methods gave different results for the guanidino derivatives. These findings may indicate that contrary to parent compounds the guanidino-terminated analogues are of mixed, agonist–antagonist character (Kosterlitz and Watt, 1968). The highest activity was shown by the guanidino derivative of [D-Met^2,Pro^5]-enkephalinamide.

With respect to analgesic activities (Table VI) the influence of amidination markedly differs from that found in the *in vitro* assays since a

Table V.   *In Vitro* Activity of Analogues of Types $(H_2N)_2\overset{+}{C}$-Tyr-Xxx-Gly-Phe-
Yyy and H-Tyr-Xxx-Gly-Phe-Yyy [a]

—Xxx— -Yyy	GPI	MVD	MVD/GPI
$(H_2N)_2\overset{+}{C}$-Tyr-D-Met-Gly-Phe-Pro-NH$_2$	1000.0(4764)[b]	136.6	0.137
H-Tyr-D-Met-Gly-Phe-Pro-NH$_2$	45.7 (46)	43.8	0.958
$(H_2N)_2\overset{+}{C}$-Tyr-D-Met-Gly-Phe-NH$_2$	457.2(1221)	114.8	0.251
H-Tyr-D-Met-Gly-Phe-NH$_2$	98.6 (99)	22.4	0.227
$(H_2N)_2\overset{+}{C}$-Tyr-D-Nle-Gly-Phe-NH$_2$	457.2 (729)	98.4	0.215
H-Tyr-D-Nle-Gly-Phe-NH$_2$	125.6 (126)	45.8	0.365
$(H_2N)_2\overset{+}{C}$-Tyr-D-Nle-Gly-Phe-Pro-NH$_2$	121.7 (457)	92.9	0.763
H-Tyr-D-Nle-Gly-Phe-Pro-NH$_2$	102.4 (102)	63.4	0.619

[a] Activities are expressed in $10^{-6}/IC_{50}$ M values taken from the data of Bajusz *et al.* (1980a).
[b] Activities in GPI determined by the single-dose method are given in parentheses (Bajusz *et al.*, 1980a).

guanidino terminal seems to be advantageous only in the tetrapeptideamides. To interpret this phenomenon it was supposed that "the shift of positive charge at the N-terminus of peptides altered the position (or spatial
arrangement) of the fifth residue, i.e. Pro-NH$_2$, to such an extent that for
analgesia its absence was more favourable than its presence" (Bajusz *et al.*,
1980a). The influence of the terminal guanidino group also seems to depend
on both the route of application and the "guest side chain" at position 2.

Table VI.   Analgesic Activity of Analogues of Types
$(H_2N)_2\overset{+}{C}$-Tyr-Xxx-Gly-Phe-Yyy and H-Tyr-Xxx-Gly-Phe-
Yyy Compared to Morphine on a Molar Basis $(= 1)$ [a]

—Xxx— -Yyy	i.c.v.	i.v.
$(H_2N)_2\overset{+}{C}$-Tyr-D-Met-Gly-Phe-Pro-NH$_2$	10.5	0.65
H-Tyr-D-Met-Gly-Phe-Pro-NH$_2$	64.6	4.2
$(H_2N)_2\overset{+}{C}$-Tyr-D-Met-Gly-Phe-NH$_2$	210.0	4.4
H-Tyr-D-Met-Gly Phe-NH$_2$	12.1	2.2
$(H_2N)_2\overset{+}{C}$-Tyr-D-Nle-Gly-Phe-NH$_2$	20.0	7.95
H-Tyr-D-Nle-Gly-Phe-NH$_2$	16.4	1.2
$(H_2N)_2\overset{+}{C}$-Tyr-D-Nle-Gly-Phe-Pro-NH$_2$	1.1	0.6
H-Tyr-D-Nle-Gly-Phe-Pro-NH$_2$	28.8	1.6

[a] Determined by the rat tail-flick test upon intracerebroventricular (i.c.v.) and
intravenous (i.v.) application (Bajusz *et al.*, 1980a).

Of the guanidino-terminated tetrapeptides the D-Met^2 analogue was 10 times more active than the D-Nle^2 congener when given i.c.v. but the D-Nle^2 peptide was twice as potent as the D-Met^2 compound upon i.v. injection.

5. Modification at the C Terminus

5.1. Truncated Analogues

Changes in activity of enkephalins due to C-terminal truncation have been investigated by comparing a series of D-Ala^2 analogues (Kosterlitz *et al.*, 1980). The potencies of these analogues in the MVD and GPI assays and those of Met-enkephalin and morphine are presented in Table VII.

Compared with Met-enkephalin, the parent compound [D-Ala^2,Leu^5]-enkephalin (I) had about the same MVD/GPI ratio and about six to seven times higher activity in both assays. Removal of the terminal COOH group (II) resulted in substantial loss of activity in the MVD and a slight increase of activity in the GPI, the MVD/GPI ratio thus being markedly diminished. When the side chain of the fifth residue was also removed (III), activity in both assays was diminished. The MVD/GPI ratio of tetrapeptide III was similar to that of des-carboxy pentapeptide II. Decarboxylation of III caused a further decrease in the activity in the MVD and a slight increase in the activity in the GPI. From these data it may be concluded that the presence of the terminal COOH group at position 5 is essential for a high activity in the MVD assay and for the maintenance of an enkephalinlike pattern of activity, while the absence of an acidic terminal can improve the activity in the GPI assay and render the activity pattern more similar to morphine. The side chain of the fifth residue is necessary for good potency in both assays.

Table VII. Changes in the *In Vitro* Activity of [D-Ala^2,Leu^5]-Enkephalin Due to C-Terminal Truncation^a

No.	Peptide	MVD	GPI	MVD/GPI
I	Tyr-D-Ala-Gly-Phe-Leu-OH	476.2	40.2	11.86
II	Tyr-D-Ala-Gly-Phe-Leu(desCOOH)	29.2	45.7	0.64
III	Tyr-D-Ala-Gly-Phe-OH	3.2	4.85	0.65
IV	Tyr-D-Ala-Gly-Phe(desCOOH)	2.3	6.45	0.36
	Met-enkephalin	68.0	6.37	10.68
	Morphine	2.0	14.38	0.14

^{a} $10^{-6}/IC_{50}$ M values have been calculated from the data of Kosterlitz *et al.* (1980).

5.2.   Analogues with Terminal Methionin-ol Residue

The highly potent Met-enkephalin derivative Tyr-D-Ala-Gly-MePhe-Met(O)-ol (Table IX, compound I) (Roemer et al., 1977) qualified as a "longer acting and orally active (i.e., more stable)" analogue has recently been further stabilized by methylation of Tyr[1] (Roemer and Pless, 1979), i.e., MeTyr-D-Ala-Gly-MePhe-Met(O)-ol (II). Studying the degradation of peptides I, II, and [D-Ala[2],Met[5]]-enkephalinamide by the enzyme complex of mouse brain ultrafiltrate, the amounts of intact peptide found in the digests after 6 hr were 46, 100, and 12%, respectively (Huguenin and Maurer, 1980). Thus, the MeTyr derivative of I (i.e., II) should be regarded as an extremely resistant peptide.

Based on the data published recently (Roemer and Pless, 1979), some structural features that seem to be of particular interest with respect to analgesic potency have been summarized in Tables VIII and IX.

As can be seen from the data of Table VIII, the more polar sulfoxide derivatives, i.e., peptides containing D-Met(O)[2] and/or Met(O)-ol[5], are more active than the corresponding thioether congeners.

The significance of the D-Xxx[2] residues in the Met(O)-ol[5] analogues is clearly indicated by the figures in Table IX. The potency order is as follows: D-Ser < D-MeAla ≃ D-Met ≤ D-Met(O) < D-Ala. Such a structure–activity relationship can hardly be explained on the basis of differences in enzyme resistance. Namely, the Tyr-D-Ala bond found in the most potent analogue (Table IX, compound I) is certainly less stable than the Tyr-D-Met bond (Bajusz et al., 1978) in compound V, and the Tyr-D-Met(O) or Tyr-D-MeAla bonds of the less active congeners III and VI, respectively, are very likely to be more resistant to enzymatic degradation. As for the MeTyr derivatives (II and IV), it is easy to explain that the most stable analogue II shows the highest potency on subcutaneous or oral administra-

Table VIII.   Analgesic Potencies of Some Met-ol[5] and Met(O)-ol[5] Analogues Compared to Morphine[a,b]

Residues		i.v.		s.c.		p.o.	
1–4	5:	Met-ol	Met(O)-ol	Met-ol	Met(O)-ol	Met-ol	Met(O) ol
Tyr-D-Ala-Gly-Phe-		0.14	0.33	0.05	1.0	i.a.[c]	0.067
Tyr-D-Met-Gly-MePhe-		0.13	1.0	0.13	1.0	i.a.	0.067
Tyr-D-Met-Gly-MePhe- (O)		0.33	1.0	1.0	1.0	0.067	0.1

[a] Data are taken from Roemer and Pless (1979).
[b] Analgesic potencies were determined by the mouse tail-flick test upon intravenous (i.v.), subcutaneous (s.c.), and oral (p.o.) administration.
[c] i.a., inactive.

Table IX. Analgesic Potencies of Analogues of Type Tyr-Xxx-Gly-MePhe-Met(O)-ol Compared to Morphine (= 1) [a,b]

No.	Xxx	i.v.	s.c.	p.o.
I (II)	D-Ala	4(3) [c]	2(10) [c]	0.2 (1) [c]
III (IV)	D-Met(O)	1(1) [c]	1 (1) [c]	0.1 (0.13) [c]
V	D-Met	1	1	0.067
VI	D-MeAla	1	1	0.067
VII	D-Ser	0.1	0.1	i.a. [d]

[a] Data are taken from Roemer and Pless (1979).
[b] Analgesic potencies were determined by the mouse tail-flick test upon intravenous (i.v.), subcutaneous (s.c.), and oral (p.o.) administration.
[c] Potency of the corresponding *N*-methyl analogue, i.e., MeTyr-D-Ala-Gly-MePhe-Met(O)-ol (II) and MeTyr-D-Met(O)-Gly-MePhe-Met(O)-ol (IV), respectively.
[d] i.a., inactive.

tion; however, it is curious that the less stable congener I is more active than its stabilized derivative II when applied intravenously. It is also worth mentioning that *N*-methylation, which markedly enhanced the subcutaneous and oral activity of I, had only marginal impact on the potency of the congener D-Met(O)^2 III.

Enzyme resistance is very likely one of the requirements that must be satisfied for high *in vivo* potency. Other factors of similar significance are favorable transport properties, suitable binding to plasma proteins, enhanced ability to cross the blood–brain barrier and to permeate other "barriers," or increased binding capacity to the receptors (Bajusz *et al.*, 1978; Morley, 1980). Thus, the enzyme resistance of analogues should not necessarily correlate with their analgesic potency.

5.3. Analogues with Terminal Methionine, Leucine, Norvaline, Proline, or Thiazolidine-4-carboxylic Acid Residue

The biological activities of analogues having the general formula Tyr-D-Xxx-Gly-Phe-Yyy-NH$_2$ (Xxx = Ala or Met, Yyy = Met, Leu, Nva, or Pro) studied by Audigier *et al.* (1980a,b,c) are shown in Table X.

In the MVD assay, the potency of D-Ala^2 analogues with terminal Met, Leu, or Nva were similar to each other and much higher than that of the Pro^5 congener. In the case of the D-Met^2 analogues, similar activity was shown by the pairs terminated by Met or Leu and Nva or Pro, respectively.

In the GPI assay, the influence of the "guest side chain" at position 2 is reversed: the D-Met^2-containing analogues show higher potency. Besides, the activity of the D-Ala^2 peptides terminated by Met or Leu were about equipotent, the Nva analogue twice as potent, while the Pro^5 analogue was

Table X.   Opioid Activities of Analogues of Type Tyr-D-Xxx-Gly-Phe-Yyy-NH$_2$[a,b]

Assay	Xxx	Yyy			
		Met	Leu	Nva	Pro
MVD	D-Ala	166.7 (4.7)	111.1 (3.8)	142.9 (2.4)	8.8 (1)
	D-Met	35.7 (1)	29.1 (1)	58.8 (1)	47.6 (5.4)
GPI	D-Ala	20.9 (1)	21.4 (1)	40.0 (1)	2.9 (1)
	D-Met	57.6 (2.3)	31.5 (1.5)	90.0 (2.3)	12.2 (4.2)
MVD/GPI	D-Ala	7.98	5.19	3.57	0.44
	D-Met	0.62	0.92	0.65	0.66
MTF	D-Ala	0.5 (1)	1.4 (1.1)	1.3 (1)	29.6 (1)
	D-Met	3.0 (6.0)	1.3 (1)	2.5 (1.9)	59.2 (2.0)

[a] *In vitro* activities are given in $10^{-6}/IC_{50}$ M values and in potency ratios refer to the corresponding less potent congeners (in parentheses) as calculated from the data of Audigier *et al.* (1980a,b,c).
[b] Analgesic potencies assessed by the mouse tail-flick test (MTF) are related to morphine as determined by Audigier *et al.* (1980a,b,c).

much weaker. In the GPI the activity of the four D-Met^2-containing analogues ranged from 31.5 to 90.9.

With respect to the MVD/GPI ratios we may recall the view (Kosterlitz *et al.*, 1980) that the free COOH terminal is essential for the maintenance of enkephalinlike activity pattern. The data presented in Table X prove that the D-amino acid residue at position 2 influences also considerably the activity pattern of analogues examined.

As for the analgesic activity, the Pro^5 substitution seems to be more advantageous than others in the C-terminal position in particular if D-Met resides at position 2. To obtain more information about the analgesic potency of Pro^5-enkephalins, Table XI presents data of such analogues with

Table XI.   Analgesic Potencies of the Analogues of Pro^5-enkephalins, Tyr-Xxx-Gly-Phe-Pro-*O*, Compared to Morphine (= 1)

Assay[a]	Xxx	*O*		
		Pro-OH	Pro-NH$_2$	Pro-NH-Et
RTF i.c.v.	D-Ala	0.24 (1)[b]	3.9 (1)	16.9 (1)
	D-Nle	1.00 (4.2)	18.8 (4.6)	4.3 (0.25)
	D-Met	1.33 (5.5)	48.8 (12.8)	16.9 (1.0)
RTF i.v.	D-Ala	0.21 (1)	0.22 (1)	0.19 (1)
	D-Nle	0.20 (1)	0.7 (3.2)	0.42 (2.2)
	D-Met	0.16 (1)	5.5 (25.0)	0.55 (2.9)

[a] Activities were determined by the rat tail-flick test (RTF) upon intracerebroventricular (i.c.v.) and intravenous (i.v.) injections (Bajusz *et al.*, 1976, 1977; Székely *et al.*, 1977).
[b] Potency ratios compared to the corresponding D-Ala^2 congeners are given in parentheses.

terminal COOH, CONH$_2$, and CONH-Et groups, respectively, and containing D-Ala, D-Nle, and D-Met at position 2 (Bajusz *et al.*, 1976, 1977; Székely *et al.*, 1977).

As for the potencies measured after i.c.v. application, the replacement of D-Ala^2 by D-Nle increased the activity of Pro^5-enkephalin and its amide by a factor of about four but decreased the potency of the ethylamide derivative by a similar extent. Changing D-Met^2 for D-Nle caused a slight increase of activity in the peptide amide (Pro-NH$_2$), while in the molecule of the ethylamide (Pro-NH-Et) the same replacement produced only the potency of the D-Ala^2 analogue.

Concerning the i.v. activities, variation of the side chain of D-Xxx^2 had practically no effect in Pro^5-enkephalin. Having Pro-NH-Et at the C terminus, D-Nle^2 and D-Met^2 were almost equally active but examining the Pro-NH$_2$ analogues D-Met^2 was some eight times more potent than D-Nle^2.

As C-terminal residue the "thia" analogue of Pro, i.e., thiazolidine-4-carboxylic acid (Thz), is also advantageous (Yamashiro *et al.*, 1977). Compared to morphine the analgesic activities of [D-Met^2,Thz^5]- and [D-Thr^2,Thz^5]-enkephalinamide were 4.2 and 4.8 (i.v.) and 7.7 and 27.1 (i.c.v.), respectively. [D-Thr^2,Thz^5]-enkephalinamide has analgesic potency also upon subcutaneous and oral application (151 and 173% respectively, as compared to morphine) (Tseng *et al.*, 1978).

A direct comparison of the analgesic potency of [D-Met^2,Pro^5]-enkephalinamide and its Thz^5 analogue is rather difficult because of the difference between the AD$_{50}$ values reported for the reference compound, morphine, in the different laboratories. Namely, 77 nmole/kg was given by Audigier *et al.* (1980c) and 1.11 nmole/mouse, that is, about 40 nmole/kg by Yamashiro *et al.* (1977) using the same method (the mouse tail-flick test, i.c.v. appliation).

5.4. Analogues with Terminal SO$_3$H and PO$_3$H$_2$ Groups

Some analogues of Nle^5-enkephalin in which the terminal COOH group is replaced by SO$_3$H and PO$_3$H$_2$, respectively, have recently been prepared (Bajusz *et al.*, 1980b). The C-terminal residue of the new analogues are α-amino-pentanesulfonic acid, -HN-CH(C$_4$H$_9$)-SO$_3$H, and α-amino-pentanephosphonic acid, -HN-CH(C$_4$H$_9$)-PO$_3$H$_2$, abbreviated as NleS and NleP, respectively. The peptides involved in the study have the general formula Tyr-Xxx-Gly-Phe-Yyy (Xxx = Gly, D-Ala, D-Nle, or D-Met; Yyy = Nle, NleS, D-Nles, NleP, or D-NleP). Substitution of Nle^5 by NleS or NleP enhanced the *in vitro* potencies but the changes in activity were markedly influeced by the Xxx^2 residue (Table XII). Accordingly, the most active compounds of this series are the [D-Met^2,Nle^5], [D-Ala^2,NleP^5], and [D-Nle^2,NleS^5] analogues, respectively. This fact reveals that a significant interdependence exists between residues 2 and 5 of enke-

Table XII. *In Vitro* Activity of Analogues of Type Tyr-
Xxx-Gly-Phe-Yyy[a]

Assay	Xxx	Yyy[b]		
		Nle	NleS	NleP
MVD	Gly	48.5 (1)[c]	235.9 (4.8)	431.0 (8.9)
	D-Ala	1000.0 (1)	1786.0 (1.8)	5130.8 (5.1)
	D-Nle	1077.2 (1)	4761.9 (4.4)	4221.2 (3.9)
	D-Met	1667.8 (1)	2041.6 (1.2)	618.2 (0.4)
GPI	Gly	2.7 (1)	17.0 (6.2)	7.8 (2.8)
	D-Ala	15.3 (1)	42.8 (2.8)	61.2 (4.0
	D-Nle	16.8 (1)	117.5 (7.0)	26.8 (1.6)
	D-Met	26.0 (1)	54.5 (2.1)	9.4 (0.4)
MVD/GPI	Gly	17.7	13.9	55.2
	D-Ala	65.4	41.7	83.9
	D-Nle	64.2	40.5	157.3
	D-Met	64.2	37.4	65.8

[a] Activities are expressed in $10^{-6}/IC_{50}$ M values (Bajusz *et al.*, 1980b).
[b] NleS and NleP denote α-amino-pentanesulfonic acid and α-amino-pentane-phosphonic acid residues, respectively.
[c] Changes in activity relative to the parent Nle[5] analogues are given in parentheses.

phalins, which might influence the peptide–receptor interactions either directly or through its contribution to the formation of preferred conformation of the given molecule. The MVD/GPI ratio of the analogues indicates that introduction of the more acidic SO_3H or PO_3H_2 did not alter the enkephalinlike activity pattern of the COOH-terminated compounds.

The peptides listed in Table XII showed no or only extremely low analgesic potency when given centrally. The only compound of the series that caused analgesia upon i.v. application was the [D-Met[2],D-NleP[5]] analogue, having a potency of 0.15 relative to morphine (= 1). The MVD/GPI ratio of this analogue was 0.7, indicating that substitution by D-NleP renders the activity pattern more similar to morphine (the corresponding NleP[5] congener had an MVD/GPI ratio of 65.8).

6. Conclusions

Modification in the molecule of Met/Leu-enkephalin could result in analogues with MVD/GPI ratios ranging through three orders of magnitude. Pentapeptide analogues of zwitterionic character generally have an enkephalinlike pattern of activity, i.e., high MVD/GPI (e.g., Tyr-D-Nle-Gly-Phe-NleP has a ratio of 157.3). Substitution and/or derivation leading to basic analogues, e.g., neutralization or removal of the terminal COOH group, result in activity pattern more or less similar to morphine, i.e., low

MVD/GPI [e.g., $(H_2N)_2\overset{+}{C}$-Tyr-D-Met-Gly-Phe-Pro-NH$_2$ has a value of 0.137]. Replacement of Gly^2 by certain D-amino acid residues generally enhances the biological potency but the activity pattern of analogues is determined primarily by the modifications applied at the C and /or N termini. It must be emphasized that general conclusions can hardly be formulated. Seemingly three main points of the enkephalin molecule can be responsible for the opioid activity, namely the terminal H$_2$N group and the side chains of Tyr^1 and Phe^4 residing in proper spatial arrangements. But the effect of these side chains is markedly influenced by the side chain of the fith residue (cf. MVD/GPI values of Met- and Leu-enkephalin in Table I). The "guest side chain" introduced in position 2 makes the picture even more complicated. There is an interdependence between residues 2 and 5 that not only influences the strength of opioid activity but also modifies the activity pattern, i.e., the ability of the molecule to interact with the δ or μ receptors. For instance, [D-Ala^2,D-Met^5]-enkephalinamide has an MVD/GPI ratio of 7.98 but its D-Met^2 congener only 0.62 (TAble X).

In spite of the difficulties in delineation of structure–activity relationships, systematic studies have led to highly potent peptide analgesics like [D-Ala^2,MePhe^4,Met(O)-ol^5]-enkephalin (Roemer *et al.*, 1977; Pless *et al.*, 1979), its MeTyr^1 congener (Roemer and Pless, 1979), [D-Met^2,Pro^5]-enkephalinamide (Bajusz *et al.*, 1977; Székely *et al.*, 1977), its Thz^5 analogue, and [D-Thr^2,Thz^5]-enkephalinamide (Yamashiro *et al.*, 1977). In addition, *in vitro* highly potent and "stable" analogues with enkephalinlike activity pattern could also be developed such as [D-Ala^2,D-Leu^5]-enkephalin (Beddell *et al.*, 1977a) or the NleS^5 and NleP^5 congeners (Bajusz *et al.*, 1980b).

DISCUSSION

KANAZIR: Dr. Bajusz, I would like to raise a question concerning the modifications of pentapeptide. Is there information that the pentapeptides are subject to modifications such as methylation, acetylation, or phosphorylation? If so, that might mean that the affinity and the capacity of pentapeptides might be regulated by these modifications.

BAJUSZ: I haven't seen any data on such modifications *in vivo;* however, we know that acetylation or methylation of the phenolic hydroxyl group, as well as acetylation of the terminal amino group, leads to inactivation. According to our own observations, opiate activity of an O-acetylderivative was regenerated parallel with deacetylation. Mono- or dimethylation of the terminal amino groups permanently lowers the opioid activity particularly in the MVD assay. I might add that acetylation of β-endorphin occurs *in vivo*.

REFERENCES

Audigier, Y., Mazarguil, H., and Cros, J., 1980a, [D-Ala^2, N-Val^5] enkephalinamide and [D-Met^2, N-Val^5] enkephalinamide: Two potent agonists of opiate and enkephalin receptors, *FEBS Lett.* **110**:88–90.

Audigier, Y., Mazarguil, H., Gout, R., and Cros, J., 1980b, Structure–activity relationships of enkephalin analogs at opiate and enkephalin receptors: Correlation with analgesia, *Eur. J. Pharmacol.* **63**:35–46.

Audigier, Y., Gout, R., Mazarguil, H., and Cros, J., 1980c, Comparative study of analgesia induced by N-Val^5- and Pro^5-enkephalin analogs, *Eur. J. Pharmacol.* **64**:187–189.

Bajusz, S., Barabás, E., Széll, E., and Bagdy, D., 1975, Peptide aldehyde inhibitors of the fibrinogen–thrombin reaction, in: *Peptides, Proceedings of the Fourth American Peptide Symposium* (R. Walter and J. Meienhofer, eds.), pp. 603–608, Ann Arbor Science, Ann Arbor.

Bajusz, S., Rónai, A. Z., Székely, J. I., Dunai-Kovács, Z., Berzétei, I., and Gráf, L., 1976, Enkephalin analogs with enhanced opiate activity, *Acta Biochim. Biophys. Acad. Sci. Hung.* **11**:305–309.

Bajusz, S., Rónai, A. Z., Székely, J. I., Gráf, L., Dunai-Kovács, Z., and Berzétei, I., 1977, A superactive antinociceptive pentapeptide, [D-Met^2,Pro^5]-enkephalinamide, *FEBS Lett.* **76**:91–92.

Bajusz, S., Patthy, A., Kenessey, Á., Gráf, L., Székely, J. I., and Rónai, A. Z., 1978, Is there correlation between analgesic potency and biodegradation of enkephalin analogs?, *Biochem. Biophys. Res. Commun.* **84**:1045–1053.

Bajusz, S., Rónai, A. Z., Székely, J. I., Miglécz, E., and Berzétei, I., 1980a, Furthur enhancement of analgesic activity: Enkephalin analogs with terminal guanidino group, *FEBS Lett.* **110**:85–87.

Bajusz, S., Rónai, A. Z., Székely, J. I., Turán, A., Juhász, A., Patthy, A., Miglécz, E., and Berzétei, I., 1980b, Enkephalin analogs containing amino sulfonic acid and amino phosphonic acid residues at position 5, *FEBS Lett.* **117**:308–310.

Beaumont, A., and Hughes, J., 1979, Biology of opioid peptides, *Annu. Rev. Pharmacol. Toxicol.* **19**:245–267.

Beddell, C. R., Clark, R. B., Follenfant, R. L., Lowe, L. A., Ubatuba, F. B., Wilkinson, S., and Miller, R. J., 1977a, Analogues of the enkephalins—Structural requirements for opioid activity, in: *Biological Activity and Chemical Structure* (J. A. Keverling Buisman, ed.), pp. 177–193, Elsevier, Amsterdam.

Beddell, C. R., Clark, R. B., Hardy, G. W., Lowe, L. A., Ubatuba, F. B., Vane, J. R., Wilkinson, S., Chang, K. J., Cuatrecasas, P., and Miller, R. J., 1977b, Structural requirements for opioid activity of analogues of the enkephalins, *Proc. R. Soc. London Ser. B* **198**:249–265.

Beluzzi, J. D., Grant, N., Garsky, V., Sarantakis, D., Wise, C. D., and Stein, L., 1976, Analgesia induced *in vivo* by central administration of enkephalin in rat, *Nature (London)* **260**:625–626.

Bradbury, A. F., Smyth, D. G., and Snell, C. R., 1975, Biosynthesis of β-MSH and ACTH, in: *Peptides, Proceeding of the Fourth American Peptide Symposium* (R. Walter and J. Meienhofer, eds.), pp. 609–615, Ann Arbor Science, Ann Arbor.

Bradbury, A. F., Feldberg, W. F., Smyth, D. G., and Snell, C. R., 1976a, Lipotropin C-fragment: An endogenous peptide with potent analgesic activity, in: *Opiates and Endogenous Opioid Peptides* (H. W. Kosterlitz, ed.), pp. 9–17, North-Holland, Amsterdam.

Bradbury, A. F., Smyth, D. G., and Snell, C. R., 1976b, Lipotropin: Precursor to two biologically active peptides, *Biochem. Biophys. Res. Commun.* **69**:950–956.

Bradbury, A. F., Smyth, D. G., and Snell, C. R., 1976c, Biosynthetic origin and receptor conformation of methionine enkephalin, *Nature (London)* **260**:165–166.

Cox, B. M., Goldstein, A., and Li, C. H., 1976, Opioid activity of a peptide, β-lipotropin-(61–91), derived from β-lipotropin, *Proc. Natl. Acad. Sci. USA* **73**:1821–1823.

Coy, D. H., Kastin, A. J., Schally, A. V., Morin, O., Caron, N. G., Labrie, F., Walker, J.

M., Felter, R., Bernston, G. G., and Sandman, C. A., 1976, Synthesis and opioid activities of stereoisomers and other D-amino acid analogs of methionine-enkephalin, *Biochem. Biophys. Res. Commun.* **73**:632–638.

D'Amour, F. E., and Smith, D. L., 1941, A method for determining loss of pain sensation, *J. Pharmacol. Exp. Ther.* **72**:74–79.

Dutta, A. S., Gormley, J. J., Hayward, C. F., Morley, J. S., Shaw, J. S., Stacey, G. J., and Turnbull, M. T., 1977, Enkephalin-like peptides: Structure–activity relationships, *Acta Pharm. Suec. Suppl.* **14**:14–15.

Feldberg, W. F., and Smyth, D. G., 1976, The C-fragment of lipotropin—a potent analgesic, *J. Physiol. (London)* **260**:30P.

Frederickson, R. C. A., 1977, Enkephalin pentapeptides—A review of current evidences for a physiological role in vertebrate neurotransmission, *Life Sci.* **21**:23–42.

Frederickson, R. C. A., Nickander, R., Smithwick, E. L., Shuman, R., and Norris, F. H., 1976, Pharmacological activity of Met-enkephalin and analogues *in vitro* and *in vivo*—Depression of single neuronal activity in specified regions, in: *Opiates and Endogenous Opioid Peptides* (H. W. Kosterlitz, ed.), pp. 239–246, North-Holland, Amsterdam.

Gacel, G., Fournie-Zaluski, M. C., Feillon, E., Roques, P. B., Senault, B., Lecomte, J. M., Malfroy, B., Swerts, J. P., and Schwartz, J. C., 1979, Conformational and biological activities of hexapeptides related to enkephalins: Respective roles of the ammonium and hydroxyl groups of tyrosine, *Life Sci.* **24**:725–732.

Goldstein, A., Tachibana, S., Lowney, L. I., Hunkapiller, M., and Hood, L., 1979, Dynorphine-(1–13), an extraordinarily potent opioid peptide, *Proc. Natl. Acad. Sci. USA* **76**:6666–6670.

Gráf, L., Székely, J. I., Rónai, A. Z., Dunai-Kovács, Z., and Bajusz, S., 1976, Comparative study on analgesic effect of Met⁵-enkephalin and related lipotropin fragments, *Nature (London)* **263**:240–242.

Gráf, L., Cseh, G., Barát, E., Rónai, A. Z., Székely, J. I., Kenessey, Á., and Bajusz, S., 1977, Structure–function relationships in lipotropins, *Ann. N.Y. Acad. Sci.* **297**:63–82.

Guillemin, R., Ling, N., and Burgus, R., 1976, Endorphines peptides d'origine hypothalamique et neurohypophysaire a activité morphinomimetique: Isolement et structure moléculair de l'α-endorphine, *C. R. Acad. Sci.* **282**:783–785.

Guillemin, R., Ling, N., Lazarus, L., Burgus, R., Minick, S., Bloom, F., Nicoll, R., Siggins, G., and Segal, D., 1977, The endorphins, novel peptides of brain and hypophyseal origin, with opiate-like activity: Biochemical and biologic studies, *Ann. N.Y. Acad. Sci.* **297**:131–156.

Hambrook, J. M., Morgan, B. A., Rance, M. J., and Smith, C. F. C., 1976, Mode of deactivation of the enkephalins by rat and human plasma and rat brain homogenates, *Nature (London)* **262**:782–783.

Huang, W. Y., Chang, R. C. C., Kastin, A. J., Coy, D. H., and Schally, A. V., 1979, Isolation and structure of pro-Met-enkephalin: Potential enkephalin precursor from porcine hypothalamus, *Proc. Natl. Acad. Sci. USA* **76**:6177–6180.

Horn, A. S., and Rodgers, J. R., 1977, The enkephalins and opiates: Structure–activity relations, *J. Pharm. Pharmacol.* **29**:257–265.

Hughes, J., 1975, Isolation of an endogenous compound from the brain with pharmacological properties similar to morphine, *Brain Res.* **88**:295–308.

Hughes, J., Kosterlitz, H. W., and Leslie, F. M., 1975a, Effect of morphine on adrenergic transmission in the mouse vas deferens: Assessment of agonist and antagonist potencies of narcotic analgesics, *Br. J. Pharmacol.* **55**:541–546.

Hughes, J., Smith, T., Morgan, B., and Fothergill, L., 1975b, Purification and properties of enkephalin—the possible endogenous ligand for the morphine receptor, *Life Sci.* **16**:1753–1758.

Hughes, J., Smith, T. W., Kosterlitz, H. W., Fothergill, L. A., Morgan, B. A., and Morris, H. R., 1975c, Identification of two related pentapeptides from the brain with potent opiate agonist activity, *Nature (London)* **258**:577–579.

Huguenin, R., and Maurer, R., 1980, Resistance of FK 33–824 and other enkephalin analogues to peptidase degradation, *Brain Res. Bull.* **5**:47–50.

Kangawa, K., Matsuo, H., and Masao, I., 1979, α-Neo-endorphin: A "big" Leu-enkephalin with potent opiate activity from porcine hypothalami, *Biochem. Biophys. Res. Commun.* **86**:153–160.

Kosterlitz, H. W., 1978, Opioid peptides and their receptors, in: *Endorphins '78* (L. Gráf, M. Palkovits, and A. Z. Rónai, eds.), pp. 205–216, Akadémiai Kiadó, Budapest.

Kosterlitz, H. W., and Watt, A. J., 1968, Kinetic parameters of narcotic agaonists and antagonists, with particular reference to *N*-allylnoroxymorphone (naloxone), *Br. J. Pharmacol.* **33**:266–267.

Kosterlitz, H. W., Lydon, R. J., and Watt, A. J., 1970, The effects of adrenaline, noradrenaline and isoprenaline on inhibitory α- and β-adrenoreceptors in the longitudinal muscle of the guinea-pig ileium, *Br. J. Pharmacol.* **39**:398–413.

Kosterlitz, H. W., Lord, J. A. H., Paterson, S. J., and Waterfield, A. A., 1980, Effects of changes in the structure of enkephalins and of narcotic analgesic drugs on their interactions with μ- and δ-receptors, *Br. J. Pharmacol.* **68**:333–342.

Law, P.-Y., and Loh, H. H., 1978, ³H-Leu⁵-enkephalin specific binding to synaptic membrane: Comparison with ³H-dihydro-morphine and ³H-naloxone, *Res. Commun. Chem. Pathol. Pharmacol.* **21**:409–434.

Lazarus, L. H., Ling, N., and Guillemin, R., 1976, β-Lipotropin as a prohormone for the morphinomimetic peptides endorphins and enkephalins, *Proc. Natl. Acad. Sci. USA* **73**:2156–2159.

Lemaire, S., Magnan, J., and Regoli, D., 1978, Rat vas deferens: A specific bioassay for endogenous opioid peptides, *Br. J. Pharmacol.* **64**:327–329.

Li, C. H., and Chung, D., 1976, Isolation and structure of an untriakontapeptide with opiate activity from camel pituitary glands, *Proc. Natl. Acad. Sci. USA* **73**:1145–1148.

Ling, N., Minick, S., Lazarus, L., Rivier, J., and Guillemin, R., 1977, Structure–activity relationships of enkephalin and endorphin analogs, in: *Peptides, Proceedings of the Fifth American Peptide Symposium* (M. Goodman and J. Meienhofer, eds.), pp. 96–99, Wiley, New York.

Ling, N., Minick, S., and Guillemin, R., 1978, Amino-terminal extension analogs of methionine-enkephalin, *Biochem. Biophys. Res. Commun.* **83**:565–570.

Loh, H. H., Tseng, L. F., Wie, E., and Li, C. H., 1976, Endorphin is a potent analgesic agent, *Proc. Natl. Acad. Sci. USA* **73**:2895–2898.

Lord, J. A. H., Waterfield, A. A., Hughes, J., and Kosterlitz, H. W., 1977, Endogenous opioid peptides: Multiple agonists and receptors, *Nature (London)* **267**:495–499.

Maryanoff, B. E., and Zelesko, M. J., 1978, Stereochemical considerations in structural comparison of enkephalins and endorphins with exogenous opiate agents, *J. Pharm. Sci.* **67**:590–591.

Miller, R. J., and Cuatrecasas, P., 1978, Enkephalins and endorphins, *Vitam. Horm. (N.Y.)* **36**:297–382.

Miller, R. J., Chang, K.-J., Cuatrecasas, P., Wilkinson, S., Lowe, L. A., Beddell, C., and Follenfant, R., 1978, Distribution and pharmacology of the enkephalins and related opiate peptides, in: *Centrally Acting Peptides* (J. Hughes, ed.), pp. 195–213, Macmillan, London.

Morgan, B. A., 1979, Enkephalins and endorphins: A review of structure–activity relationships, in: *Amino-acids, Peptides and Proteins* (R. C. Sheppard, ed.), Vol. 10, pp. 474–489, Specialist Periodical Reports, The Chemical Society, London.

Morgan, B. A., Bower, J. D., Guest, K. P., Handa, B. K., Metcalf, G., and Smith, C. F. C., 1977, Structure–activity relationships of enkephalin analogues, in: *Peptides, Proceedings of the Fifth American Peptide Symposium* (M. Goodman and J. Meienhofer, eds.), pp. 111–113, Wiley, New York.

Morley, J. S., 1980, Structure–activity relationships of enkephalin-like peptides, *Annu. Rev. Pharmacol. Toxicol.* **20**:81–110.

Nemethy, G., and Scheraga, H. A., 1962, The structure of water and hydrophobic bonding in proteins. III. The thermodynamic properties of hydrophobic bonds in proteins, *J. Phys. Chem.* **66**:1773–1789.

Pert, C. B., Bowie, D. L., Fong, B. T. W., and Chang, J.-K., 1976a, Synthetic analogues of Met-enkephalin which resist to enzymatic destruction, in: *Opiates and Endogenous Opioid Peptides* (H. W. Kosterlitz, ed.), pp. 79–86, North-Holland, Amsterdam.

Pert, C. B., Pert, A., Chang, J.-K., and Fong, B. T. W., 1976b, [D-Ala²]-Met-enkephalinamide: A potent, long lasting synthetic pentapeptide analgesic, *Science* **194**:330–332.

Pless, J., Bauer, W., Cardinaux, F., Closse, A., Hauser, D., Huguenin, R., Roemer, D., Beuscher, H. H., and Hill, R. C., 1979, Synthesis, opiate receptor binding and analgesic activity of enkephalin analogues, *Helv. Chim. Acta* **62**:398–411.

Roemer, D., and Pless, J., 1979, Structure–activity relationship of orally active enkephalin analogues as analgesics, *Life Sci.* **24**:621–624.

Roemer, D., Beuscher, H. H., Hill, R. C., Pless, J., Bauer, W., Cardinaux, F., Closse, A., Hauser, D., and Iluguenin, R., 1977, A synthetic enkephalin analogue with prolonged parenteral and oral analgesic activity, *Nature (London)* **268**:547–549.

Bónai, A. Z., and Berzétei, I., 1978, Similarities and differences of opioid receptors in different isolated organs, in: *Endorphins '78* (L. Gráf, M. Palkovits, and A. Z. Rónai, eds.), pp. 237–257, Akadémiai Kiadó, Budapest.

Rónai, A. Z., Gráf, L., Székely, J. I., Dunai-Kovács, Z., and Bajusz, S., 1977, Differential behaviour of LPH-(61–91)-peptide in different model systems: Comparison of the opioid activities of LPH-(61–91)-peptide and its fragments, *FEBS Lett.* **74**:182–184.

Rónai, A. Z., Berzétei, I., Székely, J. I., Gráf, L., and Bajusz, S., 1979a, Kinetic studies in isolated organs: Tools to design analgesic peptides and to analyze their receptor effects, *Pharmacology* **18**:18–24.

Rónai, A. Z., Székely, J. I., Berzétei, I., Miglécz, E., and Bajusz, S., 1979b, Tetrapeptide-amide analogues of enkephalin: The role of C-terminus in determining the character of opioid activity, *Biochem. Biophys. Res. Commun.* **91**:1239–1249.

Roques, B. P., Garbay-Jaureguiberry, C., Oberlin, R., Anteunis, M., and Lala, A. K., 1976, Conformation of Met⁵-enkephalin determined by high field PMR spectroscopy, *Nature (London)* **262**:778–779.

Simantov, R., and Snyder, S. H., 1976, A morphine-like peptide "enkephalin" in mammalian brain: Isolation, structure elucidation, interaction with opiate receptor, *Life Sci.* **18**:781–788.

Stern, A. S., Lewis, R. W., Kimura, S., Rosier, J., Gerber, L. D., Brink, L., Stein, S., and Udenfriend, S., 1979, Isolation of the opioid heptapeptide Met-enkephalin (Arg⁶,Phe⁷) from bovine adrenal medullary graulues and striatum, *Proc. Natl. Acad. Sci. USA* **76**:6680–6683.

Székely, J. I., Rónai, A. Z., Dunai-Kovács, Z., Miglécz, E., Berzétei, I., Bajusz, S., and Gráf, L., 1977, [D-Met²,Pro⁵]-enkephalinamide: A potent morphinelike analgesic, *Eur. J. Pharmacol.* **43**:293–294.

Terenius, L., 1978, Endogenous peptides and analgesia, *Annu. Rev. Pharmacol. Toxicol.* **18**:189–204.

Tseng, L. F., Loh, H. H., and Li, C. H., 1976, β-Endorphin as a potent analgesic by intravenous injection, *Nature (London)* **263**:240–241.

Tseng, L. F., Loh, H. H., and Li, C.H., 1978, [D-Thr^2,Thz^5]-enkephalinamide: A potent analgesic by subcutaneous and oral administration, *Life Sci.* **23**:2053–2056.

Waterfield, A. A., Smokcum, R. W. J., Hughes, J., Kosterlitz, H. W., and Henderson, G., 1977, *In vitro* pharmacology of the opioid peptides, enkephalins and endorphins, *Eur. J. Pharmacol.* **43**:107–116.

Yamashiro, D., Tseng, L. F., and Li, C. H., 1977, [D-Thr^2,Thz^5]- and [D-Met^2,Thz^5]-enkephalinamides: Potent analgesics by intravenous injection, *Biochem. Biophys. Res. Commun.* **78**:1124–1129.

2

Secondary Structure - Function Relationships in β-Endorphin

A Proposed Model of the Biologically Active Conformation

László Gráf and Miklós Hollósi

1. Introduction

Of the opioid peptides isolated from different sources and chemically identified to date, enkephalins (Hughes *et al.*, 1975) and β-endorphin (Li and Chung, 1976; Bradbury *et al.*, 1976a; Gráf *et al.*, 1976a) have received the most attention. Though the major form of enkephalins, Met-enkephalin, is apparently an N-terminal fragment of β-endorphin (for the β-endorphin amino acid sequence see Fig. 1), strong biosynthetic (Mains and Eipper, 1978; Lewis *et al.*, 1980) and morphological evidence (Bloom *et al.*, 1978; Watson *et al.*, 1978) indicate that β-endorphin is not a precursor of enkephalins, and that these two groups of opioid peptides subserve two separate neuronal systems in the brain. Evidence for the presence of at least two subtypes of opiate receptors in the brain has been reported from several laboratories (Lord *et al.*, 1977; Chang and Cuatrecasas, 1979; Smith and Simon, 1980). The receptor designated as δ prefers enkephalins to morphine, whereas the μ receptor binds morphine with higher affinity than enkephalins. β-Endorphin appears to bind to the two types of brain recep-

László Gráf and Miklós Hollósi • Institute for Drug Research, and Organic Chemistry Department of L Eötvös University, Budapest, Hungary.

László Gráf and Miklós Hollósi

<pre>
             1   10
Porcine   Tyr-Gly-Gly-Phe-Met-Thr-Ser-Glu-Lys-Ser-
Human

                                                 20
Porcine   Gln-Thr-Pro-Leu-Val-Thr-Leu-Phe-Lys-Asn-
Human

                                                 30
Porcine   Ala-Ile-Val-Lys-Asn-Ala-His-Lys-Lys-Gly-Gln
Human   Ile   Tyr   Glu
</pre>

Figure 1. The primary structure of porcine and human β-endorphin. The structures are identical with the sequence regions, residues 61–91, of porcine (Gráf *et al.*, 1971) and human β-lipotropin (Cseh *et al.*, 1972), respectively.

tors with similar affinities. Considering the structural relatedness of enkephalins to β-endorphin and the closely similar biochemical properties of the two receptor sites (Smith and Simon, 1980), it is not surprising that the two natural ligands "cross-react" with each other's receptors in *in vitro* models. The functional specificity of the neuronal mechanisms mediated by enkephalins and β-endorphin may be maintained topographically, by synaptic interactions in the brain.

Since Bradbury *et al.* (1976b) first proposed that the biological similarity of enkephalins and opiates might be due to some resemblance in their three-dimensional structures, the conformation of enkephalins has been extensively studied (for a partial listing of physicochemical methods used see Zetta *et al.*, 1979). Our investigation on the conformation of β-endorphin was initiated by the considerable protease resistance of β-endorphin as compared to its smaller N-terminal fragments (Gráf *et al.*, 1977) and by some early structure–activity data obtained in two receptor models, guinea pig ileum and mouse vas deferens (Rónai *et al.*, 1977). These latter data showed that the extension of fragment β-endorphin-(1–19) with residues 20–31 to form β-endorphin (for sequence see Fig. 1) resulted in an apparent divergence between the two *in vitro* assays suggesting that the 12 C-terminal residues may induce a conformational change in the active site of the molecule.

While the conformation of enkephalins and their analogues in solution was inferred from ^{1}H and ^{13}C nuclear magnetic resonance measurements (Zetta *et al.*, 1979), due to obvious experimental difficulties, the same method could not be used for the 31-residue β-endorphin. It is well known, however, that circular dichroism (CD) spectroscopy may furnish valuable information on the secondary structure of proteins and polypeptides. Therefore, we chose this method along with the application of the Chou–Fasman predictive rules (Chou and Fasman, 1974a,b) to explore the conformation of β-endorphin in water and organic solvents. An apparent correlation between the conformation and the biological properties of a series of β-endorphin analogues has suggested to us that some features of the conformation in organic solvents might also apply to the conformation of the molecule adopted at the functional receptor site.

We think that the way followed by us is one of the few possible approaches to uncovering the highly complex relationship between the structure and biological activity of opioid peptides.

2. Circular Dichroism Studies

Chiroptical methods, optical rotatory dispersion (ORD), and CD spectroscopy have been used for a long time to evaluate protein secondary structure (Moffitt and Yang, 1956; Shechter and Blout, 1964; Carver *et al.,* 1966). Because of the relative simplicity of CD spectra, at present this latter method is used almost exclusively.

It is generally accepted that the three main types of protein conformation are the α helix, the pleated sheet or β structure, and the unordered or aperiodic conformation. As proven by X-ray diffraction studies these three standard conformations occur in different ratios in the folded structure of protein molecules. Early chiroptical studies showed that the α helix could also exist in peptides containing as few as 5–6 amino acid residues (Goodman *et al.,* 1962). It has been a general practice to examine the CD spectra of proteins and polypeptides as a linear combination of the optical properties of α helices, β structures, and aperiodic regions.

There are two commonly used simple procedures for the evaluation of protein and polypeptide secondary structures from CD spectra. Percentage amounts of the three standard conformations can be calculated from the data published by Greenfield and Fasman (1969), which are based on the CD spectra of poly-L-lysine or poly-L-glutamic acid at different pH values. In contrast, the method of Rosenkranz and Scholtan (1971) is based on the CD spectra of poly-L-lysine in aqueous solution in the absence of salts and on the spectra of poly-L-serine in the presence of salts. The reliability of the calculated α helix, β structure, and unordered conformation of any particular polypeptide or protein may depend on the choice of the reference system. Deviation between the calculated and X-ray diffraction data may be due to the presence of structures other than the three standard conformations, e.g., 3_{10} helix or parallel β structure, in the proteins. In addition, in the far-ultraviolet (UV) region, the CD bands of the peptide bond may overlap with the far-UV bands of aromatic side chains. The extent of this

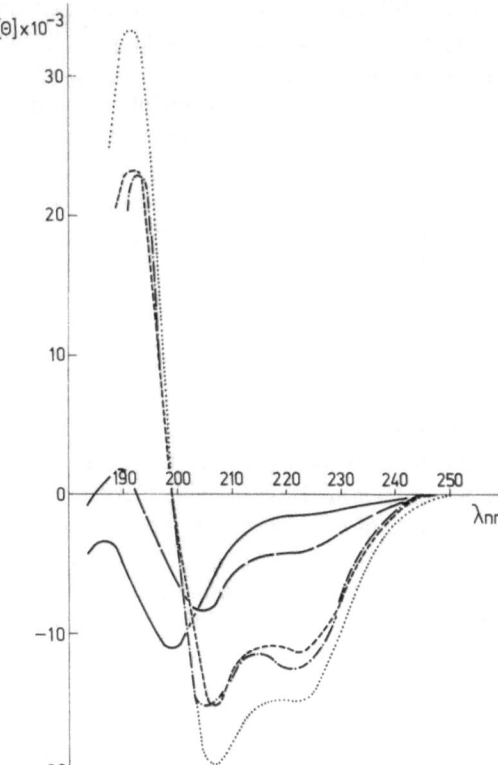

Figure 2. CD spectra of porcine β-
endorphin in mixtures of TFE–water.
TFE (· · · · ·), 60% TFE (- - - -),
40% TFE (–·–), 15% TFE (— —),
water (——). [Taken from Hollósi *et
al.*, 1977.]

contribution is determined by the ratio of aromatic to nonaromatic amino
acids. This confusing effect will also be more prominent in protein/poly-
peptide displaying low α-helix contents.

The first CD spectrum of β-endorphin in the region dominated by
amide bond absorption was reported by Hollósi *et al.* (1977). In water, the
appearance of strong negative band near 200 nm and the absence of the
double minimum at 207 and 222 nm (Fig. 2) suggest that the conformation
of β-endorphin is mostly unordered. Hollósi *et al.* (1977) have proposed,
however, that the weak but definite CD extremum at 222 nm may indicate
a small amount of α helix for β-endorphin in water. In addition, the relatively weak negative ellipticity values in the CD spectrum of the peptide,
below 195 nm, may also be due to the positive contribution of some α-helical conformation. Theoretically, the latter feature of the β-endorphin CD
spectrum might also be explained by the positive contribution of the aromatic side chain chromophores rather than α helix. In view of the more
negative CD values of β-endorphin-(1 –19) (a β-endorphin fragment with

higher aromatic amino acid to amino acid ratio than β-endorphin; see Fig. 1), however, this possibility appears to be unlikely.

In contrast to the CD spectrum in water, the CD spectrum of β-endorphin in trifluoroethanol (TFE) is dominated by one positive band at 192 nm and two negative extrema at 207 and 222 nm (Fig. 2). Cotton effects at the same wavelengths and of about the same intensity are characteristic of the CD spectra of proteins and polyamino acids with a high content of α-helical conformation (Holzwarth and Doty, 1965). The high stability of the α helix in β-endorphin in TFE was also indicated by our observation that the addition of 10% trifluoroacetic acid (known to be a helix-destroying agent) to the solution failed to affect the CD spectrum (unpublished data). Furthermore, CD studies in TFE–water mixtures have shown that β-endorphin contains a considerable amount of helical order even at relatively low TFE concentrations (Fig. 2). By contrast, addition of about 20% water to the TFE solutions of the two β-endorphin fragments (Table I) resulted in an almost complete disruption of the low helical order. The methods of Greenfield and Fasman (1969) and Rosenkranz and Scholtan (1971) were applied to estimate the percentage α helix, β structure, and unordered conformation for β-endorphin and two β-endorphin fragments in water, TFE, and TFE–water mixtures (Hollósi *et al.*, 1977). The results of the quantitative analysis of the CD data are shown in Table I.

An extensive study by Yang *et al.* (1977) on β-endorphin conformation has practically confirmed our observations outlined above. Yang *et al.*

Table I. Percentage of α Helix (α), β Structure (β), and Unordered Conformation (ρ) in the Structure of Porcine β-Endorphin and Its Plasmic Fragments as Calculated from the CD Spectra[a]

Peptide	Solvent (TFE–water)	Greenfield and Fasman (1969)			Rosenkranz and Scholtan (1971)		
		α	β	ρ	α	β	ρ
β-Endorphin	100% TFE	60	0	40	55	0	70[b]
	60% TFE	40	0	60	35	0	80[b]
	30% TFE	30	0	70	30	0	85[b]
	15% TFE	10	—[c]	—[c]	10	0	100[b]
	0% TFE	5	—[c]	—[c]	<10	0	100[b]
β-Endorphin-(1–19)	100% TFE	10	—[c]	—[c]	15	0	100[b]
	50% TFE	0	—[c]	—[c]	0	<5	95
	0% TFE	0	—[c]	—[c]	<5	0	100[b]
β-Endorphin-(20–31)	100% TFE	<10	—[c]	—[c]	10	0	100[b]
	0% TFE	0	—[c]	—[c]	<5	0	100[b]

[a] Adapted from Gráf *et al.* (1977).
[b] The sum of the calculated percentage values for the different conformations appeared to be higher than 100%.
[c] Percentage values could not be calculated by this method.

(1977) also failed to detect significant secondary structure of β-endorphin in water. In addition, the CD spectrum of β-endorphin in the region of side chain absorption indicated the lack of rigid tertiary structure with aromatic groups buried within asymmetric local environments (Yang et al., 1977). Addition of either methanol or sodium dodecyl sulfate to the aqueous solutions of β-endorphin induced α-helix formation, similarly to the effect of TFE. Though Yang et al. (1977) noted that their CD data indicate a considerable amount of helical conformation of the peptide in either methanol or sodium dodecyl sulfate solutions, "perhaps as much as one-half of the molecule," they rejected the use of any quantitative analysis of the data. The authors emphasized that the probable differences between the reference ellipticity values for the helical conformation in different solutions might considerably endanger the validity of this kind of calculation. In order to evaluate the α-helix content of β-endorphin in TFE, Hollósi et al. (1977) applied the ellipticity values of poly-γ-methyl-L-glutamate in hexafluoroisopropanol* for the α helix and those of poly-γ-morpholinyl-ethyl-L-glutamamide for the unordered conformation (Parrish and Blout, 1972). This approach gave comparable values (data are not shown in this review) to those obtained by the original methods of Greenfield and Fasman (1969) and Rosenkranz and Scholtan (1971) (Table I) suggesting that the 50–60% α-helix content of β-endorphin may be valid. In any case, the percentage values calculated for the three main conformations, in particular for α helix, may certainly serve to measure relative conformational changes of β-endorphin in response to changes of the solvent composition (see Table I) or to chemical modifications of the peptide (see below).

3.   Predictive Analysis of the Secondary Structure

It is generally accepted that the conformation of proteins is largely determined by the amino acid sequence, and several methods have been worked out for the prediction of protein conformation from analysis of the amino acid sequence. The predictive model of Chou and Fasman (1974a,b; Fasman et al., 1976) is the most widely tested and accepted of the procedures proposed so far. The method utilizes α-helix, β-sheet, and β-turn conformational parameters for the 20 amino acids quantitatively determined from a statistical analysis of 29 proteins of known X-ray structure. Chou and Fasman (1974b) have claimed 70–80% accuracy for this procedure, and indeed it has been successfully used in predicting the conformation of

---

*Hexafluoroisopropanol and TFE have very similar solvent properties as shown by the practically identical CD spectra of poly-γ-methyl-L-glutamate in the two solvents (see Parrish and Blout, 1972).

several globular proteins, e.g., adenylate kinase (Schulz *et al.,* 1974) and T4 phage lysozyme (Matthews, 1975). Furthermore, a correlation has been found between the helix-forming potentials of synthetic ribonuclease-S-(1 – 20) analogues and their abilities to bind to S-protein (Dunn and Chaiken, 1975).

On the other hand, discrepancy between the experimentally determined (CD spectroscopy) and predicted conformations of some globular polypeptide hormones (somatotropin, prolactin, and lutropin) has been reported (Jibson and Li, 1979). As also emphasized by Jibson and Li (1979), this discrepancy might be due to limitations of both CD methods and Chou–Fasman predictions. Problems involved in the quantitative analysis of the CD spectra have been briefly discussed in the previous chapter. As far as the Chou–Fasman method is concerned, some ambiguities in the predictive rules are clearly apparent and well demonstrated by the three significantly different predictions of the β-lipotropin conformation by St.-Pierre *et al.* (1976), Jibson and Li (1979), and Gráf and Hollósi (1980). The ambiguities in the predictive model, however, may have physical meaning. Particular sequence portions of a flexible polypeptide chain with comparable potentials to adopt different kinds of conformations may exist, in fact, in an equilibrium state between the conformations predicted. Upon prolonged standing in aqueous solution, a conformational transition of glucagon has been observed from a predominantly α-helical structure to a predominantly β-sheet structure (Moran *et al.,* 1977). On the other hand, X-ray studies showed the presence of mainly the helical conformation in glucagon crystals (Sasaki *et al.,* 1975). Thus, the actual conformation of a small, nonrigid peptide is highly dependent on its local environment.

While it is easy to see that the secondary structure predictions with parameters derived from the structure of globular proteins cannot be directly applied to the conformation of small, linear peptides in aqueous solutions, several lines of evidence suggest that the Chou–Fasman predictions do bear some relevance to the conformation of such peptides in environments that promote ordered structure. For both adrenocorticotropic hormone (Greff *et al.,* 1976) and β-lipotropin (Gráf and Hollósi, 1980), good correlation has been found between the α-helix contents estimated from the CD spectra in TFE and those predicted by the Chou-Fasman method.

The predictive analysis of the secondary structure of β-endorphin has been performed by two groups of investigators (Gráf *et al.,* 1977; 1980a; Jibson and Li, 1979; Li *et al.,* 1980), and the results are compared in Table II. The essential difference between the two predictions is due to the different views of the authors. While the analysis of Jibson and Li (1979) is based strictly on the predictive rules of Chou and Fasman (1974b) using the new conformational parameters of Fasman *et al.* (1976), Gráf *et al.*

Table II. Predictive Analysis of Secondary Structure in Porcine β-Endorphin According to Chou and Fasman (1974b)

Residue[a]	P_a[b]	P_β[b]	Gráf et al. (1977, 1980a)	Jibson and Li (1979)
Tyr[1]	0.69 (b)	1.47 (H)	⎤	
Gly	0.57 (B)	0.75 (b)	⎥ β turn	
Gly	0.57 (B)	0.75 (b)	⎦	
Phe	1.13 (h)	1.38 (H)		
Met	1.45 (H)	1.05 (h)		
Thr	0.83 (i)	1.19 (h)		α helix
Ser	0.77 (i)	0.75 (b)		
Glu	1.51 (H)	0.37 (B)	random	
Lys	1.16 (h)	0.74 (b)		
Ser[10]	0.77 (i)	0.75 (b)		
Gln	1.11 (h)	1.10 (h)		
Thr	0.83 (i)	1.19 (h)		
Pro	0.57 (B)	0.55 (b)		
Leu	1.21 (H)	1.30 (h)		
Val	1.06 (h)	1.70 (H)		
Thr	0.83 (i)	1.19 (h)		
Leu	1.21 (H)	1.30 (h)		
Phe	1.13 (h)	1.38 (H)		β sheet
Lys	1.16 (h)	0.74 (b)		
Asn[20]	0.67 (b)	0.89 (i)		
Ala	1.42 (H)	0.83 (i)	α helix	
Ile	1.08 (h)	1.60 (H)		
Val	1.06 (h)	1.70 (H)		
Lys	1.16 (h)	0.74 (b)		
Asn	0.67 (b)	0.89 (i)		
Ala	1.42 (H)	0.83 (i)		
His	1.00 (I)	0.87 (i)		
Lys	1.16 (h)	0.74 (b)		
Lys	1.16 (h)	0.74 (b)		β turn
Gly[30]	0.57 (B)	0.75 (b)		
Gln	1.11 (h)	1.10 (h)		

[a] Amino acids are listed in the order of occurrence in the β-endorphin structure.
[b] α helix (P_a) and β structure (P_β) potentials of the residues (Fasman et al., 1976); in parentheses, their assignments as formers (H, h), breakers (B, b), and indifferent (I, i) for helical and β-sheet regions.

(1977, 1980a) confronted the results of the predictive analysis with the CD data (Hollósi et al., 1977), and of the possible conformations consistent with the predictive rules, preference has been given to an alternative that was supported by the CD measurements. Thus, our predicted secondary structure of β-endorphin (Table II) specifically refers to the conformation of the peptide adopted in TFE rather than to secondary structure in general.

Region 13–29 of β-endorphin (Table II) represents a good example of how ambiguous the empirical rules are in predicting conformation. This region of the peptide chain satisfies all the Chou–Fasman conditions for both α helix and β sheet. The segment contains at least two clusters of helical residues (regions 14–19 and 21–26) and one cluster of strong β-forming residues at positions 14–18 to nucleate either helical or β-sheet conformations. As to the latter possibility, analysis of region 14–18 extended with 1 residue on the N-terminal side (Pro[13]) and 3 residues on the C-terminal side (residues 19–21) shows, however, that near-neighbor interactions may hinder β-structure formation in this particular fragment (Gráf *et al.*, 1980a). The helical and β-structure potentials for segment 13–29, obtained by averaging the P_α and P_β values of the individual residues (see Table II), are practically identical: $\langle P_\alpha \rangle = 1.08 \sim \langle P_\beta \rangle = 1.06$. More importantly, the β conformation could not be detected for β-endorphin in either water or TFE (Hollósi *et al.*, 1977), in methanol and sodium dodecyl sulfate solutions (Yang *et al.*, 1977). On the other hand, the 55–60% α-helix content of β-endorphin in TFE (Hollósi *et al.*, 1977) is just accounted for by this predicted helical stretch between residues 13 and 29 (Table II).

To resolve the apparent contradiction between the relatively high α-helix content of β-endorphin and the very low α-helix contents of its constituent fragments in TFE (see Tabel I), stabilizing intramolecular interactions have been postulated between the C-terminal helical and N-terminal nonhelical parts of the molecule (Hollósi *et al.*, 1977; Gráf *et al.*, 1977). Based on different considerations, originally Met[5] has been proposed to participate in nonpolar bonding with some C-terminal residues, likely Lys[28] and/or Lys[29] (Gráf *et al.*, 1977).

Recent studies of a series of C-terminal deletion analogues of β-endorphin (Gráf *et al.*, 1980b) appear to support our proposition that a C-terminal sequence portion of β-endorphin adopts the α-helical conformation in TFE. The removal of 2, 4, and 6 residues from the C-terminus of porcine β-endorphin failed to affect the percentage α-helix content of the polypeptide in TFE (Table III). These data are consistent with the prediction of α helix for residues 13–29. Further shortening β-endorphin-(1–25) with 2, 4, and 6 residues, however, resulted in a considerable and more or less gradual decrease of the α-helix content, indicating the crucial role of some residues within region 20–25 in stabilizing the helical conformation of β-endorphin. The CD spectra of β-endorphin fragments containing residues 1–31, 1–30, 1–28, and 1–21, measured in methanol (Li, 1979), are in fair agreement with our results. Thus, our previously proposed model for the β-endorphin

Table III.   Percentage α-Helix Content in the Structure of
Porcine β-Endorphin Analogues as Measured in TFE[a]

Peptide	α—Helix content (%)	
	Experimental[b]	Theoretical[c]
β-Endorphin	55 (57,52,55)	55
β-Endorphin-(1–29)	50 (52,49,50)	59
β-Endorphin-(1–27)	44 (46,44,42)	56
β-Endorphin-(1–25)	50 (49,52,50)	52
β-Endorphin-(1–23)	29 (31,27,30)	48
β-Endorphin-(1–21)	17 (19,15,17)	43
β-Endorphin-(1–19)	13 (15,12,11)	37

[a] Adapted from Gráf et al. (1980b).
[b] Average values of measurements with three different solutions. Individual values are in parentheses.
[c] 100 times the ratio of the number of theoretically helical residues in the fragment to the total number of residues (see Table II).

conformation in TFE (Gráf et al., 1977) has been modified by suggesting that the helical segment of the molecule between residues 13 and 29 may be stabilized by hydrophobic bonding of some N-terminal residue(s) to some residues at positions 20–25 rather than to $Lys^{28,29}$. Similar α-helix-stabilizing interactions have been deduced from the X-ray structure of globular proteins and also included in a proposed model of the conformation of secretin in solution (Bodanszky et al., 1969).

4.   Correlation Between the Solution Conformation and Biological Activity

All the studies on the conformation of β-endorphin in TFE (Hollósi et al., 1977), methanol and sodium dodecyl sulfate solutions (Yang et al., 1977), and lipid solutions (Wu et al., 1979), plainly or unspokenly, have addressed the question of the biologically active conformation of the peptide. The common initiative of all these attempts has been that some nonpolar solvents mimicking the natural lipophilic environment of the receptor membrane may induce a conformational change in the molecule analogous to that occurring upon interaction with the receptor. In their critical paper even Wu et al. (1979) noted that "the conformation of the complex between human β-endorphin and lipids may be related to the opiate-like function of this peptide hormone."

There has been considerable discussion concerning the receptor-bound conformation of polypeptide hormones displaying nonrigid, flexible conformations in solution (Schwyzer, 1977; Burgen *et al.*, 1975; Moran *et al.*, 1977). The "zipper" model of receptor binding (Burgen *et al.*, 1975) presumes a high degree of cooperativity in the binding process, i.e., a series of mutual conformational adjustments of both the ligand and receptor. Consequently, the receptor-bound conformation may not have common features with the conformation of the peptide in any solution. The opposite proposal is that the preferred conformation in solution would bind with the greatest affinity to the receptor site (Moran *et al.*, 1977). It is becoming increasingly apparent that the structure–function relationship data on the polypeptide hormones cannot be explained in terms of one common receptor-binding process and the peptides may interact with their receptors by mechanisms intermediate between the two main theoretical models. Due to the lack of highly purified receptor proteins, only very few possibilities are available to consider in predicting the receptor-bound conformation of the polypeptides. In our opinion, an indirect but promising approach to the problem is to correlate the biological activity of various peptide analogues to their conformation in different solutions.

The choice of proper bioassay for β-endorphin involves several problems. Two main classes of assays have been used to study structure–function relationships among the opioid peptides: *in vivo* assays for analgesia, and *in vitro* test systems (receptor binding, guinea pig ileum, mouse vas deferens, and rat vas deferens assays). Analgesic activity has been found to be a more or less unique property of β-endorphin (Loh *et al.*, 1976; Gráf *et al.*, 1976b), mediated primarily by the so-called μ receptors (Lord *et al.*, 1977). Thus, analgesia seems to offer an ideal pharmacological model to investigate the conformational requirements of a peptide for interaction with the μ receptors. It is well known, however, that the *in vivo* action of a polypeptide, like the analgesic effect of β-endorphin, is the result of several factors including transport properties and metabolic stability in addition to the receptor binding affinity of the molecule. The lack of a strict correlation between opiate receptor binding activity using tritiated β-endorphin as a primary ligand and the analgesic potency of synthetic β-endorphin analogues has recently been reported (Li *et al.*, 1980). As to the *in vitro* model systems, neither the brain membrane preparations (Lord *et al.*, 1977; Chang and Cuatrecasas, 1979; Ferrara *et al.*, 1979) nor guinea pig ileum and mouse vas deferens contain homogeneous receptor populations specific for only one of the natural ligands (Rónai *et al.*, 1977; Lord *et al.*, 1977). On the other hand, Rónai *et al.* (1977) and Lord *et al.* (1977) have suggested that the potency ratio of opiates and opioid peptides in guinea pig ileum and mouse vas deferens may serve to inidicate their relative affinity to the μ receptor. More recently, Schulz *et al.* (1979) have recognized that rat vas deferens has receptors selective for β-endorphin.

4.1. Effect of Residues 20-31 on the Conformation and Biological Activity

A comparison of some biochemical and biological properties of β-endorphin-(1-19) and β-endorphin (Table IV) (Gráf et al., 1977; Gráf, 1979) demonstrates the dramatic effect of residues 20-31 on the helical potential, enzyme resistance, and biological activities of the molecule. It has been suggested that the increased resistance to aminopeptidase action and the completely changed biological profile may both be accounted for by the same kind of conformational change at the N-terminus of the polypeptide induced by some interactions with C-terminal residues (Gráf et al., 1977; Gráf, 1979). According to this assumption, this same interaction might also stabilize the α-helical conformation in TFE. Theoretically, the conformational and receptorial effects of residues 20-31 of β-endorphin may be independent of each other, the latter being due to the presence of additional receptor binding sites within this sequence portion. β-Endorphin fragments without the enkephalin portion, however, were shown to display very weak or practically no affinity for the opiate receptors in different in vitro assay systems (Gráf, 1979; Ferrara and Li, 1980). Thus, the recent finding that β-endorphin-(6-31) and β-endorphin-(20-31) inhibit morphine-induced analgesia in vivo (Lee et al., 1980) cannot be explained simply in terms of opiate receptor binding.

To further test our hypothesis that the conformation of β-endorphin in TFE may be related to the biolgically active conformation, the α-helix contents in TFE and the in vitro biological properties of a series of β-endorphin analogues with shortened C-termini have been compared (Table V) (Gráf et al., 1980b). As seen in Table V, there is an apparent parallelism between the α-helix potential and guinea pig ileum to mouse vas deferens potency ratio changes as amino acid pairs are removed from the C-terminus in a stepwise manner. The major decrease of both the α-helix content in TFE

Table IV. Some Biochemical and Biological Properties of β-Endorphin and Its Fragments

Peptide	α-Helix content[a]	Enzyme resistance[b]	GPI/MVD[c]	Analgesic effect[d]
β-Endorphin-(1-5), Met-enkephalin	0	0	0.04	1
β-Endorphin-(1-19)	10	20	0.04	5
β-Endorphin	60	60	0.84	2500

[a] Percentage of α helix as determined by CD spectroscopy in TFE (Hollósi et al., 1977).
[b] Percentage of intact peptide in a 3-h aminopeptidase M hydrolysate (Gráf et al., 1977).
[c] Ratio of ID_{50} values determined in mouse vas deferens (MVD) and guinea pig ileum (GPI) (Rónai et al., 1977).
[d] Reciprocal value of ED_{50} as determined in the tail flick test after central administration of the peptide (Gráf et al., 1976b).

Table V.　Correlation Between the α-Helix Potential and Some Biological Properties of β-Endorphin Analogues with Shortened C-Termini

Peptide	α-Helix content[a]	GPI/MVD[b]	Rat vas deferens[c]
β-Endorphin	55	0.81	100
β-Endorphin-(1–29)	50	1.05	66
β-Endorphin-(1–27)	44	0.57	152
β-Endorphin-(1–25)	50	0.75	57
β-Endorphin-(1–23)	29	0.53	60
β-Endorphin-(1–21)	17	0.36	4
β-Endorphin-(1–19)	13	0.24	0.1

[a] Percentage content as determined by CD spectroscopy in TFE (Gráf *et al.*, 1980b).
[b] Ratio of ID_{50} vaues determined in mouse vas deferens (MVD) and guinea pig ileum (GPI).
[c] Relative potency.

and the potency ratio on guinea pig ileum and mouse vas deferens occurred in response to the removal of residues 25–20. Residues 20–23 within the same section of β-endorphin appeared to be essential for the biological activity on rat vas deferens (Albert Herz, personal communication). The lack of strict correlation between the physicochemical and biological data is clearly apparent and certainly indicates specific differences of the β-endorphin conformation in a secondary-structure-promoting solvent and at the various receptor sites. This, however, is consistent with our proposed model of β-endorphin–receptor interaction (see below).

4.2. Conformational and Biological Role of Met⁵

Extensive structure–activity studies on enkephalins have shown that the tetrapeptide Tyr-Gly-Gly-Phe has full intrinsic activity in different *in vitro* systems and that the C-terminal residue Met/Leu may function as an additional receptor binding site (see Bajusz's review in this monograph). Supporting this view, the Met^5–Pro-NH$_2$ (Bajusz *et al.*, 1976, 1977) and Met^5–D-Met substitutions (Frederickson, 1977) in the D-Met/D-Ala² enkephalin analogues considerably improved the biological properties of the pentapeptide. In contrast, the same substitutions in the β-endorphin structure led to a dramatic loss of the *in vitro* and *in vivo* biological activities (Yamashiro *et al.*, 1978). The differential biological effect of replacements of Met^5 in enkephalin and β-endorphin is in agreement with our previous proposal that Met^5 in β-endorphin may not be directly involved in receptor binding. Instead, it may participate in nonpolar bonding with some distant residues of the same molecule to stabilize a newly formed binding site at the receptor surface (Gráf, 1979).

Recently, it has been shown that modifications of Met^5 in β-endorphin

Table VI.   Effect of Modifications of Met[5] on the α-Helix Potential and
Biological Activity in Guinea Pig Ileum[a]

Peptide	α-Helix content[b]	Biological activity[c]
β-Endorphin	59	100
Met(O)[5]-β-endorphin	46	25
Met(O$_2$)[5]-β-endorphin	39	22
Met(CH$_2$COOH)[5]-β-endorphin	38	4
D-Met[5]-β-endorphin	37	0
D-Leu[5]-β-endorphin	37	0

[a] Adapted from Gráf et al. (1980a).
[b] Determined in TFE by CD spectroscopy.
[c] Relative potency.

result in a seemingly concomitant decrease of the biological activity in guinea pig ileum and the α-helix content in TFE (Table VI) (Gráf et al., 1980a). Again, this correlation suggests that Met[5] in β-endorphin may have an analogous role in stabilizing the biologically active conformation in guinea pig ileum and the conformation adopted in TFE.

4.3. Prediction of the Biologically Active Conformation

The fact that β-endorphin is able to interact with at least two kinds of opiate receptors in some peripheral tissues and the brain (Lord et al., 1977; Smith and Simon, 1980; Chang et al., 1980) is clearly a source of confusion when the structural and conformational requirements for the β-endorphin activities are discussed in the literature. Evidently, the binding of β-endorphin to the enkephalin or δ receptors (Lord et al., 1977) is due to the presence of the enkephalin sequence in the molecule. However, the C-terminal portions, residues 6–19 and also 20–31 might sterically hinder binding to this receptor site as suggested by some structure–activity data in an in vitro δ-receptor model, mouse vas deferens (Rónai et al., 1977), and also on brain membranes (Lord et al., 1977).

The biological role of residues 20–31 in β-endorphin, when tested in assays measuring predominantly μ-receptor function,* is opposite. Structure–activity studies with C-terminal deletion analogues have shown that the positive contribution of the C terminal region of the molecule to either receptor binding (Ferrara and Li, 1980) or to the potency in guinea pig ileum (Gráf et al., 1980b) is due to residues 20–28. In a similar approach, the most essential residues for activity in rat vas deferens have been localized at positions 20–23 of the β-endorphin structure (A. Herz, personal

---

*Guinea pig ileum (Lord et al., 1977), receptor binding assay with β-endorphin ligand (Ferrara et al., 1979), analgesia tests (Gráf et al., 1976b).

communication). On the other hand, C-terminal fragments of β-endorphin including region 20–31 but lacking residues 1–5 are practically inactive in the *in vitro* models. Thus, both the N-terminal (residues 1–5) and the C-terminal regions (residues 20–28) of the peptide chain appear to be intimately involved in binding to the μ receptor, which is likely the functional receptor site of β-endorphin in the brain (Fig. 3). Our model of β-endorphin–receptor interaction is substantially different from that of Li *et al.* (1980) who have suggested that the active site for the μ receptor is located at the C-terminal region, residues 20–31, of β-endorphin.

We believe that the conformation of β-endorphin at its functional receptor site is a result of a series of conformational rearrangements of the peptide upon interaction with the membrane, to which process the intrinsic conformational properties of β-endorphin significantly contribute. This view is strongly supported by the biological importance and the secondary structure stabilizing effect in organic solvents of the same entities, Met[5] and residues 22–25 of the molecule. We suggest that the conformation of β-endorphin at the receptor and in organic solvents, clearly requiring the integrity of the whole polypeptide chain, share at least one structural feature: intramolecular interaction between some N-terminal (most likely Met[5]) and C-terminal (most likely Ile[22]-Val-Lys-Asn[25]) residues (Fig. 3). Two different but related features of such a three-dimensional structure are the α-helical conformation in organic solvents and the proper orientation of

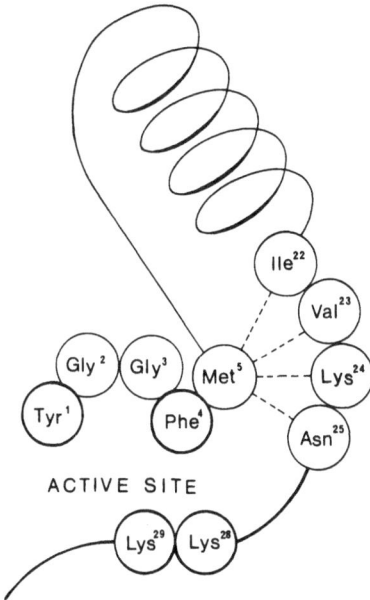

Figure 3. A proposed model of the β-endorphin conformation at the μ-receptor site.

the side chains of Tyr¹, Phe⁴, and some additional, sequentially distant residue(s), to fit into the receptor site. After this review was originally submitted, direct physicochemical evidence was provided for the existence of a tertiary structure of β-endorphin in aqueous solutions (Nicolas *et al.,* 1981). Difference spectra generated during thermolysin digestion of camel and porcine β-endorphin at both pH 8.2 and 6.5 indicated rapid blue-shifting of the near-ultraviolet absorption bands of the single N-terminal tyrosine in the molecules. A similar change was not observed for the N-terminal tyrosine in Met-enkephalin when digested under similar conditions. These results indicate that enzymatic digestion destroys or alters some structural interaction between the N-terminal tyrosine and residue(s) within the C-terminal segment of the molecule. Peptide mapping of the digests as a function of time suggests that the cleavage of the Ala²¹-Ile²² bond produces most of the observed effect (Nicolas *et al.,* 1981). These recent data are consistent with our proposed model of the β-endorphin conformation in organic solvents and probably also at the receptor site (Fig. 3).

REFERENCES

Bajusz, S., Rónai, A. Z., Székely, J. I., Dunai-Kovacs, Z., Berzetei, I., and Gráf, L., 1976, Enkephalin analogs with enhanced opiate activity, *Acta Biochim. Biophys. Acad. Sci. Hung.* **11**:305.

Bajusz, S., Rónai, A. Z., Székely, J. I., Gráf, L., Dunai-Kovacs, Z., and Berzetei, I., 1977, A superactive antinociceptive pentapeptide, (D-Met², Pro⁵)-enkephalinamide, *FEBS Lett.* **76**:91.

Bloom, F., Battenberg, E., Rossier, J., Ling, N., and Guillemin, R., 1978, Neurons containing β-endorphin in rat brain exist separately from those containing enkephalin: Immunocytochemical studies, *Proc. Natl. Acad. Sci. USA* **75**:1591.

Bodanszky, A., Ondetti, M. A., Mutt, V., and Bodanszky, M., 1969, Synthesis of secretin, IV. Secondary structure in a miniature protein, *J. Am. Chem. Soc.* **91**:944.

Bradbury, A. F., Smyth, D. G., Snell, C. R., Birdsall, N. J. M., and Hulme, E. C., 1976a, C-fragment of lipotropin has a high affinity for brain opiate receptors, *Nature (London)* **260**:793.

Bradbury, A. F., Smyth, D. G., and Snell, C. R., 1976b, Biosynthetic origin and receptor conformation of methionine enkephalin, *Nature (London)* **260**:165.

Burgen, A. S. V., Roberts, G. C. K., and Feeney, J., 1975, Binding of flexible ligands to macromolecules, *Nature (London)* **253**:753.

Carver, J. P., Shechter, E., and Blout, F. R., 1966, Analysis of the optical rotatory dispersion of polypeptides and proteins. V. A comparison of methods, *J. Am. Chem. Soc.* **88**:2562.

Chang, K.-J., and Cuatrecasas, P., 1979, Multiple opiate receptors, *J. Biol. Chem.* **254**:2610.

Chang, K.-J., Hazum, E., and Cuatrecasas, P., 1980, Possible role of distinct morphine and enkephalin receptors in mediating actions of benzomorphan drugs, *Proc. Natl. Acad. Sci. USA* **77**:4469.

Chou, P. Y., and Fasman, G. D., 1974a, Conformational parameters for amino acids in helical, β-sheet, and random coil regions calculated from proteins, *Biochemistry* **13**:210.

Chou, P. Y., and Fasman, G. D., 1974b, Prediction of protein conformation, *Biochemistry* **13**:222.

Cseh, G., Barát, E., Patthy, A., and Gráf, L., 1972, Studies on the primary structure of human β-lipotropic hormone, *FEBS Lett.* **21**:344.

Dunn, B. M., and Chaiken, I. M., 1975, Relationship between α-helical propensity and formation of the ribonuclease-S complex, *J. Mol. Biol.* **95**:497.

Fasman, G. D., Chou, P. Y., and Adler, A. J., 1976, Prediction of the conformation of the histones, *Biophys. J.* **16**:1201.

Ferrara, P., and Li, C. H., 1980, β-Endorphin: Characteristics of binding sites in rabbit spinal cord, *Proc. Natl. Acad. Sci. USA* **77**:5746.

Ferrara, P., Houghten, R., and Li, C. H., 1979, β-Endorphin: Characteristics of binding sites in the rat brain, *Biochem. Biophys. Res. Commun.* **89**:786.

Frederickson, R. C. A., 1977, Enkephalin pentapeptides, *Life Sci.* **21**:23.

Goodman, M., Listowsky, I., and Schmitt, E. E., 1962, Conformational aspects of polypeptides. V. Molar rotational model compounds for poly-γ-methyl-L-glutamate, *J. Am. Chem. Soc.* **84**:1288.

Gráf, L., 1979, Chemistry and biochemistry of pituitary endorphins, in: *Advances in Pharmacology and Therapeutics* (J. Jacob, ed.), pp. 3–14, Pergamon Press, Elmsford, N.Y.

Gráf, L., and Hollósi, M., 1980, Substrate conformation directs selective enzymic cleavage of β-lipotropin to β-endorphin, *Biochem. Biophys. Res. Commun.* **83**:1089.

Gráf, L., Barát, E., Cseh, G., and Sajgó, M., 1971, Amino acid sequence of porcine β-lipotropic hormone, *Biochim. Biophys. Acta* **229**:276.

Gráf, L., Barát, E., and Patthy, A., 1976a, Isolation of a COOH-terminal β-lipotropin fragment (residues 61–91) with morphine-like analgesic activity from porcine pituitary glands, *Acta Biochim. Biophys. Acad. Sci. Hung.* **11**:121.

Gráf, L., Székely, J. I., Rónai, A. Z., Dunai-Kovacs, Z., and Bajusz, S., 1976b, Comparative study on analgesic effect of Met-enkephalin and related lipotropin fragments, *Nature (London)* **263**:240.

Gráf, L., Cseh, G., Barát, E., Rónai, A. Z., Székely, J. I., Kenessey, A., and Bajusz, S., 1977, Structure–function relationships in lipotropins, *Ann. N.Y. Acad. Sci.* **297**:63.

Gráf, L., Hollósi, M., Patthy, A., Berzétei, I., and Rónai, A. Z., 1980a, Decrease of α-helix potential and biological activity of β-endorphin in response to modifications of Met^5, *Neuropeptides* **1**:47.

Gráf, L., Hollósi, M., Barna, I., Hermann, I., Borvendég, J., and Ling, N., 1980b, Probing the biologically and immunologically active conformation of β-endorphin: Studies on C-terminal deletion analogs, *Biochem. Biophys. Res. Commun.* **95**:1623.

Greenfield, N. J., and Fasman, G. D., 1969, Computed circular dichroism spectra for the evaluation of protein conformation, *Biochemistry* **8**:4108.

Greff, D., Toma, F., Fermandjian, S., Löw, M., and Kisfaludy, L., 1976, Conformational studies of corticotropin$_{1-32}$ and constitutive peptides by circular dichroism, *Biochim. Biophys. Acta* **439**:219.

Hollósi, M., Kajtár, M., and Gráf, L., 1977, Studies on the conformation of β-endorphin and its constituent fragments in water and trifluoroethanol by CD spectroscopy, *FEBS Lett.* **74**:185.

Holzwarth, G., and Doty, P., 1965, The ultraviolet circular dichroism of polypeptides, *J. Am. Chem. Soc.* **87**:218.

Hughes, J., Smith, T. W., Kosterlitz, H. W., Fothergill, L. A., Morgan, B. A., and Morris,

H. R., 1975, Identification of two related pentapeptides from the brain with potent opiate agonist activity, *Nature (London)* **258**:577.

Jibson, M. D., and Li, C. H., 1979, Secondary structure prediction of anterior pituitary hormones, *Int. J. Peptide Protein Res.* **14**:113.

Lee, N. M., Friedman, H. J., Leybin, L., Cho, T. M., Loh, H. H., and Li, C. H., 1980, Peptide inhibitor of morphine- and β-endorphin-induced analgesia, *Proc. Natl. Acad. Sci. USA* **77**:5525.

Lewis, R. V., Stern, A. S., Kimura, S., Rossier, J., Stein, S., and Udenfriend, S., 1980, An about 50,000-dalton protein in adrenal medulla: A common precursor of (Met)- and (Leu)enkephalin, *Science* **208**:1459.

Li, C. H., 1979, β-Endorphin: Aspects of structure–activity relationships by synthetic approach, in: *Peptides: Structure and Biological Function* (E. Gross and J. Meienhofer eds.), pp. 823–833, Pierce Chemical Company.

Li, C. H., and Chung, D., 1976, Isolation and structure of an untriakontapeptide with opiate activity from camel pituitary glands, *Proc. Natl. Acad. Sci USA* **73**:1145.

Li, C. H., Yamashiro, D., Tseng, L.-F., Chang, W.-C., and Ferrara, P., 1980, β-Endorphin omission analogs: Dissociation of immunoreactivity from other biological activities, *Proc. Natl. Acad. Sci. USA* **77**:3211.

Loh, H. H., Tseng, L.-F., Wei, E., and Li, C. H., 1976, β-Endorphin is a potent analgesic agent, *Proc. Natl. Acad. Sci. USA* **73**:2895.

Lord, J. A. H., Waterfield, A. A., Hughes, J., and Kosterlitz, H. W., 1977, Endogenous opioid peptides: Multiple agonists and receptors, *Nature (London)* **267**:495.

Mains, R. E., and Eipper, B. A., 1978, Studies on the common precursor of ACTH and endorphin, in: *Endorphins '78* (L. Gráf, M. Palkovits, and A. Z. Rónai, eds.), pp. 79–126, Akadémiai Kiadó, Budapest.

Matthews, B. W., 1975, Comparison of the predicted and observed secondary structure of T4 phage lysozyme, *Biochim. Biophys. Acta* **405**:442.

Moffitt, W., and Yang, J. T., 1956, The optical rotatory dispersion of simple polypeptides, I., *Proc. Natl. Acad. Sci. USA* **42**:596.

Moran, E. C., Chou, P. Y., and Fasman, G. D., 1977, Conformational transitions of glucagon in solution: The $\alpha-\beta$ transition, *Biochem. Biophys. Res. Commun.* **77**:1300.

Nicholas, P., Bewley, T. A., Gráf, L., and Li, C. H., 1981, β-Endorphin: Demonstration of a tertiary structure in aqueous solutions, *Proc. Natl. Acad. Sci. USA* **78**:7290.

Parrish, J. R., and Blout, E. R., 1972, The conformation of poly-L-alanine in hexafluoroisopropanol, *Biopolymers* **11**:1001.

Rónai, A. Z., Gráf, Székely, J. I., Dunai-Kovacs, Z., and Bajusz, S., 1977, Differential behaviour of LPH-(61–91)-peptide in different model systems: Comparison of the opioid activities of LPH-(61–91)-peptide and its fragments, *FEBS Lett.* **74**:182.

Rosenkranz, H., and Scholtan, W., 1971, Eine verbesserte Methode zur Konformationsbestimmung von helicalen Proteinen aus Messungen des Circular-dichroismus, *Hoppe-Seyler's Z. Physiol. Chem.* **352**:896.

Sasaki, K., Dockerrill, S., Adamiak, D. A., Tickle, I. J., and Blundell, T., 1975, X-Ray analysis of glucagon and its relationship to receptor binding, *Nature (London)* **257**:751.

Schulz, G. E., Barry, C. D., Friedman, J., Chou, P. Y., Fasman, G. D., Finkelstein, A. V., Lim, V. I., Ptitsyn, O. B., Kabat, E. A., Wu, T. T., Levitt, M., Robson, B., and Nagano, K., 1974, Comparison of predicted and experimentally determined secondary structure of adenylate kinase. *Nature (London)* **250**:140.

Schulz, R., Faase, E., Wüster, M., and Herz, A., 1979, Selective receptors for β-endorphin on the rat vas deferens, *Life Sci.* **24**:843.

Schwyzer, R., 1977, ACTH: A short introductory review, *Ann. N.Y. Acad. Sci.* **297**:3.

Shechter, E., and Blout, E. R., 1964, An analysis of the optical rotatory dispersion of polypeptides and proteins II, *Proc. Natl. Acad. Sci. USA* **51**:794.

Smith, J. R., and Simon, E. J., 1980, Selective protection of stereospecific enkephalin and opiate binding against inactivation by *N*-ethylmaleimide: Evidence for two classes of opiate receptors, *Proc. Natl. Acad. Sci. USA* **77**:281.

St.-Pierre, S., Gilardeau, C., and Chrétien, M., 1976, Circular dichroism studies of sheep β-lipotropic hormone, *Can. J. Biochem.* **54**:992.

Watson, S. J., Akil, H., Richard, C. W., III, and Barchas, J. D., 1978, Evidence for two separate opiate peptide neuronal systems, *Nature (London)* **275**:226.

Wu, C.-S. C., Lee, N. M., Loh, H. H., Yang, J. T., and Li, C. H., 1979, β-Endorphin: Formation of α-helix in lipid solutions, *Proc. Natl. Acad. Sci. USA* **76**:3656.

Yamashiro, D., Li, C. H., Tseng, L.-F., and Loh, H. H., 1978, β-Endorphin: Synthesis and analgesic activity of several analogs modified in positions 2 and 5, *Int. J. Peptide Protein Res.* **11**:251.

Yang, J. T., Bewley, T. A., Chen, G. C., and Li, C.H., 1977, Conformation of β-endorphin in methanol and sodium dodecyl sulfate solutions, *Proc. Natl. Acad. Sci. USA* **74**:3235.

Zetta, L., Cabassi, F., Tomatis, R., and Guarneri, M., 1979, 270-MHz ^{1}H nuclear-magnetic-resonance study of Met-enkephalin in water, *Eur. J. Biochem.* **95**:367.

3

Studies on the Conformation of Neuropeptides

Karl Lintner, Flavio Toma, François Piriou, Pierre Fromageot, and Serge Fermandjian

1. Introduction

A significant part of the effort directed at a better understanding of the principles governing peptide hormone action (release, transport, receptor interaction, metabolism) is devoted to studies of their precise three-dimensional structure. Most biologically active peptides have been investigated with varying intensity, thoroughness, and follow-up as to their solution conformation and its possible significance for biological activity.

So-called neuropeptides are not an exception, the more so as nowadays most peptides initially found to act in the periphery (blood vessels, intestines, uterus) are also detected in neurons and brain tissue thanks to the sensitive competitive protein binding assays such as RIA and others. Vice versa, peptides that originally were found in the brain (enkephalines, endorphins) and could have been considered genuine neuropeptides are now also known to be active at peripheral sites. The classification in peptides of the brain and peptides of the periphery is thus rapidly losing its meaning as there are no fundamental chemical or functional differences between them. This also holds for all aspects of conformational analysis.

What exactly is meant by conformational analysis and what can we expect to gain from it?

Karl Lintner, Flavio Toma, François Piriou, Pierre Fromageot, and Serge Fermandjian ● Service de Biochimie, Département de Biologie, Centre d'Etudes Nucléaires de Saclay, F-91191 Gif-sur-Yvette Cedex, France.

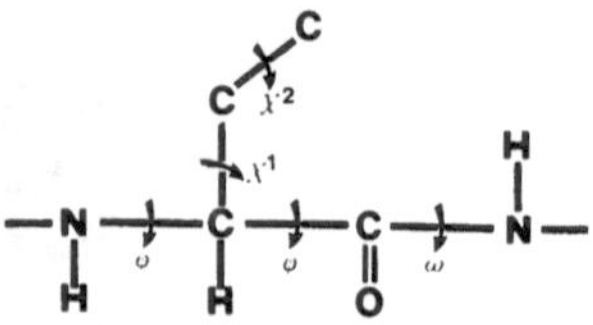

Figure 1. The peptide unit with the dihedral angles ϕ, ψ, ω, and χ.

Whereas the chemical structure of a given peptide and with it the bond lengths and bond angles are well defined, the rotational angles around the bonds (dihedral angles ϕ, ψ, ω, and χ^n; cf. Fig. 1) are not known a priori. Considerations that take into account factors of steric hindrance, van der Waals forces of attraction and repulsion, and other intramolecular interactions impose certain limits on the rotational freedom of these angles (Pullman and Pullman, 1974). Nevertheless, in a peptide the size of, say, angiotensin II with eight amino acid residues, there are a total of 49 dihedral angles. If each can adopt only six values on the average out of $360°$, we have 1.4×10^{10} or roughly 14 billion possible conformations.

Conformational analysis attempts to measure as precisely as possible the values of the dihedral angles that are *actually* adopted by the molecule, under given conditions of environment. This reference to environment is an essential restriction, as the peptide hormones in general are more or less short-chain, mostly linear compounds where the number of intramolecular interactions stabilizing a conformation may be rather small. Therefore, environmental factors (pH, ionic strength, temperature, solvent) will condition the structure of the molecule, and even in a given medium we should not expect to find one unique conformation. It is certainly the variety of protein and peptide sequences and their adaptive but well-defined flexibility that endows them with such a variety of biological action all the while retaining a high degree of specificity.

Peptide hormones act at specific target organs, at specific sites. They interact with an entity called receptor that may lie on the outer cell membrane or on the nuclear membrane, or anywhere in between. It is important that the peptide be specifically recognized and bound to its receptor before further biological effects can occur. It is this process of binding to a receptor (which may be one protein molecule, or a complex of several proteins, with or without sugar or lipid residues attached) that is conditioned by the exact alignment of the essential functional groups of the peptide.

Consider the following simplified model cases (Fig. 2):

A. The peptide possesses a number of functional groups in the side chains, some of which are destined to interact with the complementary groups in the receptor in order to assure proper binding; others provide the stimulus that triggers the biological response.

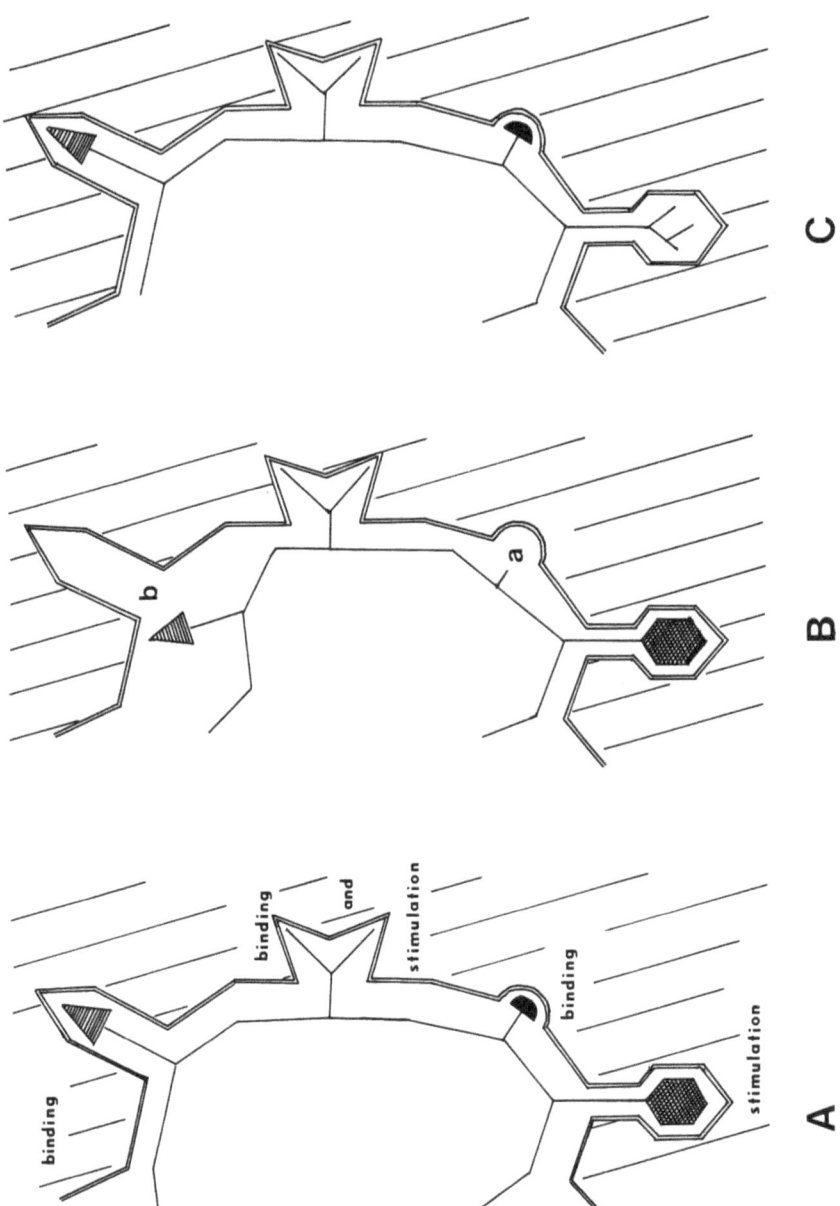

Figure 2. Simplified model of peptide–receptor interaction (see text).

In the native hormone and in structural analogues with full intrinsic activity and high affinity, this situation is ideally realized.

B. In the search for structure–activity relationships of peptides, a great number of synthetic analogues have been prepared where specific residues were either deleted or substituted by various other amino acids. Among these compounds some may turn out to have full intrinsic activity at, however, much higher doses than the hormone; in other words, their affinity for the receptor is lower. This may result from the absence of a functional group needed for binding (case a) or, if all chemically important groups are present in the molecule, from a misalignment of these groups due to a change in the peptide conformation (case b).

C. Analogues that have antagonistic properties acting as competitive inhibitors may result from modifications that do not touch the binding-related groups but affect the ones responsible for stimulation. In this case high affinity with low or no intrinsic activity may be observed.

The peptide chemist and the pharmacologist who wants to specifically design analogues of a peptide in order to potentiate, modify, or abolish its intrinsic biological activity will need to know which chemical modifications are compatible with maintaining the biologically active conformation and which will not. We are therefore looking for the correlation that ties together the terms *Structure, Conformation, Activity.*

In the following we shall examine how this aim can be approached in the case of three peptides of very different size: Thyroliberin (TRH; 3 residues), angiotensin II (A II; 8 residues), and corticotropin (ACTH; 39 residues).

2. Material and Methods

2.1. Origin of the Peptides

Our studies on the conformation of the peptide hormones are the fruit of extensive collaborations with several laboratories.

The synthesis of TRH, its isotopically enriched peptides, and most of the analogues were carried out in the laboratories of Professor K. Blaha (Prague) and Dr. J. Vičar (Olomouc) in Czechoslovakia.

Our collaboration with Drs. Khosla, Smeby, and Bumpus of the Cleveland Clinic, Ohio, concerns the study of A II. The synthesis of the innumerable analogues and isotopically enriched A II peptides and their physiological and pharmacological testing were carried out in their laboratory.

Drs. Kisfaludy and Löw of Chemical Works of G. Richter, Budapest,

Hungary, have synthesized and tested the ACTH peptides, fragments, and analogues, and have stimulated our interest in this hormone.

A few peptides (TRH and four analogues) were gifts of Dr. Studer (Hoffmann–LaRoche, Basel); [Asn^1,Val^5]-A II and ACTH-(1–24) were supplied by Dr. Riniker (Ciba-Geigy, Basel).

2.2. Methods of Investigation

Conformational analysis of peptides can be carried out by a large number of approaches. To obtain all the information necessary for reaching the goal described above, it has proven essential to combine a maximum of techniques ranging, whenever possible, from crystallographic studies of peptides in the solid state, through spectroscopic investigations of solutions, to theoretical semiempirical potential energy calculations on the isolated molecule. Each technique affords results of a specific nature; each technique also has its specific drawbacks. It is from the overlapping results and from the complementarity of the approach that a clear picture will emerge.

In our laboratory we concentrate on the spectroscopic investigation of peptides in solution, carried out by circular dichroism (CD), infrared, and nuclear magnetic resonance (NMR) spectroscopy. These techniques and their application in the field of peptides are well documented in the literature so that it will not be necessary to review them here (Leach, 1973).

It will be remarked in the following, though, that even within these techniques it is necessary to vary the approach according to the type of peptide under study.

3. ACTH

3.1. Introduction

ACTH is a peptide hormone of anterior pituitary origin and consists of 39 residues (Fig. 3). Its primary biological activity appears to be the stimulation of corticoid hormone secretion at the adrenal cortex. Recent studies on ACTH have focused more on the pathways of its biosynthesis (Roberts and Herbert, 1977) and the effort to isolate a corticotropin-releasing factor (CRF) (Sayers *et al.*, 1980) rather than on continued structure–activity relationships. However, no detailed understanding of ACTH interaction with its putative membrane receptor has emerged from the abundant data on the biological activity of ACTH fragments and analogues.

Certainly, residues 4–9 are of primary importance for steroidogenesis in *in vitro* assays; significantly, not one side chain of Met^4, Glu^5, His^6, Phe^7, Arg^8, Trp^9 turns out to be essential, irreplaceable, but rather their concerted action appears to account for the biological response (Ramachan-

1	2	3	4	5	6	7	8	9	10	11	12	13
SER	TYR	SER	MET	GLU	HIS	PHE	ARG	TRP	GLY	LYS	PRO	VAL -

14	15	16	17	18	19	20	21	22	23	24	25	26
GLY	LYS	ARG	ARG	ARG	PRO	VAL	LYS	VAL	TYR	PRO	ASN	GLY -

27	28	29	30	31	32	33	34	35	36	37	38	39
ALA	GLU	ASP	GLU	SER	ALA	GLU	ALA	PHE	PRO	LEU	GLU	PHE

Figure 3. Primary structure of human corticotropin.

dran, 1973). It has also been shown that residues 15–18 are of importance for binding affinity, that the C-terminal part of the peptide is rather unessential and seems to have mainly a stabilizing and protecting role. It is clear though that the most intense efforts of synthesizing elaborate series of analogues will not further clarify the detailed requirements of ACTH structure–function relationships.

3.2. The Conformation of ACTH in Solution

3.2.1. Earlier Studies

A few investigations of ACTH conformation in solution have been undertaken by various groups. Studies were performed on ACTH and ACTH-(1 –24) fragments in aqueous solution, where most authors did not interpret the results in terms of ordered structures. Deuterium exchange measurements, optical rotatory dispersion, CD and fluorescence spectroscopy (Edelhoch and Lippoldt, 1969; Schiller, 1972) were employed in these studies, which seemed to indicate that ACTH is of random conformation in aqueous solution at physiological pH. Only Squire and Bewley (1965) and Eisinger (1969) evoked the possibility of some α-helical content in the 1–11 segment of ACTH. In view of the progress in techniques and in the light of more recent work applying them to ACTH conformation, it is now time to revise the interpretation of the experimental results in these early studies.

Certainly, ACTH as a whole does not adopt one regular structure all along its peptide chain, such as α helix from Ser[1] to Pro[24] or Ala[32]. Short sequences of as little as four residues, however, may for instance define a β turn (Chou and Fasman, 1977; Toma *et al.*, 1980), seven or eight residues could well form an α-helix structure (Goodman *et al.*, 1971) more or less independently of what the rest of the molecule does. So it is not surprising that Schiller (1972) did not find evidence for helical conformation from fluorescence quantum yield measurements in ACTH-(1 –24): the fluorophores were in position 9 (Trp) and 21 (dansyl-lysine). α-Helical potential in ACTH, however, is located between residues 2 and 9 and is of primary importance there for efficient biological activity, as the following survey will show.

3.2.2. Statistical Sequence Analysis

It is now recognized that the three-dimensional structure of a protein or a peptide is foremost a function of its amino acid sequence. Many attempts to predict secondary structure of polypeptides based on the information content of sequence analysis alone have been published, but no single method has proven entirely successful. Nevertheless, the preferences of certain amino acids for one type of secondary structure or another can clearly be established. In this sense it is possible to apply these statistical methods also to peptides of the size of ACTH. For reasons of convenience and simplicity, the parameters of Chou and Fasman (1974, 1977) were used to delineate the conformational preferences of the residues in ACTH. The following features emerge clearly from these calculations (Löw $et\ al.$, 1975): an α-helical segment between residues 3 and 9, a β-turn structure from residues 23 to 26, and again a short helix of 4–5 residues in the 27–32 segment. The helix potential in the N-terminal part is quite pronounced (Fig. 4), strongly favored over β-pleated sheet structures, and, significantly, encompasses all residues important for biological activity.

ACTH 1·14

	Ser¹	Tyr²	Ser³	Met⁴	Glu⁵	His⁶	Phe⁷	Arg⁸	Trp⁹	Gly¹⁰	Lys¹¹	Pro¹²	Val¹³	Gly¹⁴
P_α	.77	.69	.77	1.45	1.51	1.0	1.13	.98	1.08	.57	1.16	.57	1.06	.57

$\llcorner$—— α helix ——$\lrcorner$

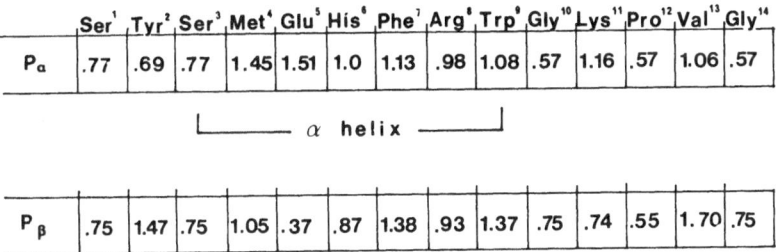

	Ser¹	Tyr²	Ser³	Met⁴	Glu⁵	His⁶	Phe⁷	Arg⁸	Trp⁹	Gly¹⁰	Lys¹¹	Pro¹²	Val¹³	Gly¹⁴
P_β	.75	1.47	.75	1.05	.37	.87	1.38	.93	1.37	.75	.74	.55	1.70	.75

	1·14	3·9
$\langle P'_\alpha \rangle$	0.95	1.13
$\langle P'_\beta \rangle$	0.96	0.96

Figure 4. Conformational parameters of ACTH-(1–14): a $\langle P_\alpha \rangle$ value > 1 indicates probable helix formation (Chou and Fasman, 1974).

3.2.3. CD Spectroscopy

The first experimental evidence confirming the helix-forming potential of this stretch of amino acids comes from CD studies on ACTH-(1 –32) and its constituent fragments (Greff *et al.*, 1976).

CD spectra were recorded in the aromatic region (320–250 nm) and in the peptide region (250–190 nm) in aqueous and in 2,2,2-trifluoroethanol (TFE) solution on the peptides ACTH-(1 –32) and ACTH-(1 –24), on the N-terminal fragments 1–14, 5–14, 5–10, and 8–14, and on the C-terminal fragments 15–32, 15–24, and 20–32. The aromatic CD signals of residues Tyr[2], Phe[7], Trp[9], and Tyr[23] in the various fragments can be used as individual probes of local variations in conformation (Greff *et al.*, 1976). The peptide region spectra (Fig. 5) are indicative of the global peptide structure. In fact, in TFE we observe CD spectra that clearly assign a certain amount of α-helix conformation to ACTH-(1 –32) and ACTH-(1 –24). Comparison of the spectra of N-terminal and C-terminal fragments shows that the helical structure is confined to the N-terminal segment, centered on the 1–14 and 5–14 peptides. The C-terminal peptides 15–32, 20–32, and 15–24 are rather devoid of regular structure.

As the CD spectra of the aqueous solution of these peptides are quite different (Greff *et al.*, 1976), suggesting a conformational transition in going from TFE to water, we examined the CD spectra of ACTH-(1 –32) in mixtures containing varying amounts of TFE and water, and we found a linear progression of the spectra: following the dichroic band at 195 nm as a function of solvent, we observe a straight line indicating that within

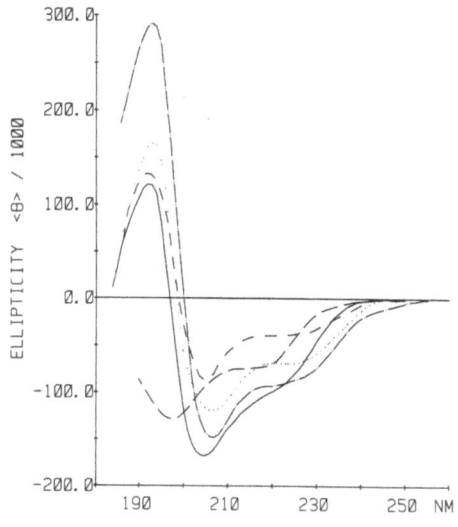

Figure 5. CD spectra of selected corticotropin fragments in TFE solution (peptide region): ———, 1–32; · · · · · ·, 1–24; –·–, 1–19; - - - -, 1–14; — —, 15–32.

these limits of environment we do not obtain pure states of conformation. Rather the α-helical content present in TFE gradually diminishes and probably converts to the extended helix conformation described in the literature (Tiffany and Krimm, 1972). The same experiment was performed on ACTH-(1–14) with exactly the same results, showing clearly that helicity is located in this section of the peptide.

In aqueous solution where some local structure seems conserved in this part of the molecule, a pH titration from pH 2 to pH 12 was performed and monitored at the wavelengths of the aromatic residues tyrosine and tryptophan. The titration curves give evidence of the spatial influence of Glu^5 and His^6 deprotonation on these aromatic side chains, an effect that would not be observed in an absolutely random structure. The energy transfer measurements by Eisinger (1969) have in fact shown that Tyr^2 and Trp^9 are separated by not more than ~ 10 Å. Arrangement of the residues 3 to 9 in helical conformation brings the side chains Tyr^2, Glu^5, His^6, and Trp^9 into positions compatible with the titration and fluorescence data.

3.2.4. Proton NMR Spectroscopy

In a way much more detailed compared to the global approach of CD, proton NMR spectroscopy confirms the above observations and goes a step further. For instance, each signal of the protons in the aromatic side chains Tyr^2, His^6, Phe^7, Trp^9, and Tyr^{23} can be followed as a function of pH and within every fragment. The behavior of the two tyrosines is clearly distinguished (Toma *et al.*, 1976). Whereas Tyr^2 signals reflect the titration of Glu^5 and His^6 in ACTH-(1–24) and ACTH-(1–32) confirming the CD results, Tyr^{23} is insensitive to them. A very different phenomenon has been observed on the signals of Tyr^{23}: the *cis–trans* isomerism of the succeeding Pro^{24} residue (Toma *et al.*, 1978). In fact, in ACTH-(1–32) and in its fragments, the peptide bond Tyr^{23}-Pro^{24} is found to be between 25 and 50% in the *cis* conformer, depending on the peptide and on the pH. In addition, the amount of *cis* isomer increases in the peptides 1–32, 15–32, and 20–32 (but not in 1–24 and 15–24) when the tyrosine side chain becomes ionized above pH 11. This suggests that around the residues Tyr^{23} and Pro^{24} there exists a special conformational feature in the longer peptides that is absent in the ones terminating in Pro^{24}. It is also significant that the other proline residues 12 and 19 in ACTH apparently do not give rise to measurable *cis–trans* isomerism. Recalling the Chou and Fasman (1977) prediction of a β turn around Tyr^{23}, Pro^{24}, Asn^{25}, Gly^{26} (Löw *et al.*, 1975) and considering the details of β-turn conformation with its defined rigidity of the backbone and side chain angles and its sensitivity to pH and solvent environment (Toma *et al.*, 1980), it seems plausible to attribute the special spectral behavior of Tyr^{23} to the presence of a β-turn folding in this site in the sequence of ACTH. Studies on the peptide Val-Tyr-Pro-Asn-Gly and derivatives are in progress, investigating the isolated sequence and its β-turn potential.

3.2.5. Carbon-13 NMR Spectroscopy

A further approach is possible by ^{13}C NMR on natural abundance and isotopically enriched compounds. The chemical shifts of carbon signals in peptides are sensitive to charge environment (pH titration), to substituent electronegativity, and, more subtly, to conformational influences. In order to study these in ACTH-(1–32) and ACTH-(1–24) it was necessary to assign most of the 171 signals in the ^{13}C spectra of ACTH-(1–32) to their respective carbons; this was achieved (with the exception of the carbonyl signals) by making use of the N- and C-terminal fragments mentioned before and by examining the specially prepared peptides [85% ^{13}C-Phe[7]]-ACTH-(1–32), [85% ^{13}C-Tyr[23]]-ACTH-(1–32), and [85% ^{13}C-Tyr[23]]-ACTH-(22–26) (Toma et al., 1981). These ^{13}C data also reflect the particularities of ACTH conformation in aqueous solution. On one hand, the increase in peptide size from ACTH-(5–14) to ACTH-(1–24) to ACTH-(1–32) causes the C^α and the C^β carbon resonances of Phe[7] to shift by -2 and -0.7 ppm, respectively, a phenomenon that has been linked to the establishment of α-helical structures in peptides (Tonelli, private communication). Certain specific chemical shifts of His[6], Glu[5], and Tyr[2] carbon resonances also indicate regular structures in this segment.

On the other hand, the chemical shift data on the residues around Pro[24] corroborate the cis–trans isomerism found by ^{1}H NMR studies and its effects on the residues Tyr[23], Asn[25], or even as far removed as Ala[27], and confirm the conspicuous absence of cis–trans isomerism at Pro[12] and Pro[19]. More detailed analysis of the ^{13}C data yields further indications in favor of a β turn at this key position in the peptide (Toma et al., 1981).

3.3. Conformation–Activity Relationship

To summarize the results described above, we can say that all physicochemical investigations of ACTH conformation—from fluorescence and ORD measurements to CD, infrared (Nabedryk-Viala et al., 1978), and ^{1}H and ^{13}C NMR spectroscopy—converge toward the conclusion that the α-helical potential of the residues 3 to 9 in the N-terminal part is well expressed in solution, more strongly in TFE, rather residually in water at neutral pH. A further element of ACTH secondary structure is the chain folding through a β turn at positions 23 to 26 for which there is also evidence in aqueous solution.

It has been stated above that no side chain of ACTH, even in the N-terminal part, is alone responsible for the expression of biological activity but that the residues 4 to 9 seem to act together. It is certain that in a well-stabilized α-helix-type structure the positioning of these side chains with respect to each other and to the backbone is quite rigidly predetermined. Perturbation of the local conformation (angles ϕ, ψ, and χ) of one residue within the helix is surely propagated along the peptide chain, thus perturbing neighboring residues and their side chain orientation.

A possibility to test this assertion is given by the series of ACTH-(1 – 19)-NH$_2$ peptides, in which successively each residue is replaced by its D-configurational isomer (Löw et al., 1980). Experimental and theoretical studies have in fact shown that all-L and all-D helices are possible and stable, but that insertion of a D residue within an L-α helix is most disruptive to this structure.

The CD spectrum of ACTH-(1 –19)-NH$_2$ in TFE was taken as a standard for comparison: it reveals an approximately 50% α-helix content in this solvent similar to what was found for the other fragments (see above). Substituting Tyr^2, then Met^4, Glu^5 one by one by their D equivalents leads to a gradual decrease of α-helical content as measured by CD; in [D-Arg^8]-ACTH-(1 –19)-NH$_2$ no more than $\sim$ 20% (a rather insignificant amount) of α helix is left. Continuing along the peptide chain, helix content rises again in D-Trp^9, D-Ala^{10}, D-Lys^{11}.

Clearly the nucleus of helical structure is found in the Arg^8 position, which by replacement with D-Arg^8 is almost entirely disrupted (Fig. 6).

Superimposing the profile of biological activity (sterioidogenesis in vitro) onto the profile of helix content in this series of analogues shows an

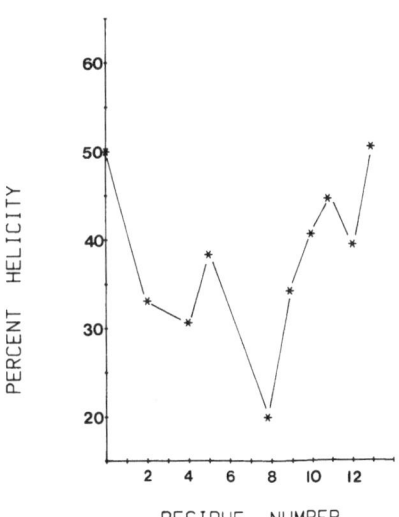

Figure 6. α-Helix content of ACTH-(1–19) amide and selected D-substituted analogues, as determined from CD spectra of TFE solution.

almost perfect correlation. As the D substitution progresses from [D-Tyr^2]-ACTH-(1–19)-NH$_2$ to [D-Arg^8]-ACTH-(1–19)-NH$_2$, steroidogenetic activity decreases continually, rapidly to a low of $\sim 5/100,000$ for [D-Arg^8]-ACTH-(1–19)-NH$_2$; it rises again in [D-Trp^9]- and [D-Ala^{10}]-ACTH-(1–19)-NH$_2$, paralleling the fall and rise in α-helix content (Fig. 6) (Löw *et al.*, 1980). With all the functional groups in the side chains present, the diminishing biological activity of these peptides is definitely due to conformational effects, on the nature of which—stabilization and disruption of α-helix—there can be no doubt in the light of everything that has been described above.

From these most recent studies the beginnings of a well-defined structure–conformation–activity correlation are emerging. It will now be necessary to investigate in detail the helical structure and the side chain orientations of residues 2 to 9 in each of the D-isomer analogues by joint use of CD, fluorescence, and ^{1}H and ^{13}C high-resolution NMR at very high fields where assignment and analysis of all spin systems and coupling patterns is possible. This work is now in progress.

4.   Angiotensin II

4.1.   Introduction

A II is a fascinating peptide to study as it is a medium-size oligopeptide of the intriguing sequence

Asp-Arg-Val-Tyr-Ile-His-Pro-Phe

with a broad spectrum of biological activity (Smeby and Fermandjian, 1978, and references therein). Its foremost effect is the regulation of arterial blood pressure either through direct smooth muscle contraction of the arteries or through the release of catecholamines and aldosterone at the adrenal medulla and the adrenal cortex, respectively. Direct effects of A II on the CNS (inducing thirst, for instance) and on the peripheral sympathetic nervous system have also been described but need further clarification as to their physiological significance. The great variety of functional side chains in A II (three aromatic residues, two carboxylate groups with a negative charge, two basic groups with a positive charge) poses the question of functional heterogeneity of the various sites of action: are the membrane receptors for A II in the different tissues of the same type or do structure–activity relationships vary from one assay to another? If so, are the conformational requirements of the hormone for receptor binding and signal stimulation different in each case?

The relative ease with which A II and its structural analogues can be synthesized has made it possible to establish quite detailed structure–activity relationships for each of the eight positions in the peptide so far, at least as regards two assay systems, the pressor and the myotropic response. These synthetic efforts have also led to the discovery of hyperactive analogues and to competitive inhibitors of A II such as [Ile⁸]-A II and [Sar¹,Ile⁸]-A II, which are used as diagnostic tools to identify renin/angiotensin-mediated hypertension.

With the availability of highly radioactive, tritiated A II it became possible to correlate binding and potency data in a way that shows that most analogues of diminished "biological activity" have in fact reduced affinity for the receptor (see Fig. 2).

To explain these effects in detail, conformational analysis of A II was undertaken as early as 1962 (Smeby *et al.*, 1962). An exhaustive and critical review of the models proposed for A II conformation in solution based on a variety of techniques, published recently (Smeby and Fermandjian, 1978), concludes that conformational analysis of A II and analogues has been refined to a degree where on the basis of the known data and the latest model derived from them (Fermandjian *et al.*, 1976), prediction of binding behavior for a given analogue can be made by comparison of certain conformational parameters, measured by CD and NMR techniques, of the analogue and of the native hormone. The following examples illustrate this point.

4.2. [Sar¹,Ile⁸]-A II

[Sar¹,Ile⁸]-A II is a very potent competitive inhibitor of A II. *In vivo* it causes a short-lived decrease in blood pressure, *in vitro* it competes for specific A II receptor sites in radioligand assays (Khosla *et al.*, 1976b). In an effort to find a longer-lasting inhibitor, two *N*-methylated derivatives were synthesized: [Sar¹,NMeTyr⁴,Ile⁸]-A II and [Sar¹,NMeIle⁵,Ile⁸]-A II. It was hoped that the *N*-methylation of the backbone would reduce enzymatic degradation of the inhibitor, thus prolonging its duration of action.

The desired resistance of chymotryptic cleavage was obtained; however, the two *N*-methylated peptides were devoid of inhibitory activity: they do not bind to the receptor (Khosla *et al.*, 1974, 1976a).

We undertook a comparative conformational study by CD and ¹H NMR spectroscopy (Piriou *et al.*, 1980) of the four peptides A II, its antagonist [Sar¹,Ile⁸]-A II, and the two derivatives thereof. Earlier studies on A II and a number of analogues have shown that certain parameters of spectral behavior, monitored for instance during a pH titration in aqueous solution, are very sensitive to conformational changes caused by amino acid substitutions (Lintner *et al.*, 1975, 1977). One such parameter is the

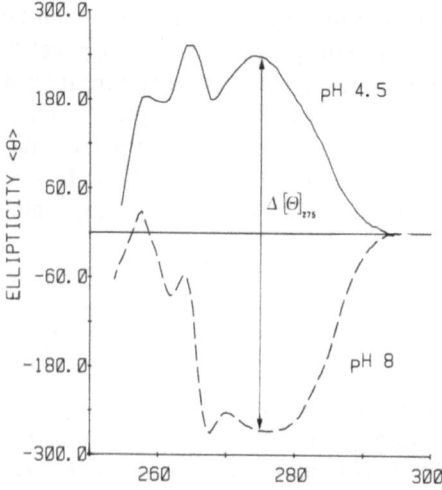

Figure 7. CD spectra of angiotensin II at pH 4.5 and pH 8: aromatic region.

amplitude of the variation of the CD signal of tyrosine at 275 nm during deprotonation of the two-residues-removed histidine side chain (Fig. 7). The difference of ellipticity $\Delta[\Theta]_{275} = [\Theta]_{275}(pH\ 4) - [\Theta]_{275}\ (pH\ 8)$ is a measure of the time-averaged distance between the two side chains tyrosine and histidine (Lintner et al., 1977; Piriou et al., 1980), this distance being a function of the rotamer distribution and the respective orientation of the phenol and imidazole rings. Figure 8 shows clearly that titration curves [Θ]$_{275}$ = f(pH) of A II and its competitive inhibitor [Sar^1,Ile^8] A-II are practically identical, whereas the ones of the modified inhibitors are distinctly

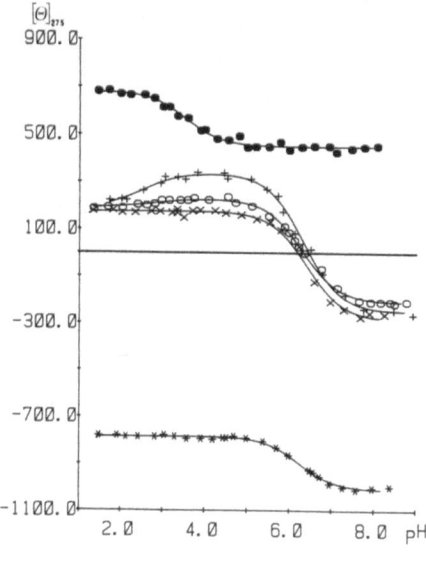

Figure 8. pH titration curves $[\Theta]_{275}$ = f(pH) of angiotensin II (o), [Sar^1,Ile^8]-A II (X), [Sar^1,NMeTyr^4,Ile^8]-A II (*), [Sar^1,NMeIle^5,Ile^8]-A II (●), and [α-MeTyr^4]-A II (+).

different with regard to sign, intensity, and amplitude of titration. In the N-methylated derivatives the two residues tyrosine and histidine, which are quite essential for binding affinity, are not aligned in the same way as in A II and [Sar^1,Ile^8]-A II. The same results but in more detail can be gleaned from the ^{1}H NMR spectra where chemical shift data, coupling constants, and titration curves are identical in the hormone and the antagonist, whereas strong perturbation of the side chain orientation of Tyr^4, Ile^5, and His^6 are evident in the spectra of the N-methylated peptides (Piriou *et al.*, 1980).

This is one of the still rare examples where modifications of biological activity of a peptide can be rigorously traced back to a definite change in conformation of essential side chains.

Following these insights, an analogue of A II was synthesized where α-MeTyr^4 was used rather than NMeTyr^4. The replacement of the α-proton with a methyl group does not have a profound effect on the peptide backbone or the side chain orientation of tyrosine, as the CD titration curve (Fig. 8) and NMR results (Khosla *et al.*, 1981) clearly show. Therefore, the peptide [α-MeTyr^4]-A II is fully active and resistant to chymotrypsin at the same time. The next step will be to synthesize [Sar^1,α-MeTry^4,Ile^8]-A II for which good inhibitory activity *and* cleavage resistance can be predicted with confidence.

4.3. [X^5]-A II

Another series of analogues of A II has enabled us to define more precisely the relationships between amino acids Tyr^4, Ile^5, and His^6 in the hormone: the series [Ile^5]-A II, [Leu^5]-A II, [Ala^5]-A II, [Gly^5]-A II. Even limiting our discussion to the three variables: CD titration curve amplitude ($\Delta[\Theta]_{275}$), side chain rotamer distribution, and $^3J_{NH-C\alpha H}$ coupling constants, we obtain a remarkably precise image of the evolution of the conformation of the central sequence 4–6 in these peptides.

The average distance between tyrosine and histidine side chain moieties increases progressively from [Ile^5]-A II to [Gly^5]-A II as the strongly diminished titration amplitude $\Delta[\Theta]_{275}$ shows (Table I). That this is not due

Table I. Analogues of Angiotensin II Modified in Position 5

Peptide	pH effect (4–8) $\Delta[\Theta]_{Tyr\ at\ 275}$	$^3J_{NH-C\alpha H}$ [a] (Hz)	Biological activity [b] (%)
[Ile^5]-A II	− 500	9.2	100
[Leu^5]-A II	− 300	8.0	25
[Ala^5]-A II	− 200	7.6	10
[Gly^5]-A II	− 30	5.6	2

[a] Coupling constant of residue 5 measured in DMSO solution.
[b] Pressor response (Smeby and Fermandjian, 1978, and references therein).

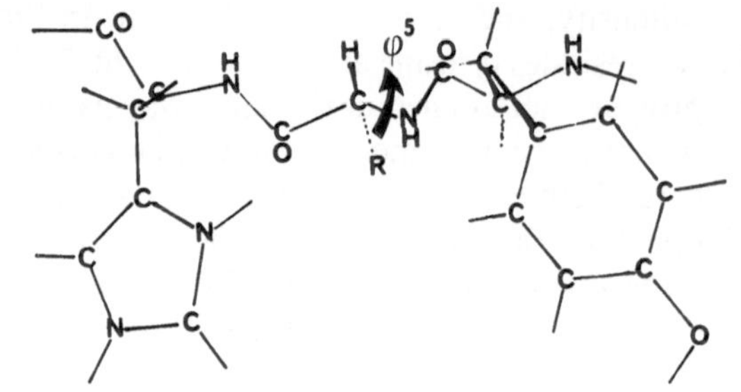

R $= CH (CH_3)_2$: Val^5
R $=H$: Gly^5

Figure 9. Side chain arrangement of tyrosine and histidine in angiotensin II.

to drastic rearrangements of the side chain distribution of tyrosine and/or histidine can be deduced from the stability of the coupling constants $J_{\alpha\beta}$ and $J_{\alpha\beta'}$ in the NMR spectra of the series. The $^3J_{NH-C^\alpha H}$ coupling constants of residue 5, however, which are tied to the dihedral angle ϕ^5 by a Karplus-type relationship (Bystrov, 1976), change gradually from 9.2 Hz in [Ile^5]-A II to 5.6 Hz in [Gly^5]-A II; therefore, ϕ^5 changes also, from a value describing a rather kinked structure at the C$^\alpha$ of [Ile^5]-A II to a more extended form in [Gly^5]-A II.

Figure 9 describes the effect better than words. Table I lists these results and confronts them with the biological activity of the series. Clearly, the diminished affinity of these peptides is correlated to the small but highly significant conformational change in the backbone structure of position 5. It will be of interest now to study the thermodynamic properties of conformational changes, of receptor binding, in order to quantitate the clearly existing qualitative structure–conformation–activity relationship in A II.

5.   Thyroliberin

5.1.   Introduction

Thyrotropin-releasing hormone (TRH, also called thyroliberin) is comprised of three amino acids in the sequence pGlu-His-Pro-NH$_2$. It was isolated (Burgus et al., 1969) on the basis of its ability to stimulate the secretion of thyroid-stimulating hormone (TSH). Since then, a number of other, apparently unrelated biological effects were found to be mediated by this peptide, such as the stimulation of synthesis and release of prolactin from pituitary cells. More striking still was the discovery of direct TRH effects on the central nervous system. There its primary effect seems to be of antidepressant nature.

Consisting of only three amino acids, the TRH molecule presents no major problem as to its chemical synthesis. A great number of analogues have therefore been prepared in the search for a clear-cut structure–activity relationship. The goal has only partially been attained as all three residues are quite sensitive to modification or substitution; even the seemingly inert side chains of proline and pyroglutamic acid cannot be modified much without drastic loss in biological activity.

The fact that TRH acts on cells of the pituitary where it definitely binds to the cell membrane receptors, but can also enter the cell to act at the nucleus, and that it exerts direct effects in neurons, makes the establishment of structure–activity relationships even more complicated. Some authors have found analogues of TRH that have highly increased CNS activity with unchanged or decreased hormonal activity (Veber *et al.*, 1976), whereas others (Björkman *et al.*, 1979) have synthesized a series of analogues for which both CNS and hormonal activity are positively correlated.

To understand TRH action in terms of its structure it is therefore necessary to be precise as to which biological assay system one is using. GH_3 clonal prolactin cells are a good model system (Pradelles *et al.*, 1972). Furthermore, binding studies with radioactively labeled TRH are essential for determining the site of the hormone activity. Tritium labeling of TRH at high specific radioactivity has made it possible to study the binding of TRH to these GH_3 cell membranes (Pradelles *et al.*, 1972).

5.2. The Solution Conformation of TRH

In these well-defined conditions it is then meaningful to investigate the molecular conformation of the peptide in order to relate conformational characteristics to binding and stimulation abilities of TRH.

The problem of TRH conformation is of different scope than the one of ACTH or A II described above. The number of possible conformations with reasonable stability is relatively small, as the rotational freedom of the dihedral angles ϕ_1, ϕ_3, χ_1^1, χ_1^2 and χ_3^1, χ_3^2, χ_3^3, χ_3^4 is severely restricted. The task consists essentially of determining ϕ_2, ψ_2, χ_2^1, and χ_2^2, the angles describing the conformation of the histidine residue, and ψ_1, ψ_3 as well as ω_2, the His-Pro peptide bond, which might exist in the *cis* or *trans* conformer. Furthermore, intramolecular hydrogen bonds that might stabilize a given conformation and the tautomeric equilibrium of the uncharged imidazole ring are subject to physicochemical investigation because of their probable biological importance.

Each of these detailed questions on TRH conformation was studied in our laboratory in a separately synthesized peptide with the appropriate isotopic enrichment: six TRH peptides were needed in addition to the natural

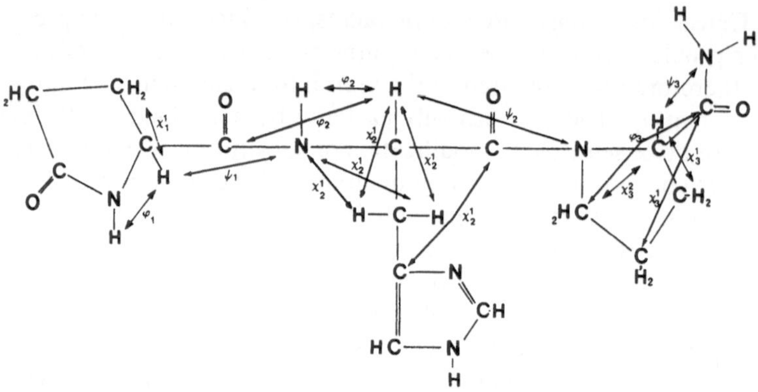

Figure 10. Primary structure of TRH: the arrows indicate possible couplings measurable by NMR in the appropriately labeled compounds.

abundance compound: [85% ^{13}C-pGlu]-TRH, [99% ^{2}H-pGlu]-TRH, [85% ^{13}C-His]-TRH, [15% ^{13}C/95% ^{15}N-His]-TRH, [85% ^{13}C-Pro]-TRH, and [15% ^{13}C/95% ^{15}N-Pro- 95% ^{15}N-NH$_2$]-TRH (Fig. 10). They allowed us to eliminate certain discrepancies of interpretation that persisted in the literature on TRH conformation.

5.2.1. The Pyroglutamic Acid Residue

The ϕ_1 angle of pGlu is comprised between 110 and 120° as determined from $^3J_{\text{NH-C}^\alpha\text{H}}$ coupling constant in ^{1}H NMR spectra, suggesting the absence of conformational constraints on this ring structure. ^{13}C NMR relaxation time studies (Deslauriers *et al.*, 1973) on the natural abundance compound confirm the high mobility of the pGlu residue atoms, in agreement with conformational energy calculations (Burgess *et al.*, 1973). This freedom, the latter results suggest, extends to the ψ_1 angle, which can rotate between -60 and 180° without interfering with the rest of the peptide. To determine the ψ_1 angle of a peptide in solution by spectroscopic means has so far been unfeasible in most cases because of the inherent limitations in the techniques of ^{1}H and ^{13}C NMR. In order to overcome these obstacles we synthesized [95% ^{15}N/15% ^{13}C-His]-TRH where it would be possible to measure ^{15}N-^{1}H coupling constants, which relate to the ψ angle by an angular dependence curve (Bystrov, 1976). The observed value of ~ 0 Hz is compatible with ψ angles of 165, 75, 15, and $-135°$. Although the spread of those values confirms the relative mobility of the pGlu moiety, it does delimit the preferred regions of conformational space of this residue (Vičar *et al.*, 1979a).

5.2.2. The Histidine Residue

1. Proton NMR spectroscopy yields three conformationally related coupling constants for this residue: $^3J_{NH-C^\alpha H}$, $^3J_{\alpha,\beta}$, and $^3J_{\alpha,\beta'}$. The ϕ_2 angles derived from the $^3J_{NH-C^\alpha H}$ value (7.5 Hz) in dimethylsulfoxide (DMSO) (Fermandjian *et al.*, 1972) are −90, −150, and 60°, the ambiguity arising from the degeneracy of the angular dependence curve (Bystrov, 1976). Although calculations on the TRH conformation yield −150° as the energetically favorable ϕ_2 value, we thought it important to narrow the choice of angles on experimental rather than on purely theoretical grounds. Synthesis of [99% ^{2}H-pGlu]-TRH allowed us to measure the interresidue coupling constant $^3J_{13C'_{pGlu}-H^\beta_{His}}$, which is also tied to the ϕ_2 angle by a different angular dependence curve (Bystrov, 1976). The value of $\sim$ 2 Hz of this coupling constant is compatible with −150 and −90°, thus eliminating the 60° value. Furthermore, this value can be measured in both DMSO and ^{2}H$_2$O solution, thus allowing access to the ϕ angle even when the NH proton is exchanged. The coupling constant does not change from one solvent to the other and indicates that, at best, interconversion between ϕ_2 = −90° and ϕ_2 = −150° takes place between the two solvent systems (Vičar *et al.*, 1979a).

2. The side chain conformation described as rotamer distribution (Fig. 11) derives from the $^3J_{\alpha,\beta}$ and $^3J_{\alpha,\beta'}$ coupling constants. However, as the protons β and β' cannot be assigned unambiguously in ^{1}H NMR spectra, rotamers I and II cannot be distinguished. In the case of TRH, this leads to a drastic discrepancy in the interpretation of TRH conformation by the authors of different laboratories. The problem is crucial as the proper orientation of the histidine side chain is certainly a factor influencing biological activity.

In aqueous solution, the rotamer population of histidine was proposed (Donzel *et al.*, 1974) to be 0.48 (I), 0.28 (II), and 0.24 (III) at acid pH and 0.49 (I), 0.32 (II), and 0.19 (III) at neutral pH. Thus, rotamer I with the imidazole ring *gauche* to the pGlu moiety dominates whatever the ionic state of the ring. Montagut *et al.* (1974), however, disagree; on the basis

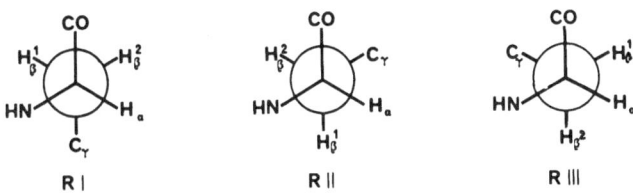

Figure 11. The three staggered rotamers of histidine in TRH.

of the δ_{Pro} resonance nonequivalence and its evolution during pH titration of imidazole, and the relative solvent inaccessibility of the $C=O$ group of histidine (Bellocq *et al.*, 1973), these authors suggest that rotamer II ($\chi_2^1 = 180°$) is preponderant, directing the side chain toward the proline ring.

In order to determine unequivocally the rotamer populations of histidine in TRH, we synthesized [85% ^{13}C-His]-TRH in which we can measure the ^{13}C–^{13}C coupling constants (Fig. 10) (inaccessible in natural abundance compounds). This coupling constant is, of course, dependent on the χ_2^1 angle and gives us the percentage of rotamer I. From the value of ~ 3 Hz we determine a rotamer I fraction of 0.46 at acid and neutral pH, which is compatible with the rotamer population proposed by Donzel *et al.* (1974).

Does changing the solvent change the conformation? According to Donzel *et al.* (1974) the rotamer distribution in DMSO solution is reversed; rotamer II predominates now and causes the proline δ-nonequivalence and the large chemical shift difference of the *syn* and *anti* carboxamide protons. This appears true only for the free base of TRH, though, as TRH · HCl in DMSO is again proposed (Donzel *et al.*, 1974) with rotamer I as the major form, although the proton nonequivalence of the proline δ-protons and some carboxamide chemical shift difference persist. To resolve these questions, we are currently studying the compounds [85% ^{13}C-His]-TRH and [95% ^{15}N/15% ^{13}C-His]-TRH in various solvents and ionic states. The coupling constants ^{13}C–^{13}C and ^{15}N–H$^\beta$, ^{15}N–H$^{\beta'}$ measured in the respective peptides will yield the unambiguous assignments of rotamers I and II.

3. Imidazole tautomerism: The isotopically enriched compound [95% ^{15}N/15% ^{13}C-His]-TRH was also used in the direct determination of the tautomeric equilibrium in the uncharged imidazole ring of TRH. In fact, the $^1J_{CN}$ coupling constants are strongly dependent on the state of hybridization sp^2 or sp^3 of the nitrogen atom. As the pH of the aqueous solution increases from 4 to 8, the nitrogen which loses its proton will become sp^2 hybridized. In agreement with the results on free histidine (Reynolds *et al.*, 1973), it is found that the N^τ tautomer is the preponderant species (by a factor of 4:1) at neutral and basic pH. This agrees well, also, with the numerous physicochemical and biological data on [N^τ MeHis]-TRH and [N^π MeHis]-TRH, the former analogue being hyperactive, the latter one having less than a thousandth of TRH activity.

The ψ_2 angle again can be approached only in selectively enriched compounds. We prepared [95% ^{15}N/15% ^{13}C-Pro- 95% ^{15}N-NH$_2$]-TRH and measured the ^{15}N–^{1}H coupling constant. The $\sim$ 0-Hz value is compatible with the ψ_2 values suggested by calculations (120 to 150° and 75 to 120°) (Vičar *et al.*, 1979a).

5.2.3. The Proline Residue

Three main features of the conformation of the proline moiety in TRH can be investigated: the *cis–trans* isomerism of ω_2; the ring puckering and flexibility; and the ψ_3 angle controlling the position of the carboxamide protons.

1. *Cis–trans* isomerism is found in TRH, but to a rather low extent. Depending on the solvent, the amount of *cis* isomer varies from 15–20% in water, 7–8% in DMSO, and 0–5% in pyridine (Deslauriers *et al.*, 1973). It is not clear if this appearance of 20% *cis* isomer in water may have a physiological role or not. Conformational studies have so far been limited to the major *trans* conformer.

2. Puckering of the pyrrolidine ring was studied in the compound [85% ^{13}C-Pro]-TRH where the measurement of $^3J_{^{13}C\alpha-^{13}C^\delta}$ coupling constants is possible (Haar *et al.*, 1975). It is found that the proline ring in TRH is somewhat restricted in its flexibility compared to free proline, with a preference for the C^γ *endo* conformer. This result is compatible with the ϕ_3 angle (-60 to $-90°$) determined similarly from $^3J_{C'-C'}$ (Vičar *et al.*, 1979b).

3. The orientation of the carboxamide NH_2 group and the possibility of a hydrogen bond between the *anti* proton to the $C=O$ group of histidine, forming a seven-membered ring, has been another point of disagreement in the interpretation of the experimental results and theoretical calculations. In the isolated molecule, the C_7 conformation is found to be energetically favored and preferred (Burgess *et al.*, 1973). The large chemical shift difference between the two protons *syn* and *anti* in DMSO and in pyridine suggest some kind of interaction involving the *anti* proton (Fermandjian *et al.*, 1972). The pyrrolidine ring puckering observed above also is in favor of a C_7-type structure (Pullman and Pullman, 1974). A hydrogen bond for the *anti* proton is detected in the model compound Z-His-Pro-NH_2 dissolved in CCl_4, a nonpolar solvent (Montagut *et al.*, 1974). CD studies on TRH and analogues in dioxane strongly suggest the of a significant amount of C_7 structure in this part of the molecule (Pradelles *et al.*, 1977). In this C_7 conformation, ψ_3 would be $\sim 80°$, which agrees well with the absence of strong coupling between $C^\alpha H_{Pro}$ and $^{15}NH_2$ measured in [95% $^{15}N/15\%$ ^{13}C-Pro-95% ^{15}N-NH$_2$]-TRH.

Only the temperature-dependence curves and hydrogen–deuterium exchange experiments on these protons do not reveal differences between the *syn* and *anti* protons and would rule out the presence of detectable hydrogen bonding in polar solvents such as DMSO and water (Feeney *et al.*, 1974; Donzel *et al.*, 1974). This discrepancy will have to be resolved as the NH_2 group and consequently its precise orientation in TRH appears to play an important role in biological activity: the one competitive inhibitor of TRH that has been found so far (Lybeck *et al.*, 1973) [cyclopentoyl-His-pyrrolidine: $(CH_2)_5$-CO-His-N-$(CH_2)_4$] demonstrates clearly that the side chain of histidine is necessary for binding to the receptor (although it can be replaced by a few other side chains with moderately diminished affinity) whereas the carbonyl of the pGlu moiety and the Pro-carboxamide group seem to be necessary for the triggering of the biological response.

The precise understanding of the conformation of TRH in a given environment (pH, solvent) is thus about to be achieved. On this basis it will now be necessary to study specific analogues such as [N^r MeHis]-TRH, [N^r MeHis]-TRH, the inhibitor mentioned above, and several other compounds of well-defined receptor affinity and biological activity. The same problems of spectral assignments, angular dependence curve degeneracy, ψ angle determination, etc., will have to be addressed; the use of specifically isotopically labeled compounds has shown the way to overcome these problems; however, the synthetic task lying ahead is formidable.

6. Conclusion

In this chapter we have tried to show how closely solution conformation and biological activity of a peptide hormone are related and how these relationships can be examined experimentally. Our understanding of how receptor affinity of a peptide is a funciton of backbone and side chain conformation and how both affinity and conformation are dependent on primary structure, is greatly enhanced by these studies. From the examples cited above it is evident, too, that detailed investigations of this kind demand a continued effort of time, synthetic material, and spectroscopic equipment in order to yield clear, rewarding results.

DISCUSSION

McKERNS: I was pleased that you have such a clear separation of cell membrane binding from biological activity in your series of peptides. In the case of gonadotropins, this is something I have been proposing for years, without much acceptance until recently.

LINTNER: Indeed, in order to correlate precise conformational features of a peptide to its biological activity parameters, it is essential to distinguish clearly the terms "recognition," "binding affinity," "signal generation," and "mechanism of action."

REFERENCES

Bellocq, A. M., Boilot, J. C., Dupart, E., and Dubien, M., 1973, Analyse conformationnelle de l'hormone hypothalamique TRF par spectroscopie Raman, *C.R. Acad. Sci. Ser. D* **276**:423.

Björkman, S., Karlsson, J. A., and Sievertsson, H., 1979, A comparison of central and hormonal effects of peptides related to thyroliberin (TRH) and melanostatin (MIF), *Acta Pharm. Suec.* **16**:95.

Burgess, A. W., Momany, F. A., and Scheraga, H. A., 1973, Conformational analysis of thyrotropin releasing factor, *Proc. Natl. Acad. Sci. USA* **70**:1456.

Burgus, R., Dunn, T. F., Ward, D. N., Vale, W., Amoss, M., and Guillemin, R., 1969, Dérivés polypeptidiques de synthèse doués d'activité hypophysiotrope TRF, *C.R. Acad. Sci. Ser. D* **268**:2116.

Bystrov, V. F., 1976, Spin–spin coupling and the conformational states of peptide systems, in: *Progress in NMR Spectroscopy* (J. W. Emsley, J. Feeney, and L. H. Sutcliffe, eds.), Vol. 10, pp. 41–81, Pergamon Press, Elmsford, N.Y.

Chou, P. Y., and Fasman, G. D., 1974, Prediction of protein conformation, *Biochemistry* **13**:222.

Chou, P. Y., and Fasman, G. D., 1977, β-Turns in proteins, *J. Mol. Biol.* **115**:135.

Deslauriers, R., Garrigou-Lagrange, C., Bellocq, A. M., and Smith, I. C. P., 1973, Carbon-13 nuclear magnetic resonance studies on thyrotropin releasing factor and related peptides, *FEBS Lett.* **31**:59.

Donzel, B., Rivier, J., and Goodman, M., 1974, Conformational studies on the hypothalamic thyrotropin releasing factor and related compounds by ^{1}H nuclear magnetic resonance spectroscopy, *Biopolymers* **13**:2631.

Edelhoch, H., and Lippoldt, R. E., 1969, Structural studies on polypeptide hormones, *J. Biol. Chem.* **244**:3876.

Eisinger, J., 1969, Intramolecular energy transfer in adrenocorticotropin, *Biochemistry* **8**:3902.

Feeney, J., Bedford, G. R., and Wessels, P. L., 1974, ^{1}H nuclear magnetic resonance studies of thyrotropin releasing factor (TRF), *FEBS Lett.* **42**:347.

Fermandjian, S., Pradelles, P., Fromageot, P., and Dunand, J. J., 1972, Proton NMR studies on thyrotropin releasing factor, *FEBS Lett.* **28**:156.

Fermandjian, S., Lintner, K., Haar, W., Fromageot, P., Khosla, M. C., Smeby, R. R., and Bumpus, F. M., 1976, Conformation–function relationship of angiotensin II and analogs, in: *Peptides 1976* (A. Loffet, ed.), pp. 339–351, Editions de l'université de Bruxelles, Bruxelles.

Goodman, M., Naider, F., and Rupp, R., 1971, Conformations of alanine oligopeptides in solution, *Bioorg. Chem.* **1**:310.

Greff, D., Toma, F., Fermandjian, S., Löw, M., and Kisfaludy, L., 1976, Conformational studies of corticotropin 1–32 and constitutive peptides by circular dichroism, *Biochim. Biophys. Acta* **439**:219.

Haar, W., Fermandjian, S., Vičar, J., Bláha, K., and Fromageot, P., 1975, ^{13}C nuclear magnetic resonance study of [85% ^{13}C-enriched proline]-thyrotropin releasing factor: ^{13}C–^{13}C vicinal coupling constants and conformation of the proline residue, *Proc. Natl. Acad. Sci. USA* **72**:4948.

Khosla, M. C., Hall, M. M., Smeby, R. R., and Bumpus, F. M., 1974, Synthesis of analogs of (8-isoleucine) angiotensin II with variations in position 1, *J. Med. Chem.* **17**:431.

Khosla, M. C., Munoz-Ramirez, H., Hall, M. M., Smeby, R. R., Khairallah, P. A., Bumpus, F. M., and Peach, M. J., 1976a, Synthesis of angiotensin II antagonists containing N- and O-methylated and other amino acid residues, *J. Med. Chem.* **19**:244.

Khosla, M. C., Bravo, E. L., Smeby, R. R., and Bumpus, F. M., 1976b, Structural requirements for stimulation or blockade of the pressor and aldosterone releasing effects of angiotensin II, in: *Peptides 1976* (A. Loffet, ed.), pp. 371–377, Editions de l'université de Bruxelles, Bruxelles.

Khosla, M. C., Stachowiak, K., Smeby, R. R., Bumpus, F. M., Piriou, F., Lintner, K., and Fermandjian ,S., 1981, Synthesis of (4-α methyltyrosine) angiotensin II and studies on its conformation, pressor activity and mode of enzymatic degradation, *Proc. Natl. Acad. Sci. USA,* **78:**757.

Leach, S. J., 1973, *Physical Principles and Techniques of Protein Chemistry,* Part C, Academic Press, New York.

Lintner, K., Fermandjian, S., Fromageot, P., Khosla, M. C., Smeby, R. R., and Bumpus, F. M., 1975, pH titration effects on the CD spectra of angiotensin II, truncated peptides and other analogues: Aromatic region, *FEBS Lett.* **56:**366.

Lintner, K., Fermandjian, S., Fromageot, P., Khosla, M. C., Smeby, R. R., and Bumpus, F. M., 1977, Circular dichroism studies of angiotensin II and analogues: Effects of primary sequence, solvent and pH on the side chain conformation, *Biochemistry* **16:**806.

Löw, M., Kisfaludy, L., and Fermandjian, S., 1975, Proposed preferred conformation of ACTH, *Acta Biochim. Biophys. Acad. Sci. Hung.* **10:**229.

Löw, M., Kisfaludy, L., Hajos, G., Mikaly, K., Makara, G. B., Toma, F., Dive, V., and Fermandjian, S., 1980, Biological and conformational properties of some corticotropin analogues containing D-amino acids, in: *Peptides 1980* (K. Brunfeldt, ed.), p. 513, Scriptor, Copenhagen.

Lybeck, H., Leppäluoto, J., Virkkunen, P., Schafer, D., Carlsson, L., and Mulder, J., 1973, Suppression of TRH-mediated thyroidal release of ^{131}I by a synthetic analog, *Neuroendocrinology* **12:**366.

Montagut, M., Lemandceau, B., and Bellocq, A. M., 1974, Conformational analysis of thyrotropin releasing factor by proton nuclear magnetic resonance spectroscopy, *Biopolymers* **13:**2615.

Nabedryk-Viala, E., Thiery, C., Calvet, P., Fermandjian, S., Kisfaludy, L., and Thiery, J. M., 1978, Conformation of corticotropin: An infrared spectrometry study of hydrogen exchange kinetics, *Biochim. Biophys. Acta* **536:**252.

Piriou, F., Lintner, K., Fermandjian, S., Fromageot, P., Khosla, M. C., Smeby, R. R., and Bumpus, F. M., 1980, Amino acid side chain conformation in angiotensin II and analogues: Correlated results of circular dichroism and ^{1}H nuclear magnetic resonance. *Proc. Natl. Acad. Sci. USA* **77:**82.

Pradelles, P., Morgat, J. L., Fromageot, P., Oliver, C., Jacquet, P., Gourdji, D., and Tixier-Vidal, A., 1972, Preparation of highly labelled ^{3}H-thyrotropin releasing hormone (pGA-His-Pro(NH$_2$)) by catalytic hydrogenolysis, *FEBS Lett.* **22:**19.

Pradelles, P., Vičar, J., Morgat, J. L., Fermandjian, S., Bláha, K., and Fromageot, P., 1977, The influence of organic solvents on the conformation of thyrotropin releasing factor (TRF) studied by circular dichroism, *Coll. Czech. Chem. Commun.* **42:**79.

Pullman, B., and Pullman, A., 1974, Molecular orbital calculations on the conformation of amino acid residues of proteins, *Adv. Protein Chem.* **28:**347.

Ramachandran, J., 1973, The structure and function of adrenocorticotropin, in: *Hormonal Proteins and Peptides* (C. H. Li, ed.), Vol. 2, pp. 1–28, Academic Press, New York.

Reynolds, W. F., Peat, I. R., Freedman, M. H., and Lyerla, J. R., Jr., 1973, Determination of the tautomeric form of the imidazole ring of L-histidine in basic solution by carbon-13 magnetic resonance spectroscopy, *J. Am. Chem. Soc.* **95:**328.

Roberts, J. L., and Herbert, E., 1977, Characterization of a common precursor to corticotropin and β-lipotropin: Identification of β-lipotropin peptides and their arrangement relative to corticotropin in the precursor synthesized in a cell free system, *Proc. Natl. Acad. Sci. USA* **74:**5300.

Sayers, G., Hanzmann, E., and Bodanszky, M., 1980, Hypothalamic peptides influencing secretion of ACTH by isolated adenohypophysial cells, *FEBS Lett.* **116:**236.

Schiller, P. W., 1972, Study of adrenocorticotropic hormone conformation by evaluation of intramolecular resonance energy transfer in N$^\epsilon$-dansyllysine-21 ACTH(1–24) tetrakosipeptide, *Proc. Natl. Acad. Sci. USA* **69:**975.

Smeby, R. R., and Fermandjian, S., 1978, Conformation of angiotensin II, in: *Chemistry and Biochemistry of Amino Acids, Peptides and Proteins* (B. Weinstein, ed.), Vol. 5, pp. 117–162, Dekker, New York.

Smeby, R. R., Arakawa, K., Bumpus, F. M., and Marsh, M. E., 1962, A proposed conformation of isoleucine-5 angiotensin II. *Biochim. Biophys. Acta* **58**:550.

Squire, P. G., and Bewley, T., 1965, Adrenocorticotropins. XXXV. The optical rotatory dispersion of sheep adrenocorticotropic hormone in acidic and basic solutions, *Biochim. Biophys. Acta* **109**:234.

Tiffany, M. L., and Krimm, S., 1972, Effects of temperature on the CD spectra of polypeptides in the extended state, *Biopolymers* **11**:2309.

Toma, F., Greff, D., Fermandjian, S., Löw, M., and Kisfaludy, L., 1976, New data on ACTH conformation, in: *Peptides 1976* (A. Loffet, ed.), pp. 625–631, Editions de l'université de Bruxelles, Bruxelles.

Toma, F., Fermandjian, S., Löw, M., and Kisfaludy, L., 1978, A proton NMR investigation of proline-24 *cis-trans* isomerism in corticotropin 1–32 and related peptides, *Biochim. Biophys. Acta* **534**:112.

Toma, F., Lam-Thanh, H., Piriou, F., Heindl, M. C., Lintner, K., and Fermandjian, S., 1980, NMR evidence for a type I β-turn in (Pro^2) tetrapeptides and interdependence of *cis-trans* isomerism, ring flexibility and backbone conformation, *Biopolymers* **19**:781.

Toma, F., Fermandjian, S., Löw, M., and Kisfaludy, L., 1981, Carbon-13 NMR studies of ACTH: Assignment of resonances and conformational features, *Biopolymers*, **20**:901.

Veber, D. F., Holly, F. W., Varga, S. L., Hirschmann, R., Nutt, F. R., Lotti, V. J., and Porter, C. C., 1976, The dissociation of hormonal and CNS effects in analogues of TRH, in: *Peptides 1976* (A. Loffet, ed.), pp. 453–461, Editions de l'université de Bruxelles, Bruxelles.

Vičar, J., Abillon, E., Toma, F., Piriou, F., Lintner, K., Bláha, K., Fromageot, P., and Fermandjian, S., 1979a, The two conformations of TRH in solution, *FEBS Lett.* **97**:275.

Vičar, J., Bláha, K., Toma, F., Piriou, F., Fromageot, P., and Fermandjian, S., 1979b, Coupling constants C–H, C–C, C–N, N–H and structural properties of TRH, in: *Peptides 1978* (I. Z. Siemion and G. Kupryszewski, eds.), pp. 489–493, Wroclaw University Press, Poland.

4

The Search for the Active Center of the Gonadotropins

Kenneth W. McKerns

1. Introduction

An earlier concept of lutropin (LH) and choriogonadotropin (hCG) action in stimulating biochemical events in the corpus luteum involves receptor binding in the cell membrane leading to the activation of adenylate cyclase and an increase in cyclic AMP. The cyclic AMP binds to a regulatory sub-unit of a cyclic AMP-dependent protein kinase, leading to the activation and release of the subunit. The activated protein kinase is believed to be the mediator of LH and hCG action. These concepts have been reviewed by Marsh (1975) and by Jungmann and Hunzicker-Dunn (1978). The concept implies that the polypeptide hormones do not enter the target cell. However, this theory cannot accommodate the newly published data showing the entry of the tropic hormones into their target cells with an activation of enzymes in the cytoplasm and chromatin. Petrusz (1978) has discussed this intracellular action of gonadotropins.

I have assumed that in the large three-dimensional folded structure of LH and hCG there is a small grouping of amino acids that become associated because of the folding of the tertiary structure, and that this constitutes the "active center." This active center is, then, the amino acids in association that initiate the biological action of these gonadotropins in their target cells. We have shown a principal site of action of the gonadotropins

Kenneth W. McKerns • International Foundation for Biochemical Endocrinology, Inc., Blue Hill Falls, Maine 04615.

LH and hCG to be a stimulus of RNA synthesis in corpus luteum chromatin (McKerns and Ryschkewitsch, 1977; McKerns, 1978). This is achieved by direct activation of all DNA-dependent RNA polymerases associated with the cell chromatin. In addition, there is an increase in nucleotide precursors (PP-ribose-P) because of activation of glucose-6-phosphate dehydrogenase in the pentose phosphate pathway. The increased pentose shunt also stimulates NADPH required for steroidogenesis. It was hoped that at least part of the active center would derive from a linear sequence of the primary structure of the beta subunits. Small peptides corresponding to a part of the active center of LH or hCG might be expected to be inhibitory to the action of these gonadotropins. A number of peptides in the range 8–12 amino acids were synthesized that were found to inhibit the LH-stimulated synthesis of RNA in isolated chromatin from corpus luteum. Some of these peptides inhibited ovulation in rats and terminated early pregnancy in mice. From these known structures it was possible to identify amino acid sequences in LH and hCG that were similar and presumably constitute part of the "active center." No such similar sequences were found in FSH beta. Subsequently, a peptide corresponding to hCG beta 133–137 was synthesized, as well as a peptide hCG alpha 34–38 coupled to hCG beta 133–137. In preliminary studies both peptides were found to be inhibitory to the action of LH or hCG *in vitro*. The amino acid sequences of the subunits of hCG are shown in Figs. 1 and 2.

ALPHA SUBUNIT

```
1                                                        10
ALA - PRO - ASP - VAL - GLN - ASP - CYS - PRO - GLU - CYS - THR - LEU -
                                          20
GLN - GLU - ASP - PRO - PHE - PHE - SER - GLN - PRO - GLY - ALA - PRO -
                                    30
ILE - LEU - GLX - CYS - MET - GLY - CYS - CYS - PHE - SER - ARG - ALA -
                  40
TYR - PRO - THR - PRO - LEU - ARG - SER - LYS - LYS - THR - MET - LEU -
      50                CHO                                          60
VAL - GLN - LYS - ASN - VAL - THR - SER - GLU - SER - THR - CYS - CYS -
                                                70
VAL - ALA - LYS - SER - TYR - ASN - ARG - VAL - THR - VAL - MET - GLY -
                              CHO         80
GLY - PHE - LYS - VAL - GLU - ASN - HIS - THR - ALA - CYS - HIS - CYS -
                              90
SER - THR - CYS - TYR - TYR - HIS - LYS - SER
```

Figure 1. Amino acid sequence of the hCG alpha subunit.

BETA SUBUNIT

```
1                                              ✝        10
SER - LYS - GLU - PRO - LEU - ARG - PRO - ARG - CYS - ARG - PRO - ILE -

CHO                                       20
ASN - ALA - THR - LEU - ALA - VAL - GLU - LYS - GLU - GLY - CYS - PRO -

                                    CHO
VAL - CYS - ILE - THR - VAL - ASN - THR - THR - ILE - CYS - ALA - GLY -

                  40
TYR - CYS - PRO - THR - MET - THR - ARG - VAL - LEU - GLN - GLY - VAL -

      50                                                        60 ✝
LEU - PRO - ALA - LEU - PRO - GLN - VAL - VAL - CYS - ASN - TYR - ARG -

                                                70
ASP - VAL - ARG - PHE - GLU - SER - ILE - ARG - LEU - PRO - GLY - CYS -

            ✝                             80
PRO - ARG - GLY - VAL - ASN - PRO - VAL - VAL - SER - TYR - ALA - VAL -

                              90                         ✝
ALA - LEU - SER - CYS - GLN - CYS - ALA - LEU - CYS - ARG - ARG - SER -

                  100
THR - THR - ASP - CYS - GLY - GLY - PRO - LYS - ASP - HIS - PRO - LEU -

      110                                                       120
THR - CYS - ASP - ASP - PRO - ARG - PHE - GLN - ASP - SER - SER - SER -

CHO                                 CHO             130             CHO
SER - LYS - ALA - PRO - PRO - PRO - SER - LEU - PRO - SER - PRO - SER -

                              CHO             140
ARG - LEU - PRO - GLY - PRO - SER - ASP - THR - PRO - ILE - LEU - PRO -

145
GLN
```

Figure 2. Amino acid sequence of the hCG beta subunit.

2. Methods in Peptide Synthesis

Peptides were synthesized by standard Merrifield solid-phase methods as described by Stewart and Young (1969), using a semiautomatic synthesizer developed by the author. The equipment, amino acids, and other chemicals necessary to perform the syntheses are readily obtainable commercially.

The following standard abbreviations are essentially those recommended by the IUPAC–IUB Commission on Biochemical Nomenclature:

Aoc: t-amyloxycarboxyl; Boc: t-butyloxycarboxyl; Bzl: benzyl; CH_2Cl_2: methylene chloride; Cl-Z: chlorocarbobenzoxy; DCCD: dicyclohexylcarbodiimide; DCHA: dicyclohexylamine; DMF: dimethylformam-

ide; MeOH: methanol; TBA: t-butylamine; TEA: triethylamine; TFA: trifluoroacetic (anhydrous); Tos: *p*-toluenesulfonyl; Z: carbobenzoxy.

Because a C-terminal amide was desired to resist enzymatic cleavage,
benzhydrylamine (BHA) resin (Beckman Bioproducts Department, Palo
Alto, California 94304) was used for the solid-phase peptide synthesis.
Alternatively, if a free C terminus is required, a resin such as the readily
available Merrifield resin can be used.

*Example: Synthesis of ⁴/₅-Ser-Arg-Tyr-Gly-Lys-Pro-Val-Gly-Lys-Lys-Val-
NH₂*

1. *Coupling of first Boc-amino acid to BHA resin (Fig. 3).* Three grams
of BHA resin was placed in a 50-ml polypropylene reaction vessel having
a filter disk. After adding 75 ml of 25%-by-vol solution of TEA in CH_2Cl_2,
the resin was stirred for 10 min to liberate the free amine. The TEA solution was then drawn off and the resin washed four times with 20-ml volumes of CH_2Cl_2. Boc-Val (571 mg or 2.5 meq/g resin) was added in 5 ml
CH_2Cl_2 and mixed briefly. The coupling agent DCCD in CH_2Cl_2 was added
in an amount equimolar to the Boc-Val, 2.5 mmoles/g resin, and mixed for
2 hr. After mixing, the solution was drawn off and the resin washed three
times with 20-ml volumes of CH_2Cl_2. To eliminate residual amino groups
on the resin, 75 ml of 25% TEA in CH_2Cl_2 was added and mixed for 10
min. Subsequently, the TEA solution was drawn off and acetic anhydride

Liberate free amine
1. 25% triethylamine (TEA) in CH_2Cl_2 (25 ml/g resin)
   Mix 10 min
   Draw off
2. *Wash* 4 × 20 ml CH_2Cl_2 ▢▢▢▢
3. *Add* 2.5 meq Boc-amino acid in CH_2Cl_2 (5 ml)
   Mix briefly
4. *Add* 2.5 mmoles dicyclohexylcarbodiimide in CH_2Cl_2
5. Mix 2 hr
6. *Draw off*
7. *Wash* 3 × 20 ml CH_2Cl_2 ▢▢▢

Terminating agents
8. 25% TEA
9. *Mix* 10 min
   Draw off
10. *Add* acetic anhydride (10 moles/mole amine) in CH_2Cl_2
11. *Mix* 10 min
12. CH_2Cl_2 (3 × 20 ml) ▢▢▢
13. MeOH (3 × 20 ml) ▢▢▢
    Dry in vacuo

Figure 3. Coupling of first Boc-L-amino acid to benzhydrylamine resin.

(10 moles/mole amine) in CH_2Cl_2 was added and mixed for 10 min. This was followed by three washings with 20-ml volumes of CH_2Cl_2 and three washings with 20-ml volumes of MeOH. After the washings, the solutions were drawn off and the resin dried *in vacuo*. An aliquot of resin was subsequently hydrolyzed with HCl. Amino acid analysis showed 0.35 meq Val coupled to the amine.

   2. *Coupling of second amino acid, Lys, obtained as Boc-Lys (Cl-Z).* *TBA.* The Boc-Lys (Cl-Z) acid was prepared from the commercially obtainable TEA salt before coupling by suspending 2 g of the powdered TBA salt in 10 ml ethyl acetate and then adding 12 meq (12 ml) of 1 N aqueous sulfuric acid. The solution was then shaken until the salt completely dissolved. The two resulting layers were separated and the aqueous layer reextracted twice with fresh ethyl acetate. The ethyl acetate extracts were combined and washed twice with water and once with saturated sodium chloride solution. The washed extracts were dried over anhydrous sodium sulfate. Concentration of the dry solution yielded an oily material, the free Boc-Lys acid, which was subsequently dissolved in CH_2Cl_2.

   Deprotection and amino acid coupling of 3 g of the resin-Val and the Boc-Lys acid were carried out by the following steps (Fig. 4): (1) three washings with 20-ml volumes of CH_2Cl_2; (2) a prewashing for 2 min with 20 ml of 25% TFA in CH_2Cl_2; (3) a washing with 20 ml TFA in CH_2Cl_2 for 30 min; (4) five washings with 20-ml volumes of CH_2Cl_2; (5) a 2-min prewashing with 20 ml of 10% TEA in CH_2Cl_2; (6) a 10-min washing with 20 ml TEA in CH_2Cl_2; (7) five washings with 20-ml volumes of CH_2Cl_2; (8) addition and mixing of 931 mg of the prepared Boc-Lys (Cl-Z) solution in approximately 5 ml CH_2Cl_2; (9) addition of approximately 2.5 meq (1.25 ml) of 2 M DCCD in CH_2Cl_2; (10) mixing for 90 min; (11) three washings with 20-ml volumes of CH_2Cl_2; (12) three washings with 20-ml volumes of MeOH; (13) three washings with 20-ml volumes of CH_2Cl_2; (14) a 2-min prewashing with 20 ml of 10% TEA in CH_2Cl_2; (15) a 10-min washing with 20 ml TEA in CH_2Cl_2; (16) five washings with 20-ml volumes of CH_2Cl_2; (17) addition and mixing of 931 mg of the prepared Boc-Lys (Cl-Z) solution in CH_2Cl_2; (18) addition of 2.5 meq DCCD in CH_2Cl_2; (19) mixing for 90 min; (20) three washings with 20-ml volumes of CH_2Cl_2; (21) three washings with 20-ml volumes of MeOH; and (22) three washings with 20-ml volumes of CH_2Cl_2.

   The above 22 steps were repeated for each of the amino acids of the compound D-Ser-Arg-Tyr-Gly-Lys-Pro-Val-Gly-Lys-Lys-Val-NH$_2$, coupling the following sequence of Boc- or Aoc-amino acids of the amounts indicated: 931 mg Boc-Lys (Z); 460 mg Boc-Gly; 570 mg Boc-Val; 538 mg Boc-Pro; 931 mg Boc-Lys (Cl-Z); 460 mg Boc-Gly; 929 mg Boc-Tyr (Bzl); 883 mg Aoc-Arg; 738 mg Boc-D-Ser (Bzl).

   If the Boc-D-Ser (Bzl) is obtained as a DCHA salt, the Boc-D-Ser

(Bzl) acid must be separated from the DCHA salt before coupling. This is accomplished by the procedure described above used to separate Boc-Lys (Cl-Z) acid from the TBA salt.

The final peptide resin was washed after step 22 with MeOH, dried, and stored in a glass desiccator in the refrigerator. The weight was 3.710 g. The protecting groups were removed and an aliquot, 1.85 g, of the above 11-amino-acid peptide-resin was cleaved in a Teflon HF apparatus by treatment with approximately 15 ml liquid hydrogen fluoride containing 5 ml anisole for 1 hr at 0°C. The hydrogen fluoride was removed by vacuum distillation and the anisole removed by ethyl acetate using filtration means.

To remove the peptide from the resin by filtration, the resin was washed several times with small volumes of 50% acetic acid. The combined filtrates were then lyophilized to obtain the peptide.

For purification purposes, the peptide was dissolved in a minimum volume of 0.5 M acetic acid and applied to a gel filtration column of Sephadex

	1. *Wash* 3 × 20 ml CH_2Cl_2 ☐☐☐
Deprotect	2. *Prewash* with 20 ml 25% trifluoroacetic (TFA) in CH_2Cl_2 (2 min) ☐
	3. *Wash* with 20 ml TFA in CH_2Cl_2 (30 min) ☐
	4. *Wash* 5 × 20 ml CH_2Cl_2 ☐☐☐☐☐
Neutralize	5. *Prewash* with 20 ml 10% triethylamine (TEA) in CH_2Cl_2 (2 min) ☐
	6. *Wash* with 20 ml TEA in CH_2Cl_2 (10 min) ☐
	7. *Wash* 5 × 20 ml CH_2Cl_2 ☐☐☐☐☐
	8. *Add* 2.5 meq Boc-amino acid in 5 ml CH_2Cl_2 ; mix (mg of Boc-) ☐
Couple	9. *Add* 2 M dichlorohexylcarbodiimide (DCCD) in CH_2Cl_2 (ml) (2.5 meq) ☐
	10. *Mix* 90 min ☐
	11. *Wash* 3 × 20 ml CH_2Cl_2 ☐☐☐
	12. *Wash* 3 × 20 ml MeOH ☐☐☐
	13. *Wash* 3 × 20 ml CH_2Cl_2 ☐☐☐
	14. *Prewash* with 20 ml 10% TEA in CH_2Cl_2 (2 min) ☐
	15. *Wash* with 20 ml TEA in CH_2Cl_2 (10 min) ☐
	16. *Wash* 5 × 20 ml CH_2Cl_2 ☐☐☐☐☐
Repeat	17. *Add* Boc-amino acid as per step 8; mix ☐
	18. *Add* DCCD as per step 9 ☐
	19. *Mix* 90 min ☐
	20. *Wash* 3 × 20 ml CH_2Cl_2 ☐☐☐
	21. *Wash* 3 × 20 ml MeOH ☐☐☐
	22. *Wash* 3 × 20 ml CH_2Cl_2 ☐☐☐

Figure 4. Deprotection and amino acid coupling.

G-25F, 1.6 cm $\times$ 190 cm long, equilibrated with 0.5 M acetic acid. The peptide was eluted with the same solvent and monitored by UV analysis. The fractions corresponding to the major peak were pooled and lyophilized to obtain 20 mg white fluffy powder.

For further purification, a partition column of Sephadex G-25F, 1.5 cm $\times$ 190 cm long, was prepared by equilibration with lower phase and then upper phase of the BAW solvent system (*n*-butanol: acetic acid: water, 4 : 1 : 5, V_H = 120 ml). The lyophilized peptide obtained by gel filtration was applied in 1.5 ml of upper phase. Elution with upper phase yielded one major peptide zone located as described above. After pooling and lyophilization, 90 mg of a white fluffy powder was obtained.

Amino acid analysis by a Beckman/Spinco analyzer of hydrolysates of this material yielded ratios close to the expected values for the following amino acids: Ser (1), Arg (1), Tyr (1), Lys (3), Pro (1), Val (2), Gly (2), NH_3 (1). Twenty-microgram loads of the peptide were homogeneous in acidic, neutral, and basic thin-layer chromatography systems when examined under UV light, iodine vapor, and Pauly reagent.

The synthesis illustrated in this example can be applied to all of the polypeptides, using the appropriate Boc- or Aoc-amino acids in the correct sequence. For example, to make the compound D-Ser-Arg-Ala-Tyr-Pro-Thr-Pro-Ala-Arg-Ser-Lys-Lys-NH$_2$, the following Boc-amino acids were used in sequence: Boc-Lys (Cl-Z); Boc-Lys (Cl-Z); Boc-Ser (Bzl); Aoc-Arg (Tos); Boc-Ala; Boc-Pro; Boc-Thr (Bzl); Boc-Pro; Boc-Tyr (Bzl); Boc-Ala; Aoc-Arg (Tos); Boc-D-Ser (Bzl). In all steps calling for addition of a Boc- or Aoc-amino acid, the amino acid is added at 2.5 meq/g resin.

3. Testing of Inhibitory Peptides

A number of peptides were examined for possible inhibitory effects of LH action *in vitro* on RNA synthesis in nuclei from corpora lutea or in chromatin prepared from corpora lutea nuclei. The incorporation of [8-^{14}C]-ATP into RNA was used to measure the activity of polymerase enzymes and the effect of LH or LH plus peptide on RNA synthesis. A high-ionic-strength incubation buffer, which favors polymerase II, was used but low-ionic-strength buffers were also studied. The methods are described by McKerns and Ryschkewitsch (1977) and McKerns (1978). It was found that the peptide D-Ser-Arg-Tyr-Gly-Lys-Pro-Val-Arg-Tyr-Lys-Lys-NH$_2$ was inhibitory to LH action. A peptide was also synthesized lacking the three positively charged Lys. This peptide was called E$_2$ and has the sequence: D-Ser-Arg-Tyr-Gly-Pro-Val-Gly-Val-NH$_2$. The effect of these peptides on LH stimulation of RNA synthesis is shown in Table I. The

Table I. Effect of Inhibitor Compounds on Lutropin
Stimulation of RNA Synthesis in Corpora Lutea Chromatin[a]

	pmoles [8-^{14}C]-ATP incorporated
Control	230
Lutropin	1290
E$_1$[b]	420
Lutropin + E$_1$	1190
E$_2$[c]	220
Lutropin + E$_2$	260
Control	1790
Lutropin	3100
Compound IV[d]	2070
Lutropin + compound IV	2400

[a] To the high-ionic-strength buffer were added chromatin (3.5 μg DNA), 1 μmole [8-^{14}C]-ATP, 1 μmole each CTP, GTP, and UTP, along with LH and peptides at 10^{-9} M. Incubation was for 5 min at 25°C. Each value is the mean of three closely agreeing values.
[b] E$_1$: D-Ser-Pro-Val-Gly-Val-NH$_2$.
[c] E$_2$: D-Ser-Arg-Tyr-Gly-Pro-Val-Gly-Val-NH$_2$.
[d] Compound IV: D-Ser-Arg-Tyr-Gly-Lys-Pro-Val-Arg-Ser-Lys-Lys-NH$_2$.

compound E$_2$ was most effective and essentially reduced the LH stimulation back to the control value. The other two compounds were less effective.

The effect of E$_2$ and three other peptides on LH-induced ovulation in the Nembutal-blocked cycling rat is shown in Table II. Virgin female Sprague–Dawley rats, weighing from 230 to 280 g, were housed in an air-conditioned room with the lights on from 5 a.m. to 7 p.m. After two consecutive 4-day vaginal cycles, the rats were injected on the early afternoon of proestrus with 32 mg Nembutal/kg body wt, or enough to render them unconscious. This blocks the endogenous release of LH. Between 1:30 and 2 p.m., 10 μg LH (NIH-LH S17) or 5 μg hCG (CR117) was injected into the jugular vein. One minute later one of the peptides in saline was injected. Additional peptide was injected 20 and 40 min later. Control rats receiving only Nembutal did not ovulate. All rats receiving LH or hCG following the Nembutal, ovulated 10–12 eggs between 10 and 11 a.m. the following morning. The rats were dissected and the eggs in the Fallopian tubes counted with the use of a low-power microscope. A minimum effective dose was not established in these experiments, but E$_2$ injected at 50–200 μg/rat, at 1, 20, and 40 min after LH, completely blocked ovulation. The other compounds were less effective.

E$_2$ was the most effective inhibitor in the two systems shown, and in other systems such as LH-induced androgen synthesis in a Leydig cell preparation. E$_2$ was also effective in terminating pregnancy in mice (data not shown). The linear sequences of the alpha and beta subunits of hCG and

Table II. Effect of Inhibitor Peptides on Lutropin-Induced Ovulation in the Nembutal-Blocked Rat

Compound	Dose (μg)	Number of animals		Average number of eggs	
		LH	LH compound	LH	LH compound
$E_2{}^a$	200	6	8	10	0
E_2	100	3	8	11	0
E_2	50	1	2	10	0
I^b	100	4	4	11	8
I	50	2	2	10	6
I	20	2	3	12	10
III^c	300	1	3	12	5
III	200	1	3	12	10
III	100	1	3	12	10
$E_1{}^d$	100	3	2	12	8

[a] E_2: D-Ser-Arg-Tyr-Gly-Pro-Val-Gly-Val-NH$_2$.
[b] I: D-Ser-Arg-Ala-Tyr-Pro-Thr-Pro-Ala-Arg-Ser-Lys-Lys-NH$_2$.
[c] III: D-Ser-Arg-Tyr-Gly-Lys-Pro-Val-Gly-Lys-Lys-Lys-NH$_2$.
[d] E_1: D-Ser-Pro-Val-Gly-Val-NH$_2$.

LH were examined for possible similar sequences to the 8-amino-acid peptide E_2. These similarities are:

residues 43–48 in hCG beta: Arg-Val-Leu-Gln-Gly-Val
residues 132–138 in hCG beta: Ser-Arg-Leu-Pro-Gly-Pro-Ser
residues 43–50 in hCG beta: Arg-Val-Leu-Gln-Gly-Val-Leu-Pro
residues 38–43 in LH alpha: Ser-Arg-Ala-Tyr-Pro-Thr
residues 34–38 in hCG alpha: Ser-Arg-Ala-Tyr-Pro

The similarities between E_2 and the subunit peptides are even more apparent when one considers the similarities in the side chains of the non-polar amino acids, Leu, Val, Ala, Pro, and Gly. To test if these compounds might be inhibitory, a peptide of amino acid residues 133–137 in hCG beta (Arg-Leu-Pro-Gly-Pro) was synthesized. Preliminary tests showed this compound to be inhibitory to the stimulatory effects of LH on androgen synthesis in an *in vitro* Leydig cell preparation.

Moreover, a peptide was synthesized consisting of hCG alpha 34–38 coupled to hCG beta 133–137 to give D-Ser-Arg-Ala-Tyr-[34] Pro-Arg-Leu-[38][133] Pro-Gly-Pro-NH$_2$[137]. This compound was also inhibitory *in vitro*. Many other alpha–beta or beta–alpha sequences are, of course, possible.

4. Discussion

The synthesis, purification, and testing of other peptides are under way. When more is known about the three-dimensional folding and other residues that might be adjacent to the identified sequences, then presumably the total active center can be assigned. However, it is interesting that at least part of the active center has been identified, if we assume that the inhibitory action of the peptides corresponding to sequences in the alpha and beta subunits of hCG and LH is at the biologically active site. This biologically active site appears to be inside the cell and to be part of the polymerase enzyme complex associated with the chromatin of the nucleus. In addition, the enzyme glucose-6-phosphate dehydrogenase in the pentose phosphate pathway is also activated by the gonadotropins LH and hCG. The active center appears to be a part of the gonadotropin structure distinct from the portion involved in binding to the cell surface.

In fact, these peptides inhibitory to the biological action of LH and hCG have no effect on the binding of FSH, LH, or PRL to cell surface receptors on Leydig cell membranes (P. A. Kelly, personal communication). Nevertheless, the inhibitory peptides are probably "recognized" at the target cell surface. The great number of cell surface receptors on target cells might simply be the mechanism for recognition and sequestration of the polypeptide hormones. The cell-recognition component of the gonado-

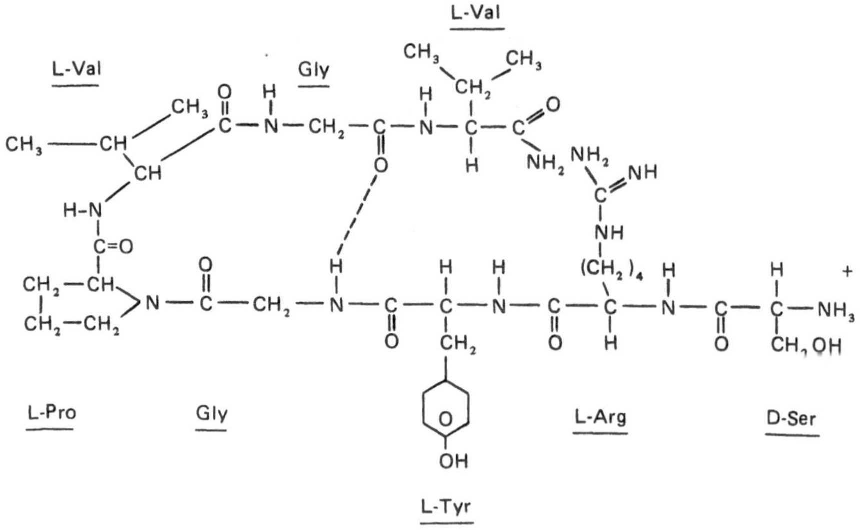

Figure 5. Possible conformation of E₂ peptide.

tropins may then be different from the biologically active component. The hormone–receptor complex invaginates (endocytosis), bringing the hormone into the cell (Szego, 1978; Tixier-Vidal, 1980; Vila-Porcile and Olivier, 1980). Regulatory processing within the cell separates the gonadotropin from the membrane receptor. The hormone, or a biologically active peptide from the hormone, then regulates cellular activity in the cytoplasm and nucleus.

A possible conformation of the 8-amino-acid peptide E_2 is shown in Fig. 5. The evidence for this derives from the work of Toma *et al.* (1980) and Piriou *et al.* (1980). There is the possibility of a Type I, beta turn in the proline residue with hydrogen bonding between Gly^4 and Gly^7. The Arg-Tyr-Gly residues are an essential component of the inhibitory activity as shown in the data of Tables I and II. The role of the rest of the molecule in binding and in antagonistic activity is unknown.

5. Summary

A number of peptides ranging from 8 to 12 amino acids were synthesized that showed inhibitory activity to the actions of LH or hCG in a number of test systems. One 8-amino-acid peptide was especially effective and appeared similar to amino acid sequences in the alpha and beta subunits of LH and hCG. It is possible that these sequences represent part of the active center of these gonadotropins. Analogues of the 8-amino-acid peptide hold promise as a once-a-month contraceptive compound and may also be of use in suppressing hormone-dependent tumors such as Leydig cell carcinoma and breast cancer.

6. Projections for the Future

A number of analogue peptides will be synthesized and tested for inhibitory action to LH and hCG, both *in vitro* and *in vivo*. The various test systems to be utilized include Leydig cell suspension, hormone-dependent mammary tumor, corpus luteum cell suspension, chromatin prepared from corpus luteum tissue, and a human prostate carcinoma in cell culture. Some *in vivo* test systems will include the inhibition of ovulation in cycling rats, the termination of pregnancy in mice during the period of their LH dependency, and the effect of inhibitory peptides on corpus luteum function and on pregnancy in baboons.

The octapeptide amide (E_2) having D-Ser at the N terminus will be resistant to aminopeptidases and carboxypeptidases. Other proteases can hydrolyze this peptide, such as chymotrypsinlike enzymes (C) at the Tyr-Gly bond. The Arg-Tyr bond can be attacked by trypsinlike proteases (T) as well as by amino dipeptidase I (ADP-I). Neutral endopeptidase (N) such as the kidney enzyme may cleave Val-NH$_2$ from the peptide:

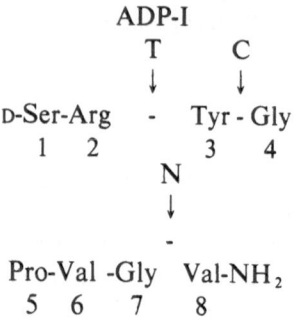

Hydrolysis of the Arg-Tyr bond can be blocked by *N*-methylation of Tyr (or alpha- or *ortho*-methyl-Tyr or -D-Tyr). In addition, a substitution of –Cl for the –OH of Tyr (*p*-Cl-Phe or *p*-Cl-D-Phe) may not only potentiate peptide binding and inhibitory activity but protect from ADP-I or T cleavage.

Replacement of Gly[4] and Gly[7] by some D-amino acid residues would prevent cleavage by C and N enzymes. It may also improve transport binding at the receptor site. This could consist of analogues containing D-Ala and D-Phe (or D-Gly) at positions 4 and/or 7. Dr. Sándor Bajusz suggests that if the D-Ala congener(s) shows higher potency than the D-Phe-containing one(s), D-Met or D-Nle analogues should be synthesized. In the opposite case the Gly residue(s) should be replaced by D-Trp or D-Tyr. Also to be considered would be the introduction of D-Thr, Trp, and Ile in place of D-Ser[1], Tyr[3], and Val[6] and/or Val[8].

The substitution of Ala for Val[6] may potentiate the beta turn at Pro.

Acknowledgments. I am grateful to Dr. Karl Lintner for helpful discussion on the conformation and activity of the inhibitory peptides at the time of the Society meeting in Dubrovnic in late September, 1980. I also wish to thank Dr. Sándor Bajusz and Dr. K. Nikolics for suggestions on analogue peptides that would be more stable to enzymatic degradation. Drs. Steven Birken and Robert Canfield provided the figures showing the linear sequences of the alpha and beta subunits of hCG. These figures were published previously in *Structure and Function of the Gonadotropins* (K. W. McKerns, ed.), Plenum Press, New York (1980).

DISCUSSION

KANAZIR: Dr. McKerns, what would be the component interacting with the chromatin? Is it a hormonally active fragment or a fragment of hormone–receptor complex?

My second question is, what determines the selectivity and specificity of transcription?

McKERNS: After the hormone–membrane receptor complex is brought into the cell (endocytosis), the membrane component is dissociated and probably recycled. For discussion of this see the chapter "Exocytosis and Related Membrane Events," by E. Vila-Porcile and L. Olivier in *Synthesis and Release of Adenohypophyseal Hormones*, edited by M. Jutisz and K. W. McKerns. I had postulated earlier that the polypeptide hormone may be processed within the cell and a fragment of the polypeptide hormone may be the biologically active component. For discussion of this, see the chapters by Clara M. Szego and by Peter Petrusz in *Structure and Function of the Gonadotropins* (K. W. McKerns, ed.), Plenum Press, 1978.

As to your second question, the gonadotropin or its biologically active fragment stimulates an increase in the rate of chain elongation, as well as an increase in the rate of initiation of all species of RNA. It does not seem to stimulate any new classes of RNA. Thus, the expression of the chromatin in a mature corpus luteum cell is already programmed and not modified by the gonadotropins.

NIKOLICS: Dr. McKerns, have you prepared radiolabeled peptide E_2, and conducted binding experiments with that? Also, what is the ratio of peptide E_2 versus LH in the experiments on inhibition of ovulation?

McKERNS: As to your first question, these experiments have not been done, but will be considered for the future. The lowest dose of peptide that would cause inhibition of ovulation has not been determined. *In vitro* experiments suggest that the peptide causes considerable inhibition at a mole to mole basis. The peptide is degraded much more rapidly than LH or hCG. One labile site for cleavage by aminopeptidase would be the Tyr-Gly bond. We propose to synthesize many analogues to try for enhanced biological activity *in vivo*. For example, the substitution of D-Ala in place of the two glycines.

PECK: I wish to comment about your assay for the inhibition of labeled hCG binding. Since this peptide is small, only 8 amino acids, it may interact with membrane receptor without interfering with the total reaction site for LH or hCG. In addition, complete kinetic analysis would be required to establish that the peptide does not interfere with hCG or LH binding. I do not find it attractive that there may be no membrane, that is, plasma membrane receptors. I would suggest that by analogy with the steroid receptor system, there may be endogenous nuclear receptors. For the estrogen Clark and I have described, Type II estrogen receptor persists in the nucleus, that is, it does not undergo translocation from the cytoplasm. It would be very interesting to label your peptide, preferably with tritium but at least with iodine and examine binding to isolated nuclei chromatin, nuclear matrix, and isolated plasma membranes. In this manner you might demonstrate specific binding sites.

McKERNS: I agree and hopefully in the future we can carry out such binding studies.

As Dr. Lintner describes in his studies with neuropeptides, there is both a binding component and a biological activating or inhibiting portion in many biologically active peptides. In the 8-amino-acid E_2 peptide, Arg^2, Tyr^3, and Gly^4 were essential to biological activity. The peptide probably has a beta turn around the Pro-Val with an H-bond between Gly^4 and Gly^7. This would present another face involved in binding or binding and biological activity.

References

Jungmann, R. A., and Hunzicker-Dunn, M., 1978. Mechanism of action of gonadotropins and the regulation of gene expression, in: *Structure and Function of the Gonadotropins* (K. W. McKerns, ed.), pp. 1–29, Plenum Press, New York.

McKerns, K. W., 1978, Regulation of gene expression in the nucleus by gonadotropins, in: *Structure and Function of the Gonadotropins* (K. W. Mc Kerns, ed.), pp. 315–328, Plenum Press, New York.

McKerns, K. W., and Ryschkewitsch, W., 1977, Lutropin stimulation of RNA synthesis in corpus luteum chromatin, *Biochim. Biophys. Acta* **478**:68–74.

Marsh, J. M., 1975, The mechanism of action of luteinizing hormone, in: *Advances in Cyclic Nucleotide Research*, Vol. 6 (P. Grungard and A. G. Robinson, eds.), p. 137, Raven Press, New York.

Petrusz, P., 1978, Gonadotropins–target cell interactions: A model based on morphological localization, in: *Structure and Function of the Gonadotropins* (K. W. McKerns, ed.), pp. 577–589, Plenum Press, New York.

Piriou, F., Lintner, K., Fermandjian, S., Fromageot, P., Khosla, M. C., Smeby, R. R., and Bumpus, F. M., 1980, Amino acid side chain conformation in angiotensin II and analogs: Correlated results of circular Dichroism and ^{1}H nuclear magnetic resonance, *Proc. Natl. Acad. Sci. USA* **77**:82–86

Stewart, J. M., and Young, J. D., 1969, *Solid Phase Peptide Synthesis*, Freeman, San Francisco.

Szego, C. M., 1978, Parallels in the modes of action of peptide and steroid hormones: Membrane effects and cellular entry, in: *Structure and Function of the Gonadotropins* (K. W. McKerns, ed.), pp. 431–472, Plenum Press, New York.

Tixier-Vidal, A., 1980 Structural basis of adenohypophyseal secretory processes, in: Synthesis and Release of Adenohypophyseal Hormones (M. Jutisz and K. W. McKerns, eds.), pp. 1–13, Plenum Press, New York.

Toma, F., Lam-Thanh, H., Piriou, F., Heindl, M.-C., Lintner, K., and Fermandjian, S., 1980, NMR evidence for a type I β-turn in (Pro^2)-tetrapeptides and interdependence of *cis:trans* isomerism, ring flexibility, and backbone conformation, *Biopolymers* **19**:781–804.

Vila-Porcile, E., and Olivier, L., 1980, Exocytosis and related membrane events, in: *Synthesis and Release of Adenohypophyseal Hormones* (M. Jutisz and K. W. McKerns, eds.), pp. 67–103, Plenum Press, New York.

5

Hormonal Antagonistic Properties of Deglycosylated Pituitary Lutropin

M. R. Sairam

1. Introduction

Detailed investigations of the structure and function of many hormones have been responsible for the design of structural analogues with desired agonistic or antagonistic activities. In the case of several small and large peptides and reasonably small proteins with hormonal activity, it has been possible to chemically synthesize analogues with desired properties using currently available synthetic technology and purification methods. The problem becomes more complicated when dealing with complex proteins such as the gonadotropins (LH, lutropin, luteinizing hormone; FSH, follitropin, follicle-stimulating hormone; hCG, human chorionic gonadotropin; PMSG, pregnant mare serum gonadotropin), which have 15–45% carbohydrate content. The last decade recorded significant developments related to the structure and function of the gonadotropic hormones (Sairam and Papkoff, 1974; Ward, 1978). Many investigations (see Ward, 1978) showed the relative importance of the α and β subunits and their amino acid side chains in the function of the hormones. Little was known concerning the role of the bulky oligosaccharide chains attached to the polypeptide backbone of the two subunits at one or more locations. This question lends itself to be studied in two possible ways in which the carbohydrates could be specifically removed by enzymatic or chemical

M. R. Sairam • Reproduction Research Laboratory, Clinical Research Institute of Montreal, Montreal, Quebec H2W 1R7, Canada.

means, without altering the polypeptide portion of the hormone. The former approach is dependent upon the availability of highly purified and specific exoglycosidases that remove susceptible sugar residues sequentially. This requires prolonged treatment and has been rather extensively studied with hCG (Moyle *et al.*, 1975). A more direct and quick approach investigated in our laboratory has resorted to the use of anhydrous hydrogen fluoride (HF) to remove oligosaccharide moieties in ovine LH in a manner similar to the process of removing protecting groups employed during the assembly of peptide chains in chemical synthesis. This has afforded us a means of obtaining modified pituitary LH that exhibits some very interesting biological properties, most notably that of antagonistic activity against the native hormone. Some of these observations are discussed in this chapter.

2. Preparation of Deglycosylated LH

Highly purified ovine LH can be deglycosylated by treatment of the dry powder with anhydrous liquid HF at 0°C for 75 min in the presence of a scavenger such as anisole. In this process the oligosaccharide moieties, which are present as three chains, are effectively removed by solvolysis. The removal of sugar residues under the conditions employed is by no means complete, but does not significantly affect the polypeptide portion of the hormone, at least as assessed by the many analytical methods employed (Sairam and Schiller, 1979). The deglycosylated hormone (DG-LH) can be recovered in powder form by lyophilization after suitable purification.

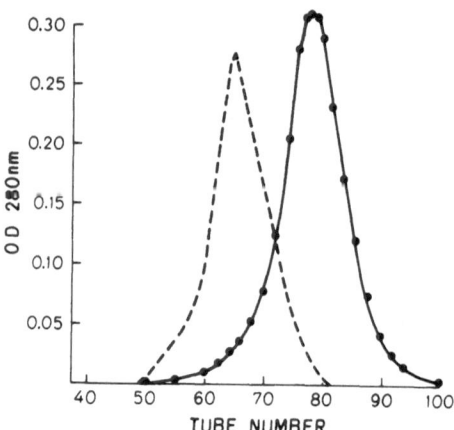

Figure 1. Chromatography of native oLH (- - - -) and DG-LH (——) on Sephadex G-100 (2.5 × 92 cm) in 0.05 M NH$_4$HCO$_3$ at 4°C, 3.2 ml/tube, 30 ml/hr flow rate. Note that DG-LH elutes with a much greater V_e/V_0 ratio.

Schematically, the procedure is as follows:

	anhydrous HF	Deglycosylated product
Native LH	→	+
	with anisole	Removed carbohydrate as adducts with anisole

$$\downarrow \text{ Sephadex G-100}$$

$$\text{Purified DG-LH}$$

The chromatographic behavior of the DG-LH is distinct from the native hormone as exemplified by the Sephadex G-100 patterns (Fig. 1) and affinity chromatography on concanavalin A-Sepharose column (Fig. 2). The DG-LH emerges much later than native LH on G-100 (Fig. 1), while on

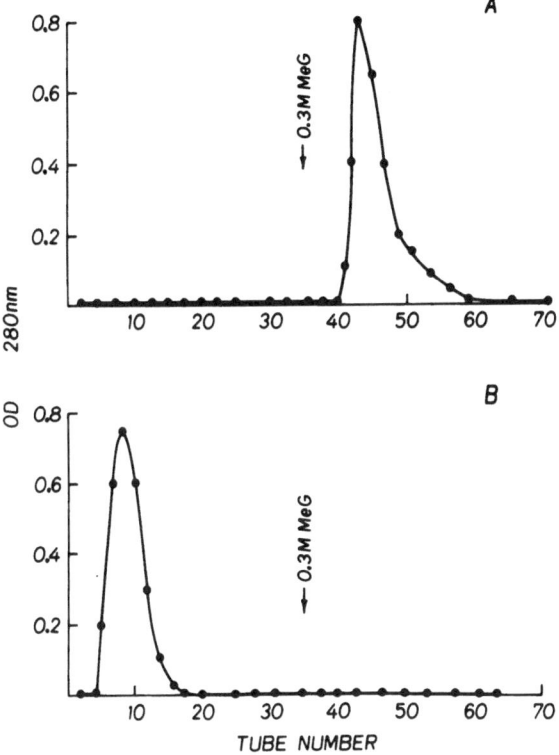

Figure 2. Behavior of native oLH (A) and DG-LH (B) on concanavalin A-Sepharose column at 4°C (1.5 × 23 cm). The column was equilibrated in 0.025 M Tris–HCl, pH 7.5, containing 1 mM MnCl$_2$, CaCl$_2$, 0.5 M NaCl, and 0.02% merthiolate; elution was carried out by 0.3 M α-methylglucopyranoside (MeG) dissolved in the starting buffer.

Table I.   Carbohydrate Composition of
Ovine LH and DG-LH (%) [a]

Component	oLH	DG-LH
Fucose	100 (2.83)	0
Hexoses	100 (10.0)	6
Galactosamine	100 (4.5)	0
Glucosamine	100 (12.8)	56

[a] The content of each residue (indicated in parentheses)
in oLH is set as 100%.

concanavalin A-Sepharose columns it is completely unadsorbed (Fig. 2).
This incidentally also reveals that there may be little or no native LH present after the HF treatment.

Of the different sugars that are present in the hormone, glucosamine is least affected by the treatment (Table I), most likely because of its close proximity to the peptide backbone to which it is linked. The overall loss of the carbohydrate moiety is approximately two-thirds.

Deglycosylation reduces the electrophoretic heterogeneity displayed by the native hormone on polyacrylamide gels (Sairam and Schiller, 1979). The DG-LH appears to be quite stable in powder form at 4°C as shown by retention of various activities over a 2- to 3-year period.

3.   Biological Properties of DG-LH

The biological and immunological properties of DG-LH are summarized in Table II and briefly discussed below.

Table II.   Biological and Immunological
Activities of Native and Deglycosylated
Ovine LH (%)

Activity	oLH	DG-LH
Receptor binding—ovary	100	80–100
Receptor binding—testis	100	80–100
cAMP—ovary	100	0
cAMP—testis	100	0
Progesterone—ovary	100	< 10 [a]
Testosterone—testis	100	< 5 [a]
Radioimmunoassay	100	100

[a] Response is nonparallel to that of native oLH and thus
estimates are only approximate.

An early step in the mechanism of action of peptide hormones is believed to be their binding to specific receptors on target cells. Consistent with this hypothesis, LH has been shown to effectively bind to membrane receptors localized on testicular and ovarian cells (Dufau and Catt, 1978). Following this an increase in adenylate cyclase activity occurs leading to the accumulation of cyclic AMP (cAMP) and formation of the end product, namely testosterone or estrogen/progesterone. Thus, the three events of hormone action—receptor binding, cAMP accumulation, steroidogenesis—have been assessed in collagenase-dispersed adult rat testicular interstitial cell suspensions or immature rat ovarian cell suspensions (Sairam, 1978).

3.1. Binding to Gonadal Receptors

For the examination of receptor binding, either rat testicular homogenates (Sairam, 1978) or pseudopregnant rat ovarian homogenates have been employed (Lee and Ryan, 1975). In both instances (Fig. 3), DG-LH effectively competed with ^{125}I-labeled ovine LH for binding to specific receptors. The calculated percentage retention of binding activity was in the range of 64–93% for several DG-LH preparations. Specificity of bind-

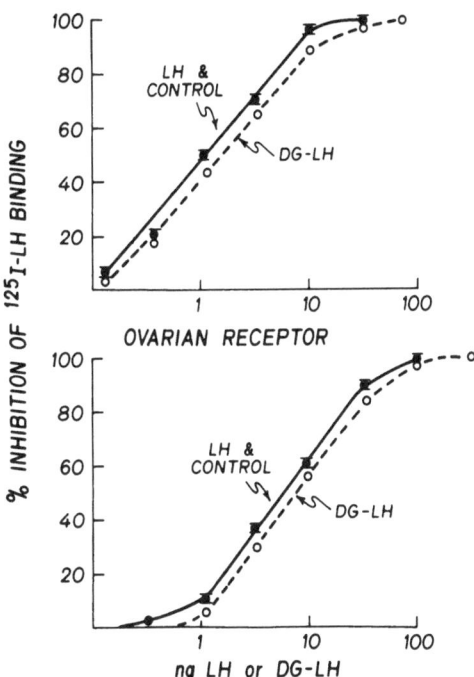

Figure 3. Comparison of the ability of oLH and DG-LH to compete with ^{125}I-labeled oLH for binding sites on adult rat testicular and pseudopregnant rat ovarian LH receptors. The binding experiments were done at 37°C for 2 hr. The specifically bound radioactivity was taken as 100% for the calculations. Control LH means that the hormone underwent the steps used in deglycosylation except for the omission of anhydrous HF.

ing of LH to its receptor was fully retained after deglycosylation as shown by the fact that a large excess of DG-LH did not affect the specific binding of ^{125}I-labeled ovine FSH to its own testicular receptor, which is distinct from that of LH.

3.2. cAMP Accumulation in Cell Suspensions

Although DG-LH preparations retained their effective capability to bind to gonadal membrane receptors as seen above, there appeared to be an almost complete loss in their ability to activate adenylate cyclase.

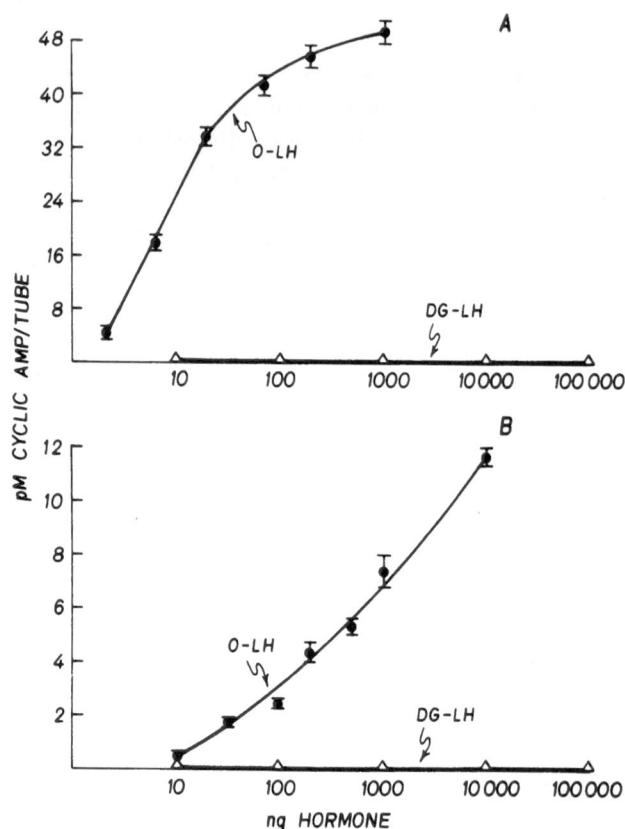

Figure 4. (A) Effect of oLH and DG-LH on cAMP accumulation in collagenase-dispersed mature rat testicular interstitial cells *in vitro*. (B) Effect of oLH and DG-LH on cAMP accumulation in immature rat ovarian cells *in vitro*. In both experiments the cells were incubated in the presence of 0.05m M IBMX (phosphodiesterase inhibitor). After 30 min at 37°C the reaction was stopped by immersion of tubes in an 80°C water bath and cAMP was then estimated by protein binding assay.

Despite the fact that the enzyme itself was not assayed, the end result can be inferred from Fig. 4. None of the DG-LH preparations stimulated cAMP accumulation in Leydig cell or ovarian cell suspensions even in the presence of the phosphodiesterase inhibitor isobutyl methyl xanthine (IBMX). This also excludes the remote possibility of an activation of phosphodiesterase which might have destroyed the accumulated cAMP in the presence of DG-LH (Sairam and Fleshner, 1981).

3.3. Steroidogenesis in Cell Suspensions

Ever since the postulate that cAMP in responsive tissues may serve as a second messenger (Robison *et al.*, 1971) there has been a continuous debate about its role in intracellular events. This argument extends to the gonadal cells as well where the exact relationship between cAMP and steroidogenesis still remains to be precisely defined (Moyle and Ramachandran, 1973; Catt and Dufau, 1973). However, recent studies (Dufau *et al.*,

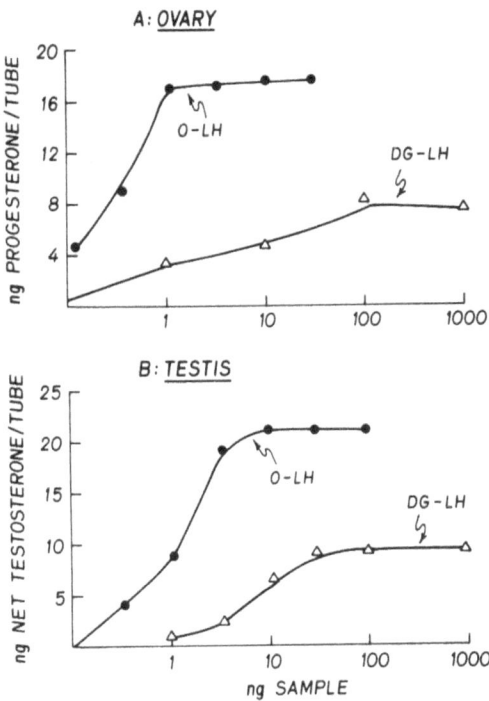

Figure 5. Steroidogenic activity of oLH and DG-LH in ovarian (A) and testicular (B) cell suspensions *in vitro*. The total volume of incubation in each case was 0.6 ml. Progesterone in (A) and testosterone in (B) were estimated by radioimmunoassay after incubation of cells for 3 and 2 hr, respectively.

1977; Ling and Marsh, 1977) suggest that indeed very minute amounts of cAMP are adequate to initiate (and maintain?) steroidogenesis in LH-responsive cells.

As no detectable amounts of cAMP accumulation were measured in our studies (see Fig. 4), it was of interest to evaluate the steroidogenic ability of the DG-LH preparations. As evident from Fig. 5, none of these elicited the characteristic steroidogenic response stimulated by the native hormone in both types of cells. In neither case did the response attain the same maximum as induced by the native hormone. However, there could be little doubt that the DG-LH preparations were indeed able to stimulate some degree of steroidogenesis, with a shallow dose–response. It is possible that the DG-LH preparations may have produced very small amounts of cAMP that were not measurable by the assay methods used in our studies (Sairam and Fleshner, 1981) but were still adequate enough to initiate some progesterone or testosterone synthesis. But the question why the steroidogenic response remains submaximal remains unresolved.

3.4. Hormonal Antagonism in Vitro

When a substance related in structure to a hormone binds to responsive cells of the target tissue, one may expect either an agonistic (in this instance, a stimulation) or an antagonistic (inhibitory) response. In view of the considerably good retention of binding activity of the DG-LH, with

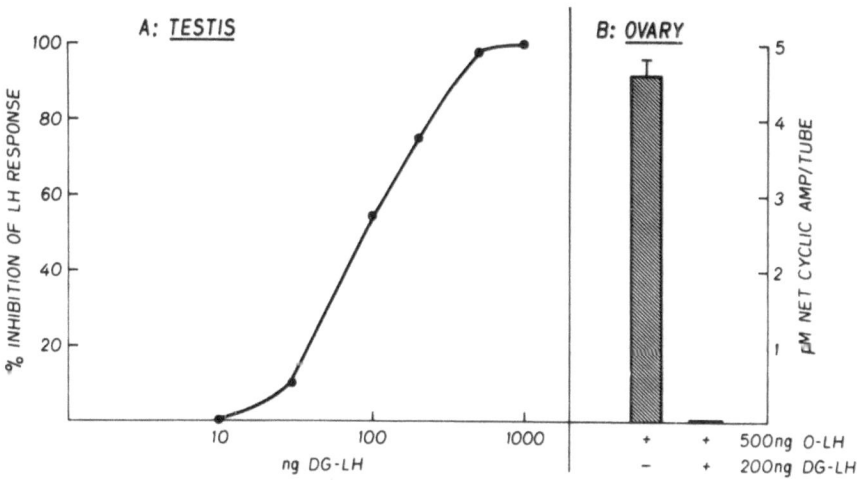

Figure 6. Inhibition of LH-induced cAMP response by DG-LH in (A) testicular and (B) ovarian cells. In (A) the net cAMP induced by 200 ng of oLH was taken as the 100% response for calculation of inhibition by DG-LH. In both experiments the oLH and DG-LH were added to the cells at the same time.

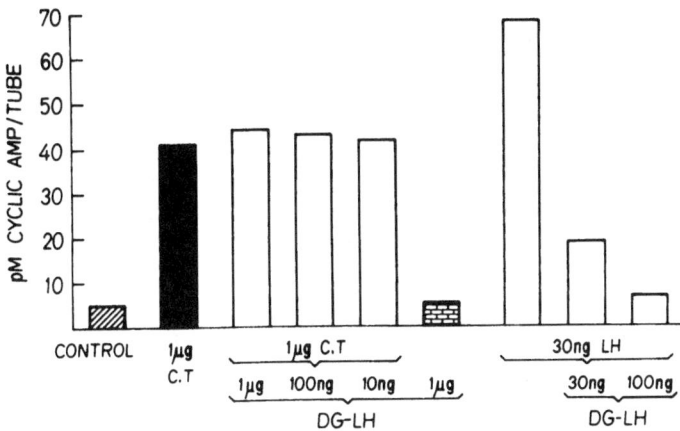

Figure 7. Inability of DG-LH to interfere with cholera toxin (CT; 1 μg) stimulation of cAMP accumulation in interstitial cells. Note that in the same set of incubations the response of oLH is inhibited by DG-LH. DG-LH was added to the cells at the same time as oLH or cholera toxin and the incubation medium contained 0.05 mM IBMX.

drastic losses in ability to induce cAMP and steroidogenic responses, their potential to inhibit hormone-induced events was investigated in gonadal cells.

It was very apparent that in both types of cells, DG-LH effectively inhibited cAMP response induced by a given concentration of ovine LH (Fig. 6). The inhibition was dose related and complete in the presence of high concentrations of DG-LH. That the inhibition of LH response in Leydig cells by DG-LH must be mediated by binding to specific receptors is revealed by the data in which the same preparation failed to alter cAMP accumulation induced by a nonhormonal agent such as cholera toxin (Sairam and Fleshner, 1981) (Fig. 7). The specificity of antagonism of LH

Table III. Failure of DG-LH to Affect cAMP Accumulation Elicited by Ovine FSH in the Testes

Treatment		pmoles cAMP/tube
oFSH	DG-LH	(mean ± S.E.M.)
0	0	24.4 ± 2.2 (4)
750 ng	0	97.6 ± 7.5 (6)
750 ng	5000 ng	96.2 ± 3.3 (4)

^aCollagenase-dispersed tubular suspensions from 20-day-old male rats were incubated at 37°C for 30 min in the presence of the indicated components. cAMP in the medium was later estimated by the protein binding assay.

action on the testes by DG-LH is clearly indicated by the fact that it is unable to affect cAMP response induced by ovine FSH in rat seminiferous tubule cell suspensions (Table III).

Accumulation of LH-induced testosterone production in Leydig cells and progesterone in ovarian cells *in vitro* was also inhibited by the concurrent presence of DG-LH (Fig. 8). Unlike inhibition of cAMP, which was usually complete, the reduction in steroidogenesis was incomplete. Some low level of steroidogenic activity, most likely due to the weak agonistic activity of the DG-LH, still persisted.

As reported elsewhere in greater detail, it is believed that the antagonistic activities of DG-LH with respect to cAMP accumulation (Sairam

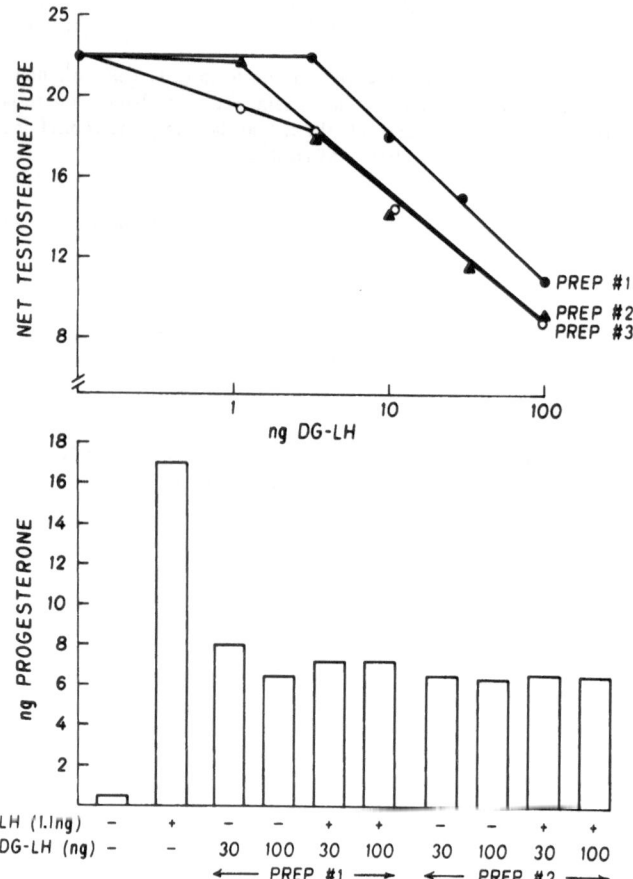

Figure 8. Inhibition of testosterone and progesterone accumulation in interstitial cells (top) and ovarian cells (bottom) by DG-LH. At the top the cells were incubated with 3.3 ng oLH, which resulted in the production of 22 ng testosterone/2 hr per tube. The three DG-LH preparations tested are noted as # 1, 2, and 3 and were each added along with the hormone.

and Fleshner, 1981) and testosterone production (Sairam and Schiller, 1979) are both competitive in nature.

3.5. Antagonistic Activity in Vivo

As known at present, the gonadotropins and thyrotropins are the only hormones that are glycoprotein in nature in the secreted form. It has been shown that removal of acidic sugar residues such as sialic acid drastically reduces the circulating half-life of gonadotropins like hCG and thus renders them inactive *in vivo* (Morell *et al.*, 1971). Hence, the presence of carbohydrate side chain (at least some of it) has been portrayed as playing an important protective role in circulation. The effects of removal of additional sugar residues on the metabolism of these hormones are not yet well understood. As the DG-LH displayed quite potent antagonistic activity in the *in vitro* studies, it was of considerable interest and importance to explore if it could exhibit similar activity *in vivo* also. We have examined this possibility in an immature rat model, wherein ovulation is induced by suitable administration of PMSG and ovine LH (Sairam, 1980). DG-LH preparations, in contrast to native LH, were unable to induce ovulation in the PMSG-primed animal when injected alone (Fig. 9). The number of ova released

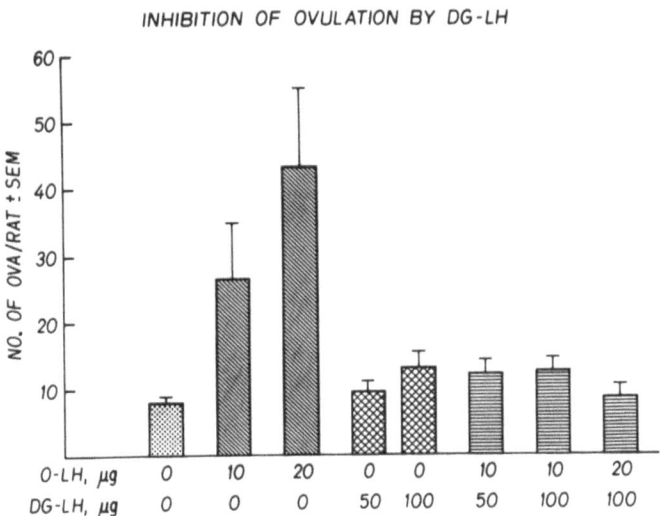

Figure 9. Inhibition of LH-induced ovulation by DG-LH. Twenty-five-day-old female rats were primed with PMSG (25 IU) on day 1. On day 3 at 12 noon, they were injected subcutaneously with 50% of the dose of DG-LH. At 3:30 p.m. of the same day, they received the appropriate dose of LH or LH + 50% of DG-LH. The animals were sacrificed on day 4 between 9 and 11 a.m. and the ova in the oviduct were counted under the microscope. Vehicle used for injection was 2% gelatin in 0.9% NaCl, pH 7.5. [Adapted from Sairam, 1980.]

into the oviduct were not significantly different from the controls. Animals treated with ovine LH released 3–6 times more ova than the controls. Prior or concurrent administration of DG-LH with the native ovine LH brought about a significant decrease ($p < 0.05$) in the ovulatory response of the animals.

These results tend to suggest that deglycosylated LH preparations can act as an inhibitor of ovulation, a process that requires the presence of LH in circulation for approximately 2 hr. The prior injection of the antagonist must hence be capable of blocking the action of the active hormone for this duration.

4. Concluding Comments

Recent studies from this laboratory have revealed that chemical deglycosylation is feasible and generates an interesting hormone derivative that has useful and interesting properties in the investigation of the various facets of the mechanism of action of the hormone on gonadal cells. The altered hormone was completely different from the native hormone in many of its physicochemical and biological properties. As far as can be ascertained, the peptide portion of the hormone remained unaffected. Specific (but incomplete) removal of the carbohydrate had drastic effects on some aspects of hormone function but not others. For example, while binding to membrane receptors was slightly decreased, other cellular responses such as cAMP increase or steroidogenesis were drastically curtailed. Immunological activity of the hormone was unaffected by partial deglycosylation. Our results at the present time indicate that apparently the conformational features required for receptor recognition are still present in DG-LH, but those that are necessary for full activation of the membrane-bound components of the adenylate cyclase and other cellular events are either lost or expressed improperly. The DG-LH was an effective antagonist of the hormone's action *in vitro*, and this phenomenon was a receptor-mediated event. The demonstration of *in vivo* antagonistic activity against the injected native hormone would lead one to believe that DG-LH may have an appreciable half-life long enough for it to bind to the ovarian receptor to block the action of native hormone. Whether its metabolism is substantially different from the native ovine LH remains to be carefully assessed by future studies.

ACKNOWLEDGMENTS. This work was supported in part by grants from the Medical Research Council of Canada and the World Health Organization in Geneva. I deeply appreciate the active participation of Messrs. P. Fleshner and M. Shevell in certain aspects of these investigations. I also thank Jayashree Sairam and Carole Chagnon-Labelle for assisting in this work.

REFERENCES

Catt, K. J., and Dufau, M. L., 1973, Spare gonadotropin receptors in rat testis, *Nature New Biol.* **244:**219–221.

Dufau, M. L., and Catt, K. J., 1978, Gonadotropin receptors and regulation of steroidogenesis in the testis ovary, *Vitam. Horm. (N.Y.)* **36:**461–492.

Dufau, M. L., Tsuruhara, T., Horner, K. A., Podesta, E., and Catt, K. J., 1977, Intermediate role of adenosine 3':5'-cyclic monophosphate and protein kinase during gonadotropin-induced steroidogenesis in testicular interstitial cells, *Proc. Natl. Acad. Sci. USA* **74:**3419–3423.

Lee, C. Y., and Ryan, R. J., 1975, Radioreceptor assay for human chorionic gonadotropin, *J. Clin. Endocrinol. Metab.* **40:**228–233.

Ling, W. L., and Marsh, J. M., 1977, Reevaluation of the role of cyclic adenosine 3',5'-monophosphate and protein kinase in the stimulation of steroidogenesis by luteinizing hormone in bovine corpus luteum slices, *Endocrinology* **100:**1571–1578.

Morell, A. G., Gregoriadis, A., Scheinbert, I. H., Hickman, J., and Ashwell, G., 1971, The role of sialic acid in determining the survival of glycoprotein in circulation, *J. Biol. Chem.* **246:**1461–1467.

Moyle, W. R., and Ramachandran, J., 1973, Effect of LH on steroidogenesis and cyclic AMP accumulation in rat Leydig cell preparations and mouse tumor Leydig cells, *Endocrinology* **93:**127–134.

Moyle, W. R., Bahl, O. P., and Marz, L., 1975, Role of the carbohydrate of human chorionic gonadotropin in the mechanism of hormone action, *J. Biol. Chem.* **250:**9163–9169.

Robison, G. A., Butcher, R. W., and Sutherland, E. W., 1971, *Cyclic AMP*, pp. 17–46, Academic Press, New York.

Sairam, M. R., 1978, Drug effects on lutropin action, in: *Structure and Function of the Gonadotropins* (K. W. McKerns, ed.), pp. 275–294, Plenum Press, New York.

Sairam, M. R., 1980, Inhibition of LH-induced ovulation in the rat by a hormonal antagonist, *Contraception* **21:**651–657.

Sairam, M. R., and Fleshner, P., 1981, Inhibition of hormone induced cyclic AMP production and steroidogenesis in interstitial cells by deglycosylated lutropin, *Mol. Cell. Endocrinol,* **22:**41–54.

Sairam, M. R., and Papkoff, H., 1974, Chemistry of pituitary gonadotropins, in: *Handbook of Physiology,* Section 7: *Endocrinology,* Vol. IV, Part 2 (E. Knobil and W. H. Sawyer, eds.), pp. 111–131, American Physiological Society, Washington, D.C.

Sairam, M. R., and Schiller, P. W., 1979, Receptor binding, biological and immunological properties of chemically deglycosylated ovine pituitary lutropin, *Arch. Biochem. Biophys.* **197:**294–301.

Ward, D. N., 1978, Chemical approaches to the structure–function relationships of luteinizing hormone (lutropin), in: *Structure and Function of the Gonadotropins* (K. W. McKerns, ed.), pp. 31–45, Plenum Press, New York.

6

Hormonal Regulation of and Ionic Requirements for in Vitro Release of Hypothalamic Peptides

Sophia V. Drouva, Jacques Epelbaum, and Claude Kordon

1. Introduction

Several neuropeptides have recently been identified within the central nervous system (see review in Snyder, 1980; Guillemin, 1978). Most of them are highly concentrated in hypothalamic areas important for neuroendocrine control (Vale *et al.*, 1980; Elde and Hökfelt, 1978). Overlapping anatomical distributions (Hökfelt *et al.*, 1980; Elde and Hökfelt, 1978; Fuxe, 1965) of several neuropeptides and neurotransmitters within these hypothalamic areas made it tempting to investigate neurotransmitter–neuropeptide interactions and their possible involvement in the adenohypophyseal regulation (Kordon *et al.*, 1980; Weiner and Ganong, 1978; Schally, 1978).

Evaluation of neurotransmitter or neuropeptide modulation of hypothalamic neurosecretory neurons can be achieved *in vitro. In vitro* preparations also make it possible to investigate cellular or ionic mechanisms underlying neuropeptide release.

The present chapter presents a critical review of the main neurotrans-

Sophia V. Drouva, Jacques Epelbaum, and Claude Kordon • Unité 159 de Neuroendocrinologie, Centre Paul Broca de l'INSERM, 75014 Paris, France.

mitter–neuropeptide or neuropeptide–neuropeptide interactions relevant for neuroendocrine control and reported to occur within the hypothalamus itself. Physiological relevancy and recent data on ionic mechanisms involved in these interactions will also be discussed.

2. Effects of Neurotransmitters or Neuropeptides on Hypothalamic Hormone Release in Vitro

Initial studies on the role of neurotransmitters in neuroendocrine control relied upon histophysiological correlation (Ahren *et al.,* 1971; Fuxe *et al.,* 1967, 1969)—that is, observations of correlated changes in the transmitter content of discrete neuronal systems and pituitary secretion rates— or upon pharmacological experiments affecting biosynthesis or release of a given neurotransmitter in the whole brain (Kordon *et al.,* 1980; McCann *et al.,* 1979; Weiner and Ganong, 1978). Each of these approaches has its advantages and its bias: histophysiological correlations permit determination of the anatomical level at which interactions take place, but do not define whether changes in the activity of a given transmitter are the cause or the consequence of correlated endocrine fluctuations. Conversely, pharmacological experiments are very global and do not localize interactions within the brain. Recently developed *in vitro* preparations—that is, incubation or superfusion systems of tissue fragments or slices, and culture of dispersed cells of pituitary or neurosecretory elements—permit both localization of interactions and assessment of their specificity with conventional pharmacological tools. *In vitro* approaches have of course their own limitations: survival time of neural tissue is limited, and data obtained on neurons separated from their afferent inputs are not necessarily applicable to the *in vivo* situation.

Assessment of the specificity of most interactions characterized *in vitro* is still incomplete. Among interactions reported so far and listed in Table I, we attempted to distinguish those that have been well characterized by pharmacological tools from those that require further confirmation. In our view, proper characterization should at least include dose–response curves fitting within the effective concentration range of other effects of the substance tested. This criterion is important, since numerous studies have shown that agonists or antagonists of neurotransmitters, for instance, can interact with recognition sites of other neurotransmitters at higher concentrations than those compatible with their specific effect (Enjalbert *et al.,* 1978). Whenever possible, further characterization with appropriate concentrations of antagonists or structural analogues represents also an important countercheck. Finally, binding studies permitting the assessment that actual binding sites are present on responsive structures and correlate with the biological effect should be included in the demonstration of a specific interaction. As far as neuroendocrine systems in the hypothalamus are concerned, the last criterion is usually lacking, due to the anatomical complexity of the region.

It is apparent from Table I that contradictory effects of neurotransmitters on neuropeptide release have been reported by different laboratories. This is probably due to the use of different preparations (i.e., synaptosomes vs. slices or fragments of hypothalamus) of different structures (i.e., median eminence, whole hypothalamus, mediobasal hypothalamus, etc.), or animals with different endocrine status.

Dopamine has been characterized as a luteinizing hormone-releasing hormone (LH-RH)-releasing transmitter *in vitro* (Negro-Vilar *et al.*, 1979; Rotsztejn *et al.*, 1977; Bennett *et al.*, 1975; Schneider and McCann, 1969); half-maximal effective doses (ED_{50}) necessary to produce the effect are comparable to those observed on other dopamine targets, and some dopamine antagonists, but not all, cancel the effect (Rotsztejn *et al.*, 1977). The action of dopamine on LH-RH is in agreement with observations that, under certain physiological conditions, dopamine can facilitate LH release (Kordon *et al.*, 1980; McCann *et al.*, 1979). In addition to this action, however, a distinct inhibitory effect of the amine on gonadotropin secretion has also been documented, in particular in castrated rodents (Drouva and Gallo, 1976, 1977; Gnodde and Schuiling, 1976). Its site of action is still unknown.

Various reports also indicate that serotonin can inhibit the release of LH-RH from the mediobasal hypothalamus. The effect has a relatively minor amplitude, but occurs for concentrations of the transmitter compatible with the hypothesis of a specific interaction (Charli *et al.*, 1978b). Its characterization by antagonists is still incomplete. Effects of other transmitters on LH-RH release have to be considered more cautiously, because they give rise to contradictory reports, were observed in the presence of high transmitter concentrations only (Fiorindo and Martini, 1975), and still lack confirmation by corresponding antagonists (Charli *et al.*, 1978b) (Table I).

Noradrenalin, dopamine, and histamine have been reported to release thyrotropin-releasing hormone (TRH) *in vitro* (Maeda and Frohman, 1980; Joseph-Bravo *et al.*, 1979; Charli *et al.*, 1978a; Hirooka *et al.*, 1978; Bennett *et al.*, 1975; Grimm and Reichlin, 1973) (see Table I); complete characterization of the effects is not fully conclusive yet.

Contradictory results have been also reported for most neurotransmitter or neuropeptide interactions on somatostatin (SRIF) release *in vitro* (Table I). For this peptide, the most extensively characterized effects appear to be the inhibitory action of the vasoactive intestinal peptide (VIP), which is dose dependent and is also observed in the presence of a structural

Table I. Modulation of the Release of Hypothalamic Neuropeptides in Vitro

Peptide	Substance	Stimulation	Inhibition	None
		Effect[a]		
SRIF	GH	*1*		
	T$_3$	*2*		
	Substance P	*3*		27
	Neurotensin	*4*		
	DA	*5*	16	28
	NA	*6*	16	29
	ACh		17	30
	GABA		18	31
	5-HT			32
	PGE$_2$			33
	Opiates			
	On basal release			34
	On induced release		19	
	TRH			35
	TSH			36
	Glucagon			37
	VIP		20	
	Secretin		21	
LH-RH	DA	*7*		38
	NA	*8*		39
	PGE$_2$	*9*		
	Melatonin	*10*		
	VIP	*11*		40
	ACh			41
	GABA			42
	5-HT		22	43
	Opiates			
	On basal release			44
	On induced release		23	
	Epinephrine			45
	Histamine			46
TRH	DA	*12*		47
	NA	*13*		48
	Histamine	*14*		
	ACh			48
	GABA			49
	5-HT		24	50
	Epinephrine			51
	Substance P			52
	Neurotensin			52
	SRIF			
	On basal release			52
	On NA-induced release		25	

Table I. Modulation of the Release of Hypothalamic Neuropeptides in Vitro
(Continued)

Peptide		Stimulation	Inhibition	None
CRF	ACh	*15*		
	5-HT	*15*		
	DA			*53*
	Histamine			*53*
	Glycine			*53*
	NA: on ACh- and 5-HT-induced release		*26*	
	GABA: on ACh- and 5-HT-induced release		*26*	
	Melatonin: on ACh- and 5-HT-induced release		*26*	

ᵃ References: *1*, Kanatsuka *et al.* (1979), Sheppard *et al.* (1978); *2*, Berelowitz *et al.* (1980); *3*, Sheppard *et al.* (1979); *4*, Maeda and Frohman (1980), Sheppard *et al.* (1979); *5*, Berelowitz *et al.* (1980), Maeda and Frohman (1980), Negro-Vilar *et al.* (1978); *6*, Epelbaum *et al.* (1981), Negro-Vilar *et al.* (1978); *7*, Bennett *et al.* (1975), Negro-Vilar *et al.* (1979), Rotsztejn *et al.* (1976, 1977), Schneider and McCann (1969); *8*, Negro-Vilar *et al.* (1979); *9*, Gallardo and Ramirez (1977), Linton *et al.* (1979); *10*, Wei Lin Kao and Weisz (1977); *11*, Samson *et al.* (1980); *12*, Maeda and Frohman (1980); *13*, Hirooka *et al.* (1978); *14*, Charli *et al.* (1978a), Joseph-Bravo *et al.* (1979); *15*, Jones *et al.* (1976); *16*, Bennett *et al.* (1979); *17*, Richardson *et al.* (1980); *18*, Gamse *et al.* (1980); *19*, Drouva *et al.* (1980, 1981b); *20*, Drouva *et al.* (1981b), Epelbaum *et al.* (1979); *21*, Epelbaum *et al.* (1979); *22*, Charli *et al.* (1978b); *23*, Drouva *et al.* (1980, 1981b), Rotsztejn *et al.* (1978); *24*, Bennett *et al.* (1975); *25*, Hirooka *et al.* (1978); *26*, Jones *et al.* (1976); *27*, Maeda and Frohman (1980); *28*, Terry *et al.* (1980); *29*, Epelbaum *et al.* (1981), Maeda and Frohman (1980), Terry *et al.* (1980); *30*, Bennett *et al.* (1979), Maeda and Frohman (1980), Terry *et al.* (1980); *31*, Epelbaum *et al.* (1981), Terry *et al.* (1980); *32*, Maeda and Frohman (1980); *33*, Terry *et al.*, (1980); *34*, Drouva *et al.* (1980, 1981b), Sheppard *et al.* (1979), Terry *et al.* (1980); *35*, Berelowitz *et al.* (1980), Maeda and Frohman (1980); *36*, Berelowitz *et al.* (1980); *37*, Epelbaum *et al.* (1979); *38*, Wei Lin Kao and Weisz (1977); *39*, Bennett *et al.* (1975), Rotsztjn *et al.*, (1977), Wei Lin Kao and Weisz (1977); *40*, Drouva *et al.* (1981b); *41*, Bennett *et al.* (1975), Charli *et al.* (1978b), Wei Lin Kao and Weisz (1977); *42*, Charli *et al.* (1978b); *43*, Bennett *et al.* (1975), Wei Lin Kao and Weisz (1977); *44*, Drouva *et al.* (1980, 1981b), Rotsztejn *et al.* (1978); *45*, Negro-Vilar *et al.* (1979); *46*, Charli *et al.* (1978a); *47*, Bennett *et al.* (1974); *48*, Bennett *et al.* (1975), Joseph-Bravo *et al.* (1979), Maeda and Frohman (1980); *49*, Joseph-Bravo *et al.* (1979); *50*, Joseph-Bravo *et al.* (1979), Maeda and Frohman (1980); *51*, Bennett *et al.* (1975); *52*, Maeda and Frohman (1980); *53*, Jones *et al.* (1976).

analogue of VIP, secretin (Epelbaum *et al.*, 1979). In a perifusion system, nanomolar concentrations of VIP also inhibited K^+-induced SRIF release (Drouva *et al.*, 1981b). At the same dose, the peptide had no effect on spontaneous or evoked LH-RH release, an observation that suggests a certain target specificity of VIP receptors presumably located on the terminals of SRIF neurons in the mediobasal hypothalamus (Drouva *et al.*, 1981b). The inhibitory effect of VIP could account for the *in vivo* stimulation of growth hormone (GH) secretion by systemic administration of the peptide (Vijayan *et al.*, 1979).

Effects of other transmitters or peptides are not as clear (Table I). For example, norepinephrine has been reported to stimulate SRIF release from median eminence fragments *in vitro* (Negro-Vilar *et al.*, 1978), while it inhibits at high concentrations the release of the peptide from hypothalamic synaptosomes (Bennett *et al.*, 1979). In addition, Epelbaum *et al.* (1981) found that the transmitter has no effect on the release of SRIF from incubated mediobasal hypothalamic slices at any dose tested, while it stimulates SRIF release in a dose-dependent manner, from incubated slices of preoptic area or amygdala (Fig. 1). Of interest is the finding that stimulatory effects of norepinephrine in preoptic area and amygdala are mediated by different types of adrenergic receptors. Phentolamine, an α-adrenergic antagonist, blocked the effect of norepinephrine only on preoptic area slices, while propanolol, a β-blocker, abolishes SRIF stimulation by norepinephrine in the amygdala, but not in the preoptic area (Fig. 2). Different populations of adrenergic receptors thus appear to be predominantly involved in the regulation of SRIF release in these two brain regions. The relevancy of these interactions in terms of GH secretion remains, however, to be elucidated.

Although GABA was found to be ineffective on SRIF release from incubated mediobasal hypothalamic slices (Epelbaum *et al.*, 1981) or perifused hypothalamic fragments (Terry *et al.*, 1980), Gamse *et al.*, (1980) reported that in micromolar concentrations, it inhibits the release of the peptide from hypothalamic cells in culture; the effect is reversed by biculline, a GABA receptor antagonist. These results are consistent with the findings that intraventricularly injected GABA leads to an increase of hypothalamic SRIF content (Takahara *et al.*, 1980), and that GABA and GABA agonists administered *in vivo* elevate plasma levels of GH (Takahara *et al.*, 1980; Vijayan and McCann, 1978).

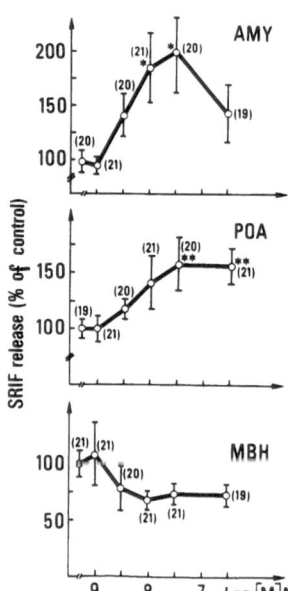

Figure 1. Effect of noradrenalin (NA) on SRIF release from mediobasal hypothalamus, preoptic area, and amygdala. [From Epelbaum *et al.*, 1981.]

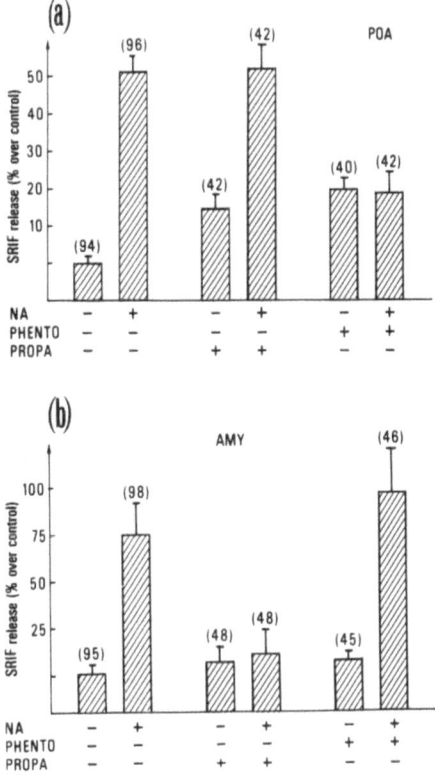

Figure 2. Effect of phentolamine (10^{-7} M) and propanolol (10^{-7} M) on SRIF release induced by NA (3×10^{-8} M). [From Epelbaum et al., 1981.]

Recently, Richardson et al. (1980) reported dose-dependent inhibition of SRIF release from rat hypothalamic fragments in culture by acetylcholine; since atropine, a muscarinic anticholinergic substance, blocks the inhibitory effect of acetylcholine, they suggest the involvement of muscarinic cholinergic mechanisms in the regulation of SRIF release and thus GH secretion. However, other authors found acetylcholine to be ineffective on SRIF release from hypothalamic fragments (Maeda and Frohman, 1980; Terry et al., 1980).

Finally, stimulation of corticotropin-releasing factor (CRF) by acetylcholine and serotonin and inhibition by GABA, noradrenaline, and melatonin have been reported (Jones et al., 1976) (Table I). Confirmation of these effects is needed, since characterization of a corticotropin-releasing factor has recently been published (Vale et al., 1981).

3.   *Interaction of Opiates with LH-RH and SRIF Release*

Several studies (Holaday and Loh, 1979; Meites *et al.*, 1979) indicate that opiate peptides stimulate GH and inhibit LH secretion. Opiates have no direct effect on GH pituitary cells (Rivier *et al.*, 1977; Shaar *et al.*, 1977) nor do they alter the response of gonadotropins to LH-RH (Cicero *et al.*, 1977). A hypothalamic site of action is thus likely, as also suggested by observations that opiate peptides, on one hand, and LH-RH and SRIF, on the other, show a well-correlated distribution in the mediobasal hypothalamus (Hökfelt *et al.*, 1980; Elde and Hökfelt, 1978). In order to evaluate whether a direct action of opiates on the release of SRIF and LH-RH could account for their *in vivo* effect on GH and LH, basal and induced release of both neurohormones from mediobasal hypothalamic slices was monitored in a superfusion system in the presence of different opiate agonists or antagonists.

As shown in Fig. 3, opiates (β-endorphin, Met-enkephalin, D-Ala^2-Met-enkephalin, Leu-enkephalin, morphine), administered from the beginning of the superfusion, did not alter the spontaneous release of LH-RH and SRIF. In contrast, they significantly inhibited maximal K$^+$-induced neuropeptide release (Figs. 3, 4) and completely blocked secretion provoked by lower concentration (28 mM) of K$^+$ (Drouva *et al.*, 1981b). The inhibitory effect was dose dependent (Fig. 5) and was reversed in all cases by a specific opiate antagonist, naloxone (10^{-7} M), which was ineffective when

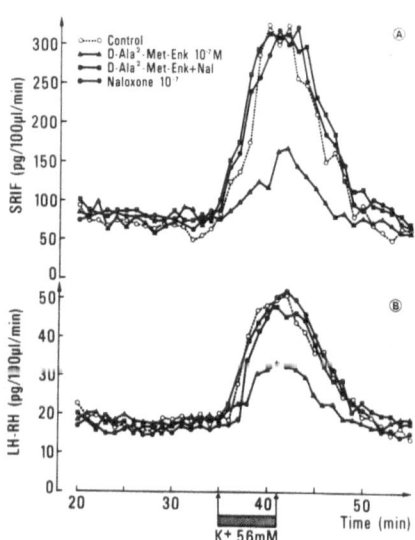

Figure 3. Effect of D-Ala^2-Metenkephalin (10^{-7} M) and naloxone (10^{-7} M) on spontaneous and K$^+$-induced LH-RH and SRIF release from mediobasal hypothalamic slices. [From Drouva *et al.*, 1981b.]

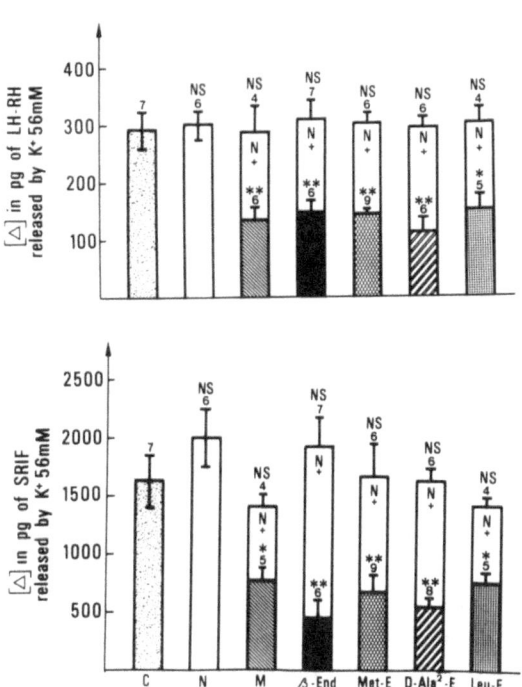

Figure 4. Inhibitory effects of opiates on K^+-induced LH-RH and SRIF release from mediobasal hypothalamic slices. Reversal effect by naloxone (N) $(10^{-7}$ M). C, control; hatched bars show the opiate agonist effect: M, morphine $(10^{-6}$ M); β-End, β-endorphin $(10^{-7}$ M); Leu-E, leucine-enkephalin $(10^{-7}$ M). Values represent the mean $\pm$ S.E. of $[\Delta]$. $[\Delta]$ represents the total amount of LH-RH and SRIF released during the whole secretory response to K^+ depolarization minus the amount of neuropeptides released over the same period of time under basal conditions. NS, not significant vs. control; *F = 5%, **F = 1% level of significance vs. control.

perifused alone (Figs. 3, 4), an observation indicating that the opiate effect could be mediated by specific opiate receptors present on LH-RH and SRIF neurons. By analogy with studies on other peptidergic neurons (Iversen et al., 1980; Iversen and Jessell, 1979; Clarke et al., 1979; Pollard et al., 1977) and since our experimental conditions permit mainly evaluation of peptide release from nerve endings, the receptors can be assumed to be presynaptic.

Opiates are also effective in antagonizing the release of LH-RH induced by dopamine; under these conditions, naloxone also prevents the effect (Rotsztejn et al., 1978).

The fact that opiates alter depolarization-induced (Drouva et al., 1980, 1981b) but not basal release of LH-RH and SRIF (Drouva et al., 1980, 1981b; Sheppard et al., 1979; Rotsztejn et al., 1978) points out a modulatory action on neuropeptide release; they appear to interfere with cellular events coupling action potentials with subsequent exocytotic response (Bloom, 1979; Kupfermann, 1979). The possibility that such uncoupling involves changes in membrane permeability to ions will be considered in the next section.

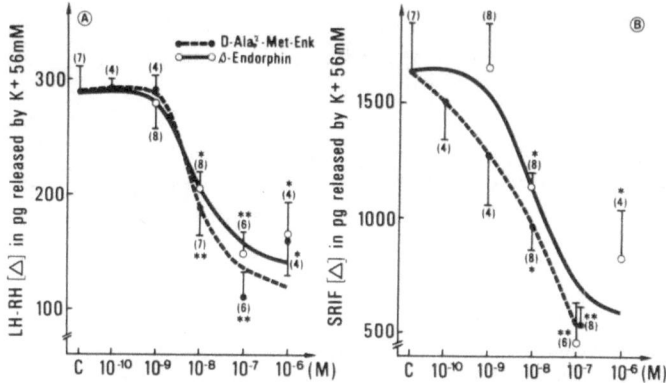

Figure 5. Dose–response of D-Ala²-Met-enkephalin and β-endorphin on K⁺-induced LH-RH and SRIF release. [From Drouva *et al.*, 1981b.]

4. Ionic Requirements of Neuropeptide Release

Ever since the experiments of Douglas and Poisner (1964a) established the stimulus–secretion coupling hypothesis, there has been wide agreement that peptide secretion is a Ca^{2+}-dependent process (Kordon, 1979; Iversen and Jessell, 1979). Treatments resulting in membrane depolarization—such as induction of action potentials, use of high extracellular concentrations of K^+ under *in vitro* conditions, or addition of depolarizing drugs such as veratridine—all result directly or indirectly in opening of voltage-sensitive Ca^{2+} channels and peptide exocytosis, provided extracellular Ca^{2+} is present. A Ca^{2+}-dependent release of several neuropeptides has been reported by different laboratories using one of the *in vitro* techniques, i.e., the release of LH-RH (Drouva *et al.*, 1981a; Dyer *et al.*, 1980; Harter and Ramirez, 1980; Negro-Vilar *et al.*, 1979; Gallardo and Ramirez, 1977; Warberg *et al.*, 1977; Rotsztejn *et al.*, 1976), SRIF (Drouva *et al.*, 1981a; Gamse *et al.*, 1980; Maeda and Frohman, 1980; Richardson *et al.*, 1980; Terry *et al.*, 1980; Epelbaum *et al.*, 1979; Berelowitz *et al.*, 1978; Iversen *et al.*, 1978), TRH (Maeda and Frohman, 1980; Joseph-Bravo *et al.*, 1979; Charli *et al.*, 1978a; Warberg *et al.*, 1977), substance P (Iversen *et al.*, 1976), neurotensin (Iversen *et al.*, 1978), opiates (Iversen and Jessell, 1979; Henderson *et al.*, 1978; Osborne *et al.*, 1978), and CCK-8 (Dodd *et al.*, 1980), VIP (Emson *et al.*, 1978), CRF (Bennett and Edwardson, 1975).

In vitro preparations are not all equally sensitive to Ca^{2+}-dependent depolarization-induced release of peptide. In particular, under conventional conditions of purified nerve ending (synaptosome) preparations, K^+ depolarization releases only small amounts of peptides; both induced and basal peptide secretion are only slightly inhibited by Ca^{2+} removal. This has been taken as an indication that integrity of granular peptide pools and/or membrane permeability may be disrupted in the fractionation process (Ramirez and Kordon, 1978).

In contrast, data obtained from minced cerebral tissue are in better agreement with the stimulus–secretion coupling hypothesis. On this material, use of perifusion systems offers the advantage of evaluating the time-dependent dynamics of the neurosecretory process, while it eliminates possible feedback effects on the tissue of other substances released simultaneously; in addition, it permits an accurate analysis of ionic requirements of peptide release, as shown by the following examples.

Spontaneous release of LH-RH and SRIF from mediobasal hypothalamic slices stabilizes after 20 min of superfusion (Fig. 6). The onset of the

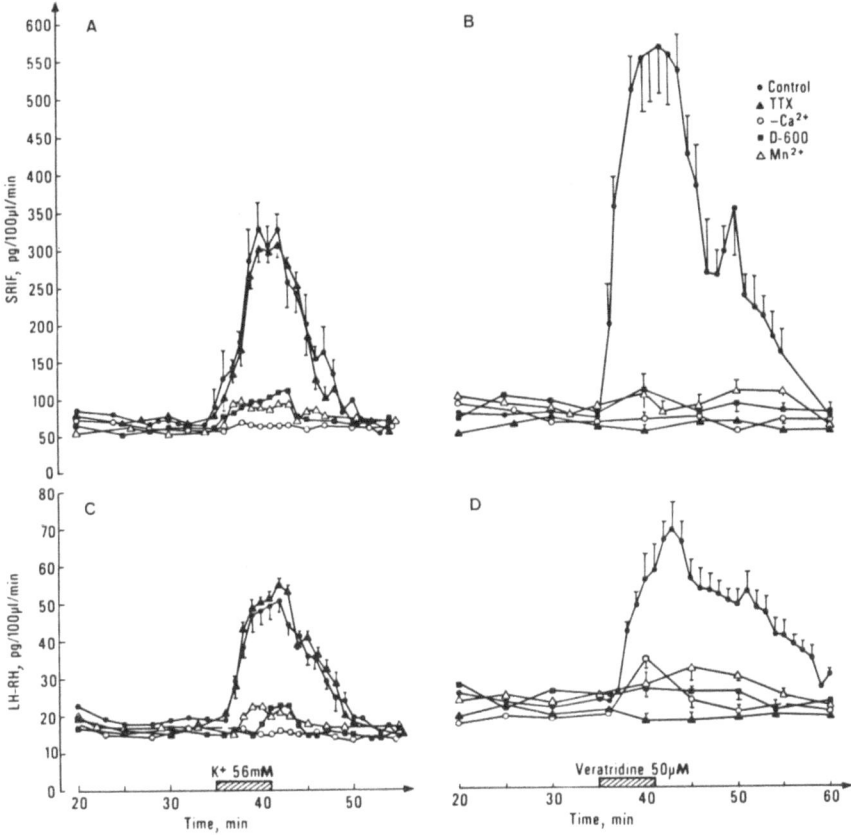

Figure 6. Spontaneous and K^+- or veratridine-induced LH-RH and SRIF release from mediobasal hypothalamic slices superfused with either control Locke medium or Locke medium containing 5×10^{-7} M tetrodotoxin, 10^{-4} M D-600, or 3 mM Mn^{2+}. Values represent the mean $\pm$ S.E. from four experiments. [From Drouva *et al.*, 1981a.]

response to depolarization by K^+ or veratridine is rapid and dose dependent. Maximal release is obtained with 56 mM K^+ and 50 μM veratridine
(Fig. 7). Removal of Ca^{2+} (with or without replacement with Mg^{2+}) blocks
neuropeptide release induced by both agents (Fig. 6). In addition, the magnitude of the secretory response to K^+ is related to the Ca^{2+} concentration
of the medium (Fig. 8). According to the stimulus–secretion coupling
hypothesis and as verified from $^{45}Ca^{2+}$ uptake studies (Nordmann and
Dyball, 1978; Blaustein, 1975; Douglas and Poisner, 1964a,b), depolarization stimulates secretion by promoting influx of Ca^{2+} into neurosecretory
terminals. Kinetics of LH-RH and SRIF response to depolarization (Figs.
6, 7), sustained release after successive pulses of K^+ (56 mM; Fig. 9),
independence of release profiles of both neuropeptides from depolarization
duration (Fig. 10), and Ca^{2+} dependency of the effect (Figs. 6, 8) are all
consistent with Ca^{2+} uptake measurements (Nordmann, 1976; Blaustein,
1975; Baker *et al.*, 1973) and strongly indicate that Ca^{2+} is a prerequisite
for the neurosecretory response. In contrast, neither Mg^{2+} nor chlorine is
needed for spontaneous or evoked release of LH-RH and SRIF (Drouva *et
al.*, 1981a).

In order to further define the type of ionic channels involved in Ca^{2+}
influx during depolarization, different blockers of ionic channels were used.

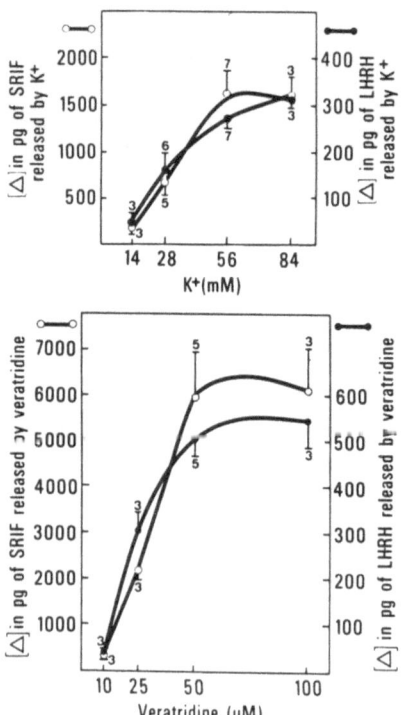

Figure 7. Dose–response effect of K^+ and
veratridine on LH-RH and SRIF release from
superfused mediobasal hypothalamic slices.
[From Drouva *et al.*, 1981a.]

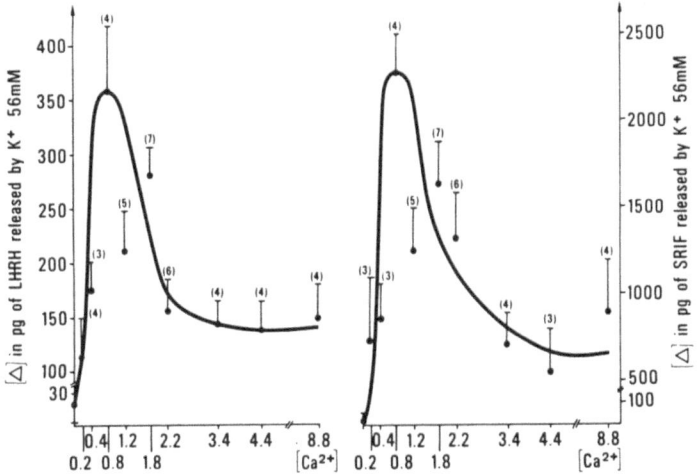

Figure 8. Effect of different Ca²⁺ concentrations on secretory response of LH-RH and SRIF to 56 mM K⁺ from superfused mediobasal hypothalamic slices. [From Drouva *et al.*, 1981a.]

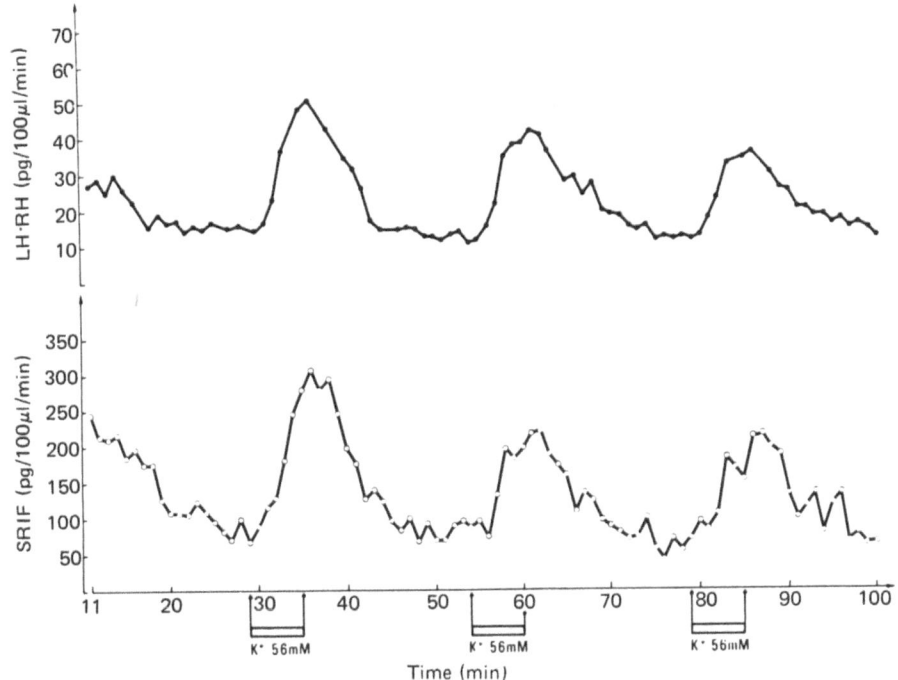

Figure 9. Secretory response of LH-RH and SRIF to pulses of 56 mM K⁺. [From Drouva *et al.*, 1981a.]

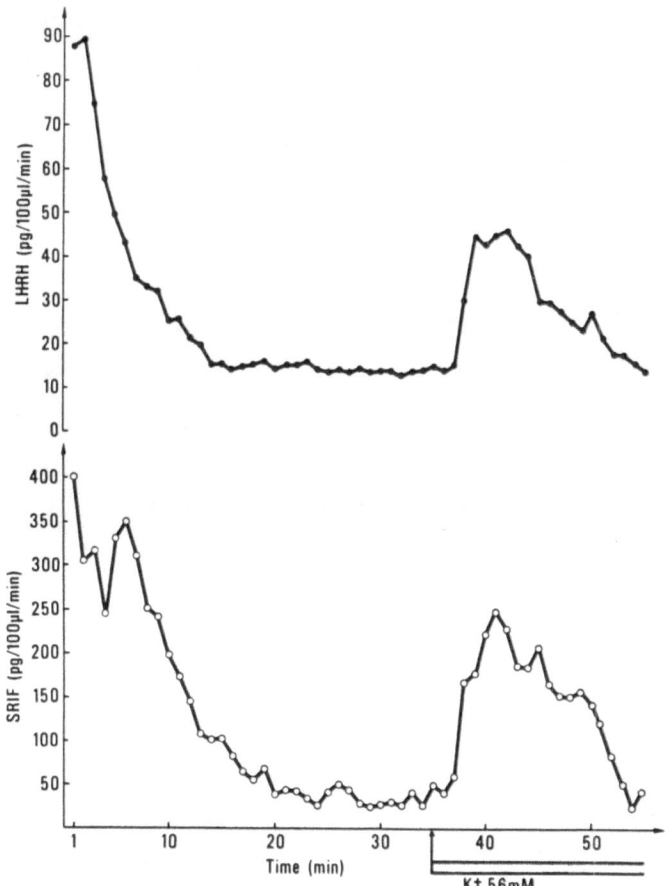

Figure 10. LH-RH- and SRIF-induced release by continuous stimulation of 56 mM K⁺. [From Drouva *et al.*, 1981a.]

Administration of voltage-sensitive Ca^{2+} channel blockers, such as D-600 $(10^{-4}M)$, Mn^{2+} (3 mM), completely abolished K^+- and veratridine-induced neuropeptide release (Fig. 6). Moreover, high Mg^{2+} concentrations could also block the K^+-evoked release (Drouva *et al.*, 1981a). Blockade of "fast" sodium channels and "early" calcium channels (Rasmussen and Goodman, 1977) by tetrodotoxin $(5 \times 10^{-7} \text{ M})$ abolishes the stimulatory effect of veratridine, whereas the stimulation of K^+ is unaffected (Fig. 6). Under veratridine stimulation, Na^+ channels present on LH-RH and SRIF nerve endings thus seem to induce an opening of voltage-dependent Ca^{2+} channels, following their prior activation (Drouva *et al.*, 1981a).

That the stimulatory effect of both depolarizing agents is Ca^{2+} dependent and is abolished by voltage-sensitive Ca^{2+} channel blockade indicates that Ca^{2+} influx through voltage-dependent Ca^{2+} channel is a prerequisite to the neurosecretory process and that the activation state of this Ca^{2+} channel regulates the exocytotic release of LH-RH and SRIF in response to a depolarizing stimulus.

In addition, K^+ stimulation is not abolished by tetrodotoxin. This suggests that Na^+ and early Ca^{2+} channels are not essential for the K^+-evoked neuropeptide release, a conclusion further supported by maintenance of the induced release in the absence of Na^+. In fact, K^+-induced LH-RH and SRIF release was even potentiated at low (50 or 6 mM) Na^+ concentration, an observation possibly due to intervention of the Na–Ca exchange system (Dipolo and Beaugé, 1980; Rubin, 1970) present in nerve terminals (Fig. 11). Similar conclusions have been derived from studies of neurohypophyseal hormone release (Dreifuss *et al.*, 1978).

The spontaneous release of LH-RH and SRIF is not altered by omission of Ca^{2+} or by administration of Ca^{2+} or Na^+ channel blockers (Fig. 6). The mechanisms coupling Ca^{2+} influx with exocytotic process are not yet elucidated. Of potential interest, however, may be the report by Greengard (1979) that Ca^{2+} influx and phosphorylation of specific proteins are induced in parallel by K^+ and veratridine. The two phenomena were also potentiated by administration of calmodulin (DeLorenzo *et al.*, 1979), a specific Ca^{2+}-binding protein involved in intracellular regulatory mechanisms of Ca^{2+}-dependent nerve ending functions (Klee *et al.*, 1980; Roufogalis, 1980).

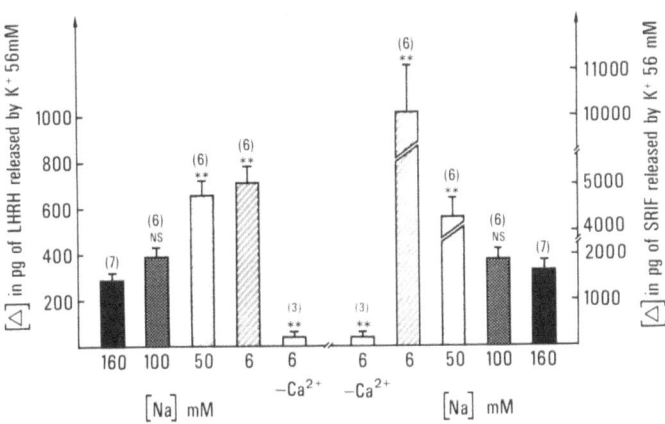

Figure 11. K^+-induced LH-RH and SRIF release from mediobasal hypothalamic slices at different Na^+ concentrations. [From Drouva *et al.*, unpublished.]

In conclusion, Na$^+$ and Ca^{2+} voltage-dependent channels are present on LH-RH and SRIF nerve endings. Their activation leads to membrane depolarization with subsequent Ca^{2+} entry through Ca^{2+} voltage-dependent channels and consequently to neuropeptide release. Neither magnesium nor chlorine is needed for spontaneous or evoked release of LH-RH and SRIF (Drouva *et al.*, 1981a).

5. Are Neuropeptide-Neurotransmitter Interactions Coupled with Ionic Channels?

Progress in our understanding of ionic process involved in peptide release makes it tempting to analyze the possible involvement of discrete ionic channels in a given peptide–peptide or neurotransmitter–peptide interaction. Modulatory effects, such as for example those described in the preceding sections of this chapter, may indeed be due to specific coupling between transmitter receptors and ionic influx. This hypothesis could account for observations already discussed showing that some transmitter interactions affect peptide release under discrete experimental or physiological conditions only; the difference between effective and ineffective interactions would thus depend upon the "open" or "closed" state of corresponding channels.

Coupling of transmitter receptors to channels involved in neuropeptide release has mainly been investigated so far in the case of ACh (Marty, 1980; Kirpekar and Prat, 1979), GABA (Kupferman, 1979), and opiate peptides (Drouva *et al.*, 1981b; Iversen *et al.*, 1980; Clarke *et al.*, 1979; Kupfermann, 1979; Mudge *et al.*, 1979; Ross and Cardenas, 1979). Effects of GABA are assumed to be mediated by the chlorine channels (Kupfermann, 1979) while the ACh action is associated with the activation of Ca^{2+} and Na$^+$ or Ach channels (Kirpekar and Prat, 1979; Marty, 1980).

There are strong indications that opiate receptors modulate Ca^{2+} channels (Drouva *et al.*, 1981b). Voltage-dependent Ca^{2+} channels and opiates are able to abolish K$^+$ stimulation of LH-RH and SRIF release; opiates can also inhibit ^{45}Ca^{2+} uptake from brain synaptosomes, as has been reported by Guerrero-Munoz *et al.* (1979a,b). In addition, we have found (Drouva *et al.*, 1981b) the inhibitory effect of opiates to be more pronounced in the presence of lower than maximal depolarizing concentrations of K$^+$ or when Ca^{2+} concentrations in the medium are reduced from 1.8 mM to 0.8 mM (in the latter case, inhibition rose from 50% to 70%). These findings further support the opiate–Ca^{2+} interaction for the inhibitory effect on LH-RH and SRIF release. Relationships between morphine action and Ca^{2+} levels have also been reported in different systems (Opmeer and Van Ree, 1979; Ross and Cardenas, 1979; Jhamandas *et al.*,

1978; Harris *et al.*, 1976). One can thus postulate that activation of opiate receptors initiates changes in the activation state of voltage-dependent Ca^{2+} channels and, hence, alters Ca^{2+} membrane permeability and neuropeptide release in response to a stimulus.

Electrophysiological data on the effect of opiates on vasopressin and oxytocin release are also consistent with the hypothesis of an opiate coupling to Ca^{2+} channels (Iversen *et al.*, 1980; Clarke *et al.*, 1979).

6. Hormonal Effect on Neuropeptide Release in Vitro

It is well established from pharmacological experiments that the hormonal status of the animal can determine the pituitary hormone response to a given neurotransmitter (Gallo and Drouva, 1979; Labrie *et al.*, 1979; McCann *et al.*, 1979; Rotsztejn *et al.*, 1976). Data already available concern the influence of steroids on LH-RH release and of thyroid hormones and GH on SRIF.

6.1. LH-RH

Earlier studies from Edwardson's group have demonstrated that gonadal steroids, 17β-estradiol and progesterone, administered either *in vivo* or *in vitro* could influence basal LH-RH release from synaptosomal preparations (Bennett *et al.*, 1975), suggesting that the effect of steroids may result from their action on the synthesis and/or the release process of this neuropeptide. In addition, it has been shown that the dopamine-(Rotsztejn *et al.*, 1976; Schneider and McCann, 1969) or the electrical stimulation (Dyer *et al.*, 1980)-induced LH-RH release from incubated hypothalamic fragments is dependent upon prior exposure of animals to steroids.

Recently, Ramirez and co-workers (1980), in very elegant studies, showed that the release of LH-RH from mediobasal hypothalamic–anterior hypothalamic–preoptic area of prepubertal female rats measured in a perifusion system varied between intact, ovariectomized, ovariectomized/estrogen implanted, and ovariectomized rats primed with estradiol and injected with progesterone. The basal release of LH-RH from ovariectomized animals was significantly lower than that of the intact females. However, it was increased when the animals were treated with estradiol and progesterone. Furthermore, data from our laboratory (Drouva *et al.*, unpublished) showed that in both male and female rats gonadectomized 4 weeks prior to the superfusion of mediobasal hypothalamic slices, there is no significant increase in LH-RH release after K^+ depolarization. In addition, mediobasal hypothalamic slices from estrogen-implanted animals for

5 days could release significant amount of LH-RH. By contrast, there is no complete recovery of LH-RH release to the level of intact male rats, in castrated animals treated with testosterone (Fig. 12). These findings suggest that the presence of steroids can modify the releasable pool(s) of LH-RH from nerve endings.

6.2. SRIF

It has been shown that SRIF levels decrease significantly following hypophysectomy (Kanatsuka *et al.*, 1979; Fernandez-Durango *et al.*, 1978; Hoffman and Baker, 1977), while administration of GH prevents this depletion (Kanatsuka *et al.*, 1979; Hoffman and Baker, 1977). In addition, in GH-treated rats, hypothalamic SRIF content is high (Kanatsuka *et al.*, 1979), suggesting that circulating levels of GH may exert a "short-loop" feedback control on SRIF release in hypothalamus (Hoffman and Baker, 1977). This is consistent with the finding of Sheppard *et al.* (1978) that rGH causes dose-dependent release of SRIF from incubated rat hypothalamus.

The involvement of the activity of hypothalamic–thyroid axis on hypothalamic SRIF release, and vice versa, is not yet clear (Berelowitz *et al.*, 1980; Gillioz *et al.*, 1979; Hirooka *et al.*, 1978).

Thyroidectomy does not seem to modify SRIF release into the hypothalamo–hypophyseal portal system (Gillioz *et al.*, 1979). However, SRIF release from incubated hypothalamic fragments under basal conditions or

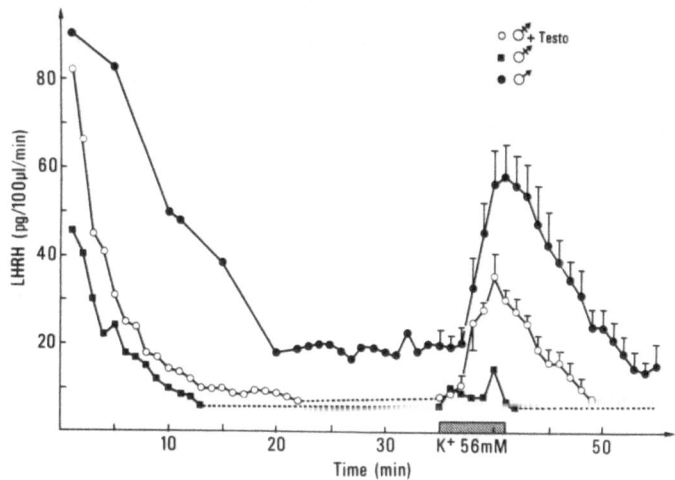

Figure 12. Spontaneous and K^+-induced LH-RH release from mediobasal hypothalamic slices from (♂) intact, (♂) castrated, and ♂ + testosterone treated male rats. [From Drouva *et al.*, unpublished.]

after stimulation by K^+ or dopamine is lower in hypothyroid than in euthyroid controls or hypothyroid T_3-treated rats (Berelowitz *et al.*, 1980). In addition, somatostatin release is stimulated after exposure to T_3 (Berelowitz *et al.*, 1980).

The observation that SRIF can inhibit basal and norepinephrine-induced thyrotropin-releasing factor release from rat hypothalamus in organ culture (Hirooka *et al.*, 1978) further supports the existence of functional relationships between hypothalamic–thyroid axis and hypothalamic SRIF levels.

7. Conclusion

Modulation by transmitters or peptides of neurosecretion into the hypothalamo–hypophyseal portal system seems to play an important role in regulating pituitary function. This modulation is presumably mediated by specific recognition sites located in the vicinity of neurosecretory nerve endings. In those cases, specificity of the interaction can be characterized by *in vitro* methods; *in vitro* data must, however, always be interpreted with great care, in view of the limited concentration range at which given agonists or antagonists of a transmitter are specifically recognized by the corresponding receptors. Consistent peptide effects obtained independently under *in vitro* and *in vivo* conditions can be assumed physiologically relevant. However, final demonstration of the functional role of endogenous peptides requires availability of specific peptide antagonists, a condition only met so far for opiate peptides.

Presynaptic peptide receptors appear coupled to ionic channels involved in the regulation of neuronal exocytosis. This is in particular the case for opiates, which have been shown to modulate the activity of voltage-dependent Ca^{2+} channels. Coupling processes with ionic channels may account for the modulatory action of several peptides, and explain why their effect often depends upon the initial activation state of neuronal channels. Interestingly, peripheral hormones also appear able to interfere with ionic fluxes into neurosecretory neurons. Further work is required to complete our knowledge of peptide and hormone interactions in the hypothalamus and of their coupling mechanisms.

REFERENCES

Ahren, K., Fuxe, K., Hamberger, C., and Hökfelt, T., 1971, Turnover changes in the tuberoinfundibular dopamine neurons during the ovarian cycle of the rat, *Endocrinology* **88**:1415.

Baker, P. F., Meves, H., and Ridgway, E. B., 1973, Calcium entry in response to maintained depolarization of squid axons, *J. Physiol. (London)* **231**:527.

Bennett. G. W., and Edwardson, J. A., 1975, Release of corticotrophin releasing factor and other hypophysiotropic substances from isolated nerve endings (synaptosomes), *J. Endocrinol.* **65**:33.

Bennett, G. W., Edwardson, J. A., Holland, D. T., Jeffcoate, S. L., and White, N., 1975, Release of immunoreactive luteinizing hormone and thyrotropin-releasing hormone from hypothalamic synaptosomes, *Nature (London)* **257**:323.

Bennett, G. W., Edwardson, J. A., Marcano de Cotte, D., Berelowitz, M., Pimstone, B. L., and Kronheim, S., 1979, Release of somatostatin from rat brain synaptosomes, *J. Neurochem.* **32**:1127.

Berelowitz, M., Kronheim, S., Pimstone, B., and Sheppard, M., 1978, Potassium stimulated dependent release of immunoreactive somatostatin from incubated rat hypothalamus, *J. Neurochem.* **31**:1537.

Berelowitz, M., Maeda, K., Harris, S., and Frohman, A., 1980, The effect of alterations in the pituitary–thyroid axis on hypothalamic content and *in vitro* release of somatostatin-like immunoreactivity, *Endocrinology* **107**:24.

Blaustein, M. P., 1975, Effects of potassium, veratridine and scorpion venom on calcium accumulation and transmitter release by nerve terminals *in vitro, J. Physiol. (London)* **247**:617.

Bloom, F. E., 1979, Contrasting principles of synaptic physiology: Peptidergic and non-peptidergic neurons, in: *Central Regulation of the Endocrine System* (K. Fuxe, T. Hökfelt, and R. Luft, eds.), pp. 173–187, Plenum Press, New York.

Charli, J. L., Joseph-Bravo, P., Palacios, J. M., and Kordon, C., 1978a, Histamine-induced release of thyrotropin releasing hormone from hypothalamic slices, *Eur. J. Pharmacol.* **52**:401.

Charli, J. L., Rotsztejn, W. H., Pattou, E., and Kordon, C., 1978b, Effect of neurotransmitters on *in vitro* release of luteinizing hormone releasing hormone from the mediobasal hypothalamus of male rats, *Neurosci. Lett.* **10**:159.

Cicero, T. J., Badger, T. M., Wilcox, C. E., Bell, R. D., and Meyer, E. R., 1977, Morphine decreases luteinizing hormone by an action on the hypothalamic–pituitary axis, *J. Pharmacol. Exp. Ther.* **203**:548.

Clarke, G., Wood, P., Merrick, L., and Lincoln, D. W., 1979, Opiate inhibition of peptide release from the neurohumoral terminals of hypothalamic neurons, *Nature (London)* **282**:746.

DeLorenzo, R. J., Freedman, S. D., Yohe, W. B., and Maurer, S. C., 1979, Stimulation of Ca^{2+}-dependent neurotransmitter release and presynaptic nerve terminal protein phosphorylation by calmodulin and a calmodulin-like protein isolated from synaptic vesicles, *Proc. Natl. Acad. Sci. USA* **76**:1838.

Dipolo, R., and Beaugé, L., 1980, Mechanisms of calcium transport on the giant axon of the squid and their physiological role, *Cell Calcium* **1**:147.

Dodd, P. R., Edwardson, J. A., and Dockray, G. J., 1980, The depolarization-induced release of cholecystokinin C-terminal octapeptide (CCK-8) from rat synaptosomes and brain slices, *Regul. Pept.* **1**:17.

Douglas, W. W., and Poisner, A. M., 1964a, Stimulus–secretion coupling in a neurosecretory organ: The role of calcium in the release of vasopressin from the neurohypophysis, *J. Physiol. (London)* **172**:1.

Douglas, W. W., and Poisner, A. M., 1964b, Calcium movement in the neurohypophysis of the rat and its relation to the release of vasopressin, *J. Physiol. (London)* **172**:19.

Dreifuss, J. J., Theodosis, D. T., and Zingg, H. H., 1978, Ionic influences on neurohypophyseal secretion *in vitro*, in: *Biologie Cellulaire des Processus Neurosécrétoires Hypothalamiques* (C. Kordon and J. D. Vincent, eds.), pp. 563–578, CNRS, Paris.

Drouva, S. V., and Gallo, R. V., 1976, Catecholamine involvement in episodic luteinizing hormone release in adult ovariectomized rats, *Endocrinology* **99:**651.

Drouva, S. V., and Gallo, R. V., 1977, Further evidence for inhibition of episodic luteinizing hormone in ovariectomized rats by stimulation of dopamine-receptors, *Endocrinology* **100:**792.

Drouva, S. V., Epelbaum, J., Tapia-Arancibia, L., Laplante, E., and Kordon, C., 1980, Met-enkephalin inhibition of K$^+$-induced LHRH and SRIF release from rat mediobasal hypothalamic slices, *Eur. J. Pharmacol.* **61:**411.

Drouva, S. V., Epelbaum, J., Héry, M., Tapia-Arancibia, L., Laplante, E., and Kordon, C., 1981a, Ionic channels involved in the LHRH and SRIF release from rat mediobasal hypothalamus, *Neuroendocrinology* **32:**155.

Drouva, S. V., Epelbaum, J., Tapia-Arancibia, L., Laplante, E., and Kordon, C., 1981b, Opiate receptors modulate LHRH and SRIF release from mediobasal hypothalamic neurons, *Neuroendocrinology* **32:**163.

Dyer, R. G., Mansfield, S., and Yates, S. O., 1980, Discharge of gonadotrophin-releasing hormone from the mediobasal part of the hypothalamus: Effect of stimulation frequency and gonadal steroids, *Exp. Brain Res.* **39:**453.

Elde, R., and Hökfelt, T., 1978, Distribution of hypothalamic hormones and other peptides in the brain, in: *Frontiers in Neuroendocrinology,* Vol. 5 (W. F. Ganong and L. Martini, eds.), pp. 1–33, Raven Press, New York.

Emson, P. C., Fahrenkrug, J., Schaffalitzky de Muckadell, O. B., Jessell, T. M., and Iversen, L. L., 1978, Vasoactive intestinal peptide (VIP) vesicular localization and potassium evoked release from rat hypothalamus, *Brain Res.* **143:**174.

Enjalbert, A., Bourgoin, S., Hamon, M., Adrien, J., and Bockaert, J., 1978, Postsynaptic serotonin sensitive adenylate cyclase in the central nervous system. I. Development and distribution of serotonin and dopamine-adenylate-cyclase in rat and guinea pig, *Brain Mol. Pharmacol.* **14:**2.

Epelbaum, J., Tapia-Arancibia, L., Besson, J., Rotsztejn, W. H., and Kordon, C., 1979, Vasoactive intestinal peptide inhibits release of somatostatin from hypothalamus *in vitro, Eur. J. Pharmacol.* **58:**493.

Epelbaum, J., Tapia-Arancibia, L., and Kordon, C., 1981, Noradrenalin stimulates somatostatin release from incubated slices of the amygdala and the hypothalamic preoptic area, *Brain Res.* **215:**393.

Fernandez-Durango, R., Arimura, A., Fishback, J., and Schally, A. V., 1978, Hypothalamic somatostatin and LHRH after hypophysectomy, in hyper or hypothyroidism and during anesthesia in rats, *Proc. Soc. Exp. Biol. Med.* **157:**235.

Fiorindo, R. P., and Martini, L., 1975, Evidence for a cholinergic component in the neuroendocrine control of luteinizing hormone (LH) secretion, *Neuroendocrinology* **18:**322.

Fuxe, K., 1965, Evidence for the existence of monoamine neurons in the central nervous system. IV. Distribution of monoamine nerve terminals in the central nervous system, *Acta Physiol. Scand.* **64:**37.

Fuxe, K., Hökfelt, T., and Nilsson, O., 1967, Activity changes in the tuberoinfundibular DA neurons of the rat during various states of the reproductive cycle, *Life Sci.* **6:**2057.

Fuxe, K., Hökfelt, T., and Nilsson, O., 1969, Castration, sex hormones and tuberoinfundibular dopamine neurons, *Neuroendocrinology* **5:**107.

Gallardo, E., and Ramirez, V. D., 1977, A method for the superfusion of rat hypothalamus: Secretion of luteinizing hormone-releasing hormone (LHRH) *Proc. Soc. Exp. Biol. Med.* **155:**79.

Gallo, R. V., and Drouva, S. V., 1979, Effect of intraventricular infusion of catecholamines on luteinizing hormone release in ovariectomized and ovariectomized steroid-primed rats, *Neuroendocrinology* **29:**149.

Gamse, R., Vaccaro, D. E., Gamse, G., Di Pace, M., Fox, T. O., and Leeman, S. E., 1980, Release of immunoreactive somatostatin from hypothalamic cells in culture: Inhibition by γ-aminobutyric acid, *Proc. Natl. Acad. Sci. USA* **77:**5552.

Gillioz, P., Giraud, P., Conte-Derolx, B., Jaquet, P., Codaccioni, J. L., and Oliver, C., 1979, Immunoreactive somatostatin in rat hypophysial portal blood, *Endocrinology* **104:**1407.

Gnodde, H. P., and Schuiling, G. A., 1976, Involvement of catecholaminergic and cholinergic mechanisms in the pulsatile release of LH in the long-term ovariectomized rat, *Neuroendocrinology* **20:**212.

Greengard, P., 1979, Cyclic nucleotide and protein phosphorylation mechanisms in the central nervous system, in: *Central Regulation of the Endocrine System* (K. Fuxe, T. Hökfelt, and R. Luft, eds.), pp. 157–172, Plenum Press, New York.

Grimm, Y., and Reichlin, S., 1973, TRH: Neurotransmitter regulation of secretion by mouse hypothalamic tissue *in vitro, Endocrinology* **93:**626.

Guerrero-Munoz, F., De Lourdes Guerrero, M., and Way, E. L., 1979a, Effect of morphine on calcium uptake by lysed synaptosomes, *J. Pharmacol. Exp. Ther.* **211:**370.

Guerrero-Munoz, F., De Lourdes Guerrero, M., Way, E. L., and Li, C. H., 1979b, Effects of β-endorphin on calcium uptake in the brain, *Science* **206:**89.

Guillemin, R., 1978, Peptides in the brain: The new endocrinology of the neuron, *Science* **202:**390.

Harris, R. A., Yamamoto, H., Loh, H. H., and Way, E. L., 1976, Alterations in brain calcium localization during the development of morphine tolerance and dependence, in: *Opiates and Endogenous Opioid Peptides* (H. W. Kosterlitz, ed.), pp. 361–368, Elsevier/North-Holland, Amsterdam.

Harter, D. S., and Ramirez, V. D., 1980, The effect of ions, metabolic inhibitors and colchicine on luteinizing hormone releasing hormone release from superfused rat hypothalamus, *Endocrinology* **107:**375.

Henderson, G., Hughes, J., and Kosterlitz, W. H., 1978, *In vitro* release of Leu and Met enkephalin from the corpus striatum, *Nature (London)* **271:**677.

Hirooka, Y., Hollander, G. S., Suzuki, S., Ferdinand, P., and Juan, S. J., 1978, Somatostatin inhibits release of thyrotropin releasing factor from organ cultures of rat hypothalamus, *Proc. Natl. Acad. Sci. USA* **75:** 4509.

Hoffman, D. L., and Baker, B. L., 1977, Effect of treatment with growth hormone on somatostatin in median eminence of hypophysectomized rats, *Proc. Soc. Exp. Biol. Med.* **156:**265.

Hökfelt, T., Johansson, O., Ljungdahl, A., Lundberg, M. J., and Schultzberg, M., 1980, Peptidergic neurons, *Nature (London)* **284:**515.

Holaday, J. W., and Loh, H. H., 1979, Endorphin–opiate interactions with neuroendocrine systems, in: *Advances in Biochemical Psychopharmacology,* Vol. 20 (H. H. Loh and D. H. Ross, eds.), pp. 227–258, Raven Press, New York.

Iversen, L. L., and Jessell, T. M., 1979, Regulation of neuropeptide release in rat brain, in: *Central Regulation of the Endocrine System* (K. Fuxe, T. Hökfelt, and R. Luft, eds.), pp. 189–207, Plenum Press, New York.

Iversen, L. L., Jessell, T., and Kanazawa, J., 1976, Release and metabolism of substance P in rat hypothalamus, *Nature (London)* **264:**81.

Iversen, L. L., Iversen, S. D., Bloom, F., Douglas, C., Brown, M., and Vale, W., 1978, Calcium dependent release of somatostatin and neurotensin from rat brain *in vitro, Nature (London)* **273:**161.

Iversen, L. L., Iversen, S. D., and Bloom, F. E., 1980, Opiate receptors influence vasopressin release from nerve terminals in rat neurohypophysis, *Nature (London)* **284**:350.

Jhamandas, K., Sawynok, J., and Sutak, M., 1978, Antagonisms of morphine action on brain acetylcholine release by methylxanthines and calcium, *Eur. J. Pharmacol.* **49**:309.

Jones, M. T., Hillhouse, E., and Burden, J., 1976, Secretion of corticotropin releasing hormone *in vitro,* in: *Frontiers in Neuroendocrinology,* Vol. 4 (L. Martini and W. F. Ganong, eds.), pp. 195–226, Raven Press, New York.

Joseph-Bravo, P., Charli, J. L., Palacios, J. M., and Kordon, C., 1979, Effect of neurotransmitters on the *in vitro* release of immunoreactive thyrotropin-releasing hormone from rat mediobasal hypothalamus, *Endocrinology* **104**:801.

Kanatsuka, A., Makino, H., Matsushima, Y., Osegawa, M., Yamamoto, M., and Kumagai, A., 1979, Effect of hypophysectomy and growth hormone administration on somatostatin content in the rat hypothalamus, *Neuroendocrinology* **29**:186.

Kirpekar, S. M., and Prat, J. C., 1979, Release of catecholamines from perifused cat adrenal gland by veratridine, *Proc. Natl. Acad. Sci. USA* **76**:2081.

Klee, C. B., Crouch, T. H., and Richman, P. G., 1980, Calmodulin, *Annu. Rev. Biochem.* **49**:289.

Kordon, C., 1979, Role and regulation of neuropeptide neurons, in: *Central Regulation of the Endocrine System* (K. Fuxe, T. Hökfelt, and R. Luft, eds.), pp. 473–487, Plenum Press, New York.

Kordon, C., Enjalbert, A., Héry, M., Joseph-Bravo, P., Rotsztejn, W. H., and Ruberg, M., 1980, Role of neurotransmitters in the control of adenohypophyseal secretion, in: *Handbook of the Hypothalamus* (J. P. Morgane and J. Panskepp, eds.), Vol. 2, pp. 253–306, Dekker, New York.

Kupfermann, J., 1979, Modulatory actions of neurotransmitters, *Annu. Rev. Neurosci.* **2**:447.

Labrie, F., Drouin, J., Lagace, L., Ferland, L., Beaulieu, M., Raymond, V., and Mascicotte, J., 1979, Interactions between hypothalamic and peripheral hormones at the anterior pituitary level, in: *Central Regulation of the Endocrine System* (K. Fuxe, T. Hökfelt, and R. Luft, eds.), pp. 85–107, Plenum Press, New York.

Linton, E. A., Bennett, G. W., and Whitehead, S. A., 1979, Prostaglandins and the release of LHRH from hypothalamic synaptosomes, *Neuroendocrinology* **28**:394.

McCann, S. M., Krulich, L., Ojeda, S. R., Negro-Vilar, A., and Vijayan, E., 1979, Neurotransmitters in the control of anterior pituitary function, in: *Central Regulation of the Endocrine System* (K. Fuxe, T. Hökfelt, and R. Luft, eds.), pp. 329–347, Plenum Press, New York.

Maeda, K., and Frohman, L. A., 1980, Release of somatostatin and thyrotropin releasing hormone from rat hypothalamic fragments *in vitro, Endocrinology* **106**:1837.

Marty, A., 1980, Action of calcium ions on acetylcholine-sensitive channels in aplysia neurons, *J. Physiol. (Paris)* **76**:523.

Meites, J., Bruni, J. F., Van Vugt, D. A., and Smith, A. F., 1979, Relation of endogenous opioid peptides and morphine to neuroendocrine functions, *Life. Sci.* **24**:1325.

Mudge, A. N., Leeman, S. E., and Fishbach, G. D., 1979, Enkephalin inhibits release of substance P from sensory neurons in culture and decreases action potential duration, *Proc. Natl. Acad. Sci. USA* **76**:526.

Negro-Vilar, A., Ojeda, S. R., Arimura, A., and McCann, S. M., 1978, Dopamine and norepinephrine stimulate somatostatin release by median eminence fragments *in vitro, Life Sci.* **23**:1493.

Negro-Vilar, A., Ojeda, S. D., and McCann, S. M., 1979, Catecholaminergic modulation of luteinizing hormone releasing hormone release by median eminence terminals *in vitro, Endocrinology* **104**:1749.

Nordmann, J. J., 1976, Evidence for calcium inactivation during hormone release in the rat neurohypophysis, *J. Exp. Biol.* **65**:669.

Nordmann, J. J., and Dyball, R. E. J., 1978, Effects of veratridine on Ca fluxes and the release of oxytocin and vasopressin from the isolated rat neurohypophysis, *J. Gen. Physiol.* **72**:297.

Opmeer, F. A., and Van Ree, J. M., 1979, Competitive antagonism of morphine action *in vitro* by calcium, *Eur. J. Pharmacol.* **53**:395.

Osborne, H., Höllt, V., and Herz, A., 1978, Potassium-induced release of enkephalins from rat striatal slices, *Eur. J. Pharmacol.* **48**:219.

Pollard, H., Llorens-Cortes, C., and Schwartz, J. C., 1977, Enkephalin receptors on dopaminergic neurons in rat striatum, *Nature (London)* **268**:745.

Ramirez, V. D., and Kordon, C., 1978, *in vitro* release and regulation of neurohormones from hypothalamic fragments on isolated nerve endings, in: *Biologie Cellulaire des Processus Neurosécrétoires Hypothalamiques* (C. Kordon and J. D. Vincent, eds.), pp. 579–594, CNRS, Paris.

Ramirez, V. D., Dluzen, D., and Lin, D., 1980, Progesterone administration *in vivo* stimulates release of luteinizing hormone-releasing hormone *in vitro, Science* **208**:1037.

Rasmussen, H., and Goodman, D. P. B., 1977, Relationships between calcium and cyclic nucleotides in cell activation, *Physiological. Rev.* **57**:421.

Richardson, S. B., Hollander, G. S., D'Eletto, R., Greenleaf, P. W., and Thaw, G., 1980, Acetylcholine inhibits the release of somatostatin from rat hypothalamus *in vitro, Endocrinology* **107**:122.

Rivier, C., Vale, W., Ling, N., Brown, M., and Guillemin, R., 1977, Stimulation *in vivo* of the secretion of prolactin and growth hormone by β-endorphin, *Endocrinology* **100**:238.

Ross, D. H., and Cardenas, H. L., 1979, Nerve cell calcium as a messenger for opiate and endorphin actions: Neurochemical mechanisms of opiates and endorphins, in: *Advances in Biochemical Psychopharmacology,* Vol. 20 (H. H. Loh and D. H. Ross, eds.), pp. 301–336, Raven Press, New York.

Rotsztejn, W. H., Charli, J. L., Pattou, E., Epelbaum, J., and Kordon, C., 1976, *In vitro* release of luteinizing hormone-releasing hormone (LHRH) from rat mediobasal hypothalamus: Effects of potassium, calcium, and dopamine, *Endocrinology* **99**:1663.

Rotsztejn, W. H., Charli, J. L., Pattou, E., and Kordon, C., 1977, Stimulation by dopamine of luteinizing hormone-releasing hormone (LHRH) release from mediobasal hypothalamus in male rats, *Endocrinology* **101**:1475.

Rotsztejn, W. H., Drouva, S. V., Pattou, E., and Kordon, C., 1978, Metenkephalin inhibits *in vitro* dopamine-induced LHRH release from mediobasal hypothalamus of male rats, *Nature (London)* **274**:281.

Roufogalis, B. G., 1980, Calmodulin: Its role in synaptic transmission, *Trends Neurosci.* **3**:238.

Rubin, R. P., 1970, The role of calcium in the release of neurotransmitter substances and hormones, *Pharmacol. Rev.* **22**:389.

Samson, W. K., Koening, J., Reeves, J., and McCann, S. M., 1980, Vasoactive intestinal peptide stimulates LHRH release from hypothalamic synaptosomes, in: *Proceedings of the 62nd Annual Endocrine Society Meeting,* abstract 746.

Schally, A. V., 1978, Aspects of hypothalamic regulation of the pituitary gland, *Science* **202**:18.

Schneider, H. P. G., and McCann, S. M., 1969, Possible role of dopamine as transmitter to promote discharge of LH-releasing factor, *Endocrinology* **85**:121.

Shaar, C. J., Frederickson, R. C. A., Dininger, N. B., and Jackson, L., 1977, Enkephalin analogues and naloxone modulate the release of growth hormone and prolactin: Evidence for regulation by an endogenous opioid peptide in brain, *Life Sci.* **21**:853.

Sheppard, M. C., Kronheim, S., and Pimstone, B. L., 1978, Stimulation by growth hormone of somatostatin release from the rat hypothalamus *in vitro, Clin. Endocrinol.* **9**:583.

Sheppard, M. C., Kronheim, S., and Pimstone, B. L., 1979, Effect of substance P, neurotensin and the enkephalins on somatostatin release from the rat hypothalamus *in vitro, J. Neurochem.* **32**:647.

Snyder, S. H., 1980, Brain peptides as neurotransmitters, *Science* **209**:976.

Takahara, J., Yunoki, S., Hosogi, H., Yakushiji, W., Kageyama J., and Ofuji, T., 1980, Concomitant increases in serum growth hormone and hypothalamic somatostatin in rats after injection of γ-aminobutyric acid, amino-oxyacetic acid, or γ-hydroxybutyric acid, *Endocrinology* **106**:343.

Terry, L. C., Rostad, O. P., and Martin, J. B., 1980, The release of biologically and immunologically reactive somatostatin from perifused hypothalamic fragments, *Endocrinology* **107**:794.

Vale, W., Rivier, C., and Brown, M., 1980, Physiology and pharmacology of hypothalamic regulatory peptides, in: *Handbook of the Hypothalamus* (J. P. Morgane and J. Panskepp, eds.), Vol. 2, pp. 165–252, Dekker, New York.

Vale, W., Spiess, J., Rivier, C., and Rivier, J., 1981, Characterization of a 41-residue ovine hypothalamic peptide that stimulates secretion of corticotropin and β-endorphin, *Science* **213**:1394.

Vijayan, E., and McCann, S. M., 1978, Effects of intraventricular injection of γ-aminobutyric acid (GABA) on plasma growth hormone and thyrotropin in conscious ovariectomized rats, *Endocrinology* **103**:1883.

Vijayan, E., Samson, W. K., Said, S. J., and McCann, S. M., 1979, Vasoactive intestinal peptide: Evidence for a hypothalamic site of action to release growth hormone, luteinizing hormone, and prolactin in conscious ovariectomized rats, *Endocrinology* **104**:53.

Warberg, S., Eskay, R. L., Barnea, A., Reynolds, R. C., and Porter, J. C., 1977, Release of luteinizing hormone releasing hormone and thyrotropin releasing hormone from a synaptosome enriched fraction of hypothalamic homogenates, *Endocrinology* **100**:814.

Wei Lin Kao, L., and Weisz, J., 1977, Release of gonadotropin-releasing hormone (GnRH) from isolated perifused medial-basal hypothalamus by melatonin, *Endocrinology* **100**:1723.

Weiner, R. I., and Ganong, W. F., 1978, Role of brain monoamines and histamine in regulation of anterior pituitary secretion, *Pharmacol. Rev.* **68**:905.

7

In Vitro Release of Luteinizing Hormone-Releasing Hormone

Influence of Estrogen and a Mammary Carcinogen

E. J. Peck, Jr., C. Pasqualini, and B. Kerdelhué

1. Introduction

The brain has been recognized as an organ responsive to gonadal steroids since the experiments of Berthold (1849) on castration and gonadal transplantation in roosters. Implant, lesion, and stimulation experiments in the mid-20th century established certain areas of the hypothalamus, limbic system, and midbrain as specific targets for steroids. More recent investigations employing technology for the assay of specific steroid receptors have confirmed and extended the previous findings. These topics have been the subject of several recent reviews (Feder *et al.,* 1978; McEwen, 1978; Clark and Peck, 1979) and are not discussed as extensively or intensively here as there. In this report we use conclusions from these reviews to develop a model for the control of gonadotropin secretion, one aspect of gonadal steroid action in the brain, and to extend our knowledge of this system by studying potential perturbants of the neuroendocrine axis.

Endocrine control of gonadotropin secretion may include a series of

E. J. Peck, Jr. • Department of Cell Biology, Baylor College of Medicine, Houston, Texas 77030.   *C. Pasqualini and B. Kerdelhué* • Laboratoire des Hormones Polypeptidiques, CNRS, 91190 Gif-sur-Yvette, France.

feedback mechanisms involving steroid, peptide, and protein hormones. "Long" feedback loops may utilize gonadal steroids as regulators of neural and pituitary function while "short" and "ultrashort" feedback systems may utilize luteinizing hormone (LH) or follicle-stimulating hormone (FSH) and luteinizing hormone-releasing hormone (LH-RH) respectively as regulators or modulators within the hypothalamic–hypophyseal axis. The study of these feedback systems is complicated by a number of circumstances. First, they are not independent of one another; rather each may have direct and/or indirect effects on the others. Second, they are influenced by (and perhaps dependent upon) neural imputs arising from sensory stimuli (olfaction, photoperiod, and, in some instances, somatosensation). In addition to these complexities, the precise mechanism(s) employed in the control of gonadotropin secretion varies with the species (and perhaps strain) under study. Thus, in the primate, "long" and "short" feedback loops may function primarily at the level of the anterior pituitary (Hoff *et al.*, 1977; Nakai *et al.*, 1978; Plant *et al.*, 1978). In the rodent, on the other hand, it seems certain that steroids have profound influences on both the pituitary and the hypothalamus. Castration of the female rat reduces hypothalamic content of LH-RH and, at the same time, stimulates its release, while estrogen replacement results in LH-RH accumulation within the hypothalamus (Piacsek and Meites, 1966; Araki *et al.*, 1975; Kalra, 1976; Kobayashi *et al.*, 1978; Simpkins *et al.*, 1980; Tytell *et al.*, 1980a,b,c,d).

In previous studies we have examined the effects of steroids on the levels, storage form, and *in vitro* release of LH-RH using nerve terminal preparations from the mediobasal hypothalamus. Levels of LH-RH in nerve terminals increase with estrogen replacement (Peck *et al.*, 1979; Tytell *et al.*, 1980c,d). Attendant to this accumulation of LH-RH within neurosecretory terminals is a reduction in release (Peck *et al.*, 1979; Tytell *et al.*, 1980b) and an apparent change in its storage form (Peck *et al.*, 1979; Tytell *et al.*, 1980c,d). Certainly these data, as well as those of others (see above), suggest either direct or indirect involvement of estrogens at the hypothalamic level in the control of gonadotropin secretion in the rodent.

Our investigations of the hypothalamic content, subcellular distribution, and *in vitro* release of LH-RH as a function of hormonal manipulation have employed the subcellular fractionation techniques of De Robertis *et al.* (1961) and Whittaker (1959). These methods were developed for the study of neurotransmitter systems, especially acetylcholine, and only recently have been extended for the study of neuropeptides. The subcellular distributions of corticotropin-releasing hormone (Mulder, 1970; Fink *et al.*, 1972), thyrotropin-releasing hormone (TRH) (Barnea *et al.*, 1975; Winokur *et al.*, 1977; Peck *et al.*, 1979), somatostatin (Styne *et al.*, 1977; Peck *et al.*, 1979), and LH-RH (Ishii, 1970; Fink *et al.*, 1972; Shin *et al.*,

1974; Taber and Karavolas, 1975; Barnea *et al.*, 1975, 1976, 1977; Ramirez *et al.*, 1975; Gautron *et al.*, 1977; Peck *et al.*, 1979; Tytell *et al.*, 1980c,d) have been studied with this technique in a number of species. Three observations are generally made: (1) the major portion of the releasing hormone is contained in the equivalent of Whittaker's synaptosome-containing P_2 fraction; (2) hypotonic shock of P_2 (or synaptosomes separated from P_2 via centrifugation) results in a loss of peptide from that fraction and its attendant appearance in a P_3 (or vesicle) fraction; (3) little peptide is observed in soluble or microsomal fractions. We certainly do not question the first two generalizations–neuropeptides are contained in vesicles or secretory granules within neurosecretory terminals that can be isolated as synaptosomes (Peck *et al.*, 1979; Tytell *et al.*, 1980c,d). However, considerable amounts of neuropeptide may exist in soluble or microsomal fractions. To observe this form the investigator must inactivate the degrading system present in hypothalamic extracts. For instance, the inclusion of bacitracin (McKelvy *et al.*, 1976) at 100 μM in a hypotonic medium prevents the degradation of somatostatin, TRH, and LH-RH following synaptosomal rupture and allows the demonstration of these peptides in a free, or soluble, form (Peck *et al.*, 1979). Previous reports that failed to demonstrate free peptide invariably did not inhibit endogenous degradative systems. Using this methodology we have demonstrated an osmotically sensitive pool of LH-RH within synaptosomes, the size of which is influenced by estrogens (Peck *et al.*, 1979). However, as reported by others, the bulk of the LH-RH within synaptosomes is contained within vesicles or granules that are presumed to be important for neurosecretion.

Given that neuropeptides are stored in vesicles or granules within neurosecretory terminals, we have considered these terminals as one potential site for steroid regulation of the neuroendocrine axis. As mentioned above, we have used equilibrium and rate sedimentation centrifugation in combination with hypotonic shock to study the subcellular and subsynaptosomal distribution of LH-RH as a function of castration and gonadal steroid replacement (Peck *et al.*, 1979; Tytell *et al.*, 1980c,d). In addition, P_2 or synaptosomal fractions have been employed to examine the *in vitro* release of LH-RH. *In vitro* release can be induced by depolarizing influences, as electrical stimulation or elevated extracellular K^+ (Bennett *et al.*, 1975; Warberg *et al.*, 1977; Tytell *et al.*, 1980a,b) and such release is primarily a Ca^{2+}-coupled process (Bennett and Edwardson, 1973; Tytell *et al.*, 1980a).

Using the P_2 system derived from the mediobasal hypothalamus of castrate females, we have shown that LH-RH is released *in vitro* in two forms, soluble and particulate (or matrix bound). Soluble LH-RH is released under basal, that is, nondepolarizing conditions, and this release is not dependent on Ca^{2+}. The release of soluble LH-RH appears to be a function of alterations in the tonicity of the media. Thus, 10 mM Mn^{2+}, 60 mM K^+, or 120 mM sucrose will stimulate a 2-fold increase in the release of soluble LH-RH, and this release is unaffected by the Ca^{2+} antagonist, La^{3+} (Tytell et al., 1980a). The release of particulate LH-RH, on the other hand, is strictly dependent on Ca^{2+}, is not affected to any great extent by changes in tonicity, and represents a minor fraction of release under basal conditions. With the addition of depolarizing agents, release of particulate LH-RH is rapid and marked, with K^+-stimulated release of LH-RH (particulate) exceeding that of basal release by 5- to 10-fold under most circumstances. Our investigation of these two forms of LH-RH release (soluble and particulate) has led us to suggest (Tytell et al., 1980a) that soluble LH-RH may be analogous to the cytoplasmic pool of acetylcholine that is seen in synaptosomes (Whittaker et al., 1964) and that exchanges readily with exogenous acetylcholine (Marchbanks, 1968). Particulate LH-RH, on the other hand, may be analogous to "stable bound" or vesicular acetylcholine (Whittaker et al., 1964) and may be released as a particle, or bound to a matrix. In view of our observation that castration and estrogen replacement changes the relative size of these putative pools, the regulation of LH-RH secretion at the median eminence by gonadal steroids might involve changes in soluble versus particulate stores of the peptide. The proof of such a hypothesis certainly will require considerable research. At present the hypothesis serves to design experiments, as those discussed below, for the study of endocrine alterations in the subcellular distribution and release of LH-RH.

In the studies to be reported here, we have examined the effect of estradiol and a mammary carcinogen, 7,12-dimethylbenz(a)anthracene (DMBA), on the P_2 content and in vitro release of hypothalamic LH-RH. It was previously reported that DMBA inhibits preovulatory gonadotropin surges in the cycling Sprague–Dawley female (Kerdelhué and El Abed, 1979), a strain of rat that is susceptible to mammary tumor induction following a single dose of DMBA (Huggins et al., 1961). Alteration of gonadotropin surges in Wistar females, a strain resistant to tumor induction, was not observed (Kerdelhué and El Abed, 1979). A possible direct effect of DMBA on the hypothalamus was previously suggested by Stern et al. (1968). Stereotaxic implantation of DMBA in the preoptic region of the hypothalamus resulted in alterations in vaginal cyclicity. If implanted in neonatally androgenized females, DMBA immediately released the sterile females from constant estrus and sporadic cyclicity ensued. These studies, together with those of Kerdelhué and El Abed (1979), suggest that DMBA may have an effect on the hypothalamus and, further, that alteration of the neuroendocrine axis may play a role in the induction of mammary tumors by DMBA.

We have utilized the DMBA-treated female rat and the in vitro release of LH-RH from synaptosomes to test whether the effect of DMBA is at the level of the hypothalamus. In this study both estrogen-primed castrate females of the Sprague–Dawley strain and cycling females of Sprague–Dawley and Wistar strains have been used to examine the effect of DMBA on LH-RH release.

2. Methods

For the estrogen-primed, castrate model, Sprague–Dawley female rats (Hormone Assay, Chicago, Ill.) were employed. They were housed under controlled conditions (22 °C; 12-hr light cycle with lights on from 0700 to 1900) with food and water *ad libitum*. Between 55 and 60 days after birth they were given sesame oil with or without DMBA (20 mg) by gastric intubation (Huggins *et al.*, 1961). All animals were ovariectomized 5 days after DMBA treatment. At noon 2 weeks after ovariectomy, animals from each treatment group (oil or DMBA in oil) were given a single subcutaneous injection of 50 μg of estradiol benzoate (EB) in 0.5 ml of sesame oil. Groups of three animals per treatment group were decapitated without anesthesia at 1500, 1700, and 1900 hr on the next day, trunk blood collected, and hypothalami dissected for the preparation of synaptosomes (see below).

For the cycling female model, female Sprague–Dawley (IFFA-CREDO, Lyon, France) and Wistar (lab strain, Gif-sur-Yvette, France) rats were housed under controlled conditions (22 °C; 14-hr light cycle with lights on from 0530 to 1930) and given food and water *ad libitum*. DMBA in sesame oil or sesame oil alone was administered by gastric intubation as above. Vaginal smears were taken every day and only rats with regular 4-day estrous cycles were used. Groups of five or six rats in proestrus were decapitated at 1600 and 1800 hr 3–4 weeks after treatment with DMBA or oil. As above, trunk blood was collected and hypothalami removed for the preparation of median eminence synaptosomes.

The procedure for the study of LH-RH release from a synaptosome-enriched P_2 fraction of hypothalami was essentially as described by Tytell *et al.* (1980a). A schematic representation of this procedure is shown in Fig. 1. Briefly, 3-mm-diameter punches of mediobasal hypothalamus were homogenized in isotonic (0.32 M) sucrose containing 10 μM $CaCl_2$ and 100 μM bacitracin. A P_2 fraction was prepared by differential centrifugation and resuspended in incubation medium (0.32 M sucrose, 10 mM glucose, 2 mM $CaCl_2$, 100 μM bacitracin). An aliquot of this suspension was extracted with 0.1 N acetic acid in 50% ethanol for the assay of LH-RH content by radioimmunoassay (see below). One-milliliter aliquots of P_2 suspension equivalent to 0.45 hypothalamus were preincubated at 25 °C for 7

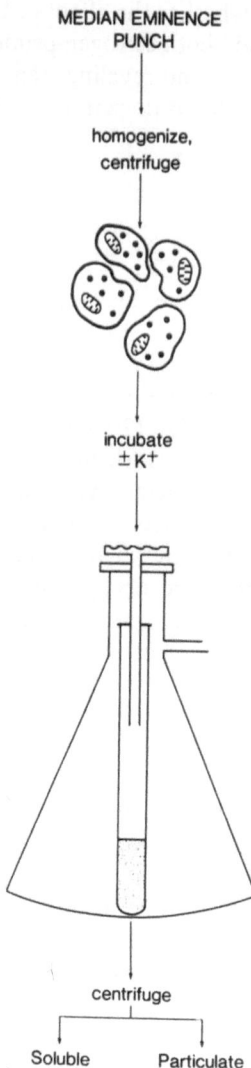

MEDIAN EMINENCE
PUNCH

homogenize,
centrifuge

incubate
± K+

centrifuge

Soluble
(supernatant)

Particulate
(pellet)

Figure 1. A schematic of the protocol for the study of *in vitro* release of neural peptides. Details are given in the text and in Tytell *et al.* (1980a).

min and subsequently incubated at the same temperature for 5 min in the presence or absence of 55 mM KCl. Incubation was terminated by filtration on Whatman GF/C filters and the filtrate kept on ice until used. Aliquots of filtrate were taken for the determination of total LH-RH release, and the remaining filtrate was centrifuged at 48,000g for 20 min to separate soluble and particulate (sedimentable) LH-RH. Aliquots of supernatant were taken for the assay of soluble LH-RH, and pellets were extracted with

0.1 N acetic acid in 50% ethanol for the determination of particulate LH-RH. Particulate LH-RH was also calculated by difference as described by Tytell *et al.* (1980a).

For studies involving the ovariectomized, estrogen-primed model, LH-RH was measured by radioimmunoassay using a solid-phase second-antibody procedure as described by Tytell *et al.* (1980a). For studies involving cycling female rats, LH-RH was measured by the radioimmunoassay described by Kerdelhué *et al.* (1969). All values for LH-RH content and release were adjusted to a per hypothalamic equivalent basis.

Samples of serum were assayed for LH in duplicate using the radioimmunoassay previously described (Kerdelhué *et al.* 1973). Results were expressed as nanograms per milliliter serum in terms of a laboratory rat LH preparation (1.4 × NIH LH S1).

For statistical analysis Student's *t* test was used. A *p* value of 0.05 or less was considered significant.

3. Results

Figure 2 summarizes the results of a study of oil and DMBA-treated estrogen-primed, ovariectomized Sprague–Dawley females. A complete report of the results of this study is currently in press (Kerdelhué and Peck, 1981). In the upper panel are shown serum levels of LH at 3, 5, and 7 p.m.

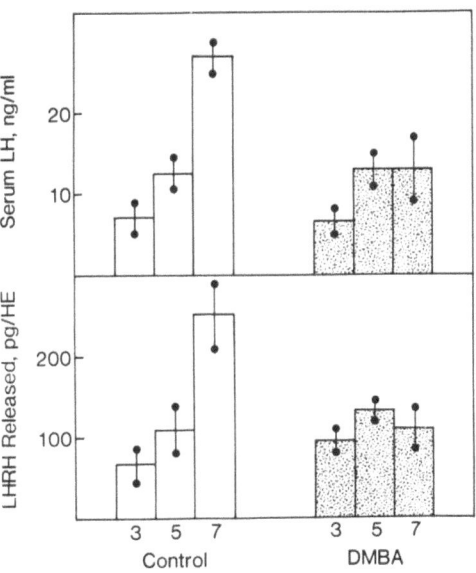

Figure 2. Effect of DMBA on the neuroendocrine axis of ovariectomized, estrogen-primed Sprague–Dawley rats. Upper panel: values for serum LH are given as a function of time (3, 5, or 7 p.m.). Values are the mean ± S.E.M. of three independent experiments with five to six animals assayed per time point per experiment. Lower panel: values for net K^+-stimulated release (K^+-stimulated release − basal release) of particulate LH-RH as a function of time of sacrifice (3, 5, or 7 p.m.). Values are the mean ± S.E.M. of three independent experiments with three to four determinations per experiment.

in control (oil) and DMBA-treated castrate rats primed with 50 mg EB for 1 day. In control estrogen-primed, castrate animals, as previously reported by Legan et al. (1975), a surge in serum LH is noted. However, this surge is blunted in estrogen-primed, castrate animals previously treated one time with DMBA. These results are in agreement with those of Kerdelhué and El Abed (1979) for the cycling female model.

In the lower panel of Fig. 2, net K^+-stimulated release of *particulate* LH-RH is shown for P_2 fractions prepared from hypothalami of the same animals. In control (oil treated) animals, *in vitro* release of particulate LH-RH by elevated K^+ increases as a function of time of day in a fashion parallel to that of the LH surge. However, the release of particulate LH-RH is suppressed in the case of DMBA-treated animals. The release of soluble LH-RH was not significantly different between experimental and control systems (data not shown). In addition, basal release of particulate LH-RH was not different from synaptosomes of oil- or DMBA-treated rats at any time (Kerdelhué and Peck, 1981).

To compare the effects of DMBA on a sensitive (Sprague–Dawley) and a resistant (Wistar) strain, cycling female rats of each strain were sacrificed at 1600 and 1800 hr on the day of proestrus. Serum samples were taken and P_2 fractions of mediobasal hypothalamus were prepared as described in Methods. Serum LH as well as P_2 content and *in vitro* release of LH-RH was examined in three independent experiments. In each case hypothalami from five to six animals were pooled and release from P_2 was studied under basal conditions as well as with elevation of K^+.

Table I shows the values for serum LH in oil- and DMBA-treated animals of both strains at 1600 and 1800 hr for one representative experiment. Note that between 1600 and 1800 hr there is a surge in serum LH in every case. However, in the case of DMBA-treated Sprague–Dawley females, the LH surge is blunted (50% decrease, $p < 0.05$ when compared with oil-treated control). A comparison of oil- and DMBA-treated Wistar females does not reveal a significant depression upon DMBA treatment. These results are in agreement with those of Kerdelhué and El Abed (1979) for cycling females of the Sprague–Dawley and Wistar strains.

Table I.   Serum LH in Oil- and DMBA-Treated Animals[a]

Time	Sprague–Dawley		Wistar	
	Oil	DMBA	Oil	DMBA
4 p.m.	2.9 ± 0.5	2.9 ± 0.6	3.1 ± 0.4	1.3 ± 0.1
6 p.m.	21.0 ± 0.6	11.1 ± 2.8[b]	15.9 ± 1.0	11.3 ± 3.6

[a] Animals were sacrificed on the afternoon of the fifth or sixth estrous cycle after treatment. Values are the mean $\pm$ S.E.M. of five determinations.
[b] $p < 0.05$ when compared with control Sprague–Dawley.

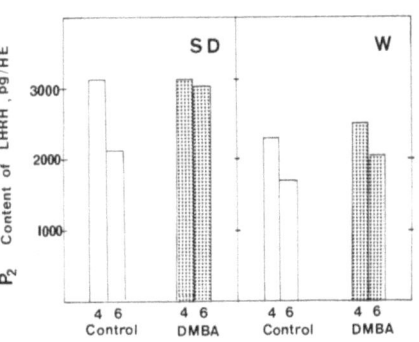

Figure 3. P$_2$ content of LH-RH in cycling females at 1600 and 1800 hr on proestrus. Results are presented for one experiment comparing the levels of LH-RH in neurosecretory terminals of Sprague–Dawley (SD) and Wistar (W) females with (DMBA) or without (control) previous exposure to DMBA. Each value is the average of two to three measurements on a single pool of P$_2$.

Figure 3 shows the LH-RH content in the P$_2$ fraction of the same animals as studied in Table I. Since all hypothalami were pooled for the preparation of P$_2$ fractions, these values represent a single determination and no statistical analysis of the population is possible. However, it is interesting to note that P$_2$ content of LH-RH decreases between 1600 and 1800 hr in all cases except the DMBA-treated Sprague–Dawley females, that group which shows a suppression in the LH surge.

Figure 4 shows soluble and particulate release of LH-RH from P$_2$ fractions prepared at 1800 hr, under basal and K$^+$-stimulated conditions for the same experiment as in Table I and Fig. 3. Release of soluble LH-RH is almost similar in all cases, representing about 3–6% of total P$_2$ content

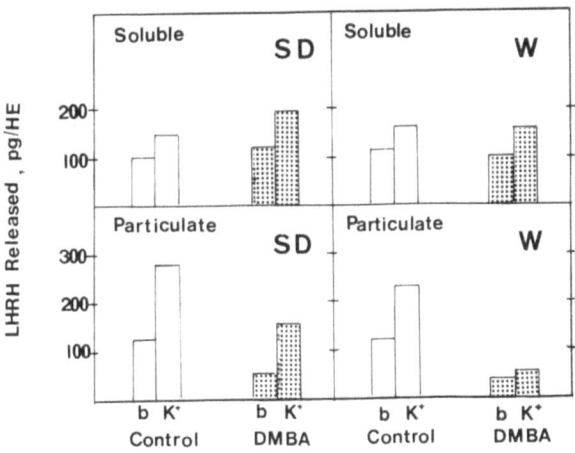

Figure 4. *In vitro* release of LH-RH from P$_2$ fractions of cycling females. Basal (b) and K$^+$-stimulated (K$^+$) release of LH-RH are presented for Sprague-Dawley (SD) and Wistar (W) females sacrificed at 1800 h on the day of proestrus. Each value is the average of two to three determinations.

of LH-RH. Release of particulate LH-RH, however, is different between treatment groups (oil vs. DMBA) and strains (Sprague–Dawley vs. Wistar). In every instance, less LH-RH is released in DMBA-treated animals when compared with controls. In addition, the release of LH-RH from P_2 fractions of Wistar animals is generally less than that of the Sprague–Dawley strain.

Table II summarizes the results of three experiments on cycling females of the Sprague–Dawley and Wistar strains. All values for release (1600 and 1800 hr) have been pooled since for each individual experiment variation in the timing of the afternoon LH surge existed. Only net K^+-stimulated LH-RH release is shown; however, the inhibition effects of DMBA shown in Table II were generally observed for basal or K^+-stimulated release (as in Fig. 4). Thus, in agreement with the experiment presented in Fig. 4, release of LH-RH from the P_2 fraction of Sprague–Dawley females sacrificed on the afternoon of proestrus is greater than that of P_2 fractions from Wistar females. This difference appears to be most evident when total release or particulate (total − soluble) is examined. In addition, treatment with DMBA results in a suppression of LH-RH release. The effect of DMBA is marked for total and particulate release. Release of soluble LH-RH is not affected in general (see Fig. 4 as well). Unfortunately the cycling female model presents variation in the timing of the LH surge and so no significant differences exist in Table II. However, for three independent experiments the trend was always as shown in Fig. 4 and Table II. DMBA treatment resulted in a suppression of LH-RH release (especially particulate) in both Sprague–Dawley and Wistar females with, however, the effect more marked in the Sprague–Dawley

Table II.   Net K^+-Stimulated LH-RH Release from P_2 Fractions of Oil- and DMBA-Treated Cycling Females[a]

	Sprague–Dawley	Wistar
Total release	pg/HE	
Control	367 ± 116	211 ± 37
DMBA-treated	244 ± 61	153 ± 68
Release of soluble LH-RH		
Control	149 ± 70	123 ± 38
DMBA-treated	131 ± 39	102 ± 37
Release of particulate LH-RH		
Control	218 ± 90	88 ± 39
DMBA-treated	113 ± 50	51 ± 30

[a] Values are the mean ± S.E.M. of three independent experiments with 10 animals per experiment. All values for release experiments (1600 and 1800) were pooled.

strain. Additional experiments will be necessary to establish the significance of these strain and treatment differences.

4. Discussion

The results contained in this report, when taken together with those of others (Bennett *et al.*, 1975; Warberg *et al.*, 1977; Tytell *et al.*, 1980a,b), establish the synaptosome-containing P_2 fraction as a useful preparation for the study of neurosecretory systems. The study of *in vitro* release of LH-RH from these terminals has proven a useful model for the study of the release of hypothalamic LH-RH, especially since at this time our methodology does not allow the direct measurement of LH-RH in serum on a reliable basis. The results shown in Fig. 2 suggest that the *in vitro* release of LH-RH from P_2 fractions is tightly coupled with one *in vivo* event, the LH surge.

Of considerable interest to us is the existence of two pools of LH-RH, soluble and particulate. In both the ovariectomized, estrogen-primed model and the cycling female it would appear that the most important, or at least physiological, pool is the particulate fraction. This pool has proven to be tightly coupled with the *in vivo* LH surge while the soluble pool appears to change very little.

We have previously demonstrated that the release of particulate LH-RH is Ca^{2+}-coupled while the release of soluble is not (Tytell *et al.*, 1980a). Perhaps the particulate pool is a reflection of that pool released via neurotransmitter-mediated processes at the median eminence. Only additional research can answer this question. Certainly it appears that the size and/or releasability of this pool varies with steroid manipulation (Tytell *et al.*, 1980b; Kerdelhué and Peck, 1981) and with the estrous cycle. These previous studies have shown that *in vitro* release of LH-RH is greatest in the afternoon (Tytell *et al.*, 1980b). In the present report our examination of the ovariectomized, estrogen-primed model shows a progressive increase in the *in vitro* release of particulate LH-RH as a function of the time of day *and* a strong coupling with the *in vivo* level of serum LH. Thus, the *in vitro* release of particulate LH-RH from P_2 fractions of the median eminence parallels the *in vivo* situation. It was our intention to demonstrate this fact in the cycling female. However, the variability in this physiological situation has not permitted this conclusion as yet. More experiments at frequent time intervals will be necessary to establish this relationship in the cycling animal.

Our studies with DMBA and the estrogen-primed, ovariectomized female support our contention that *in vitro* release of particulate LH-RH reflects the *in vivo* situation. Thus, when the LH surge is blunted, as with DMBA treatment, the release of particulate LH-RH *in vitro* is similarly depressed. The surprising result comes from our study of the cycling system. It would seem that DMBA treatment can affect LH-RH release at all levels, except perhaps the soluble pool. Thus, a single exposure to DMBA results in a reduction in basal as well as K^+-stimulated release of LH-RH (Fig. 4) and a marked reduction in net K^+-stimulated release (Table II). The mechanism by which DMBA produces its effect is not known. We are currently investigating its effect on estrogen receptor mechanisms in the hypothalamus and pituitary. In addition, we are examining the effect of prolactin (PRL) on the releasability of LH-RH from P_2 fractions. In fact, in DMBA-treated Sprague–Dawley cycling rats the reduced preovulatory LH surge is accompanied by an enhanced PRL surge (Kerdelhué and El Abed, 1979). Furthermore, the administration of PRL does block preovulatory gonadotropin surges in Sprague–Dawley but not in Wistar rats (Pasqualini *et al.*, 1980).

Regardless, it must be kept in mind that the effect of DMBA on the neuroendocrine axis and LH-RH release may be separated from its carcinogenic action. This conclusion arises from the observation reported here that Wistar as well as Sprague–Dawley females apparently are affected by DMBA treatment, at least with respect to *in vitro* release of LH-RH.

In summary, we have studied the release of LH-RH *in vitro* and attempted to correlate this release with the *in vivo* LH surge. For the ovariectomized, estrogen-primed female such a correlation is possible. Thus, the releasability of LH-RH parallels serum levels of LH on a temporal basis and both are suppressed by DMBA treatment. Unfortunately, the cycling female rat is more variable and, to date, has not afforded us with statistically significant data.

ACKNOWLEDGMENTS. This work was supported by a joint NCI–INSERM exchange program and grants from the NIH (HD08389), INSERM (650-7378), the Ligue Nationale contre le Cancer, and the Fondation de Recherche en Hormonologie. E. J. Peck, Jr. is a USPHS Career Development Awardee (HD0022).

We wish to thank the NIAMDD, Rat Pituitary Hormone Distribution Program, for the radioimmunoassay kit for LH. We are grateful to our many collaborators who helped in the implementation of this research.

DISCUSSION

KRAICER: I am intrigued by the particulate nature of LH-RH released from the synaptosomes. What are the "particles"? What do they look like at the EM level? What is the physiological significance of LH-RH, in the particulate form, in the portal circulation?

PECK: We do not yet know the nature of the particles released. They sediment at 48,000 g in low-viscosity media. We have not yet succeeded in visualizing them with the electron microscope. We believe that the particles are a protective mechanism which makes LH-RH less available to degrading enzymes. Thus, particles can be exposed to high-speed supernatants from brain tissue and the half-life of LH-RH is longer than that for soluble LH-RH. Such studies are preliminary, and require confirmation, but we view the particles as protecting LH-RH within the perivascular space.

KRAICER: What is the ionic constitution of your incubation medium? I ask this in respect to whether you know that the synaptosomes have a transmembrane potential and that high K^+ really does depolarize.

PECK: Our medium is isotonic sucrose containing about 2 M $CaCl_2$, 100 μM bacitracin, and 10 nM glucose. If we attempt to use more "physiologic" media, the synaptosomes release most of their LH-RH within a short time. Thus, in our hands they are not very stable in the presence of ions such as phosphate, sulfate, etc.

KRAICER: What is the time base for release with high K^+? Is it constant, reversible, and repeatable? Does the release "spike" while maintaining high K^+ or is it maintained during high K^+ stimulation?

PECK: The release process is a spike. If one stops release by filtration at various times, release is almost complete within 1 minute. By 5 minutes the process is over. Thus, it is not constant. We have been unable to demonstrate uptake of LH-RH, suggesting (but not proving) that release is not reversible. We have not tried multiple stimulation of the same synaptosomes.

KANAZIR: Since the estrogen effects are mediated via receptor, what would be the role of the receptor in the regulation of release? Is the effect of calcium coupled with the n¹ ᴐs-phorylation of synaptosomes? Is the release of soluble LH-RH altered?

PECK: In my opinion the estrogen receptor is acting at the genomic level to alter the protein composition of neurosecretory elements. We do not know what or how many proteins are involved. Analysis via denaturing gel electrophoresis is currently in progress to examine the question.

We have no conclusive evidence for a role of phosphorylation in the release process. We are currently examining calmodulin in this respect.

Yes, the amount of LH-RH in the soluble pool is altered by estrogen treatment; however, we do not understand the significance of the shift between soluble and particulate pools.

REFERENCES

Araki, S., Ferin, M., Zimmerman, A., and Vande Wiele, R. L., 1975, Ovarian modulation of immunoreactive gonadotropin-releasing hormone (Gn-RH) in the rat brain: Evidence for a differential effect on the anterior and mid-hypothalamus, *Endocrinology* **96**:644.

Barnea, A., Ben-Jonathan, N., Colston, C., Johnston, J. M., and Porter, J., 1975, Differential subcellular compartmentalization of thyrotropin releasing hormone (TRH) and gonadotropin releasing hormone (LRH) in hypothalamic tissue, *Proc. Natl. Acad. Sci. USA* **72**:3152.

Barnea, A., Ben-Jonathan, N., and Porter, J. C., 1976, Characterization of hypothalamic subcellular particles containing luteinizing hormone releasing hormone and thyrotropin releasing hormone, *J. Neurochem.* **27**:477.

Barnea, A., Neaves, W. B., and Porter, J. C., 1977, Ontogeny of the subcellular compartmentalization of thyrotropin releasing hormone and luteinizing hormone releasing hormone in the rat hypothalamus, *Endocrinology* **100**:1068.

Bennett, G. W., and Edwardson, J. A., 1973, Calcium dependent release of hypophysiotropic substances from nerve-endings (synaptosomes) isolated from the hypothalamus, *J. Endocrinol.* **59**:XV–XVI.

Bennett, G. W., Edwardson, J. A., Holland, D., Jeffcoate, S. L., and White, N., 1975, Release of immunoreactive luteinizing hormone-releasing hormone and thyrotrophin-releasing hormone from hypothalamic synaptosomes, *Nature (London)* **257**:323.

Berthold, A. A., 1849, Transplantation der Hoden, *Arch. Anat. Physiol. Wiss. Med.* **16**:42.

Clark, J. H., and Peck, E. J., Jr., 1979, *Female Sex Steroids: Receptors and Function*, Springer-Verlag, Berlin.

De Robertis, E., De Iraldi, A. P. , Rodriguez, G., and Gomez, C. J., 1961, On the isolation of nerve endings and synaptic vesicles, *J. Biophys. Biochem. Cytol.* **9**:229.

Feder, H. H., Landau, I. T., and Walker, W. A., 1978, Anatomical and biochemical substrates of the action of estrogens and antiestrogens on brain tissues that regulate female sex behavior of rodents, in: *Endocrine Control of Sexual Behavior* (C. Beyer, ed), Raven Press, New York.

Fink, G., Smith, G. C., Tiballs, J., and Lee, V. W. K., 1972, LRF and CRF release in subcellular fractions of bovine median eminence, *Nature New Biol.* **239**:57.

Gautron, J. P., Pattou, E., and Kordon, C., 1977, New data on the subcellular distribution of LH-RH in various structures of the rat hypothalamus, *Mol. Cell. Endocrinol.* **8**:81.

Hoff, J. D., Lashley, B. L., Wang, C. F., and Yen, S. S. C., 1977, The two pools of pituitary gonadotropin: Regulation during the menstrual cycle, *J. Clin. Endocrinol. Metab.* **44**:302.

Huggins, C., Grand, L. C., and Brillantes, F. P., 1961, Mammary cancer induced by a single feeding of polynuclear hydrocarbons and its suppression, *Nature (London)* **189**:204.

Ishii, S., 1970, Association of luteinizing hormone-releasing factor with granules separated from equine hypophysial stalk, *Endocrinology* **86**:207.

Kalra, S., 1976, Tissue levels of luteinizing hormone-releasing hormone in the preoptic area and hypothalamus, and serum concentrations of gonadotropins following anterior hypothalamic deafferentation and estrogen treatment of the female rat, *Endocrinology* **99**:101.

Kerdelhué, B., and El Abed, A., 1979, Inhibition of preovulatory gonadotropin secretion and stimulation of prolactin secretion by 7,12-dimethylbenz(*a*)anthracene in Sprague–Dawley rats, *Cancer Res.* **39**:4700.

Kerdelhué, B., and Peck, E. J., Jr., 1981, *In vitro* LH-RH release: Correlation with the LH surge and alteration by a mammary carcinogen, *Peptides* **2**:219.

Kerdelhué, B., Bérault, A., Courte, C., and Jutisz, M., 1969, Mise au point d'un dosage radioimmunologique de l'hormone lutéinisante (LH) de rat au moyen d'un immunserum anti-LH de mouton et d'une préparation marquée de LH de rat, *C. R. Acad Sci. Ser. D* **269**:2413.

Kerdelhué, B., Jutisz, M., Gillessen, D., and Studer, R. O., 1973, Obtention of antisera against the decapeptide which stimulates the release of pituitary gonadotropins and development of its radioimmunoassay, *Biochim. Biophys. Acta* **297**:540.

Kobayashi, R. M., Lu, K. H., Moore, R. Y., and Yen, S. C. C., 1978, Regional distribution of hypothalamic luteinizing hormone-releasing hormone in proestrous rats: Effects of ovariectomy and estrogen replacement, *Endocrinology* **102**:98.

Legan, S. J., Coon, G. A., and Karsch, F. J., 1975, Role of estrogen as initiator of daily LH surges in the ovariectomized rat, *Endocrinology* **96**:50.

McEwen, B. S., 1978, Gonadal steroid receptors in neuroendocrine tissues, in: *Receptors for Hormones* (B. W. O'Malley and L. Birnbaumer, eds.), Vol. II, pp. 453–500, Academic Press, New York.

McKelvy, J. F., Leblanc, P., Laudes, C., Perrie, S., Grimm-Jorgensen, Y., and Kordon, C., 1976, The use of bacitracin as an inhibitor of the degradation of thyrotropin releasing factor and luteinizing hormone releasing factor, *Biochem. Biophys. Res. Commun.* **73**:507.

Marchbanks, R. M., 1968, Exchangeability of radioactive acetylcholine with the bound acetylcholine of synaptosomes and synaptic vesicles, *Biochem. J.* **106**:87.

Mulder, A. H., 1970, On the subcellular localization of corticotropin-releasing factor (CRF) in the rat median eminence, in: *Progress in Brain Research,* Vol. 32 (D. De Weid and J. A. W. M. Weijnen, eds.), pp. 31–41, Elsevier, Amsterdam.

Nakai, Y., Plant, T. M., Hess, D. L., Keogh, E. J., and Knobil, E., 1978, On the sites of the negative and positive feedback actions of estradiol in the control of gonadotropin secretion in the rhesus monkey, *Endocrinology* **102**:1008.

Pasqualini, C., El Abed, A., and Kerdelhué, B., 1980, L'inhibition de la sécrétion préovulatoire des gonadotropines par la prolactine chez la ratte semble dépendre de la souche utilisée, *C.R. Acad. Sci. Ser. D*, **291**:1055.

Peck, E. J., Jr., Tytell, M., Boyd, A. E., and Barr, G., 1979, Effects of steroid hormones on neural peptides, in: *Brain Peptides: A New Endocrinology* (A. M. Gotto, Jr., E. J. Peck, Jr., and A. E. Boyd, III, eds.), pp. 241–258, Elsevier/North-Holland, Amsterdam.

Piacsek, B. E., and Meites, J., 1966, Effects of castration and gonadal hormones on hypothalamic content of luteinizing hormone releasing factor (LRF), *Endocrinology* **79**:432.

Plant, T. M., Nakai, Y., Belchetz, P., Keogh, E., and Knobil, E., 1978, The sites of action of estradiol and phentolamine in the inhibition of the pulsatile, circhoral discharge of LH in the rhesus monkey *(Macaca mulatta), Endocrinology* **102**:1015.

Ramirez, V. O., Gautron, J. P., Epelbaum, J., Pattou, E., Zamora, A., and Kordon, C., 1975, Distribution of LH-RH in subcellular fractions of the basomedial hypothalamus, *Mol. Cell. Endocrinol.* **3**:339.

Shin, S. H., Morris, A., Snyder, J., Hymer, W. C., and Milligan, J. V., 1974, Subcellular localization of LH releasing hormone in the rat hypothalamus, *Neuroendocrinology* **16**:191.

Simpkins, J. W., Kalra, P. S., and Kalra, S. P., 1980, Temporal alterations in luteinizing hormone-releasing hormone concentrations in several discrete brain regions: Effects of estrogen–progesterone and norepinephrine synthesis inhibition, *Endocrinology* **107**: 573.

Stern, E., Mickey, R. M., and Gorski, R. A., 1968, Neuroendocrine factors in experimental carcinogenesis, *Ann. N.Y. Acad. Sci.* **164**:494.

Styne, D. M., Goldsmith, P. C., Burnstein, S. R., Kaplan, S. L., and Grumbach, M. M., 1977, Immunoreactive somatostatin and luteinizing hormone releasing hormone in median eminence synaptosomes of the rat: Detection by immunohistochemistry and quantification by radioimmunoassay, *Endocrinology* **101**:1099.

Taber, C. A., and Karavolas, H. J., 1975, Subcellular localization of LH releasing activity in the rat hypothalamus, *Endocrinology* **96**:446.

Tytell, M., Clark, J. H., and Peck, E. J., Jr., 1980a, Properties of LHRH release from a hypothalamic synaptosomal fraction of estrogen-primed ovariectomized rats, *Neurochem. Res.* **5**:479.

Tytell, M., Clark, J. H., and Peck, E. J., Jr., 1980b, Effects of estrogen and progesterone on LHRH release from a hypothalamic synaptosomal fraction of ovariectomized rats, *Neurochem. Res.* **5**:493.

Tytell, M., Clark, J. H., and Peck, E. J., Jr., 1980c, Effect of estrogen on the subcellular distribution of luteinizing hormone releasing hormone in the ovariectomized rat, *Endocr. Res. Commun.* **25**:213.

Tytell, M., Clark, J. H., and Peck, E. J., Jr., 1980d, Estrogen and the subcellular distribution of luteinizing hormone releasing hormone, *Peptides* **1**(4):301.

Warberg, J., Eskay, F. L., Barnea, A., Reynolds, R. C., and Porter, J. C., 1977, Release of luteinizing hormone and thyrotropin releasing hormone from a synaptosome-enriched fraction of hypothalamic homogenates, *Endocrinology* **100**:814.

Whittaker, V. P., 1959, The isolation and characterization of acetylcholine-containing particles from brain, *Biochem. J.* **72**:694.

Whittaker, V. P., Michaelson, J. A., and Kirland, R. J. A., 1964, The separation of synaptic vesicles from nerve ending particles (synaptosomes), *Biochem. J.* **90**:293.

Winokur, A., Davis, R., and Utiger, R. D., 1977, Subcellular distribution of thyrotropin-releasing hormone (TRH) in rat brain and hypothalamus, *Brain Res.* **120**:423.

8

Mode of Opioid and Catecholamine Involvement in Regulating LH Secretion

Satya P. Kalra

1. Introduction

Since the recent discovery of opioid peptides, numerous investigators have demonstrated that these peptides occur in the pituitary as well as in several discrete regions of the diencephalon and that they are secreted into the hypophyseal portal veins (Hughes *et al.*, 1975; Guillemin *et al.*, 1976; Bloom *et al.*, 1978; Goldstein *et al.*, 1979; Larsson *et al.*, 1979; Snyder, 1980; Uhl *et al.*, 1978; Wamsley *et al.*, 1980; Wardlaw *et al.*, 1980). Recent research efforts have been focused on demonstrating the endocrine effects of opiates and endogenous ligands of opiate receptors—the β-endorphins and enkephalins (Meites *et al.*, 1979; Kalra *et al.*, 1981b; Van Vugt and Meites, 1980). Although these peptides alter the secretion of several pituitary hormones, few studies have attempted to understand their mode of action (Meites *et al.*, 1979; Muraki *et al.*, 1978; Cicero *et al.*, 1980; Kalra *et al.*, 1981b). In this paper, we will review our recent investigations on the effects of opiates and opiate antagonists on pituitary luteinizing hormone (LH) release. We have attempted to examine the nature of the functional link between steroid feedback and opiate action on LH release and how catecholaminergic transmission may serve as an important mediator for these interactions.

Satya P. Kalra • Department of Obstetrics and Gynecology, University of Florida College of Medicine, Gainesville, Florida 32610.

2. Effects of Opiates on LH Release

Morphine (M) and related narcotics have long been known to impair
reproductive function in rodents and man (Azizzi *et al.*, 1973; Thomas and
Dombrosky, 1975; Cicero *et al.*, 1974). When administered prior to the
critical period on proestrus, M blocked ovulation the following day (Barraclough and Sawyer, 1955). This antiovulatory effect of M was due to
blockade of the preovulatory LH discharge and appeared to involve the
hypothalamo–pituitary axis (Pang *et al.*, 1977; Cicero *et al.*, 1977). Basal
gonadotropin secretion in intact rats was also suppressed by M treatment
(Grandison *et al.*, 1980; Cicero *et al.*, 1977; Meites *et al.*, 1979). However,

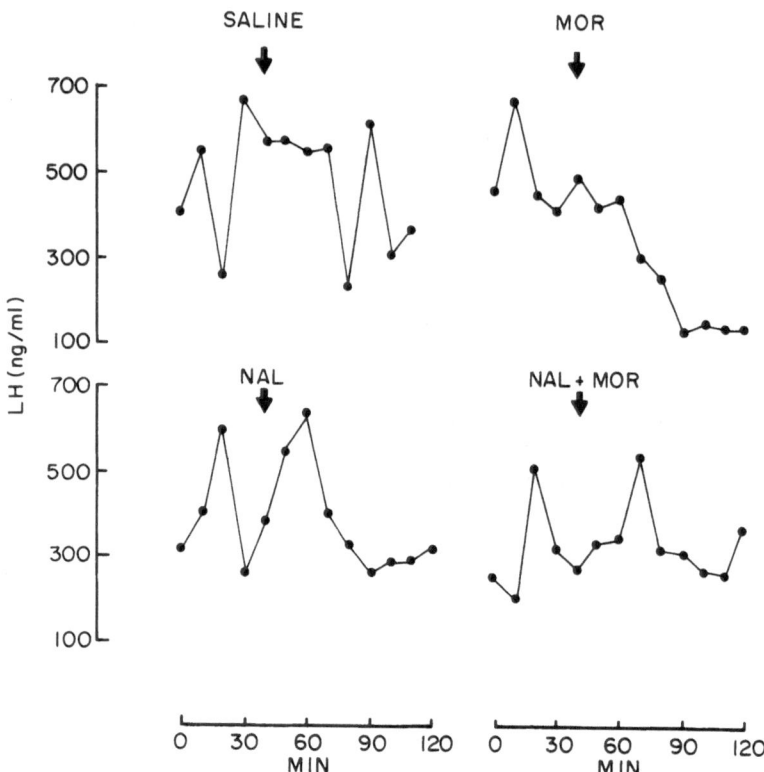

Figure 1. The effects of morphine (MOR, 30 mg/kg) and naloxone (NAL, 2 mg/kg) on episodic
LH secretion in a long-term ovariectomized rat. Blood samples for LH analysis were withdrawn
at 10-min intervals through an indwelling jugular cannula. MOR treatment suppressed episodic
LH secretion. While NAL failed to influence LH release in the ovariectomized rat, NAL treatment counteracted the suppression of LH release by MOR.

the most dramatic effects of M on LH release were demonstrable in ovariectomized rats (Fig. 1). Secretion of LH in these rats occurs episodically at 20- to 30-min intervals (Gallo, 1980); an injection of M inhibited this episodic LH pattern within 1 hr (Fig. 1). In short-term orchidectomized rats, M similarly suppressed LH release; surprisingly though, M was not effective in long-term orchidectomized rats (Cicero *et al.*, 1980). The cause of this intriguing refractoriness, which apparently develops only in castrated male rats, is not understood.

Evidence has accumulated to show that inhibition of LH release may result from the interaction of M with brain opiate receptors. The opiate antagonist naloxone completely reversed the LH-inhibiting effects of M in ovariectomized rats (Fig. 1). Similar observations were made by Cicero *et al.* (1979) and Meites *et al.* (1979) in intact and castrated male rats. These results, taken together with the observation that intracerebral injection of β-endorphins reduced serum LH levels in ovariectomized rats (Grandison *et al.*, 1980), agree with the view that endogenous opioid peptides may play a physiological role in regulation of pituitary LH secretion.

3. Effects of Opiate Receptor Antagonists on Pituitary LH Release

During evaluation of the effects of opiates and opioid peptides on pituitary hormone secretion it became evident that the opiate receptor antagonists themselves were potent stimulators of LH release from the pituitary gland (Cicero *et al.*, 1979; Blank *et al.*, 1979; Kalra *et al.*, 1981b). The stimulatory effects of these antagonists, naloxone and naltrexone, have been studied in detail (Cicero *et al.*, 1979). Generally, naltrexone produced a greater LH response than naloxone. It is of interest to note that the naloxone-induced LH response was found to be remarkably different during sexual maturation of male and female rats (Blank *et al.*, 1979; Cicero *et al.*, 1979; Ieiri *et al.*, 1979). While the hypothalamo–pituitary axis of the prepubertal and pubertal male rats could not be activated to release LH, intact and castrated adult rats responded readily to naloxone treatment (Cicero *et al.*, 1979; Blank *et al.*, 1979). However, in prepubertal and pubertal female rats between 10 and 50 days old, naloxone treatment readily elicited LH release but in adult cycling rats this response was not apparent (Blank *et al.*, 1979). In marked contrast to these results of Blank *et al.* (1979) in cycling female rats, we have observed that naloxone stimulated LH release when administered just prior to the critical period on proestrus (unpublished). In general, thus, there is convincing evidence to suggest that the steroid hormonal milieu may modulate the hypothalamic influence of opioid peptides on LH secretion.

4. Effects of Estrogen on LH Release Induced by Naloxone

It is well established that estrogens suppress pituitary LH release in ovariectomized rats (Kalra *et al.*, 1973; Negro-Vilar *et al.*, 1973). Blank *et al.* (1979) observed that estrogen treatment for two consecutive days inhibited naloxone-induced LH release in immature ovariectomized rats. These investigators reported that even multiple injections of naloxone failed to reverse the LH inhibition that occurred shortly after a single injection of estradiol benzoate (Blank *et al.*, 1980). However, our investigations suggest that estrogens may not completely inhibit the naloxone-induced LH release but may modulate the magnitude of LH response and this may well depend upon the time allowed for estradiol to act prior to naloxone injection (Fig. 2). Naloxone injection 60–90 min after estrogen, when serum LH levels are partially suppressed, stimulated LH release lasting only for 10 min (Fig. 2). In contrast, 2 days after estrogen injection similar naloxone treat-

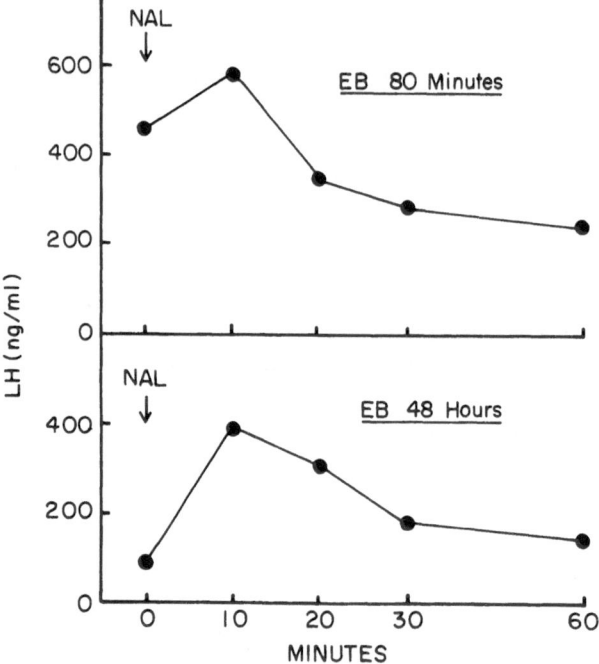

Figure 2. The effects of estradiol benzoate (EB, 10 μg/rat) on naloxone (NAL)-induced LH release. NAL (2 mg/kg) injection 60–90 min after EB treatment produced a small rise in serum LH levels, whereas after similar NAL treatment 48 h after EB injection there was a marked increase in LH release that persisted for 60 min.

ment produced a relatively greater LH response that persisted for 60 min (Fig. 2). There are two possibilities to explain these differences in the naloxone-induced LH response produced by prior estrogen treatment. Evidence exists to suggest that naloxone acts centrally to promote LH-releasing hormone (LH-RH) release from peptidergic neurons in the hypothalamus. In that case, it is likely that estrogen treatment may acutely restrain release of the neurohormone induced by naloxone. However, a more likely possibility is the one that takes into account estrogen-induced changes in pituitary sensitivity to LH-RH. It is noteworthy that the differences in the magnitude of LH response after naloxone treatment (Fig. 2) paralleled the estrogen-induced biphasic alterations in pituitary responsiveness to LH-RH as reported previously (Kalra, 1978; Negro-Vilar *et al.*, 1973). A comparison of these results with those depicted in Fig. 2 revealed that after estrogen treatment the naloxone-induced LH release was attenuated at the time when pituitary responsiveness decreased while the relatively greater LH release in response to naloxone was coincident with enhanced pituitary responsiveness to LH-RH. Thus, while estrogen-induced changes in pituitary sensitivity to LH-RH may explain the variable LH release after naloxone injection in these rats, we cannot completely rule out the possibility that estrogen may modulate the release of LH-RH by naloxone.

5. Mode of Opiate Involvement in Pituitary LH Release

5.1 Effects on Hypothalamic LH-RH Secretion

While there is growing evidence to show that opiates and opiate receptor antagonists act centrally rather than at the level of the pituitary gland to influence LH secretion, few attempts have been made to directly evaluate the influence of opiates on LH-RH secretion in the rat (Muraki *et al.*, 1978; Cicero *et al.*, 1980; Simpkins and Kalra, 1980). Recently, we have shown that progesterone administration to estrogen-primed ovariectomized rats activated those secretary processes in peptidergic neurons that enhanced LH-RH elaboration in the mediobasal hypothalamus (MBH) (Kalra *et al.*, 1978; Simpkins *et al.*, 1980; Simpkins and Kalra, 1980; Kalra *et al.*, 1981b). Acceleration in the accumulation and/or synthesis of LH-RH in the MBH occurred prior to its discharge to elicit the afternoon surge of LH (Levine and Ramirez, 1980). Activation of opiate receptors with M blocked the stimulatory effects of progesterone on the MBH LH-RH and serum LH response (Simpkins and Kalra, 1980; Kalra *et al.*, 1981b). The specificity of brain opiate receptor involvement in progesterone action was afforded by the fact that prior blockade of opiate receptors with naloxone overcame the inhibitory effects of M on LH-RH secretion. These experimental data not only documented the influence of brain opiate receptors on the central action of gonadal steroids but further revealed the pivotal role of opioids in the chain of events triggered by progesterone to stimulate hypersecretion of LH-RH in ovariectomized, estrogen-primed rats.

5.2. Neurotransmitter Mediation of Opiate Effects on LH Release

5.2.1. Dopamine

Morphine analgesia has long been known to significantly change the level of activity in several monoaminergic neurons (Iwamoto and Way, 1979; Korf et al., 1974). Demonstration of opiate receptors both in the regions innervated by monoaminergic neurons and on aminergic projections has led to the realization that these two systems may interact to regulate several behavorial and neuroendocrine functions (Watson et al., 1980; Snyder, 1980; Wamsley et al., 1980; Sar et al., 1978; Herkenham and Pert, 1980; Pert et al., 1976). The distribution patterns of neurons containing β-endorphin and enkephalins and of monoaminergic neurons overlap in the diencephalon (Watson et al., 1980; Elde and Hökfelt, 1978; Ungerstedt, 1971). Ferland et al. (1977) showed that Met-enkephalin decreased turnover in the tuberoinfundibular dopamine (TIDA) neurons in the hypothalamus. Gudelsky and Porter (1979) reported decreased DA levels in pituitary stalk portal plasma of rats treated with morphine or β-endorphin. These and several other studies strongly imply a regulatory role of endogenous opioid peptides on the release of DA from the TIDA terminals in the median eminence. It is quite possible that opioid peptide neurons may affect LH-RH release by regulating DA release from TIDA neurons in the median eminence. On the other hand, Rotsztejn et al. (1978a,b), based on their observation that opiates and opioid peptides inhibited DA-induced LH-RH release from in vitro hypothalamic cultures, believe that opioid peptide neurons may act at the level of LH-RH neurons to modulate DA action. In contrast to these in vitro studies, we failed to find any evidence of dopaminergic participation in LH release elicited by naloxone. As shown in Fig. 3, prior blockade of central DA receptors with pimozide had little effect on LH release stimulated by naloxone.

Furthermore, the physiological significance of DA participation in the opiate effects on LH release is unclear. First, an excitatory role of DA on LH release has been seriously questioned (Gallo, 1980; Kalra, 1977). Second, there is compelling evidence to show that opiates and opioid peptides suppress turnover and release of DA into the portal circulation and this effect may in fact be solely involved with prolactin release from the pituitary (Shaar and Clemens, 1980; Gudelsky and Porter, 1979).

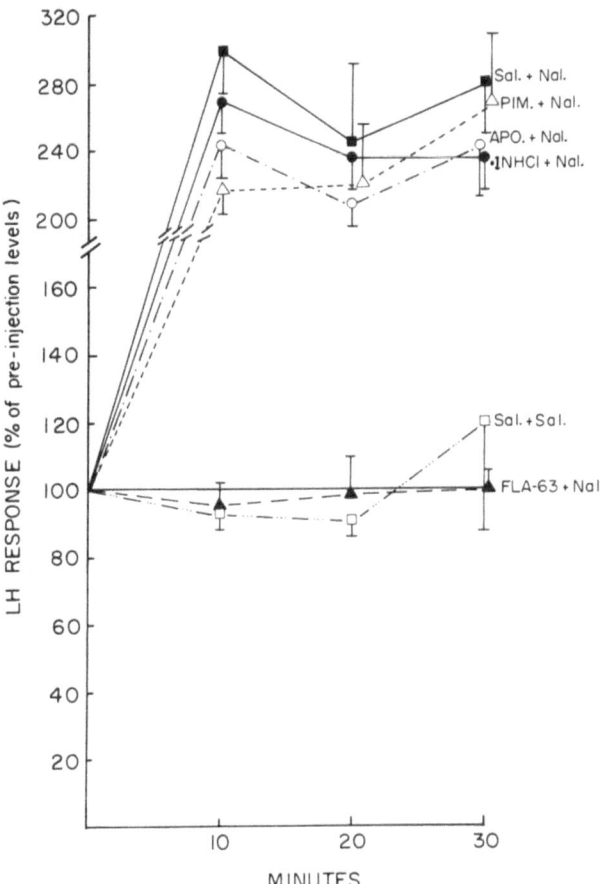

Figure 3. Ineffectiveness of pimozide (PIM, a dopamine receptor antagonist) and apomorphine (APO, a dopamine agonist) in suppressing LH release induced by naloxone (Nal, 2 mg/kg) in ovariectomized, estrogen/progesterone-treated rats. FLA-63, a norepinephrine synthesis inhibitor, blocked the stimulation of LH release by Nal.

5.2.2. Norepinephrine

Norepinephrine (NE) has been identified as an excitatory neurotransmitter for LH release (Kalra *et al.*, 1972; Kalra, 1977; Gallo, 1980). It is, therefore, quite possible that central NE neurons may mediate the effects of opiates, opiate antagonists, and possibly of the endogenous opioid peptides on LH release. Evidence showing that M can inhibit NE release from presynaptic nerve terminals supports such a mode of opiate action in inhibiting LH release (Stark *et al.*, 1977; Korf *et al.*, 1974; Bird and Kuhar, 1977). We found that prior blockade of central α-adrenergic receptors with phenoxybenzamine abolished the naloxone-induced LH release in estro-gen–progesterone-treated rats (Simpkins and Kalra, 1980; Kalra *et al.*, 1981b). Furthermore, diethyldithiocarbamate (DDC) and bis-(4-methyl-l-homopiperanzinyl-thiocarbanyl) disulfide (FLA-63), which suppressed hypothalamic NE levels (Estes *et al.*, 1981), also completely inhibited LH release induced by naloxone (Fig. 3).

To further delineate the mechanistic aspects of opiate and opiate antagonists actions, a possible anatomical interrelationship between opioid peptide, LH-RH, and NE containing neurons is presented in Fig. 4. The highlight of this interaction is that opioid peptide neurons by axo-axonic links modulate the release of NE, which in turn either directly or indirectly through another intervening neuronal system may influence LH-RH secre-tion. Accordingly, decrements in serum LH levels observed after M treat-ment would reflect a decrease in excitatory NE input to LH-RH neurons. Indeed, several studies show decreased brain NE turnover in M-treated rats (Iwamoto and Way, 1979; Korf *et al.*, 1974; Watson *et al.*, 1980). Our successful attempts to stimulate LH release in M-pretreated NE-deficient rats with the α-agonist clonidine were in line with this view (Simpkins and Kalra, 1980; Kalra *et al.*, 1981b). With respect to the action of opiate antagonists, presumably by displacing endogenous opioid peptides from receptors located on NE terminals, naloxone facilitated a transient release on NE that in turn promoted LH-RH release. In accordance with this anatomical relationship, blockade with phenoxybenzamine of α-adrenergic receptors localized either on LH-RH neurons or on another intervening neuronal system(s) would prevent the action of NE, which may be released by naloxone treatment. Also, failure of naloxone to stimulate LH release in FLA-63- and DDC-treated rats may be due to inadequate release of NE from an already depleted releasable pool.

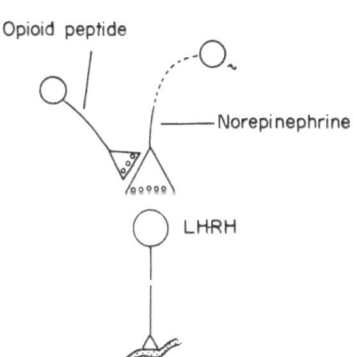

Figure 4. Conceptual depiction of anatomical inter-relationship between opioid peptide, norepinephrine, and LH-RH containing neuron in the brain to explain the blockade of naloxone-induced LH release by α-adrenergic receptor blocker phenoxybenzamine and norepinephrine synthesis inhibitors DDC and FLA-63 (for detail see text).

6.   Site(s) of Opiate Modulation of LH Release

The probable brain loci where opioid peptide neurons interact with NE neurons to modulate LH release would be those where opiate receptors and NE projections are coextensively distributed (Snyder, 1980; Bloom *et al.,* 1978; Uhl *et al.,* 1978; Elde and Hökfelt, 1978; Herkenham and Pert, 1980; Ungerstedt, 1971; Pert *et al.,* 1976). Several distinct regions in the brain, innervated by NE neurons of the locus coeruleus and where opiate receptors also have been demonstrated, are the locus coeruleus, amygdala, septum, preoptic area, and several distinct nuclei in the hypothalamus. These loci previously have been shown to exert modulatory influence on LH release (Kalra and McCann, 1975; Kalra *et al.,* 1971; Piva *et al.,* 1979). Interestingly, however, implantation of minute quantities of naloxone in dorsal raphae nucleus, medial lemniscus, amygdaloid region, medial septum, diagonal band of Broca, or dorsal and ventral medial hypothalamus of estrogen/progesterone-treated rats failed to alter LH release. In contrast, similar implantation of naloxone crystals or injection of naloxone solution in the medial preoptic area or the median eminence–arcuate nucleus rapidly stimulated LH release in these rats (Fig. 5). The onset of LH release was apparent within minutes of intracranial implantation of naloxone into these sites, and pretreatment with M or DDC blocked these stimulatory effects of naloxone (Fig. 6). There are two noteworthy outcomes of these observations. First, these results reaffirm our view that NE projections in the diencephalon may mediate the effects of opiates and opiate antagonists on LH release. Second, since distribution of LH-RH perikarya and their projections more or less correspond to naloxone-responsive

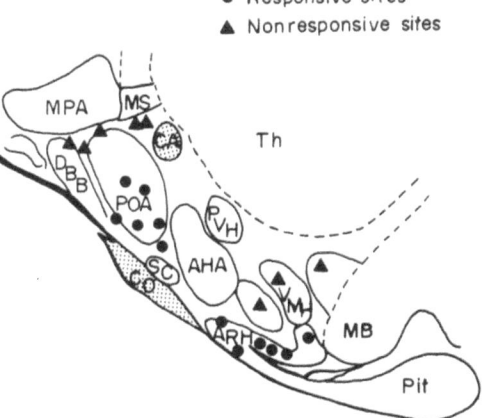

Figure 5. Location of naloxone implant sites in the diencephalon of rat brain. Responsive sites = naloxone implants stimulated LH release; nonresponsive sites = naloxone did not stimulate LH release. MPA, area parolfactoria medialis; MS, medial septum; DBB, diagonal band of Broca; POA, preoptic area; CA, anterior commissures; CO, optic chiasma; SC, suprachiasmatic nucleus; AHA, anterior hypothalamic area; PVH, paraventricular nucleus; VMH, ventral medial hypothalamus; MB, mamillary body; Th, thalamus; ARH, arcuate nucleus; Pit, pituitary.

regions, we suggest that the site(s) of linkage between opioid peptides and NE neurons might exist in close proximity of LH-RH neurons (Kalra, 1976; Barry, 1979). A close functional association between the two classes of these peptidergic neurons–opioid peptide and LH-RH-containing neurons—was also evident from deafferentation studies (Kalra *et al.*, 1981a). Complete surgical isolation of the MBH from the rest of the brain or transection of anterior links of the MBH alone produced a drastic fall in the MBH Met-enkephalin levels (Fig. 7). There were concurrent increments in Met-enkephalin levels of the preoptic area, a region that lies in front of the transections. It appeared that the MBH may be innervated by Met-enkephalin perikarya located in the MBH and more rostrally in the septal-preoptic region. Immunocytochemical localization of enkephalin-containing neurons in these sites is in accord with this view. Interestingly, a similar innervation of the MBH by two distinct populations of LH-RH perikarya

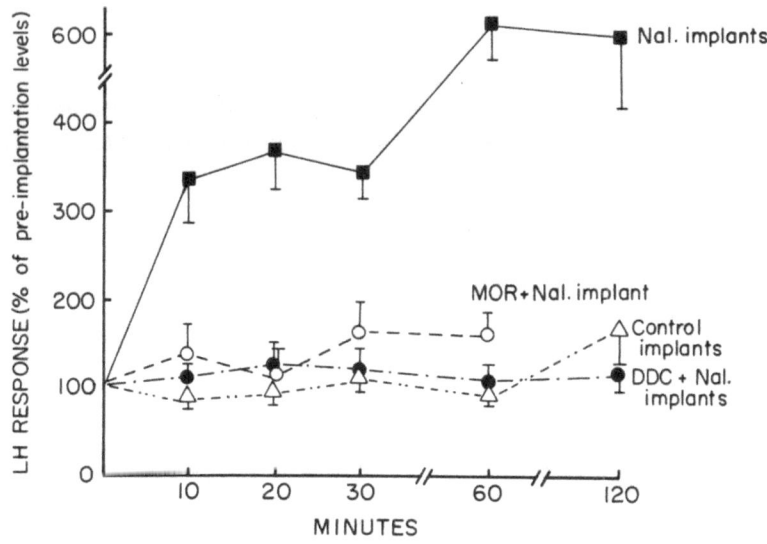

Figure 6. Pattern of LH release after intracranial implantation of naloxone (Nal) in the median eminence–arcuate region (ME-ARC) of steroid-pretreated rats. Naloxone-induced LH release was blocked by pretreatment with morphine (MOR) or a norepinephrine synthesis inhibitor, diethyldithiocarbamate (DDC).

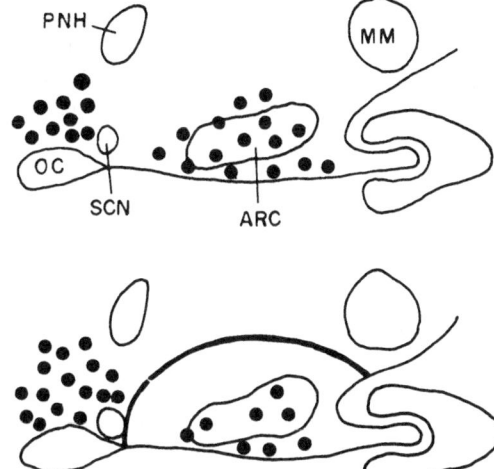

Figure 7. Marked reduction in the hypothalamic island and concurrent increase in the rostral preoptic area of Met-enkephalin levels after complete hypothalamic deafferentation of ovariectomized rats. PVN, paraventricular nucleus; ARC, arcuate nucleus; OC, optic chiasma; SCN, suprachiasmatic nucleus; MM, median mamillary nucleus.

MET-ENKEPHALIN ● = 500 pg/mg protein

was postulated by us and later confirmed by extensive immunocytochemical studies (Kalra, 1976; Barry, 1979). Based on the fact that naloxone-sensitive sites were found only in those areas that were richly innervated by LH-RH and Met-enkephalin neurons, we suggest that the functional linkage between these peptidergic neurons may exist within the preoptic–tuberal pathway (Kalra *et al.,* 1971; Everett, 1977). However, the possibility that β-endorphin neurons, whose perikarya are exclusively found in the basal tuberal region, may influence the secretion of LH-RH neurons cannot be discounted.

7. Summary

A new level of inquisition in the central control of gonadotropin secretion has been made possible with the discovery of opioid peptides in the brain. This new class of peptides exert primarily an inhibitory influence on LH secretion that can be modified by gonadal steroid milieu. By utilizing an opiate and an opiate antagonist it has been possible to recognize the specificity of opiate receptor involvement in the progesterone-induced changes in hypothalamic LH-RH secretion in estrogen-primed, ovariectomized rats. Several lines of evidence developed in this report indicate the apparent necessity of intact noradrenergic system in mediation of the effects of opiate agonist and antagonist on LH release. This action of opiate can be attributed to a functional link between opioid peptides and noradrenergic neurons occurring in close vicinity of the LH-RH neurons in the preoptic–tuberal pathway.

ACKNOWLEDGMENTS. This work was supported by NIH Grants HD 08634 and HD 14006. Thanks are due to Mrs. Sandra Barnawell for assistance in preparing the manuscript.

DISCUSSION

KANAZIR: How can the progesterone effect on the secretion of LH be incorporated into the model you propose? In addition, is there any information on the polymorphism of the opiate receptor? How are these differences discriminated?

KALRA: On the basis of the model proposed in Fig. 4, we speculate that progesterone treatment may temporarily remove the inhibitory effect of endogenous opioid peptides on noradrenergic terminals in the hypothalamus. The facilitatory effects of noradrenergic neurons may, thus, elicit hypersecretion of LH-RH. I am not able to answer your second question.

PECK: What doses of phenoxybenzamine, clonidine, and DDC were employed in your studies of the blockage of naloxone-induced release of LH? Secondly, how long after deafferentation were you able to see a decrease in Met-enkephalin in the island?

KALRA: The doses employed were: phenoxybenzamine, 20mg/kg; clonidine, 0.3 mg/kg; and DDC, 500 mg/kg. As to your second question, rats were sacrificed 7 days after hypothalamic deafferentation. The Met-enkephalin concentration was significantly decreased at this time.

NIKOLICS: Did you implant naloxone in other areas containing LH-RH cell bodies? Secondly, how do you explain the differential time course in LH release after implants into POA and ME?

KALRA: We have implanted naloxone in the medial septum region. These implants failed to change LH secretion in the estrogen/progesterone treated rats. As to your second question, we did not observe a significant difference in either the onset or the duration of LH response after naloxone implants in the median eminence or preoptic areas. The total amount of LH stimulated by ME naloxone implants was greater than produced by POA naloxone implants. We do not know the cause of this difference other than to suggest that this may be due to clustering of LH-RH neurons in the MBH.

REFERENCES

Azizzi, F., Vagenakis, A. G., Longcope, C., Ingbar, S. H., and Braverman, L. E., 1973, Decreased serum testosterone concentration in male heroin and methadone addicts, *Steroids* **22**:467.
Barraclough, C. A., and Sawyer, C. H., 1955, Inhibition of the release of pituitary ovulatory hormone in the rat by morphine, *Endocrinology* **57**:329.

Barry, J., 1979, Immunohistochemistry of luteinizing hormone-releasing hormone producing neurons of the vertebrates, *Int. Rev. Cytol.* **60**:179.

Bird, S. J., and Kuhar, M. J., 1977, Iontophorectic application of opiates to the locus coeruleus, *Brain Res.* **122**:523.

Blank, M. S., Paneri, A. E., and Friesen, H. G., 1979, Opioid peptides modulate luteinizing hormone secretion during sexual maturation, *Science* **203**:1129.

Blank, M. S., Paneri, A. E., and Friesen, H. G., 1980, Effects of naloxone on luteinizing hormone and prolactin in serum of rats, *J. Endocrinol.* **85**:307.

Bloom, F., Battenberg, E., Rosier, J., Ling, N., and Guillemin, R., 1978, Neurons containing β-endorphin in rat brain exist separately from those containing enkephalin: Immunocytochemical studies, *Proc. Natl. Acad. Sci. USA 75:1591.*

Cicero, T. J., Meyer, E. R., Bell, R. D. and Weist, D., 1974, Effects of morphine on the secondary sex organs and plasma testosterone levels of rats, *Res. Commun. Chem. Pathol. Pharmacol.* **7**:17.

Cicero, T. J., Bell, R. D., Meyer, E. R., and Schweitzer, J., 1977, Narcotics and the hypothalamic pituitary gonadal exis: Acute effects on luteinizing hormone, testosterone and androgen-dependent systems, *Pharmacol. Exp. Ther.* **201**:76.

Cicero, T. J., Schainker, B. A., and Meyer, E. R., 1979, Endogenous opioids participate in the regulation of the hypothalamic pituitary-luteinizing hormone axis and testosterone's negative feedback control of luteinizing hormone, *Endocrinology* **104**:1286.

Cicero, T. J., Meyer, E. R., Gabrial, S. M., Bell, R. D., and Wilcox, C. E., 1980, Morphine exerts testosterone-like effects in the hypothalamus of the castrated male rat, *Brain Res.* **202**:151.

Elde, R., and Hökfelt, T., 1978, Distribution of hypothalamic hormones and other peptides in the brain, in: *Frontiers in Neuroendocrinology,* Vol. 5 (W. F. Ganong and L. Martini, eds.), pp. 1–33, Raven Press, New York.

Estes, K., Simpkins, J. W., and Kalra, S. P., 1981, Resumption with clonidine of pulsatile LH release following acute norepinephrine depletion in ovariectomized rats, *Neuroendocrinology,* in press.

Everett, J. W., 1977, The timing of ovulation, *J. Endocrinol.* **75**:1P

Ferland, L., Fuxe, K., Eneroth, P., Gustafsson, J., and Skelt, P., 1977, Effects of methionine-enkephalin on prolactin release and catecholamine levels and turnover in the median eminence, *Eur. J. Pharmacol.* **43**:89.

Gallo, R. V., 1980, Neuroendocrine regulation of pulsatile luteinizing hormone release in the rat, *Neuroendocrinology* **30**:122.

Goldstein, A., Tachibana, S., Lowney, L. I., Hunkapiller, M., and Hood, L., 1979, Dynorphin (1–13), an extraordinary potent opioid peptide, *Proc. Natl. Acad. Sci. USA* **76**:6666.

Grandison, L., Fratta, W., and Guidotti, H., 1980, Location and characterization of opiate receptors regulating pituitary secretion, *Life Sci.* **26**:1633.

Gudelsky, G., and Porter, J., 1979, Morphine and opioid-peptide-induced inhibition of the release of dopamine from the tuberoinfundibular neurons, *Life Sci.* **25**:1697.

Guillemin, R., Ling, N., and Burgus, R., 1976, Endorphins, peptides d'origine hypothalaminque et neurophysaire a activité morphionmimetique: Isolement et structure moléculaire de L'2-endorphin, *C.R. Acad. Sci. Ser. D* **282**:783.

Herkenham, M., and Pert, C. B., 1980, *In vitro* autoradiography of opiate receptors in rat brain suggests loci of "opiatergic" pathway, *Proc. Natl. Acad. Sci. USA* **77**:5532.

Hughes, J., Smith, T. W., Kosterlitz, H. W., Fothergill, L. A., Morgan, B. A., and Morris, H. R., 1975, Identification of two related pentapetides from the brain with potent opiate agonist activity, *Nature (London)* **258**:577.

Ieiri, T., Chen, H. T., and Meites, J., 1979, Effects of morphine and naloxone on serum levels of luteinizing hormone and prolactin in prepubertal male and female rats, *Neuroendocrinology* **29**:288.

Iwamoto, E. T., and Way, L., 1979, Opiate actions and catecholamines, in: *Neurochemical Mechanisms of Opiates and Endorphins* (H. H. Loh and D. H. Ross, eds.) Raven Press, New York.

Kalra, P. S., and McCann, S. M., 1975, The stimulatory effect on gonadotropin release of implants of estradiol or progesterone in certain sites in the central nervous system, *Neuroendocrinology* **19**:289.

Kalra, P. S., Kalra, S. P., Krulich, L., Fawcett, C. P., and McCann, S. M., 1972, Involvement of norepinephrine in transmission of the stimulatory influence of progesterone on gonadotropin release, *Endocrinology* **90**:1168.

Kalra, P. S., Fawcett, C. P., Krulich, L., and McCann, S. M., 1973, The effect of gonadal steroid on plasma gonadotropins and prolactin in the rat, *Endocrinology* **92**:1256.

Kalra, S. P., 1976, Tissue levels of luteinizing hormone releasing hormone in the preoptic area and hypothalamus, and serum concentrations of gonadotropins following anterior hypothalamic deafferentation and estrogen treatment of female rat, *Endocrinology* **99**:101.

Kalra, S. P., 1977, Neuroamines in gonadotropin secretion, *Excerpta Med. Int. Congr. Ser.* **42**:152.

Kalra, S. P., 1978, Effects of estrogen on ^{3}H-leucine uptake by the hypothalamus and pituitary: Correlation of hypothalamic and serum LHRH and LH, *J. Endocrinol. Invest.* **2**:131.

Kalra, S. P., Ajika, K., Krulich, L., Fawcett, C. P., Quijada, M., and McCann, S. M., 1971, Effects of hypothalamic and preoptic electrochemical stimulation on gonadotropin and prolactin release in proestrous rats, *Endocrinology* **88**:1150.

Kalra, S. P., Kalra, P. S., Chen, C. L., and Clemens, J. A., 1978, Effects of norepinephrine synthesis inhibitors and a dopamine agonist on hypothalamic LHRH, serum gonadotropins and prolactin levels in gonadal steroid treated rats, *Acta Endocrinol. (Copenhagen)* **89**:1.

Kalra, S. P., Chen, C. L., and Kumar, M. S. A., 1981a, Differential response of methionine-enkephalin in basal hypothalamus and preoptic area following hypothalamic deafferentation, *Brain Res.*, **215**:410.

Kalra, S. P., Simpkins, J. W., and Chen, C. L., 1981b, The opioid–catecholamine modulation of gonadotropin and prolactin secretion, in: *Functional Correlates of Hormone Receptors in Reproduction* (V. Mahesh, Muldoon, T., Saxena, B., and Saddler, W., eds.), pp. 135–150, Elsevier, Amsterdam, 1981.

Korf, J., Bunney, B. S., and Aghajanian, G. K., 1974, Noradrenergic neurons: Morphine inhibition of spontaneous activity, *Eur. J. Pharmacol.* **25**:165.

Larsson, L., Childres, S., and Snyder, S. H., 1979, Met- and Leu-enkephalin immunoreactivity in separate neurons, *Nature (London)* **282**:407.

Levine, J. E., and Ramirez, V. D., 1980, *In vivo* release of luteinizing hormone releasing hormone estimated with pushpull cannulae from mediobasal hypothalami of ovariectomised, steroid primed rats, *Endocrinology* **107**:1782.

Meites, J., Bruni, J. R., Van Vugt, D. A., and Smith, A. F., 1979, Relation of endogenous opioid peptides and morphine to neuroendocrine functions, *Life Sci.* **24**:1325.

Muraki, T., Tokunaga, Y., Matsumoto, S., and Makino, T., 1978, Time course of effects of morphine on hypothalamic content of LHRH and serum testosterone and LH levels of morphine tolerant and nontolerant male rats, *Arch. Int. Pharmacodyn. Ther.* **233**:290.

Negro-Vilar, A., Orias, R., and McCann, S. M., 1973, Evidence for a pituitary site of action for the acute inhibition of LH release by estrogen in the rat, *Endocrinology* 92:1680.

Pang, C. N., Zimmerman, E., and Sawyer, C. H., 1977, Morphine inhibition of the preovulatory surges of plasma luteinizing hormone and follicle stimulating hormone in the rat, *Endocrinology* 101:1726.

Pert, C. B., Kuhar, J. J., and Snyder, S. H., 1976, The opiate receptor: Autoradiographic localization of rat brain, *Proc. Natl. Acad. Sci. USA* 73:3729.

Piva, F., Borrell, J., Limonta, P., Gavazzi, G., and Martini, L., 1979, Role of amygdala and the organum vasculosum laminae terminalis in the control of ovarian function in the female rat, *J. Steroid Biochem.* 11:1007.

Rotsztejn, W., Drouva, S., Pattou, E., and Kordon, C., 1978a, Met-enkephalin inhibits *in vitro* dopamine-induced LHRH release from medio-basal hypothalamus of male rats, *Nature (London)* 274: 281.

Rotsztejn, W., Drouva, S., Pattou, E., and Kordon, C., 1978b, Effect of morphine on the basal and dopamine induced release of LHRH from medio-basal hypothalamic fragments *in vitro, Eur. J. Pharmacol.* 50:285.

Sar, M., Stumpf, W. E., Miller, R. J., Chang, K. J., and Cuatrecasas, P., 1978, Immunohistochemical localization of enkephalin in rat brain and spinal cord, *J. Comp. Neurol.* 182:17.

Sawyer, C. H., and Clifton, L., 1980, Aminergic innervation of the hypothalamus, *Fed. Proc.* 39:2889.

Shaar, C. J., and Clemens, J. A., 1980, The effects of opiate agonists on growth hormone and prolactin release in rats, *Fed. Proc.* 38:2539.

Simpkins, J. W., and Kalra, S. P., 1980, Evidence for noradrenergic (NE) mediation of opioid effects on LH secretion, in: Proceedings of the 62nd Annual Endocrine Society Meeting, Washington, D.C., p. 643.

Simpkins, J. W., Kalra, P. S., and Kalra, S. P., 1980, Temporal alternation in LHRH concentrations in several discrete brain regions: Effects of estrogen-progesterone and norepinephrine synthesis-inhibition, *Endocrinology* 107:573.

Snyder, S. H., 1980, Brain peptides as neurotransmitters, *Science* 209:976.

Stark, K., Taube, H. D., and Borowski, E., 1977, Presynaptic receptor system in catecholaminergic transmission, *Biochem. Pharmacol.* 26:259.

Thomas, J. A., and Dombrosky, J. T., 1975, Effects of methadone on the male reproductive system, *Arch. Int. Pharmacodyn. Ther.* 215:215.

Uhl, G. R., Childres, S. R., and Snyder, S. H., 1978, Opioid peptides and the opiate receptors, in: *Frontiers in Neuroendocrinology,* Vol. 5 (W. F. Ganong and L. Martini, eds.), Raven Press, New York.

Ungerstedt, W., 1971, Stereotaxic mapping of monoamine pathways in the rat brain, *Acta Physiol. Scand. Suppl.* 367:1.

Van Vugt, D. A., and Meites, J., 1980, Influence of endogenous opiates on anterior pituitary function, *Fed. Proc.* 39:2533.

Wamsley, J., Young, W. S., and Kuhar, M. J., 1980, Immunohistochemical localization of enkephalin in rat forebrain, *Brain Res.* 190:153.

Wardlaw, S. L., Warenbert, W. B., Ferin, M., Carmel, P. W., and Franz, A. G., 1980, High level of β-endorphin in hypophysial portal blood, *Endocrinology* 106:1323.

Watson, S., Richards, G. W., Ciarvanello, R. D., and Barchas, J. D., 1980, Interaction of opiate peptides and noradrenalin systems: Light microscopic studies, *Peptides* 1:30.

9

Stress, Corticoliberin (CRF), and Glucocorticoids in the Regulation of ACTH Release

Ervin Stark and Gábor B. Makara

1. Introduction

The main components in the control of adrenocorticotropin (ACTH) secretion have been identified during the last 25 years and include: a hypothalamic releasing factor (corticoliberin, corticotropin releasing factor, CRF), vasopressin, the nervous circuits controlling the release of CRF, the afferent neural pathways conveying information to the controlling circuits about the "milieu interne," and the conditions in the world outside the organism. In addition, feedback loops, both hormonal and neural, convey information about the performance of the various components of the hypothalamo–pituitary–adrenocortical system. Recently, a great deal of *new* information became available about the various components and the mechanisms involved in the organization of their interrelationships.

The present chapter attempts to review recent knowledge and to reevaluate some of the older data in the literature in the light of recent developments.

Some 30 years ago the first theory of the regulation of ACTH secretion was beautifully simple since it was thought that under both resting conditions and during stressful stimulation the negative feedback action of the

Ervin Stark and Gábor B. Makara • Institute of Experimental Medicine, Hungarian Academy of Sciences, Budapest, Hungary.

adrenal glucocorticoid hormones, acting on the anterior pituitary, would provide the signal for ACTH secretion (Sayers, 1950). With the advent of the neurohumoral theory of the control of the anterior pituitary (Harris, 1955), the emphasis shifted from the pituitary to the hypothalamus as the site of action of glucocorticoid feedback. It took a long time before the multitude of components and their intricate relationship became appreciated (Yates and Maran, 1974).

Contemporary methods are capable of measuring both adrenal and pituitary hormone levels in the plasma with sufficient sensitivity and precision and an array of bioassays are available for CRF measurements in tissue extracts; this methodological inventory allows us to take a look at old questions from a new viewpoint and to set up experiments to study the details of the controlling mechanisms.

2. ACTH-Secreting Cells

With the help of advanced immunocytochemical techniques and the combination of biochemical and immunochemical methods, the ACTH-containing cells have been identified in the anterior and the intermediate lobes of the pituitary gland (Nakane *et al.,* 1977), and in the hypothalamic arcuate nucleus (Watson *et al.,* 1978). In the rat, practically all glandular cells of the intermediate lobe contain, among numerous other peptides, ACTH-like immunoreactive substance(s), whereas in the anterior lobe only 7 to 15% of all parenchymal cells can be stained with antibodies to ACTH. It is also known that ACTH extracted from the two lobes is biologically active. Most authors agree that ACTH-producing cells also contain β-lipotropin (β-LPH) and β-endorphin. In these stellate cells the bulk of immunoreactive material is packed in rather small granules (diameter 150–200 nm) that line up along the internal face of the cell membrane.

Only the ACTH-secreting cells of the anterior lobe seem to participate to a large extent in the control of adrenocortical function. The main product of the intermediate lobe cells appears to be α-MSH, corticotropinlike intermediate lobe peptide (CLIP), and β-endorphin. Physiological studies also suggest that in anterior lobectomized rats adrenal function is as seriously damaged as in animals with complete hypophysectomy (Greer *et al.,* 1975).

3. The Nature of CRF

It is a general view that the ACTH-producing cells of the anterior pituitary are mainly controlled by CRF-producing neuroendocrine cells projecting to the hypothalamic median eminence and the pituitary stalk. The CRF is thought to be released at these sites into the portal capillaries of the hypothalamo–pituitary portal circulation, which transports the neurohormone to the anterior pituitary.

The ACTH-releasing effect of hypothalamic extracts was first described by Guillemin and Rosenberg (1955) and Saffran *et al.* (1955). Despite the considerable amount of work, the structure of CRF is still unknown. Therefore, physiologists developed a variety of advanced *in vivo* and *in vitro* bioassay techniques for CRF measurements. The sensitivity of these assays range from 0.05 to 0.005 part of an extract made from a single rat hypothalamus, and thus allows study of the variations in tissue CRF content.

There seems to be a general agreement (1) that CRF specifically triggers ACTH secretion, (2) that arginine-8-vasopressin is not identical with hypothalamic CRF, and (3) that vasopressin is in some way involved in the regulation of ACTH secretion.

There is, however, some disagreement concerning the role of vasopressin in CRF release, the size and number of ACTH-releasing peptides, and the origin of CRF-containing structures that are present in the neurohemal regions of the median eminence and the pituitary stalk. It is also debated whether CRF is released only from the median eminence, or also from the pituitary stalk and the neural lobe of the pituitary, and whether these latter structures take part in the control of adenohypophyseal ACTH secretion.

Most earlier work on isolation of CRF resulted in multiple chromatographic fractions of CRF activity the nomenclature of which was various: α- and β-CRF (Schally *et al.*, 1960), or CRF A and B (Jones *et al.*, 1976). Recently the group of Schulster (Bristow *et al.*, 1980) found a single peak of CRF activity and suggested that multiple peaks in earlier studies might have arisen from enzymatic breakdown during collection and/or acid extraction of hypothalamic tissue. When the tissue was extracted under conditions designed to minimize or eliminate protease effects, a better recovery of activity was found yielding a single peak using either conventional ion-exchange chromatography or high-pressure liquid chromatography.

The earlier debate (see Fortier, 1966; Martini, 1966) whether arginine-8-vasopressin is the only, dominant CRF appears to lead nowadays to the conclusion that it is not, since CRF activity has been found in Brattleboro rats genetically lacking the ability to synthesize vasopressin; moreover, some laboratories have succeeded in a clear-cut chromatographic separation of CRF from vasopressin. What is debated now is the extent to which vasopressin may participate in ACTH release under various conditions. Gillies and Lowry (1979) suggested that CRF is essentially modulated vasopressin. They drew this conclusion from experiments using a CRF bioassay (based on cell columns of freshly dispersed anterior pituitary cells) that is rather sensitive to AVP. It was shown that the AVP-free fraction of hypothalamic extract efficiently potentiated the action of AVP and steepened the slope of the AVP dose–response curve. Buckingham (1979) and Buckingham and Leech (1980) also found that the hypothalamic content as well as the release of CRF from incubated hypothalami of homozygous Brattleboro rats was greatly reduced. By contrast, in experiments performed in our laboratory (Kárteszi *et al.,* 1981), using a bioassay insensitive to vasopressin, the CRF activity of stalk–median eminence or nueral lobe extracts from Brattleboro animals was found to be the same as in extracts from the same tissues of Sprague–Dawley rats. We could also provide evidence that exogenous arginine-8-vasopressin failed to potentiate the effect of hypothalamic extracts from homozygous Brattleboro rats (Table I).

The controversial findings may be explained by the different characteristics of the various bioassays. The cell column system of Gillies and Lowry (1978) is very sensitive to both CRF and vasopressin, while the sensitivity of the assays using incubated pieces of anterior pituitary (see Buckingham, 1979) is intermediate and depends very much on experimental conditions (e.g., length of preincubation which eliminates endogenous vasopressin; priming with vasopressin; etc.). It is to be noted that in all these systems vasopressin acts like a partial agonist. In contrast the cell culture bioassay is almost completely insensitive to vasopressin but quite sensitive to hypothalamic extract. It is likely, therefore, that the outcome of an experiment aiming to elucidate the role of vasopressin is very much dependent on the assay method selected for measuring CRF activity.

In our work on the nature and origin of CRF in the neurohemal regions we relied on either the AVP-insensitive method based on monolayer cell cultures or on evidence from *in vivo* studies. Monolayer cell cultures in

Table I.   ACTH-Releasing Potency of Stalk–Median Eminence and Neurohypophyseal Extracts[a]

Test materials	Net release of immunoreactive ACTH, pmoles/0.6 ml, due to equivalent amounts of brain extracts		
	0.033	0.1	0.33
Cerebral cortical extract (in NL equivalents)	3.40 ± 0.27 (4)	3.24 ± 0.09 (2)	5.18 + 0.86 (3)
Brattleboro SME	6.33 ± 0.32 (4)	6.90 ± 1.19 (2)	11.00 ± 1.19 (3)
Brattleboro SME + 3.3 ng AVP	7.06 ± 0.41 (4)	7.92 ± 0.02 (2)	9.84 ± 0.91 (3)
Brattleboro NL	6.54 ± 0.18 (4)	10.68 ± 0.34 (2)	29.37 ± 3.76 (3)
Brattleboro NL + 333 ng AVP	5.85 ± 0.33 (4)	10.27 ± 1.30 (2)	26.57 ± 2.55 (3)

[a] From Kárteszi *et al.* (1981). Reproduced with permission of the publisher.

multiwell plates (24 or 96 cultures) form an easy, reproducible experimental model for CRF bioassay. The test solutions are mixed with the culture medium and the ACTH released in the course of 2 hr was measured by a direct radioimmunoassay validated with a bioassay (Makara *et al.,* 1979).

It should also be noted that the sampling method we used is different from that generally used in other laboratories. We dissected the anatomically defined median eminence + the pituitary stalk (SME) quickly after decapitation under a stereomicroscope so that the samples did not include any gray matter of the tuberal hypothalamus. In contrast, most other laboratories obtain median eminence that contains substantial amounts of hypothalamic gray matter.

4. The Site of CRF Release

Since the advent of the neurohumoral theory of anterior pituitary control it is customary to regard the median eminence as the main or even the only release site for the various hypophysiotropic hormones. This seems, however, an oversimplified concept, for biochemical analysis has shown that high concentrations of these hormones are present not only in the ME but also in the neural lobe (NL) and the pituitary stalk. These latter two subdivisions of the neurohypophysis have a capillary bed contiguous with that of the ME and thus the angioarchitectonics is such that any material released from nerve endings in the stalk and the NL may reach the anterior pituitary and act there as a releasing or an inhibiting hormone.

This concept seems to be applicable to the case of TRH, somatostatin, and possibly CRF, which neurohormone is present in the pituitary stalk and the NL in high concentration (Yasuda *et al.,* 1977; Kárteszi *et al.,* 1978). Whether CRF present in these regions can be released *in vivo* in response to appropriate stimuli and may participate in the control of the ACTH-secreting cells is a subject worth studying since the quantity of CRF-like activity stored in the latter two subdivisions of the neurohypophysis is much larger than the amount stored in the ME itself.

Direct electrical stimulation seems to be a suitable experimental approach for selectively activating the CRF-containing structures in the NL. Since ACTH release may be strongly enhanced by undesired stressful side effects of the surgical procedure associated with the placement of electrodes into the NL, pretreatment with dexamethasone, morphine, and pentobarbitone was used to prevent stressful activation. Passing a train of electrical pulses through a pair of electrodes placed into the NL via the parapharyngeal approach elicited marked ACTH release as shown by the elevation of plasma corticosterone level 20–25 min after the stimulation (Makara, 1979; Makara *et al.,* 1980a) (Fig. 1).

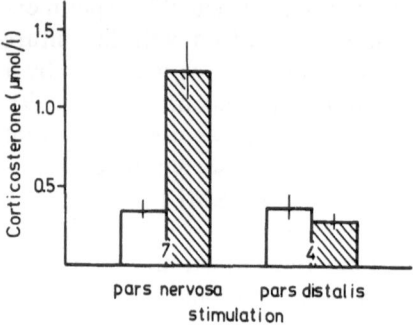

Figure 1. Effect of electrical stimulation of the pars distalis or the pars nervosa of the pituitary gland on plasma corticosterone level of dexamethasone, morphine, and pentobarbitone pretreated rats. Open columns: before stimulation; cross-hatched columns: 20–30 min after stimulation. Number of rats shown at the base of the columns. [Reproduced from Makara *et al.*, 1980a, with permission.]

However, the possibility exists that electrical stimulation of the NL activates CRF-containing nerve fibers that give off collaterals to the palisade zone of the ME and the CRF released in the ME is responsible for ACTH release elicited by electrical stimulation of the neural lobe. Anatomical basis for such a possibility has been shown to exist by the electrophysiological experiments of Pittman *et al.* (1978). However, Baertschi *et al.* (1980) tested the response to NL stimulation after placing a cut across the origin of the pituitary stalk at the hypothalamus and succeeded in eliciting ACTH release even after severing the neural connections between the NL and the ME; these data point to the NL as a possible site of CRF release; further experiments should, however, elucidate under physiological conditions or stressful stimulation how much of CRF is contributed by the various parts of the neurohypophysis: the ME, the stalk, and the pars nervosa of the pituitary.

Since vasopressin is a potent releaser of ACTH in the normal animal and human and is known to be highly concentrated in the NL, it is logical to ask whether mainly vasopressin or a nonvasopressin CRF is the agent responsible for the ACTH-releasing effect of NL stimulation. One way to answer this question is to test the pituitary–adrenal response to NL stimulation both in normal and in Brattleboro rats with hereditary diabetes insipidus. In the experiments of Baertschi *et al.* (1980), electrical stimulation of the NL of homozygous Brattleboro rats failed to elevate plasma ACTH level; however, the prestimulus plasma ACTH levels were already exceedingly high in these experiments. The results of these experiments were difficult to reconcile in the light of our finding of a high ACTH-releasing potency in the NL of homozygous Brattleboro rats; therefore, we also performed NL stimulation in Brattleboro rats, taking care to reduce the unwanted interference of surgical stress by using both the relatively mild parapharyngeal approach and pharmacological blockade with dexamethasone, morphine, and pentobarbitone rather than exposing the full hypo-

thalamus + the pituitary and using urethane, a potent stimulus for the
pituitary–adrenal system, as in the experiments of Baertschi *et al.* (1980).
We have shown (Kárteszi *et al.*, 1981) that the placement of electrodes in the NL without stimulation fails to increase ACTH release but
electrical pulses through NL electrodes significantly increase plasma corticosterone levels in Sprague–Dawley rats, and in heterozygous or homozygous Brattleboro rats as well (Fig. 2). This finding clearly indicates that
even in animals lacking vasopressin synthesis, stimulation of the NL is
capable of releasing CRF (and consequently ACTH and corticosterone);
however, the relatively small response may indicate that either vasopressin
itself plays a role *in vivo* in the ACTH release caused by electrical stimulation of the NL or that the general responsiveness of the pituitary–adrenocortical system of Brattleboro rats is less than that of normal rats. This
decreased responsiveness is not necessarily the direct consequence of the
lack of vasopressin, since the secondary consequences of diabetes insipidus
may also cause this phenomenon.

Taking into account that in the homozygous diabetes insipidus rat
basal indices of pituitary–adrenal system are either near normal or
depressed and their response to adrenalectomy and to a number of (but not
all) stressful stimuli is normal, we may conclude that a nonvasopressin

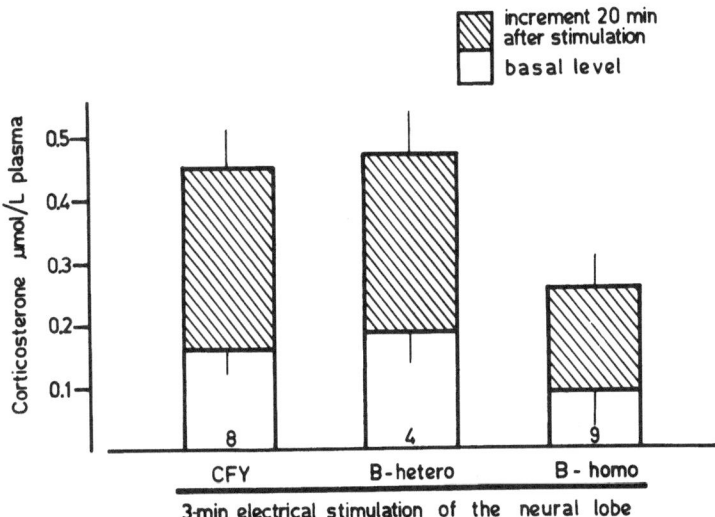

Figure 2. Effect of electrical stimulation of the neural lobe on plasma corticosterone level. Rats
were pretreated with dexamethasone, morphine, and pentobarbitone. B-hetero: heterozygous
Brattleboro rats; B-homo: homozygous Brattleboro rats. [Reproduced from Kárteszi *et al.*, 1981,
with permission.]

CRF certainly exists in both the SME and the NL of such rats and that it is sufficient for maintaining an almost normal pituitary–adrenal function. The deficits observed by other authors (McCann *et al.*, 1966; Arimura *et al.*, 1967), however, suggest that the absence of vasopressin, some physiological change secondary to diabetes insipidus, or the situation as a chronic stressor may impair the pituitary–adrenal axis.

5. The Location of CRF-Producing Cells

Between 1950 and 1970 it was widely accepted that the hypophysiotropic hormones are produced in that part of the hypothalamus that immediately surrounds the ME. The best candidates as the releasing-hormone-producing cells were the tuberoinfundibular neurons, mainly located in the hypothalamic arcuate nucleus. Some authors suggested that separate gonadotropic, thyrotropic, somatotropic areas can be localized in the hypothalamus, but this refinement of the concept was not accepted widely. Although several lines of indirect evidence seemed to be consistent with this idea, on closer scrutiny most of this evidence could also be reconciled with possible alternative locations outside the mediobasal hypothalamus (MBH). Until immunohistochemical methods make possible the direct tracing of the axons between terminals and cell bodies, studies with hypothalamic islands seem to provide the best evidence for deciding whether or not the tuberoinfundibular parvocellular neuronal system produces the various releasing and inhibiting hormones.

The use of biochemical micromethods alone or in conjunction with surgical disruption of possible axonal pathways confirmed the conclusions of the histochemical approaches in the case of the identified hypophysiotropic hormones: LH-RH, somatostatin (Brownstein *et al.*, 1976, 1977), and dopamine (see Palkovits, 1979). While dopamine-containing projections to the neurohemal regions appear to mainly arise from the MBH where dopamine-containing tuberoinfundibular neurons are located, recent data on somatostatin and LH-RH provide evidence that the bulk of axons transporting these hormones reach the SME from cell bodies located in the anterior periventricular hypothalamus and in the region lying between the optic chiasma, anterior commissure, and the septum, respectively; the tuberoinfundibular neurons send little or no such projections.

Based on slightly increased plasma corticosterone and pituitary ACTH levels and a statistically nonsignificant difference in response to stress in rats with complete hypothalamic deafferentations, it has been claimed that the CRF-containing cells are contained within the boundaries of the hypothalamic cut (Halász *et al.*, 1967). Studies with intrahypothalamic grafts in or around the ME (Csernus *et al.*, 1975) pointed to a location near or within the ME itself. Because these ideas were well in line with neuroen-docrinological thinking of the time, it was a long period before some incon-sistencies of details in these reports and some seemingly irregular findings in our studies forced us to test some of the underlying assumptions and to reevaluate those former experiments that gave an almost unequivocal basis to the idea that CRF-containing nerve cells are within the MBH and main-tain functional projections to the primary capillary bed of the ME even after complete hypothalamic deafferentation.

The very first inconsistency was that in spite of increased basal plasma corticosterone levels and increased anterior pituitary ACTH content, the adrenal glands were slightly atrophic (Halász *et al.,* 1967). However, the alleged increase in plasma corticosterone or ACTH level (Siegel *et al.,* 1980) was not confirmed in a number of laboratories, and it is our opinion (Stark *et al.,* 1978; Makara *et al.,* 1981b) that whenever conditions are optimized for verifying the completeness of hypothalamic cuts and for excluding extraneous stress, the basal plasma corticosterone and ACTH levels are less than or equal to that of the controls. Thus, hypothalamic deafferentation fails to "disinhibit" the pituitary–adrenal system as it was originally proposed, and this is well in line with the slight atrophy of the adrenals.

The second inconsistency was in the reports suggesting that ether stress stimulates corticosterone secretion in rats bearing complete hypotha-lamic islands made with the Halász knife. The majority of the authors (for references see Makara *et al.,* 1980b) reported near-normal response of the plasma corticosterone or ACTH level at several time intervals after hypo-thalamic deafferentation. Halász *et al.* (1967) found only a nonsignificant elevation of plasma corticosterone after the severe stress of surgery plus ether inhalation, and Voloschin *et al.* (1968) failed to find a significant response of plasma corticosterone to ether inhalation in one group of rats with fairly large hypothalamic islands. These inconsistencies remained unnoticed for a few years, and it was only after other discordant findings had already emerged that we set out to reevaluate the response of the pitu-itary–adrenal system to ether inhalation in rats with various hypothalamic cuts around the MBH. These studies (Makara *et al.,* 1980b; Kárteszi *et al.,* 1980) revealed that out of six series of experiments only one showed a slight increase of plasma corticosterone in response to ether anesthesia, and in the five other experiments there was no significant change. In normal rats 2 hr after placing an anterolateral hypothalamic cut or in adrenalec-tomized rats 7 to 8 days after anterolateral isolation of the MBH, plasma ACTH measurements failed to reveal any change in response to ether inhalation (Fig. 3). Interestingly, only those rats that had at least some fibers uncut in the lateral retrochiasmatic area of the hypothalamus showed a measurable rise of either plasma corticosterone level or ACTH level.

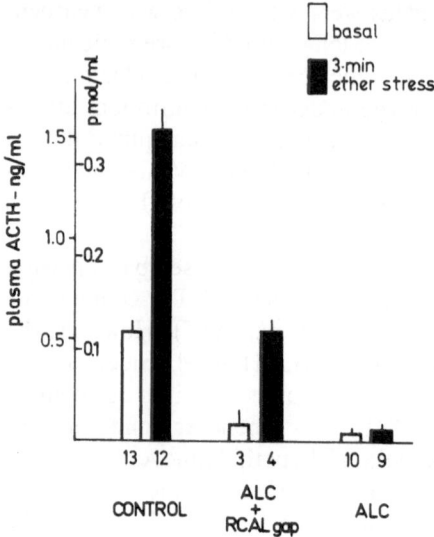

Figure 3. Effect of ether on plasma ACTH in adrenalectomized rats with an anterolateral cut (ALC) around the mediobasal hypothalamus. RCAL: lateral retrochiasmatic area. Columns and bars represent means and S.E.M. [Reproduced from Kárteszi et al., 1980, with permission.]

These findings suggested (but in no way proved) that the earlier reports claiming significant ether-induced increments in pituitary–adrenal function following complete hypothalamic cuts around the MBH might have included a good number of animals in which a portion of the nerve fibers reaching the MBH in the basal part of the lateral retrochiasmatic area might have been left uncut during the operation and remained unnoticed on examination.

Standing alone, the lack of pituitary–adrenal response to ether inhalation cannot be used as evidence against the intra-MBH location of the CRF-producing neurons, since it is possible that intact neural pathways reaching the CRF-producing neurons from a site of action outside the MBH may be required for a normal response. To exclude that most of the CRF-containing neurons are within the MBH, we have to use either a form of stimulation that certainly reaches the isolated tissue of the MBH or direct measurement of CRF activity within this tissue.

Electrical stimulation via electrodes within the MBH 7 to 8 days after placing various hypothalamic cuts around the MBH was used in an attempt to reveal the existence of intra-MBH cells producing CRF and releasing it near the primary capillaries of the portal circulation (Makara et al., 1978). We found no rise in plasma corticosterone whenever the groups of rats used had all nerve fibers via the lateral retrochiasmatic area severed on the side of electrical stimulation (Fig. 4). Conversely, if electrical stimulation was contralateral to unilateral transections or was in MBH tissue only incompletely separated from the rest of the brain (with a slab of uncut tissue in the lateral basal part of the retrochiasmatic area), a significant response to stimulation was often observed. In agreement with these findings, chemical stimulation with intraventricular infusion of acetylcholine or the powerful neuroexcitant sodium glutamate failed to elicit a rise of corticosterone in rats with complete hypothalamic cuts around the MBH. It is well known that at least acetylcholine is a potent stimulus release of CRF by incubated hypothalamic blocks (Jones *et al.*, 1976; Buckingham and Hodges, 1977) or synaptosomes (Edwardson and Bennett, 1974). Thus, attempts to directly stimulate postulated intra-MBH CRF cells with different stimuli failed, which strongly suggested to us that CRF-containing axons and terminals of the SME complex as well as in the NL might arise from cell bodies lying outside the areas usually included in a "deafferented" MBH.

Bioassay for CRF content of extracts made from tissue taken from rats with cuts around the MBH also gave divergent results. Yasuda and Greer (1976) failed to find a change in CRF activity in their extract made from MBH tissue, whereas Krieger *et al.* (1977) found a slightly increased CRF activity in MBH samples. However, the sampling methods of these studies precluded a thorough verification of the completeness of the hypothalamic cuts. In our studies (Makara *et al.*, 1979) we have sampled the SME or the NL of rats from fresh unfrozen brain and retained the hypothalamus for a rigorous histological study. We used only extracts from rats in which we found no neural connections between the hypothalamic island and the rest of the brain. In such extracts we found a decrease to less than one-tenth of the normal CRF content of the SME and NL obtained 7 to 8 days after the complete or anterolateral cut around the MBH (Makara *et al.*, 1979, 1981b). In the case of the SME, a marked decrease was already evident 3 days after placing an anterolateral cut (Table II). In the case of the NL, the marked decrease of CRF content following anterolateral hypothalamic cut was also verified with the pituitary microinjection method

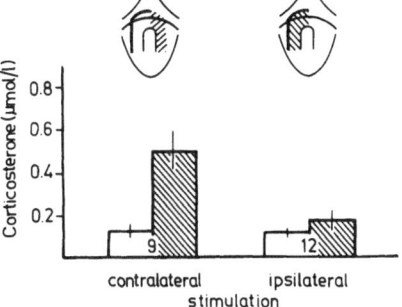

Figure 4. Effect of electrical stimulation on plasma corticosterone level in pharmacologically blocked rats with unilateral cuts around the mediobasal hypothalamus. The drawings above the bars show the ventral aspect of the hypothalamus with the contour of the cut (thick line) and the stimulated area (cross-hatching). Open column: before stimulation; cross-hatched column: 20 min after stimulation. [Reproduced from Makara *et al.*, 1978, with permission.]

for measuring CRF activity and the decreased neurohormone content was paralleled by an almost complete disappearance of the neural elements from the NL as seen under the electron microscope (Makara *et al.*, 1980a).

Another way to demonstrate that the CRF-containing neural projections of the SME and the NL originate from outside the MBH would be to try to locate an area the destruction of which prevents or markedly decreases pituitary–adrenal function even if challenged by stimuli reaching the MBH and/or the terminals in the SME or the NL. The best place to look for CRF cells seemed to be the hypothalamic paraventricular nuclei (PVN) since the ME projection of these nuclei show morphophysiological changes in parallel with changes in activity of the pituitary–adrenal system (Bock and Jurna, 1977; Lobo Antunes *et al.*, 1977; Vandesande *et al.*, 1977). It has also been shown that electrical stimulation of the PVN is followed by a marked pituitary–adrenal activation (Dunn and Critchlow, 1973; Maran *et al.*, 1978). Destroying the PVN and its immediate surroundings with a special rotating knife caused the following changes: the response to electrical stimulation of the MBH or the NL is significantly diminished, the plasma ACTH response to adrenalectomy and adrenalectomy + ether inhalation is reduced, the plasma corticosterone response to surgical trauma and ether inhalation + venesection stress is impaired and the weight of the adrenals is slightly reduced (Makara *et al.*, 1981b). It is interesting to note that at least the response to ether inhalation returns to normal by 4 weeks after removal of the PVN (Fig. 5). All these findings are consistent with the idea that CRF-containing cells are either in or near the PVN or at least their axons go through this region and thus PVN damage is impairing pituitary–adrenal function via a partial removal of the CRF-containing system. It should be noted, however, that some CRF cells should be located outside this region, as a small but significant amount of

Table II.   CRF Activity in SME Extract Decrease after Placing an Anterolateral Cut around the Mediobasal Hypothalamus[a]

Addition to culture	Fraction of SME added/ml medium	
	00.25	0.25
Cerebral cortical extract[b]	0.53 ± 0.04[c]	0.61 ± 0.05
SME from sham-operated rats	0.89 ± 0.09[d]	1.31 ± 0.09[d]
SME 1 day after anterolateral cut	0.90 ± 0.04[d]	1.21 ± 0.13[d]
SME 3 days after anterolateral cut	0.83 ± 0.05[d]	0.98 ± 0.07[d,e]
SME 7 days after anterolateral cut	0.57 ± 0.05	0.71 ± 0.05[e]

[a] From Makara *et al.* (1979). Reproduced with permission of the publisher.
[b] In equivalents of SME (average protein content of one SME: 30.6 μg).
[c] Values are means ± S.E.M.; $n = 5$.
[d] $p < 0.05$: significantly different from the effect of the same dose of cerebral cortical extract.
[e] $p < 0.05$: significantly less than the response to SME from sham-operated control animals.

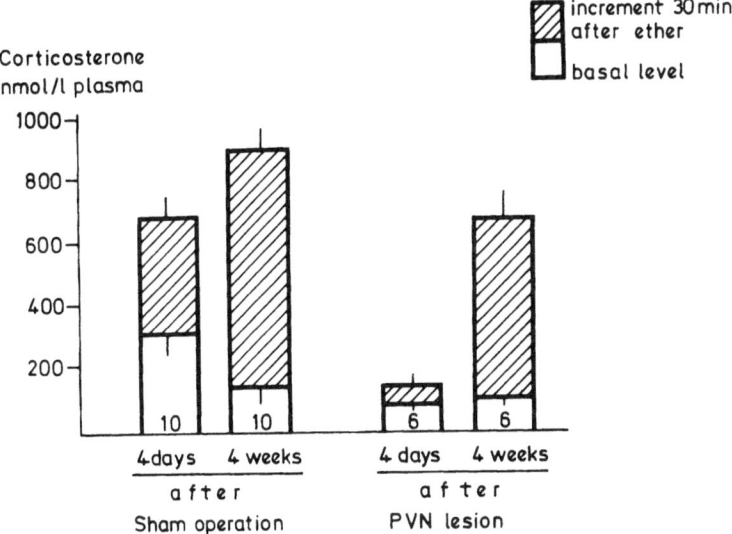

Figure 5. Effect of ether + venesection stress on plasma corticosterone level in rats with a lesion of the hypothalamic paraventricular nucleus. [Reproduced from Makara *et al.*, 1981b, with permission.]

CRF remains in the SME shortly after PVN lesioning and might be responsible for the small but significant response to ether+venesection stress at this time point.

Also, the return of normal response to ether suggests that compensatory CRF mechanisms are brought into operation some weeks after the destruction of the cells in or near the PVN. The evidence for CRF cells or axons in the paraventricular region of the hypothalamus gives indirect support also to our hypothesis that most CRF-producing cells are not within the MBH, since the PVN cannot be regarded as basal hypothalamic structure and in most cases is excluded from it by hypothalamic "deafferentation."

6.   Afferent Pathways of Stressful Stimuli

When the organism encounters a stressful situation or a stressor agent, the impact of the stressor should give rise to some signals, reaching either the CRF-producing cells of the hypothalamus or the ACTH-secreting cells of the anterior pituitary and activating their target. The large variety of stressors capable of stimulating CRF and ACTH release makes it very likely that a number of different mechanisms might, independently or in concert, elicit the endocrine response to stress. The links between the site(s) of impact and the CRF (or ACTH)-producing cells may be considered as the afferent pathways of the various stressful stimuli.

Theoretically, stressful stimuli may be subdivided according to the pathways by which they activate the hypothalamo–pituitary–adrenal system (Makara *et al.*, 1969b). Stressors with "neural" pathway(s) would activate peripheral and/or central neural pathways, which, ultimately, reach and activate the CRF-containing cells of the hypothalamus. CRF cells might also bear receptors sensitive to changes in some components of the blood and/or the interstitial fluid and, thus, it is possible that some "humoral" pathways may convey information (i.e., by an increase or a decrease in the amount of specific components) about the impact of some stressors to the CRF-producing cells. Unfortunately, there is no direct evidence for the existence of such a mechanism. However, there is evidence suggesting that some stressors stimulate ACTH release bypassing the usual hypothalamic transduction mechanism, since their effect is present even after complete destruction of the MBH, i.e., after elimination of the nerve fibers and terminals normally secreting CRF into the portal circulation. Such stimuli, not requiring the intimate links between the hypothalamus and the pituitary for their ACTH-releasing action, have been called "systemic" since they are supposed to be capable of stimulating ACTH secretion at the level of the anterior pituitary by a transduction mechanism involving the systemic circulation.

Unfortunately, most studies of the stress-activated pathways in the literature used partial or complete Halász-type cuts around the MBH, and in the light of our recent findings cited in the earlier part of this chapter, such cuts inflict a substantial and usually uncontrolled amount of damage to the CRF-producing cells and thus such cuts are hardly producing information about the afferent pathways abutting the CRF cells. We think hypothalamic deafferentations are of doubtful value in studying the pathways to the hypothalamic CRF cells. The value of those experiments that employ MBH lesions has also been questioned recently (Yasuda and Greer, 1978) because of the possibility that CRF in small quantities may reach the systemic circulation and may reach the transplanted or *in situ* pituitary even if it does not have direct vascular links with the MBH.

The only undisputed information about the pathways of the stressful stimuli came from sectioning of peripheral nerves or central nervous system pathways at extrahypothalamic sites. Peripheral denervation of the site of impact of a localized stressor such as surgical incision, injection of dilute formaldehyde subcutaneously, electrical stimulation of somatosensory nerves, or leg fracture prevented the pituitary–adrenal activation normally elicited by these stimuli. Furthermore, it has also been shown that the neural pathways mediating the effects of traumatic stress and the injection of dilute formaldehyde, ascend in the spinal cord most likely via the contralateral lateral column (Makara *et al.,* 1969b, 1970). Thus, these stimuli are certainly neural in that their presence is signaled in a large part, if not entirely, by activation of a neural mechanism. Interestingly, however, increasing the amount of formaldehyde injected subcutaneously into a denervated hindlimb led to an activation of ACTH secretion, suggesting that blood-borne changes occur with increased dosage and may activate ACTH release by some nonneural mechanism(s) (Stark *et al.,* 1970).

The pathways are largely neural and involve the median forebrain bundle as the main afferent pathway to the hypothalamus in the case of somatosensory, photic, and acoustic stimulation (Feldman *et al.,* 1972, 1975).

7. Feedback Regulation of ACTH Secretion

Suppression of glucocorticoid secretion and adrenal atrophy are well-known consequences of long-term administration of corticosteroid hormones in therapy. These phenomena formed the basis of the first theory of pituitary–adrenal regulation via negative feedback (Sayers, 1950). The objection to the importance of feedback was that no decrease of plasma corticosteroids could be found before the stress-induced increase occurred. In response to these experimental findings the "variable set point" theory of feedback regulation was formulated (Yates *et al.,* 1961). This ingenious version postulated that stressful stimuli rapidly induce a resetting of the feedback controller to a higher set point, thereby producing a virtual shortage of corticosteroids, which stimulated ACTH release. During testing the predictions of the variable set point concept we attempted to produce very high blood corticosteroid levels by cortisol and studied the stress-induced rise of corticosterone in rats. During the period of high cortisol levels the stress-induced corticosterone rise was present but it disappeared later when cortisol decreased to undetectable levels (Stark and Fachet, 1963). Similar conclusions were reached by Smelik (1963) and Hodges and Jones (1963) in experiments using exogenous corticosterone as the feedback stimulus. In further studies the analysis of the time course of the inhibitory effect revealed that there are two components in the corticosteroid feedback (Dallman and Yates, 1969): a fast, rate-sensitive feedback is induced by the rapid rise of blood corticosteroid level and this feedback signal will inhibit ACTH secretion for only a few minutes. If steroid levels are high after a latent period, a delayed phase of feedback inhibition can be observed; this inhibition is dependent upon the magnitude and the duration of corticosteroid increase and is effective hours later when the previous high levels might have already disappeared. It has long been debated whether the pituitary or the hypothalamus is the site at which corticosteroids inhibit the activity of the hypothalamo–pituitary–adrenal system. The debate was centered on the question whether or not the pituitary can be the site of action. The arguments against a pituitary site of action rested upon negative findings obtained shortly after corticosteroid administration (see Stark *et al.,* 1968b). When it became known that a relatively long time is necessary to demonstrate steroid inhibition at the level of the pituitary, many authors reported a decreased release of immunoreactive ACTH after adding corticosterone or dexamethasone to anterior pituitary fragments or cultures (e.g., Vale and Rivier, 1977).

This finding has been confirmed also in our laboratory. In vivo ACTH release can also be inhibited by dexamethasone in rats with pituitary islands prepared either by removing the medial hypothalamus (Stark *et al.,* 1973/74) or in rats with pituitary grafts implanted under the kidney capsule (Yasuda *et al.,* 1978). Thus, one of the sites of feedback inhibition is at the anterior pituitary.

It is now clear that a fast and a delayed feedback occur at both the hypothalamus and the pituitary. The hypothalamic fast feedback appears to be the one important under conditions of surgical stress. A complex and integrated system of the two types of feedback may form an economical means for the control of the pituitary–adrenal system (Jones *et al.,* 1979).

Another question of some importance is whether high blood ACTH levels themselves give rise to a negative feedback signal. The existence of such a "short feedback" has been postulated several years ago (see Szentágothai *et al.,* 1968). In acute experiments, Kendall *et al.* (1975) have shown that injection of ACTH-(1 –24) failed to decrease the secretion of immunoreactive ACTH-(1 –39), a finding not supporting the above concept. The short feedback inhibition of ACTH release might have been a cause of the peculiar stress-nonresponsive state seen after prolonged ACTH treatment: in rats receiving 3 IU of long-acting ACTH daily for 14 days, the adrenals were two to three times their normal size, the resting plasma corticosterone level was low 24 hr after the last injection, the sensitivity of the adrenals to a test dose of ACTH was increased (Stark *et al.,* 1963; Hodges and Mitchley, 1971), but there was no rise of plasma corticosterone in response to a number of stressful stimuli (Stark *et al.,* 1963; Hodges and Mitchley, 1971) with the exception that the response to bacterial endotoxin was maintained (Stark *et al.,* 1968a).

The stress-nonresponsive period lasts for 2 to 4 days after concluding the 14 days of ACTH treatment (Stark *et al.,* 1968a, 1969; Hodges and Mitchley, 1971) and when the response returned toward normal it was somewhat increased compared to the appropriate saline-treated group. In order to determine the site of action where prolonged ACTH treatment may inhibit pituitary–adrenal function, first the responsiveness of the various components of the system was studied. The adrenal sensitivity of ACTH was maintained both *in vitro* and *in vivo*. Thus, the nonresponsive components may be only at the hypophyseal or suprahypophyseal level.

Intracarotid injection of a hypothalamic extract was used to test the sensitivity of the anterior pituitary. This extract elicited a significantly smaller rise of plasma corticosterone in the ACTH-treated rats than in the controls (Stark *et al.*, 1968). Taking into account the enhanced adrenal response to ACTH, this finding suggested a marked decrease in pituitary responsiveness. Using both a bioassay and a radioimmunoassay the ACTH content of the anterior pituitary gland was found to be about one-half of the normal. When incubated in a Krebs–Ringer glucose solution, the pituitary segments released significantly less ACTH both under resting conditions and in the presence of various amounts of control SME extract (Fig. 6). The marked decrease observed in CRF-induced ACTH release appears to be an important factor in preventing stress-induced activation of the pituitary–adrenal axis in ACTH-treated rats.

The functional state of the hypothalamic CRF-producing cells was tested by hypothalamic electrical stimulation, the response to which was blocked in rats after prolonged ACTH treatment. This finding, however, might be explained by a decreased production and release of hypothalamic CRF and/or the decreased sensitivity of the pituitary toward CRF. Further evidence for a hypothalamic site of action has been provided by bioassay of CRF activity in SME extracts taken from ACTH-treated rats. The dose–

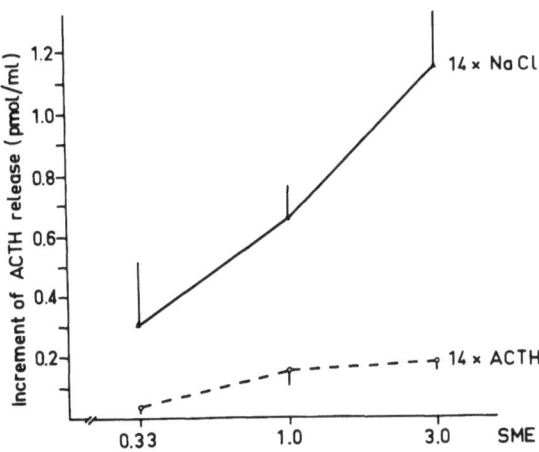

Figure 6. Responsiveness to CRF-containing stalk–median eminence (SME) extract of hemipituitary glands obtained from rats receiving either daily injection of 3 IU ACTH (dashed line) or 50 μl 0.15 M NaCl (solid line) for 14 days. Net release was calculated as the increment of ACTH caused by SME above the release by the control hemipituitary. For each point $n = 7$. Vertical lines indicate S.E.M. [Reproduced from Stark *et al.*, 1981, with permission.]

response curve for the SME taken from ACTH-treated animals ran significantly lower than that of the controls. Potency was about 30% of the controls.

The mechanism by which prolonged ACTH treatment may inhibit the pituitary–adrenal system may involve either the primary effect of the exogenous ACTH (possibly via the postulated short feedback loop) or an effect secondary to ACTH-stimulated adrenal steroidogenesis. Although the changes observed in the ACTH-treated animals were considered compatible with an explanation invoking corticosteroid feedback due to the summated effects of the 12- to 16-hr daily periods of high blood corticosteroids level, the inhibitory effect of episodes of high blood ACTH level could not be excluded either. The short-loop feedback regulation of ACTH secretion has been postulated (see Szentágothai et al., 1968) but received little if any direct experimental support (see Motta et al., 1965; Kendall et al., 1975). According to Hodges and Mitchley (1971), the inhibition of endogenous ACTH secretion is probably due to the hormone itself, because a small dose of tetracosactin, which produced only slight and transient rises in plasma corticosterone, caused an inhibition of the stress-induced ACTH release, and because tetracosactin given after chronic betamethasone treatment increased the delay in recovery of function of the pituitary–adrenal axis without producing any marked elevation in plasma corticosterone.

In an attempt to elucidate the role of ACTH per se in inhibiting pituitary ACTH release, the effect of prolonged ACTH treatment was studied in adrenalectomized rats receiving corticosterone supplementation. We reasoned that the marked decrease in pituitary ACTH content and CRF-

Table III. Effect of SME Extract on Release of ACTH (pmoles/Hemipituitary Gland) from Incubated Hemipituitary Glands of Rats Receiving Daily Injections of ACTH (3 IU) or 50 μl 0.15 M NaCl for 2 Weeks[a]

Treatment	Replacement with corticosterone (μmoles/liter)	SME equivalent added to the medium			
		None (basal)	0.33	1.0	3.0
Intact + NaCl	None	0.22 ± 0.08[b] (6)		1.54 ± 0.35 (6)	
Intact + ACTH	None	0.07 ± 0.02[c] (6)		0.29 ± 0.03[c] (6)	
Adrenalectomy + NaCl	462	0.15 ± 0.06[c] (18)	0.43 ± 0.06 (5)	0.88 ± 0.22 (6)	1.04 ± 0.14 (7)
Adrenalectomy + ACTH	462	0.10 ± 0.22[c] (16)	0.56 ± 0.14 (5)	0.94 ± 0.29 (6)	1.41 ± 0.28 (5)

[a]From Stark et al. (1981). Reproduced with permission of the publisher.
[b]Values are means ± S.E.M. Number of rats given in parentheses.
[c]$p < 0.05$ compared with intact saline-treated control values in the same column (Dunn's test).

induced ACTH release should be present also in adrenalectomized, corticosterone + ACTH-treated rats if ACTH itself, and not the sustained daily stimulation of corticosterone secretion, was the dominant component of the effect.

Corticosterone supplements were dissolved in drinking water. Decreasing corticosterone supply increased content and *in vitro* release of pituitary ACTH, and the dose at which these parameters were near the control values was 231 μmoles/liter. In the adrenalectomized and corticosterone-supplemented rats, ACTH treatment failed to inhibit *in vitro* ACTH release or to decrease anterior pituitary stores of ACTH (Table III). These findings imply that the adrenals are necessary for ACTH-induced inhibition of pituitary (ACTH) function, and it seems likely that the inhibition results from several factors such as summation of the delayed feedback signals repeated daily for 14 days, the amplification of the effect of each ACTH injection resulting in slightly higher and longer plasma corticosterone level, and a further increase in the biologically active free corticosterone level by a concomitant decrease in the binding capacity of plasma transcortin during prolonged ACTH treatment (Ács *et al.,* 1967). That the inhibition is connected with plasma free corticosterone levels is indicated by the finding that two injections of transcortin-rich plasma at the end of the prolonged ACTH treatment caused a partial reappearance of responsiveness to at least some stressful stimuli (Ács and Stark, 1973).

The above data pertinent to feedback regulation suggest that glucocorticoid feedback acts at both hypothalamic and pituitary levels and, at least in the general circulation, the level of glucocorticoids is a much more powerful feedback signal than that of ACTH. Whether ACTH postulated to reach the brain via the retrograde blood flow from the pituitary (Mezey *et al.,* 1978; Page and Bergland, 1977) or ACTH-like molecules released from neural elements play a role in the regulation of pituitary–adrenal function has at present no supporting experimental evidence.

REFERENCES

Ács, Z., and Stark, E., 1973, The role of transcortin in the regulation of corticotrophin secretion, *J. Endocrinol.* **56:**317.

Ács, Z., Stark, E., and Csáki, L., 1967, The effect of long-term corticotrophin treatment on the corticosteroid binding capacity of transcortin, *J. Endocrinol.* **39:**565.

Arimura, A., Saito, T. M., Bowers, C. Y., and Schally, A. V., 1967, Pituitary–adrenal activation in rats with hereditary hypothalamic diabetes insipidus, *Acta Endocrinol. (Copenhagen)* **54:**155.

Baertschi, A. J., Vallet, P., Baumann, J. B., and Girard, J., 1980, Neural lobe of pituitary modulates corticotropin release in the rat, *Endocrinology* **106:**878.

Bock, R., and Jurna, I., 1977, Ipsilateral diminution of CRF-granules after unilateral hypothalamic lesions, *Cell Tissue Res.* **185:**215.

Bristow, A. F., Montague, D., Synetos, D., Jenkins, G., Cockayne, D., and Schulster, D., 1980, An improved methodology for the extraction and partial purification of porcine hypothalamic corticotrophin releasing factor, *J. Endocrinol.* **84**:189.

Brownstein, M. J., Arimura, A., Schally, A. V., Palkovits, M., and Kizer, J. S., 1976, The effect of surgical isolation of the hypothalamus on its luteinizing hormone-releasing hormone content, *Endocrinology* **98**:662.

Brownstein, M. J., Arimura, A., Fernandez-Durango, R., Schally, A. V., Palkovits, M., and Kizer, J. S., 1977, The effect of hypothalamic deafferentation on somatostatin-like activity in the rat brain, *Endocrinology* **100**:246.

Buckingham, J. C., 1979, Corticotropin releasing factor, *Pharmacol. Rev.* **31**:253.

Buckingham, J. C., and Hodges, J. R., 1977, Production of corticotrophin releasing hormone by the isolated hypothalamus of the rat, *J. Physiol. (London)* **272**:469.

Buckingham, J. C., and Leach, J. H., 1980, Hypothalamo–pituitary–adrenocortical function in rats with inherited diabetes insipidus, *J. Physiol. (London)* **305**:397.

Csernus, V., Lengvári, I., and Halász, B., 1975, Further studies on ACTH secretion from pituitary grafts in the hypophysiotrophic area, *Neuroendocrinology* **17**:18.

Dallman, M. F., and Yates, F. E., 1969, Dynamic asymmetries in the corticosteroid feedback path and distribution–metabolism–binding elements of the adrenocortical system, *Ann. N.Y. Acad. Sci.* **156**:696.

Dunn, J., and Critchlow, V., 1973, Electrically stimulated ACTH release in pharmacologically blocked rats, *Endocrinology* **93**:835.

Edwardson, J. A., and Bennett, G. W., 1974, Modulation of corticotrophin-releasing factor release from hypothalamic synaptosomes, *Nature (London)* **251**:425.

Feldman, S., Conforti, N., and Chowers, I., 1972, Neural pathways mediating adrenocortical responses to photic and acoustic stimuli, *Neuroendocrinology* **10**:316.

Feldman, S., Conforti, N., and Chowers, I., 1975, Subcortical pathways involved in the mediation of adrenocortical responses following sciatic nerve stimulation, *Neuroendocrinology* **18**:359.

Fortier, C., 1966, Nervous control of ACTH secretion, in: *The Pituitary Gland* (G. W. Harris and B. T. Donovan, eds.), Vol. 2, pp. 195–234, Butterworths, London.

Gillies, G., and Lowry, P. J., 1978, Perfused rat isolated anterior pituitary cell column as a bioassay for factor(s) controlling release of adrenocorticotropin: Validation of a technique, *Endocrinology* **103**:521.

Gillies, G., and Lowry, P., 1979, Corticotrophin releasing factor may be modulated vasopressin, *Nature (London)* **278**:463.

Greer, M. A., Allen, C. F., Panton, P., and Allen, J. P., 1975, Evidence that the pars intermedia and pars nervosa of the pituitary do not secrete functionally significant quantities of ACTH, *Endocrinology* **96**:718.

Guillemin, R., and Rosenberg, B., 1955, Humoral hypothalamic control of anterior pituitary: A study with combined tissue cultures, *Endocrinology* **57**:599.

Halász, B., Slusher, M., and Gorski, R. A., 1967, Adrenocorticotrophic hormone secretion in rats after partial or total deafferentation of the medial basal hypothalamus, *Neuroendocrinology* **2**:43.

Harris, G. W., 1955, *Neural Control of the Pituitary Gland*, Arnold, London.

Hodges, J. R., and Jones, M. T., 1963, The effect of injected corticosterone on the release of adrenocorticotrophic hormone in rats exposed to acute stress, *J. Physiol. (London)* **167**:30.

Hodges, J. R., and Mitchley, S., 1971, Effect of prolonged treatment with tetracosactin on hypothalamo–pituitary–adrenal function in the rat, *Br. J. Pharmacol.* **43**:804.

Jones, M. T., Hillhouse, E., and Burden, J., 1976, Secretion of corticotropin-releasing hormone in vitro, in: *Frontiers in Neuroendocrinology,* Vol. 4 (L. Martini and W. F. Ganong, eds.), pp. 195–226, Raven Press, New York.

Jones, M. T., Gillham, B., Mahmoud, S., and Homes, M. C., 1979, The characteristics and mechanism of action of corticosteroid negative feedback at the hypothalamus and anterior pituitary, in: *Interaction within the Brain–Pituitary–Adrenocortical System* (M. T. Jones, B. Gillham, M. F. Dallman, and S. Chattopadhyay, eds.), pp. 163–180, Academic Press, New York.

Kárteszi, M., Stark, E., Makara, G. B., Fazekas, I., and Rappay, G., 1978, Corticoliberin (CRF) activity of the rat neurohypophysis, *Endocrinol. Exp.* **12**:204.

Kárteszi, M., Makara, G. B., and Stark, E., 1980, Rise of plasma ACTH induced by ether is mediated through neural pathways entering the medial basal hypothalamus, *Acta Endocrinol. (Copenhagen)* **93**:129.

Kárteszi, M., Stark, E., Rappay, G., László, F. A., and Makara, G. B., 1981, Corticoliberin (CRF) activity of the rat neurohypophysis is distinct from vasopressin, *Am. J. Physiol. Endocrinol. Metab.,* **240**:E689.

Kendall, J. W., Tang, L., and Cook, D. M., 1975, Sites of feedback control in the pituitary–adrenocortical system, in: *Anatomical Neuroendocrinology* (W. E. Stumpf and L. D. Grant, eds.), pp. 276–283, Karger, Basel.

Krieger, D. T., Liotta, A., and Brownstein, M. J., 1977, Corticotropin releasing-factor distribution in normal and Brattleboro rat brain, and effect of deafferentation, hypophysectomy and steroid treatment in normal animals, *Endocrinology* **100**:227.

Lobo Antunes, J., Carmel, P. W., and Zimmerman, E. A., 1977, Projections from the paraventricular nucleus to the zona externa of the median eminence of the rhesus monkey: An immunohistochemical study, *Brain Res.* **137**:1.

McCann, S. M., Antunes-Rodriguez, J., Naller, R., and Valtin, H., 1966, Pituitary–adrenal function in the absence of vasopressin, *Neuroendocrinology* **79**:1058.

Makara, G. B., 1979, The site of origin of corticoliberin (CRF), in: *Interaction within the Brain–Pituitary–Adrenocortical System* (M. T. Jones, B. Gillham, M. F. Dallman, and S. Chattopadhyay, eds.), pp. 97–113, Academic Press, New York.

Makara, G. B., Stark, E., and Mihály, K., 1969a, Corticotrophin release induced by injection of formalin in rats with hemisection of the spinal cord, *Acta Physiol. Acad. Sci. Hung.* **35**:331.

Makara, G. B., Stark, E., Palkovits, M., Révész, T., and Mihály, K., 1969b, Afferent pathways of stressful stimuli: Corticotrophin release after partial deafferentation of the medial basal hypothalamus, *J. Endocrinol.* **44**:187.

Makara, G. B., Stark, E., and Mihály, K., 1970, Corticotrophin release induced by traumatic stress in rats with unilateral spinal cord lesion, *Acta Physiol. Acad. Sci. Hung.* **38**:199.

Makara, G. B., Stark, E., and Palkovits, M., 1978, ACTH release after tuberal electrical stimulation in rats with various cuts around the medial basal hypothalamus, *Neuroendocrinology* **27**:109.

Makara, G. B., Stark, E., Rappay, G., Kárteszi, M., and Palkovits, M., 1979, Changes in corticotrophin releasing factor of the stalk median eminence in rats with various cuts around the medial basal hypothalamus, *J. Endocrinol.* **83**:165.

Makara, G. B., Stark, E., Kárteszi, M., Fellinger, E., Rappay, G., and Szabó, D., 1980a, Effect of electrical stimulation of the neurohypophysis on ACTH release in rats with hypothalamic lesions, *Neuroendocrinology* **31**:237.

Makara, G. B., Stark, E., and Palkovits, M., 1980b, Reevaluation of the pituitary–adrenal response to ether in rats with various cuts around the medial basal hypothalamus, *Neuroendocrinology* **30**:38.

Makara, G. B., Stark, E., Kárteszi, M., Palkovits, M., and Rappay, G., 1981a, Effects of paraventricular lesions on stimulated ACTH release and CRF in stalk–median eminence of the rat, *Am. J. Physiol. Endocrinol. Metab.* **240**:E441.

Makara, G. B., Stark, E., Kárteszi, M., Palkovits, M., and Rappay, G., 1981b, Hypothalamic organization of corticoliberin (CRF) producing structures, in: *Advances in Physiological Sciences,* Volume 14: *Endocrinology, Neuroendocrinology, Neuropeptides* Part II (E. Stark, G. B. Makara, B. Halász, and G. Rappay, eds.), Akadémiai Kiadó/Pergamon Press, Budapest.

Maran, J. W., Carlson, D. E., Grizzle, W. E., Ward, D. G., and Gann, D. S., 1978, Organization of the medial hypothalamus for control of adrenocorticotropin in the cat, *Endocrinology* **103**:957.

Martini, L., 1966, Neurohypophysis and anterior pituitary activity, in: *The Pituitary Gland* (G. W. Harris and B. T. Donovan, eds.), Vol. 3, pp. 535–577, Butterworths, London.

Mezey, É., Palkovits, M., deKloet, E. R., Verhoef, J., and deWied, D., 1978, Evidence for pituitary–brain transport of a behaviorally potent ACTH analog, *Life Sci.* **22**:831.

Motta, M., Mangili, G., and Martini, L., 1965, A "short" feedback loop in the control of ACTH secretion, *Endocrinology* **77**:392.

Nakane, P. K., Sétáló, G., and Mazurkiewicz, J. E., 1977, The origin of ACTH cells in rat anterior pituitary, *Ann. N.Y. Acad. Sci.* **297**:201.

Page, R. B., and Bergland, R. M., 1977, The neurohypophyseal capillary bed. I. Anatomy and arterial supply, *Am. J. Anat.* **148**:345.

Palkovits, M., 1979, Effect of surgical deafferentation on the transmitter and hormone content of the hypothalamus, *Neuroendocrinology* **29**:140.

Pittman, Q. J., Blume, H. W., and Renaud, L. P., 1978, Electrophysiological indications that individual hypothalamic neurons innervate both median eminence and neurohypophysis, *Brain Res.* **157**:364.

Saffran, M., Schally, A. V., and Benfey, B. G., 1955, Stimulation of the release of corticotropin from the adenohypophysis by a neurohypophysial factor, *Endocrinology* **57**:439.

Sayers, G., 1950, The adrenal cortex and homeostasis, *Physiol. Rev.* **30**:241.

Schally, A. V., Anderson, R. N., Lipscomb, H. S., Long, J. M., and Guillemin, R., 1960, Evidence for the existence of two corticotrophin-releasing factors, alpha and beta, *Nature (London)* **189**:1192.

Siegel, R., Chowers, I., Conforti, N., and Feldman, S., 1980, Corticotrophin and corticosterone secretory patterns following acute neurogenic stress and in variously hypothalamic deafferentated male rats, *Brain Res.* **188**:399.

Smelik, P. G., 1963, Relation between blood level of corticoids and their inhibiting effect on the hypophyseal stress response, *Proc. Soc. Exp. Biol. Med.* **113**:616.

Stark, E., and Fachet, J., 1963, Effect of blood corticosteroid levels on ACTH release caused by stress, *Acta Med. Acad. Sci. Hung.* **9**:366.

Stark, E., Fachet, J., and Mihály, K., 1963, Pituitary and responsiveness in rats after prolonged treatment with ACTH, *Can. J. Biochem. Physiol.* **41**:1772.

Stark, E., Ács, Z., Makara, G. B., and Mihály, K., 1968a, The hypophyseal–adrenocortical response to various different stressing procedures in ACTH-treated rats, *Can. J. Physiol. Pharmacol.* **46**:567.

Stark, E., Gyével, A., Ács, Z., Szalay, S. K., and Varga, B., 1968b, The site of the blocking action of dexamethasone on ACTH secretion: *In vivo* and *in vitro* studies, *Neuroendocrinology* **3**:275.

Stark, E., Ács, Z., and Szalay, K. S., 1969, Further studies on the hypophyseal–adrenocortical response to various stressing procedures in ACTH-treated rats, *Acta Physiol. Acad. Sci. Hung.* **36**:55.

Stark, E., Makara, G. B., Palkovits, M., and Mihály, K., 1970, Afferent pathways of stressful stimuli: Their dependence on strength and the time elapsed after the onset of stimulation, *Acta Physiol. Acad. Sci. Hung.* **38**:43.

Stark, E., Makara, G. B., Marton, J., and Palkovits, M., 1973/74, ACTH release in rats after removal of the medial hypothalamus, *Neuroendocrinology* **13**:224.

Stark, E., Makara, G. B., Palkovits, M., Kárteszi, M., and Mihály, K., 1978, Basal levels of pituitary ACTH and plasma corticosterone after complete or frontal cuts around the medial basal hypothalamus, *Endocrinol. Exp.* **12**:209.

Stark, E., Kárteszi, M., Rappay, G., and Makara, G. B., 1981, Effects of treatment with adrenocorticotrophin on the hypothalamo–pituitary–adrenal system, *J. Endocrinol.* **88**:131.

Szentágothai, J., Flerkó, B., Mess, B., and Halász, B., 1968, *Hypothalamic Control of the Anterior Pituitary: An Experimental–Morphological Study,* Akadémiai Kiadó, Budapest.

Vale, W., and Rivier, C., 1977, Substances modulating the secretion of ACTH by cultured anterior pituitary cells, *Fed. Proc.* **36**:2094.

Vandesande, F., Dierickx, K., and DeMey, J., 1977, The origin of the vasopressinergic and oxytocinergic fibres of the external region of the median eminence of the rat hypophysis, *Cell Tissue Res.* **180**:443.

Voloschin, L., Joseph, S. A., and Knigge, K. M., 1968, Endocrine function in male rats following complete and partial isolations of the hypothalamo–pituitary unit, *Neuroendocrinology* **3**:387.

Watson, S. J., Richard, C. W., III, and Barchas, J. D., 1978, Adrenocorticotropin in rat brain: Immunocytochemical localization in cells and axons, *Science* **200**:1180.

Yasuda, N., and Greer, M. A., 1976, Rat hypothalamic corticotropin-releasing factor (CRF) content remains constant despite marked acute or chronic changes in ACTH secretion, *Neuroendocrinology* **22**:48.

Yasuda, N., and Greer, M. A., 1978, Evidence that the hypothalamus mediates endotoxin stimulation of adrenocorticotropic hormone secretion, *Endocrinology* **102**:947.

Yasuda, N., Greer, M. A., Greer, S. E., and Panton, P., 1977, Distribution of corticotrophin releasing factor activity within the hypothalamic–pituitary complex of rats and cattle, *J. Endocrinol.* **75**:293.

Yasuda, N., Greer, M. A., Greer, S. E., and Panton, P., 1978, Studies on the site of action of vasopressin in inducing adrenocorticotropin secretion, *Endocrinology* **103**:906.

Yates, F. E., and Maran, J. W., 1974, Stimulation and inhibition of adrenocorticotropin release, in: *Handbook of Physiology,* Section 7: *Endocrinology,* Vol. IV, (R. O. Greep and E. B. Astwood, eds.), pp. 367–404, American Physiological Society, Washington, D.C.

Yates, F. E., Leeman, S. E., Glenister, D. W., and Dallman, M. F. 1961, Interaction between plasma corticosterone concentration and adrenocorticotropin-releasing stimuli in the rat: Evidence for the reset of an endocrine feedback control, *Endocrinology* **69**:67.

10

The Role of Steroid Receptors in the Regulation and Integration of Steroid and Peptide Hormone Actions in Common Target Cells

Facts and Speculations

Dušan T. Kanazir

1. Introduction

In this survey I shall attempt to review the results of 10 years' research of my colleagues from the Institute Boris Kidrič-Vinča and Institute for Biological Research, Belgrade, and to update the results of several recent pertinent experiments. Based on some new ideas and concepts, a new model is proposed for the integrating action of both steroid and peptide hormones in a common target brain cell. Consequently, my efforts will be concentrated on some new aspects of the regulatory and modulatory role of glucocorticoid (steroid) receptors. The basic concepts of our model are derived from our extensive and comparative studies of the structure and function of glucocorticoid receptors in anabolic and catabolic target organs (Ribarac-Stepić *et al.*, 1973, 1979a, b; Trajković *et al.*, 1973, 1974, 1980; Kanazir,

Dušan T. Kanazir • Department of Biochemistry and Molecular Biology, Faculty of Natural Sciences and Mathematics, University of Belgrade, Belgrade, Yugoslavia. *Present address:* The Serbian Academy of Sciences and Arts, Belgrade, Yugoslavia.

1974; D.T. Kanazir *et al.,* 1978, 1979, 1980; Djordjević-Marković, 1980), as well as from numerous data from other laboratories that had not been taken into consideration because they were not compatible with today's two-step model for steroid action. Usually, those are the events observed very early after steroid hormone administration. They have been considered as side effects, not directly related to hormone action, such as for example: changes in cell permeability (Le Cam and Freychet, 1977), in polyribosomal distribution (Burridge *et al.,* 1976), etc. However, it is now evident that the steroid hormones trigger in target cells a series of fast (immediate) effects. A satisfactory hypothesis to account for these immediate, short-term, acute extragenomic effects, mediated by steroids, has not yet been formulated. An enormous body of data supports the view that various steroids regulate in their target cells the gene expression, at the level of transcription, by a common two-step molecular mechanism (for details, see references in Jensen and DeSombre, 1972; Muller and Cowan, 1974; O'Malley and Means, 1974; Kanazir, 1974; Chan *et al.,* 1978). This model has already acquired the status of a "dogma" and has become a generally accepted theoretical framework for the experimental research currently being conducted. Recently, this "dogma" has been questioned, because it does not give a coherent, overall picture on the mode of steroid hormone action. Steroids trigger in respective target cells a series of rapid molecular events in which the genome of target cells is not involved. These rapid extragenomic effects represent the fast cell responses to hormones and have not as yet been integrated into the two-step model (D. T. Kanazir *et al.,* 1978, 1979, 1980). The present model is concerned only with the relatively slow responses caused by changes in gene expression. The problem becomes still more complex when we try to understand the interplay of steroids on the one hand and peptides, protein hormones, and neurotransmitters on the other, occurring in a common target cell such as a brain cell. The main, still unanswered, question is how these biochemical processes mediated by different hormones in common target cells are integrated? The brain cells are common targets for steroids, peptides, protein hormones, and neuro-transmitters. Consequently, it is accepted that the brain is an important target organ for steroids, in which they mediate neuroendocrine and behavioral effects (McEwen, 1979).

On the other hand, it was found that hypothalamic hypophysiotropic peptide hormones can mediate in target cells three different functions: hormonal (pituitary), behavioral, and neural (brain) (Barker, 1976; Hökfelt *et al.,* 1980). Therefore, they can be considered as multisignal integrators in the brain of man and animals. It is also evident that homeostasis results from the integrated functions of steroid and peptide hormones. The peptides act as hormones and simultaneously as possible neurotransmitters and are involved in the initiation of specific CNS-dependent behaviors and in neuronal excitability of hypothalamic as well as extrathalamic neurons. They induce short-term and long-term effects. These peptides, having different sizes, shapes, and electrical charge distributions, could, as well as steroids, function in brain, by interaction with appropriate specific receptors, an important role in information transfer and storage (Barker, 1976; Hökfelt *et al.*, 1980; Jessell and Kelly, 1980; Hinkle and Tashijian, 1976; Matusik and Rosen, 1978).

The molecular mechanism(s) of coordination and integration of multiple actions and responses of common target cells to different hormones is as yet still obscure. However, the rapid extragenomic events seem to be a feature common to all peptide, polypeptide, and steroid hormones. The most important events of this rapid response are the change in intracellular cyclic nucleotides, the alteration of membrane transport, and the changes of the intracellular pools of membrane transport, and the changes of the intracellular pools of precursors and ions (Tata, 1976; Greengard, 1979) and, in the case of steroids, the phosphorylation of soluble proteins (Blečić *et al.*, 1980), steroid receptors (Trajković *et al.*, 1980; Ribarac-Stepić *et al.*, 1980), and ribosomal proteins (D. T. Kanazir *et al.*, 1980; Stefanović, 1980). The end results of these relatively rapid changes in target cells would be to modify the intracellular milieu so as to promote and control transcriptional and translational machinery. It is important to emphasize that virtually all the rapid responses of pituitary tropic hormones, brain peptides, and conventional neurotransmitters can be mimicked by cAMP (Robinson *et al.*, 1971; Nathanson, 1977) whereas those of glucosteroids can be potentiated by cAMP (Jost and Averner, 1975; Baxter, 1979).

Although our limited knowledge of regulation of one cellular function by another in *in vivo* systems does not allow us to draw a definite, well-defined, and comprehensive model concerning multiple effects of different hormones in common target cells, some "links" between steroid and peptide hormone action in common target cells seem to be apparent (D. T. Kanazir *et al.*, 1979, 1980).

In this review article I shall attempt to make a theoretical approach to the problem, based partly on experimental facts from our and other laboratories and free speculations and putative assumptions. I will try to elaborate a highly speculative model in which steroid receptors might serve as integrating systems of many diverse and seemingly unrelated molecular events caused by steroid and peptide hormones in common target cells. The model we propose might serve as a device for the information transfer, switching, and integration in corresponding target cells. It is assumed that peptide and steroid hormone actions are "linked" or "coupled" via specific receptor systems operating in common target cells. It is very likely that glucocorticoid (steroid) receptors are involved in modulating the number of surface receptors specific for peptide hormones and neurotransmitters and responsiveness of common target cells to peptide hormones. Although the data I will present here are, at the present moment, insufficient to support unequivocally the model, nevertheless it might speed up the change of our thinking patterns on the mode of different hormone actions in common target cells.

2. The Basic (Hypothetical) Concepts of the Proposed Model for Integrating the Actions of Steroid and Peptide Hormones in Common Target Cells

We propose that the native steroid receptor is a heteromultimer of a specific association of "steroidophilic" subunit with some regulatory proteins. The receptor seems to be a molecular system "encoded" with a defined metabolic "code" for multiple, sequential, and cooperative biochemical and physiological functions.

1. It is very likely that the keystone of molecular organization of the native receptor system is a steroidophilic subunit(s), which exhibits an extremely high affinity and specificity toward steroids whereas other subunits may have regulatory functions.

2. The "activation" of the receptor caused by a specific binding of appropriate steroid seems to be a phosphorylation-dependent process that results in the disaggregation of the steroid receptor into monomeric subunits or association components. Consequently, the "activation" of the receptor would be a key event in steroid hormone action.

3. Some of the released subunits might be involved in regulations at posttranscriptional and translational levels—underlying the extragenomic events—and seem to be involved or linked with the phosphorylation/dephosphorylation regulatory mechanism operative in target cells. These would be "kinaselike" subunits.

4. Steroidophilic subunits control and modulate gene expression at the transcriptional level.

5. The primary role of steroids is to "activate" metabolic "code" "encoded" into the specific receptor molecule.

6. Extragenomic events seem to be the prerequisites for gene expression (late events).

7. The role of glucocorticoids (steroids) in brain neuroendocrine and behavioral functions is mediated by brain structure-specific receptors (McEwen, 1979).

8. The phosphorylation of regulatory subunits of cAMP-dependent phosphoprotein kinase mediated and modulated by the putative kinaselike subunit of steroid receptor (released upon the activation of steroid receptor) and the effect of peptide hormones via cAMP on the same substrate

(cAMP-dependent protein kinase) might provide a molecular basis for the link of steroid on one hand and peptide hormone and conventional neurotransmitter actions on another hand. This might serve as a molecular mechanism(s) for potentiating, synergistic, and antagonistic action of steroid and peptide hormones.

9. Glucocorticoids modulate the number of plasma membrane receptors for peptide hormones and catecholamines (for references see Wolfe *et al.*, 1976).

10. The endogenous protease activity in cell plasma membrane seems to be involved in the regulation of basal adenylate cyclase and its hormonal, neurotransmitter responsiveness (Richert and Rayan, 1977). Most of the membrane protease would exist as inactive precursors that could be activated by steroid receptor kinaselike subunit (upon the activation of steroid receptor) and/or by peptide hormone binding (Constantopoulos and Najjar, 1973; Jusit *et al.*, 1976; Catt *et al.*, 1979).

The modulation of membrane protease activity may ultimately be responsible for both adenylate cyclase stimulation and loss of peptide hormone–receptor complexes, i.e., for down-regulation or desensitization of peptide hormone receptors (Catt *et al.*, 1979).

It may be assumed that glucocorticoids via the regulatory subunit of the receptor might, by phosphorylation/dephosphorylation of inhibitors, modulate the endogenous membrane-bound phosphatase and other enzymes' activities.

Membrane protease and phosphatase activities (Killilea *et al.*, 1979) may be cooperatively modulated. This might regulate the transport of ions such as Ca^{2+} and Na^+ (Kometiani *et al.*, 1978; Schulman and Greengard, 1978).

3. Experimental Facts Supporting the Basic Concepts of the Proposed Model

Is native steroid receptor a multimeric molecule or molecular system? How is this assumption supported by experimental data?

It should be emphasized that the actual organization and composition of native cytoplasmic steroid holoreceptor is still unknown. Bearing in mind diverse experimental data, it appears logical to us to propose that the native steroid holoreceptor is a specific multimolecular system consisting of various subunits. The steroidophilic subunit, which is today generally considered as holoreceptor, seems to be isoprotein (tissue specific) and may aggregate with several different subunits giving rise to different macromolecular forms of native steroid receptor. This would imply that the receptor may have "encoded" in it different metabolic "programs" and that the structure of a given receptor might vary from tissue to tissue. This might offer one of the plausible explanations for heterogeneity and tissue-specific polymorphism of steroid receptors. The alternative is that holoreceptor is a tissue-specific association of receptor protein, steroidophilic subunit, and several regulatory proteins, which dissociate from the receptor upon binding of the corresponding steroid hormone to receptor protein. Therefore, in both cases the composition of native holoreceptor would depend on differentiation, specificity, and diversity of target cells.

3.1. The Structure of Steroid Receptors: Experimental Facts

It was demonstrated that the progesterone receptor in the chick oviduct is a dimer composed of A and B subunits "encoded" with different functions (Chan *et al.*, 1978; Schrader *et al.*, 1980; Liao, 1976).

Estrogen-specific receptor seems also to be a multimer (Jensen and DeSombre, 1972; Chan *et al.*, 1978; Jungblut *et al.*, 1976).

Our data suggest that the receptor for glucocorticoids is also a multimer, which can be resolved into three fractions by DEAE-cellulose chromatography. Only one of these fractions is taken up by isolated nuclei, and stimulates the rate of [^{3}H]-UTP incorporation into nuclear RNAs. Two other components, which bind glucocorticoids and are fractionated after activation, remain in the cytoplasm and do not, at least under *in vitro* conditions, translocate in nuclei (Ribarac-Stepić *et al.*, 1973, 1979b; Kanazir *et al.*, 1979; Radojčić and Kanazir, 1980; Djordjević-Marković, 1980).

Our comparative studies of the glucocorticoid receptors from various tissues suggest very strongly the existence of tissue-specific polymorphism of receptors. Following thermal activation of glucocorticoid receptors from liver and kidney, a difference in the elution patterns of activated complexes was observed (Kanazir *et al.*, 1979; Marković and Litwack, 1980; Djordjević-Marković, 1980). The results on receptors from various tissues are presented in Table I. By comparing the ratios of IB fractions as well as the percentage of binding of the activated receptor–hormone complexes to DNA, one might conclude that the structure of the receptor varies from tissue to tissue (Djordjević-Marković, 1980; Marković *et al.*, 1980). Taken together, our data suggest that glucocorticoid receptors might be multimers and probably heteromultimers or heteromolecular associations. However, other alternative explanations of our data are not excluded.

3.2. Glucocorticoid Binding Proteins of Brain

It is widely accepted that the brain is an important target tissue for glucocorticoids (steroids) in mediating their neuroendocrine and behavioral actions (McEwen, 1979). The glucocorticoid binding proteins are differentially localized in the brain regions (structures) such as hippocampus, septum, amygdala, hypophysis (McEwen, 1979; Mileusnić *et al.*, 1980; Ribarac-Stepić *et al.*, 1980; S. Kanazir *et al.*, 1980). This pattern of distribution was observed in rats and in rhesus monkey brain (Gerlach *et al.*, 1976; S. Kanazir *et al.*, 1980). These regions of brain and pituitary retain labeled estradiol, corticosterone, and cortisol differentially (Gerlach *et al.*, 1976). Few results are available on the glucocorticoid receptors in human brain (Tsuboi *et al.*, 1979). The anterior pituitary showed the highest binding capacity followed by the hypothalamus and hippocampus. This pattern of hormone distribution in human brain differs from that observed in rat brain cytosol (McEwen, 1979). The binding capacity in rat pituitary was rather low. These differences in cytosol binding capacities between human and rat brain structures might suggest differences in the functions of these particular brain structures between these two species. The recent data from

Table I. Distribution of Activated Forms of the Glucocorticoid Receptor from Different Tissues Assayed by DEAE-Sephadex Minicolumn Before (4°C) and after (25°C) Activation[a]

Tissue cytosol	[^{3}H]-TA concn (final) (nM)	Protein II (%)	Protein IB (%)	Ratio Between Protein II and IB after activation	Ratio between IB forms before and after activation	% bound (CD assay)	DNA (% of bound complex)
Liver							
4°C	31	10.7	3.7	11.2	2.2	68	1.5
25°C		91.8	8.2				54.0
Kidney							
4°C	50	—	11.3		8.1	29	0.9
25°C		—	91.5				20.1
Brain							
Septum							
4°C	55	7.0	9.8	7.3	1.2	4	4.0
25°C		88.0	12.0				13.0
Hippocampus							
4°C	47.9	10.0	3.3	10.7	2.6	12	1.9
25°C		91.5	8.5				12.4
Hypothalamus							
4°C	50	44.0	7.1	3.9	1.8	36	11.0
25°C		51.0	13.0				11.0

[a] Cytosols were labeled *in vitro* with 30–50 nM [^{3}H]triamcinolone acetonide ([^{3}H]-TA) for 90 min at 4°C, and analyzed by DEAE-Sephadex A-50 ion-exchange chromatography before (4°) or after (25°) heat activation as previously described (Marković *et al.*, 1980). The radioactivity associated with different receptor forms (which correspond to the activated forms Protein II and IB) was calculated as a percentage of total recovered radioactivity from minicolumns. All cytosols were examined by charcoal-dextran assay and bound radioactivity was measured. DNA-cellulose binding study was performed with all cytosols.

our laboratory suggest that the capacity of receptors to bind natural (cortisol) and artificial (TA) glucocorticoids varies from structure to structure in brain. Thus, for example, many more binding sites were detected in hippocampus than in other brain structures as shown in Fig. 1. The open question remains whether the polymorphism of receptors occurs in brain. Recent evidence suggests the presence of more than one population of receptors for TA and cortisol in the brain structures (de Kloet and McEwen, 1976). Our data on thermal activation could suggest that the putative receptors for TA in hippocampus, hypothalamus, and septum seem to be somehow different from each other (Fig. 1, Table I).

Preliminary results of Mileusnić et al. (1980) and S. Kanazir et al. (1980) suggest very strongly that cortisol is involved in the regulation of synthesis of some proteins in hippocampus and septum. The identification of these proteins is now in progress and might be of importance for better understanding of steroid action(s) in brain structures. Especially the data on the hippocampus receptor and its functions may be of biological (neuroendocrine) and behavioral interest, as it is known that the hypersecretion of cortisol can cause anxiety, suicidal impulses, etc. (Endröczi, 1975).

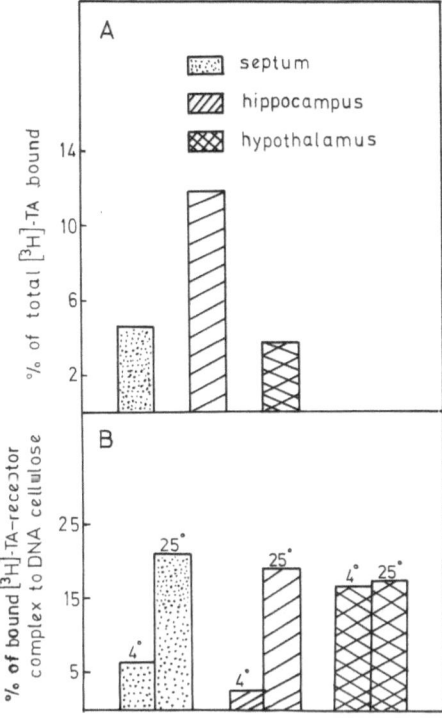

Figure 1. Cytosols obtained from septum, hippocampus, and hypothalamus were incubated in vitro with 50 nM [³H]-TA (triamcinolone acetonide) for 90 min at 4°C. (A) Bound radioactivity measured using charcoal-dextran assay. (B) DNA-cellulose binding study performed with unactivated (4°C) and with activated (30 min at 25°C) cytosols; percentage of bound complex to DNA presented.

3.3. The Activation of the Specific Cytoplasmic Receptor Seems to Be a Key Event in Mediating Steroid Hormone Action

We propose (Kanazir *et al.*, 1978, 1979) that the activation of the receptor caused by binding of appropriate glucocorticoid (steroid) hormone is a process resulting in disaggregation of native holoreceptor into monomeric subunits. Upon the activation of the receptor, steroidophilic subunit is, as a complex with hormone, translocated to nuclei where it interacts with acceptor sites of the chromatin and regulates the rate of initiation of transcription of a few specific structural genes (O'Malley and Means, 1974; Chan *et al.*, 1978; Schrader *et al.*, 1980). The other released subunit(s) apparently remains in cytoplasm and seems to be involved in modulating the rates of regulatory mechanism(s) operating in differentiated cells. Data from our laboratory (D. T. Kanazir *et al.*, 1979, 1980) suggest strongly that one or more of the subunits released upon activation of the receptor are involved in modulating the rates of phosphorylation/dephosphorylation mechanism. Consequently, bearing in mind the tissue-specific polymorphism of the glucocorticoid (steroid) receptors and variations in the constitution of various differentiated cells, one would expect that the activation of the receptor may cause different metabolic and molecular events in respective target cells, as it seems to be the case with anabolic and catabolic target cells. The primary function of glucocorticoid (steroid) hormones would be to trigger the activation of the appropriate specific receptor, but the nature and the patterns of metabolic changes induced by steroids are dependent not only on the steroid hormone, but on the structure of the receptor(s) and the intracellular milieu (composition) of the corresponding differentiated target cell(s). The data from our laboratory indicate a very rapid increase in the phosphorylation levels of the receptor 5 to 10 min after intraperitoneal cortisol administration. This phosphorylation increases the binding capacity of the receptor by factor of 2 to 3. The uptake of the phosphorylated complex by isolated liver (target) cell nuclei is about 2 to 3 times greater than that of unphosphorylated complexes (Kanazir *et al.*, 1979; Trajković *et al.*, 1980). It is very likely, therefore, that the phosphorylation of the receptor increases both the binding capacity of the receptor and the rate of its translocation to nuclei, as well as the rate of interaction with chromatin of the target cell.

3.4. The Extragenomic Events — Rapid Effects Dependent on Glucocorticoid (Steroid) Hormones

The steroid hormones trigger in respective target cells a series of rapid effects that are not caused by modulation of gene activity, and are relatively insensitive to inhibitors of *de novo* synthesis of mRNAs such as actinomycin D. Androgens may exert immediate effect on the posttranscriptional level in rat ventral prostate. Orchiectomy of rats prevents the methionine-tRNA (initiator tRNA) binding to prostate ribosomes. This is reversed by injection of a potent androgen into the castrated animals. Androgens also stimulate, in the presence of GTP, the binding of methionine-tRNA to initiation factors (Liang *et al.*, 1977). This effect seems to be dependent on receptor, for antiandrogens that antagonize receptor binding of androgen can block this effect (Fang and Liao, 1971). The rapidity of the effect and the fact that this effect cannot be prevented by actinomycin D and cycloheximide suggest that new RNA or protein synthesis may not be needed for this extragenomic steroid hormone action (Liang *et al.*, 1977).

Among the rapid extragenomic effects of the cortisol, we observed a rapid synthesis of cortisol-specific receptor protein occurring, very likely, on the preformed specific mRNA (Kanazir *et al.*, 1979).

In addition to the early synthesis of proteins, an increased level of phosphorylation of some ribosomal proteins of small and large subunits and increased *in vitro* translational capacity of ribosomes were reported (D. T. Kanazir *et al.*, 1980; Stefanović and Kanazir, 1980). The level of phosphorylation of S_6 basic protein of small (40 S) subunit, and one of acidic pro-

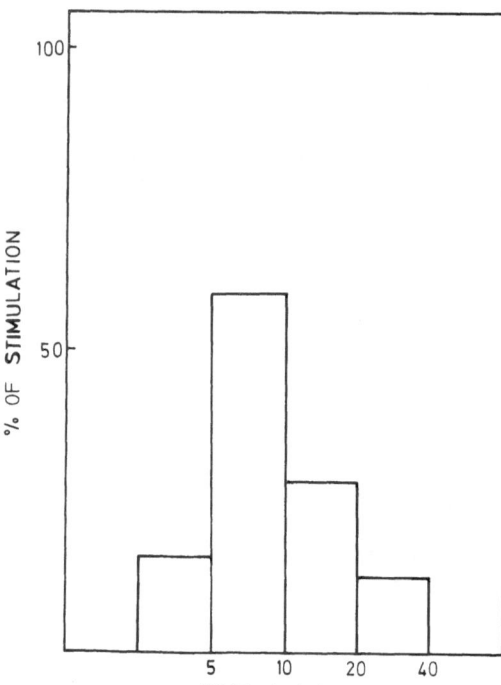

Figure 2. Pulse labeling of ribosomal proteins following cortisol administration. Cortisol (10 mg/200 g body wt) was given i.p. and animals were sacrificed at the indicated intervals of time. The labeling of ribosomal proteins was performed by 10-min pulse of ^{32}P (4 mCi/200 g body wt). Polysomes were extracted from livers and analyzed for incorporation of ^{32}P into total ribosomal proteins as previously described (D. T. Kanazir *et al.*, 1980).

Table II.   The Specific Activity of S6
Ribosomal Protein[a]

Cortisol-treated (cpm/A_{605})	Control (cpm/A_{605})	Percentage stimulation
121	32	278
98	61	61
119	49	143

[a] Liver ribosomal proteins were prepared from rats that had received i.p. cortisol (20 mg/200 g body wt) and ^{32}P (4 mCi/200 g body wt) 10 min before sacrificing. Proteins were separated by two-dimensional electrophoresis and the spot corresponding to S6 was treated by SDS-urea in order to extract protein together with bound dye. Radioactivity was determined and expressed as radioactivity/605-nm absorbance ratio (D. T. Kanazir *et al.*, 1980). The data from three different experiments are presented.

teins, not yet identified, of large (60 S) subunit was markedly increased 10 min after intraperitoneal administration of cortisol (Fig. 2, Table II). This is in agreement with data previously published (D. T. Kanazir *et al.*, 1980). The labeled phosphate was bound to serine and threonine (Table III) (Stefanović, 1980). This phosphorylation seems to depend on the activation of cortisol-specific receptor, for the binding of the antihormone cortexolone decreases the level of phosphorylation of ribosomal proteins (Table IV). Cortexolone, by binding to cortisol-specific receptor, prevents its activation in a proper way. After thermal activation, cortexolone was found to be preferentially bound (Fig. 3) to the IB fraction of the receptor (Djordjević-Marković and Kanazir, 1980), i.e., the fraction that very likely does not translocate into the liver cell nuclei. This observation is of interest as it suggests that this IB fraction could be involved in the posttranscriptional regulation of gene expression.

Table III.   Radioactivity of Phosphoserine
Separated from Ribosomal Protein S6[a]

Cortisol-treated (cmp/fraction)	Control (cpm/ fraction)	Percentage stimulation
90	29	233
87	40	118

[a] Liver ribosomal proteins extracted as described in Table II were hydrolyzed and subjected to paper electrophoresis. Spots representing phosphoserine were cut out after staining and radioactivity was determined by liquid scintillation counting. The data from two different experiments are presented.

Table IV.   The Influence of Cortexolone on the
Phosphorylation of Basic and Acidic Rat Liver Ribosomal
Proteins

Cortexolone-treated (cpm/mg protein)	Control (cpm/mg protein)	Percentage inhibition
	Basic	
5820	16,862	65
1100	1,300	15
485	2,402	80
318	606	48
	Acidic	
692	2,977	77
3480	8,428	59
1038	1,352	23
271	1,120	76

[a] Rats received i.p. injection of cortexolone (5 mg/100 g body wt) and ^{32}P (4 mCi/200 g body wt) and were sacrificed 10 min later. Basic and acidic rat liver ribosomal proteins were extracted as previously described (D. T. Kanazir et al., 1980). The data from four different experiments are presented, where given specific activities represent the mean value for proteins isolated from three to five collected livers, corrected for specific radioactivity of corresponding ATP pools.

Ribosomes prepared from rat liver 10 min after cortexolone treatment had lower capacity to translate poly(U) *in vitro* (Fig. 4) and the fidelity of poly(U) translation was lower (Stefanović, 1980).

Among the extragenomic effects of glucocorticoids, we observed *in vivo* an increased translational capacity. This increase in the translational capacity might be attributed to the immediate modifications of elongation and initiation factors and/or phosphorylation of ribosomal proteins critical for the rate and fidelity of translation (Metlaš *et al.*, 1979; Kanazir *et al.*, 1979; Stefanović and Kanazir, 1980).

Our results pertaining to cortisol-induced early increase of translational capacity of polyribosomes may suggest that this increase is correlated with the level of phosphorylation of ribosomal proteins, which seems to be dependent on the activation of the receptor. For several reasons we favor the hypothesis that immediate increase of the phosphorylation levels of ribosomal S 6 and other proteins (Metlaš *et al.*, 1979; Stefanović and Kanazir, 1980), as well as that of subunits of eiF2 and eiF3 (Traugh and Lundak, 1978), is dependent and mediated by the activation of steroid receptor and may be involved in modulating the translation efficiency, specificity (discrimination), and accuracy. The steroid-dependent phosphorylation of

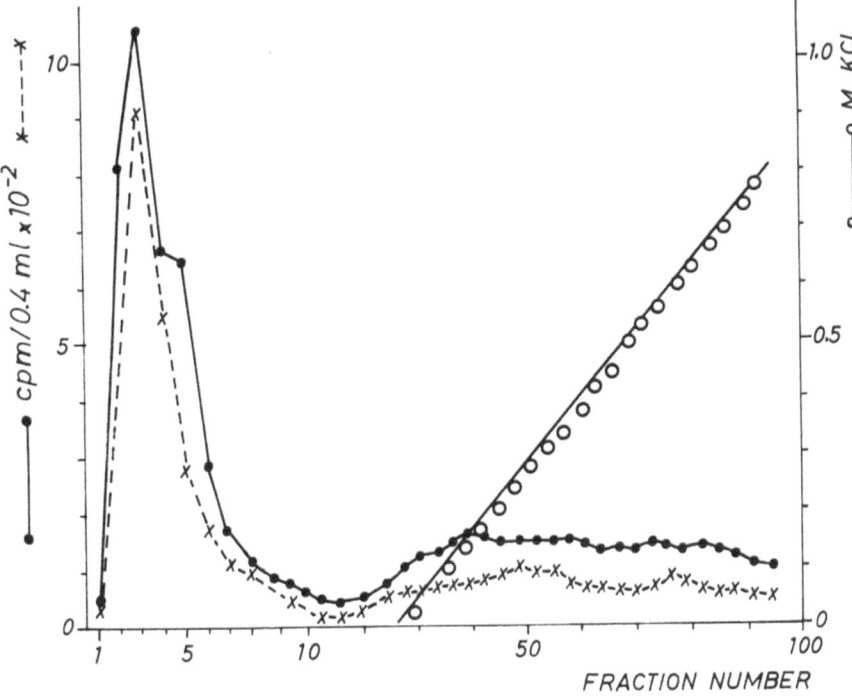

Figure 3. DEAE-Sephadex A-50 chromatography of unactivated and activated liver cytosol labeled *in vitro* with [³H]cortexolone. Liver cytosol was labeled *in vitro* with 30 nM [³H]cortexolone for 90 min at 4°C and analyzed by DEAE-Sephadex A-50 minicolumn before (o) and after (X) heat activation as previously described (Marković *et al.*, 1980).

the ribosomal proteins may serve as a regulator of qualitative rather than quantitative protein synthesis (D. T. Kanazir *et al.*, 1980).

Our hypothesis is also supported by data from other laboratories: an increase in the clustering of ribosomes for mRNA was observed after cortisol administration (Kulkarno *et al.*, 1976); heavier polysomes induced by estradiol were shown in chick oviduct (Palmiter *et al.*, 1970); the capacity for protein synthesis of prostatic ribosomes was reduced by castration and restored by testosterone administration (Liao and Williams-Ashman, 1962; Liang *et al.*, 1977); and the enhanced maturation of oocytes under steroid action was described (Smith and Ecker, 1971).

For the sake of our discussion it should be noted that peptide hormones also evoke in their respective target cells the rapid extragenomic events, such as the changes in the balance between intracellular concentrations of cAMP and cGMP (ratios), in the size of the pools of precursors, and in the

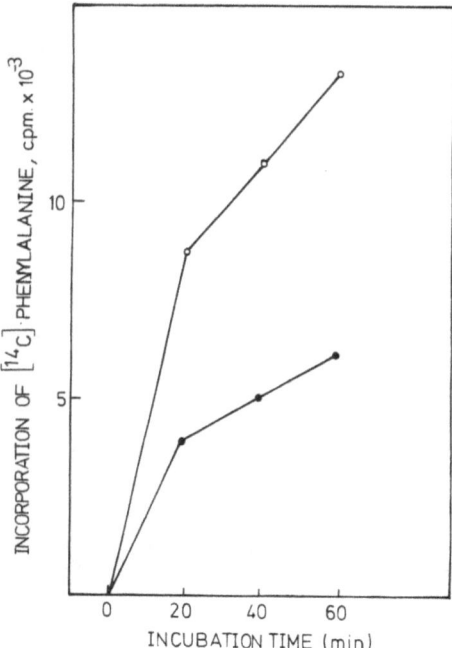

Figure 4. The effect of cortexolone on the *in vitro* translation directed by poly(U). Animals were sacrificed 10 min following cortexolone administration (10 mg/200 g body wt). The ribosomal subunits were prepared as previously described (D. T. Kanazir *et al.*, 1980). The *in vitro* translation was followed by [^{14}C]phenylalanine (0.5 μCi/0.95 nmole) incorporation into the polyphenylalanine. The reaction mixture was composed as previously described (D. T. Kanazir *et al.*, 1980). To each reaction mixture 0.3 optical density (O.D.) of small (40 S) and 0.6 O.D. of large (60 S) ribosomal subunits were added. The incubation was carried out at 37°C for the time intervals indicated on the abscissa. The radioactivities of the precipitates obtained by hot TCA were counted. (o) Control incorporation of [^{14}C]phenylalanine; (●) the incorporation following cortexolone treatment.

transport of ions such as Ca^{2+} and Na^+ (membrane events). It seems that polypeptide hormones rapidly alter the translational capacity of ribosomes and the rate of translation and induce the early, actinomycin D-insensitive and late, actinomycin D-sensitive bursts of protein synthesis (for complete review of references see Tata, 1976; Clemans and Korner, 1970).

With respect to steroid and/or peptide hormone action one of the most important questions that remains unanswered is whether late genomic events caused by these hormones in respective target cells or in common target cells are secondary to rapid extragenomic events or not? The answer to this question may be of importance for better understanding of fundamental events involved in gene expression. The available data are more compatible with the idea that rapid extragenomic responses to hormones might play an important role in facilitating selective gene expression (D. T. Kanazir *et al.*, 1979, 1980).

4. Coordination and Integration of Steroid and Peptide Hormone Actions in Common Target Cells

The brain and endocrine cells are common targets for steroid and peptide hormones. Many steroid actions in common target cells are either synergistic with or antagonistic to the effect of peptide hormones and neurotransmitters whose actions are mediated through cyclase systems, although steroid hormone action cannot be explained by the same mechanism (Jost and Averner, 1975; Greengard, 1976, 1978, 1979). However, it should be noted that representatives of all major classes of steroid as well as peptide hormones affect the protein phosphorylation system in their target cells (Greengard, 1978, 1979). It is evident that both peptide and steroid hormones are involved in the phosphorylation/dephosphorylation of proteins in common target cells but via different mechanisms whose molecular basis is still unclear.

4.1. Possible Link between Steroid Receptor Activation and Phosphoprotein Kinase Activity

Data from our laboratory (D. T. Kanazir *et al.*, 1979, 1980; Stefanović, 1980; Blečić *et al.*, 1980; Trajković *et al.*, 1980) indicate that the activation of cortisol receptor and increased level of phosphorylation of ribosomal and other proteins seem to be time-linked series of events. Consequently, within the frame of our model we propose that cortisol (steroid) receptor in anabolic target cells might contain as subunit(s) either cAMP-independent kinase or catalytic subunit of cAMP-dependent phosphoprotein kinase [kinase C-like subunit(s)] or phosphokinase modifying proteins, whereas in catabolic target (thymus) cells one would expect phosphoprotein phosphatase or its modifying (regulatory) protein as receptor subunit due to the putative tissue-specific polymorphism of the steroid receptor and/or to the differences in the state of differentiation of various target cells. Upon activation of the native receptor the released putative kinase C-like subunit(s) might affect the phosphorylation/dephosphorylation rate of regulatory (R) subunit of cAMP-dependent phosphoprotein kinases (RC). This might be a molecular link between steroid receptor activation and adenylate cyclase system through which peptide hormones and neurotransmitters act on the cell regulation. The proposed putative link might provide a molecular basis for the "permissive," "potentiating," synergistic, and antagonistic actions of steroids with peptide hormones and neurotransmitters in common target cells.

The concept I favored above is supported by the following experimental facts. The adenylate cyclase activity, cAMP levels, and polyamine levels decrease in the ventral prostate of the rat after antiandrogen administration. The biosynthesis of cAMP is dependent on the gonadal status of the animal (Singhal *et al.*, 1976). A rapid decline of type I protein kinase activity parallels the postcastration decay of receptors (Liao, 1976; Fuller *et al.*, 1978). The action of cAMP is impaired in the absence of glucocorticoids (Rousseau, 1977; Fuller *et al.*, 1978). The phosphorylation of ribosomal proteins is, as shown earlier, impaired by cortexolone (Stefanović, 1980). The steroid hormones control the phosphorylation of the R subunits of cAMP-dependent kinase (Liu and Greengard, 1976; Greengard, 1978, 1979). The phosphorylation of R subunits of protein kinase alters the state of dissociation and reassociation and the activity of cAMP-dependent protein kinase (Rangel-Aldao and Mandelsohn-Rosen, 1976). Nuclear translocation of cytoplasmic cAMP-dependent kinase seems to be dependent on phosphorylation and is a critical step in the regulation of gene expression (Costa *et al.*, 1976), induction of enzymes (Byus and Russell, 1976), and RNA polymerase activity (Jungmann and Kranias, 1977).

We propose, therefore, that steroid hormones on the one hand and peptide hormones and neurotransmitters on the other hand regulate in the common target cells the activity of the same enzyme, cAMP-dependent protein kinase, but do so by different mechanisms. It is very likely that this event may serve as a possible molecular basis for the integration of steroid and peptide hormone actions.

4.2.   Recent Experimental Data and Hypothetical Concepts

Our concepts concerning the structure of steroid receptors could be extended to brain cells. The steroid receptors in these cells seem to be involved in the regulation of rapid extragenomic effects in neurons (neurotransmission and neural discharges, the release of brain peptide hormones, and transport of ions) as well as in the regulation of late gene expression. Although we still do not know enough about the regulatory functions of steroid receptors, nevertheless we do know that: (1) steroid receptors are prevalent in several brain regions (structures) such as hypothalamus, thalamus, hippocampus, septum, pituitary, etc. (McEwen, 1979; S. Kanazir *et al.*, 1980); (2) glucocorticoids (steroids) exert both extragenomic and genomic regulatory influence over the production and release of brain–pituitary peptides such as for example growth hormone, gonadotropins, etc. (Martial *et al.*, 1977; Bratin and Porta Nova, 1977; Yu *et al.*, 1977).

That would mean that the acute, immediate release of neurotransmitters and releasing hormones might be due to modifications such as phosphorylation (Greengard, 1976, 1979), methylation, acetylation of preformed protein, mediated by the activation of steroid receptors, whereas the late effects (e.g., *de novo* synthesis of brain peptides) arise from the selective action of steroid hormone–receptor complex on the expression of series (batteries) of genes (Bratin and Porta Nova, 1977; Yu *et al.*, 1977) whose activities are also modulated by peptide hormones and neurotransmitters in brain target cells.

Peptide hormones evoke the acute responses of target cells following binding to the cell surface receptors. The adenylate cyclase is a part of a complex regulatory system that mediates the action of peptide hormones and neurotransmitters on their respective target cells. (For more detailed information on the role of hormone receptors in membrane transduction see the review article of Rodbell, 1980.) Recent data led to the notion that membrane receptors are mobile (Singer and Nicolson, 1972). But their numbers seem to be regulated by the concentrations of hormones (self-regulation). Regulation of the receptor concentration by homologous hormone and down-regulation were demonstrated for several peptide hormones (Kahn *et al.,* 1973; Lesniak and Roth, 1976; Hinkle and Tashijian, 1976; Mukherjee *et al.,* 1975; Shear *et al.,* 1976). Certain peptide receptors such as those for prolactin (Posner *et al.,* 1972) and angiotensin II (Hauger *et al.,* 1978) may increase in number after exposure to elevated hormone concentration, but the more common response is a loss of receptors from the cell surface. Hormonal down-regulation of receptors is accompanied by cooperative desensitization. This represents an additional form of receptor regulation, in which partial occupancy of receptor sites leads to decreased affinity of the residual receptors with the acute reduction of target cell sensitivity in the presence of increased concentrations of homologous hormones. Hormone-induced receptor loss has sometimes been found to depend on protein synthesis and sometimes not, but the return (replenishment) of receptors seems to require the synthesis of new proteins (see review by Catt *et al.,* 1979). The internalization and desensitization of receptors may protect target cells from prolonged and excessive stimulation by peptide hormones and neurotransmitters whereas the intracellular fragments of peptide hormones may exert long-term specific effects on the genome of target cells. The loss of receptor and sensitization are accompanied by the loss of cyclase activity (Mukherjee *et al.,* 1975; Shear *et al.,* 1976). It may be of interest for our discussion to stress several very important points resulting from the relevant recent research. The molecular nature of events through which binding of conventional neurotransmitters or brain peptides to membrane receptors of pituitary and extrapituitary targets (Scharpe, 1980; Clayton *et al.,* 1979), is coupled with the alteration of cell membrane is not yet understood. There are at least two general classes of mechanisms by which this process may occur. Receptor may be coupled with the ionophore. The binding to the receptor would induce a conformational change of ionophore causing an alteration in the permeability of the membrane to specific ions. The permeability change resulting from one-step, nonenzymatic process is expected to have very short latency and duration of the order of several milliseconds (ultrarapid changes). The second would be "receptor–second messenger" mechanism. The model for that mechanism predicts that electrophysiological changes and/or alterations of membrane permeability are caused by biochemical modification of membrane proteins involved in the regulation of permeability (Greengard, 1978, 1979). These changes might be expected to have longer latency and duration (rapid changes). However, slow genomic responses will be caused either by second-messenger cyclic nucleotides (cAMP and/or cGMP) and/ or by internalized peptide fragments found in the cytoplasm and in perinuclear regions of target cells (Catt *et al.*, 1979). Although the peptide hormones and conventional neurotransmitters act on target cells via "cyclase–kinases system," the precise role or roles of cyclic nucleotides formed during physiological activity of target cells as well as the regulation of cyclase activity are less well understood. Recent data indicate that endogenous protease activity in the cell membrane is involved in the regulation of basal adenylate cyclase activity and its hormonal (neurotransmitter) responsiveness. In the absence of hormones, a low level of membrane protease activity would maintain basal adenylate cyclase (Richert and Rayan, 1977; Jusit *et al.*, 1976). The mechanism of protease stimulation is consistent with a limited proteolysis. The majority of membrane protease would exist as inactive precursors that could be activated by hormone binding to membrane receptors. I am proposing in addition that phosphorylation/dephosphorylation might be involved in the regulation of membrane protease activity via a putative mechanism similar to that already described (Constantopoulos and Najjar, 1973). A membrane enzyme with protease properties has been described in liver (Jusit *et al.*, 1976). It was also shown that *in vivo* injection of pituitary gonadotropins increases ovarian cell protease activity (Reichert, 1962).

If steroid hormone receptor activation is involved in modulating the activity of membrane protease, then steroid hormones might via membrane protease activity regulate the number of membrane receptors for peptide hormones and neurotransmitters as well as the adenylate cyclase activity. The protease activity could ultimately be responsible for both adenylate cyclase activity and the loss of the hormone receptor (Catt *et al.*, 1979). Recently a membranal proteinas that brings about a specific, limited degradation of catalytic subunit of cAMP-dependent kinase was described (Alhanaty *et al.*, 1981). Several reports have documented an increased response of adenylate cyclase to catecholamines in adrenalectomized rats. A three- to fivefold increase in the number of β receptors was found after adrenalectomy. Cortisone reversed both the increased responsiveness of adenylate cyclase and the increased binding of catecholamines in adrenalectomized rats (Leray *et al.*, 1973; Wolfe *et al.*, 1976; Jost and Averner, 1975). It has also been reported that corticosteroids (both glucocorticoids and mineralocorticoids) play a role as modulators of the formation of cAMP by altering the activity of adenylate cyclase in cerebral cortex of rats (Nakagawa and Kuriyama, 1976; Bloom, 1979) and in tumor cells in culture (see references in Nakagawa and Kuriyama, 1976). Thus, steroid hormones could, upon activation of steroid receptor, via modulation of membrane protease activity regulate the responsiveness of the common target cells to peptide hormones and neurotransmitters. In addition to the kinases, phosphoprotein phosphatases are also involved in the regulatory phosphorylation/dephosphorylation mechanism. The holoenzyme contains inhibitory (or regulatory) and catalytic subunit. When phosphorylated, this inhibitory subunit is a potent inhibitor of phosphatase. This phosphorylation is catalyzed by cAMP-dependent protein kinase. Trypsin-controlled proteolysis causes a partial dissociation of catalytic subunit and partial activation of the enzyme. All peptide hormones that cause a rise of cAMP may activate the inhibitor and inactivate protein phosphatase (Killilea *et al.*, 1979).

Brain and many other tissues possess a protein kinase (or kinases) that requires both Ca^{2+} and calcium-dependent regulator (CDR), suggesting that (1) this enzyme(s) may be of general importance as a mediator in regulatory actions of calcium (Schulman and Greengard, 1978; Greengard, 1979) and (2) that Ca^{2+} is involved in the cell regulatory mechanism(s) (Brostrom *et al.*, 1975; Hammerschlag *et al.*, 1975). A most significant feedback relationship is the ability of cAMP to modulate Ca^{2+} homeostasis, acting in some systems to increase Ca^{2+} influx across the plasma membrane or to increase calcium release from internal stores. Ca^{2+} activates a Ca^{2+}-dependent protein kinase in nerve terminals that phosphorylates protein I (calmodulin) in presynaptic terminals, which in turn control the release and synthesis of neurotransmitters (Greengard, 1978, 1979).

5. Potentiation

Glucocorticoids play a permissive and synergistic role in regulating peptide hormone action in common target cells. Thyroid and glucocorticoids stimulate growth hormone (GH) production in cultured rat pituitary cells (Tsai and Samuels, 1974; Yu *et al.*, 1977; Bratin and Porta Nova, 1977; Baxter, 1979). The hormonal control occurs at the level of gene expression, i.e., by influencing the levels of GH mRNA (Martial *et al.*, 1977). Although these results showed an increase in the number of copies of GH mRNA in response to glucocorticoids, the data do not exclude the possibility that the primary effect is at the level of translation efficiency of preexisting mRNA which causes later enhancement of GH gene transcription. It was reported that adrenalectomy results in a marked decrease (50%) of GH synthesis and that treatment of adrenalectomized rats with either natural or synthetic glucocorticoids restores GH synthesis to normal or above-normal levels (Bratin and Porta Nova, 1977; Martial *et al.*, 1977). This effect is due to the increased synthesis of GH mRNA (Yu *et al.*,

1977). These data suggest that glucocorticoids are not required for GH synthesis but rather that they act to modulate the basic synthesis rate determined by thyroid hormone levels. Cortisol, however, increases *in vitro* the number of thyroid hormone receptors per cell (GH₃ pituitary cell line) (Perrone *et al.*, 1980; Martial *et al.*, 1977). Corticosterone potentiates to some extent the effects of locally secreted adrenergic substances (Barsegnian and Levine, 1980; Greengard, 1978). A potentiating effect of cortisone on the cAMP-mediated induction of the enzyme tyrosine aminotransferase (TAT) in rat liver was observed by administering a low dose of glucocorticoids (cortisone acetate), which alone does not induce the enzyme (Hoshino *et al.*, 1975; Baxter, 1979). This induction has been shown to be completely dependent upon coordinate action of both the glucocorticoid hormone and cyclic AMP (Ernest *et al.*, 1977). Hydrocortisone seems to be required for *in vitro* maximal induction of casein mRNA by prolactin. This induction was inhibited by progesterone added to organ cultures (Matusik and Rosen, 1978). As it is known that progesterone prevents the activation of glucocorticoid receptors (Rousseau *et al.*, 1973; Baxter, 1979), the data mentioned above might suggest that glucocorticoids via activation of the receptor are involved in modulating the responsiveness to prolactin. For the explanation of that glucocorticoid effect, several alternative mechanisms are possible such as modulation of the number of prolactin-specific receptors in organ cultures of mammary glands mediated by kinase C-like subunit upon activation of steroid receptor and/or via action of that subunit(s) at various levels of translation or transcription causing the facilitation of specific gene expression. It should be noted that the corticosteroids are also involved in modulating the activities of some enzymes involved in the brain cell metabolism of neurotransmitters such as tyrosine and tryptophan hydroxylases and *N*-methyltransferase (Azmitia and McEwen, 1974; McEwen, 1979).

These experimental facts support the predictions of our model. We propose that the physiological role of glucocorticoids (steroids) in the cAMP-mediated enzyme induction might be explained by assuming that one of the steroid receptor's subunits is the catalytic subunit of cAMP-dependent kinase whose concentration is modulated by the activation of steroid receptor.

6.   The Model—Description, Postulates, and Speculations

Within the frame of the proposed model the native steroid receptor—a multimer—would contain in its structure either cAMP-independent kinase or catalytic subunit of cAMP-dependent kinase (RC) (kinase C-like subunit of the receptor) which, released from the holoreceptor, upon binding of the corresponding hormone, will be involved in the cell regulatory phosphorylation/dephosphorylation mechanism. The key events (see Fig. 5) would be the activation of the native holoreceptor and consequent phosphorylation of the R subunit of cAMP-dependent cytoplasmic kinase (R_2 C_2) (Greengard, 1978, 1979; Liu and Greengard, 1976) by the kinase C-like subunit(s) released from the steroid receptor. This would result in the increased capacity of the R subunit to bind cAMP and in enhancement of the kinase dissociation on R and C subunits (Rangel-Aldao and Mandelsohn-Rosen, 1976; Majumder, 1977; Fuller *et al.*, 1978). The phosphorylation of the R subunit of kinase suppresses the rate of reassociation of R and C subunits of kinase (Rangel-Aldao and Mandelsohn-Rosen, 1976) and thereby intensifies the kinase migration (translocation) to nuclei (Costa *et al.*, 1976; Jungmann and Kranias, 1977) and extends the kinase activity time. This event, the activation of kinase by glucocorticoid (steroid) receptor activation, would amplify and potentiate the effects mediated via adenylate cyclase system by peptide hormones and conventional neurotransmitters (Jost and Averner, 1975; Greengard, 1978). According to our model the glucocorticoid (steroid) hormones and peptide hormones as well as neurotransmitters would regulate the activity of the same (key) enzyme, cAMP-dependent protein kinase (type II), but would do so by different mechanisms. The receptor subunit involved in the regulation might, like kinase I (independent of cAMP), represent the catalytic subunit (C) which could be, as claimed by Majumder, converted to the cAMP-dependent (RC) form of kinase after being associated with kinase regulatory subunit (R) to form holoenzyme (RC) cAMP-dependent kinase (Majumder, 1977). The fact that both classes of hormones affect the same substrate, cAMP-dependent protein kinase, provides a possible molecular basis for integrating metabolic and physiologic actions of these two classes of different hormones in the common target cells. Tissue cell-specific differences in the activation of cAMP-dependent protein kinases caused by two classes of hormones might be due on the one hand to the nature of activated protein kinase and on the other hand to the nature of the protein(s) that is (are) phosphorylated and might depend on the cell differentiation. The concepts mentioned above are supported very strongly by experimental facts discussed in Section 4.1.

Within the frame of our model, we thus propose that the activation of steroid receptor serves as an additional regulatory mechanism, which is, via phosphoprotein kinase system, coupled with phosphorylation/dephosphorylation physiological regulatory mechanism operating in target cells (Greengard, 1978, 1979; Robinson *et al.*, 1971). It therefore seems very likely that the steroid hormones in their target cells regulate the rate of phosphorylation of kinase regulatory subunit and translocation of kinase to nuclei and the cytoplasmic "pool" size of cAMP-dependent kinase (Fuller

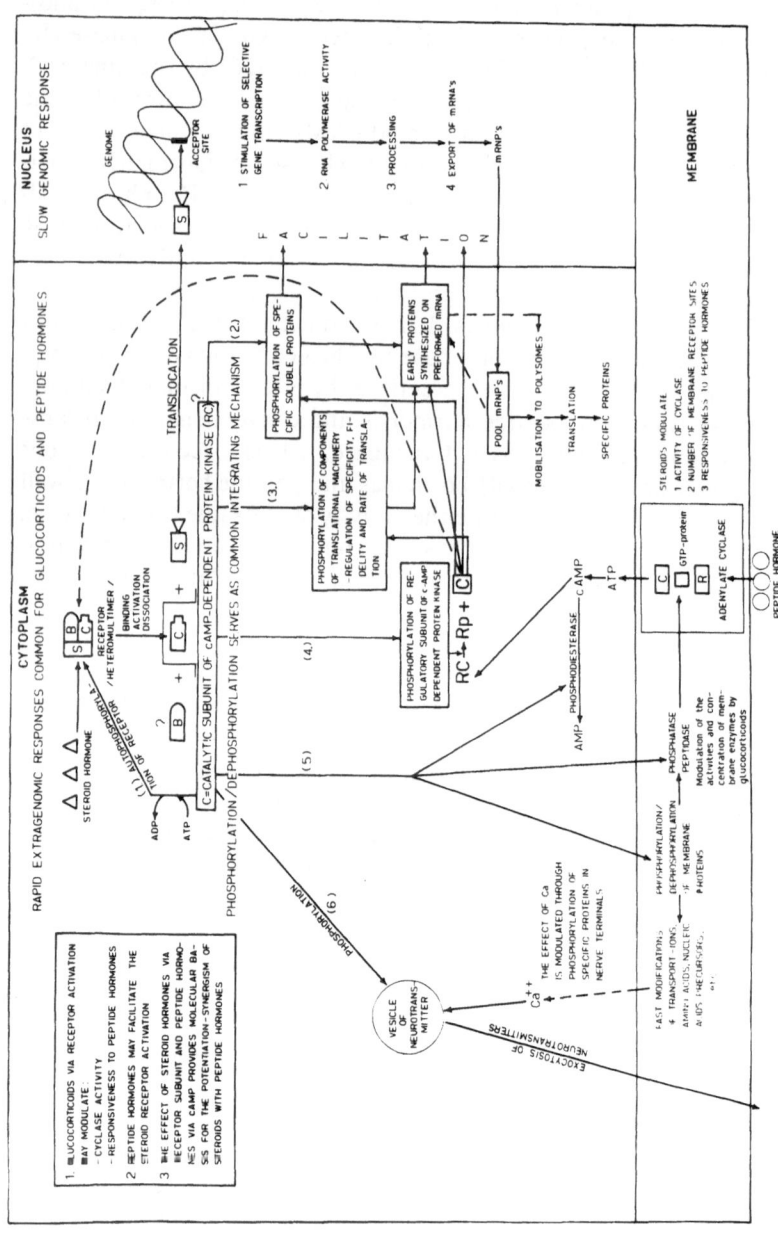

Figure 5. A diagrammatic presentation of the basic concepts of the proposed highly speculative model for integrating actions of steroid and peptide hormones. The concepts of the model have appeared in part in *J. Steroid Biochem.* (1979) **11**:389–400.

et al., 1978; Singhal *et al.,* 1976). This action of steroids on phosphoprotein kinase might provide the molecular basis for integrating the molecular (physiological) actions of steroid and peptide hormones in common target cells (D. T. Kanazir *et al.,* 1979, 1980).

The steroid hormones, via activation of specific receptor, might be, through released kinase C-like subunit, involved in modulating the adenylate cyclase activity and the number of cell membrane receptor sites (Leray *et al.,* 1973; Mukherjee *et al.,* 1975; Wolfe *et al.,* 1976; Catt *et al.,* 1979) for protein and peptide hormones as well as for conventional neurotransmitters.

The experimental results presented in Section 4 suggest a very fundamental role of membrane protease in regulating basal adenylate cyclase activity and its hormonal responsiveness (Reichert, 1962; Richert and Rayan, 1977; Jusit *et al.,* 1976; Alhanaty *et al.,* 1981). In the absence of peptide hormones, a low level of membrane protease activity would maintain basal adenylate cyclase activity. The majority of membrane protease would exist as inactive precursors that could be activated by peptide hormone or neurotransmitter binding to receptors. We propose that in common target cells, kinase C-like subunit, released upon the activation of steroid receptor, might be involved in modulating protease activity by direct phosphorylation/dephosphorylation of protease or through phosphorylation of an inhibitor or regulator of protease activity. Membrane protease may ultimately be responsible for both adenylate cyclase stimulation and the loss of the peptide hormone receptors (peptide hormone–receptor complex) (Catt *et al.,* 1979; Mukherjee *et al.,* 1975). Membrane protease activity may underlie the internalization of peptide hormone–receptor complexes, i.e., hormonal down-regulation of membrane receptors or desensitization of peptide hormone receptors (Catt *et al.,* 1979). Thus, the steroid hormones through activation of respective receptor might, via released kinase C-like subunit, modulate the membrane protease activity (Richert and Rayan, 1977; Reichert, 1962) and thereby regulate in common target cells the adenylate cyclase activity (Fuller *et al.,* 1978), number of surface receptors (Wolfe *et al.,* 1976; Shear *et al.,* 1976), and cell responsiveness to peptide hormones and neurotransmitters (Mukherjee *et al.,* 1975; Catt *et al.,* 1979; Rodbell, 1980; Shear *et al.,* 1976). Thus, for example, it is known that cortisol is involved in the modulation of the TRH receptor loss caused by thyroid hormone in GH_3 pituitary cells (Perrone *et al.,* 1980; Catt *et al.,* 1979). There are data supporting the view that corticoids may regulate the number of plasma membrane peptide hormone receptors and responsiveness to peptide hormones and catecholamines (Wolfe *et al.,* 1976) and the activity of adenylate cyclase (Leray *et al.,* 1973; Nakagawa and Kuriyama, 1976; Greengard, 1979). Glucocorticoids seem also to be involved in specific modifications of concentration and activity of some membrane enzymes (Fuller *et al.*, 1978). Glucocorticoid (steroid) hormones might be involved, via receptor activation, in modulating the membrane protein phosphatase and ATPase activities which seem to be a part of a system of active transport of ions (Kometiani *et al.*, 1978; Bloom, 1979; Greengard, 1979). Membrane protein phosphatase is composed of a regulatory (inhibitory) and a catalytic subunit, and it is known that phosphorylation of the regulatory subunit prevents phosphatase activity, whereas a limited proteolysis may activate phosphoprotein phosphatase (Constantopoulos and Najjar, 1973; Killilea *et al.*, 1979).

AMP-dependent kinase that phosphorylates endogenous membrane substrate as well as exogenous histone is abundant in plasma membrane synaptosomes and membrane synapses (Constantopoulos and Najjar, 1973; Kometiani *et al.*, 1978; Greengard, 1979; Schulman and Greengard, 1978). The membrane protein kinase apparently phosphorylates proteins that modulate the activity of enzymes (Na^+/K^+-dependent magnesium- or calcium-activated ATPase or phosphatase) responsible for active transport of ions in the membrane of excitable cells and synapses (Kometiani *et al.*, 1978; Greengard, 1979).

Among agents that have been implicated in adenylate cyclase modulation is Ca^{2+} but the nature of its involvement remains obscure. Generally, Ca^{2+} has been regarded as an inhibitor of adenylate cyclase. However, low concentrations (micromolar range) have been reported to increase the accumulation of cAMP in intact cells and to stimulate adenylate cyclase (Lefkowitz *et al.*, 1970; Brostrom *et al.*, 1975; Greengard, 1979). Ca^{2+}-dependent regulation is conferred upon brain adenylate cyclase by Ca^{2+}-binding protein (CDR) (Brostrom *et al.*, 1975). This protein seems also to be involved in the regulation of brain cyclic nucleotide phosphodiesterase activity (Kakiuchi and Yamazaki, 1970; Wolff and Brostrom, 1974). A low concentration of Ca^{2+} is required for activation of the adenylate cyclase and the cyclic nucleotide phosphodiesterase, and therefore it is assumed that their activities *in vivo* are regulated by variations of intracellular Ca^{2+} concentrations. It is assumed that the Ca^{2+}-dependent cyclic nucleotide phosphodiesterase preferentially hydrolyzes cGMP rather than cAMP at substrate concentrations in the micromolar range. An attractive hypothesis consistent with this proposal is that Ca^{2+} influx results in the formation of a Ca^{2+}–CDR complex. The ensuing activation of adenylate cyclase would increase the intracellular cAMP whereas concomitant activation of cyclic phosphodiesterase would decrease cGMP and thereby would increase the levels of GMP indispensable to maintain the enhanced activity of adenylate cyclase. CDR dependency may not be a characteristic of all forms of adenylate cyclase. Ca^{2+} activation of adenylate cyclase and cGMP phosphodiesterase may underlie important changes in cAMP and cGMP concentrations and physiological functions. Recent studies indicate that, in addition to brain, many other tissues possess a protein kinase that requires both Ca^{2+} and CDR for the activity, suggesting that this enzyme (kinase) may be of general importance as a mediator of the regulatory action of calcium (Greengard, 1979). It is therefore apparent that calcium plays an important role in physiological homeostasis via modulation of adenylate cyclase and cyclic nucleotide phosphodiesterase activities as well as by regulation of neurotransmission (Nathanson, 1977; Bloom, 1979; Hommerschlag *et al.*, 1975) and secretion of hormones from endocrine glands (Barsegnian and Levine, 1980). The model we propose predicts that steroid hormones may play a regulatory role in the homeostasis of calcium and that the transport of calcium might be modulated by the activation of steroid receptors via the kinase C-like subunit of activated receptors, which might be involved in the regulation of membrane kinase activity (see Fig. 5).

However, our model predicts that peptide hormones and neurotransmitters via cAMP and cAMP-dependent kinase may modulate the state (level) of phosphorylation of steroid receptors and thereby may increase affinity and capacity of receptors to specific steroids and facilitate the activation of receptor. If this is the case, it would mean that peptide hormones and neurotransmitters may potentiate the action of steroid hormones in common target cells. It is now established that several enzymes inducible by glucocorticoids may also be induced by cAMP (Jost and Averner, 1975; Byus and Russell, 1976; Baxter, 1979). For example, in rats, hepatic tyrosine aminotransferase and ornithine decarboxylase can be induced both by glucocorticoids and by cAMP (Lin and Knox, 1957; Greengard, 1969; Beck *et al.*, 1972; Ernest *et al.*, 1977). Furthermore, it was shown that in seminal vesicles, key glucolytic enzymes can be induced either by testosterone or by cAMP (Singhal *et al.*, 1970). These experimental facts are in agreement with the predictions of our model. Namely the role of peptide hormones, neurotransmitters, and/or cAMP is to modulate (facilitate) the activation of steroid receptor via phosphorylation caused by the catalytic subunit of activated cAMP-dependent phosphoprotein kinase, which results in disaggregation of the steroid receptor even in the presence of suboptimal concentrations of steroid hormones. Our model predicts that the activated cAMP-dependent protein kinase might, in common target cells, bring about maximal induction of enzymes even in the presence of low concentrations of glucocorticoids (steroids), which otherwise alone would not induce enzymes. The data mentioned above support strongly the view that the activation of glucocorticoid (steroid) receptors and that of cAMP-dependent phosphoprotein kinase are cooperative potentiating processes in common target cells and key events in the putative integrating mechanism of steroid and peptide hormone actions in common target cells. Both the activation of the steroid receptor and that of cAMP-dependent protein kinase may modulate the translation and induce the synthesis of early proteins on preformed mRNAs—extragenomic effects that seem to be essential for modulating the rate of transcription and for the amplification of genomic expression, i.e., late synthesis of RNAs and specific proteins (late genomic effects). Only early extragenomic plus late genomic effects represent the full course of hormone action in target cells. One of the key postulates of our model is that both glucocorticoid (steroid) and peptide hormones modulate in their common target cells the efficiency of phosphorylation/dephosphorylation regulatory mechanism that underlies the ultrarapid and rapid extragenomic as well as the late (slow) genomic physiological responses of target cells to hormones. This seems to be, also, a dynamic integrator mechanism for steroid and peptide hormone actions in common target cells.

In summary we believe that the putative kinase C-like subunit of the steroid receptor might have several functions as diagrammatically presented in Fig. 5. It might be involved in the following processes: (1) autophosphorylation of the steroid receptor modulating the binding capacity of the receptor; (2) phosphorylation of one or several preformed regulatory proteins involved in some metabolic functions not yet identified; (3) phosphorylation of ribosomal proteins and other components of translational machinery and thereby it might be involved in the regulation of the rate, specificity, and fidelity of the translation as well as in the control of early protein synthesis (translations on the preformed mRNAs) and in the regulation of mobilization of preformed mRNAs from the cytoplasmic pool into the polysomes; (4) phosphorylation of regulatory subunit of cAMP-dependent phosphoprotein kinase, and thereby kinase C-like subunit might be involved in modulating the activation of cAMP-dependent kinase, the rate of dissociation and reassociation of subunits R and C of kinase, and the duration of the activity of C subunit. This event is the key event in integrating the actions of both steroid and peptide hormones. Probably this is the most important function of kinase C-like subunit, as it represents the molecular "link" with the adenylate cyclase system. The receptor kinase C-like subunit might be converted to RC form after being associated with kinase R subunit. (5) Phosphorylation and modulation of membrane protein kinase, phosphatase, and peptidase—activities that seem to be part of the mechanism(s) regulating (i) activity of adenylate cyclase, (ii) membrane permeability (transport of ions and precursors), and (iii) process of internalization of membrane receptor–peptide hormone complexes (downregulation and desensitization of membrane receptors), and thereby steroids may modulate the cell responsiveness to peptide hormones. (6) Finally, this subunit of steroid receptor might be involved in brain cells (common target cells) in the regulation of calcium homeostasis.

However, our model, as presented in Fig. 5, predicts that peptide and protein hormones, as well as conventional neurotransmitters, acting via adenylate cyclase system may potentiate, in common target cells, the action of steroid hormones by modulating the phosphorylation of steroid receptor by C subunit of RC kinase, upon the activation of kinase, i.e., binding of cAMP to kinase. The model, as presented, therefore predicts that peptide hormones, neurotransmitters, and cAMP may play, in common target cells, an important role in modulating the binding capacity of the steroid receptor and the rate of its activation. Consequently, the activation of steroid receptors and that of cAMP-dependent kinase(s) seem to be the key events in the cooperative regulation and integration of the actions of both classes of hormones, as described in detail in this and the two preceding sections.

7. Summary and Prospects for the Future

Much information is yet to be obtained concerning the exact mechanism by which the steroid and peptide hormones carry out their regulatory functions in common target cells such as for example brain cells. The key components of the integrating mechanism seem to be (1) steroid receptor, tissue-specific multimeric structure "encoded" with specific metabolic "code," and (2) phosphoprotein kinase systems. The activation of the steroid receptor by steroid hormones on the one hand and the activation of kinase systems by peptide hormones and neurotransmitters or cAMP on the other hand represent the key events in the operation of the integrating mechanism. These events seem to be coupled in common target cells.

The mechanism(s) by which steroid and peptide hormones induce in common target cells the sequence of biochemical events underlying tissue-specific responsiveness will be an area of active research in the immediate future. New techniques, approaches, ideas, and imagination will stimulate the efforts of investigators in the next decade. One could expect and hope to see the unraveling of many of the mysteries of gene activity (expression) and regulation in eukaryotic cells. We believe that the model proposed here, although highly speculative, will stimulate the imagination and provide a new approach to the study of the role of both steroid and peptide hormones in gene expression in common target cells as well as the study of the molecular nature of integrator mechanism(s) of steroid and peptide hormone actions, especially in brain cells.

ACKNOWLEDGMENTS. Work from the author's laboratory was supported by grants from the Serbian Research Council and Serbian Academy of Sciences and Arts. I am indebted to all of my colleagues working on the project for their invaluable contributions to the studies described herein. I thank Dr. Nevena Ribarac-Stepić, Dr. Radmila Djordjević-Marković, and

Mr. S. Dragana Stefanović for their kind permission to use their recent data not yet published; Mrs. Mirjana Stojanović and Miss Emina Ljutić for preparation of the manuscript; and Mrs. Bojka Kesegić for drawing the figures.

DISCUSSION

PECK: In your model, the first step is the dissociation of a multimeric complex to produce a "steroidophilic" subunit and several other regulatory subunits. Do you have evidence for this first step?

KANAZIR: We have some experimental evidence. Using ion-exchange chromatography (DEAE-cellulose and DEAE-Sephadex A-50 minicolumns) we were able to resolve unactivated form of receptor from activated forms, i.e., following the activation we were able to see two glucocorticoid receptor complexes which differ in molecular weight, charge, and nuclear uptake, from unactivated form as well as among themselves. Only one of those two complexes is able to transfer to nuclei and induce the incorporation of UTP into nuclear RNA. The other form (called IB) does not translocate to the nuclei and stays in the cytoplasm. The regulatory role of this "subunit" is not known and our assumption is that it could be involved in regulation of early, immediate effects which appear following the administration of glucocorticoids. Our results suggest that the early effects are caused by phosphorylation/dephosphorylation regulatory mechanism and since this process is impaired by cortexolone (antihormone) and since it was found that this antihormone is bound to IB "subunit" of receptor, we are inclined to believe that IB "subunit" is involved in phosphorylation/dephosphorylation mechanism. Since under *in vitro* conditions one can see only one form of receptor–hormone complex before activation, and two following the activation, while all three forms appear *in vivo*, it seems very likely that our assumption about subunit structure holds. On the other hand, we have not been able to purify "subunits" and to analyze their detailed structure, because, as it is known, the structure of receptor is very unstable. We are now working on that subject.

PECK: If only sedimentation profiles are employed, one must be cautious. Modification of ionic strength can cause aggregation and dissociation of multiple protein components.

KANAZIR: You are absolutely right; the higher ionic strength can cause aggregation and dissociation of receptor, and can affect sedimentation profiles. For these reasons we did not use sedimentation studies which can provide very useful data about the structure and specially the size of nonactivated receptor and its hypothetical subunits. Unfortunately, our hypothesis was based only on the results obtained using ion-exchange chromatography and gel filtration.

PECK: Are the affinities of the various forms of receptor the same or different? I mean, among the multiple forms that you observe in a given tissue, do they have the same or different affinities for steroid?

KANAZIR: We assume that the three components resolved on DEAE-cellulose are the components of the given tissue-specific receptor. We do know that the capacity of binding is not the same but we did not determine their affinity.

HERBERT: Do you think that the receptor is regenerated after each induction event by *de novo* synthesis? Is there any evidence for turnover of the receptor? (What is turnover time?)

KANAZIR: I cannot give you a direct answer to this question, since we did not study this problem. But I do know that there is early synthesis of receptor 10 min after intraperitoneal hormone administration, and we believe that the pool size of free receptors is low and that the synthesis of receptor is under control of the homologous hormone itself (self-regulation of the pool size of receptor). However, in the literature there is evidence indicating that the regeneration of receptor might take place in the cytoplasm via its phosphorylation. We also observed the increase of phosphorylation of receptor but at the present moment we don't know if this phosphorylation is related to the regeneration of preexisting (used) receptor molecules or to phosphorylation of *de novo* synthesized receptor molecules.

REFERENCES

Alhanaty, E., Patinkin, J., Tauber-Finkelstein and Shaltiel, S., 1981, Degradative inactivation of cyclic AMP-dependent protein kinase by a membranal proteinase is restricted to the free catalytic subunit in its native conformation, *Proc. Natl. Acad. Sci. USA* **78**:3492–3495.

Azmitia, E. C., and McEwen, B. S., 1974, Adrenocortical influence on rat brain tryptophan hydroxylase activity, *Brain Res.* **78**:291–302.

Barker, J. L., 1976, Peptides role in neuronal excitability, *Physiol. Rev.* **56**:435–452.

Barsegnian, G., and Levine, R., 1980, Effect of corticosterone on insulin and glucagon secretion by the isolated perfused rat pancreas, *Endocrinology* **106**:547–552.

Baxter, J. D., 1979, Glucocorticoid hormone action, in *International Encyclopedia of Pharmacology and Therapeutics* (B. L. Bowman, ed.), pp. 67–133, Pergamon Press, Elmsford, N.Y.

Beck, W. T., Bellantone, R. A., and Canellakis, E. S., 1972, The *in vivo* stimulation of rat liver ornithine decarboxylase activity by dibutyryl cyclic adenosine 3′, 5′-monophosphate, theophylline and dexamethasone, *Biochem. Biophys. Res. Commun.* **48**:1649–1655.

Blečić, G., Trajković, D., and Kanazir, D. T., 1980, Early effects of cortisol on protein phosphorylation in rat liver cytosol, *Eur. J. Cell. Biol.* **22**(1):71.

Bloom, E. F., 1979, Cyclic nucleotides in central synaptic function, *Fed. Proc.* **38**:2203–2207.

Bratin, J. W., and Porta Nova, R., 1977, Effect of glucocorticoid levels *in vivo* on growth hormone biosynthesis, *Mol. Cell. Endocrinol.* **15**:19–27.

Brostrom, O. C., Huang, C. Y., Breckenridge, M. B., and Wolff, J. D., 1975, Identification of a calcium-binding protein as a calcium-dependent regulator of brain adenylate cyclase, *Proc. Natl. Acad. Sci. USA* **72**:64–68.

Burridge, M. V., Farmer, S. R., Green, D. C., Henshaw, E. S., and Tata, J. R., 1976, Characterisation of polysomes from *Xenopus laevis* liver synthesising vitellogenin and *in vitro* translation of vitellogenin and albumin mRNA's, *Eur. J. Biochem.* **62**:161–171.

Byus, C. V., and Russell, D. H., 1976, Possible regulations of ornithine decarboxylase activity in the adrenal medulla of the rat by a cAMP dependent mechanism, *Biochem. Pharmacol.* **25**:1595–1600.

Catt, K. J., Harwood, J. P., Aguilera, G., and Dufau, M. L., 1979, Hormonal regulation of peptide receptors and target cell responses, *Nature (London)* **280**:109–116.

Chan, L., Means, R. A., and O'Malley, B. W., 1978, Steroid hormone regulation of specific gene expression, in: *Vitamins and Hormones* (P. L. Munson, E. Diczfalusy, J. Glover, R. E. Olson, eds.), pp. 259–295, Academic Press, New York.

Clayton, J. R., Harwood, J. P., and Catt, K. J., 1979, Gonadotropin-releasing hormone analogue binds to luteal cells and inhibits progesterone production, *Nature (London),* **282**:90–92.

Clemans, M. J., and Korner, A., 1970, Amino acid requirement for the growth hormone stimulation of incorporations of precursors into protein and nucleic acids of liver slices, *Biochem. J.* **119**:629–634.

Constantopoulos, A., and Najjar, V. A., 1973, The activation of adenylate cyclase, *Biochem. Biophys. Res. Commun.* **53**:794–805.

Costa, E., Kurosawa, A., and Guidotti, A., 1976, Activation and nuclear translocation of protein kinase during transsynaptic inductions of tyrosine 3-monooxygenase, *Proc. Natl. Acad. Sci. USA* **73**:1058–1062.

de Kloet, E. R., and McEwen, B. S., 1976, Differences between cytosol receptor complexes with corticosterone and dexamethason in hippocampal tissue from rat brain, *Biochim. Biophys. Acta* **421**:124–133.

Djordjević-Marković, R., 1980, Ph.D. thesis, University of Belgrade.

Djordjević-Marković, R., and Kanazir, D. T., 1980, personal communication.

Endröczi, E., 1975, Mechanism of steroid hormone actions on motivated behaviour, *Prog. Brain Res.* **42**:125–134.

Ernest, M. J., Chen, Ch. L., and Feigelson, P., 1977, Induction of tyrosine aminotransferase synthesis in isolated liver cell suspensions, *J. Biol. Chem.* **252**:6783–6791.

Fang, S., and Liao, S., 1971, Androgen receptors, *J. Biol. Chem.* **246**:16–24.

Fuller, J. I. D., Byus, C. V., and Russell, D. H., 1978, Specific regulation by steroid hormones of the amount of type I cAMP-dependent protein kinase holoenzyme, *Proc. Natl. Acad. Sci. USA* **75**:223–227.

Gerlach, J. L., McEwen, B. S., Pfaff, D. W., Moskowitz, S., Ferin, M., Carmel, F. W., and Zimmerman, E. A., 1976, Cells in regions of rhesus monkey brain and pituitary retain radioactive estradiol, corticosterone and cortisol differentially, *Brain Res.* **103**:603–612.

Greengard, P., 1969, The hormonal regulation of enzymes in prenatal and postnatal rat liver, *Biochem. J.* **115**:19–24.

Greengard, P., 1976, Possible role for cyclic nucleotides and phosphorylated membrane proteins in postsynaptic actions of neurotransmitters, *Nature (London)* **260**:101–108.

Greengard, P., 1978, Phosphorylated proteins as physiological effectors, *Science* **199**:146–152.

Greengard, P., 1979, *Cyclic Nucleotides Phosphorylated Proteins, and Neuronal Function,* Raven Press, New York.

Hauger, R. L., Aguilera, G., and Catt, K. J., 1978, Angiotensin II regulates its receptor sites in the adrenal glomerulosa zone, *Nature (London)* **281**:176–178.

Hinkle, P. M., and Tashijian, A. H., Jr., 1976, Thyrotropin-releasing hormone regulates the number of its own receptors in the GH$_3$ strain of pituitary cells in culture, *Biochemistry* **14**:3845–3851.

Hökfelt, T., Johansson, O., Ljungdahl, A., Lundberg, J. M., and Schultzberg, M., 1980, Peptidergic neurones, *Nature (London)* **284**:515–521.

Hommerschlag, R., Dravid, A. R., and Chiu, A. Y., 1975, Mechanism of axonal transport: A proposed role for calcium ions, *Science* **188**:273–275.

Hoshino, J., Kühne, U., Filjak, B., and Kröger, H., 1975, Potentiating effect of a physiological dose of cortisone acetate on the dibutyryl cyclic AMP-mediated induction of tyrosine aminotransferase in rat liver, *FEBS Lett.* **56**:62–65.

Jensen, E. V., and De Sombre, E. R., 1972, Mechanisms of the female sex hormones, *Annu. Rev. Biochem.* **41**:789–821.

Jessell, T. M., and Kelly, Z. S., 1980, Peptide pathology of mice and men, *Nature (London)* **285**:131–132.

Jost, J. P., and Averner, M., 1975, Gene regulation in mammalian cells: A model for the interaction of steroids and 3',5'-cyclic AMP, *J. Theor. Biol.* **49**:337–344.

Jungblut, P. W., Gaues, J., Hughes, A., Kallweit, E., Sierralta, W., Szendro, P., and Wagner, R. K., 1976, Activation of transcription-regulating proteins by steroids, *J. Steroid Biochem.* **7**:1109–1116.

Jungmann, R. A., and Kranias, E. G., 1977, Nuclear phosphoprotein kinases and the regulation of gene transcription, *Int. J. Biochem.* **8**:819–830.

Jusit, M., Seifert, S., Weiss, E., Haac, R., and Heinrith, P. K., 1976, Isolation and characterisation of a membrane bound proteinase from rat liver, *Arch. Biochem. Biophys.* **177**:355–363.

Kahn, C. R., Naville, D. M., Jr., and Roth, J., 1973, A model of insulin-resistance, insulin-receptor interaction in the obese-hyperglycemic mouse, *J. Biol. Chem.* **384**:244–250.

Kakiuchi, S., and Yamazaki, R., 1970, Calcium dependent phosphodiesterase activity and its activating factor (PAF) from brain, *Biochem. Biophys. Res. Commun.* **41**:1104–1110.

Kanazir, D. T., 1974, The mechanism of steroid hormone action, *Prakt. Akad. Athenon* **10**:463–471.

Kanazir, D. T., Trajković, D., Ribarac-Stepić, N., Popić, D. S., and Metlaš, R., 1978, Cortisol dependent acute metabolic responses in rat liver cells, *J. Steroid Biochem.* **9**:467–476.

Kanazir, D. T., Ribarac-Stepić, N., Trajković, D., Blečić, G., Radojčić, M., Metlaš, R., Stefanović, D., Katan, M., Perišić, O., Popić, S., and Marković-Djordjević, R., 1979, Structure and regulatory functions of cortisol receptor. I. Extragenomic effects dependent on the cortisol receptor activation, *J. Steroid Biochem.* **11**:389–400.

Kanazir, D. T., Stefanović, D., Metlaš, R., Popić, S., Katan, M., Trajković, D., Ribarac-Stepić, N., and Djordjević-Marković, R., 1980, Cortisol mediated control at the translation level in anabolic target liver cells. I. Modulation of translation and ribosomal proteins phosphorylation, in: *Enzyme Regulation and Mechanism of Action,* Vol. 60 (P. Mildner and B. Ries, eds.), pp. 245–257, Pergamon Press, Elmsford, N.Y.

Kanazir, S., Mileusnić, R., Ruždijić, S., and Rakić, M. L., 1980, Translation *in vitro* of rat brain poly (A) mRNA's after cortisol treatment, Abstracts 13th FEBS Meeting, Jerusalem, p. 289.

Killilea, D. S., Mell Gren, L. R., Aylward, H. J., Metieh, M. E., and Lee, E. Y. C., 1979, Studies of the presumptive native forms of phosphorylase phosphatase in liver extracts and their dissociation to a catalytic subunit, *Arch. Biochem. Biophys.* **193**:130–134.

Kometiani, P., Kometiani, Z., and Mikeladze, D., 1978, 3',5'-AMP-dependent protein kinase and membrane ATP-ase of the nerve cell, *Prog. Neurobiol.* **11**:233–247.

Kulkarno, B. S., Netrawali, S. M., Pradhan, D. C., and Sreenivasan, A., 1976, Action of hydrocortisone at a translation level in the rat liver, *Mol. Cell. Endocrinol.* **4**:195–203.

Le Cam, A., and Freychet, P., 1977, Effect of glucocorticoids on amino acid transport in isolated hepatocytes, *Mol. Cell. Endocrinol.* **9**:205–214.

Lefkowitz, R. J., Roth, J., and Pastan, I., 1970, Effects of calcium on ACTH stimulation of the adrenal: Separation of hormone binding from adenyl cyclase activation, *Nature (London)* **228**:864–866.

Leray, F., Chambaut, A. M., Perrenoud, M. L., and Hanoune, J., 1973, Adenylate-cyclase activity of rat liver plasma membranes: Hormonal stimulations and effects of adrenalectomy, *Eur. J. Biochem.* **38**:185–192.

Lesniak, M. A., and Roth, J., 1976, Regulation of receptor concentration by homologous hormone: Effect of human growth hormone on its receptor in IM-9 lymphocytes, *J. Biol. Chem.* **251**:3720–3729.

Liang, T., Castaneda, E., and Liao, S., 1977, Androgen and initiation of protein synthesis in the prostate, *J. Biol. Chem.* **252**:5692–5700.

Liao, S., 1976, in: *Receptors and Mechanisms of Action of Steroid Hormones,* (J. R. Pasqualini, ed.), Part I, Receptors and the mechanism of action of androgens, pp. 159–214, Dekker, New York.

Liao, S., and Williams-Ashman, H. G., 1962, An effect of testosterone on amino acid incorporation by prostatic ribonucleoprotein particles, *Proc. Natl. Acad. Sci. USA* **48**:1956–1964.

Lin, E. C. C., and Knox, W. E., 1957, Adaptation of the rat liver tyrosine α-ketoglutarate transaminase, *Biochim. Biophys. Acta* **26**:85–88.

Liu, A. Y.-C., and Greengard, P., 1976, Regulation by steroid hormones of phosphorylation of specific protein common to several target organs, *Proc. Natl. Acad. Sci. USA* **73**:568–572.

McEwen, B. S., 1979, Steroid hormone interactions with the brain: Cellular and molecular aspects, in: *Reviews of Neurosciences* (D. M. Schneider, ed.), Vol. 4, pp. 1–30, Raven Press, New York.

Majumder, L. G., 1977, Protein kinase activity in mouse mammary carcinoma, *Biochem. Biophys. Res. Commun.* **74**:1140–1145.

Marković, D. R., and Litwack, G., 1980, Activation of liver and kidney glucocorticoid-receptor complexes occurs *in vivo,* *Arch. Biochem. Biophys.* **202**:374–379.

Marković, D. R., Eisen, J. H., Parchman, L. G., Barnet, C. H., and Litwack, G., 1980, Evidence for physiological role of corticosteroid binder IB, *Biochemistry* **19**:4556–4564.

Martial, J. A., Baxter, D. J., Goodman, M. H., and Seeburg, H. P., 1977, Regulation of growth hormone messenger RNA by thyroid and glucocorticoid hormones, *Proc. Natl. Acad. Sci. USA* **74**:1816–1820.

Matusik, R. J., and Rosen, J. M., 1978, Prolactin induction of casein mRNA in organ culture: A model system for studying peptide hormone regulation of gene expression, *J. Biol. Chem.* **253**:2343–2347.

Metlaš, R., Stefanović, D., Popić, S., and Kanazir, D. T., 1979, Cortisol dependent phosphorylation of rat liver ribosomal proteins, *Proc. Serbian Acad. Sci. (Beograd)* **1979**:205–210.

Mileusnić, R., Kanazir, S., and Rakić, M. L., 1980, Tubulin in developing rat brain: Regional distribution and effect of glucocorticords, in: *Circulatory and Developmental Aspects of Brain Metabolism* (M. Spatz, B. B. Mršulja, M. L. Rakić, and W. D. Lust, eds.), pp. 247–260, Plenum Press, New York.

Mukherjee, C., Carong, M., and Lefkowitz, J. R., 1975, Catecholamine-induced subsensitivity of adenylate cyclase associated with loss of β-adrenergic receptor binding sites, *Proc. Natl. Acad. Sci. USA* **72**:1945–1949.

Muller, C. G., and Cowan, A. R., 1974, Current molecular insights into the mechanism of estrogen action, in: *Advances in the Biosciences* (G. Raspe, ed.), pp. 57–73, Pergamon Press, Elmsford, N.Y.

Nakagawa, K., and Kuriyama, K., 1976, Modulating role of pituitary adrenal axis in cerebral metabolism of adenosine 3',5' monophosphate, *J. Neurochem.* **27**:609–612.

Nathanson, J. A., 1977, Cyclic nucleotides and nervous system function, *Physiol. Rev.* **57**:157–256.

O'Malley, B. W., and Means, R. A., 1974, Female steroid hormones and target cell nuclei, *Science* **183**:610–620.

Palmiter, R. D., Christensen, A. K., and Schimke, R. T., 1970, Organization of polysomes from pre-existing ribosomes in chick oviduct by secondary administration of either estradiol or progesterone, *J. Biol. Chem.* **245**:833–845.

Perrone, H. M., Greer, L. T., and Hinkle, M. P., 1980, Relationships between thyroid hormone and glucocorticoid effects in GH3 pituitary cells, *Endocrinology* **106**:600–605.

Posner, B. I., Kelly, P. A., and Friesen, H. G., 1975, Prolactin receptors in rat liver: Possible induction by prolactin, *Science* **188**:57–59.

Radojčić, M., and Kanazir, D. T., 1980, Some aspects of structural organisation of glucocorticoid receptor in rat liver cell cytosol, *Eur. J. Cell. Biol.* **22**(1):70.

Rangel-Aldao, R., and Mandelsohn-Rosen, O., 1976, Dissociation and reassociation of the phosphorylated forms of cAMP-dependent protein kinase from bovine cardiac muscle, *J. Biol. Chem.* **251**:3375–3380.

Reichert, L. E., 1962, Endocrine influences on rat ovarian proteinase activity, *Endocrinology* **70**:697–700.

Ribarac-Stepić, N., Trajković, D., and Kanazir, D. T., 1973, Some properties of cortisol receptor complex isolated from cytoplasm of rat liver and its effect on biosynthesis of nuclear RNA, *Steroids* **22**:155–169.

Ribarac-Stepić, N., Trajković, D., and Kanazir, D. T., 1979a, Cortisol binding components of cytosol receptor in normal and adrenalectomized rats, *Arch. Int. Physiol. Biochim.* **87**:543–555.

Ribarac-Stepić, N., Trajković, D., Stanković, J., and Kanazir, D. T., 1979b, Specific binding of estradiol and cortisol in rat liver and uterine cytosol, *Proc. Serbian Acad. Sci. (Beograd)* **1979**:181–189.

Richert, D. N., and Rayan, J. R., 1977, Protease inhibitors block hormonal activation of adenylate cyclase, *Biochem. Biophys. Res. Commun.* **78**:799–805.

Robinson, G. A., Butcher, R. W., and Sutherland, E. W., 1971, *Cyclic AMP,* Academic Press, New York.

Rodbell, M., 1980, The role of hormone receptors and GTP-regulatory proteins in membrane transduction, *Nature (London)* **284**:17–21.

Rousseau, G. G., 1977, Activity of protein kinases dependent on adenosine 3′ : 5′ monophosphate and its thermostable protein inhibitor in rat hepatoma cells, *Eur. J. Biochem.* **76**:309–316.

Rousseau, G. G., Baxter, J. D., Higgins, S. J., and Tomkins, G. M., 1973, Steroid-induced nuclear binding of glucocorticoid receptors in intact hepatoma cells, *J. Mol. Biol.* **79**:539–554.

Scharpe, R. M., 1980, Extra-pituitary actions of LHRH and its agonists, *Nature (London)* **286**:12–13.

Schrader, W. T., Selesnev, Y., Vedeckis, W. V., and O'Malley, B. W., 1980, Steroid receptor subunit structure, in: *Gene Regulation by Steroid Hormones* (A. K. Roy and J. H. Clark, eds.), Springer-Verlag, Berlin.

Schulman, H., and Greengard, P., 1978, Stimulation of brain membrane protein phosphorylation by calcium and an endogenous heat stable protein, *Nature (London)* **271**:478–479.

Shear, M., Insel, P. A., Melmon, K. L., and Coffino, P., 1976, Agonist-specific refractoriness induced by isoproterenol: Studies with mutant cells, *J. Biol. Chem.* **251**:7572–7576.

Singer, S. J., and Nicolson, G. L., 1972, The fluid mosaic model of the structure of cell membranes, *Science* **175**:720–731.

Singhal, R. L., Vijjayvargiya, R., and Ling, G. M., 1970, Cyclic adenosine monophosphate: Andromimetic action on seminal vesicular enzymes, *Science* **168**:261–263.

Singhal, R. L., Tsang, B. K., and Sutherland, D. J., 1976, in: *Advances in Sex Hormone Research* (R. L. Singhal and J. A. Thomas, eds.), Vol. 2, pp. 325–424, University Park Press, Baltimore.

Smith, L. D., and Ecker, R. E., 1971, The interaction of steroids with *Rana pipiens* oocytes in the induction of maturation, *Dev. Biol.* **25**:232–247.

Stefanović, D., 1980, in preparation.

Stefanović, D., and Kanazir, T. D., 1980, Structural and functional modifications of ribosomes induced by steroids, Abstracts 13th FEBS Meeting, Jerusalem, p. 285.

Tata, J. R., 1976, Growth-promoting actions of peptide hormones, in: *Polypeptide Hormones: Molecular and Cellular Aspects,* CIBA Foundation Symposium 41, pp. 297–308, Elsevier, Amsterdam.

Trajković, D., Ribarac-Stepić, N., and Kanazir, D. T., 1973, The effect of cortisol on biosynthesis and phosphorylation of histones in rat liver, *Arch. Int. Physiol. Biochim.* **81**:859–869.

Trajković, D., Ribarac-Stepić, N., and Kanazir, D. T., 1974, The effect of cortisol on the phosphorylation of rat liver nuclear acidic proteins and role of these proteins in biosynthesis of nuclear RNA, *Arch. Int. Physiol. Biochim.* **82**:211–216.

Trajković, D., Blečić, G., and Kanazir, D. T., 1980, Modulation of glucocorticoid receptor binding capacity by ATP and phosphokinase and phosphatase inhibitors, *Eur. J. Cell. Biol.* **22**(1):71.

Traugh, J. A., and Lundak, T. S., 1978, Phosphorylation of translational initiation factor 3 (eiF3) by cyclic AMP regulated protein kinase, *Biochem. Biophys. Res. Commun.* **83**:379–384.

Tsai, J. S., and Samuels, H. H., 1974, Thyroid hormone action: Stimulation of growth hormone and inhibition of prolactin secretion in cultured GH1 cells, *Biochem. Biophys. Res. Commun.* **59**:420–428.

Tsuboi, S., Kowashima, R., Tomioka, O., Kanata, M., Sakamoto, N., and Fujita, T., 1979, Glucocorticoid binding proteins of human brain cytosol, *Brain Res.* **179**:181–185.

Yu, L.-Y., Tushinski, J. R., and Bancroft, F. C., 1977, Glucocorticoid induction of growth hormone synthesis in a strain of rat pituitary cells, *J. Biol. Chem.* **252**:3870–3875.

Wolfe, B. B., Harden, K. T., and Molinoff, B. R., 1976, Adrenergic receptors in rat liver: Effect of adrenalectomy, *Proc. Natl. Acad. Sci. USA* **73**:1343–1347.

Wolff, D. J., and Brostrom, C. O., 1974, Calcium binding phosphoprotein from pig brain: Identification as a calcium dependent regulator of brain cyclic nucleotide phosphodiesterase, *Arch. Biochem. Biophys.* **163**:349–358.

11

Effects of Estrogens on Receptor "Nucleotropy" and "Activation"

Walter Sierralta, Jürgen Gaues, Linda Görlich, Alun Hughes, Erhard Kallweit, Janet Kielhorn, Melvin Little, Itzhak Maschler, Sharon McCann, Heinrich H. D. Meyer, Fritz Parl, Gary C. Rosenfeld, Grant Stone, Pablo Szendro, Cecilia Terán, Anne J. Truitt, Rüdiger K. Wagner, and Peter W. Jungblut

1. Introduction

The experiments to be discussed in this chapter have been performed with steroid-sensitive uterine cells. The question then immediately arises whether steroid-sensitive cells of other organs, as for example the brain, display the same basic molecular mechanisms in reacting to hormonal stimuli. Although backed only by circumstantial evidence, the answer is soundly affirmative: gonadal steroids—and all other steroid hormones—act intracellularly by interacting with specific proteins called "receptor." The complexes formed are noncovalent, but surprisingly stable at low temperatures. Equilibrium constants of association are in the 10^9 M^{-1} range. Receptors bind "their" steroid with high specificity. These properties are shown by all

Walter Sierralta, Jürgen Gaues, Linda Görlich, Alun Hughes, Erhard Kallweit, Janet Kielhorn, Melvin Little, Itzhak Maschler, Sharon McCann, Heinrich H. D. Meyer, Fritz Parl, Gary C. Rosenfeld, Grant Stone, Pablo Szendro, Cecilia Terán, Anne J. Truitt, Rüdiger K. Wagner, and Peter W. Jungblut • Max-Planck-Institut für experimentelle Endokrinologie, D-3000 Hannover 61, FRG.

receptors, independent of their tissue of origin. Furthermore, the accumulation of steroid-receptor complexes in the nucleus, the site of action, is common to all target cells.

It is understandable that in the past, emphasis has been placed on the steroid as the active component, while a merely supportive role was assigned to the receptor. The noncovalent attachment, which by "heat exchange" is reversible *in vitro,* also suggested repeated usage of the intracellular steroid "carrier," hence a slow turnover rate of the protein and a steady concentration level. Neither of these assumptions was correct. The essential element is the receptor, not the steroid, which "activates" the protein by inducing favorable conformational changes. Receptors do not shuttle between cytoplasm and nucleus: They are degraded after acting. Their action consists in an enhancement of transcription, the extent of which is not only receptor- but also cell-specific, meaning that the scale of messages transcribed with the aid of an individual receptor depends on cell differentiation. At least one receptor governs the transcription of its own message, the estradiol-receptive, transcription-regulating protein (Jungblut *et al.,* 1976). The study of its turnover under controlled physiological conditions is thus a promising approach for unveiling the unknown.

2. Subcellular Origins and Properties of Uterine Estradiol Receptors

Estradiol receptors have been found in three homogenate fractions of the uterus: The soluble phase ("cytosol") (Toft and Gorski, 1966), the nuclei (Jungblut *et al.,* 1967), and the microsomal particles (Little *et al.,* 1972). Of these, only one can be appropriately defined.

Nuclei can be isolated to a high degree of purity and, after removing the outer layer of the nuclear envelope, allow for a quantitative assessment of their receptor content, when optimal extraction procedures are applied.

The particle-free supernatant of homogenates is of course not identical with the cytosol, but rather represents a mixture of extra- and intracellular fluids, part of the latter possibly arising from cytoplasmic containments, ruptured during homogenization.

Microsomes are fragments of rough and smooth endoplasmic reticulum and other membranous structures of the cell and cannot be quantitatively recovered from homogenates.

This hampers distribution studies, which, for the "cytosol" and the "microsomes," have to rely on reference markers. Fortunately, some physicochemical differences between the receptors originating from the three subcellular sites are helpful discriminators.

The differences do not concern the steroid-binding site. Judging from their indistinguishably high affinity and steroid-binding specificity, it must be identical for the cytosol, the nuclear, and the microsomal estradiol receptor (Jungblut *et al.*, 1979). Rather, they are due to the absence or presence of structural features adjacent to a common core, resulting in characteristic sedimentation velocities and electrophoretic mobilities.

(It ought to be mentioned here that the early observed difference between the 9 S cytosol receptor and the 5 S nuclear receptor is likely a puzzling, still poorly understood artifact. The former can be produced by homogenizing tissue with low-ionic-strength buffer; the latter is extracted from nuclei by salt. Subsequent addition of salt to freshly prepared low-ionic-strength cytosols disperses the 9 S or larger receptor aggregates to a 4 S subunit.)

Since the sedimentation velocity of the microsomal receptor is lowest of all and also its electrophoretic mobility at pH 8.2 is markedly reduced as compared to that of the cytosol and the nuclear receptor, it could be the common protein core awaiting some unknown posttranslational finishing. This "primordial" receptor is not only capable of steroid binding, but already contains the structures required for a phenomenon interpreted as receptor "activation."

When estradiol-containing microsomal extracts are warmed, the macromolecular-bound labeled steroid is gradually shifted from the 3.5 S to the 4.5 S position in density gradient analyses (left-hand panel Fig. 1). The

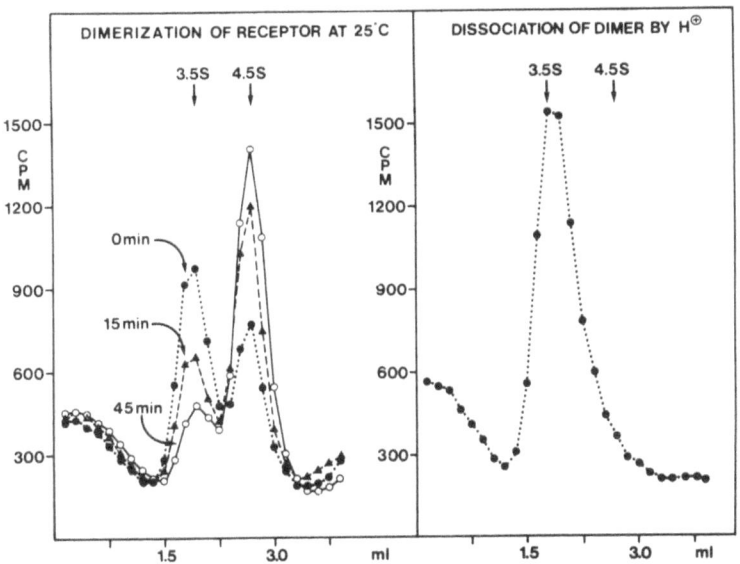

Figure 1. Estradiol-facilitated conversion of microsomal 3.5 S ⇌ 4.5 S receptor.

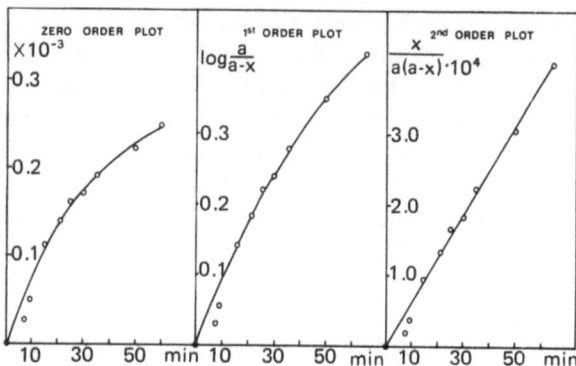

Figure 2. Kinetics of estradiol-facilitated conversion of microsomal 3.5 S → 4.5 S receptor.

transition follows second-order kinetics for dimerization (Fig. 2) (Little *et al.*, 1973). Stability studies (Fig. 3) indicate the participation of histidyl and tyrosyl residues, provided their "microscopic" pKs in the polypeptide are close to those of the free amino acids. While the decrease of 4.5 S receptor concentration above pH 10 is not matched by an increase of the 3.5 S complex (progressing denaturation), the dimer can be reversibly dissociated between pH 7.0 and 6.5 (right-hand panel Fig. 1). The back-to-back/head-to-toe model of the receptor dimer (Fig. 4) is fictitious. Its actual shape might be more elongated, should the polar groups exist in

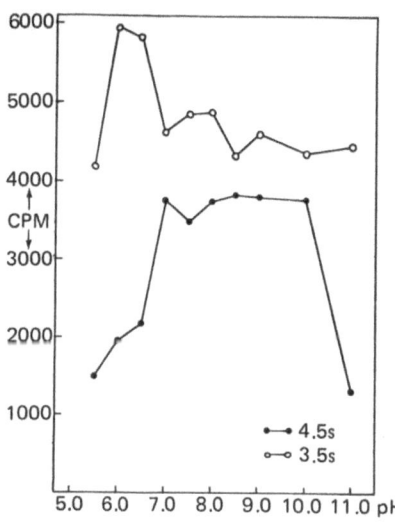

Figure 3. Stability of 3.5 S and 4.5 S microsomal estradiol receptor complexes to pH.

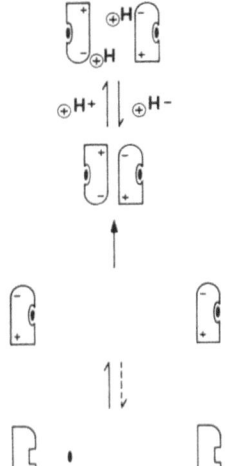

Figure 4. Model of steroid-facilitated receptor dimerization/dissociation.

closer vicinity, and the arrangement could be front-to-front/head-to-toe instead. For steric reasons, however, all possible forms of the homodimer must feature inversely arranged surface areas of its constituent (identical) monomers.

With some more difficulty than with the microsomal receptor, which lacks the tendency of forming uncontrollable homo- and heteroaggregates, the same dimerization/dissociation processes are demonstrable for the cytosol receptor (Fig. 5) (Little *et al.*, 1975).

Receptor extracted from highly purified nuclei after *in vivo* adminis-

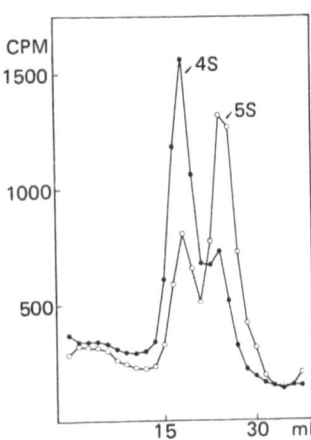

Figure 5. Estradiol-facilitated dimerization of cytosol 4 S → 5 S receptor: ●—●: unheated extract; ○—○: heated extract

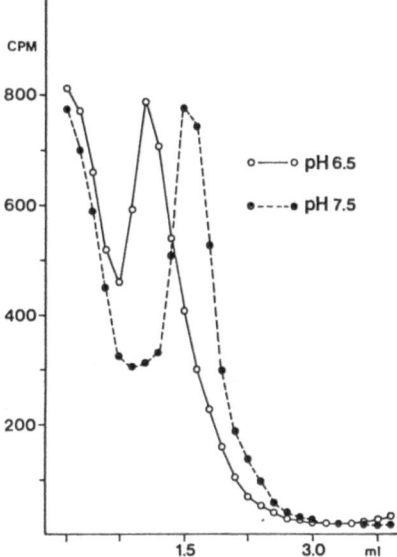

Figure 6. Dissociation of nuclear 5 S estradiol receptor dimer by protonation.

tration of estradiol is apparently a dimer, indistinguishable from the *in vitro*-produced dimer of the dispersed cytosol receptor, as its reversible dissociation (Jungblut *et al.*, *1978*) by proton addition and withdrawal suggests (Fig. 6).

Whatever accounts for the difference between the microsomal and the cytosol/nuclear receptors then is beyond the two functions of steroid binding and dimer formation. Besides those already mentioned, some more distinctive properties are listed in Table I. They further add to the apparent identity of the cytosol and the nuclear receptor and, considering the total absence of microsomal-type receptor in purified nuclei (Fig. 7), allow for

Table I. Properties of Estrogen Receptors

	Site of extraction		
	"Microsomes"	"Cytosol"	Nucleus
K_A estradiol	$\sim 5 \times 10^9 \ M^{-1}$	$\sim 5 \times 10^9 \ M^{-1}$	$\sim 5 \times 10^9 \ M^{-1}$
Velocity (S)	$3.5 \rightleftharpoons 4.5$ (low salt)	$4 \rightleftharpoons 5$ (high salt)	$5 \rightleftharpoons 4$ (high salt)
Mobility (pH 8.2)	$\ominus$	$\oplus$	$\oplus$
Heparin-Sepharose	*not* adsorbed	adsorbed	adsorbed
Protamine chloride	*not* precipitated	precipitated	precipitated

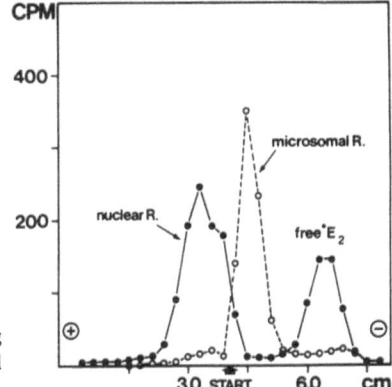

Figure 7. Comparison of mobilities demonstrating absence of microsomal-type receptor in purified and stripped uterine nuclei.

a tentative functional assignment to the "finishing" entity: a participation in either nuclear uptake or retention or in both.

The removal of this entity from cytosol and nuclear receptor was a chance observation. Attempts to disperse suspected glycosaminoglycan-receptor aggregates in cytosol by hyaluronidase treatment produced "microsomal" receptor. The same happened with receptor from purified

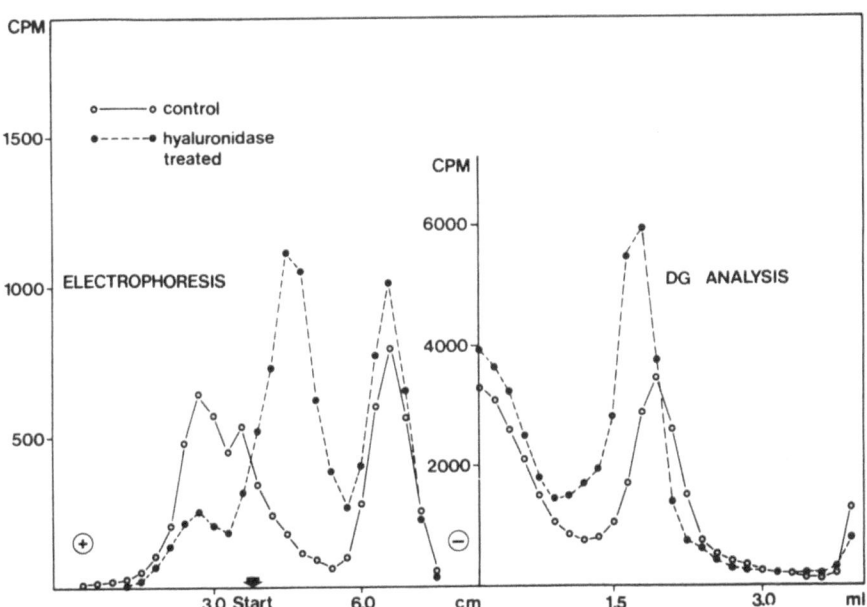

Figure 8. Effect of hyaluronidase on extractability, electrophoretic mobility, and sedimentation velocity of nuclear receptor.

Table II. Improvement of Receptor Yields from Pig Uterus
Nuclei by Hyaluronidase

| | Number of binding sites/nucleus extracted by | |
Ablative treatment	Salt only	Salt + hyaluronidase
Ovariectomy	2690	12,630
	3010	8,420
	2910	9,220
	2170	11,100
	2230	8,500
Ovariectomy + adrenalectomy	2230	4,760

nuclei (Fig. 8) (Jungblut *et al.*, 1980) accompanied by three- to fivefold
higher receptor yields (Table II). Since the "clipping" proceeds both with
testes hyaluronidase and with "eliminase" from *Streptomyces hyaluronol-
yticus,* a contamination of both endoglycosidases with an identical, highly
specific endopeptidase-"clipase" is a remote explanation. The tryptic fission
product of cytosol/nuclear receptor, moreover, differs from that of hyaluronidase action. Whether "finished" receptor is indeed a hitherto unknown
intracellular proteoglycan traveling along unfamiliar routes remains to be
proven. The data available on this peculiar macromolecule are interpreted
in Fig. 9.

3. Reactions of Uterine Cells to Pulse-Administered Estradiol

The discovery of the small and less acidic microsomal estradiol receptor prompted the development of an experimental design for delineating its
physiological role (Hughes *et al.*, 1977). We use German Landrace pigs
for two reasons: (1) for securing sufficiently large quantities of tissue and
(2) because the necessary manipulations can be carried out on wake,
trained animals, which minimizes the complicating interference of adrenal-
derived estradiol. In brief: 3-month-old (premature) animals are ovariectomized and the left horn is detached from the corpus uteri. A silastic tub-
ing containing a suspension of crystalline estradiol is subcutaneously
implanted 4–6 days prior to the operation and left *in situ* until 8–10 days
before the experiment, 2 months later. This treatment facilitates the operation, reduces the formation of adhesions, and helps to standardize the size
of the uterus. The chronically castrated pigs are then trained for tolerating
a simulated insemination, which naturally proceeds in the sow via tran-
scervical filling of the uterus with some 200 ml of ejaculate by the boar. In

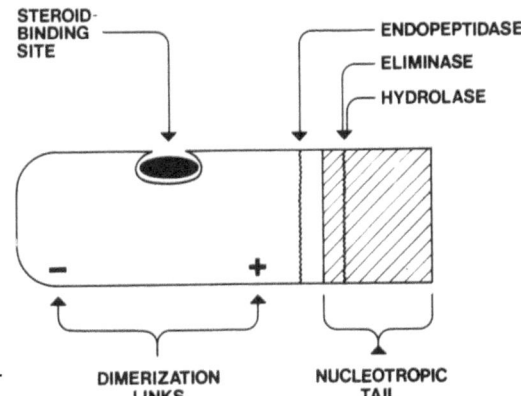

Figure 9. Model of "finished" estradiol receptor monomer.

a typical experiment, instead, 20 ml of a 1×10^{-6} M solution of (unlabeled) estradiol in buffered saline is injected into the open horn. The animals are stunned and exsanguinated at various times after the injection, treated and untreated horns are excised, chilled, and processed.

The macroscopic effect seen at 90 min after the estradiol injection is shown in Fig. 10. This so-called water imbibition of the treated horn, still missing at 60 min postinjection, is a late phenomenon as compared to the swiftness of other events. It is preceded by a rapid uptake of estradiol from the injectant, the steroid concentration of which drops within minutes below the 1‰ level. Most of the estradiol passes through the uterine wall into the systemic circulation and is metabolized before reaching the "blind" control horn. We have thus the benefit of a pulse-exposure of target cells to a hormonal stimulus under optimal physiological conditions, to be compared with untreated cells of the same "walking incubator."

Although we have been improving the physiological technique over the past 7 years to almost perfection, the originally intended purpose still suffers from biochemical shortcomings. As already mentioned, microsomes cannot be quantitatively harvested from homogenates, nor can we safely extract all microsomal receptor from the obtained particulate fraction. If, however, receptor extracted by standardized procedures and referred to protein concentration and marker enzyme activity should be representative, the "basic" microsomal 3.5 S estradiol receptor qualifies as the precursor of the cytosol/nuclear receptor. Its "specific" concentration rises within 30–60 min after an estradiol pulse to over fourfold.

Compared to this problem, the pursuit of estradiol and receptor uptake by the nucleus is an easy task, provided the nuclei analyzed have been carefully purified and stripped off the outer layer of the nuclear envelope, which

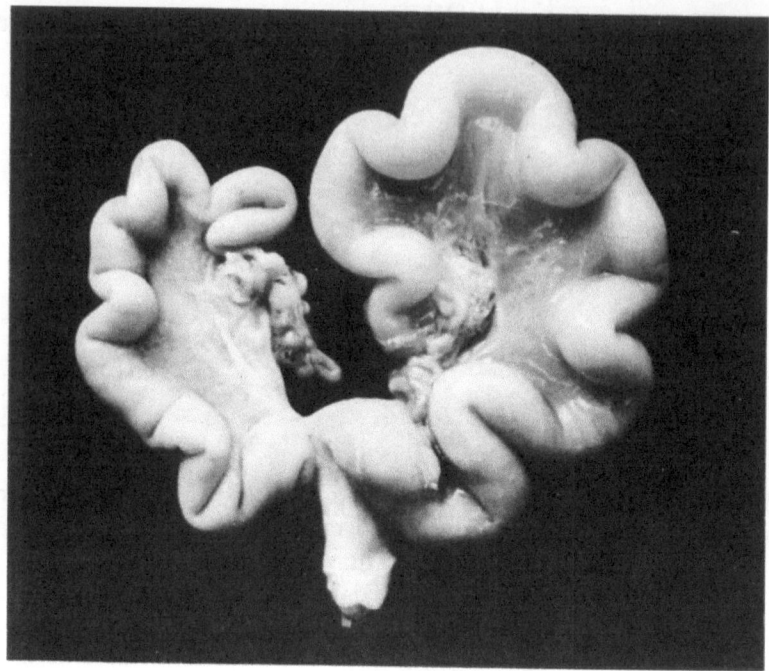

Figure 10. Uterus of chronically castrated pig 90 min after injection of 20 ml of 1 × 10⁻⁶ M
estradiol into open horn (right side of picture.)

is rough ER and contains microsomal receptor. (We contend that experiments with crude, $600xg$ nuclear sediments are hazardous. The sediments trap up to 60% of the microsomal fraction of uterine homogenates and exchange experiments—"cold" for "hot" estradiol with concomitant receptor extraction—are unavoidably confused by the admixture of extranuclear receptors.) We split the nuclei isolated at various times after estradiol injection for parallel receptor and estradiol assays. The extracted receptor is labeled by exchange with tritiated estradiol at elevated temperature and quantitated by density gradient centrifugation or electrophoresis; the accumulated ("cold") estradiol is measured in the other aliquot by radioimmunoassay. Both values are expressed per nucleus.

    The time course (Fig. 11) shows a sudden and equimolar increase of estradiol and receptor content in the nucleus proper, albeit arising from different starting levels (Jungblut *et al.,* 1979). This is followed by a slow and parallel decline of both concentrations until some 8 hr after the pulse, whereafter estradiol continues fading away from the nucleus, while the receptor content levels off to the starting concentration in the vicinity of 10,000/nucleus. The increase in nuclear receptor content coincides with the disappearance of receptor from the cytosol (upper panel Fig. 12). (The cytosol receptor concentration per cell is calculated from the DNA concentration of the homogenate divided by the DNA content per nucleus $= 7 \times 10^{-12}$ g.) The timing of the shift and the close absolute figures are convincing evidence for an estradiol-mediated receptor translocation between the two compartments of the cell. This translocation is irreversible. Receptor leaving the nucleus does not reappear as a steroid-binding entity in the cytosol. The late rise in cytosol receptor concentration, which is preceded by that of its presumable microsomal precursor, can be inhibited by (locally administered) actinomycin D or puromycin and therefore depends on *de novo* synthesis.

It is worth noting that major morphological responses of the stimulated endometrial epithelium are missing until the eighth hour after estradiol injection and are then sequentially patterned (lower panel Fig. 12) (Jungblut *et al.,* 1979). At this time, nuclear estradiol and receptor contents have returned to control levels. The requirement of a prolonged presence of active steroid–receptor complexes for promoting the late events including cell division cannot be concluded from these data. Rather, it appears that the hormone–receptor complex merely has a trigger-pulling function.

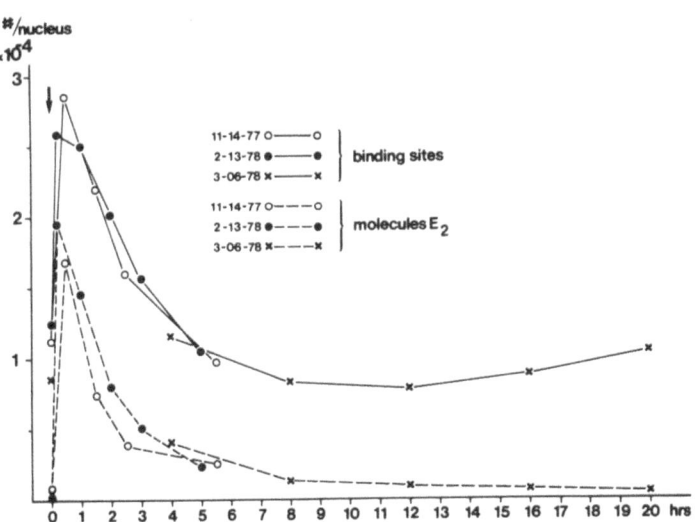

Figure 11. Comparison of nuclear estradiol and receptor contents.

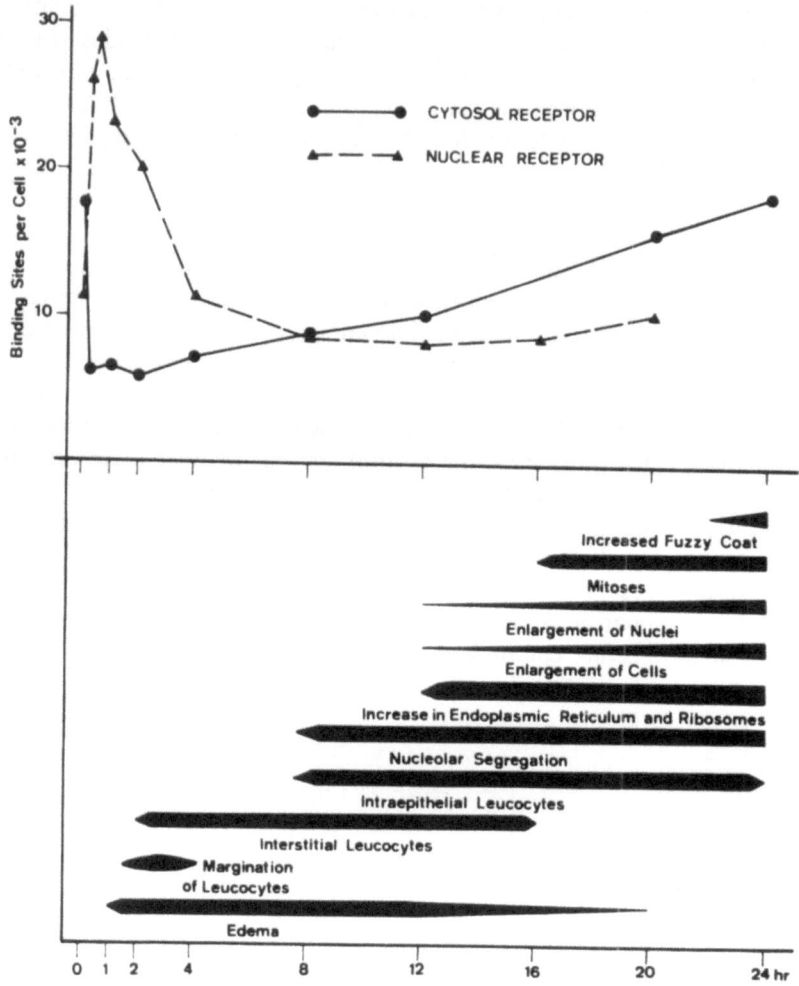

Figure 12. Correlation of biochemical and morphological events after estradiol pulse.

4. Receptor "Nucleotropy" and "Activation"

Estradiol receptors are synthesized in the cytoplasm and act inside the nucleus. How do they get there? In which way are they conditioned for nuclear uptake? The *in vitro* formation of nuclear-type 5 S receptor from 4 S cytosol receptor suggested that the underlying dimerization would "activate" the receptor for its "pas de deux" to the nucleus. The estradiol-facilitated dimerization of the 3.5 S microsomal receptor (which is never found inside the nucleus) did cast some doubt on this notion, although logistical restraint might be the physiological cause. There was, however, one observation at hand for testing the issue: estrone, in contrast to estradiol, does not facilitate the formation of receptor dimers; estriol is somewhat less effective than estradiol. The relative potencies of the three estrogens are equally demonstrable for the microsomal receptor (Fig. 13) and—with adapted experimental conditions—for the cytosol receptor. If then only receptor twins should be allowed into the nucleus, estrone would be an ineffective usher. It is by no means! Estrone, like estradiol, promotes a rapid translocation of receptor from the cytosol into the nucleus (upper panel Fig. 14). But the estrone-mediated translocation does not last; it is quickly followed by a receptor reshuttle. We must thus conclude that monomeric receptor can enter the nucleus and that the enhancement of its "nucleotropy" is a process different from dimerization. The nuclear retention of receptor, however, appears to be related to the estrogen involved. It is longest with estradiol, shorter if estriol is administered (not shown), and missing after estrone. This order is the same as that for facilitating receptor dimerization and, provided retention time correlates to activity, receptor dimers should then be the active element in the nucleus.

Two aspects of the estrone experiments remain to be discussed: first, does estrone—and all other estrogens—enhance receptor "nucleotropy" by unfolding a "sticky tail," like that portion of the cytosol (and nuclear) receptor, which can be removed by hyaluronidase (converting it to microsomal-type receptor)? The thought is intriguing. All but 100 sec would be needed for draining freely diffusible receptor from the cytoplasm to the

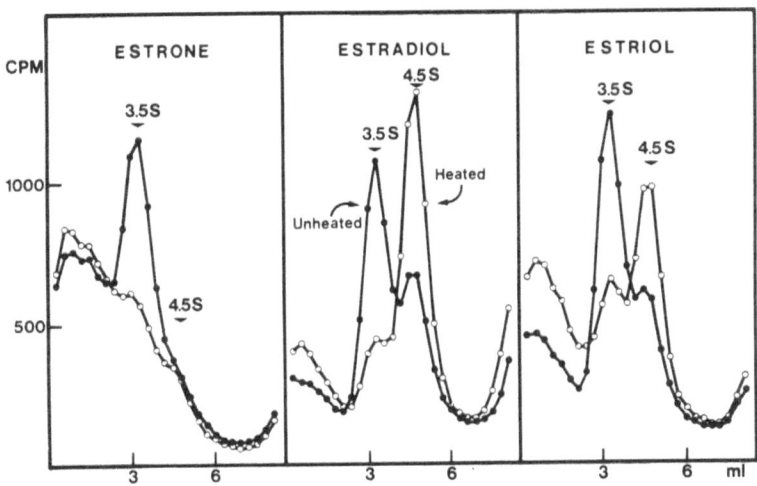

Figure 13. Relative potencies of estrone, estradiol, and estriol in facilitating receptor dimerization.

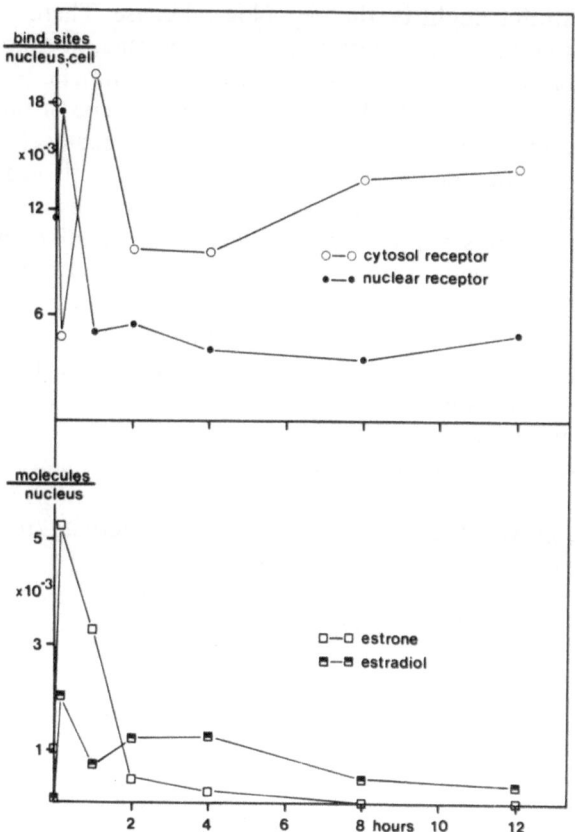

Figure 14. Reshuttle of receptor after estrone-facilitated translocation.

nucleus by postulating an effector-enhanced nucleotropy. The model resembles in essence the interactions of hormones and antigens with plasma membrane receptors! Occurring within the confinement of the cytoplasm, it could practically achieve a unidirectional flux.

The second aspect concerns the fate of receptor, which is quickly "replenished" into the cytosol after estrone-facilitated translocation into the nucleus, but then vanishes before the onset of receptor synthesis, brought about by the complicating action of estradiol-1 1β-ol-oxidoreductase (lower panel Fig. 14). The disappearance of the nucleus-released receptor might well be considered the second limb of a unidirectional migration, ending in receptor degradation. The site at which the degradation takes place is as yet unknown. Some observations indicate an involvement of lysosomes, which in porcine endometrial cells also contain a highly active estradiol-17β-ol-dehydrogenase (Entenmann *et al.,* 1980). Unless the fate of receptor translocated with estrone support is exceptional, the nucleus itself does not "process" the receptor to non-steroid-binding fragments. Whether it "conditions" the receptor for extranuclear degradation remains to be elucidated.

5. Ovarian- and Steroid-Independent Fluctuations of Uterine Estradiol Receptor Levels

We and others have been plagued by seasonal variations of receptor levels in immature animals and postmenopausal women (Hughes *et al.,* 1976). These can be attributed to a changing influence of the adrenal cortex, which in summertime and in stress releases increased quantities of estrogen precursors. A not too pronounced circadian rhythm of rat uterine estradiol receptor concentrations should have the same cause. But this explanation fails for the 7- to 12-day up-and-down periods of cytosol receptor concentration in the uteri of ovariectomized/hypophysectomized rats. We suspected a slow turnover of the receptor in the absence of steroid, likely without the implication of a nuclear passage. It was not until the pulse-administration experiments with pigs had arrived at the pure nuclei perfection, that we reconsidered. Like everyone else, we clung to the idea that possibly the steroid, but not the receptor, could independently enter the nucleus. Considering the lability of the receptor and the excellent extractability of the stable, noncovalently bound steroid, each analysis of nuclei should then have shown an apparent excess of steroid. The failure to meet this expectation has already been stated. In retrospect, the later discovered difference between receptor "nucleotropy" and the formation of active dimers might have appealed as a handy explanation. "Finished" cytosol receptor could have a somewhat less effective "sticky tail" without the steroid and gradually enter the nucleus. It would remain there for some time as "unfilled site," in the inactive, monomeric form.

This retrospective hypothesis is only in part a true reflection. Uterine nuclei of ovariectomized and successfully adrenalectomized pigs are completely devoid of estradiol (Jungblut *et al.,* 1978). They do contain some monomeric receptor, but, as late as 33 days after removing the second adrenal, the majority of receptor present is in its dimeric form (Fig. 15). Both processes, nuclear entrance and dimerization, are therefore essentially steroid-independent. It still could be disputed whether this is sufficient evidence for a steroid-independent action of the receptor. But, since the estradiol receptor is good only for one nuclear passage and part of its action is

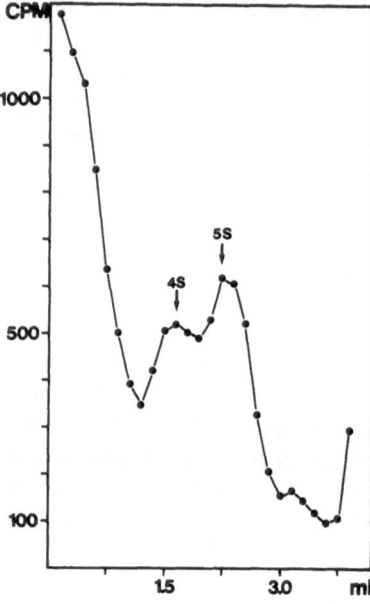

Figure 15. Presence of monomer and dimer estradiol receptor in uterine nuclei from ovariectomized/ adrenalectomized pigs.

the provision of its own message for resynthesis, admittedly proven only with steroidal support, an entirely different mechanism in the absence of estradiol can hardly be envisaged. A quantitative difference in receptor turnover, and all related processes, rather than a qualitative one in the presence or absence of steroid is the more reasonable alternative.

6. The Basic Mechanism of Estradiol Receptor Action

Recognizing the receptor protein as the essential element does not curtail the importance of the steroid. Called for only is substituting the "all or nothing" theorem by the less absolute "very much or very little" in explaining the physiological reality. (This might impress as a hair-splitting exercise, but there is at least one situation possible, in which the steroid-independent turnover of the estradiol and other steroid receptors could exceed the leisurely basal rate of cellular activity: the cancerous growth of "derailed" target cells. Should they retain *receptor* dependency as an essential mechanism, withdrawal of the steroidal conformation catalysts might not suffice for a total blockade, which would require an effective "poisoning" of the transcription-regulating proteins, called receptor.)

Our present view of estradiol/receptor action is outlined in Fig. 16. The diagram emphasizes the unidirectional voyage of the receptor, many stations of which are still in a haze. The sites of core-protein synthesis and receptor "finishing" remain to be pinpointed, as is the site of receptor degradation. The structure of the "nucleotropic tail" must be identified before the penetration of the nuclear envelope can be understood. Where exactly inside the nucleus the receptor lodges for action is unknown. The dimer structure of the receptor, however, allows for a reasonable assumption. Its toppled Janus-head features can only align to stretches of double-stranded DNA with fitting nucleotide sequences or to other inversely arranged macromolecular twin structures, hitherto unknown. Although our old hypothesis, that the receptor dimer would act as an unwinding protein for transcription, retains its appeal, direct proof is hard to come by. The "sticky" nucleotropic tail complicates experiments with isolated systems, which cannot copy the intricate arrangement of macromolecules in the intact nucleus whereby unspecific adsorptions are obviously avoided, as the shedding of (monomeric) estrone receptor complex shows.

There are many indications that the estrogen receptor's mode of action exemplifies those of all other steroid hormone receptors. Understandably, their sites of interaction and hence the products are different. At least one of them, the progestagen receptor, does not enhance the transcription of its own message, which is taken care of by the estrogen receptor. This is indicated by ③ in the diagram; ① = cytosol estradiol receptor and ② = estradiol in the cell nucleus are additional parameters we recommend be analyzed as criteria of persisting hormone-responsiveness in mammary cancer biopsies. (Wavefunctions cannot be described by a single point!)

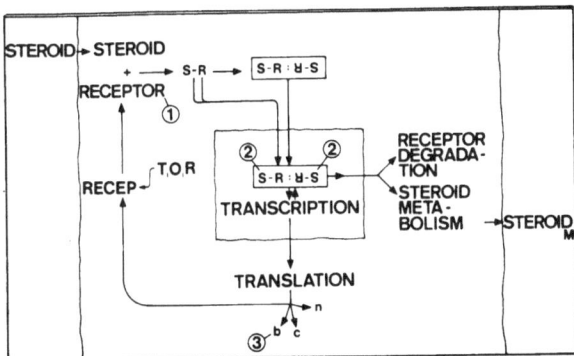

Figure 16. Basic mechanism of estradiol/receptor action.

7. How to Succeed in Mapping the Steroid-Sensitive Areas of the Brain

The last remarks connect to the difficulties encountered in mapping the steroid-sensitive areas of the brain. The plight of the neurophysiologist is twofold: the scarcity of steroid-responsive neural cells and their variable states of response. The collection of sufficient amounts of topically well-defined cells already is a painstaking exercise, not to speak of the various biochemical microanalyses necessary for monitoring the response situation.

What is the alternative? Recalling that the receptor, but not its specific steroid-"catalyst," can enter the cell nucleus independently, Stumpf's technique of injecting labeled hormone and tracking it by autoradiography is of convincing safety (Stumpf, 1968). Whether the silver grains mirror the injected steroid or some active metabolite can with some certainty be checked on other, more abundant target cells, for as long as hypothetical *cell*-specific steroid receptors do not materialize. Proper timing for optimal accumulation must be taken care of by time-course experiments.

The Stumpf technique, however, can only identify "resting" target cells, containing high concentrations of cytoplasmic receptor, which swiftly takes the marker steroid along into the nucleus. It "blacks out" those cells, which have recently been exposed to endogenous steroid and are in a state of "saturation" or "recovery." The immunohistochemical procedure of Nenci *et al.,* (1976) for locally arresting steroids after their release from denatured receptor could fill the gap left by "occupied" nuclei.

Used in combination under well-defined physiological conditions, the two methods certainly will add to our knowledge. They are not only first choice for exploring the topography of steroid-responsive perikarya, but also could reveal whether all cells of the organism containing identical steroid-receptor systems are "in phase" or not.

REFERENCES

Jungblut, P. W., Gaues, J., Hughes, A., Kallweit, E., Sierralta, W., Szendro, P., and Wagner, R. K., 1976, Activation of transcription-regulating proteins by steroids, *J. Steroid Biochem.* 7:1109–1116.

Toft, D. O., and Gorski, J., 1966, A receptor molecule for estrogens: Isolation from rat uterus and preliminary identification, *Proc. Natl. Acad. Sci. USA* 55:1574–1581.

Jungblut, P. W., Hätzel, I., DeSombre, E., and Jensen, E. V., 1967, Die oestrogenbindenden Prinzipien der Erfolgsorgane, in: *18th Coll. Ges. phys. Chemie* (Mosbach), pp. 58–86, Springer-Verlag, Berlin.

Little, M., Rosenfeld, G. C., and Jungblut, P. W., 1972, Cytoplasmic estradiol receptors associated with the microsomal fraction of pig uterus, *Hoppe-Seyler's Z. Physiol. Chem.* 353:231–242.

Jungblut, P. W., Hughes, A., Gaues, J., Kallweit, E., Maschler, I., Parl, F., Sierralta, W., Szendro, P., and Wagner, R. K., 1979, Mechanisms involved in the regulation of steroid receptor levels, *J. Steroid Biochem.* **11**:273–278.

Little, M., Szendro, P., and Jungblut, P. W., 1973, Hormone-mediated dimerization of microsomal estradiol receptor, *Hoppe-Seyler's Z. Physiol. Chem.* **354**:1599–1611.

Little, M., Szendro, P., Terán, C., Hughes, A., and Jungblut, P. W., 1975, Biosynthesis and transformation of microsomal and cytosol estradiol receptors, *J. Steroid Biochem.* **6**:493–500.

Jungblut, P. W., Kallweit, E., Sierralta, W., Truitt, A., and Wagner, R. K., 1978, The occurrence of steroid-free "activated" estrogen receptor in target cell nuclei, *Hoppe-Seyler's Z. Physiol. Chem.* **359**:1259–1268.

Jungblut, P. W., Meyer, H., and Wagner, R. K., 1980, The interrelationship of estrogen receptors extracted from various subcellular compartments, in: *Perspectives in Steroid Receptor Research* (F. Bresciani, ed.), Raven Press, New York.

Hughes, A., Szendro, P., Terán, C., Kielhorn, J., Sierralta, W., Stone, G., Little, M., and Jungblut, P. W., 1977, Biosynthesis of steroid-hormone receptors, in: (A. Vermeulen, P. Jungblut, A. Klopper, L. Lerner, and F. Sciarra., eds.) *Research on Steroids,* Vol. VII, North-Holland, Amsterdam.

Entenmann, A., Sierralta, W., and Jungblut, P. W., 1980, Studies on the involvement of lysosomes in estrogen action. III. The dehydrogenation of estradiol to estrone by porcine endometrial lysosomes, *Hoppe-Seyler's Z. Physiol. Chem.* **361**:959–968.

Hughes, A., Jacobson, H. J., Wagner, R. K., and Jungblut, P. W., 1976, Ovarian independent fluctuations of estradiol receptor levels in mammalian tissues, *J. Cell. Mol. Endocrinol.* **5**:379–388.

Stumpf, W., 1968, Estradiol-concentrating neurons: Topography in the hypothalamus by dry mount autoradiography, *Science* **162**:1001–1003.

Nenci, I., Beccati, D., Piffanelli, A., and Lanza, G., 1976, Detection and dynamic localization of estradiol–receptor complexes in intact target cells by immunofluorescence technique, *J. Steroid Biochem.* **7**:505–510.

12

Steroids and Membrane-Associated Events in Neurons and Pituitary Cells

B. Dufy, L. Dufy-Barbe, E. Arnauld, and J. D. Vincent

1. Introduction

Classically we are accustomed to differentiating between the mechanisms of action of sex steroids on one hand, and those of peptide hormones and neurotransmitters on the other. Peptide hormones were shown to first interact with membrane receptor sites, activating membrane enzymes and/or channels to various ions, whereas steroid hormones were supposed to passively cross the lipid bilayer of the cell membrane to bind to a cytosolic receptor, as the first step in a number of processes that lead to the genetic machinery of protein synthesis. Evidence supporting this concept of estrogen action on peripheral target organs has been extensively reviewed (Baulieu et al., 1975; Jensen and De Sombre, 1973; O'Malley and Means, 1974; Mueller et al., 1972). A number of clues were also provided to support such a mechanism at the level of brain and pituitary (McGuire and Lisk, 1969; Stumpf, 1970; Whalen and Luttge, 1971; Mowles et al., 1971; Wade and Feder, 1972; for review see Warembourg, 1978). Sex steroids have been shown to control or modulate a number of functions in the central nervous system (CNS). A fundamental argument against a single mechanism of action for sex steroids is the various delays observed for their different phys-

B. Dufy, L. Dufy-Barbe, E. Arnauld, and J. D. Vincent. • INSERM U176, 33077 Bordeaux Cedex, France.

iological effects. Indeed, some of the effects take place in the range of seconds while others take hours and even days. Moreover the effects of sex steroids are generally biphasic, i.e., an early inhibitory effect that is followed by a stimulatory effect some hours later or vice versa. Up to now there have not been sufficient studies of whether the short-term biological effects of steroids are possible in the classical scheme of genetic protein synthesis and no alternative has been provided.

There is a current conceptualization for differentiating between nervous elements, which are supposed to communicate using neurotransmitters in specific and localized pathways, and endocrine cells, which use hormones for a widespread mode of communication. Such a rigid dichotomy is no longer valid. Indeed, signs of secretory activity of the type encountered in endocrine cells were observed a long time ago in neurons of basal brain areas, and recent works have shown definite neuronal characteristics in endocrine cells of various origin; this sytem was named APUD (acronym for amine precursor uptake and decarboxylase system) by Pearse (1968). As well as the cytochemical and ultrastructural features that they share with neurons, most of these endocrine cells are excitable and are able to display action potentials (Kidokoro, 1975; Tischler et al., 1976; Taraskevich and Douglas, 1977; Dufy et al., 1979a). Finally, results from various types of investigations have raised the possibility that steroid hormones may interact directly or indirectly with the receptors for neurotransmitters (Euvrard et al., 1979; Dufy et al., 1979b; Schaeffer and Hsueh, 1979; Hruska and Silbergeld, 1980a,b; Arnauld et al., 1981). There is no doubt that these new observations will lead to a more generalized but also more complex concept of neurohormonal transfer of information.

The purpose of the present essay is to put forward evidence for a membrane involvement of steroid hormone effects in the brain and pituitary and to examine the possibility of interactions at the level of the cell membrane for various substances involved in neurohormonal communication.

2.   The Electrophysiological Approach

Compelling evidence to indicate that steroid hormones influence brain function has been available for more than a century (for review see Beach, 1948); however, the physiological processes by which these substances exert their effects are at present poorly understood. It is important to realize that, whereas long-term effects of steroids are well substantiated, crucial experimental evidence, other than those obtained by electrophysiological techniques, is lacking for short-term effects. Indeed, electrophysiological techniques have been the only techniques to show short-term effect of steroids on brain central processes. Electrophysiology is an extremely useful approach for visualizing membrane electrical properties in nervous and other tissues. Modifications in the structure of the membrane such as opening of ionic channels, interactions of various substances with membrane receptors leading to movement of ionic particles across the cell membrane are suitable to be studied by electrophysiology. Moreover, such powerful techniques as voltage clamp and power spectra analysis of membrane fluctuations allow precise analysis of opening and closing of individual channels. Such techniques have been widely used for the study of the mechanisms of action of neurotransmitters, peptides, and various amino acids, and their interaction with specific receptor sites. However, these techniques have not yet been used for the study of a putative action of steroids on target cell membrane, certainly because steroids are supposed to cross passively the lipid bilayer of target cells and were not supposed to affect the movement of ions across the cell membrane.

3. Electrophysiological Approach to the Study of Sex Steroid Effects on Neurons

3.1. Sex Steroids Directly Affect the Electrical Excitability of Central Neurons

The studies of Barraclough and Cross (1963) constituted one of the earliest attempts to determine whether steroid hormones directly influence brain electrical activity. These authors succeeded in recording single unit activity in the hypothalamus of the cyclic female rat. They observed that some stimuli accelerated the firing rate of the majority of neurons tested, and that progesterone was able to induce a selective depression of the response of lateral hypothalamic neurons to cervical probing. This technique has since been applied to demonstrate changes in excitability of discrete brain regions following the administration of hormonal steroids.

Effects on electrical activity, having latencies of a few minutes or less, were reported in various areas known for their sensitivity to steroids, i.e., the hypothalamus (Dufy *et al.*, 1976; Kelly *et al.*, 1977) and the hippocampus (Teyler *et al.*, 1980). Testosterone has also been reported to elicit early responses in hypothalamic neurons (Orsini and Mei, 1979) and to alter the refractory period of stria terminalis neurons (Kendrick and Drewett, 1979) (for review and discussion of these rapid effects of steroid in neurons see M. Kelly, this volume).

Many of the functional characteristics of hormonal steroid target tissues have been elucidated by biochemical techniques. Our knowledge of the mechanisms of action of steroid hormones has greatly progressed, yet our understanding is still extremely limited as to the role of steroid hormones in brain functions and their mechanisms of action. Besides the now classical process leading to synthesis of new proteins, several other mechanisms of action for steroid hormones are possible. It has been suggested that the cell membrane can be a target for steroids (Dufy *et al.*, 1975; Kelly *et al.*, 1976; Pietras and Szego, 1977; Baulieu, 1978; Kelly *et al.*, 1980). Such a hypothesis is particularly attractive in view of the short latency of some responses. This would require the presence of high-affinity steroid receptors in the brain; however, to date, no membrane receptors for steroid hormones have been identified in the CNS. Without ruling out such a possibility, an alternative explanation, involving early interactions with receptors to other substances, can be advanced.

3.2. *Rapid Interactions between Estrogen and Dopamine on Caudate Neurons*

Evidence for interaction between estrogens and striatal DA receptors is still indirect. Clinically, estrogen has been reported to improve symptoms of tardive dyskinesia following chronic treatment with neuroleptics (Bedard *et al.*, 1977; Gordon *et al.*, 1980), while elevated concentration of estrogens during pregnancy or oral contraceptive use can induce chorea (Nausieda *et al.*, 1979). Experimentally, it has been observed that estrogens can influence the circling behavior induced by amphetamine and apomorphine in rats with unilateral lesion of the nigro-striatal pathway (Hruska and Silbergeld, 1978a; Bedard *et al.*, 1978).

In a series of experiments we tested the electrical activity of caudate neurons and their response to iontophoretically applied DA at different time intervals before and after intramuscular injection of estradiol benzoate (EB) in rats ovariectomized 1 week previously (Arnauld *et al.*, 1981).

The experiments were performed on female adult rats anesthetized with urethane (1.25 g/kg, i.p.). The animals were ovariectomized 1 week

Table I. Electrical Activity and Response to DA of Caudate Neurons in Ovariectomized Rats

	Number of neurons	Electrical activity		Response to DA	
		Spontaneously active neurons (%)	Glutamate-driven neurons (%)	Positive (%)	Negative (%)
Control	103	25	75	86 (i:98; e:2)[a]	14
After EB					
0–2 hr	25	20	80	88 (i:100; e:0)	12
2–4 hr	23	48	52	39 (i:57; e:43)	61
4–6 hr	21	76	24	57 (i:11; e:89)	43
6–8 hr	15	87	13	73 (i:73; e:27)	27
8–12 hr	18	83	17	83 (i:75; e:25)	17

[a] i: neurons inhibited by DA in percentage of cells affected by DA; e: neurons excited by DA in percentage of cells affected by DA.

before the experiment. All animals received an intramuscular injection of 20 µg EB dissolved in sesame oil. Unit activity was recorded from the frontal part of the caudate nucleus; using a glass micropipet (5–7 M Ω) filled with a saturated solution of pontamine blue dye in 2 M NaCl permitted localization of the tip. A seven-barreled micropipet, glued to the side of the recording pipet, allowed for controlled application in the close vicinity of the cell of DA (0.2 M, pH 4) and of the following test compounds: sodium glutamate (0.2 M, pH 8); GABA (0.3 M, pH 4.2); and NaCl (2 M) for current effect control. Of the 103 neurons recorded before estradiol, 25% were spontaneously active and 75% were driven by glutamate. DA affected most of the neurons (86%), inducing an inhibition of their firing activity (Table I). Between 2 and 6 hr after the estradiol injections, the proportion of spontaneously active neurons was elevated to 87% and their firing rate

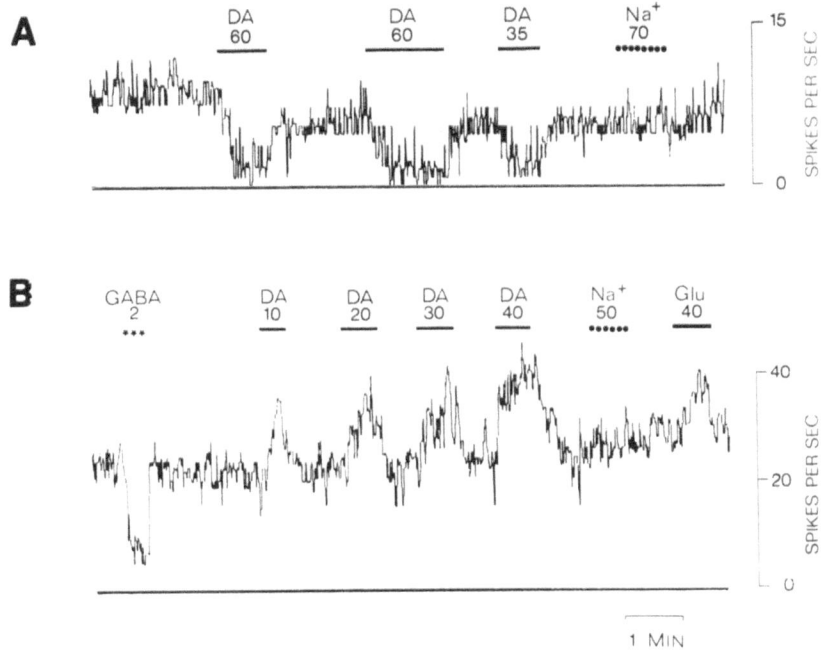

Figure 1. Effect of microiontophoretic administration of dopamine (DA), γ-aminobutyric acid (GABA), and glutamate (Glu) on spontaneously firing neurons in the anterior caudate nucleus. Each trace is a pen recording of cell discharge expressed in spikes per second. The bars indicate the duration of ejection of the drugs and the numbers give the currents used expressed in nA. (A) Depression of spontaneous firing by DA in a control neuron. Current has no effect by itself as checked by ejection of Na^+. (B) Increase in spontaneous firing evoked by DA in a neuron recorded 5 hr after estradiol benzoate injection (20 µg, i.m.). GABA is still active in depressing electrical activity. The responses to Na^+ and glutamate are not altered.

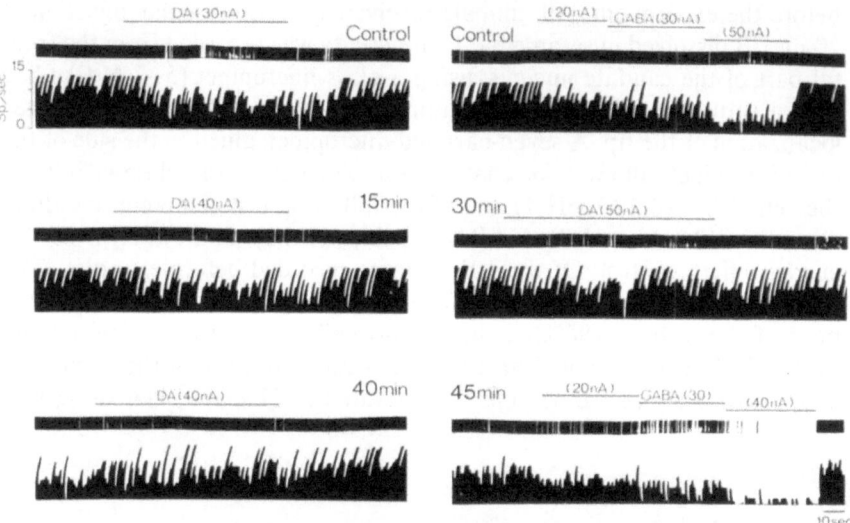

Figure 2. Time course of responses to DA in a caudate neuron following administration of estradiol benzoate (20 µg, i.m.; solvent: sesame oil). Each trace is a pen recording of cell firing
expressed in spikes (sp) per second. During the control period DA and GABA depress action
potential firing. Thirty minutes following estradiol administration DA is no longer active; after
40 min the response of the same cell is reversed since DA becomes excitatory whereas GABA
(45 min) is still active in inhibiting neuronal discharge.

dramatically increased. Concomitantly, the proportion of cells affected by
DA decreased to 39% and most of the DA-responsive cells were now
excited (Fig. 1).

Therefore, following systemic administration of estrogen the response
of caudate neurons shifted from an inhibition to an excitation (Fig. 2). DA
was found to be less effective 30 min after administration of estrogen. The
fact that estradiol reverses the response from an inhibitory to an excitatory
one may be explained by the intervention of a different class of DA receptors normally masked or inactivated in the absence of estrogen. Such a supposition is not too unlikely in view of the numerous data describing the
existence of different types of DA receptors (Kebabian and Calne, 1979).

A possible action of estrogen on central DA receptors has also been
suggested by different groups (Di Paolo et al., 1979; Euvrard et al., 1979;
Hruska and Silbergeld 1980a,b; Chiodo et al., 1979). In these experiments,
removal of the pituitary generally blocked the effects of estrogens, suggesting that such effects may be indirect. Since estrogen is known to stimulate
prolactin secretion (Chen and Meites, 1970) and since prolactin, in turn,
can act on central DA receptors (Scapagnini et al., 1980), it is possible that
the antagonist effect of estrogen on DA receptors in the striatum is
mediated by an elevation of the plasma level of prolactin.

4. Electrophysiological Approach to the Study of Sex Steroid Effects on Pituitary Cells

4.1. Electrophysiological Properties of Pituitary Cells

Classically, a major difference between neurons and gland cells is the regenerative voltage-dependent conductance changes (action potentials) observed in the former and the passive membrane characteristics of the latter. In neurons, including neuroendocrine cells, action potentials provoke secretion by the opening of voltage-dependent calcium channels. However, the increase in intracellular Ca^{2+} concentration that accompanies secretion from gland cells has not been considered to be associated with action potentials. To date some exceptions to this classical division have already been found: the β cells of the endocrine pancreas (Matthews and Sakamoto, 1975), the chromaffin cells of the adrenal medulla (Biales *et al.*, 1976; Brandt *et al.*, 1976), and pituitary cells from pars distalis (Taraskevich and Douglas, 1977) and pars intermedia (Douglas and Taraskevich, 1978).

The ionic requirements for action potentials in endocrine cells appear to be different from those for neurons and even between the various types of endocrine cells: β cells and pituitary cells of the pars distalis only require calcium, whereas chromaffin cells and cells of the pars intermedia seem to require both Na^+ and Ca^{2+} to sustain spiking activity. Moreover, spiking activity always results in activation of Ca^{2+} channels and hence an increase in intracellular Ca^{2+} concentration.

The general opinion is that calcium ions play a major role in the secretion process of pituitary hormones (for review see Trifaro, 1977; Moriarty, 1978). The stimulus–secretion coupling concept, proposed by Douglas (1968) to explain the release of catecholamines and posterior pituitary hormones, may also apply to the release of several anterior pituitary hormones including prolactin.

Kidokoro (1975) was the first to report Ca^{2+}-dependent action potentials in a tumoral line of anterior pituitary cells (GH3). Dissociated cells from rat anterior pituitary were also found to generate action potentials that were insensitive to tetrodotoxin (Taraskevich and Douglas, 1977). Biales *et al.* (1977) reported a Na^+ component in the action potentials of GH3 but such cells seem to have a prominent Ca^{2+} component, at least under the experimental culture conditions used (Dufy *et al.*, 1979b; Taraskevich and Douglas, 1980). Human pituitary cells derived from pituitary adenomas were also found to generate Ca^{2+}-dependent action potentials (Biales *et al.*, 1977; Vincent *et al.*, 1980). All the authors agree that TRH is able to induce Ca^{2+}-dependent spike activity in silent cells or to increase the firing rate of spontaneously active cells (Dufy *et al.*, 1979a; Ozawa and Kimura, 1979; Taraskevich and Douglas, 1980; Sand *et al.*, 1980). These findings are consistent with the observation that Ca^{2+} is essential for the secretion of adenohypophyseal hormones and provide a valuable clue for a direct involvement of Ca^{2+}-dependent action potentials in the secretory process of some anterior pituitary cells.

4.2. Rapid Changes in Membrane Conductance and Permeability Following Administration of Estrogens

We have observed that 17β-estradiol ($17\beta E_2$) when added to the recording medium (50 pg/ml) increased the percentage of spontaneously firing GH3/B6 cells and potentiated the electrical response to TRH (Dufy *et al.*, 1979a). Direct administration of $17\beta E_2$ to the cell being recorded elicited a series of transient changes in membrane potential followed by a sustained discharge of action potentials (Fig. 3). These action potentials were demonstrated as Ca^{2+}-dependent spikes (Dufy *et al.*, 1979a). They were observed within 1 minute following $17\beta E_2$ administration and lasted for 3 to 4 minutes; the cell was then completely desensitized to further application of the steroid. $17\alpha E_2$ was much less effective (Fig. 3).

These results reveal a rapid effect of $17\beta E_2$ on the membrane of GH3/ B6 cells. The electrical activity elicited by $17\beta E_2$ was similar in time course and Ca^{2+}-dependence to that induced by TRH. This rapid effect of $17\beta E_2$ on membrane properties implies recognition sites for the steroid at the membrane surface and probably reflects conformational changes in membrane components. Recent evidence for a membrane site of action of estro-

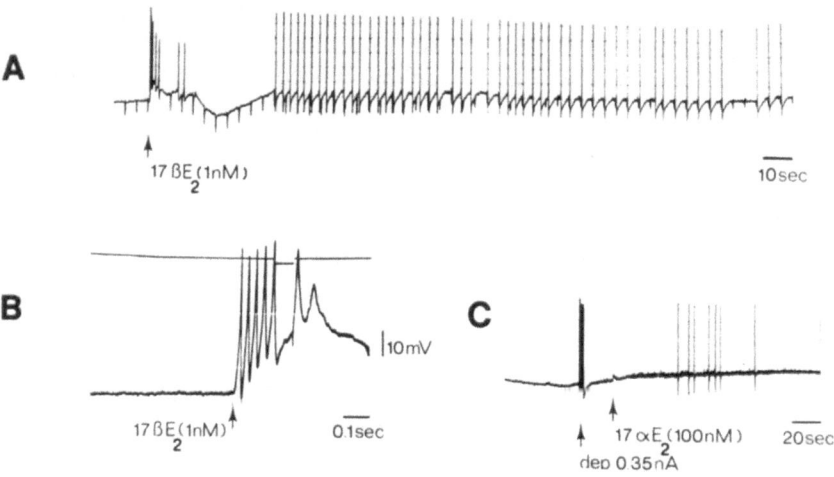

Figure 3. Effects of estrogen on the clonal pituitary cell line GH3/B6. (A) $17\beta E_2$ directly administered onto the membrane of an excitable cell induces action potentials. (B) Early effect of $17\beta E_2$. (C) $17\alpha E_2$ is much less efficient than $17\beta E_2$ in promoting action potentials.

gen has been reported for uterine cells (Pietras and Szego, 1977), and an early intracellular accumulation of calcium has been clearly demonstrated following estrogen administration to endometrial cells (Pietras and Szego, 1979). However, the physiological implication of such early effects of estradiol on the passive membrane properties and excitability of pituitary cells remains unclear.

Estrogens have long been considered as important regulators of PRL secretion *in vivo* and *in vitro*. *In vitro* parameters of cell culture such as cell morphology, growth, responsiveness to TRH and DA are under the influence of estrogen (Brunet *et al.*, 1977). $17\beta E_2$ increased the number of membrane receptors for TRH (Gershengorn *et al.*, 1979; Gourdji, 1980) and stimulated both basal and PRL response to TRH in different systems (Gourdji, 1980; Tixier-Vidal *et al.*, 1978; Vician *et al.*, 1979). Estradiol stimulates PRL production in GH3 cells (Haug and Gautvik, 1976). However, this stimulatory effect was only significant after 4 days and reached a maximum by 10 days. The short-term effect of $17\beta E_2$ on the electrical activity of GH3/B6 cells is hardly compatible with the long-term effect of the steroid on PRL production. We have therefore investigated the possibility that PRL release is stimulated within minutes after $17\beta E_2$ injection, which would be consistent with the Ca^{2+}-dependent electrical effect we observed. Using the GH3/B6 cell line we found a significant increase of PRL in the medium (83% increase from control values; $n = 4$) 10 min following administration of estradiol ($17\beta E_2$, 10^{-8} M). This increase was only 63% at 15 min and not different from controls at 40 min; a second increase of PRL in estrogen-treated samples is observed several hours later. Thus, the Ca^{2+}-dependent spike activity observed in the minutes following estradiol administration may account for the early transient release of PRL induced by the steroid in the GH3/B6 cell line.

It is therefore worthwhile to note that the early effect of estradiol on calcium spiking activity, which we observed, is consistent with previous observations by Pietras and Szego (1977) showing a rapid uptake of Ca^{2+} following administration of estrogen to isolated endometrial cells. These observations strongly suggest a membrane site of action for estrogen.

4.3. Modulation by Estrogen of the Inhibitory Effect of DA on PRL-Secreting Cells

PRL-secreting cells are under the stimulatory influence of TRH and of estrogen, but they are mainly controlled by inhibitory substances among which DA is the most potent. In the GH3/B6 cell line, DA administered directly onto a cell provokes an inhibition of firing within 30 to 60 sec (Fig. 4). Similar results were observed in teleostean PRL cells (Taraskevich and Douglas, 1978) and also in human cells cultured from pituitary adenomas

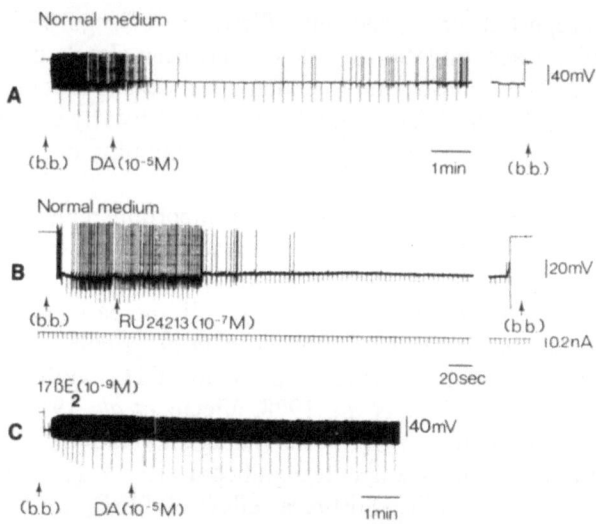

Figure 4. Effect of dopamine (DA) and a dopamine agonist RU 24213 (from Roussel-UCLAF) on action potential firing of spontaneously active GH3/B6 cells. b.b. indicates bridge balance of the amplifier. Membrane resistance is measured by injecting currents of 0.2 nA. (A, B) DA and RU 24213 inhibit action potential firing when the cells are recorded in an estrogen-free medium. (C) When $17\beta E_2$ (10^{-9} M) is added to the recording medium DA is no longer effective. [Reprinted from Dufy *et al.*, 1979b, with permission.]

(Vincent *et al.*, 1980). In addition, DA inhibits TRH-induced action potentials. In the GH3/B6 cell line this inhibitory effect is concomitant with a rapid decrease of the input resistance without any detectable change in the resting membrane polarization (Dufy *et al.*, 1979b), whereas in human cells DA induced a clear hyperpolarization (Vincent *et al.*, 1980). These responses were specific since the DA antagonists haloperidol and chlorpromazine were effective in blocking the response to DA (Dufy *et al.*, 1979). The DA inhibition of both spontaneous and TRH-induced action potentials may be related to the effect of DA on PRL secretion. It has previously been shown that the DA agonist CB 154 decreased TRH-induced PRL release in GH3/B6 cells without modifying the binding of [³H]-TRH (Gourdji *et al.*, 1973). Similarly, a significant decrease of PRL release in GH3/B6 cells was induced within 30 min of exposure to [³H]-DA. Simultaneously, the cells showed a binding capacity for [³H]-DA and this was partially inhibited by an excess of unlabeled DA (Dufy *et al.*, 1979b). These observations were not confirmed by Cronin *et al.* (1980) and Faure *et al.* (1980), for in their experiments "the basal release of PRL from *GH3* cells was unaffected when incubated for 6 h with DA." We do not know where the discrepancies between their results and ours come from. It is possible that the repetitive trypsinizations necessary for culture replication alter the responsiveness of GH3 cells to dopaminergic inhibition. In support of such a hypothesis, we currently observe a loss of responsiveness to TRH after numerous trypsinizations. Similar loss of response to estradiol has been reported on the same preparation by Dannies *et al.* (1977).

Finally, we have observed that the inhibitory effect of DA on PRL-secreting cells was modulated by estrogen. The percentage of cells inhibited by DA was different according to the presence or absence of estrogen in the medium in which the cells were grown. When the cells were grown for at least 24 hr in an estrogen-depleted medium, the percentage of DA-inhibited cells reached 80% (n = 43) instead of the 40% in normal medium. When $17\beta E_2$ (10^{-8} M) was added to the recording solution, DA did not affect the action potential firing nor the input resistance of all the cells tested (n = 30) (Fig. 4). We have reported above that $17\beta E_2$ when directly applied to the cell provokes action potential firing; DA did not inhibit this estrogen-induced electrical activity.

Using different methodologies several other groups have already reported interactions between estradiol and DA at the level of pituitary. Raymond *et al.* (1978) have shown a potent antidopaminergic effect of estradiol on rat primary cultures of anterior pituitary cells. This effect may be mediated through catecholestrogens since 2-hydroxy-estrogen is a competitive inhibitor of binding of spiroperidol to DA receptors in the anterior pituitary (Schaeffer and Hsueh, 1979). Moreover, the enzymes necessary for conversion have been found in pituitary tissues (Paul *et al.*, 1977). Another important contribution to the same problem comes from the observations of Quijada *et al.* (1980) that tamoxifen, an antiestrogenic compound, rapidly enhances the sensitivity of dispersed PRL-secreting pituitary cells to DA. This action of tamoxifen occurred within 2.5–4 hr (Quijada *et al.*, 1980). These observations are consistent with those reported, here above, for striatal neurons. This is not surprising since several classes of DA receptors have been identified and similarities between DA receptors found in CNS and pituitary have been described (Kebabian and Calne, 1979).

5. Concluding Remarks

Taken together, results of these investigations suggest that some effects of steroid hormones occur too rapidly to be accounted for by genomic activation. A great variety of latencies have been reported for the effects of hormonal steroids and it is likely that different modes of action are used. An important question posed by the above findings relates to the mechanisms by which hormonal steroids can rapidly affect the function of target cells. Do steroids directly interact with the membrane and, if so, how is this performed [since, as lipophilic agents, they are supposed to diffuse passively through the lipid bilayer of the cell membrane (Clark *et al.*, 1978; Mueller *et al.*, 1979)]?

Accumulating evidence suggests a membrane site of action in different systems. High-affinity membrane binding sites have been shown in the uterus (Pietras and Szego, 1979). Progesterone acts at the level of the cell surface in *Xenopus laevis* oocytes (Baulieu *et al.*, 1978), to induce a rapid increase in Ca^{2+} influx (Wasserman *et al.*, 1980; Kostellow and Morrill, 1980), which in turn triggers a number of cytoplasmic events leading to meiotic maturation. Estrogen has also been shown to rapidly induce CA^{2+} influx in endometrial cells (Pietras and Szego, 1977) and Ca^{2+}-dependent action potentials followed shortly after administration of estradiol to PRL-secreting pituitary cells (Dufy *et al.*, 1979a). As yet, the molecular understanding of the activation of cells by steroids which should use a membrane site of action lags behind our knowledge of the classical steroid receptor system. It is now clear that steroids enter indifferently target and nontarget tissues with equal facility (for review see Clark and Peck, 1978), which suggests a simple diffusion. Without ruling out the possibility of a carrier-mediated process (Milgrom *et al.*, 1973; Pietras and Szego, 1979), it should be fruitful to investigate effects of steroids on membrane lipids. Obviously, the properties of membrane proteins are highly sensitive to the physical state and biochemical composition of membrane lipids; changes in membrane fluidity can affect membrane properties in a number of ways (for review see Jackson, 1978; Barnett, 1978). The properties of ionic channels and the efficiency of membrane receptors may be affected by changes in the configuration of lipids and by modification of fluidity of the membrane. Na^+/K^+-ATPase, which maintains the sodium potassium gradients in the cell, is inhibited when the fluidity of phospholipids is reduced (Jackson, 1978); phospholipid methylation is rapidly followed by Ca^{2+} influx and histamine release in mast cells (Axelrod and Hirata, 1980). The acetylcholine receptor purified from *the Torpedo* is also extremely sensitive to lipid structure (Giraudat and Changeux, 1980). The ability of the hormone-receptor activation of adenylate cyclase is altered by changes in membrane fluidity (for review see Barnett, 1978; Hollenberg, 1979). This last observation may account for the decreased sensitivity of DA in the presence of estrogen in striatal neurons and pituitary cells reported above.

Unfortunately, only a limited number of experiments have been done to test this new possibility of steroid action at the membrane level. However, significant changes in phospholipid composition have been reported following glucocorticoid treatment (Johnston *et al.*, 1980).

In summary, rapid effects of hormonal steroids can be mediated by:

1. The classical steroid receptor system leading to genomic activation and protein synthesis, since steroids have been reported to rapidly alter chromatin structure (Johnson *et al.,* 1979); also, significant rise in nuclear RNA polymerase II activity has been observed as early as 30 min following steroid treatment (Clark *et al.,* 1978). Rapid involvement of cytoplasmic receptors without nuclear participation may also be invoked.

2. Membrane receptor for steroid hormones as suggested by Pietras and Szego (1975, 1979).

3. Direct alterations of the properties of ionic channels or modification of efficiency of receptors to other substances (peptides, neurotransmitters) by changes in the physical state or biochemical composition of membrane lipids.

ACKNOWLEDGMENTS. We thank Dr. A. F. Parlow and the NIAMDD for providing the reagents for rat PRL assay, and J. Seal for editorial assistance. This work was supported by grants from INSERM (CRL 78.1.2656), CNRS (ERA 493), and the Philippe Foundation.

DISCUSSION

PECK: Using biochemical techniques, we have attempted many times to demonstrate membrane receptors for estrogens. These have been without success. Since you have $17\beta E$ versus $17\alpha E$ differences, does this demand a receptor or could these differences result from physical-chemical changes (or differences in lipid–lipid interactions) in the membrane?

DUFY: I know that few people have succeeded in showing membrane receptors for estrogen and that bothers me a little. This is the reason why I think that membrane receptors, in a classical meaning, are not perhaps necessary. A given kind of modification of membrane lipids may account for the effects we observed; this can only be a working hypothesis; at the present time, we have no data on that.

KRAICER: What is the range of TMPs (transmembrane potentials) in the GH3 cells?

DUFY: The membrane potential depends on the type of chemical used to fill the recording pipets. With 3 M KCl, the membrane potential ranges from -30 to -60 mV. The cells are more polarized when using potassium sulfate or potassium citrate.

KRAICER: Does dopamine alter TMP?

DUFY: In the GH3/B6 cell line we did not find any alteration of TMP following administration of DA. The inhibition of firing was accompanied by a decrease in membrane resistance. Things seem to be very different in human pituitary cells since we observed that DA induced an important hyperpolarization.

KRAICER: Does estrogen desensitize the GH3/B6 cells to TRH?

DUFY: I don't know; we have not done this experiment.

KRAICER: You report experiments done with a high concentration of Ca^{2+} in the recording medium. Why?

DUFY: It has been reported in the literature that it is difficult to keep an intracellular recording with low Ca^{2+}. So, we have done as others and recorded with 10 mM Ca^{2+}. Recently, we have used a much lower concentration (2 mM Ca^{2+}) and did not find any differences in the responses compared to 10 mM; so it works with 2mM Ca^{2+}. However, with Ca^{2+}-free medium the cells have a very low resting potential.

REFERENCES

Arnauld, E., Dufy, B., Pestre, M., and Vincent, J. D., 1980, Effects of estrogen on the responses of caudate neurons to microiontophoretically applied dopamine, *Neuroscience,* **21**:325–331.

Axelrod, J., and Hirata, F., 1980, Lipids and the transduction of biological signals through membranes, in: *Progress in Psychoneuroendocrinology* (F. Brambilla, G. Racagni, and D. de Wied, eds.), pp. 1–12, Elsevier, Amsterdam.

Barnett, R. E., 1978, Fluidity in membranes, in: *Receptors and Hormone Action,* (B. W. O'Malley and L. Birnbaumer, eds.), Vol. I, pp. 427–446, Academic Press, New York.

Barraclough, C., and Cross, B., 1963, Unit activity in the hypothalamus of the cyclic female rat: Effect of genital stimuli and progesterone, *J. Endocrinol.* **26**:339.

Baulieu, E. E., 1975, Some aspects of the mechanism of action of steroids hormones, *Mol. Cell. Biochem.* **7**:157.

Baulieu, E. E., 1978, Cell membrane, a target for steroids hormones, *Mol. Cell. Endocrinol.* **12**:247.

Baulieu, E. E., Godeau, F., Schorderet, M., and Schorderet-Slatkine, S., 1978, Steroid induced meiotic division in *Xenopus laevis* oocytes: Surface and calcium, *Nature (London)* **275**:593.

Beach, F., 1948, *Hormones and Behavior,* Harper & Row (Hoeber), New York.

Bedard, P., Langelier, P., and Villeneuve, A., 1977, Estrogens and extrapyramidal system, *Lancet* **2**:1367.

Bedard, P., Dankoud, J., Boucher, R., and Langelier, P., 1978, Effects of estrogens on apomorphine-induced circling behavior in the rat, *Can. J. Physiol.* **56**:538.

Biales, B., Dichter, M., and Tischler, A., 1976, Electrical excitability of cultured adrenal chromaffin cells, *J. Physiol. (London)* **262**:743.

Biales, B., Dichter, M. A., and Tischler, A., 1977, Sodium and calcium action potential in pituitary cells, *Nature (London)* **267**:172.

Brandt, B. L., Hagiwara, S., Kidokoro, Y., and Miyasaki, S., 1976, Action potentials in the rat, chromaffin cell and effects of acetylcholine, *J. Physiol. (London)* **263**:417.

Brunet, N., Gourdji, D., Moreau, M. F., Grouselle, D., Bournaud, F., and Tixier-Vidal, A., 1977, Effect of 17β-estradiol on prolactin secretion and thyroliberin responsiveness in two rats prolactin continuous cell lines: Definition of an experimental model, *Ann. Biol. Anim. Biochim. Biophys.* **17**:413.

Chen, C. L., and Meites, J., 1970, Effects of estrogen and progesterone on serum and pituitary prolactin levels in ovariectomized rats, *Endocrinology* **86**:503.

Chiodo, L. A., Caggiula, A. R., and Saller, C. F., 1979, Estrogen increases both spiperone-induced catalepsy and brain levels of 3H spiperone in rat, *Brain Res.* **172**:360.

Clark, J. H., and Peck, E. J., 1978, Steroid hormone receptors: Basic principles and measurement, in: *Receptors and Hormone Action* (B. W. O'Malley and L. Birnbaumer, eds.), Vol. I, pp. 383–410, Academic Press, New York.

Clark, J. H., Peck, E. J., Hardin, J. W., and Eriksson, H., 1978, The biology and pharmacology of estrogen receptor binding: Relationship to uterine growth, in: *Receptors and Hormone Action* (B. W. O'Malley and L. Birnbaumer, eds.), Vol. II, pp. 1–31, Academic Press, New York.

Cronin, M. J., Faure, N., Martial, J. A., and Weiner, R. I., 1980, Absence of high affinity dopamine receptors in GH3 cells: A prolactin-secretory clone resistant to the inhibitory action of dopamine, *Endocrinology* **106**:718.

Dannies, P. J., Yen, P. M., and Tashjian, A. H., 1977, Anti-estrogenic compounds increase prolactin and growth hormone synthesis in clonal strains of rat pituitary cells, *Endocrinology* **101**:1151.

Di Paolo, T., Carmichael, R., Labrie, F., and Raymond, J. P., 1979, Effects of estrogen on the characterization of ^{3}H spiroperidol and ^{3}H RU 24213 binding in rat anterior pituitary gland and brain, *Mol. Cell. Endocrinol.* **16**:99.

Douglas, W. W., 1968, Stimulus–secretion coupling: The concept and clues from chromaffin and other cells, *Br. J. Pharmacol.* **34**:451.

Douglas, W. W., and Taraskevich, P. S., 1978, Action potentials in gland cells of rat pituitary pars intermedia: Inhibition by dopamine, an inhibitor of MSH secretion, *J. Physiol. (London)* **285**:171.

Dufy, B., Partouche, C., Dufy-Barbe, L., and Vincent, J. D., 1975, Effects of oestrogen on the electrical activity of hypothalamic units: Correlations with gonadotrophic hormone levels, *Proc. Hung. Acad. Sci. (Budapest)* **1975**:303.

Dufy, B., Partouche, C., Poulain, D., Dufy-Barbe, L., and Vincent, J. D., 1976, Effects of estrogen on the electrical activity of identified and unidentified hypothalamic units, *Neuroendocrinology* **22**:38.

Dufy, B., Vincent, J. D., Fleury, H., Du Pasquier, P., Gourdji, D., and Tixier-Vidal, A., 1979a, Membrane effects of thyrotropin-releasing hormone and estrogen shown by intracellular recording from pituitary cells, *Science* **204**:509.

Dufy, B., Vincent, J. D., Fleury, H., Du Pasquier, P., Gourdji, D., and Tixier-Vidal, A., 1979b, Dopamine inhibition of action potentials in a prolactin secreting cell line is modulated by estrogen, *Nature (London)* **282**:855.

Euvrard, C., Labrie, F., and Boissier, J., 1979, Effects of estrogen on changes in the activity of striatal cholinergic neurons induced by DA drugs, *Brain Res.* **169**:215.

Faure, N., Cronin, M. J., Martial, J. A., and Weiner, R. I., 1980, Decreased responsiveness of GH3 cells to the dopaminergic inhibition of prolactin, *Endocrinology* **107**:1022.

Gershengorn, M. C., Marcus Samuels, B. E., and Geras, E., 1979, Estrogens increase the number of thyrotropin releasing hormone receptors on mammotropic cells in culture, *Endocrinology* **105**:171.

Giraudat, J., and Changeux, J.-P., 1980, The acetylcholine receptor, *Trends Pharmacol. Sci.* **1**:198.

Gordon, J. H., Borison, R. L., and Diamond, B. I., 1980, Estrogen in experimental tardive dyskinesia, *Neurology* **30**:551.

Gourdji, D., 1980, Characterization of thyroliberin (TRH) binding sites and coupling with prolactin and GH secretion in rat pituitary cell lines, in: *Synthesis and Release of Adenohypophyseal Hormones* (K. W. McKerns and M. Jutisz, eds.), pp. 463–493, Plenum Press, New York.

Gourdji, D., Tixier-Vidal, A., Morin, A., Pradelles, P., Morgat, J. L., Fromageot, P., and Kerdelhué, B., 1973, Binding of tritiated thyrotropin-releasing factor to a prolactin secreting clonal cell line (GH3), *Exp. Cell Res.* **82**:39.

Haug, E., and Gautvik, K. M., 1976, Effects of sex steroids on prolactin secreting rat pituitary cells in culture, *Endocrinology* **99**:1982.

Hollenberg, M. D., 1979, Hormone–receptor interactions at the cell membrane, *Pharmacol. Rev.* **30**:393.

Hruska, R. E., and Silbergeld, E. K., 1980a, Increased dopamine receptor sensitivity after estrogen treatment using the rat rotation model, *Science* **208**:1466.

Hruska, R. E., and Silbergeld, E. K., 1980b, Estrogen treatment enhances dopamine receptor sensitivity in the rat striatum, *Eur. J. Pharmacol.* **61**:397.

Jackson, R. L., 1978, Current views on the organization of lipids and proteins in plasma membranes, in: *Receptors and Hormone Action* (B. W. O'Malley and L. Birnbaumer, eds.), Vol. I, pp. 411–426, Academic Press, New York.

Jensen, E. V., and De Sombre, E. R., 1973, Estrogen–receptor interaction, *Science* **182**:126.

Johnson, L. K., Lan, N. C., and Baxter, J. D., 1979, Stimulation and inhibition of cellular functions by glucocorticoids: Correlations with rapid influences on chromatin structure, *J. Biol. Chem.* **254**:7785.

Johnston, D., Matthews, E. R., and Melnykovich, G., 1980, Glucocorticoid effects on lipid metabolism in Hela cells: Inhibition of cholesterol synthesis and increased sphingomyelin synthesis, *Endocrinology* **107**:1482.

Kebabian, J. W., and Calne, D. B., 1979, Multiple receptors for dopamine, *Nature (London)* **277**:93.

Kelly, M. J., Moss, R. L., and Dudley, C. A., 1976, Differential sensitivity of preoptic septal neurons to microelectrophoresed estrogen during the estrous cycle, *Brain Res.* **114**:152.

Kelly, M., Moss, R., and Dudley, C., 1977, The effects of microelectrophoretically applied estrogen, cortisol and acetylcholine on medialpreopticoseptal unit activity throughout the estrous cycle of the female rat, *Exp. Brain Res.* **30**:53.

Kelly, M. J., Kuhnt, U., and Wuttke, W., 1980, Hyperpolarisation of hypothalamic parvocellular neurons by 17β-estradiol and their identification through intracellular staining with procion yellow, *Exp. Brain Res.* **40**:440.

Kendrick, K. M., and Drewett, R. F., 1979, Testosterone reduces refractory period of stria terminalis neurones on the rat brain, *Science* **204**:877.

Kidokoro, Y., 1975, Spontaneous calcium action potentials in a clonal pituitary cell line and their relationship to prolactin secretion, *Nature (London)* **258**:741.

Kostellow, A. B., and Morrill, G. A., 1980, Calcium dependence of steroid and guanine 3′,5′ monophosphate induction of germinal vesicle breakdown in *Rana pipiens* oocytes, *Endocrinology* **106**:1012.

McGuire, J. L., and Lisk, R. D., 1969, Localization of estrogen receptors in the rat hypothalamus, *Neuroendocrinology* **4**:289.

Martin, S., York, D. H., and Kraicer, J., 1973, Alternations in transmembrane potential of adenohypophyseal cells in elevated potassium and calcium-free media, *Endocrinology* **92**:1084.

Matthews, E. K., and Sakamoto, Y., 1975, Electrical characteristics of pancreatic islet cells, *J. Physiol. (London)* **246**:421.

Milgrom, E., Atger, M., and Baulieu, E. E., 1973, Studies on estrogen entry into uterine cells and on estradiol–receptor complex attachment to the nucleus: Is the entry of estrogen into uterine cells a protein-mediated process? *Biochim. Biophys. Acta* **320**:267.

Moriarty, C. M., 1978, Role of calcium in the regulation of adenohypophysial hormone release, *Life Sci.* **23**:185.

Mowles, T. F., Ashkanazy, B., Mix, E., and Sheppard, H., 1971, Hypothalamic and hypophyseal estradiol-binding complexes, *Endocrinology* **89**:484.

Mueller, G. C., Vonderhaar, B., Kim, U. H., and Le Mathieu, M., 1972. Estrogen action: An inroad to cell biology, *Recent Prog. Horm. Res.* **28**:1.

Mueller, R. E., Johnston, T. C., and Wotiz, H. H., 1979, Binding of estradiol to purified uterine plasma membranes, *J. Biol. Chem.* **254**:7895.

Nausieda, P. A., Koller, W. C., Weiner, W. J., and Klawans, L., 1979, Chorea induced by oral contraceptives, *Neurology* **29**:1605.

O'Malley, B. W., and Means, A. R., 1974, Female steroid hormones and target cell nuclei, *Science* **183**:610.

Orsini, J. C., and Mei, N., 1979, Reponse precoce des neurones hypothalamiques à une injection de testosterone chez le rat mâle, *C. R. Soc. Biol.* **173**:96.

Ozawa, S., and Kimura, N., 1979, Membrane potential changes caused by thyrotropin-releasing hormone in the clonal GH3 cell and their relationship to secretion of pituitary hormone, *Proc. Natl. Acad. Sci. USA* **76**:6017.

Paul, S. M., Axelrod, J., and Diliberto, E. J., 1977, Catechol estrogen-forming enzyme of brain: Demonstration of a cytochrome P-450 monooxygenase, *Endocrinology* **10**:1604.

Pearse, A. G. E., 1968, Common cytochemical and ultrastructural characteristics of cells producing polypeptide hormones (the APUD series) and their relevance to thyroid and ultimobranchial C cells and calcium, *Proc. R. Soc. London Ser. B.* **170**:71.

Pietras, R. J., and Szego, C. M., 1975, Endometrial cell calcium and oestrogen action, *Nature (London)* **253**:357.

Pietras, R. J., and Szego, C. M., 1977, Specific binding sites for oestrogen at the outer surfaces of isolated endometrial cells, *Nature (London)* **265**:69.

Pietras, R. J., and Szego, C. M., 1979, Estrogen receptors in uterine plasma membrane, *J. Steroid Biochem.* **11**:1471.

Quijada de, M., Timmermans, H. A. T., Lamberts, S. W. J., and McLeod, R. M., 1980, Tamoxifen enhances the sensitivity of dispersed prolactin secreting pituitary tumor cells to dopamine and bromocryptine, *Endocrinology* **106**:702.

Raymond, V., Beaulieu, M., and Labrie, F., 1978, Potent antidopaminergic activity of estradiol at the pituitary level on prolactin release, *Science* **200**:1173.

Sand, O., Haug, E., and Gautvik, K. M., 1980, Effects of thyroliberin and 4-aminopyridine on action potentials and prolactin release and synthesis in rat pituitary cells in culture, *Acta Physiol. Scand.* **108**:247.

Scapagnini, U., Rizza, V., Drago, F., Canonico, P. L., Pelligrini-Quarantotti, B., Ragusa, M., Clementi, G., Patro, A., Marchetti, A., and Gessa, G. L., 1980, Prolactin effects on the brain, in: *Central and Peripheral Regulation of Prolactin Function* (R. M. McLeold and U. Scapagnini, eds.), pp. 293–309, Raven Press, New York.

Schaeffer, J. M., and Hsueh, A. J. W., 1979, 2-Hydroxy-estradiol interaction with dopamine receptor binding in rat anterior pituitary, *J. Biol. Chem.* **254**:5606.

Stumpf, W., E., 1970, Estrogen neurons and estrogen neuron systems in the periventricular brain, *Am. J. Anat.* **129**:207.

Taraskevich, P. S., and Douglas, W. W., 1977, Action potentials occur in cells of the normal anterior pituitary gland and are stimulated by the hypophysiotropic peptide thyrotropin-releasing hormone, *Proc. Natl. Acad. Sci. USA* **74**:4064.

Taraskevich, P. S., and Douglas, W. W., 1978, Catecholamines of supposed inhibitory hypophysiotropic function suppress action potentials in prolactin cells, *Nature (London)* **276**:832.

Taraskevich, P. S., and Douglas, W. W., 1980, Electrical behaviour in a line of anterior pituitary cells (GH cells) and the influence of the hypothalamic peptide, thyrotrophin releasing factor, *Neuroscience* **5**:421.

Teyler, T. J., Vardaris, R. M., Lewis, D., and Rawitch, A. B., 1980, Gonadal steroids: Effects on excitability of hippocampal pyramidal cells, *Science* **209**:1017.

Tischler, A. S., Dichter, M. A., Biales, B., Delellis, R. A., and Wolfe, H., 1976, Neuronal properties of human endocrine tumor cells of proposed neural crest origin, *Science* **192**:902.

Tixier-Vidal, A., Brunet, N., and Gourdji, D., 1978, Morphological and molecular aspects of the regulation of prolactin secretion by rat pituitary cell lines, in: *Progress in Prolactin Physiology and Pathology* (C. Robyn and M. Harter, eds.), pp. 29–43, Elsevier, Amsterdam.

Trifaro, J. M., 1977, Common mechanisms of hormones secretion, *Annu. Rev. Pharmacol. Toxicol.* **17**:27.

Vician, L., Shupnik, M. A., and Gorski, J., 1979, Effects of estrogen on primary ovine pituitary cell cultures: Stimulation of prolactin secretion synthesis and preprolactin messenger ribonucleic acid activity, *Endocrinology* **104**:736.

Vincent, J. D., Dufy, B., Israel, J. M., Zyzek, E., Dufy-Barbe, L., Guerin, J., Gourdji, D., and Tixier-Vidal, A., 1980, Neurohormonal communication: An electrophysiological study of the membrane properties of anterior pituitary cells, in: *Progress in Psychoneuroendocrinology* (F. Brambilla, G. Racagni, and D. de Wied, eds.), pp. 25–37, Elsevier/North-Holland, Amsterdam.

Wade, G. N., and Feder, H., 1972 [1,2-^{3}H]-Progesterone uptake by guinea pig brain and uterus: Differential localization, time-course of uptake and metabolism effects of age, sex, estrogen-priming and competing steroids, *Brain Res.* **45**:525.

Warembourg, M., 1978, Distribution of steroid receptor in the CNS, in: *Cell Biology of Hypothalamic Neurosecretion* (J. D. Vincent and C. Kordon, eds.), pp. 221–337, CNRS, Paris.

Wasserman, W. J., Pinto, L. H., O'Connor, C. M., and Smith, L. D., 1980, Progesterone induces a rapid increase in [Ca^{2+}] in *Xenopus laevis* oocytes, *Proc. Natl. Acad. Sci. USA* **77**:1534.

Whalen, R. E., and Luttge, W. F., 1971, Differential localization of progesterone uptake in brain: Role of sex, estrogen pretreatment and adrenalectomy, *Brain Res.* **33**:147.

13

Electrical Effects of Steroids in Neurons

Martin J. Kelly

1. Introduction

Neuroendocrinologists have recognized for over three decades that gonadal steroids participate in CNS (hypothalamic) control of pituitary hormone secretion via a feedback mechanism (Sawyer *et al.*, 1949). It is known that removal of the gonads results in enhanced synthesis and release of the gonadotropins. However, restoration of circulating levels of gonadal steroid(s) through systemic administration of the steroid(s) inhibits this augmentation. Therefore, this phenomenon has been designated "negative feedback" of gonadal steroids (see McCann, 1974, for review). The mechanism of steroid feedback has been thought to involve DNA-dependent RNA transcription for the hypothalamus as well as the pituitary gland (McEwen and Luine, 1978). The nuclear mediation of new protein synthesis has been a well-documented mode of action of steroids in peripheral target tissues such as estrogen's actions in the uterus (Jensen *et al.*, 1969). Indirect evidence for this mode of action in the CNS came from the early autoradiographic localization of estrogen, progesterone, and testosterone in hypothalamic neurons (Pfaff, 1968; Stumpf, 1968; Sar and Stumpf, 1973; Pfaff and Keiner, 1973; Stumpf and Sar, 1976) and glucocorticoids in hippocampal neurons (McEwen *et al.*, 1972a; Stumpf and Sar, 1976). Furthermore, it was found that these areas contain specific cytosol and nuclear receptors for estrogen (Zigmond and McEwen, 1970), progesterone (Kato and Onouchi, 1977), and cortisol (McEwen *et al.*, 1972b). However, these anatomical and biochemical data are only indirect evidence for the physiological action

---

Martin J. Kelly ● Oregon Health Sciences University, Portland, Oregon 97201.

of steroids on neurons. In fact, it was noted quite early that there are more rapid changes produced by steroids that cannot be explained by a nuclear-mediated mechanism of action, i.e., inhibition of luteinizing hormone release in the ovariectomized rat within 15 min following a systemic injection of estrogen (Negro-Vilar *et al.,* 1973). The data reviewed below will illustrate that steroids can directly modify the excitability of CNS neurons through specific membrane events, and these actions are relevant to the physiological role of steroids in "feedback" regulation of the CNS.

2.   Effects of Estrogen on Hypothalamic Neurons

This review will concentrate on the effects of the gonadal steroids on the electrical activity of hypothalamic neurons because there is compelling evidence that these changes in electrical excitability are involved in the negative feedback actions of these steroids on the hypothalamic–pituitary axis.

Sawyer and colleagues (Sawyer *et al.,* 1949; Everett and Sawyer, 1949) postulated that steroids might exert their feedback effects at the CNS level when they found that dibenamine (α-adrenergic blocker) or atropine (muscarinic cholinergic blocker) blocked estrogen-induced ovulation in pregnant rats and progesterone-advanced ovulation in the 5-day cyclic rat. However, it was not until 10 years later that Kawakami and Sawyer (1959) demonstrated that progesterone when given systemically had a biphasic effect on the threshold (initially attenuating and then augmenting) of the coitus-induced hypothalamic–rhinencephalic EEG activity in the estrogen-primed female rabbit. Lateral hypothalamic neurons were shown to exhibit increased activity with cervical probing, but this activity was attenuated by large doses of progesterone (Cross and Silver, 1965). Later investigations (Lincoln and Cross, 1967; Lincoln, 1969a) demonstrated that some of these effects were probably nonspecific in that they were linked to a general increase in the frontal EEG arousal. These early studies also tended to use large doses of steroids (e.g., 400 mg of progesterone, Barraclough and Cross, 1963), which made the results questionable in terms of a physiological response.

Of physiological significance, Lincoln and Cross (1967) recorded from single neurons in the preoptic, septal, and anterior hypothalamus that did show a specific increase in activity after cervical stimulation. These investigators were confident that these cells were under estrogen-feedback control (Lincoln, 1967; Lincoln and Cross, 1967). Moreover, Cross and Dyer (1971b) recorded from preoptic–anterior hypothalamic cells throughout the estrous cycle of the rat and found a prominent increase in the activity of these cells on the afternoon of proestrus, and Moss and Law (1971) corroborated these findings. Multiunit recordings also showed similar increases on the afternoon of proestrus in the arcuate nucleus (Terasawa and Sawyer, 1969a,b; Kawakami *et al.,* 1970) and the medial preoptic (Wuttke, 1974). Wuttke correlated these increments in activity with elevated plasma levels of luteinizing hormone (LH) in alert, freely moving rats. It should be emphasized that during the 4-day estrous cycle of the rat plasma estrogen levels peak during the afternoon of proestrus (Nequin *et al.,* 1975). Cross and Dyer (1971a) extended their previous observations by documenting that the cyclic fluctuations in medial preoptic–anterior hypothalamic (mPOA–AH) unit activity were not dependent on afferent inputs, for they recorded the activity in rats with diencephalic islands. Dyer (1973) noted further that these mPOA–AH neurons did not project to the mediobasal hypothalamus, for they were not driven antidromically by median eminence stimulation, a technique that allows one to identify the axonal projection of a neuron. Cross (1973) concluded from the aforementioned findings that the cyclicity of mPOA–AH activity must be related to the estrous cycle and is ovarian (steroid) dependent. Furthermore, it was more likely that an interneuron was involved since antidromically identified cells (putative peptide neurosecretory neurons) did not show similar fluctuations.

The next logical step was to look for changes in activity after estrogen administration. Yagi (1970, 1973) demonstrated that mPOA neurons, which were recorded from ovariectomized rats, would change their firing frequency in response to estrogen given intravenously (i.v.). Fifty percent of the responsive mPOA neurons increased their firing frequency with a mean latency to response of 16 min. Bueno and Pfaff (1976) found that estrogen given i.v. decreased the number of cells with spontaneous activity in the mPOA of ovariectomized rats. Dufy and colleagues (Dufy *et al.,* 1975, 1976) reported similar findings in the ovariectomized rabbit but also measured serum LH and follicle-stimulating hormone (FSH). Furthermore, Dufy and co-workers antidromically identified (AI) these mPOA neurons as having their axons projecting to the median eminence. Essentially all of the AI cells that were responsive to estradiol benzoate ($n = 15$) showed a decrease in their firing rate following an i.v. injection of 20 μg of estradiol benzoate (Fig. 1). This decrease in the firing frequency preceded the precipitous drop in LH and FSH. A sequel study (Dufy *et al.,* 1978) demonstrated that actinomycin D or cycloheximide, potent RNA or protein synthesis inhibitors respectively, could not prevent the inhibitory effects of estrogen on LH and FSH secretion. In this study the authors did not measure electrical activity but assumed that there was a causal relationship between the actions of estrogen on the electrical activity and the secretion of the gonadotropins. Dufy and colleagues proposed at this time that a "membrane-mediated process" at the hypothalamic level was involved in the negative feedback of estrogen.

It is important to note that the latencies for the effects of estrogen

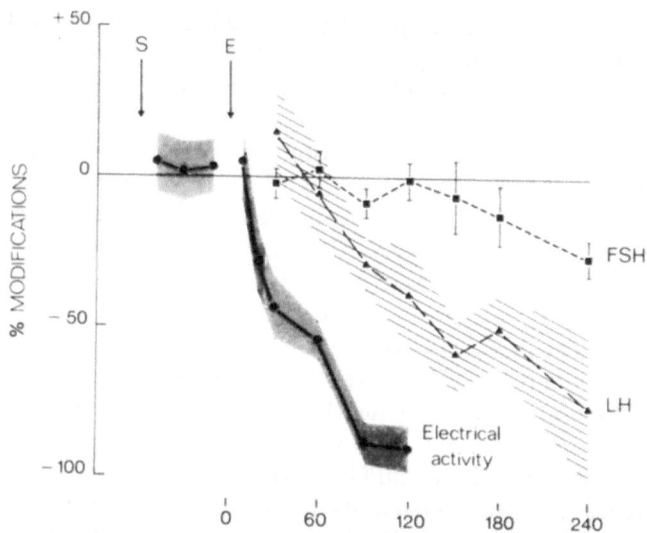

Figure 1. The effects of estradiol benzoate injection (E) or solvent injection (S) on the spontaneous activity of 14 antidromically identified hypothalamic neurons in the anesthetized ovariectomized rabbit. Mean serum LH and FSH were also monitored. Percent modification in electrical activity, serum LH and FSH are expressed relative to controls along the ordinate. The time in minutes is represented along the abscissa. Standard deviations are also expressed (bars for FSH and shaded areas for LH and spontaneous activity). [From Dufy et al., 1975, with permission.]

following systemic administration of the steroid do not eliminate the possibility of a nuclear-mediated event (i.e., DNA-dependent RNA synthesis). For example, Means and Hamilton (1966) reported a rapid incorporation of [³H]uridine into nuclear RNA in the uterus 2 min after administration of estrogen. However, Dufy's observations using protein synthesis inhibitors argued against such a nuclear-mediated event being responsible for the negative feedback of estrogen.

We began our studies in Dallas about 1974 to look at the effects of estrogen on mPOA single unit activity. At this time, the question was, does estrogen act via a nuclear mechanism or does the steroid have direct effects on the membrane to bring about changes in electrical excitability? To begin to answer this we utilized the technique of microelectrophoresis* to deposit small quantities of steroids (femtomoles) into the immediate vicinity of a cell while simultaneously measuring unit activity (Kelly et al., 1977a). This technique was used first by Ruf and Steiner (1967) to study the effects of

---

*Microelectrophoresis and iontophoresis will be used interchangeably in this chapter to denote the movement of a solute from a micropipet as a result of passing current.

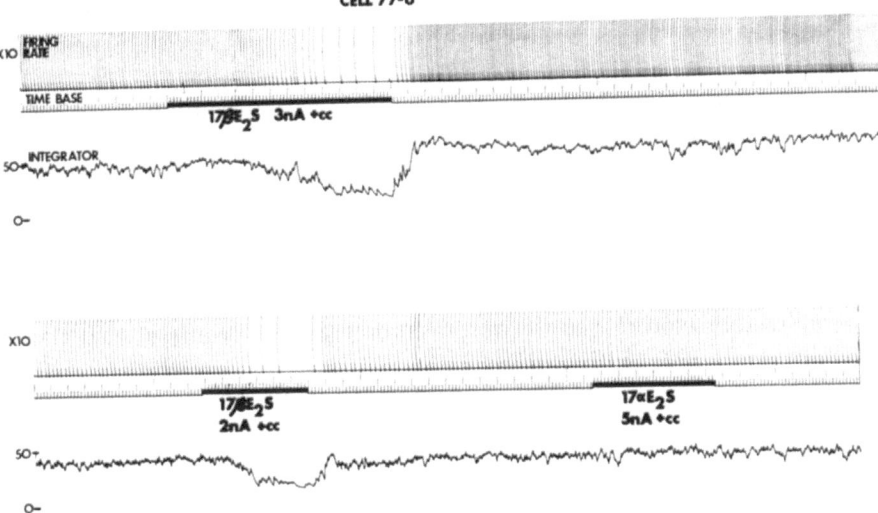

Figure 2. A strip-chart recording of the spontaneous activity of an mPOA neuron and the effects of 17β-estradiol hemisuccinate and 17α-estradiol hemisuccinate applied microelectrophoretically (see text). The 17β-estradiol ester caused a direct inhibition of the cell's activity that was not mimicked by 17α-estradiol even at higher ejection currents (higher "doses"). The baseline activity of the cell was 8 Hz. Each set of recordings shows pulses of the activity (upper trace), the time base in seconds (middle trace), and an integration of the activity over 5-sec epochs in the lower trace. [From Kelly *et al.*, 1977b, with permission.]

dexamethasone (phosphate) on hypothalamic and mesencephalic neurons (see below). We showed that 17β-estradiol hemisuccinate (17βE$_2$S) was released via microelectrophoresis in great enough quantity to affect neuronal activity (Kelly *et al.*, 1977b). Cycling female rats were used in our study, and we were able to demonstrate that estrogen had a rapid action on mPOA neurons with a latency of milliseconds to seconds (Fig. 2). A total of 161 cells were tested with estrogen. A number of these cells ($n = 23$) were identified through antidromic activation as having their axons projecting to the arcuate–median eminence. We used this technique to identify putative peptidergic neurons (Moss, 1976). Four of the cells were inhibited almost "immediately" by microelectrophoresed 17βE$_2$S. The remaining cells were unresponsive, but the majority of these ($n = 17$) lacked spontaneous activity so we were not able to measure inhibition without pharmacological intervention. Of the total 161 cells, 138 could not be driven antidromically, but they also were tested with 17βE$_2$S. Fifty-four of these responded within milliseconds or seconds to the steroid. Forty showed a decrease in firing rates and 14 showed increased activity following 17βE$_2$S. These effects were both reversible (terminated with the end of the microelectrophoresis) and reproducible with several applications (Fig. 2). In order to demonstrate the specificity of the response, an additional 36 neurons were tested with $17\beta E_2 S$ and 17α-estradiol hemisuccinate ($17\alpha E_2 S$). Twelve neurons (including two cells that were antidromically identified) showed decreased firing with $17\beta E_2 S$, and none of these, or the other 24 neurons, were responsive to $17\alpha E_2 S$ even at higher "doses" (higher ejection currents) (Fig. 2). Therefore, we demonstrated that estrogen could rapidly modify the electrical activity of mPOA neurons, that these effects were specific and that they were not mediated by the cell nucleus. Cortisol hemisuccinate was also applied microelectrophoretically, and these results will be discussed below. Since we were recording extracellularly at that time (Kelly et al., 1977a), we could not say that the cell from which we were recording was the "target" cell of the steroid's action, but we hypothesized that the inhibitory actions of $17\beta E_2 S$ could reflect an inhibition of inhibitory interneurons that resulted in the activation of luteinizing hormone-releasing hormone (LH-RH) neurons on proestrus (Kelly et al., 1977a).

Yamada (1975) reported that iontophoresed estrone (sodium estrone sulfate) inhibited hypothalamic neurons in the male rat and inhibited mPOA–AH neurons in the castrated female rat (Yamada and Nishida, 1978). Although they used higher ejection currents (high doses) than in our studies (Kelly et al., 1975, 1977a,b), the inhibition was similarly rapid in onset and completely reversible. A lower percentage of responsive neurons in Yamada's study (16 out of 145) could be attributed to the differences in the estrogen compounds used or the effects of long-term castration on the response (Kelly et al., 1978a).

We did not know at this point in our studies whether estrogen was acting presynaptically or postsynaptically to alter the spontaneous activity of rat mPOA neurons. We did have evidence that the effects were exerted postsynaptically, since $17\beta E_2 S$ could antagonize the effects of acetylcholine (Ach) excitation (Kelly et al., 1977a). However, Whitehead and Ruf (1974) concluded from their studies with intravenously administered estrogen that the steroid was acting presynaptically, since the dose–response curve of dopamine, norepinephrine, or Ach for antidromically identified mPOA neurons was not altered after the steroid. They measured the response 30 min postinfusion even though estrogen inhibited the spontaneous activity of some mPOA neurons within 5 min.

Another question that arose from our aforementioned studies was the identity of the estrogen-responsive neurons. The majority of these were not found to have projections to the mediobasal hypothalamus. Furthermore, the time course of the action of $17\beta E_2 S$ suggested that its mode of action was similar to that of a neurotransmitter rather than the classical nuclear-mediated mode of action. In order to begin to answer these questions, it was necessary to perform intracellular recordings, which would allow us to

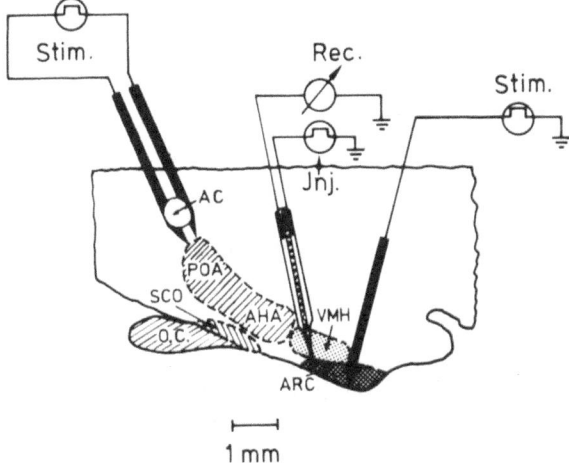

├──────┤
1 mm

Figure 3. Schematic drawing of 400-μm-thick sagittal hypothalamic slice. Slices were prepared from cycling (diestrous) female guinea pigs and maintained in oxygenated medium for 8–10 hr. This figure shows the slice with stimulating electrodes (Stim) placed in the stria terminalis and median eminence (ventral surface of slice). Arcuate (ARC), ventromedial (VMH), and cell-poor-zone neurons were recorded (Rec) through single-barrel micropipets filled with procion yellow, or theta glass micropipets (as illustrated) via conventional intracellular recording techniques. A bridge circuit was utilized to inject current (Inj) and measure voltage changes in order to determine the input resistance of a cell. AC, anterior commissure; AHA, anterior hypothalamus; POA, preoptic; SCO, suprachiasmatic; OC, optic chiasm. [From Kelly *et al.*, 1981, with permission.]

measure changes in membrane excitability and to label the neurons intracellularly. It has been virtually impossible to obtain stable intracellular recordings in the small parvocellular hypothalamic neurons *in vivo*. Therefore, we developed an *in vitro* slice preparation for the hypothalamus to obtain long-term intracellular recordings (Kelly *et al.*, 1978b, 1979c). The hypothalamic slice also allowed us to apply 17β-estradiol (E_2) in the bathing medium (Kelly *et al.*, 1979a,b). [The methodology for the preparation of the sagittal hypothalamic slice has been published in detail (Kelly *et al.*, 1979c, 1980).] Figure 3 illustrates the basic recording scheme. In these studies, we chose to record from the mediobasal hypothalamus of female guinea pigs because of the presence of LH-RH-containing neurons and their documented role in control of the tonic secretion of LH from the pituitary gland (Silverman and Krey, 1978; Krey and Silverman, 1978). A total of 85 neurons were recorded in the arcuate nucleus (ARC), ventromedial (VM), and cell-poor zone (CPZ) between the ARC and VM. Twenty-eight cells were tested with E_2, and Fig. 4 shows a representative response from one of the 11 responsive neurons. Eleven neurons (ARC:$n = 9$; CPZ:$n =$

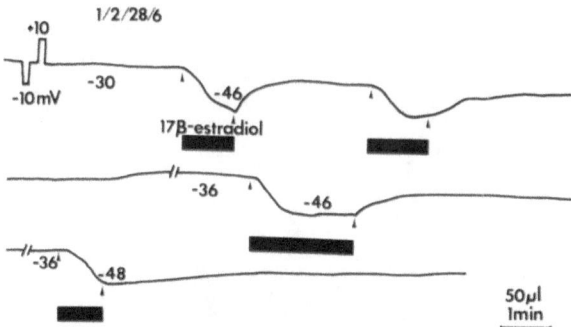

Figure 4. Continuous script-chart recording of changes in the membrane potential of an arcuate neuron during exposure to 17β-estradiol (E₂) (see text). Solid bars indicate time of influx of E₂-medium into the recording chamber. Note the significant hyperpolarization of 16 mV which occurred at concentrations of 10^{-10} M E₂. Breaks in the recording indicate that times of 5 and 6.5 min have passed respectively. The cell was stabilized at -30 mV ("resting membrane potential"). Calibration bar represents a flow rate of 50 µl/min. 10-mV calibration pulses are at the beginning of the record. [From Kelly *et al.,* 1980, with permission.]

2) were hyperpolarized 2 to 24 mV by physiological concentrations of E₂ (10^{-10} M). It was impossible to accurately measure response latency since the steroid-containing medium was perfused through the chamber reservoir. However, the time course for this response is similar to that of putative neurotransmitters applied in a slice perfusion system (Pittman *et al.,* 1980). Figure 4 also illustrates that the hyperpolarization is only reversed upon washing-out the E₂. Estrone was applied to a few of these neurons at concentrations of 10^{-8} M and was unable to produce any changes. In neurons that demonstrated a spontaneous activity there was a decrease in firing frequency of the cell accompanying the hyperpolarization. Finally, the input resistance, a measure of membrane ion conductances, was found to decrease, which could indicate that K^+ and Cl^- conductance increases are involved.

As a first step toward identifying these cells, procion yellow, a fluorescent intracellular marker, was injected into these neurons. Three of the E₂-responsive neurons (ARC cells) were identified antidromically as having axonal projections to the median eminence; this was verified histologically by tracing the labeled axon. Figure 5 contrasts the differences between the E₂-responsive neurons (Fig. 5a) and the nonresponsive neurons (Fig. 5b). All of the E₂-responsive cells were small fusiform neurons with little branching of the primary dendrites, and with the exception of one cell, they had no spine-like appendages. None of the pyramidal-like neurons of Fig. 5b were shown to be responsive to E₂. We reasoned that since spines are indicative of synaptic contact (Palay and Chan-Palay, 1977), the estrogen-responsive neurons are more likely to be under humoral control because of their lack of spines (Kelly *et al.*, 1981). Also, these estrogen-responsive neurons lie within the hypophysiotropic area originally described by Halász and colleagues (1962). In the future, we plan on identifying these procion yellow-labeled neurons immunocytochemically to see if they are LH-RH neurons.

Presently, we hypothesize that estrogen's negative feedback effect, which is known to be relatively rapid (McCann, 1974), is exerted through a membrane-mediated event, i.e., the hyperpolarization of the cell membrane. The findings of Dufy *et al.* (1975, 1976) are congruent with this hypothesis. Positive feedback, which is known to have a much longer latency (McCann, 1974), is probably nuclear-mediated. Although some assumptions have been made based on temporal correlations, this hypothesis offers a clear cellular separation of these two distinct phenomena. Future research in my laboratory will reveal the underlying mechanism for the changes in membrane excitability and with which cell types this is associated.

3. Effects of Other Steroids (Progesterone, Testosterone, and Cortisol) on Hypothalamic Neurons

3.1. Progesterone

Progesterone was one of the first steroids to be investigated in conjunction with hypothalamic electrical recording (Kawakami and Sawyer, 1959). Kawakami and Sawyer (1959) found progesterone to have a biphasic effect (initially decreasing and then increasing) on the coitus-induced activity (EEG) in the hypothalamus of the rabbit. Barraclough and Cross (1963) found that large doses of progesterone (400 mg) depressed the response of lateral hypothalamic neurons to vaginal stimulation. However, progesterone, when given intravenously, causes nonspecific anesthetic-like effects on hypothalamic unit activity and EEG activity (Ramirez *et al.*, 1969; Komisaruk *et al.*, 1967; Lincoln, 1969b). Furthermore, Beyer *et al.* (1967) found that 100–200 μg of progesterone caused synchronization of the cortical EEG (spindling) and a decrease in hypothalamic multiunit activity. These effects of progesterone were always associated with an increase in blood pressure.

Terasawa and Sawyer (1970) recorded multiunit activity in the arcuate nucleus–median eminence of the rat and found specific changes in ativity after progesterone administration. In the estrogen-primed ovariectomized rat, they found an increase in activity several hours after progesterone was given subcutaneously. In the intact rat the response varied

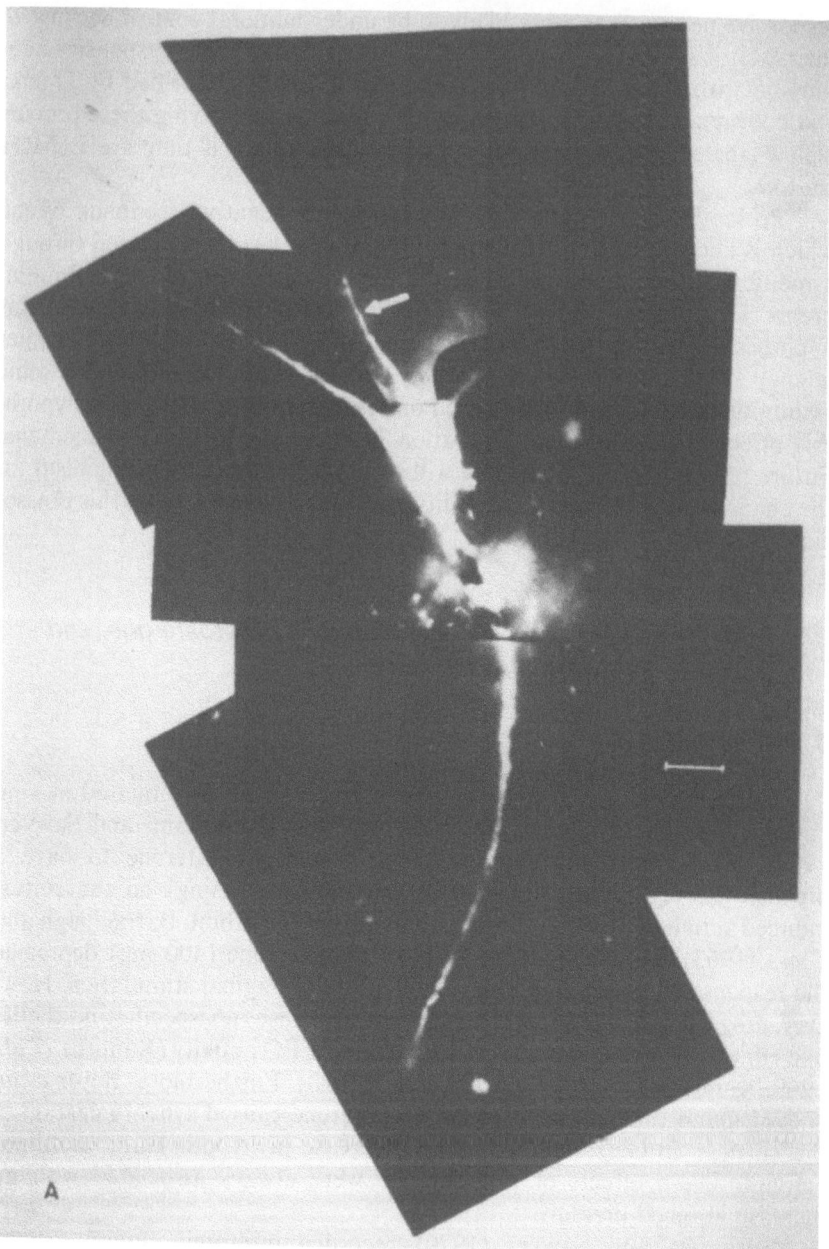

Figure 5. (a) Montage of procion yellow-stained arcuate neuron of Fig. 4. The recording electrode was situated in a primary dendrite (hole surrounded by "halo" of fluorescence). The cell exhibits three primary dendrites without branches (traced for at least 200 μm) and an axon (arrow) exiting from the soma and projecting rostrally. This morphology was typical for estrogen-sensitive neurons. Calibration bar = 10 μm. (b) Montage of a procion yellow-labeled arcuate

according to when in the cycle the steroid was given, but all of these changes were independent of EEG and blood pressure. Most of the animals showed a phasic response (increase–decrease–increase or decrease–increase). Dufy and Dufy-Barbe (1979) measured the effects of 20α-hydroxy-progesterone given intravenously on rabbit hypothalamic neurons and found some cells to increase their spontaneous activity (5 out ot 50) and some to decrease their firing frequency (12 out of 50) with latencies to the response of several minutes.

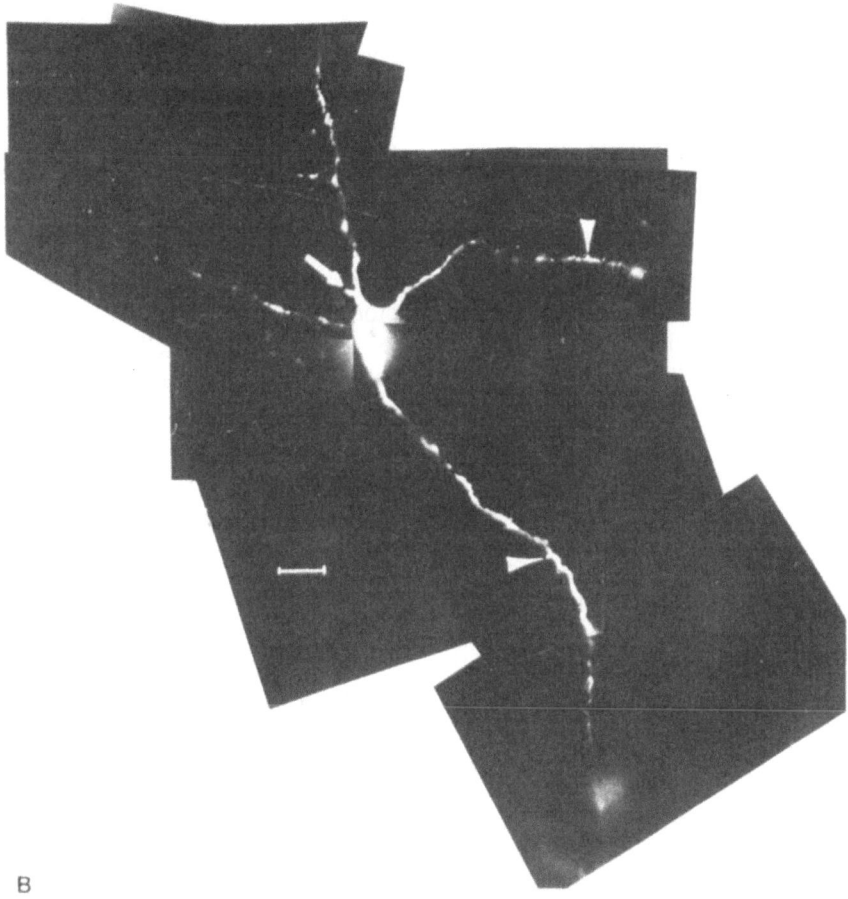

B

neuron which was not estrogen-sensitive. Spinelike appendages (arrowheads) are numerous and axon (arrow) exits from primary dendrite and takes a tortuous ventral course toward the median eminence (to the right of the montage). This cell had four primary dendrites, one of which branched after 30 μm. Resting membrane potential was 55 mV. Calibration bar = 10 μm. [Both from Kelly *et al.*, 1980, with permission.]

Although an attempt has been made to relate the changes in hypothalamic neuronal excitability produced by progesterone to specific neuroendocrine events, there seems to be no temporal correlation, i.e., progesterone is not involved in an acute feedback regulation of neurosecretory events. With higher doses of progesterone, one sees an anesthetic-like effect similar to barbiturates' action, which cannot be differentiated from general changes in the EEG. Therefore, progesterone's acute effects on hypothalamic electrical activity appear to be more pharmacological than physiological.

3.2. Androgens

Testosterone exerts a negative feedback effect on LH secretion at the hypothalamus in the male rat similarly to the effects of estrogen in the female but apparently with a longer latency (Cheung and Davidson, 1977). Also, testosterone can lead to increased sexual behavior in the male rat (Davidson, 1966; Lisk, 1967). However, at least some of the effects of testosterone are believed to be a result of aromatization of testosterone to estradiol (Naftolin et al., 1975). Pfaff and Pfaffmann (1969b) showed that testosterone applied directly to the preoptic area via a cannula enhanced the response of preoptic neurons to olfactory bulb stimulation. Also, testosterone could modify the spontaneous activity of preoptic cells (Pfaff and Pfaffmann, 1969b). A recent study investigated the effects of testosterone on the absolute refractory period following antidromic stimulation of the corticoamygdaloid neurons that project to the medial preoptic area (Kendrick and Drewett, 1979). In the castrated male rat, the refractory period increased by 50% when compared to the intact control. Testosterone treatment for 18 to 22 days restored the absolute refractory period to its initial value (mean 1 msec). In a sequel paper using the same model (Kendrick and Drewett, 1980), the investigators found that dihydrotestosterone was not able to decrease the refractory period following antidromic stimulation in the castrated male rat. The authors concluded that testosterone and not dihydrotestosterone directly modifies activity of brain neurons. However, another interpretation could be that the dihydrotestosterone is not aromatized to estrogen, and it is the estrogen that is the active compound.

Recently, Yamada (1979) demonstrated that the iontophoresis of testosterone directly onto anterior hypothalamic and septal neurons causes an increase in their spontaneous activity. Moreover, estrone did not cause any changes in firing frequency in any of these cells ($n = 13$). Yamada (1979) felt that he had identified specific testosterone-sensitive neurons. An earlier study (Yamada and Nishida, 1978) showed that neurons responsive to estrone were not responsive to dehydroepiandrosterone, which, together with the aforementioned studies, would suggest that aromatase activity is not involved in the neuronal-sensitivity to testosterone. Therefore, there appears to be direct effect of testosterone on membrane excitability, but its role in negative feedback regulation and/or behavior needs to be ascertained.

3.3. Adrenocortical Steroids

Early studies documented the role of glucocorticoids in negative feedback on adrenocorticotropic hormone (ACTH) secretion using the implantation of steroids into the hypothalamus (Davidson and Feldman, 1963). Moreover, corticoids were shown to affect both evoked potentials and the unit activity of the hypothalamus (Feldman and Dafny, 1970). Ruf and Steiner (1967) were the first to utilize the technique of microelectrophoresis to apply dexamethasone (phosphate), a synthetic corticosteroid and potent inhibitor of ACTH secretion, to identify hypothalamic and mesencephalic neurons that were sensitive to small quantities of cortical steroids. Fifteen out of 115 neurons were inhibited, and the inhibition persisted for 20–30 sec after steroid application. In a more extensive study of hypothalamic and mesencephalic neurons, Steiner and colleagues (1969) identified 61 dexamethasone-responsive neurons. Four of these were excited by dexamethasone, but the majority (93%) were inhibited. These responsive neurons were distributed throughout the hypothalamus. Interestingly, Steiner *et al.* (1969) noted that some of the neurons' activity was not depressed immediately, but their spontaneous activity was "modulated" (greater fluctuation in phasic activity). Steiner (1972) in a later study demonstrated that corticosterone (phosphate), when applied microelectrophoretically, was an even more potent inhibitor of spontaneous activity.

Mandelbrod *et al.* (1974) investigated the effects of cortisol sodium succinate, applied microelectrophoretically, on the spontaneous activity of tuberal hypothalamic neurons of male rats. A total of 356 neurons were tested with the steroid and 177 of these (49%) exhibited a response. The major response (145 out of 177) was inhibition of spontaneous activity. The effects of the cortisol ceased with the switching-off of the iontophoretic current. Also, the pretest level of spontaneous activity of a neuron did not determine its response. This is consistent with our findings for estrogen (Kelly *et al.*, 1977a). We also applied cortisol (hemisuccinate) to preoptic–septal neurons. None of the AI neurons (see Section 2) responded to cortisol, and a different population of neurons from those that responded to the estrogen, responded to the cortisol. The major effect was inhibition of spontaneous activity (31 out of 45 responsive neurons). Interestingly, the response characteristics to the steroid were similar for preoptic–septal neurons (Kelly *et al.*, 1977a) and arcuate neurons (Mandelbrod *et al.*, 1974). It should be emphasized that the steroids are iontophoresed into the "immediate" vicinity of a neuron and a slight difference in response latencies (a few seconds) might reflect a difference in diffusion time to the cell membrane. Also, an increase (excitation) in neuronal firing using extracellular recording techniques could reflect an inhibition of an adjacent inhibitory interneuron that synapses on the neuron from which one is recording.

In summary, the major effect of cortisol (dexamethasone) on hypothalamic neurons, when applied directly to the cell membrane by microelectrophoresis, is inhibition of spontaneous activity. This direct inhibition is perhaps related to the negative feedback exerted by corticoids on the hypothalamus, which is known to attenuate the secretion of ACTH from the pituitary gland (Mangili *et al.*, 1966).

4. Effects of Steroids on Behavior-Related Electrophysiological Events

4.1. Olfactory Bulb

Probably one of the more prominent actions of steroids is on behavior-related neuronal activity in the olfactory bulb. The urine of estrous female rats (odor) has been shown to increase the sexual behavior in intact male rats but not in castrated rats (Pfaff and Pfaffmann, 1969a). Moreover, testosterone will enhance the response of olfactory bulb units of male rats to these urine odors (Pfaff and Pfaffmann, 1969b). As previously mentioned, testosterone also enhances the response of preoptic neurons to olfactory bulb stimulation, presumably an important pathway for sexual behavior (Pfaff and Pfaffmann, 1969a). Recently, Macleod *et al.* (1979) have tested both testosterone and 5α-androst-16-en-3-one for their ability to stimulate mitral cell activity in the olfactory bulb of male and female pigs. Both these steroids are manufactured in the testis, stored in the submaxillary gland, and transmitted into the environment during excitement and sexual arousal. Mitral cells were recorded extracellularly and antidromically driven by stimulation of the lateral olfactory tract. Thirty-seven cells were tested with the two steroids, and 25 of these responded to one or both of the steroids when they were blown into the nasal pharynx by a constant air flow system (various other organic substances were also tested). There were prominent excitations of cell activity, within seconds of application, by both substances, which could not be mimicked by amylacetate, pyridine, or benzene. The excitation persisted for periods of up to a minute following cessation of the application of the steroid. A few of the neurons ($n = 3$) were inhibited by one or both of the substances.

The mitral cells have been identified autoradiographically to concentrate labeled testosterone (Pfaff, 1968), but the effects of the steroids are much too rapid to be explained by a nuclear-mediated event. Clearly, the response of olfactory bulb neurons (mitral cells) to gonadal steroids is related to the induction of sexual arousal in mammalian species such as the rat and pig.

4.2. Hippocampus

The hippocampus is an area that has been found from autoradiographic studies to concentrate labeled corticosterone (Gerlach and McEwen, 1972) and to have specific soluble corticosterone-binding macromolecules (McEwen *et al.*, 1972b). The hippocampus as a target site for corticosteroid feedback has both endocrine and behavioral significance. Its role in ACTH secretion is believed to be an inhibitory one, and it is an integral structure involved in different learning behaviors (Bohus, 1975).

Changes in hippocampal activity have been observed following adrenal cortical hormone administration (Feldman and Dafny, 1970; Pfaff *et al.*, 1971). Pfaff and co-workers (1971) demonstrated that an intraperitoneal injection of corticosterone can inhibit the spontaneous activity of hippocampal neurons with a latency of 10–40 min in freely moving animals. These effects were also observed in hypophysectomized animals, which eliminated the possibility that corticosterone's actions were via changes in ACTH secretion. Also, Pfaff's group utilized a cannula system to apply corticosterone directly into the hippocampus and were able to see a depression of pyramidal cell activity (Pfaff *et al.*, 1971). Because of the longer time course of the actions of corticosteroid, it is not known if the steroid is acting directly on the cell membrane or via cellular metabolic pathways to change membrane excitability. However, Teyler and co-workers (Teyler *et al.*, 1979; Vardaris *et al.*, 1980) have found an enhancement of synaptic input to the hippocampal CA1 neurons following application of 10^{-8} M testosterone (in the female) or 10^{-10} M estrogen (in the male) in the medium bathing hippocampal slices. Since the hippocampus is a limbic structure and is believed to be an integral part of emotional behavior (Papez, 1937), the role of the gonadal steroids could be to enhance synaptic input related to this behavior.

Kawakami and co-workers (Kawakami and Kubo, 1971; Kubo *et al.*, 1975) have found that the hippocampal input into the hypothalamus is inhibitory relative to the control of LH secretion in the rat. The threshold for activation (stimulation) of this inhibitory input is reduced in proestrous, estrous, or ovariectomized rats pretreated with estradiol benzoate. The hippocampus then appears to be a target site for the adrenal, and possibly the gonadal steroids. The significance of the actions of steroids on electrical activity of the pyramidal neurons has not been fully elucidated, but the role of the hippocampus in the mediation of learning and emotional behavior and its anatomical connection with the hypothalamus emphasize the importance of this structure as a site of steroid action.

4.3. Estrogen and the Nigrostriatum

The discussion so far has centered on the hypothalamus and other limbic structures that exhibit changes in electrical activity induced by steroids. An argument has been made that these rapid changes are an integral part of the feedback (negative) regulation by the steroids. Also, there are behavior-related electrophysiological events throughout the limbic system that are induced by steroids. The following discussion will look at extralimbic sites where the steroids have been shown convincingly to have some "modulatory" role in synaptic transmission.

Several groups have begun to study the effects of estrogen on the nigrostriatal dopamine system. These studies have been prompted by the clinical observations that women are more prone than men to develop basal ganglia disorders in response to neuroleptic drugs (Donlon and Stenson, 1976). Estrogen has been shown to increase apomorphine- and amphetamine-induced stereotypy (gnawing, licking, etc.) in ovariectomized female rats (Chiodo et al., 1980c). Moreover, female rats show a higher incidence of chlorpromazine-induced catalepsy than intact males (Mislow and Friedhoff, 1973). Recently, Chiodo et al. (1980a) have identified two types of dopamine neurons within zona compacta of the substantia nigra of rats according to their response to sensory stimuli. Type A neuron, as it is called by the investigators, responds to activating tactile or olfactory stimuli with an acceleration of its spontaneous activity. Type B neuron shows a decrease in its activity following the same stimuli. Both types of neurons were shown by the investigators to fit all the electrophysiological and pharmacological criteria for dopamine cells (Bunney and Aghajanian, 1976). Chiodo et al. (1980b) have also demonstrated that an intravenous injection of E_2 can enhance Type A neuron activity and attenuate Type B activity (Fig. 6). Moreover, E_2 pretreatment is able to "modulate" the response of dopamine neurons to iontophoretically applied dopamine, which causes an activation of the autoreceptors on the dopamine cells (Chiodo et al., 1980b). The doses of estrogen utilized do not affect the metabolism of amphetamine or apomorphine by the brain (Chiodo et al., 1980c).

Interestingly, substantia nigra dopamine neurons (A9) have not been identified autoradiographically to concentrate labeled estrogen. Although it cannot be stated that estrogen is involved in the "normal" electrophysiology of nigrostriatal dopamine neurons, the aforementioned studies do give compelling evidence that estrogen has a role (modulatory?) in the pathophysiology of the basal ganglia disorders.

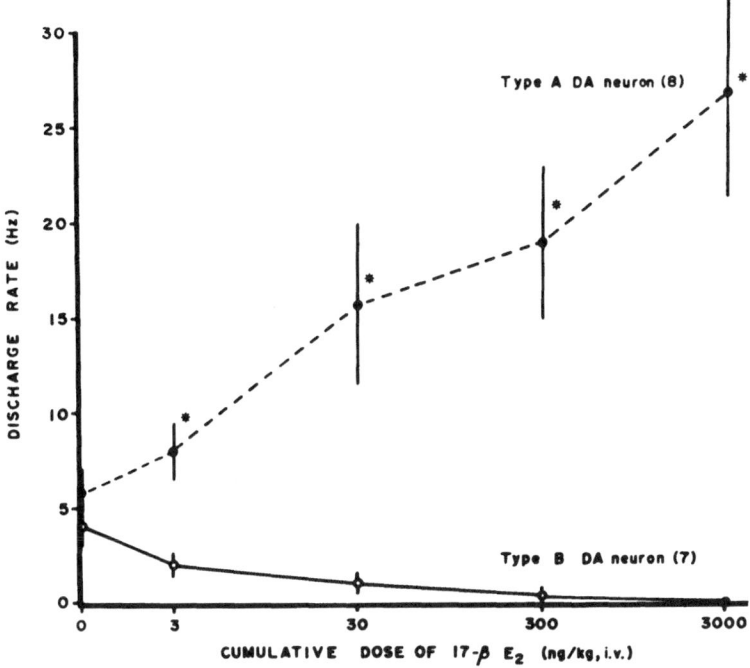

Figure 6. The effects of 17β-estradiol on Type A and Type B nigrostriatal dopaminergic neurons located in the zona compacta of the substantia nigra of ovariectomized female rats (see text for classification). Estrogen was given intravenously at 2-min intervals and total accumulative dosage is shown along the abscissa. The frequency of discharge (4- to 6-Hz resting level for nigrostriatal dopaminergic neurons) is shown along the ordinate. Estrogen was able to markedly attenuate the resting discharge of Type B neurons but increased the spontaneous activity of Type A neurons, presumably via actions on the dopamine autoreceptors with a latency of 5–10 sec after intravenous administration (L. A. Chiodo and A. R. Caggiula, unpublished observations).

4.4. Estrogen ''Modulation'' of Central Gray Neurons

The lordosis "reflex," dorsiflexion of the vertebral column, is an essential element of female copulatory behavior in the rodent and is dependent on estrogen levels (Pfaff *et al.*, 1974). Furthermore, central gray (CG) neurons are known to be an integral part of this "reflex" pathway (Sakuma and Pfaff, 1980a). These CG neurons receive afferents from the preoptic and ventromedial hypothalamus (Conrad and Pfaff, 1976; Kreiger *et al.*, 1979), and stimulation of the preoptic area suppresses (Moss *et al.*, 1974) and stimulation of the ventromedial area facilitates lordosis (Pfaff and Sakuma, 1979). Moreover, Sakuma and Pfaff (1979a,b) have demonstrated that stimulation of the central gray area facilitates this behavior in the estrogen-primed ovariectomized female rat and that lesions of this area diminish the lordosis response. Sakuma and Pfaff (1980b) also have found that CG neurons, which were antidromically identified as having projections to the medullary reticular formation (presumably involved in the descending pathway for lordosis), exhibited a higher resting discharge following estrogen treatment of ovariectomized female rats. Moreover, estrogen alone or stimulation of the ventromedial nucleus alone, increased the likelihood of antidromic action potentials to invade the soma (somatodendritic spike) of CG neurons in these ovariectomized female rats (Sakuma and Pfaff, 1980b). Estrogen has been shown by autoradiographic studies to be taken up by preoptic and ventromedial neurons (Pfaff and Keiner, 1973), and long-term treatment of ovariectomized female rats with estrogen does increase the resting discharge of ventromedial neurons and decreases the discharge of preoptic neurons (Bueno and Pfaff, 1976). The authors conclude that the actions of estrogen on the lordosis "reflex" are probably via new protein synthesis in preoptic and/or ventromedial neurons, whose projections impinge directly on CG neurons and "modulate" their excitability.

5. Conclusion and Prospectus

The concept of steroid action in the CNS being mediated by protein synthesis has gained widespread acceptance based on autoradiographic and biochemical data. However, McEwen et al. (1972a) recognized in their review 10 years ago that not all actions of steroids can be explained in this classical framework of events for steroid actions. Indeed, many neuroendocrine and behavioral events occur much too rapidly to be explained by new protein synthesis.

There are effects of estrogens on monoamine metabolism that can be attributed to the indirect actions of the steroid on the genome (McEwen and Luine, 1978). Recent reports have shown also that estrogen can increase the efflux of catecholamines from short-term cultures of rat hypothalamus (Paul et al., 1979); and estrogen can cause the accumulation of cyclic AMP in hypothalamic slices in vitro, which is blocked by adrenergic and dopaminergic blocking agents (Gunaga et al., 1974; Weissman et al., 1975). Weissman et al. (1975) proposed that estrogen was acting presynaptically to cause the release of catecholamines via a nuclear-mediated mechanism and that the catecholamines act postsynaptically via cyclic AMP as in the cerebellum (Siggins et al., 1971). However, the steroid could act directly on the presynaptic terminal to cause the release of catecholamines. Bennett et al. (1975) have found that in vitro application of E_2 to a hypothalamic synaptosomal preparation causes augmentation of the release of LH-RH into the medium following electrical stimulation. In conflict to this report, however, Tytell *et al.* (1980) did not find any effect of estrogen when applied directly to mitochondrial synaptosomal fractions from the mediobasal hypothalamus of ovariectomized rats.

The only real direct action of the steroids that has been measured in a physiological context is on the electrical activity, which has been the subject of this review. Although these actions of the steroids appears to be "transmitterlike," we do not have any information on membrane receptor sites. Two groups (Szego, 1978; Jungblut, 1981) have identified microsomal-bound receptors for E_2 in pig and rat uteri, and Szego (1978) has identified receptors on the plasma membrane of uterine epithelial cells. It would not be unreasonable to assume that similar membrane receptors exist in the CNS for estrogen as in the case for cytoplasmic (nuclear) receptors. Once these receptors have been isolated and characterized, we can begin to explore the physiological significance of Szego's (Jungblut's) finding—that is, does membrane receptor blockade prevent the electrophysiological effects of estrogen? Indeed, Carette *et al.* (1979) have found that estradiol-7α-butyric acid applied microelectrophoretically inhibits preoptic neurons within seconds, yet it does not cross the cell membrane (Carette *et al.*, 1979).

A final question on the electrophysiological effects concerns the site of action of the steroid. Are the steroid actions on the same cells that have been identified autoradiographically to concentrate steroids? Estrogen "inhibits" the activity of a similar percentage (30%) of mediobasal hypothalamic neurons (Kelly *et al.*, 1980) as the percentage of neurons that take up labeled estrogen (Pfaff, personal communication). Moreover, other adjacent neurons, which are not directly "inhibited" by estrogen, might be electrically coupled to steroid-responsive neurons. Andrew *et al.* (1980) have presented recent evidence for the existence of gap junctions in the paraventricular nucleus, which would explain our observation of dye coupling between arcuate neurons (Kelly *et al.*, 1979c). Therefore, a change in a single neuron's activity could ramify throughout other neurons and create an electrophysiological syncytium. This would explain how changes in a few neurons could reflect a gross change in LH-RH output from the slice. Finally, we do not know if these estrogen-responsive cells are LH-RH neurons, but immunocytochemistry in combination with intracellular injection of procion yellow will yield this information.

ACKNOWLEDGMENTS. Portions of the work described in this review were carried out in the laboratory of Dr. R. L. Moss (Dallas), and in the laboratory of Drs. W. Wuttke and U. Kuhnt (Germany) as an NINCDS postdoctoral fellow. Current portions of this work are being supported by the author's research grants, HRSF W-22 and NIH NS16419.

I would like to thank Drs. B. Dufy and L. A. Chiodo for providing figures, and Drs. D. W. Pfaff and S. M. Smith for their comments on the manuscript. Also I am grateful to Ms. Gerry Camp for preparation of the manuscript.

REFERENCES

Andrew, R. D., MacVicar, B. A., Dudek, F. E., and Hatton, G. I., 1980, Dye-coupling by gap junctions in magnocellular neuroendocrine cells of rat hypothalamus: Evidence for electrotonic coupling, *Soc. Neurosci. Abstr.* Vol. 6, p. 456.

Barraclough, C. A., and Cross, B. A., 1963, Unit activity in the hypothalamus of the cyclic female rat: Effect of genital stimuli and progesterone, *J. Endocrinol.* **26:**339–359.

Bennett, G. W., Edwardson, J. A., Holland, D., Jeffcoate, S. L., and White, N., 1975, Release of immunoreactive luteinizing hormone-releasing hormone and thyrotropin-releasing hormone from hypothalamic synaptosomes, *Nature (London)* **257:**323–325.

Beyer, C., Ramirez, V. D., Whitmoyer, D. I., and Sawyer, C. H., 1967, Effects of hormones on the electrical activity of the brain in the rat and rabbit, *Exp. Neurol.* **18:**313–326.

Bohus, B., 1975, The hippocampus and the pituitary–adrenal hormones, in: *The Hippocampus,* Vol. 1: *Structure and Development* (R. L. Isaacson and K. H. Pribram, eds.), pp. 323–354, Plenum Press, New York.

Bueno, J., and Pfaff, D. W., 1976, Single unit recording in hypothalamus and preoptic area of estrogen-treated and untreated ovariectomized female rats, *Brain Res.* **101:**67–78.

Bunney, B. S., and Aghajanian, G. K., 1976, The effects of antipsychotic drugs on the firing rate of dopaminergic neurons: A reappraisal, in: *Antipsychotic Drugs: Pharmacodynamics and Pharmacokinetics* (G. Sedvall, B. Uvans, and V. Zotterman, eds.), pp. 305–318, Pergamon Press, Elmsford, N.Y.

Carette, B., Barry, J., Linkie, D., Ferin, M., Mester, J., and Beaulieu, E. E., 1979 Effets de l'aestradiol-7-α-acide burtyrique au niveau de cellules hypothalamiques, *C. R. Acad. Sci. Ser. D* **288:**631–634.

Cheung, C. V., and Davidson, J. M., 1977, Effects of testosterone implants and hypothalamic lesions on luteinizing hormone regulation in the castrated male rat, *Endocrinology* **101:**920–928.

Chiodo, L. A., Antelman, S. M., Caggiula, A. R., and Lineberry, C. G., 1980a, Sensory stimuli alter the discharge of dopamine (DA) neurons: Evidence for two functional types of DA cells in the substantia nigra, *Brain Res.* **189:**544–549.

Chiodo, L. A., Caggiula, A. R., and Lucik, R. R., 1980b, Estrogen-induced subsensitivity of dopamine autoreceptors, *Soc. Neurosci. Abstr.* Vol. 6, p. 451.

Chiodo, L. A., Caggiula, A. R., and Saller, C. F., 1980c, Estrogen potentiates the stereotypy induced by dopamine agonists in the rat, *Life Sci.* **28:**827–835.

Conrad, L. C. A., and Pfaff, D. W., 1976, Efferents from medial basal forebrain and hypothalamus in the rat. I. An autoradiographic study of the medial preoptic area, *J. Comp. Neurol.* **169:**221–262.

Cross, B. A., 1973, Towards a neurophysiological basis for ovulation, *J. Reprod. Fertil.* Suppl. 20, pp. 97–117.

Cross, B. A., and Dyer, R. G., 1971a, Unit activity in rat diencephalic islands: The effect of anaesthetics, *J. Physiol. (London)* **212:**467–481.

Cross, B. A., and Dyer, R. G., 1971b, Cyclic changes in neurons on the anterior hypothalamus during the rat estrous cycle and the effect of anaesthesia, in: *Steroid Hormones and Brain Function* (C. H. Sawyer and R. Gorski, eds.), pp. 95–102, University of California Press, Los Angeles.

Cross, B. A., and Silver, I. A., 1965, Effect of luteal hormone on the behaviour of hypothalamic neurons in pseudopregnant rats, *J. Endocrinol.* 31:251–263.

Davidson, J. M., 1966, Activation of the male rat's sexual behavior by intracerebral implantation of androgen, *Endocrinology* 79:783–794.

Davidson, J. M., and Feldman, S., 1963, Cerebral involvement in the inhibition of ACTH secretion by hydrocortisone, *Endocrinology* 72:936–946.

Donlon, P. T., and Stenson, R. L., 1976, Neuroleptic induced extrapyramidal symptoms, *Dis. Nerv. Syst.* 37:629–635.

Dufy, B., and Dufy-Barbe, L., 1979, Effects of gonadal steroids on the electrical activity of hypothalamic neurons, in: *Biologie Cellulaire des Processus Neurosécrétoires Hypothalamique* (J. D. Vincent and C. Kordon, eds.), pp. 207–220, CNRS, Paris.

Dufy, B., Partouche, C. H., Dufy-Barbe, L., and Vincent, J. D., 1975, Effects of oestrogen on the electrical activity of hypothalamic units: Correlations with gonadotrophic hormone levels, *International Congress on Psychoneuroendocrinology*, Budapest, pp. 303–312.

Dufy, B., Partouche, C. H., Poulain, D., Dufy-Barbe, L., and Vincent, J. D., 1976, Effects of oestrogen on the electrical activity of identified and unidentified hypothalamic units, *Neuroendocrinology* 22:38–47.

Dufy, B., Dufy-Barbe, L., and Vincent, J. D., 1978, Effects of protein synthesis inhibitors on the negative feed back effect of estrogen on LH release, *Horm. Res.* 9:279–291.

Dyer, R. G., 1973, An electrophysiological dissection of the hypothalamic regions which regulate the pre-ovulatory secretion of luteinizing hormone in the rat, *J. Physiol. (London)* 234:421–442.

Everett, J. W., and Sawyer, C. H., 1949, A neural timing factor in the mechanism by which progesterone advances ovulation in the cyclic rat, *Endocrinology* 45:581–595.

Feldman, S., and Dafny, N., 1970, Effects of cortisol on unit activity in the hypothalamus of the rat, *Exp. Neurol.* 27:375–387.

Gerlach, J. L., and McEwen, B. S., 1972, Rat brain binds adrenal steroid hormone: Radioautography of hippocampus with corticosterone, *Science* 175:1133–1136.

Gunaga, K. D., Kawano, A., and Menon, K. M. J., 1974, *In vivo* effect of estradiol benzoate on the accumulation of adenosine 3' 5'-cyclic monophosphate in the rat hypothalamus, *Neuroendocrinology* 16:273–281.

Halász, B., Pupp, L., and Uhlarik, S., 1962, Hypophysiotropic area in the hypothalamus, *J. Endocrinol.* 25:147–154.

Jensen, E. V., Suzuki, T., Numata, M., Smith, S., and De Sombre, E. R., 1969, Estrogenbinding substances of target tissues, *Steroids* 14(4):417–427.

Jungblut, P., Intracellular action of gonadal steroids, *Exp. Brain Res.*, **Suppl. 3**: 37–60.

Kato, J., and Onouchi, T., 1977, Specific progesterone receptors in the hypothalamus and anterior hypophysis of the rat. *Endocrinology* 101:920–928.

Kawakami, M., and Kubo, K., 1971, Neuro-correlate of limbic–hypothalamo–pituitary–gonadal axis in the rat: Changes in limbic–hypothalamic unit activity induced by vaginal and electrical stimulation, *Neuroendocrinology* 7:65–89.

Kawakami, M., and Sawyer, C. H., 1959, Induction of behavioral and electroencephalographic changes in the rabbit by hormone administration or brain stimulation, *Endocrinology* 65:631–643.

Kawakami, M., Terasawa, E., and Ibuki, T., 1970, Changes in multiple unit activity of the brain during the estrous cycle, *Neuroendocrinology* 6:30–48.

Kelly, M. J., Dudley, C. A., and Moss, R. L., 1975, Identification of estrogen sensitive neurons in the preoptic–septal area of the normal cyclic female rat, *Soc. Neurosci. Abstr.* Vol. 1, p. 457.

Kelly, M. J., Moss, R. L., and Dudley, C. A., 1977a, The effects of microelectrophoretically applied estrogen, cortisol, and acetylcholine on medial preoptic septal unit activity throughout the estrous cycle of the female rat, *Exp. Brain Res.* **30**:53–64.

Kelly, M. J., Moss, R. L., Dudley, C. A., and Fawcett, C. P., 1977b, The specificity of the response of preoptic septal area neurons to estrogen: 17α-estradiol versus 17β-estradiol and the response of extrahypothalamic neurons, *Exp. Brain Res.* **30**:43–52.

Kelly, M. J., Moss, R. L., and Dudley, C. A., 1978a, The effect of ovariectomy on the responsiveness of preoptic–septal neurons to microelectrophoresed estrogen, *Neuroendocrinology* **25**:204–211.

Kelly, M. J., Kuhnt, U., and Wuttke, W., 1978b, *In vitro* electrophysiological studies on hypothalamic neurons, *Pfluegers Arch. Eur. J. Physiol.* **373**(Suppl.):R54.

Kelly, M. J., Kuhnt, U., and Wuttke, W., 1979a, Effects of 17β-estradiol on hypothalamic parvocellular neurons as revealed by intracellular recordings and staining with procion yellow, The Endocrine Society, 61st Annual Meeting, p. 282.

Kelly, M. J., Kuhnt, U., and Wuttke, W., 1979b, Intracellular electrophysiological studies on the parvocellular neurons of the hypothalamus, *Fed. Proc.* **38**:2557.

Kelly, M. J., Kuhnt, U., and Wuttke, W., 1979c, Morphological features of physiologically identified hypothalamic neurons as revealed by intracellular marking, *Exp. Brain Res.* **34**:107–116.

Kelly, M. J., Kuhnt, U., and Wuttke, W., 1980, Hyperpolarization of hypothalamic parvocellular neurons by 17β-estradiol and their identification through intracellular staining with procion yellow, *Exp. Brain Res.* **40**:440–447.

Kelly, M. J., Kuhnt, U., and Wuttke, W., 1981, Electrophysiological and morphological properties of E_2-responsive neurons, *Exp. Brain Res., Suppl.* **3**:294–308.

Kendrick, K. M., and Drewett, R. F., 1979, Testosterone reduces refractory period of stria terminalis neurons in the rat brain, *Science* **204**:877–879.

Kendrick, K. M., and Drewett, R. F., 1980, Testosterone sensitive neurons respond to oestradiol but not to dihydrotestosterone, *Nature (London)* **286**:67–68.

Komisaruk, B. R., McDonald, P. G., Whitmoyer, D. I., and Sawyer, C. H., 1967, Effects of progesterone and sensory stimulation on EEG and neuronal activity in the rat, *Exp. Neurol.* **19**:494–507.

Kreiger, M. S., Conrad, L. C. A., and Pfaff, D. W., 1979, An autoradiographic study of the efferent connections of the ventromedial nucleus of the hypothalamus, *J. Comp. Neurol.* **183**:785–816.

Krey, L. C., and Silverman, A. J., 1978, The luteinizing hormone-releasing hormone (LH-RH) neuronal networks of the guinea pig brain. II. Projections to the median eminence and the regulation of gonadotropin secretion, *Brain Res.* **157**:247–255.

Kubo, K., Gorski, R. A., and Kawakami, M., 1975, Effects of estrogen on neuronal excitability in the hippocampal-septal hypothalamic system, *Neuroendocrinology* **18**:176–191.

Lincoln, D. W., 1967, Unit activity in the hypothalamus, septum and preoptic area of the rat: Characteristics of spontaneous activity and the effects of oestrogen, *J. Endocrinol.* **37**:171–189.

Lincoln, D. W., 1969a, Responses of hypothalamic units to stimulation on the vaginal cervix: Specific versus nonspecific effects, *J. Endocrinol.* **43**:683–684.

Lincoln, D. W., 1969b, Effects of progesterone on the electrical activity of the forebrain, *J. Endocrinol.* **45**:585–596.

Lincoln, D. W., and Cross, B. A., 1967, Effect of oestrogen on the responsiveness of neurons in the hypothalamus, septum and preoptic area of rats with light-induced persistent oestrus, *J. Endocrinol.* **37**:191–203.

Lisk, R. D., 1967, Neural localization of androgen activation of copulatory behavior in the male rat, *Endocrinology* **80**:754–761.

McCann, S. M., 1974, Regulation of secretion of follicle stimulating hormone and luteinizing hormone, in: *Handbook of Physiology*, Section 7: *Endocrinology*, Vol. V, Part 2 (R. O. Greep and E. B. Astwood, eds.), pp. 489–517, American Physiological Society, Washington, D.C.

McEwen, B. S., and Luine, V. N., 1978, Specificity, mechanisms and functional significance of steroid-receptor interactions in the brain and pituitary, in: *Biologie Cellulaire des Processus Neurosécrétoires Hypothalamique* (J. D. Vincent and C. Kordon, eds.), pp. 239–267, CNRS, Paris.

McEwen, B. S., Zigmond, R. E., and Gerlach, J. L., 1972a, Sites of steroid binding and action in the brain, in: *The Structure and Function of Nervous Tissue*, Vol. V (G. H. Bourne, ed.), pp. 205–291, Academic Press, New York.

McEwen, B. S., Magnus, C., and Wallach, G., 1972b, Soluble corticosterone-binding macromolecules extracted from rat brain, *Endocrinology* **90**:217–226.

Macleod, N., Reinhardt, W., and Ellendorff, F., 1979, Olfactory bulb neurons of the pig respond to an identified steroidal pheromone and testosterone, *Brain Res.* **164**:323–327.

Mandelbrod, I., Feldman, S., and Werman, R., 1974, Inhibition of firing is the primary effect of microelectrophoresis of cortisol to units in the rat tuberal hypothalamus, *Brain Res.* **80**:303–315.

Mangili, G., Motta, M., and Martini, L., 1966, Control of adrenocorticotrophic hormone secretion, in: *Neuroendocrinology* (L. Martini and W. F. Ganong, eds.), Vol. 1, pp. 298–370, Academic Press, New York.

Means, A. R., and Hamilton, T. H., 1966, Early estrogen action: Concomitant stimulations within two minutes of nuclear RNA synthesis and uptake of RNA precursor by the uterus, *Proc. Natl. Acad. Sci. USA* **56**:1596–1598.

Mislow, J. F., and Friedhoff, A. J., 1973, A comparison of chlorpromazine-induced extrapyramidal syndrome in male and female rats, in: *Hormones and Brain Function* (K. Lissak, ed.), pp. 315–326, Plenum Press, New York.

Moss, R. L., 1976, Unit responses in preoptic and arcuate neurons related to anterior pituitary function, in: *Frontiers in Neuroendocrinology*, Vol. 4 (L. Martini and W. F. Ganong, eds.), pp. 95–128, Raven Press, New York.

Moss, R. L., and Law, O. T., 1971, The estrous cycle: Its influence on single unit activity in the forebrain, *Brain Res.* **30**:435–438.

Moss, R. L., Paloutizian, R. F., and Law, O. T., 1974, Electrical stimulation of forebrain structures and its effect on copulatory as well as stimulus-bound behavior in ovariectomized hormone-primed rats, *Physiol. Behav.* **12**:997–1004.

Naftolin, F., Ryan, K. J., Davies, I. J., Reddy, V. V., Flores, F., Petro, Z., and Kuhn, M., 1975, The formation of estrogens by central neuroendocrine tissues, *Recent Prog. Horm. Res.* **31**:295–315.

Negro-Vilar, A., Orias, R., and McCann, S. M., 1973, Evidence for a pituitary site of action for the acute inhibition of LH release by estrogen in the rat, *Endocrinology* **92**:1680–1684.

Nequin, L. G., Alvarez, J., and Schwartz, N. B., 1975, Steroid control of gonadotropin release, *J. Steroid Biochem.* **6**:1007–1012.

Palay, S. L., and Chan-Palay, V., 1977, Morphology of neurons and neuroglia, in: *Handbook of Physiology*, Section 1: *The Nervous System*, Vol. 1: *Cellular Biology of Neurons*, Part 1 (J. M. Bookhart and V. B. Mountcastle, eds.), pp. 5–37, American Physiological Society, Washington, D.C.

Papez, J. W., 1937, A proposed mechanism of emotion, *Arch. Neurol. Psychiatry* **38:**725–744.

Paul, S. M., Axelrod, J., Saavedra, J. M., and Skolnick, D., 1979, Estrogen-induced efflux of endogenous catecholamines from the hypothalamus *in vitro, Brain Res.* **178:**499–505.

Pfaff, D. W., 1968, Autoradiographic localization of radio-activity in rat brain after injection of tritiated sex hormones, *Science* **161:**1355–1356.

Pfaff, D. W., and Keiner, M., 1973, Atlas of estradiol-concentrating cells in the central nervous system of the female rat, *J. Comp. Neurol.* **151:**121–158.

Pfaff, D., and Pfaffmann, C., 1969a, Behavioral and electrophysiological responses of male rats to female rat urine odors, in: *Olfaction and Taste* (C. Pfaffmann, ed.), pp. 258–267, Rockefeller University Press, New York.

Pfaff, D. W., and Pfaffmann, C., 1969b, Olfactory and hormonal influences on the basal forebrain of the male rat, *Brain Res.* **15:**137–156.

Pfaff, D. W., and Sakuma, Y., 1979, Facilitation of the lordosis reflex of female rats from the ventromedial nucleus of the hypothalamus, *J. Physiol. (London)* **288:**189–202.

Pfaff, D. W., Gregory, E., and Silva, M. T. A., 1971, Testosterone and corticosterone effects on single unit activity in the rat brain, in: *Influence of Hormones on the Nervous System* (O. H. Ford, ed.), pp. 269–281, Karger, Basel.

Pfaff, D. W., Diakow, C., Zigmond, R. E., and Kow, L.-M., 1974, Neural hormonal determinants of female mating behavior in rats, in: *The Neurosciences: Third Study Program* (F. O. Schmitt and F. G. Worden, eds.), pp. 621–646, MIT Press, Cambridge, Mass.

Pittman, Q. J., Hatton, J. D., and Bloom, F. E., 1980, Morphine and opioid peptides reduce paraventricular neuronal activity: Studies on the rat hypothalamic slice preparation, *Proc. Natl. Acad. Sci. USA* **77:**5527–5531.

Ramirez, V., Komisaruk, B. R., Whitmoyer, D. I., and Sawyer, C. H., 1969, Effects of hormones and vaginal stimulation on the EEG and hypothalamic units in rats, *Am. J. Physiol.* **212:**1376–1384.

Ruf, K., and Steiner, F. A., 1967, Steroid-sensitive single neuron in rat hypothalamus and midbrain: Identification by microelectrophoresis, *Science* **165:**667–669.

Sakuma, Y., and Pfaff, D. W., 1979a, Facilitation of female reproductive behavior from mesencephalic central gray in the rat, *Am. J. Physiol.* **237:**R278–R284.

Sakuma, Y., and Pfaff, D. W., 1979b, Mesencephalic mechanisms for integration of female reproductive behavior in the rat, *Am. J. Physiol.* **237:**R285–R290.

Sakuma, Y., and Pfaff, D. W., 1980a, Cells of origin of medullary projections in central gray of rat mesencephalon, *J. Neurophysiol.* **44:**1002–1011.

Sakuma, Y., and Pfaff, D. W., 1980b, Excitability of female rat central gray cells with medullary projections: Changes produced by hypothalamic stimulation and estrogen treatment, *J. Neurophysiol.* **44:**1012–1023.

Sar, M., and Stumpf, W. E., 1973, Audioradiographic localization of radioactivity in the rat brain after injection of 1,2-^{3}H-testosterone using dry mount autoradiography, *Endocrinology* **92:**251–256.

Sawyer, C. H., Everett, J. W., and Markee, J. E., 1949, A neural factor in the mechanism by which estrogen induces the release of luteinizing hormone in the rat, *Endocrinology* **44:**218–233.

Siggins, G. R., Hoffer, B. J., and Bloom, F. E., 1971, Studies on norepinephrine-containing afferents to Purkinje cells of rat cerebellum. III. Evidence for mediation of norepinephrine effects by cyclic 3',5'-adenosine monophosphate, *Brain Res.* **25:**535–539.

Silverman, A., and Krey, L. C., 1978, The luteinizing hormone-releasing hormone (LH-

RH) neuronal networks of the guinea pig brain. 1. Intra- and extra-hypothalamic projections, *Brain Res.* **157**:233–246.

Steiner, F. A., 1972, Effects of locally applied hormones and neurotransmitters in hypothalamic neurons, in: *Proceedings, 4th International Congress on Endocrinology*, pp. 202–207.

Steiner, F. A., Ruf, K., and Akert, K., 1969, Steroid sensitive neurones in rat brain: Anatomical localization and responses to neurohumours and ACTH, *Brain Res.* **12**:74–85.

Stumpf, W. E., 1968, Estradiol-concentrating neurons: Topography in the hypothalamus by dry-mount autoradiography, *Science* **162**:1001–1003.

Stumpf, W. E., and Sar, M., 1976, Steroid hormone target sites in the brain: The differential distribution of estrogen, progestin, androgen and glucocorticoid, *J. Steroid Biochem.* **7**:1163–1170.

Szego, C. M., 1978, Parallels in the modes of action of peptide and steroid hormones: Membrane effects and cellular entry, in: *Structure and Function of the Gonadotropins* (K. W. McKerns, ed.), pp. 431–472, Plenum Press, New York.

Terasawa, E., and Sawyer, C. H., 1969a, Electrical and electrochemical stimulation of the hypothalamo–adenohypophysial system with stainless steel electrodes, *Endocrinology* **84**:918–925.

Terasawa, E., and Sawyer, C. H., 1969b, Changes in electrical activity in the rat hypothalamus related to electrochemical stimulation of adenohypophysial function, *Endocrinology* **85**:143–149.

Terasawa, E., and Sawyer, C. H., 1970, Diurnal variation in the effects of progesterone on multiple unit activity in the rat hypothalamus, *Exp. Neurol.* **27**:359–374.

Teyler, T. J., Foy, M., and Vardaris, R. M., 1979, Modulation of hippocampal excitability in adult castrated and ovariectomized rats, *Soc. Neurosci. Abstr.* Vol. 5, p. 465.

Tytell, M., Clark, J. H., and Peck, E. J., Jr., 1980, Effects of estrogen and progesterone on LHRH release from a hypothalamic synaptosomal fraction of ovariectomized rats, *Neurochem. Res.* **5**:493–504.

Vardaris, R. M., Teyler, T. J., and Reiheld, C. T., 1980, Effects of neonatal alterations in gonadal steroids in CA 1 hippocampal pyramidal cells: Challenge with sex hormones to *in vitro* slice preparations from adult rats, *Soc. Neurosci. Abstr.* Vol. 6, p. 458.

Weissman, B. A., Daly, J. W., and Skolnick, P., 1975, Diethylstilbestrol-elicited accumulation of cyclic-AMP in incubated rat hypothalamus, *Endocrinology* **97**:1559–1566.

Whitehead, S. A., and Ruf, K. B., 1974, Response of antidromically identified preoptic neurons in the rat to neurotransmitters and to estrogen, *Brain Res.* **79**:185–198.

Wuttke, W., 1974, Preoptic unit activity and gonadotropin release, *Exp. Brain Res.* **19**:205–216.

Yagi, K., 1970, Effects of estrogen on the unit activity of the rat hypothalamus, *J. Physiol. Soc. Japan* **32**:692–693.

Yagi, K., 1973, Changes in firing rates of single preoptic and hypothalamic units following an intravenous administration of estrogen in the castrated female rat, *Brain Res.* **53**:343–352.

Yamada, Y., 1975, Effects of iontophoretically applied prolactin on unit activity of the rat brain, *Neuroendocrinology* **18**:263–271.

Yamada, Y., 1979, Effects of testosterone on unit activity in rat hypothalamus and septum, *Brain Res.* **172**:165–168.

Yamada, Y., and Nishida, E., 1978, Effects of estrogen and adrenal androgen on unit activity of the rat brain, *Brain Res.* **142**:187–190.

Zigmond, R. E., and McEwen, B. S., 1970, Selective retention of oestradiol by cell nuclei in specific brain regions of the ovariectomized rat, *J. Neurochem.* **17**:889–899.

14

Recent Data on Neuropeptide Mapping in the Central Nervous System

M. Palkovits

1. Introduction

Mapping of neuropeptides in the brain began in the past decade. In the 1950s only oxytocin and vasopressin were known as far as cellular topography was concerned and even this knowledge proved to be rather superficial. Following the bioassays establishing the existence of hypothalamic releasing and inhibiting hormones by the early 1970s, three neuropeptides of such effect (LH-RH, TRH, and somatostatin) were chemically identified, whereas the chemical structure of growth hormone-releasing hormone, corticotropin-releasing hormone, and many others remain unknown. In the mid-1970s it was shown that pituitary hormones are present in the central nervous system, although in much lower concentrations. Furthermore, the number of newly discovered neuropeptides that are neither "pituitary" nor "hypothalamic" is rapidly increasing. A classical example is substance P, whose existence was first reported in 1931 by von Euler and Gaddum but chemically characterized only much later (Chang and Leeman, 1970). Several neuropeptides such as vasoactive intestinal polypeptide (VIP), gastrin, cholecystokinin, glucagon, secretin were first identified in the gastrointestinal tract and later verified to be present in cerebral neurons.

In this review neuropeptides will be considered in three groups: hypothalamic, pituitary, and other (Table I). It has to be emphasized, however,

M. Palkovits • 1st Department of Anatomy, Semmelweis University Medical School, Budapest, Hungary. Present address: Laboratory of Clinical Science, National Institute of Mental Health, Bethesda, Maryland 20205.

Table I. Immunocytochemistry and Radioimmunoassay Studies for the Topographical Localization of Neuropeptide-Containing Neurons in the Central Nervous System

Neuronal peptides	Immunocytochemistry	Radioimmunoassays
"Hypothalamic" neuronal peptides		
LH-RH	Barry et al. (1973), Barry and Dubois (1976), Silverman (1976), Barry (1978), Hoffman et al. (1978), Vigh et al. (1978), Phillips et al. (1980)	Palkovits et al. (1974), Kizer et al. (1976), Samson et al. (1980), Selmanoff et al. (1980)
TRH	Hökfelt et al. (1975)	Brownstein et al. (1974), Jackson and Reichlin (1974), Oliver et al. (1974), Kizer et al. (1976)
Somatostatin	Sétáló et al. (1975), Parsons et al. (1976), Krisch (1978), Bennett-Clarke et al. (1980)	Brownstein et al. (1975), Palkovits et al. (1976), Kobayashi et al. (1977), Epelbaum et al. (1979), Palkovits et al. (1980)
CRF		Lang et al. (1976),[a] Krieger et al. (1977a)[a]
Vasopressin	Vandesande and Dierickx (1975), Vandesande et al. (1975), Buijs (1978), Sofroniew and Weindl (1978), Sofroniew et al. (1979), Morris et al. (1980)	George and Jacobowitz (1976), Dogterom et al. (1978), Hawthorn et al. (1980)
Oxytocin	Vandesande and Dierickx (1975), Vandesande et al. (1975), Buijs (1978), Sofroniew et al. (1979), Morris et al. (1980)	George et al. (1976), Dogterom et al. (1978)
"Pituitary" neuronal peptides		
β-LPH	Watson et al. (1977b), Akil et al. (1978), Zimmerman et al. (1978), Bloch et al. (1979)	Krieger et al. (1977b)
β-END	Bloom et al. (1978), Bloch et al. (1979), Gramsch et al. (1979)	Krieger et al. (1977b), Palkovits et al. (1978), Dupont et al. (1980)

Table I. Immunocytochemistry and Radioimmunoassay Studies for the Topographical Localization of Neuropeptide-Containing Neurons in the Central Nervous System (*Cont.*)

Neuronal peptides	Immunocytochemistry	Radioimmunoassays
ACTH	Akil *et al.* (1978), Watson *et al.* (1978), Bloch *et al.* (1979), Pelletier and Leclerc (1979)	Krieger *et al.* (1977b), Palkovits *et al.* (1978)
α-MSH	Dubé *et al.* (1978), Jacobowitz and O'Donohué (1978), Bloch *et al.* (1979), Eskay *et al.* (1979), O'Donohué *et al.* (1979)	Dubé *et al.* (1978), O'Donohué *et al.* (1979)[b]
Growth hormone	Pacold *et al.* (1978)	Pacold *et al.* (1978)[a]
Thyrotropin		Moldow and Yalow (1978)
Prolactin	Fuxe *et al.* (1977)	
"Other" neuronal peptides		
Substance P	Cuello and Kanazawa (1978), Ljungdahl *et al.* (1978)	Brownstein *et al.* (1976a), Kanazawa and Jessell (1976)
Neurotensin	Uhl *et al.* (1977, 1979b)	Uhl and Snyder (1976), Kobayashi *et al.* (1977)
Enkephalins	Elde *et al.* (1976), Watson *et al.* (1977), Johansson *et al.* (1978), Gramsch *et al.* (1979), Simantov *et al.* (1977), Uhl *et al.* (1979a), Wamsley *et al.* (1980)	Simantov *et al.* (1976), Hong *et al.* (1977), Kobayashi *et al.* (1978), Palkovits *et al.* (1978), Dupont *et al.* (1980), Palkovits *et al.* (1981b)
Renin-angiotensin I	Changaris *et al.* (1977)	Changaris *et al.* (1977), Fuxe *et al.* (1980)
Angiotensin II	Fuxe *et al.* (1976)	
Bradykinin	Corrêa *et al.* (1979)	
DSIP		Kastin *et al.* (1978)
VIP	Silverman (1976), Lorén *et al.* (1979), Sims *et al.* (1980)	Fahrenkrug and Schaffalitzky de Muckadell (1978), Besson *et al.* (1979), Samson *et al.* (1979), Roberts *et al.* (1980), Palkovits *et al.* (1981a)
Gastrin		Rehfeld (1978)
Cholecystokinin	Larsson and Rehfeld (1979), Vanderhaeghen *et al.* (1980)	Innis *et al.* (1979), Larsson and Rehfeld (1979)
Bombesin		Brown *et al.* (1978), Walsh *et al.* (1979)

[a] Bioassays.
[b] High-pressure liquid chromatography.

that all known neuropeptides may occur in any part of the brain, and therefore classification from other aspects may equally be justified.

A short chapter cannot cover all relevant details as data are available on the neuroanatomical topography of 25 neuropeptides and the number of brain areas studied is over 150. Data are tabulated and for details the reader is referred to the original publications (Table I) or reviews (Brownstein et al., 1976b; Elde and Hökfelt, 1978; Hökfelt et al., 1978a,b; Palkovits, 1978; Sétáló et al., 1978; Fuxe et al., 1979; Palkovits, 1979a,b; Phillips et al., 1979; Dierickx, 1980; Palkovits, 1980a,b).

In the tables herein, the biochemical data represent, unless otherwise indicated, results obtained by radioimmunoassay of anatomically well-defined and correctly localized areas. Immunocytochemistry is reviewed covering light and electron microscopic findings. It should be noted that critical evaluation of specificity of immune reactions is beyond the scope of this review. Discrimination between findings is made only in case of convincing refutation. Biochemical values will be expressed in relative terms, i.e., high, moderate, and low concentrations will be mentioned. Numerical values obtained by different laboratories differ substantially, and thus, when possible, concentrations were classified within the results of each laboratory. Immunocytochemical data were not evaluated quantitatively; only axonal and perikaryal reactions were distinguished. With light microscopic immunocytochemistry, axonal and terminal reactions cannot be safely distinguished, and thus they are referred to herein as axonal.

The topography of neuropeptides has been described in several species (Palkovits, 1978), but this review shall be concerned only with that in the rat as these species have been the most extensively investigated.

2. Distribution of Neuropeptides in the Major Brain Areas

2.1. Rhinencephalon (Table II)

Relatively low concentrations of a number of neuropeptides were found in the olfactory bulb and tubercle. Their localization is mainly axonal although some LH-RH-, somatostatin-, and VIP-immunopositive perikarya are present. The nucleus of the diagonal band, thought to be a cholinergic center, contains at least five types of neuropeptide-synthesizing perikarya (LH-RH, somatostatin, substance P, enkephalin, bradykinin). In the hippocampus almost all neuropeptides were found to be present. Of them VIP, cholecystokinin, DSIP, and β-LPH concentrations match the cerebral mean. The dentate gyrus has been scarcely investigated.

2.2. Cerebral Cortex (Table III)

All known neuropeptides have been identified in the cerebral cortex. VIP and cholecystokinin show outstandingly the highest values in the brain.

Table II. Topographical Distribution of Neuronal Peptides in the *Rhinencephalon* (Limbic Lobe) Determined by Radioimmunoassays and Immunocytochemistry[a]

	LH-RH	TRH	Somatostatin	CRF	Vasopressin	Oxytocin	β-LPH	β-END	ACTH	α-MSH	Growth hormone	Thyrotropin	Substance P	Neurotensin	Enkephalins	Renin-angiotensin I	Bradykinin	DSIP	VIP	Gastrin	Cholecystokinin	Bombesin
Bulbus olfactorius	CF	□	□				□	⊠	□	□			□	□	□				□CF	F	■	
Nucl. olfact. ant.	F		CF							F									□C	F	F	F
Nucl. olfact. post.	CF		CF							F			F		□						F	
Tuberculum olfact.	F		⊠ F		—	F		F		F			⊠CF CF		□ ⊠C	F	CF		□ F	CF	CF	
Nucl. tractus diag.	—CF		CF		F					□ F			F	F	F	F			CF		CF	
Hippocampus ant.					□	□		□														
Hippocampus CA1*			□C		□	F — F	⊠	□ F	□	□	□	—	□	□	□ □	□		⊠	⊠CF	⊠CF	⊠CF	□
Hippocampus CA3						F F	F	F	F	F					F	F			⊠CF	⊠CF	CF	
Gyrus dentatus						—																

a *Symbols:* Biochemistry: ■ = high, ⊠ = moderate, □ = low concentrations, — = nondetectable. Immunocytochemistry: C = immunoreactive cell bodies, F = immunoreactive nerve fibers and terminals.
*Biochemical data referred to the hippocampus *in general.* (For references see Table I.)

Table III. Topographical Distribution of Neuronal Peptides in the *Cerebral Cortex* Determined by Radioimmunoassays and Immunocytochemistry[a]

Cortex	LH-RH	TRH	Somatostatin	CRF	Vasopressin	Oxytocin	β-LPH	β-END	ACTH	α-MSH	Growth hormone	Thyrotropin	Substance P	Neurotensin	Enkephalins	Renin-angiotensin I	Bradykinin	DSIP	VIP	Gastrin	Cholecystokinin	Bombesin
Frontal*	—	□	□	□	□	—	□	□	□	□	□		□	□	□	□		⊠ F	■ CF	—	■ CF	⊠
Frontopolar			C																■ CF		■ CF	
Parietal			C							□			F	□					■ CF		F	
Insular			C							□			F	□	□			F	■ CF		■ CF	
Temporal					□								F	□ □	□			F	■ CF		CF	
Occipital			C		□			F					F	□ □	□ □			F	■ CF		⊠ F	
Cingulate	—		C		—	—		F		—			F		□			F	■ CF		F	
Pyriform						F				□			— □	⊠	F		F	⊠	CF		CF	
Entorhinal						F				□					F				CF		CF	
Subiculum						F				□					F				⊠ CF		CF	

[a] *Symbols:* Biochemistry: ■ = high, ⊠ = moderate, □ = low concentrations, — = nondetectable. Immunocytochemistry: C = immunoreactive cell bodies, F = immunoreactive nerve fibers and terminals.

* Biochemical data referred to the cerebral cortex *in general.* (For references see Table I.)

Both peptides were shown with immunocytochemistry to be contained by perikarya and to form an abundant neuronal network. Other neuropeptides appeared to be axonal except for somatostatin, which was shown in some cells.

2.3. Basal Ganglia (Table IV)

The nucleus accumbens is rich in neuropeptides. Particularly high is its VIP concentration, but in addition 10 other neuropeptides were demonstrated. Several neuropeptides are found in the caudate-putamen (somatostatin, substance P, enkephalin, VIP) but in low concentrations. Prominently high is the enkephalin concentration of the globus pallidus; however, not a single neuropeptide-containing perikaryon has been shown as yet.

2.4. Septum (Table V)

The lateral septal nucleus is rich in neuropeptides, dozens of which were shown in axons. It is of interest that this nucleus is rich also in biogenic amines (Palkovits, 1978). Although other septal nuclei were found to contain fewer neuropeptides, it must be noted that these nuclei have been less investigated. With immunocytochemistry LH-RH, somatostatin, substance P, enkephalin, and VIP perikarya were detected in the septum.

2.5. Amygdala (Table VI)

The amygdala is relatively rich in neuropeptides. Its somatostatin and VIP concentrations exceed the cerebral mean. These two neuropeptides are present in all amygdaloid nuclei in axons and perikarya. In all nuclei, immunoreactive substance P-fibers were described; in the central and medial nuclei, cells were also found. In the amygdala the highest CNS concentration of growth hormone was measured, while neurotensin and enkephalin concentrations also reach the cerebral mean. LH-RH-, vasopressin-, enkephalin-, and cholecystokinin-positive axons were seen in several amygdaloid nuclei.

2.6. Preoptic Region (Table VII)

This area is rich in neuropeptides with a prominent neurotensin level but other neuropeptides (somatostatin, β-END, ACTH, α-MSH, substance P, and VIP) also match the cerebral mean. The preoptic region contains a colorful assembly of immunoreactive perikarya: LH-RH-, somatostatin-, substance P-, enkephalin-, VIP-, and cholecystokinin-positive cells are present in great abundance. Many neuropeptides are found in the bed nucleus of the stria terminalis, which contains a high number of aminergic terminals (Palkovits, 1978).

Table IV. Topographical Distribution of Neuronal Peptides in the *Basal Ganglia* Determined by Radicimmunoassays and Immunocytochemistry[a]

	LH-RH	TRH	Somatostatin	CRF	Vasopressin	Oxytocin	β-LPH	β-END	ACTH	α-MSH	Growth hormone	Substance P	Neurotensin	Enkephalins	Renin-angiotensin I	Angiotensin II	Bradykinin	DSIP	VIP	Cholecystokinin	Bombesin
Nucl. caudatus	—						□	□	□	□		⊠CF	⊠	⊠CF				□	□	F	F
Caudatus-putamen		□	□CF	—	□			□	□	⊠		□CF	□	□CF					□CF	⊠	F
Nucl. accumbens			F □CF		—			□	□	□		⊠CF	□	⊠CF		F	F	□	■	F	F
Globus pallidus								F	F	F	—	F	⊠	⊠	□				F	F	F
Claustrum			C									F	F	F		F	F		CF	F	F

[a]*Symbols:* Biochemistry: ■ = high, ⊠ = moderate, □ = low concentrations, — = nondetectable. Immunocytochemistry: C = immunoreactive cell bodies, F = immunoreactive nerve fibers and terminals.

*Biochemical data referred to the hippocampus *in general*. (For references see Table I.)

Table V. Topographical Distribution of Neuronal Peptides in the *Septal Nuclei* Determined by Radioimmunoassays and Immunocytochemistry[a]

Nuclei	LH-RH	TRH	Somatostatin	CRF	Vasopressin	Oxytocin	β-LPH	β-END	ACTH	α-MSH	Growth hormone	Substance P	Neurotensin	Enkephalins	Renin-angiotensin I	Angiotensin II	Bradykinin	DSIP	VIP	Cholecystokinin	Bombesin
Septum*	F	□										⊠		□					⊠		
Dorsal septal		□	□			□	⊠	□	□	⊠		⊠	□	—		F			□		
Intermedial septal		□			F							F									
Medial septal		□	CF		F	F				□		CF							F		
Lateral septal	□CF	F	F					F	F	F		CF		□CF				F	CF	F	
Fimbrial septal		□										CF		C					F		
Triangular septal	—				—							F								F	

[a]*Symbols:* Biochemistry: ■ = high, ⊠ = moderate, □ = low concentrations, — = nondetectable. Immunocytochemistry: C = immunoreactive cell bodies, F = immunoreactive nerve fibers and terminals.
* Biochemical data referred to the hippocampus *in general.* (For references see Table 1.)

Table VI. Topographical Distribution of Neuronal Peptides in the *Amygdaloid Nuclei* Determined by Radioimmunoassays and Immunocytochemistry[a]

Amygdala[*]	LH-RH	TRH	Somatostatin	CRF	Vasopressin	Oxytocin	β-LPH	β-END	ACTH	α-MSH	Growth hormone	Thyrotropin	Substance P	Neurotensin	Enkephalins	Angiotensin II	Bradykinin	DSIP	VIP	Cholecystokinin	Bombesin
Anterior area			⊠		□	—	□	□	□	□	■F	□	⊠	⊠C	⊠	F			□F	CF	
N. tract. olf. lat.			⊠CF														F		⊠	F	
Lateral nucl.	F		⊠CF										F						⊠CF	F	
Lat. post. nucl.			⊠CF		F			□					F		□				F	F	
Medial nucl.			□CF					□					F		□				⊠CF	F	
Med.-post. nucl.	F		□CF		F			F	F	F			CF		□	C			F		
Central nucl.	□		□CF		F		F	□	F	F			CF	F	□	F			⊠—F	CF	
Cortical nucl.	□		⊠CF							□□□			CF		□				⊠CF	CF	
Basal-med. nucl.	F		□CF		F		F	□		□			F		—	F			⊠CF		
Basal-lat. nucl.			⊠CF										F		□				⊠CF		
Posterior nucl.			□CF					□					F		F				F		F

[a] *Symbols:* Biochemistry: ■ = high, ⊠ = moderate, □ = low concentrations, — = nondetectable. Immunocytochemistry: C = immunoreactive cell bodies, F = immunoreactive nerve fibers and terminals.
[*] Biochemical data referred to the hippocampus *in general*. (For references see Table I.)

Table VII. Topographical Distribution of Neuronal Peptides in the *Preoptic Region* Determined by Radioimmunoassays and Immunocytochemistry[a]

	LH-RH	TRH	Somatostatin	CRF	Vasopressin	Oxytocin	β-LPH	β-END	ACTH	α-MSH	Thyrotropin	Prolactin	Substance P	Neurotensin	Enkephalins	Angiotensin II	Bradykinin	DSIP	VIP	Cholecystokinin
Preoptic region*	□CF		⊠		—			⊠	⊠	⊠			⊠	■	□				⊠	
Med. preoptic nucl.	CF	□ □	⊠	□	—	□		⊠		□ ⊠	□		⊠CF	■CF						CF
Periventr. nucl.	□CF		F	CF	—			F	F	F			F	F	C	F	F		CF	CF
Suprachiasm. nucl.	□CF		CF		—								⊠CF	CF	C	F				
Lateral nucl.	□CF		□	□	—	—		□		⊠			⊠CF	⊠CF	⊠CF	⊠			⊠CF	
NIST			⊠					⊠	⊠	⊠				CF					□	F
NISM	□CF		⊠CF			—		F	F	F			F	F		F			F	CF

[a] *Symbols:* Biochemistry: ■ = high, ⊠ = moderate, □ = low concentrations, — = nondetectable. Immunocytochemistry: C = immunoreactive cell bodies, F = immunoreactive nerve fibers and terminals.
* Biochemical data referred to the hippocampus *in general.* (For references see Table I.)

Table VIII. Topographical Distribution of Neuronal Peptides in the *Hypothalamus* and *Mamillary Body* Determined by Radioimmunossays and Immunocytochemistry[a,b]

	LH-RH	TRH	Somatostatin	GH-RH***	CRF	Vasopressin	Oxytocin	β-LPH	β-END	ACTH	α-MSH	Growth hormone	Thyrotropin	Prolactin	Substance P	Neurotensin	Enkephalins	Renin-angiotensin I	Angiotensin II	Bradykinin	DSIP	VIP	Gastrin	Cholecystokinin	Bombesin
Hth*	—	■	■		□			■	□		■	□	■		⊠	C	□	□	F		□	⊠	⊠	⊠	■
Med. Hth**	□	□	■						■ ■	⊠ ■	⊠ ■				⊠ F	⊠ F	⊠		F	CF		□ F	F	CF F	
Lat. Hth	—	⊠	⊠		⊠	⊠	□		■ □	F F	F F			F	F F	F F	– CF F	C	F	CF		⊠ F		CF F	
MBH	—	F	F			CF	CF		CF CF	CF CF	CF CF			F	CF CF	CF CF	C C	C	CF			CF F		CF F	
ME	■	■	■ ■		⊠	⊠ CF	⊠ CF		■ □ ■	⊠ ■ ■	□ ⊠ ■			F	⊠ CF F	⊠ ⊠	⊠ CF F	C C	CF			■ CF	F		
NPE	□	F F	F F	□	⊠	⊠ CF	⊠ CF		F F	F F	F F			F	□ CF F	□ CF	□ CF C	CF C				□ F			
NSC	—	⊠	⊠		⊠	⊠ CF	⊠ CF			F F	F F			F	F CF	F	– CF F					⊠ F			
NSO	—	⊠ CF	⊠		□	⊠ CF	⊠ CF			F F	F F			F	F CF	⊠	⊠ CF					□ F			
NPV	F	F CF	F		□	⊠ CF	⊠ CF		CF	F F	F F			F	⊠ CF	⊠	⊠ CF					□ F			
NHA	F	F	F		□	–	–		F	F F	F F			F	□ CF	□	F CF					□ F			
NPV	F	F	F		□	–	–		F	CF	CF			F	⊠ CF	⊠ CF	⊠ CF					□ F			
AR	□	⊠	⊠		□	⊠	⊠		■	CF F	CF F			F	□ CF	□	□ C					■ CF F			
NA		F	⊠		□					F F	F F			F	□ CF	⊠	□ CF		CF			□ F			
NVM		⊠ CF	⊠		□				CF F	CF F	⊠				⊠ CF	⊠	⊠ CF		CF			⊠ F		CF	
NDM	⊠ CF	⊠ CF	⊠		⊠	⊠ CF	⊠ CF		■	F F	■ F				⊠ CF	⊠	⊠ CF		F	CF		■ CF		CF	
NPF	□	F	□		□				CF						F CF		F CF			F		□ F			
NPMD	□	□	□		□										F CF		CF					□ F			
NPMV	□	□	□		□				CF						F CF		□ CF					□ F			
NHP	□	□			□	F	–			F	F			F	F CF	□	□ C					□ F			
NSM						–	–							F	F CF		□ C					F			
MB																									
NMM	F	F	F		□	F	F		⊠	□	□				⊠	⊠	□ F					□	⊠	⊠	
NML		–													F		F C								
NMP															F		C								

a *Symbols*: Biochemistry: ■ = high, ⊠ = moderate, □ = low concentrations, — = nondetectable. Immunocytochemistry: C = immunoreactive cell bodies, F = immunoreactive nerve fibers and terminals.
* Biochemical data referred to the *whole* hypothalamus or **to the *whole medial* hypothalamus. ***Bioassay data (Krulich et al., 1977) (For references see Table L)
b *Abbreviations*: Hth, hypothalamus; Med. Hth, medial hypothalamus; Lat. Hth, lateral hypothalamus; MBH, mediobasal hypothalamus; ME, median eminence; NPE, periventricular nucleus; NSC, suprachiasmatic nucleus; NSO, supraoptic nucleus; NPV, paraventricular nucleus; NHA, anterior hypothalamic nucleus; AR, retrochiasmatic area; NA, arcuate nucleus; NVM, ventromedial nucleus; NDM, dorsomedial nucleus; NPF, perifornical nucleus; NPMD, dorsal premamillary nucleus; NPMV, ventral premamillary nucleus; NHP, posterior hypothalamic nucleus; NSM, supramamillary nucleus; MB, mamillary body; NMM, medial mamillary nucleus; NML, lateral mamillary nucleus; NMP, posterior mamillary nucleus.

2.7. Hypothalamus (Table VIII)

The hypothalamus is the richest source of brain neuropeptides, all known neuropeptides being present. It has been most extensively studied both biochemically and immunocytochemically so that its neuropeptide mapping is fairly well known.

Both "hypothalamic" and "pituitary" neuropeptides are found in the highest concentrations (except for growth hormone, whose peak concentration occurs in the amygdala). The concentration of neurotensin and bombesin is particularly high in the hypothalamus while that of other neuropeptides reaches the central mean.

Within the hypothalamus the highest levels of neuropeptides are present in the median eminence (cf. Palkovits, 1979a, 1980a). The concentration of some exceeds the cerebral or hypothalamic mean with one (TRH, CRF, α-MSH, neurotensin) or more (LH-RH, somatostatin, oxytocin, vasopressin) orders of magnitude. In the median eminence, perikarya are practically absent; neuropeptides occur in axons and axonal varicosities. (There are no axo-axonal synapses in the median eminence; chemical substances are contained by axonal varicosities.) In this region 16 types of immunoreactive neuropeptidergic axons have been described to date. Their role is most probably neurohormonal, exerting their effects on the pituitary through the portal circulation. A further role attributable to these peptides is to act as local neurotransmitters stimulating or inhibiting the release of another neurotransmitter from other fibers. This is not necessarily realized by synaptic contacts. Moreover, there is a theoretical possibility of influencing the permeability of the fenestrated capillaries, thereby facilitating or blocking the uptake of other biologically active substances.

Among the cell groups of the hypothalamus, the medial-basal ones (mainly the arcuate nucleus) contain particularly numerous neuropeptide-containing cells and axons. Here are found the opiocortin (β-LPH, β-END, ACTH, α-MSH) cell bodies, whose axons enmesh the whole CNS. Apart from them (and from dopamine and acetylcholine cells), LH-RH-, substance P-, and enkephalin-containing perikarya were shown in the arcuate nucleus. Several peptidergic cells occur in the periventricular (somatostatin, enkephalin, bradykinin, cholecystokinin), supraoptic and paraventricular (vasopressin, oxytocin, TRH, substance P, enkephalin, angiotensin II, cholecystokinin), and dorsomedial (TRH, substance P, enkephalin, bradykinin, cholecystokinin) nuclei. Two or three cell types were shown in many other hypothalamic nuclei. Almost all nuclei contain ACTH-, α-MSH-, substance P-, and VIP-immunopositive fibers. VIP is present in prominent concentration in the suprachiasmatic nucleus, the only one within the hypothalamus containing VIP-synthesizing perikarya.

The mamillary body is much poorer in neuropeptides than the hypothalamus, but it must be noted that this area has been scarcely investigated for neuropeptide topography.

Table IX. Topographical Distribution of Neuronal Peptides in the *Thalamus*, *Epithalamus* (Habenula), *Metathalamus*, and *Subthalamus* Determined by Radioimmunoassays and Immunocytochemistry[a]

	LH-RH	TRH	Somatostatin	CRF	Vasopressin	Oxytocin	β-LPH	β-END	ACTH	α-MSH	Thyrotropin	Prolactin	Substance P	Neurotensin	Enkephalins	Angiotensin I	Bradykinin	DSIP	VIP	Cholecystokinin	Bombesin
Thalamus*		□	□	□	□		□	□	⊠	□	⊠		□	□	□	□		■	⊠	⊠	⊠
Ant.-dors. nucl.																					
Ant.-vent. nucl.																					
Ant.-med. nucl.					—					□			□		F				F		
Reticular nucl.										□			□	□	F				F		
Ventral nucl.										□			□	□	F				F		
Vent.-dors. nucl.								□		□			□		F						
Lateral nucl.								□		□			□□		F						
Lat.-post. nucl.								□					□□		F						
Posterior nucl.										—			F	F	F				F		
Parataenial nucl.						F			F	F			F	F	F						
Med.-dors. nucl.								□		□			F	□	F				F	F	
Med.-vent. nucl.										F			F		F		F		F		
Parafasc. nucl.					□	□				□			F		CF						
Reuniens nucl.					□	□				□F			F	F	F		CF		F		
Rhomboideus nucl.										⊠			F	F	F						
Periventr. nucl.					—	F	F	F	F	□F		F	⊠	F	CF		F		F	F	
Habenula**																					
Habenula med.	□				—	F	F			□			C	F	F				⊠		
Habenula lat.	F					—		□							F				□		
Corp. gen. med.										□			□□	□	□CF				□F		
Corp. gen. lat.													□	□□	□CF				□F		
Area pretect.														□	□						
Entopeduncl. nucl.			C										□		F				F		
Subthalamic nucl.		F								—			⊠	F	—CF				□	F	
Zona incerta	F	F	C				F		F	—F			F	F	—CF				F		
Forel's fields										F		F	F		F						

[a]*Symbols:* See Table VIII. *Biochemical data referred to the whole thalamus or **to the whole habenula. (For references see Table I.)

2.8.   Thalamus (Table IX)

The thalamus belongs to the territories relatively poor in neuropeptides, despite the fact that the highest DSIP concentrations are measured here. This statement refers to the concentration rather than the number of neuropeptides, 16 of which are present. The levels of two rarely encountered neuropeptides (bombesin, thyrotropin) reach the cerebral mean. Concerning thalamic nuclei biochemical or immunocytochemical data are sporadic. A fairly comprehensive mapping is available for α-MSH, and an immunocytochemical one for substance P, neurotensin, enkephalin, and VIP. The periventricular thalamic nucleus is outstandingly rich in neuropeptides as compared to other thalamic nuclei. In this small nucleus situated rostrocaudally on the top of the thalamus, 11 neuropeptides were identified in axons. In addition, this nucleus is rich in noradrenaline, dopamine, and serotonin.

2.9.   Habenula (Table IX)

The habenula is poor in neuropeptides. Both the medial and the lateral nucleus contain a few kinds of peptidergic fibers, and substance P cells were described in the medial habenular nucleus.

2.10.   Metathalamus (Table IX)

Geniculate bodies contain enkephalin perikarya and enkephalin, substance P, VIP, cholecystokinin (in the lateral only) and β-LPH (in the medial only) axons. Concentrations hitherto measured (α-MSH, substance P, neurotensin, enkephalin, VIP) are low.

2.11.   Subthalamus (Table IX)

In the zona incerta several types of neuropeptide-containing axons (LH-RH, TRH, β-LPH, ACTH, α-MSH, substance P, enkephalin, VIP) were described, but only somatostatin and enkephalin perikarya.

2.12.   Mesencephalon (Table X)

Nearly 20 neuropeptides have been demonstrated in the mesencephalon to date. Their concentrations, except for the "hypothalamic" ones, match the cerebral mean. They are localized mainly in axons; only five (substance P, neurotensin, enkephalin, VIP, cholecystokinin) types of neuropeptide-synthesizing perikarya are encountered, in low numbers.

Of the midbrain regions, the highest neuropeptide content is found in the central gray matter. All neuropeptides present in the midbrain are

Table X. Topographical Distribution of Neuronal Peptides in the *Mesencephalon* (Midbrain Nuclei) Determined by Radioimmunoassays and Immunocytochemistry[a]

	LH-RH	TRH	Somatostatin	CRF	Vasopressin	Oxytocin	β-LPH	β-END	ACTH	α-MSH	Thyrotropin	Prolactin	Substance P	Neurotensin	Enkephalins	Angiotensin II	Bradykinin	DSIP	VIP	Cholecystokinin	Bombesin
Mesencephalon*	□		□				⊠	⊠	□	□			⊠	□	⊠	F		⊠	□CF	CF	□
SGC	F		⊠F		F	F	⊠F	⊠F	□F	F			⊠CF	⊠CF	⊠CF		F		F	CF	F
Colliculus sup.	F		⊠F		F	F	F	F	□	□	□		□CF	□CF	□C				□CF	F	
Colliculus inf.			F										□	□	□				□CF	F	
Vent. tegm. area	F						□F	F		□			⊠CF	CF	⊠CF				□CF	F	
Interpedunc. nucl.	□				□	F	F	⊠	F	□			F/CF	□	F				F	CF	
Substantia nigra	—	F	□		—	F	F	⊠	F	F			■	⊠	□CF				F	CF	
Nucleus ruber	—	F							□	F			□	□	□				F	F	
Cuneiform nucl.										F			CF	CF	CF	F			F	CF	
Dorsal raphe nucl.		F	□		F	F	F	F	F	F			□	F	F	F			F	CF	
Cent. sup. (mid. ra.)		F											□	F	□				□	CF	
Nucl. lineares													F	F	□				□	CF	
Nucl. lemn. lat.	F	F	F		F		F				□		F	F	CF		F		F	F	

[a]*Symbols:* Biochemistry: ■ = high, ⊠ = moderate, □ = low concentrations, — = nondetectable. Immunocytochemistry: C = immunoreactive cell bodies, F = immunoreactive nerve fibers and terminals.
*Biochemical data referred to the *whole* mesencephalon. (For references see Table 1.)

found here. The central gray is followed by the substantia nigra, known formerly as containing dopamine cells, and the dorsal raphe nucleus, where serotonin cells occur. Both nuclei contain in addition numerous neuropeptidergic cells and axons. The substance P content of the substantia nigra is the highest in the CNS. Nine neuropeptides were demonstrated in the interpeduncular nucleus.

Several mesencephalic areas contain substance P-, neurotensin-, enkephalin-, VIP-, cholecystokinin-, β-LPH-, and vasopressin-immunopositive fibers but some containing "hypothalamic" peptides (LH-RH, TRH, somatostatin) and opiocortins also occur.

2.13. Pons (Table XI)

Dozens of neuropeptides are found in the pons but their concentrations, except for β-LPH, cholecystokinin, and DSIP, are low. Two cell groups were shown to be rich in neuropeptides: the locus coeruleus and the parabrachial nucleus. In the locus coeruleus, in addition to its high number of noradrenaline cells, substance P and neurotensin neurons were also described and some 11 different neuropeptidergic axons were shown with immunocytochemistry. In the parabrachial nucleus, substance P, neurotensin, enkephalin, and cholecystokinin cell bodies and five more types of axons were described.

2.14. Cerebellum (Table XI)

Immunocytochemical data are scarce. With radioimmunoassay low concentrations of 13 neuropeptides were measured. No intracerebellar topographical details are known.

2.15. Medulla Oblongata (Table XII)

Three neuropeptides (oxytocin, ACTH, DSIP) are present in moderate concentrations. Thirteen others were measured in low concentrations. Very few topographical data are available; only the mapping of substance P, neurotensin, enkephalin, and VIP have been performed in detail. Within the medulla the nucleus of the solitary tract is prominently the richest in neuropeptides. Besides its known aminergic cells, substance P, neurotensin, and enkephalin perikarya have recently been demonstrated. Eight additional peptides were found in axons. The nucleus ambiguus is also rich in neuropeptide-containing axons. In addition, several cell groups are found in the medulla oblongata containing at least four to five neuropeptides. To these contribute the vasopressin, oxytocin, ACTH, and α-MSH axons descending from the hypothalamus.

Table XI. Topographical Distribution of Neuronal Peptides in the *Pons* and the *Cerebellum* Determined by Radioimmunoassays and Immunocytochemistry[a]

	TRH	Somatostatin	CRF	Vasopressin	Oxytocin	β-LPH	β-END	ACTH	α-MSH	Growth hormone	Thyrotropin	Substance P	Neurotensin	Enkephalin	Renin-angiotensin I	Angiotensin II	Bradykinin	DSIP	VIP	Gastrin	Cholecystokinin	Bombesin
Pons*	□	□	—	□	—	⊠	□	□	□	—	□	□	□	□				⊠	□		⊠	□
Nuclei pontis																						
N. ret. tegm. p.																						
N. ret. pontis or.																						
N. ret. pont. caud.						F								F								
Raphe pontis						F						F	F			F						
Parabrach. nucl.		F			F	F	F	F	F			CF	CF	CF					F	F	CF	
Locus coeruleus					F	F	F	F	F			CF	CF	F		F			F	F	F	
Nucl. tegm. dors.								F	□ F			CF	CF	□ F					□	F	F	—
Nucl. tegm. vent.							F		F			F	F	F								
Oliva superior																CF						
Corpus trapezoid.												F	F	F					F			
Nucl. mot. N. Vth	F						F		□			F	□	F	C							
Nucl. sens. N. Vth									F			F	□	F							□	□
Cerebellum**	□	□	—	□		□	□	□	—	—	□	□	□	□				⊠	□		—	□
Cortex												□	□	□	C						—	
Nuclei												F	F	F							—	
Med. nucl.																						
Interpositus nucl.																						
Lateral nucl.																						

[a]*Symbols:* Biochemistry: ■ = high, ⊠ = moderate, □ = low concentrations, — = nondetectable. Immunocytochemistry: C = immunoreactive cell bodies, F = immunoreactive nerve fibers and terminals.
*Biochemical data referred to the hippocampus *in general.* (For references see Table I.)

Table XII. Topographical Distribution of Neuronal Peptides in the *Medulla Oblongata* Determined by Radioimmunoassays and Immunocytochemistry[a]

	TRH	Somatostatin	CRF	Vasopressin	Oxytocin	β-LPH	β-END	ACTH	α-MSH	Growth hormone	Thyrotropin	Substance P	Neurotensin	Enkephalins	Angiotensin I	Angiotensin II	Bradykinin	DSIP	VIP	Gastrin	Cholecystokinin	Bombesin
Medulla*	□	□	—	□	⊠	□	□	⊠	□		□	□	□	□	□			⊠	□		□	□
Nucl. ret. gig. cell.		F						F	F			F	F									
Nucl. ret. med. obl.												F	F								C	
Nucl. ret. parvocell.												F	F	F								
Nucl. ret. paramed.		F		F	F		F	F				F	CF	CF					F	F	F	
Nucl. tract. s. Vth		F		F	F		F	F				■	CF	CF		F			F	F	F	
Nucl. ret. lat.	F			F	F	F	F					F	F	F						F	F	
Nucl. mot. VIIth						F	F		F			CF	F	CF								
Oliva inferior						F	F					CF	CF	CF								
Nucl. raphe magn.									F			CF	CF	CF				□	F	F	F	
Nn. cochleares												CF	CF	C								
Nucl. raphe pall.												CF	F	CF								
Nucl. raphe obsc.												CF	F	CF								
Nucl. vest. lat.															C							
Nn. vestibulares													CF	CF								
Nucl. tract. solit.	F	F		F	F				F			CF	CF	⊠CF		F			⊠	F	F	
Nucl. ambiguus									F			CF	CF	—F						F □		
Nucl. prepositus																						
Nucl. mot. N. XIIth	F																					
Nucl. gracilis		F										F	F	—F					F	F		
Nucl. cuneat. lat.																						
Nucl. cuneatus														F							CF	

[a] *Symbols:* Biochemistry: ■ = high, ⊠ = moderate, □ = low concentrations, — = nondetectable. Immunocytochemistry: C = immunoreactive cell bodies, F = immunoreactive nerve fibers and terminals.
*Biochemical data referred to the hippocampus *in general.* (For references see Table I.)

Table XIII. Topographical Distribution of Neuronal Peptides in the *Spinal Cord* Determined by Radioimmunoassays and Immunocytochemistry[a]

	LH-RH	TRH	Somatostatin	CRF	Vasopressin	Oxytocin	β-LPH	β-END	ACTH	α-MSH	Thyrotropin	Prolactin	Substance P	Neurotensin	Enkephalins	Angiotensin I	Angiotensin II	Bradykinin	VIP	Cholecystokinin
Spinal cord																				
Fun. post.																				
Fun. lat.					□ F	F			F				□	□	F	□		—	□	
Fun. med.			□		□	F	□	□		□			□	□	□	□			□	□
Cornu post.			F		F	F							■CF	CF	CF					
Cornu lat.													CF		F		F			F
Cornu ant.		F											F		F		F		□	F

[a] *Symbols:* Biochemistry: ■ = high, ⊠ = moderate, □ = low concentrations, — = nondetectable. Immunocytochemistry: C = immunoreactive cell bodies, F = immunoreactive nerve fibers and terminals.
• Biochemical data referred to the spinal cord in general. (For references see Table I.)

2.16. Spinal Cord (Table XIII)

There are still gaps in the neuropeptide mapping of the spinal cord. Biochemical measurements usually show low levels. With immunocytochemistry substance P, neurotensin, and enkephalin perikarya were detected. Other neuropeptides (TRH, somatostatin, oxytocin, vasopressin, ACTH, angiotensin II, cholecystokinin) are present in axons.

3. Peptidergic Axons in Major Anatomical Pathways (Table XIV)

There are only a few CNS tracts whose chemical composition is elucidated. It seems likely that no pathways are chemically homogeneous and that most of them are bidirectional. Neuropeptide-containing axons often run together with aminergic, cholinergic, or amino acid-containing fibers. It is also frequently seen that neuropeptide innervation does not join the well-known anatomical pathways but forms a separate bundle or reaches its target area by means of solitary fibers.

Between the forebrain and the lower brain stem and spinal cord two long tracts are identified comprising ascending aminergic and descending peptidergic fibers. The periventricular pathway is gathered at the sides of the third ventricle and runs through the diencephalon. It follows the ventricular system: in the midbrain it passes along the central canal; in the pons–medulla and tegmentum it is found near the floor of the fourth ventricle. The other pathway is situated more laterally intermingled with the fibers of the medial forebrain bundle and passes through the ventrolateral part of the midbrain tegmentum. In the pons and medulla it runs ventrolaterally near the surface.

From the data summarized in Table XIV it appears that some bundles (for example the stria terminalis, supraoptic decussation, etc.) contain a surprisingly high number of neuropeptide-immunoreactive axons in addition to their aminergic ones. Other bundles are also of heterogeneous composition. It should be noted, however, that the chemical composition of most large tracts is unknown.

4. Concluding Remarks

The large number of neuropeptides occurring in the CNS gives at first a chaotic impression rather than the feeling of increased understanding. No doubt, the classical views based on the assumption of homogeneous brain centers and their unidirectional interconnections are challenged by recent neurochemical findings. However, taking into consideration the thousands

Table XIV. Neuronal Peptide-Containing Nerve Fibers in the Major Neuronal Pathways in the Central Nervous System[a]

	LH-RH	Somatostatin	Vasopressin	β-LPH	β-END	ACTH	α-MSH	Substance P	Neurotensin	Enkephalins	Angiotensin I	Angiotensin II	Bradykinin	DSIP	VIP	Gastrin	Cholecystokinin	Bombesin
Tractus olfact. med.	+																	
Tractus olfact. lat.	+																	
Corpus callosum																		
Tractus diagonalis	+						+	+		+							+	
Fornix		+	+	+				+		+					+		+	
Internal capsule			+	+											+			
Stria medullaris			+		+		+	+		+					+			
Stria terminalis	+	+	+		+	+	+	+		+							+	
Fimbria hippocampi								+	+	+								
Supraoptic decussation					+	+	+	+		+								
Medial forebrain bundle	+	+	+	+	+	+	+	+	+	+			+		+		+	
Ansa lenticularis					+		+	+		+								
Fasciculus retroflexus							+	+		+					+			
Lemniscus medialis																		
Pyramidal tract																		
Pedunculus cerebell. sup.										+					+			
Tractus spin. n. trigemini								+		+								

[a] + = immunocytochemically identified fibers.

of axons terminating on one single neuron, great variety of substances acting on a neuron becomes more feasible. Each neuron is monosynaptically linked with dozens of brain areas. Between these areas close connection is established by the dominance of one or another neurotransmitter. This, however, does not exclude the possibility that other areas may have a modulatory influence on this connection by other transmitters acting on the general responsiveness of the neurons. Neuropeptides seem to have such a neuromodulatory effect in the CNS.

Mapping has currently become a widely applied approach to neuropeptide research. Its up-to-date results are summarized in the tables of this review. However, it should be borne in mind that mapping is only one prerequisite to the elucidation of the function of neuropeptides. The "neuropeptides here and there" attitude must be kept within its own limits or, what is more desirable, must be complemented with experimental studies. In other words, mapping is not an aim but an aid of neuropeptide research.

REFERENCES

Akil, H., Watson, S. J., Berger, P. A., and Barchas, J. D., 1978, Endorphins, β-LPH, and ACTH: Biochemical, pharmacological and anatomical studies, *Adv. Biochem. Psychopharmacol.* **18**:125.

Barry, J., 1978, The distribution, organization and connections of neurons containing luteinizing hormone releasing hormone in the vertebrate brain, in: *Biologie Cellulaire des Processus Neurosécrétoires Hypothalamiques* (J. D. Vincent and C. Kordon, eds.), pp. 399–413, CNRS, Paris.

Barry, J., and Dubois, M. P., 1976, Immunoreactive neurosecretory pathways in mammals, *Acta Anat.* **94**:497.

Barry, J., Dubois, M. P., and Poulain, P., 1973, LRF producing cells of the mammalian hypothalamus, *Z. Zellforsch. Mikrosk. Anat.* **146**:351.

Bennett-Clarke, C., Romagnano, M. A., and Joseph, S. A., 1980, Distribution of somatostatin in the rat brain: Telencephalon and diencephalon, *Brain Res.* **188**:473.

Besson, J., Rotsztejn, W., Laburthe, M., Epelbaum, J., Beaudet, A., Kordon, C., and Rosselin, G., 1979, Vasoactive intestinal peptide (VIP): Brain distribution, subcellular localization and effect of deafferentation of the hypothalamus in male rats. *Brain Res.* **165**:79.

Bloch, B., Bugnon, C., Fellmann, D., Lenys, D., and Gouget, A., 1979, Neurons of the rat hypothalamus reactive with antisera against endorphins, ACTH, MSH and β-LPH, *Cell Tissue Res.* **204**:1.

Bloom, F., Battenberg, E., Rossier, J., Ling, N., and Guillemin, R., 1978, Neurons containing β-endorphin in rat brain exist separately from those containing enkephalin: Immunocytochemical studies, *Proc. Natl. Acad. Sci. USA* **75**:1591.

Brown, M., Allen, R., Villarreal, J., Rivier, J., and Vale, W., 1978, Bombesin-like activity: Radioimmunology assessment in biological tissues, *Life Sci.* **23**:2721.

Brownstein, M. J., Palkovits, M., Saavedra, J. M., Bassiri, R. M., and Utiger, R. D., 1974, Thyrotropin-releasing hormone in specific nuclei of the brain, *Science* **185**:267.

Brownstein, M., Arimura, A., Sato, H., Schally, A. V., and Kizer, J. S., 1975, The regional distribution of somatostatin in the rat brain, *Endocrinology* **96**:1456.

Brownstein, M. J., Mroz, E. A., Kizer, J. S., Palkovits, M., and Leeman, S. E., 1976a, Regional distribution of substance P in the brain of the rat, *Brain Res.* **116**:299.

Brownstein, M. J., Palkovits, M., Saavedra, J. M., and Kizer, J. S., 1976b, Distribution of hypothalamic hormones and neurotransmitters within the diencephalon, in: *Frontiers in Neuroendocrinology*, Vol. 4 (L. Martini and W. F. Ganong, eds.), pp. 1–23, Raven Press, New York.

Buijs, R. M., 1978, Intra- and extrahypothalamic vasopressin and oxytocin pathways in the rat. Pathways to the limbic system, medulla oblongata and spinal cord, *Cell Tissue Res.* **192**:423.

Chang, M. M., and Leeman, S. E., 1970, Isolation of a sialogic peptide from bovine hypothalamic tissue and its characterization as substance P, *J. Biol. Chem.* **245**:4784.

Changaris, D. G., Demers, L. M., Keil, L. C., and Severs, W. B., 1977, Immunopharmacology of angiotensin I in brain, in: *Central Actions of Angiotensin and Related Hormones* (J. P. Buckley, C. M. Ferrario, and M. F. Lokhandwala, eds.), pp. 233–243, Pergamon Press, Elmsford, N.Y.

Corrêa, F. M. A., Innis, R. B., Uhl, G. R., and Snyder, S. H., 1979, Bradykinin-like immunoreactive neuronal systems localized histochemically in rat brain, *Proc. Natl. Acad. Sci. USA* **76**:1489.

Cuello, A. C., and Kanazawa, I., 1978, The distribution of substance P immunoreactive fibers in the rat central nervous system, *J. Comp. Neurol.* **178**:129.

Dierickx, K., 1980, Immunocytochemical localization of the vertebrate nonapeptide neurohypophyseal hormones and neurophysins, *Int. Rev. Cytol.* **62**:119.

Dogterom, J., Snijdewint, F. G. M., and Buijs, R. M., 1978, The distribution of vasopressin and oxytocin in the rat brain, *Neurosci. Lett.* **9**:341.

Dubé, D., Lissitzky, J. C., Leclerc, R., and Pelletier, G., 1978, Localization of α-melanocyte-stimulating hormone in rat brain and pituitary, *Endocrinology* **102**:1283.

Dupont, A., Barden, N., Cusan, L., Mérand, Y., Labrie, F., and Vaudry, H., 1980, β-Endorphin and met-enkephalins: Their distribution, modulation by estrogens and haloperidol, and role in neuroendocrine control, *Fed. Proc.* **39**:2544.

Elde, R., and Hökfelt, T., 1978, Distribution of hypothalamic hormones and other peptides in the brain, in: *Frontiers in Neuroendocrinology*, Vol. 5 (W. F. Ganong and L. Martini, eds.), pp. 1–33, Raven Press, New York.

Elde, R., Hökfelt, T., Johansson, O., and Terenius, L., 1976, Immunohistochemical studies using antibodies to leucine-enkephalin: Initial observations of the nervous system of the rat, *Neuroscience* **1**:349.

Epelbaum, J., Tapia-Arancibia, L., Kordon, C., Ottensen, O. P., and Ben-Ari, Y., 1979, Regional distribution of somatostatin within the amygdaloid complex of the rat brain, *Brain Res.* **174**:172.

Eskay, R. L., Giraud, P., Oliver, C., and Brownstein, M. J., 1979, α-Melanocyte stimulating hormone in the rat brain, evidence that α-MSH containing cells in the arcuate region send projections to extrahypothalamic sites, *Brain Res.* **178**:55.

Fahrenkrug, J., and Schaffalitzky de Muckadell, O. B., 1978, Distribution of vasoactive intestinal polypeptide (VIP) in the porcine central nervous system, *J. Neurochem.* **31**:1445.

Fuxe, K., Ganten, D., Hökfelt, T., and Bolme, P., 1976, Immunohistochemical evidence for the existence of angiotensin II-containing nerve terminals in the brain and spinal cord in the rat, *Neurosci. Lett.* **2**:229.

Fuxe, K., Hökfelt, T., Eneroth, P., Gustafsson, J.-Å., and Skett, P., 1977, Prolactin-like immunoreactivity: Localization in nerve terminals of rat hypothalamus, *Science* **196**:899.

Fuxe, K., Andersson, K., Hökfelt, T., Mutt, V., Ferland, L., Agnati, L. F., Ganten, D.,

Said, S., Eneroth, P., and Gustafsson, J.-Å., 1979, Localization and possible function of peptidergic neurons and their interactions with central catecholamine neurons, and the central actions of gut hormones, *Fed. Proc.* **38**:2333.

Fuxe, K., Ganten, D., Hökfelt, T., Locatelli, V., Poulsen, K., Stock, G., Rix, E., and Taugner, R., 1980, Renin-like immunocytochemical activity in the rat and mouse brain, *Neurosci. Lett.* **18**:245.

George, J. M., and Jacobowitz, D. M., 1976, Localization of vasopressin in discrete areas of the rat hypothalamus, *Brain Res.* **93**:363.

George, J. M., Staples, S., and Marks, B. M., 1976, Oxytocin content of microdissected areas of the hypothalamus, *Endocrinology* **98**:1430.

Gramsch, C., Höllt, V., Mehraein, P., Pasi, A., and Herz, A., 1979, Regional distribution of methionine-enkephalin and beta-endorphin-like immunoreactivity in human brain and pituitary, *Brain Res.* **171**:261.

Hawthorn, J., Ang, V. T. Y., and Jenkins, J. S., 1980, Localization of vasopressin in the rat brain, *Brain Res.* **197**:75.

Hoffman, G. E., Melnyk, V., Hayes, T., Bennett-Clarke, C., and Fowler, E., 1978, Immunocytology of LHRH neurons, in: *Brain–Endocrine Interaction, Vol. III: Neural Hormones and Reproduction* (D. E. Scott, G. P. Kozlowski, and A. Weindl, eds.), pp. 67–82, Karger, Basel.

Hökfelt, T., Fuxe, K., Johansson, O., Jeffcoate, S., and White, N., 1975, Distribution of thyrotropin-releasing hormone (TRH) in the central nervous system as revealed with immunocytochemistry, *Eur. J. Pharmacol.* **34**:389.

Hökfelt, T., Elde, R., Fuxe, K., Johansson, O., Ljungdahl, Å., Goldstein, M., Luft, R., Efendić, S., Nilsson, G., Terenius, L., Ganten, D., Jeffcoate, S. L., Rehfeld, J., Said, S., Perez de la Mora, M., Possani, L., Tapia, R., Teran, L., and Palacios, R., 1978a, Aminergic and peptidergic pathways in the nervous system with special reference to the hypothalamus, in: *The Hypothalamus* (S. Reichlin, R. J. Baldessarini, and J. B. Martin, eds.), pp. 69–135, Raven Press, New York.

Hökfelt, T., Elde, R., Johansson, O., Ljungdahl, Å., Schultzberg, M., Fuxe, K., Goldstein, M., Nilsson, G., Pernow, B., Terenius, L., Ganten, D., Jeffcoate, S. L., Rehfeld, J., and Said, S., 1978b, Distribution of peptide-containing neurons, in: *Psychopharmacology: A Generation of Progress* (M. A. Lipton, A. DiMascio, and K. F. Killam, eds.), pp. 39–66, Raven Press, New York.

Hong, J. S., Yang, H.-Y., Fratta, W., and Costa, E., 1977, Determination of methionine enkephalin in discrete regions of rat brain, *Brain Res.* **134**:383.

Innis, R. B., Corrêa, F. M. A., Uhl, G. R., Schneider, B., and Snyder, S. H., 1979, Cholecystokinin octapeptide-like immunoreactivity: Histochemical localization in rat brain, *Proc. Natl. Acad. Sci. USA* **76**:521.

Jackson, I. M. D., and Reichlin, S., 1974, Thyrotropin-releasing hormone (TRH): Distribution in hypothalamic and extrahypothalamic brain tissues of mammalian and submammalian chordates, *Endocrinology* **95**:854.

Jacobowitz, D. M., and O'Donohué, T. L., 1978, α-Melanocyte stimulating hormone: Immunohistochemical identification and mapping in neurons of rat brain, *Proc. Natl. Acad. Sci. USA* **75**:6300.

Johansson, O., Hökfelt, T., Elde, R. P., Schultzberg, M., and Terenius, L., 1978, Immunohistochemical distribution of enkephalin neurons, *Adv. Biochem. Psychopharmacol.* **18**:51.

Kanazawa, I., and Jessell, T., 1976, Post mortem changes and regional distribution of substance P in the rat and mouse nervous system. *Brain Res.* **117**:362.

Kastin, A. J., Nissen, C., Schally, A. V., and Coy, D. H., 1978, Radioimmunoassay of DSIP-like material in rat brain, *Brain Res. Bull.* **3**:691.

Kizer, J. S., Palkovits, M., Tappaz, J., Kebabian, J., and Brownstein, M., 1976, Distribution of releasing factors, biogenic amines and related enzymes in the bovine median eminence, *Endocrinology* **98**:685.

Kobayashi, R. M., Brown, M., and Vale, W., 1977, Regional distribution of neurotensin and somatostatin in rat brain, *Brain Res.* **126**:584.

Kobayashi, R. M., Palkovits, M., Miller, R. J., Chang, K.-J., and Cuatrecasas, P., 1978, Distribution of enkephalin in the brain is unaltered by hypophysectomy, *Life Sci.* **22**:527.

Krieger, D. T., Liotta, A., and Brownstein, M. J., 1977a, Corticotropin releasing factor distribution in normal and Brattleboro rat brain, and effect of deafferentation, hypophysectomy and steroid treatment in normal animals, *Endocrinology* **100**:227.

Krieger, D. T., Liotta, A., Suda, T., Palkovits, M., and Brownstein, M. J., 1977b, Presence of immunoassayable β-lipotropin in bovine brain and spinal cord: Lack of concordance with ACTH concentrations, *Biochem. Biophys. Res. Commun.* **76**:930.

Krisch, B., 1978, Hypothalamic and extrahypothalamic distribution of somatostatin-immunoreactive elements in the rat brain, *Cell Tissue Res.* **195**:499.

Krulich, L., Quijada, M., Wheaton, J. E., Illner, P., and McCann, S. M., 1977, Localization of hypophysiotrophic neurohormones by assay of sections from various brain areas, *Fed. Proc.* **36**:1953.

Lang, R. E., Voight, K.-H., Fehm, H. L., and Pfeiffer, E. F., 1976, Localization of corticotropin-releasing activity in the rat hypothalamus, *Neurosci. Lett.* **2**:19.

Larsson, L.-I., and Rehfeld, J. F., 1979, Localization and molecular heterogeneity of cholecystokinin in the central and peripheral nervous system, *Brain Res.* **165**:201.

Ljungdahl, Å., Hökfelt, T., Nilsson, G., and Goldstein, M., 1978, Distribution of substance P-like immunoreactivity in the central nervous system of the rat. I. Cell bodies and nerve terminals, *Neuroscience* **3**:861.

Lorén, I., Emson, P. C., Fahrenkrug, J., Björklund, A., Alumets, J., Håkanson, R., and Sundler, F., 1979, Distribution of vasoactive intestinal polypeptide in the rat and mouse brain. *Neuroscience* **4**:1953.

Moldow, R. L., and Yalow, R. S., 1978, Extrahypophyseal distribution of thyrotropin as a function of brain size. *Life Sci.* **22**:1859.

Morris, R., Salt, T. E., Sofroniew, M. V., and Hill, R. G., 1980, Actions of microiontophoretically applied oxytocin, and immunohistochemical localization of oxytocin, vasopressin and neurophysin in the rat caudal medulla, *Neurosci. Lett.* **18**:163.

O'Donohué, T. L., Miller, R. L., and Jacobowitz, D. M., 1979, Identification, characterization and stereotaxic mapping of intraneuronal α-melanocyte stimulating hormone-like immunoreactive peptides in discrete regions of the rat brain, *Brain Res.* **176**:101.

Oliver, C., Eskay, N. L., Ben-Jonathan, N., and Porter, J. C., 1974, Distribution and concentration of TRH in the rat brain, *Endocrinology* **95**:540.

Pacold, S. T., Kirsteins, L., Hojvat, S., Lawrence, A. M., and Hagen, T. C., 1978, Biologically active pituitary hormones in the rat brain amygdaloid nucleus, *Science* **199**:804.

Palkovits, M., 1978, Topography of chemically identified neurons in the central nervous system: A review, *Acta Morphol. Acad. Sci. Hung.* **26**:211.

Palkovits, M., 1979a, Microchemistry of microdissected hypothalamic nuclear areas, *Int. Rev. Cytol.* **56**:315.

Palkovits, M., 1979b, Dopamine levels in individual brain regions: Biochemical aspects of DA distribution in the central nervous system, in: *The Neurobiology of Dopamine* (A. S. Horn, J. Korf, and B. H. C. Westerink, eds.), pp. 343–356, Academic Press, London.

Palkovits, M., 1980a, Mapping of neurotransmitters and hypothalamic hormones, in: *Neuroactive Drugs in Endocrinology* (E. E. Müller, ed.), pp. 35–48, Elsevier/North-Holland, Amsterdam.

Palkovits, M., 1980b, Topography of chemically identified neurons in the central nervous system: Progress in 1977–1979, *Med. Biol.* **58**:188.

Palkovits, M., Arimura, A., Brownstein, M., Schally, A. V., and Saavedra, J. M., 1974, Luteinizing hormone-releasing hormone (LH-RH) content of the hypothalamic nuclei in rat, *Endocrinology* **96**:554.

Palkovits, M., Brownstein, M., Arimura, A., Sato, H., Schally, A. V., and Kizer, J. S., 1976, Somatostatin content of the hypothalamic ventromedial and arcuate nuclei and the circumventricular organs in the rat, *Brain Res.* **109**:430.

Palkovits, M., Gráf, L., Hermann, I., Borvendég, J., Ács, Z., and Láng, T., 1978, Regional distribution of enkephalins, endorphins and ACTH in the central nervous system of rats determined by radioimmunoassay, in: *Endorphins '78* (L. Gráf, M. Palkovits, and A. Z. Rónai, eds.), pp. 187–195, Akadémiai Kiadó, Budapest.

Palkovits, M., Kobayashi, R. M., Brown, M., and Vale, W., 1980, Somatostatin levels following various hypothalamic transections in rat, *Brain Res.* **195**:499.

Palkovits, M., Besson, J., and Rotsztejn, W., 1981a, Distribution of vasoactive intestinal polypeptide in intact, stria terminalis transected and cerebral cortex isolated rats, *Brain Res.* **213**:455.

Palkovits, M., Epelbaum, J., and Gros, C., 1981b, Met-enkephalin concentrations in individual brain nuclei of ansa lenticularis and stria terminalis transected rats, *Brain Res.* **216**:203.

Parsons, J. A., Erlandsen, S. L., Hegre, O. D., McEvoy, R. C., and Elde, R. P., 1976, Central and peripheral localization of somatostatin immunoenzyme immunocytochemical studies, *J. Histochem. Cytochem.* **24**:872.

Pelletier, G., and Leclerc, R., 1979, Immunohistochemical localization of adrenocorticotropin in the rat brain, *Endocrinology* **104**:1426.

Phillips, M. I., Weyhenmeyer, J., Felix, D., Ganten, D., and Hoffman, W. E., 1979, Evidence for an endogenous brain renin-angiotensin system, *Fed. Proc.* **38**:2260.

Phillips, H. S., Hostetter, G., Kerdelhué, B., and Kozlowski, G. P., 1980, Immunocytochemical localization of LH-RH in central olfactory pathways of hamster, *Brain Res.* **193**:574.

Rehfeld, J. F., 1978, Localization of gastrins to neuro- and adenohypophysis, *Nature (London)* **271**:771.

Roberts, G. W., Woodhams, P. L., Bryant, M. G., Crow, T. J., Bloom, S. R., and Polak, J. M., 1980, VIP in the brain: Evidence for a major pathway linking the amygdala and hypothalamus via the stria terminalis, *Histochemistry* **65**:103.

Samson, W. K., Said, S. I., and McCann, S. M., 1979, Radioimmunologic localization of vasoactive intestinal polypeptide in hypothalamic and extrahypothalamic sites in the rat brain, *Neurosci. Lett.* **12**:265.

Samson, W. K., McCann, S. M., Chud, L., Dudley, C. A., and Moss, R. L., 1980, Intra- and extrahypothalamic luteinizing hormone-releasing hormone (LHRH) distribution in the rat with special reference to mesencephalic sites which contain both LHRH and single neurons responsive to LHRH, *Neuroendocrinology* **31**:66.

Selmanoff, M. K., Wise, P. M., and Barraclough, C. A., 1980, Regional distribution of luteinizing hormone-releasing hormone (LH-RH) in rat brain determined by microdissection and radioimmunoassay, *Brain Res.* **192**:421.

Sétáló, G., Vigh, S., Schally, A. V., Arimura, A., and Flerkó, B., 1975, GH-RH-containing neural elements in the rat hypothalamus, *Brain Res.* **90**:352.

Sétáló, G., Flerkó, B., Arimura, A., and Schally, A. V., 1978, Brain cells as producers of releasing and inhibiting hormones, in: *Neuronal Cells and Hormones* (G. H. Bourne, J. F. Danielli, and K. W. Jeon, eds.), pp. 1–52, Academic Press, New York.

Silverman, A. J., 1976, Distribution of luteinizing hormone-releasing hormone (LHRH) in the guinea pig brain, *Endocrinology* **99**:30.

Simantov, R., Kuhar, M. J., Pasternak, G. W., and Snyder, S. H., 1976, The regional distribution of a morphine-like factor enkephalin in monkey brain, *Brain Res.* 106:189.

Simantov, R., Kuhar, M. J., Uhl, G. R., and Snyder, S. H., 1977, Opioid peptide enkephalin: Immunohistochemical mapping in the rat central nervous system, *Proc. Natl. Acad. Sci. USA* 74:2167.

Sims, K. B., Hoffman, D. L., Said, S. I., and Zimmerman, E. A., 1980, Vasoactive intestinal peptide (VIP) in mouse and rat brain: An immunocytochemical study, *Brain Res.* 186:165.

Sofroniew, M. V., and Weindl, A., 1978, Projections from the parvocellular vasopressin- and neurophysin-containing neurons of the suprachiasmatic nucleus, *Am. J. Anat.* 153:391.

Sofroniew, M. V., Weindl, A., Schinko, I., and Wetzstein, R., 1979, The distribution of vasopressin-, oxytocin-, and neurophysin-producing neurons in the guinea pig brain. I. The classical hypothalamo–neurohypophyseal system, *Cell Tissue Res.* 196:367.

Uhl, G. R., and Snyder, S. H., 1976, Regional and subcellular distributions of brain neurotensin, *Life Sci.* 19:1827.

Uhl, G. R., Kuhar, M. J., and Snyder, S. H., 1977, Neurotensin: Immunohistochemical localization in rat central nervous system, *Proc. Natl. Acad. Sci. USA* 74:4059.

Uhl, G. R., Goodman, R. R., Kuhar, M. J., Childers, S. R., and Snyder, S. H., 1979a, Immunohistochemical mapping of enkephalin-containing cell bodies, fibers and nerve terminals in the brain stem of the rat, *Brain Res.* 166:75.

Uhl, G. R., Goodman, R. R., and Snyder, S. H., 1979b, Neurotensin-containing cell bodies, fibres and nerve terminals in the brain stem of the rat: Immunohistochemical mapping, *Brain Res.* 167:77.

Vanderhaeghen, J. J., Lotstra, F., De Mey, J., and Gilles, C., 1980, Immunohistochemical localization of cholecystokinin- and gastrin-like peptides in the brain and hypophysis of the rat, *Proc. Natl. Acad. Sci. USA* 77:1190.

Vandesande, F., and Dierickx, K., 1975, Identification of the vasopressin producing and of the oxytocin producing neurons in the hypothalamic magnocellular neurosecretory system of the rat, *Cell Tissue Res.* 164:153.

Vandesande, F., Dierickx, K., and De Mey, J., 1975, Identification of the vasopressin–neurophysin II and the oxytocin–neurophysin I producing neurons in the bovine hypothalamus, *Cell Tissue Res.* 156:189.

Vigh, S., Sétáló, G., Schally, A. V., Arimura, A., and Flerkó, B., 1978, LH-RH-containing nerve fibres in the brain of rats treated with sulpiride or reserpine, *Brain Res.* 152:401.

von Euler, U. S., and Gaddum, J. H., 1931, An unidentified depressor substance in certain tissue extracts, *J. Physiol. (London)* 72:74.

Walsh, J. H., Wong, H. C., and Docray, G. J., 1979, Bombesin-like peptides in mammals, *Fed. Proc.* 38:2315.

Wamsley, J. K., Young, W. S., III, and Kuhar, M. J., 1980, Immunohistochemical localization of enkephalin in rat forebrain, *Brain Res.* 190:153.

Watson, S., Akil, H., Sullivan, S., and Barchas, J., 1977a, Immunocytochemical localization of methionine enkephalin: Preliminary observations, *Life Sci.* 21:733.

Watson, S., Barchas, J., and Li, C. H., 1977b, β-Lipotropin: Localization of cells and axons in rat brain by immunocytochemistry, *Proc. Natl. Acad. Sci. USA* 74:5155.

Watson, S. J., Richard, C. W., and Barchas, J. D., 1978, Adrenocorticotropin in rat brain: Immunocytochemical localization in cells and axons, *Science* 200:1180.

Zimmerman, E. A., Liotta, A., and Krieger, D. T., 1978, β-Lipotropin in brain: Localization in hypothalamic neurons by immunoperoxidase technique, *Cell Tissue Res.* 186:393.

15

Distribution and Changes in Peptides in the Brain

Nicholas Barden and André Dupont

1. Introduction

For many years biogenic amines were the only established and investigated neurotransmitters in the central nervous system. More recently, certain amino acids such as glutamate, γ-aminobutyrate, and glycine have been recognized as major neurotransmitter systems and within the past few years certain small peptide molecules have been proposed as neurotransmitter candidates. The list of peptides possibly involved in neurotransmitter systems is rapidly increasing and currently consists of more than a dozen different compounds ranging from tripeptides to molecules of about 100 amino acids. Although not all of these are as yet firmly established as classical neurotransmitters, the physiological roles of certain peptides, in particular the opiate peptides, and their interactions with other systems in the brain are beginning to be understood.

This chapter will attempt to summarize our findings to date on the modification of concentrations of certain peptides in discrete brain nuclei during both physiological states and pharmacological treatments. These studies include distributions of enkephalins, endorphins, somatostatin, cholecystokinin, neurotensin, and thyrotropin-releasing hormone and their interactions with biogenic amines or peripheral hormonal systems, in particular estrogens and thyroid hormones.

Nicholas Barden and André Dupont • Department of Molecular Endocrinology, Le Centre Hospitalier de l'Université Laval, Quebec GIV 4G2, Canada.

2. Distribution of Neurotransmitter Peptides in Brain

2.1. Opiate Peptides

Following reports of the presence of endogenous opiate activity in the brain (Pasternak *et al.*, 1977; Terenius and Wahlstrom, 1975), two pentapeptides having the following structures: H-Tyr-Gly-Gly-Phe-Met-OH (Met-enkephalin) and H-Tyr-Gly-Gly-Phe-Leu-OH (Leu-enkephalin) were isolated from porcine (Hughes *et al.*, 1975) and calf brain (Simantov and Snyder, 1976a,b) and shown to possess high opiate agonist activity. The sequence of Met-enkephalin is the same as the N terminus of β-endorphin, the C-fragment [β-LPH-(61–91)] of β-lipotropin first isolated from sheep pituitary gland (Li *et al.*, 1965). β-Endorphin has been isolated and characterized from camel pituitary glands (Li and Chung, 1976) as well as from many other species including man, and shown to be an analgesic agent as measured by both *in vitro* and *in vivo* methods. Both Met-enkephalin and β-endorphin have potent morphinelike activity (Hughes *et al.*, 1975; Li and Chung, 1976) in many biological assays and bind to the opiate receptor (Bradbury *et al.*, 1976; Morin *et al.*, 1976).

These findings suggested that β-LPH may be the precursor of endorphins and enkephalins. However, increasing evidence supports the existence of two separate systems: an enkephalin system and a β-LPH/β-endorphin/ACTH system (Bloom *et al.*, 1978; Krieger *et al.*, 1977; Pelletier *et al.*, 1977).

We have measured the distribution of enkephalin- and β-endorphin-like immunoreactivities in 42 discrete areas of rat and bovine brain. In all of these studies, rat brains were rapidly removed after decapitation, frozen on dry ice, and serial coronal slices of 300 μm were cut in the stereotaxic plane of König and Klippel (1963) in a cryostat at −10°C as described by Palkovits (1973). Samples were then punched from the sections with small stainless-steel needles (inside diameter 0.5 or 1 mm) and homogenized in 2 N acetic acid for extraction of peptides.

Bovine brains were obtained from 30 castrated males (steers) chilled on ice and dissected within 1 hr after the death of animal. The brains were cut into coronal slices (approximately 1 cm thick) and the areas dissected were immediately frozen on dry ice before homogenization in 2 N acetic acid. Peptides were measured by specific radioimmunoassays as previously described (Dupont *et al.*, 1980; Gros *et al.*, 1978; Miller *et al.*, 1978).

As shown in Table I, the highest concentrations of β-endorphin are found in the median eminence and the hypothalamic nuclei followed by the medial preoptic nucleus, the nucleus interstitialis striae terminalis, periaqueductal gray, nucleus medialis thalami, and nucleus amygdaloideus centralis. The highest concentration of Met-enkephalin immunoreactivity was

Table I. Regional Distribution of Met-Enkephalin and β-Endorphin in Rat Brain[a]

	Met-enkephalin (ng/mg protein)	β-Endorphin (ng/mg protein)
Frontal cortex	0.36 ± 0.07	ND[b]
Septum	1.43 ± 0.36	0.77 ± 0.08
Tractus diagonalis	2.07 ± 0.33	1.06 ± 0.33
Medial preoptic nucleus	10.99 ± 0.54	21.93 ± 0.96
Lateral preoptic nucleus	6.38 ± 1.66	2.90 ± 0.13
Nucleus interstitialis striae terminalis	16.22 ± 1.11	15.46 ± 1.57
Nucleus accumbens	1.88 ± 0.29	ND
Tuberculum olfactorium	1.41 ± 0.33	ND
Striatum A 9000	2.33 ± 0.68	ND
Striatum A 7000	0.85 ± 0.44	—
Globus pallidus	18.25 ± 3.07	—
N. amygdaloideus medialis A 4250	1.28 ± 0.35	4.28 ± 1.14
N. amygdaloideus medialis A 3500	1.35 ± 0.30	3.47 ± 0.29
N. amygdaloideus centralis A 4250	7.13 ± 1.81	5.76 ± 1.25
N. amygdaloideus lateralis, pars post	0.50 ± 0.13	3.09 ± 0.91
N. amygdaloideus basalis, pars lat	0.50 ± 0.06	4.43 ± 1.73
N. amygdaloideus corticalis	0.49 ± 0.05	ND
Nucleus anterior hypothalami	7.21 ± 0.62	21.17 ± 1.80
Nucleus lateralis hypothalami	9.16 ± 2.95	5.81 ± 1.32
Nucleus suprachiasmaticus	4.80 ± 0.95	25.76 ± 2.83
Nucleus periventricularis	1.88 ± 0.11	30.79 ± 3.16
Paraventricular nucleus	6.42 ± 0.83	13.18 ± 0.22
Median eminence	2.69 ± 0.51	41.78 ± 0.77
Arcuate nucleus I–II–III	2.45 ± 0.61	29.67 ± 0.37
Nucleus ventromedialis hypothalami	5.44 ± 1.79	21.84 ± 1.80
Nucleus dorsomedialis hypothalami	3.26 ± 0.65	25.68 ± 4.92
Nucleus arcuate IV–V	3.6 ± 0.84	26.39 ± 8.01
Nucleus premamillaris ventralis	4.00 ± 0.77	9.60 ± 0.40
Nucleus premamillaris dorsalis	0.67 ± 0.03	4.59 ± 0.95
Nucleus mamillaris medialis, pars medialis	0.80 ± 0.12	6.32 ± 0.94
Nucleus lateralis thalami	0.16 ± 0.04	1.84 ± 0.21
Nucleus medialis thalami	0.33 ± 0.01	6.16 ± 1.41
Nucleus ventralis thalami	0.13 ± 0.03	1.32 ± 0.41
Nucleus paraventricularis thalami	—	30.47 ± 5.29
Hippocampus	0.17 ± 0.02	1.47 ± 0.20
Nucleus habenulae lateralis	1.51 ± 0.07	6.32 ± 0.94
Substantia nigra	0.37 ± 0.07	ND
Area ventralis tegmenti	0.95 ± 0.15	ND
Formatio reticular + lemniscus medialis	0.38 ± 0.07	ND
Interpeduncular nucleus	2.55 ± 0.27	1.33 ± 0.15
Periaqueductal gray	1.93 ± 0.39	8.19 ± 1.77
Nucleus medianus raphes	0.50 ± 0.10	ND

[a] The brains were frozen and specific areas were punched as described in text. Values are the mean ± S.E.M. of three groups of four rats each (pooled).
[b] ND, not detectable.

found in the globus pallidus, followed by the nucleus interstitialis striae terminalis, medial preoptic nucleus, nucleus lateralis hypothalami, nucleus amygdaloideus centralis, and paraventricular nucleus.

The regional distribution of β-endorphin, Met- and Leu-enkephalin-like material in bovine brain is shown in Table II. β-Endorphin-like immunoreactivity appears to be mainly concentrated in the hypothalamic area, preoptic area, habenula, lateral pontine nuclei, nucleus interpeduncularis, and septum. While the highest concentration of enkephalinlike material

Table II. Regional Distribution of Met-enkephalin, Leu-enkephalin, and β-Endorphin in Bovine Brain[a]

	β-Endorphin (ng/mg protein)	Met-enkephalin (ng/mg protein)	Leu-enkephalin (ng/mg protein)	Ratio Met-enk Leu-enk
Median eminence	1.916 ± 2.500	5.340 ± 0.087	0.560 ± 0.013	9.54
Polus frontalis	—	0.570 ± 0.037	—	—
Cortex temporalis	—	0.430 ± 0.047	0.198 ± 0.06	2.17
N. caudatus	—	15.640 ± 0.400	2.004 ± 0.061	7.80
Putamen	—	11.100 ± 1.330	1.557 ± 0.013	7.13
Tuberculum olfactorium	—	14.990 ± 2.340	2.231 ± 0.267	6.72
N. accumbens septi	—	18.500 ± 1.640	2.036 ± 0.384	9.09
Area preoptica	3.752 ± 0.480	3.280 ± 0.127	0.677 ± 0.110	4.84
Septum	1.036 ± 0.250	4.433 ± 0.395	0.655 ± 0.110	6.77
Globus pallidus	—	37.980 ± 4.070	6.173 ± 1.402	6.15
Cortex parietalis	—	0.274 ± 0.033	—	—
Hypothalamus anterior	2.120 ± 0.060	0.908 ± 0.130	0.362 ± 0.031	2.51
Thalamus anterior	—	0.422 ± 0.036	0.369 ± 0.038	1.14
Corpus amygdaloideum	—	1.387 ± 0.147	0.409 ± 0.010	3.39
Hippocampus	—	0.184 ± 0.020	—	—
Hypothalamus median	3.924 ± 0.032	2.392 ± 0.329	0.516 ± 0.034	4.64
Thalamus median	—	1.067 ± 0.146	0.364 ± 0.069	2.93
Habenula	3.503 ± 0.338	2.764 ± 0.223	0.319 ± 0.055	8.66
Hypothalamus posterior	1.374 ± 0.400	2.401 ± 0.083	1.161 ± 0.250	2.07
N. interpeduncularis	1.653 ± 0.264	1.034 ± 0.192	0.223 ± 0.115	4.64
Tectum mesencephali	—	0.784 ± 0.054	0.276 ± 0.080	1.75
S. grisea centralis	—	1.045 ± 0.085	0.220 ± 0.038	4.75
S. nigra	—	1.112 ± 0.062	0.564 ± 0.300	1.97
Tegmentum	—	2.009 ± 0.132	—	—
Medial pontine area	—	1.429 ± 0.208	0.570 ± 0.250	2.51
Lateral pontine area	1.850 ± 0.040	0.978 ± 0.034	0.216 ± 0.080	4.53
Area postrema	—	0.868 ± 0.080	—	—
Anterior medulla	—	0.547 ± 0.170	—	—
Cerebellar vermis	—	0.057 ± 0.008	—	—
Cerebellar hemisphere	—	0.060 ± 0.01	—	—

[a] Peptides were extracted from discrete structures pooled from 30 brains. Values are the mean ± S.E.M. of triplicate determinations.

was found in the basal ganglia, certain structures of the olfactory brain, namely the nucleus accumbens septi and the tuberculum olfactorium, also contain high amount of Leu- and Met-enkephalin-like immunoreactivity. These results support the view that enkephalins and endorphins form distinct systems in the brain. This mutual exclusivity of β-endorphinergic and enkephalinergic systems has been previously suggested by immunohistochemical data (Bloom *et al.*, 1978; Watson *et al.*, 1978).

Parallels can be seen between the distribution of the peptides in bovine and rat brains. As observed in the rat, the β-endorphin system in the cattle is also concentrated within the hypothalamic area. The high amounts of β-endorphin in the area preoptica, habenula, lateral pontine nuclei, nucleus interpeduncularis, and septum suggest, as for the rat, the existence of an extrahypothalamic endorphinergic neural network.

In all structures of bovine brain studied Met-enkephalin was always present in higher concentrations than was Leu-enkephalin, with the ratio of Met-enkephalin/Leu-enkephalin ranging from 1.14 in anterior thalamus to 9.09 in nucleus accumbens. These results differ from those of Snyder's group (Simantov and Snyder, 1976b), who reported a predominance of Leu-enkephalin, and may result from the decreased degradation during extraction when acetic acid is used as in the present studies. Recently, Larsson *et al.* (1979) have indicated the existence of separate neuronal systems for Met-enkephalin and Leu-enkephalin and report that Leu-enkephalin-containing neurons are only 20–25% as numerous as are Met-enkephalin-containing neurons in the striatum. In light of this report, the different ratios of Met-enkephalin/Leu-enkephalin found in various structures may reflect differences in innervation by the two distinct neuronal systems.

2.2. Neurotensin

In bovine brain, neurotensin was found in very high concentrations in globus pallidus followed by nucleus accumbens and tuberculum olfactorium (Table III), while low levels are present in hypothalamic areas, habenula, tegmentum, and amygdala. The distribution of neurotensin in bovine brain is strikingly similar to that of the enkephalins (Table II). Although similarities between the distributions of Met-enkephalin and neurotensin of rat brain, particularly with respect to an amygdalo-fugal pathway (Uhl and Snyder, 1979), are evident, differences can also be seen (Table IV). Moreover, in animals pretreated with colchicine, Metenkephalin was significantly increased in medial preoptic nucleus and nucleus ventromedialis hypothalami, while it was decreased in the globus pallidus and median eminence. Neurotensin was significantly increased after pretreatment with colchicine in the medial preoptic nucleus, nucleus interstitialis

Table III.   Regional Distribution of Neurotensin in Bovine
Brain

	Neurotensin (ng/mg protein)
Median eminence	2.690 ± 0.100
Polus frontalis	0.059 ± 0.010
Cortex temporalis	—
N. caudatus	1.504 ± 0.880
Putamen	1.340 ± 0.240
Tuberculum olfactorium	2.964 ± 0.018
N. accumbens septi	3.220 ± 0.300
Area preoptica	2.250 ± 0.540
Septum	2.342 ± 0.225
Globus pallidus	5.580 ± 1.420
Cortex parietalis	—
Hypothalamus anterior	0.610 ± 0.125
Thalamus anterior	—
Corpus amygdaloideum	0.330 ± 0.031
Hippocampus	0.186 ± 0.012
Hypothalamus median	1.300 ± 0.072
Thalamus median	—
Habenula	0.956 ± 0.047
Hypothalamus posterior	1.541 ± 0.079
N. interpeduncularis	2.190 ± 0.049
Tectum mesencephali	—
S. grisea centralis	0.200 ± 0.029
S. nigra	0.070 ± 0.002
Tegmentum	0.630 ± 0.025
Medial pontine nuclei	0.035 ± 0.009
Lateral pontine nuclei	0.193 ± 0.009
Area postrema	—
Anterior medulla	—
Cerebellar vermis	0.034 ± 0.404
Cerebellar hemisphere	0.075 ± 0.038

striae terminalis, and globus pallidus (Dupont and Barden, unpublished observations), indicating that these two neuropeptides exist in distinct cell populations.

2.3.   Cholecystokinin

Cholecystokinin (CCK) was first recognized in duodenum (Jorpes *et al.,* 1964) and subsequently shown to occur primarily as the C-terminal octapeptide in the central nervous system (Vanderhaeghen *et al.,* 1975, 1980; Müller *et al.,* 1977a; Dockray *et al.,* 1978). While the distribution of CCK in rat brain, as determined using immunohistochemical techniques, has been described (Innis *et al.,* 1979), this has not as yet been corroborated

Table IV.   Regional Distribution of Neurotensin in Rat
Brain[a]

	Neurotensin (ng/mg protein)
Nucleus accumbens	0.72 ± 0.15
Tuberculum olfactorium	0.52 ± 0.03
Striatum A 9000	0.59 ± 0.3
Frontal cortex A 9000	0.31 ± 0.05
Septum	0.97 ± 0.20
Tractus diagonalis	1.17 ± 0.18
Medial preoptic nucleus	3.87 ± 0.23
Lateral preoptic nucleus	2.85 ± 0.13
N. interstitialis striae terminalis	2.79 ± 0.18
Striatum A 7000	ND[b]
Nucleus anterior hypothalami	2.12 ± 0.41
Nucleus lateralis hypothalami	1.96 ± 0.34
Globus pallidus	0.86 ± 0.08
Striatum A 6000	ND
Periventricular nucleus	1.42 ± 0.18
Hippocampus	ND
Nucleus lateralis thalami	ND
Nucleus medialis thalami	0.67 ± 0.03
Nucleus ventralis thalami	ND
Nucleus dorsomedialis hypothalami	2.91 ± 0.05
Nucleus ventromedialis hypothalami	1.13 ± 0.3
Arcuate nucleus I-II-III	1.57 ± 0.12
Median eminence	4.19 ± 0.88
Nucleus habenulae lateralis	1.05 ± 0.09
Nucleus habenulae median	1.28 ± 0.17
Nucleus entopeduncularis	1.30 ± 0.15
N. amygdaloideus lateralis, pars post	0.57 ± 0.06
N. amygdaloideus basalis, pars lat	0.89 ± 0.07
N. amygdaloideus medialis A 4250	1.03 ± 0.17
N. amygdaloideus centralis A 4250	4.48 ± 0.69
N. amygdaloideus medialis A 3500	ND
N. amygdaloideus corticalis	0.69 ± 0.04
Cortex entorhinalis	0.85 ± 0.04
Nucleus arcuate IV-V	2.01 ± 0.05
Substantia nigra	0.48 ± 0.05
Area ventralis tegmenti	1.72 ± 0.21
Interpeduncular nucleus	0.82 ± 0.22
Formatio reticularis	0.69 ± 0.08
Gray periaqueductal	1.56 ± 0.32
N. medianus raphes	0.73 ± 0.09

[a] Values are the mean ± S.E.M. of three groups of six animals each.
[b] ND, not detectable.

by radioimmunoassay studies. The highest concentrations of CCK in bovine brain were found in frontal and temporal cortex, closely followed by the basal ganglia and mesolimbic structures (Table V). Relatively low amounts of CCK were present in median eminence and habenula as well as in structures of the mesencephalon. In rat, the highest amount of CCK was found in the nuclei of amygdala, then in polus frontalis, nucleus accumbens, and cortex entorhinalis. Medial preoptic nucleus, median eminence, and nucleus arcuatus appear to be the richest hypothalamic structures.

Table V. Regional Distribution of
Immunoreactive Cholecystokinin in Bovine
Brain[a]

	Cholecystokinin (ng/mg protein)
Median eminence	0.208 ± 0.0065
Polus frontalis	2.172 ± 0.56
Cortex temporalis	1.024 ± 0.016
N. caudatus	1.044 ± 0.05
Putamen	0.568 ± 0.032
Tuberculum olfactorium	0.716 ± 0.028
N. accumbens septi	0.368 ± 0.037
Area preoptica	0.174 ± 0.002
Septum	0.081 ± 0.002
Globus pallidus	0.167 ± 0.008
Cortex parietalis	0.256 ± 0.011
Hypothalamus anterior	0.087 ± 0.009
Thalamus anterior	ND[b]
Corpus amygdaloideum	1.267 ± 0.045
Hippocampus	0.07 ± 0.008
Hypothalamus median	0.03 ± 0.001
Thalamus median	ND
Habenula	0.044 ± 0.003
Hypothalamus posterior	ND
N. interpeduncularis	0.07 ± 0.007
Tectum mesencephali	ND
S. grisea centralis	0.168 ± 0.008
S. nigra	ND
Tegmentum	0.05 ± 0.0005
Medial pontine area	0.048 ± 0.002
Lateral pontine area	0.055 ± 0.003
Area postrema	ND
Anterior medulla	0.05 ± 0.002
Cerebellar vermis	ND
Cerebellar hemisphere	ND

[a] Values are the mean ± S.E.M. of three determinations of cholecystokinin concentrations in tissue pooled from 30 animals.
[b] ND, not detectable.

Other brain structures contain low or very low amounts of CCK-like immunoreactivity (Table VI).

Recent study suggests that most of the CCK-like immunoreactivity in rat brain consists of CCK8-like peptide (Dockray, 1980). In both rat and bovine species, CCK is predominant in the cortex, the rhinencephalon, and the structures of the telencephalic basal ganglia. Nuclei amygdalae, so-called archistriatum, show the highest amount of CCK-like activity, while a low concentration of the peptide is seen in the neostriatum of the rat. However, in bovine brain CCK-like substance is approximately equally distributed between these two structures of the telencephalic basal ganglia.

3. Physiological Changes in Opiate Peptide Distribution

3.1. Estrous Cycle

A role for endogenous opiate peptides in the control of prolactin secretion is suggested by the prevention of stress- or suckling-induced increases in serum prolactin by naloxone administration (Ferland *et al.*, 1978). In order to obtain further evidence for the involvement of endogenous opiate peptides, we have measured the β-endorphin content of discrete brain nuclei during the estrous cycle of rats and have found changes to occur in certain regions simultaneously with the peak of prolactin release. Measurement of the β-endorphin content of 16 specific hypothalamic and extrahypothalamic nuclei during all stages of the estrous cycle showed that statistically significant variations occurred only in suprachiasmatic and arcuate nuclei and the median eminence (Fig. 1). In the arcuate nucleus, which contained the highest β-endorphin content of all brain nuclei, the afternoon of proestrus coincided with a 50% decrease in β-endorphin content compared to the mean content on all other days of the cycle. In contrast, the β-endorphin content of suprachiasmatic nucleus and median eminence were increased by 100 and 65%, respectively, on the afternoon of proestrus when compared to the mean content on other days. No significant changes in the β-endorphin content of these three regions were apparent at 1500 hr on diestrus. In the suprachiasmatic nucleus, the β-endorphin content remained elevated throughout estrus, whereas in both arcuate nucleus and median eminence by the day of estrus, values returned to those seen on other days of the cycle. Since opiate peptides have been demonstrated to reduce the turnover of dopamine in the median eminence (Ferland *et al.*, 1977) and inhibit release of dopamine from tuberoinfundibular neurons (Gudelsky and Porter, 1979) and since dopamine levels in the median eminence are decreased on the afternoon of proestrus (Crowley *et al.*, 1978), it is possible that the increased β-endorphin content of median eminence seen on the

Table VI. Regional Distribution of Immunoreactive Cholecystokinin in Rat Brain[a]

	Immunoreactive CCK (ng/mg protein)
Polus frontalis	5.37 ± 0.87
Cortex temporalis	1.66 ± 0.25
Cortex pyriformis	2.82 ± 0.41
Cortex entorhinalis	4.49 ± 1.19
Septum	1.16 ± 0.25
Tractus diagonalis	0.84 ± 0.13
Medial preoptic nucleus	1.99 ± 0.25
Lateral preoptic nucleus	0.59 ± 0.11
Nucleus interstitialis striae terminalis	1.05 ± 0.08
Commissura anterior	1.17 ± 0.13
Nucleus accumbens	4.62 ± 1.20
Nucleus supraopticus	0.89 ± 0.09
Tubercle olfactorium	1.99 ± 0.22
Striatum A 9000	0.97 ± 0.11
Striatum A 7000	0.84 ± 0.08
Globus pallidus	0.20 ± 0.01
N. amygdaloideus medialis A 4250	4.43 ± 1.25
N. amygdaloideus medialis A 3500	8.18 ± 0.55
N. amygdaloideus centralis A 4250	2.88 ± 0.28
N. amygdaloideus corticalis	10.85 ± 0.52
N. amygdaloideus basalis, pars lateralis	9.71 ± 1.35
Nucleus anterior hypothalami	0.87 ± 0.23
Nucleus lateralis hypothalami	0.41 ± 0.05
Nucleus suprachiasmaticus	1.20 ± 0.14
Nucleus periventricularis stellatocellularis	1.23 ± 0.13
Nucleus periventricularis sotundocellularis	1.77 ± 0.57
Paraventricular nucleus	1.37 ± 0.17
Median eminence	2.31 ± 0.30
Arcuate nucleus I–II–III	1.79 ± 0.50
Nucleus ventromedialis hypothalami	0.90 ± 0.16
Nucleus dorsomedialis hypothalami	0.23 ± 0.05
Nucleus arcuate IV–V	0.64 ± 0.07
Nucleus premamillaris ventralis	1.05 ± 0.25
Nucleus premamillaris dorsalis	0.63 ± 0.02
Nucleus mamillaris medialis, pars medialis	0.57 ± 0.07
Nucleus lateralis thalami	0.21 ± 0.02
Nucleus medialis thalami	0.87 ± 0.05
Nucleus ventralis thalami	0.23 ± 0.05
Gyrus dentatus	0.97 ± 0.26
Hippocampus	1.75 ± 0.38
Nucleus habenulae lateralis	0.22 ⊥ 0.03
Substantia nigra	0.18 ± 0.02
Zona incerta	0.15 ± 0.03
Area ventralis tegmenti	0.16 ± 0.03
Interpeduncular nucleus	1.31 ± 0.27
Periaqueductal gray	1.42 ± 0.41
Nucleus medianus raphes	0.13 ± 0.01
Nucleus dorsalis raphes	0.29 ± 0.03

[a] The brains were frozen and specific areas were punched as described in text. Values are the mean ± S.E.M. of three groups of four rats each.

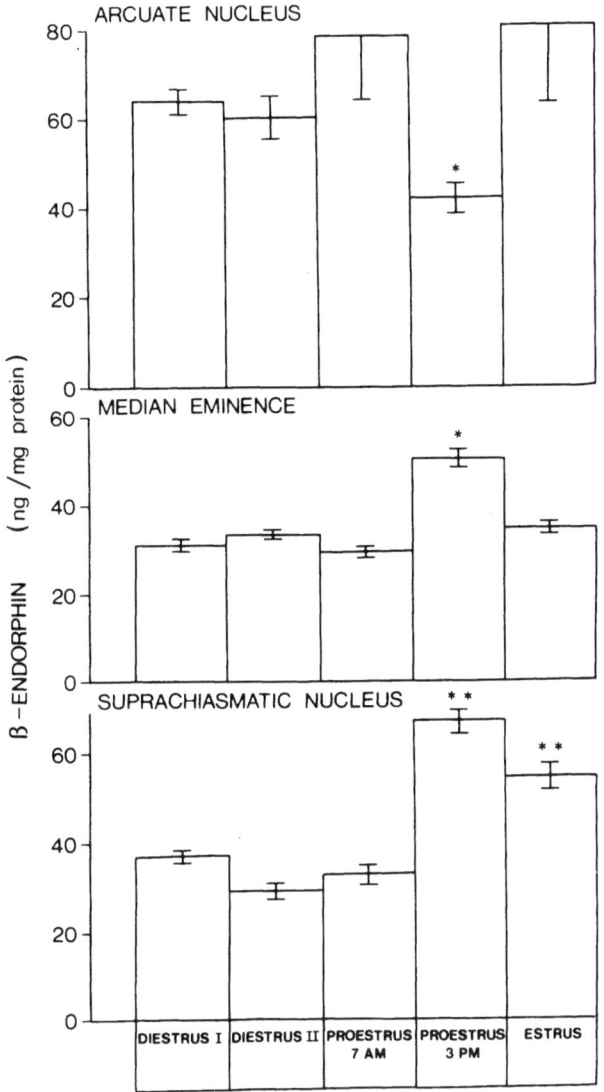

Figure 1. Changes in the β-endorphin content of certain hypothalamic nuclei during the estrous cycle of the rat. Groups of 18 four-day cycling rats were sacrificed at 7 a.m. each day of the estrous cycle and also at 3 p.m. on the days of proestrus and diestrus. β-Endorphin content of specific brain nuclei removed by 0.5- or 1.0-mm punches of frozen brain slices was measured by radioimmunoassay using synthetic β-endorphin as standard and a specific antiserum as described (Linnitsky *et al.*, 1978). Results shown are the mean $\pm$ S.E.M. of three separate determinations of the β-endorphin content of tissue pooled from six animals. On the afternoon of diestrus, the β-endorphin content of suprachiasmatic and arcuate nuclei and the median eminence were respectively 120 $\pm$ 18, 120 $\pm$ 17, and 100 $\pm$ 20% of the concentrations at 7 a.m. diestrus. Significance of differences in the means was tested by the Duncan–Kramer multiple range test following analysis of variance (* $p < 0.01$ vs. means on diestrus I, II, proestrus a.m., and estrus; ** $p < 0.01$, *** $p < 0.05$ vs. means on diestrus I, II, and proestrus a.m.).

afternoon of proestrus is related to this effect, and the subsequent increase in prolactin secretion results from removal of the dopamine inhibition at the level of the anterior pituitary. It is to be noted that the changes in β-endorphin concentrations seen on the afternoon of proestrus do not appear to be related to the diurnal rhythm since corresponding changes were not evident on the afternoon of diestrus. The cell bodies of the brain β-endorphinergic system are believed to be located exclusively in the arcuate nucleus (Bloom et al., 1978), and the decreased β-endorphin content of this nucleus on the afternoon of proestrus may reflect increased axonal transport to the median eminence and suprachiasmatic nucleus. We are not able to relate the increased β-endorphin content of suprachiasmatic nucleus on the afternoon of proestrus and during estrus to any physiological effects. However, it is interesting to recall the importance of this nucleus in the regulation of diurnal rhythms and, presumably, the estrous cycle in view of its acute sensitivity to light/dark cycles. Moreover, a role for endogenous opioid peptides in the regulation of LH secretion has been indicated (Cicero et al., 1979; Dupont et al., 1977a,b), and recently Arendash and Gallo (1979) reported that electrical stimulation of the suprachiasmatic nucleus can modify the episodic LH release in untreated and estrogen-primed ovariectomized rats. It is becoming increasingly clear that peripheral hormones can influence central nervous system neurotransmitter systems, and the changes in β-endorphin seen during the estrous cycle are likely due to fluctuations in peripheral gonadal steroids, particularly estrogens, the level of which is markedly increased immediately prior to the regional changes in β-endorphin levels on the afternoon of proestrus.

3.2. Aging

The influence of the central nervous system on anterior pituitary function is well established (Kordon et al., 1976; Müller et al., 1977b), and in old male rats the plasma levels of LH and FSH are lower and that of PRL is higher (Shaar et al., 1975) than in young animals. The observation that intracerebral injection of opiate peptides leads to modification of PRL, GH, and LH/FSH release (Dupont et al., 1977a,b; Meites et al., 1979) while naloxone inhibits the increase of serum PRL induced by suckling and during the afternoon of proestrus (Barden et al., 1981) indicates a role for endogenous opiate peptides in the control of PRL release. For these reasons, we have measured the β-endorphin content of different brain regions of young and old male rats.

The basal plasma PRL levels of old rats were significantly higher ($p < 0.05$) than those of three-month-old rats (95.2 ± 22.4 vs. 38.6 ± 3.7 ng/ml). The β-endorphin content was measured in brain areas previously shown to contain substantial amounts of this peptide. As shown in Fig. 2,

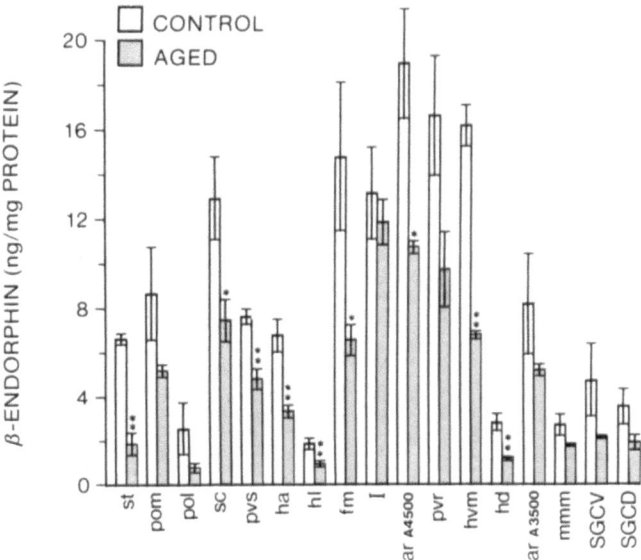

Figure 2. Distribution of β-endorphin in brain of young (3-month) and old (24-month) Sprague–Dawley rats. Tissue was punched from brains of six rats and pooled for measurement of β-endorphin by radioimmunoassay. The results shown are the mean ± S.E.M. of measurements performed in three groups of six animals. *$p < 0.05$; **$p < 0.01$. Abbreviations used in this and subsequent figures: st, nucleus interstitialis striae terminalis; pom, nucleus preopticus medialis; pol, nucleus preopticus lateralis; sc, nucleus suprachiasmaticus; pvs, nucleus periventricularis (hypothalami); ha, nucleus anterior hypothalami; hl, nucleus lateralis hypothalami; fm, nucleus paraventricularis; I, infundibulum; ar, nucleus arcuatus; pvr, nucleus periventricularis (thalami); hvm, nucleus ventromedialis (hypothalami); hd, nucleus dorsomedialis (hypothalami); mmm, nucleus mamillaris medialis; SGCV, substantia grisea centralis, pars ventralis; SGCD, substantia grisea centralis, pars dorsalis, ZI, zona incerta.

highest concentrations of β-endorphin are found in arcuate nucleus and median eminence followed by other hypothalamic nuclei, suprachiasmatic nucleus, medial preoptic nucleus, nucleus interstitialis striae terminalis, and periaqueductal gray. A significant decrease in the β-endorphin content of old rat brain to a mean of 50.3 ± 2.8% of control (young animals) is noted in all nuclei studied, with the exception of median eminence, where no difference between young and old animals could be detected.

The cell bodies of the brain β-endorphin system are believed to be located exclusively in the arcuate nucleus (Bloom *et al.*, 1978). In accordance with this, we found the highest concentration of β-endorphin in the arcuate nucleus. Moreover, the 55% decrease of β-endorphin content in this nucleus in old rats as compared to young controls is paralleled by a corresponding mean decrease of 50% in all other brain nuclei. We have noted

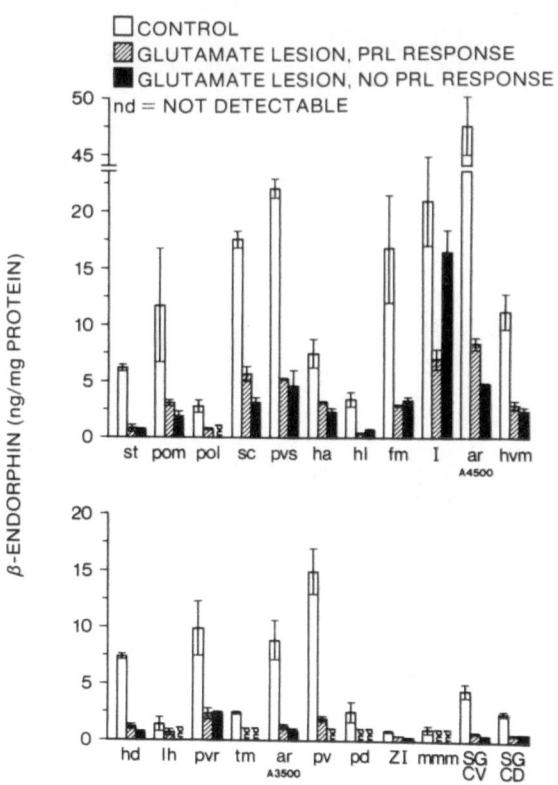

Figure 3. Effect of monosodium glutamate (MSG)-induced brain lesions of neonatal rats on adult rat brain β-endorphin concentrations. Neonatal rats were injected intraperitoneally with MSG (2 mg/g body wt on days 2 and 4, and 4 mg/g body wt on days 6, 8, and 10 postpartum). At 90 days of age, animals were tested for increased serum PRL levels in response to cold stress and divided into two groups on the basis of positive or negative results. β-Endorphin concentrations shown are the mean ± S.E.M. of three groups of six animals each, and in all regions of glutamate-lesioned animals (except median eminence of lesioned group without PRL response to cold stress) were significantly ($p < 0.01$) below concentrations found in corresponding brain nuclei of normal littermate controls. Abbreviations used in addition to those in the legend to Fig. 2 are: lh, nucleus habenulae lateralis; tm, nucleus medialis thalami; pv, nucleus premamillaris ventralis; pd, nucleus premamillaris dorsalis.

similar decreases in brain β-endorphin content parallel to that of the arcuate nucleus following lesion of this structure with glutamate (Fig. 3). Since opiate peptides injected intracerebrally stimulated PRL secretion, it is difficult to reconcile the decrease in brain β-endorphin levels with the increased plasma PRL levels of old rats. However, old animals have been reported to have decreased hypothalamic dopamine levels (Pradhan, 1980; Robinson, 1975; McGeer and McGeer, 1976), and it is possible that under chronic conditions such as aging, some refractoriness develops to the inhibitory action of opiates on the tuberoinfundibular dopaminergic system (Ferland *et al.,* 1977). In view of the high incidence of pituitary tumors in old rats, it is also possible that undetected PRL-secreting microadenomas could cause elevated plasma PRL levels.

4. Effects of Hypothyroidism on Brain Peptides

4.1. Thyrotropin-Releasing Hormone

Little is known of the effects of hypothyroid or hyperthyroid states on classical neurotransmitters, and effects on recently discovered brain peptides are unknown. The effects of postnatal thyroidectomy and thyroxine replacement therapy on the development of TRH-containing neurons are shown in Fig. 4. Hypothyroid animals showed a 75% decrease in the TRH

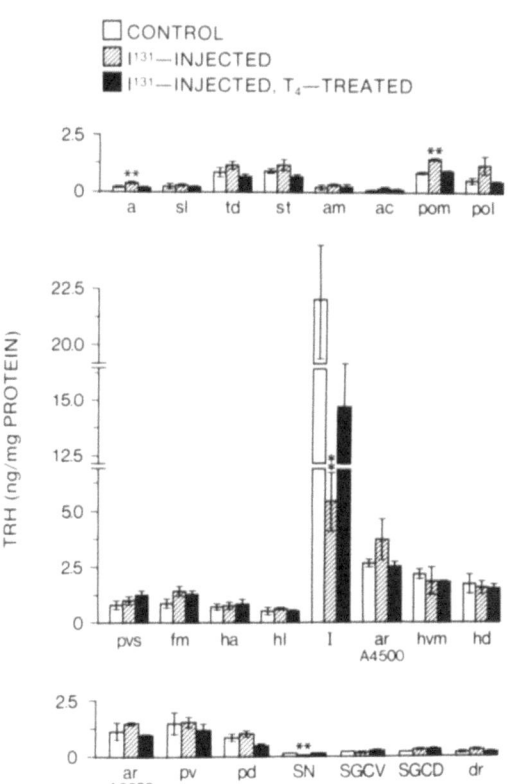

Figure 4. Changes in the distribution of thyrotropin-releasing hormone in the brain following neonatal thyroidectomy and T₄ replacement therapy. A group of 120 newborn male rats were injected with 125 μCi of ^{131}I on the first day after birth. Half of these animals were subsequently injected daily with L-thyroxine (T₄) (1 μg/100 g body wt). A further group of 60 littermates were permitted to mature normally. All animals were sacrificed after 45 days. Results shown are the mean ± S.E.M. of three groups of six animals each. *$p < 0.05$; **$p < 0.01$. Additional abbreviations used are: a, nucleus acumbens; sl, nucleus septum lateralis; td, nucleus tractus diagonalis; am, nucleus amygdaloideus medialis; ac, nucleus amygdaloideus centralis; pv, nucleus premamillaris ventralis; pd, nucleus premamillaris dorsalis; SN, substantia nigra; dr, nucleus Raphes dorsalis.

content of median eminence when compared to control animals. A significant decrease was also noted in the substantia nigra of hypothyroid animals, while increases in the TRH content of nucleus accumbens and median preoptic nucleus were seen. All of these changes, both increases and decreases in TRH content of particular brain nuclei of hypothyroid animals, were completely reversed by T_4 replacement therapy. Thyroid hormones exert a negative feedback action at the hypothalamic level (Belchetz et al., 1977), and the high plasma TRH levels of hypothyroid animals result from excessive TRH secretion (Jackson and Reichlin, 1979). This high rate of TRH release is probably reflected in the markedly decreased TRH content of median eminence seen in hypothyroid animals. Although TRH is distributed throughout the extrahypothalamic central nervous system, its physiological relevance remains obscure. TRH, acting as a neurotransmitter, may serve in functionally different pathways and the changes in TRH concentrations in the few brain nuclei of hypothyroid animals seen may reflect specific interactions of thyroid hormones with these particular pathways rather than global effects on TRH metabolism.

4.2. Substance P

Highest concentrations of substance P are found in substantia nigra, pars reticulata, and the peptide is widely distributed throughout the central nervous system (Fig. 5). Very marked changes in the distribution of substance P were present in brain of hypothyroid animals and were completely reversed by chronic T_4 replacement therapy. Hypothyroid animals exhibited large increases in substance P up to $2\frac{1}{2}$ times the concentrations seen in control animals in 19 of the 32 brain nuclei studied (Fig. 5). These widespread stimulatory effects of hypothyroidism on substance P concentrations are in contrast to the very localized effects seen on TRH levels, and imply a major role for substance P in the development of hypothyroid-induced central nervous system malfunction. Although substance P may be responsible for transmission of nociceptive stimuli in primary afferent neurons, its function in the brain is not clearly defined (Nicoll et al., 1980). In the major striatonigral and habenulo-interpeduncular tracts, substance P acts as excitatory neurotransmitter. The increased substance P content of hypothyroid animal brain nuclei could result from increased synthesis or decreased turnover. Since similar increases are seen in both soma and terminal regions of substance P-containing neurons, modification of axonal transport is unlikely. Although increased protein synthesis in the brain has been reported following thyroidectomy (Gonzalez and Geel, 1980), it appears unlikely that the increased substance P content of hypothyroid animals is a reflection of an overall increase in protein synthesis at the translational level, since we do not see similar increases in the concentrations of

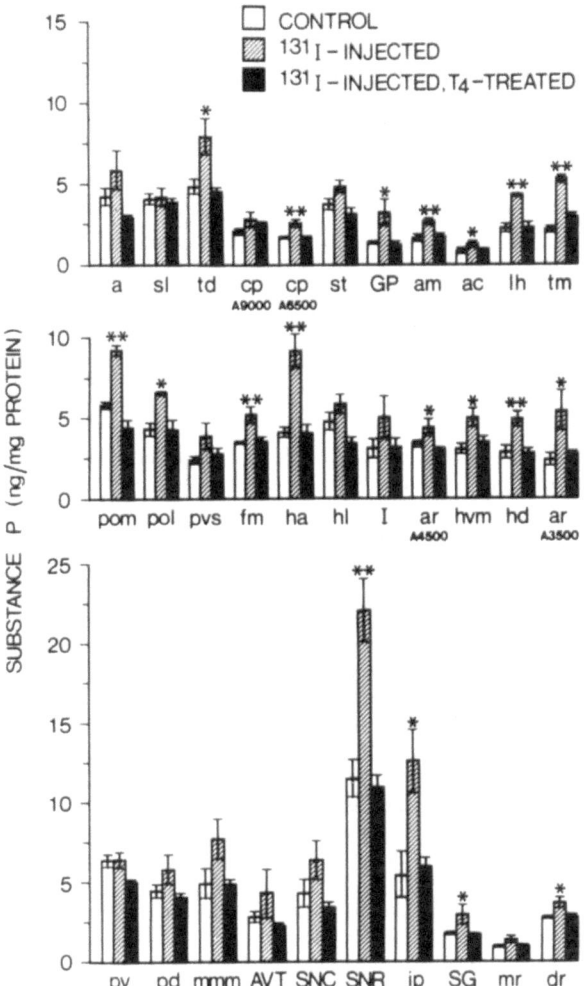

Figure 5. Changes in the distribution of substance P following neonatal thyroidectomy and T_4 replacement therapy. Results shown are the mean ± S.E.M. of three groups of six animals each. The significance of differences between means was tested by the multiple range test of Kramer, $*p < 0.05$, $**p < 0.01$. Animals were treated as described in the legend to Fig. 4. Additional abbreviations used are: cp, caudate putamen; GP, globus pallidus; lh, nucleus habenulae lateralis; tm, nucleus medialis thalami; ip, nucleus interpedumicularis; SNC, substantia nigra, zona compacta; SNR, substantia nigra, zona reticulata; AUT, area ventralis tegmenti; SG, substantia grisea centralis; mr, nucleus medianus raphes.

other peptides. Opiate peptides inhibit release of substance P from slices of trigeminal nucleus (Jessell and Iversen, 1977) or cultured neurons (Mudge et al., 1979), but we have found no generalized effect of thyroid hormone status on opiate peptides in the brain (Barden and Dupont, unpublished observations). Thyroid hormone status appears to be intimately related to effects of other growth-promoting factors including nerve growth factor and growth hormone (Walker et al., 1979; Roger and Fellows, 1979), and their combined effects on central nervous system peptide neurotransmitters is a field ripe for investigation.

REFERENCES

Arendash, G. W., and Gallo, R. V., 1979, Regional differences in response to electrical stimulation within the median suprachiasmatic region on blood luteinizing hormone levels in ovariectomized and ovariectomized, estrogen-primed rats, *Endocrinology* **104**:333.

Barden, N., Mérand, Y., Rouleau, D., Garon, M., and Dupont, A., 1981, Changes in β-endorphin content of discrete hypothalamic nuclei during the estrous cycle of the rat, *Brain Res.,* **204**:441.

Belchetz, P. E., Gredley, G., Bird, D., and Himsworth, R. H., 1977, Regulation of thyrotropin secretion by negative feedback of triiodothyronine on the hypothalamus, *J. Endocrinol.* **76**:439.

Bloom, F., Battenberg, E., Rossier, J., Ling, N., and Guillemin, R., 1978, Neurons containing β-endorphin in rat brain exist separately from those containing enkephalin: Immunocytochemical studies, *Proc. Natl. Acad. Sci. USA* **75**:1591.

Bradbury, A. F., Smyth, P. G., and Snell, C. R., 1976, C-fragment of lipotropin has a high affinity for brain opiate receptors, *Nature (London)* **260**:793.

Cicero, T. J., Schainker, B. A., and Mayer, E. R., 1979, Endogenous opioids participate in the regulation of the hypothalamic–pituitary–luteinizing hormone axis and testosterone negative feedback control of luteinizing hormone, *Endocrinology* **104**:1286.

Crowley, W. R., O'Donohue, T. L., and Jacobowitz, D. M., 1978, Changes in catecholamine content in discrete brain nuclei during the estrous cycle of the rat, *Brain Res.* **147**:315.

Dockray, G. J., 1980, Cholecystokinins in rat cerebral cortex: Identification, purification and characterization by immunochemical methods, *Brain Res.* **188**:155.

Dockray, G. J., Gregory, R. A., Hutchinson, J. B., Harris, J. I., and Runswick, M. J., 1978, Isolation, structure and biological activity of two cholecystokinin octapeptides from sheep brain, *Nature (London)* **274**:711.

Dupont, A., Cusan, L., Garon, M., Labrie, F., and Li, C. H., 1977a, β-Endorphin: Stimulation of growth hormone release *in vivo*, *Proc. Natl. Acad. Sci. USA* **73**:358.

Dupont, A., Cusan, L., Labrie, F., Coy, D. H., and Li, C. H., 1977b, Stimulation of prolactin release in the rat by intraventricular injection of β-endorphin and methionine-enkephalin, *Biochem. Biophys. Res. Commun.* **75**:76.

Dupont, A., Lépine, J., Langelier, P., Mérand, Y., Rouleau, D., Vaudry, H., Gros, C., and Barden, N., 1980, Differential distribution of β-endorphin and enkephalins in rat and bovine brain, *Regul. Pept.* **1**:43.

Ferland, L., Fuxe, K., Eneroth, P., Gustaffson, J. A., and Skett, P., 1977, Effects of methionine-enkephalin on prolactin release and catecholamine levels and turnover in the median eminence, *Eur. J. Pharmacol.* **43**:89.

Ferland, L., Kledzik, G. S., Cusan, L., and Labrie, F., 1978, Evidence for a role of endorphins in stress- and suckling-induced prolactin release in the rat, *J. Mol. Cell. Endocrinol.* **12**:267.

Gonzales, L. W., and Geel, S. E., 1980, Thyroid hormone state and the incorporation of [^{14}C]-leucine by brain microsomes in developing rats, *Brain Res. Bull.* **5**:1.

Gros, C., Pradelles, P., Rouget, C., Bepoldin, O., Dray, F., Fournie-Zaluwski, M. C., Roques, B. P., Pollard, H., Llorens-Cortes, C., and Schwartz, J. C., 1978, Radioimmunoassay of methionine- and leucine-enkephalins in regions of rat brain and comparison with endorphins estimated by radioreceptor assay, *J. Neurochem.* **31**:29.

Gudelsky, G. A., and Porter, J. C., 1979, Morphine and opioid peptide-induced inhibition of the release of dopamine from tuberoinfundibular neurons, *Life Sci.* **25**:1697.

Hughes, J., Smith, T. W., Kosterlitz, L. A., Fothergill, B. A., Morgan, B. A., and Morris, H. R., 1975, Identification of two related pentapeptides from the brain with potent opiate activity, *Nature (London)* **258**:577.

Innis, R. B., Correa, F. M. A., Uhl, G. R., Schneider, B., and Snyder, S. H., 1979, Cholecystokinin octapeptide-like immunoreactivity: Histochemical localization in rat brain, *Proc. Natl. Acad. Sci. USA* **76**:521.

Jackson, I. M. D., and Reichlin, S., 1979, Distribution and biosynthesis of TRH in the nervous system, in: *Central Nervous System Effects of Hypothalamic Hormones and Other Peptides* (R. Collu, A. Barbeau, J. G. Rochefort, and J. R. Ducharme, eds.), pp. 3–54, Raven Press, New York.

Jessell, T. M., and Iversen, L. L., 1977, Opiate analgesics inhibit substance P release from rat trigeminal nucleus, *Nature (London)* **268**:549.

Jorpes, E., Mutt, V., and Toczko, K., 1964, Further purification of cholecystokinin and pancreozymin, *Acta Chem. Scand.* **18**:2408.

Krieger, D. T., Liotta, A., and Li, C. H., 1977, Hormone plasma immunoreactive β-lipotropin: Correlation with basal and stimulated plasma ACTH concentrations, *Life Sci.* **21**:1771.

Köenig, J. F. R., and Klippel, R. A., 1963, The Rat Brain, R. E. Krieger, New York.

Kordon, C., Mevy, M., and Enjalbert, A., 1976, Neurotransmitters and control of pituitary function, in: *Hypothalamus and Endocrine Functions* (F. Labrie, G. Pelletier, and J. Meites, eds.), pp. 51–61, Plenum Press, New York.

Larsson, L. I., Childers, S., and Snyder, S., 1979, Met- and Leu-enkephalin immunoreactivity in separate neurons, *Nature (London)* **282**:407.

Li, C. H. and Chung, D., 1976, Isolation and structure of an Untriarontapeptide with opiate activity from camel pituitary glands, *Proc. Natl. Acad. Sci. USA*, **73**:1145.

Li, C. H., Barnafi, L., Chrétien, M., and Chung, D., 1965, Isolation and amino acid sequence of β-LPH from sheep pituitary gland, *Nature (London)* **208**:1093.

Linnitsky, J. C., Morin, O., Dupont, A., Labvie, F., Seidah, H. G., Chrétien, M., Lis, M., and Coy, D. H., 1978, Content of β-LPH and its fragments (including endorphins) in anterior and intermediate lobes of bovine pituitary gland, *Life Sci.* **22**:1715.

McGeer, E. G., and McGeer, P. L., 1976, Neurotransmitter metabolism in the aging brain, in: *Neurobiology of Aging* (R. D. Terry and S. Gershon, eds.), pp. 389–403, Raven Press, New York.

Meites, M., Bruni, J. F., Van Vugt, D. A., and Smith, A. F., 1979, Relation of endogenous opioid peptides and morphine to neuroendocrine functions, *Life Sci.* **24**:1325.

Miller, R. J., Chang, K. J., Cooper, B., and Cuatrecasas, P., 1978, Radioimmunoassay and characterization of enkephalins in rat tissues, *J. Biol. Chem.* **253**:531.

Morin, O., Caron, M. G., De Léan, A., and Labrie, F., 1976, Binding of the opiate pentapeptide methionine-enkephalin to a particulate fraction from rat brain, *Biochem. Biophys. Res. Commun.* **73**:940.

Mudge, A. W., Leeman, S. E., and Fischback, G., 1979, Enkephalin inhibits release of substance P from sensory neurons in culture and decreases action potential duration, *Proc. Natl. Acad. Sci. USA* **76**:526.

Müller, J. E., Straus, E., and Yalow, R. S., 1977a, Cholecystokinin and its COOH-terminal octapeptide in the pig brain, *Proc. Natl. Acad. Sci. USA* **74**:3035.

Müller, J. E., Nisticu, G., and Scapagnini, U. (eds.), 1977b, *Neurotransmitter and Anterior Pituitary Function,* Academic Press, New York.

Nicoll, R. A., Schenkev, C., and Leeman, S., 1980, Substance P as a transmitter candidate, *Annu. Rev. Neurosci.,* **3**:227.

Palkovits, M., 1973, Isolated removal of hypothalamic or other brain nuclei of the rat, *Brain Res.* **59**:449.

Pasternak, G. W., Goodman, R., and Snyder, S. H., 1977, An endogenous morphine-like factor in mammalian brain, *Life Sci.* **16**:1765.

Pelletier, G., Leclerc, R., Labrie, F., Côté, J., Chrétien, M., and Lis, M., 1977, Immunohistochemical localization of β-lipotropic hormone in the pituitary gland, *Endocrinology* **100**:770.

Pradhan, S. N., 1980, Central neurotransmitters and aging, *Life Sci.* **26**:1643.

Robinson, D. S., 1975, Changes in monoamine oxidase and monoamines with human development and aging, *Fed. Proc.* **34**:103.

Roger, L. J., and Fellows, R. E., 1979, Evidence for thyroxine–growth hormone interaction during brain development, *Nature (London)* **282**:414.

Shaar, C. J., Euker, J. S., Reigle, G. D., and Meites, J., 1975, Effects of castration and gonadal steroids on serum luteinizing hormone and prolactin in old and young rats, *J. Endocrinol.* **66**:45.

Simantov, R., and Snyder, S., 1976a, Morphine-like peptides in mammalian brain: Isolation, structure, elucidation and interactions with the opiate receptor, *Proc. Natl. Acad. Sci. USA* **73**:2515.

Simantov, R., and Snyder, S., 1976b, Isolation and structure identification of a morphine-like enkephalin in bovine brain, *Life Sci.* **18**:781.

Terenius, L., and Wahlstrom, A., 1975, Research for an endogenous ligand for opiate receptor, *Acta Physiol. Scand.* **94**:74.

Uhl, G. R., and Snyder, S. H., 1979, Neurotensin: A neuronal pathway projecting from amygdala through striae terminalis, *Brain Res.* **161**:522.

Vanderhaeghen, J., Signeau, J., and Gepts, W., 1975, New peptide in the vertebrate CNS reacting with antigastrin antibodies, *Nature (London)* **257**:604.

Vanderhaeghen, J. J., Lotstra, F., DeMey, J., and Gilles, C., 1980, Immunohistochemical localization of cholecystokinin- and gastrin-like peptides in the brain and hypophysis of the rat, *Proc. Natl. Acad. Sci. USA* **77**:1190.

Walker, P., Weichsel, M. E., Jr., Guo, S. M., Fisher, D. A., and Fisher, D. A., 1979, Radioimmunoassay for mouse nerve growth factor (NGF): Effects of thyroxine administration on tissue NGF levels, *Brain Res.* **186**:331.

Watson, S. J., Akil, H., Richard, C. W., and Barchas, J. D., 1978, Evidence for two separate opiate peptide neuronal systems, *Nature (London)* **275**:226.

16

In Vitro Synthesis of Hypothalamic Neurophysin Precursors

Irwin M. Chaiken, Ernst A. Fischer, Linda C. Giudice, and Christopher J. Hough

1. Introduction

Biochemical study of the neurophysins, hypothalamo–neurohypophyseal protein carriers of the neuropeptide hormones oxytocin and vasopressin (Breslow, 1979), has focused on fundamental questions about the biosynthetic and molecular events by which these species are formed and stabilized into noncovalent complexes (see Fig. 1). While interacting complexes can be formed between native protein and hormone in their isolated forms, it has become evident from several properties of these molecules that biosynthetic assembly is more complex. For example, unlike most proteins that remain close to their intact biosynthesized states after translation, neurophysins do not themselves contain sufficient amino acid sequence information to code for spontaneous folding to functional forms. When these proteins, which are highly disulfide cross-linked (seven disulfides per protein molecule of 10,000 daltons), are treated with "catalytic" (substoichiometric) amounts of disulfide-reducing agent, they are inactivated readily,

Irwin M. Chaiken, Ernst A. Fischer, Linda C. Guidice, and Christopher J. Hough • Laboratory of Chemical Biology, National Institute of Arthritis, Metabolism and Digestive Diseases, National Institutes of Health, Bethesda, Maryland 20205. *Present address of E.A.F.:* Hoffman–LaRoche & Co. AG, 4002 Basel, Switzerland. *Present address of L.C.G.:* Department of Neurobiology, Stanford University School of Medicine, Stanford, California 94305.

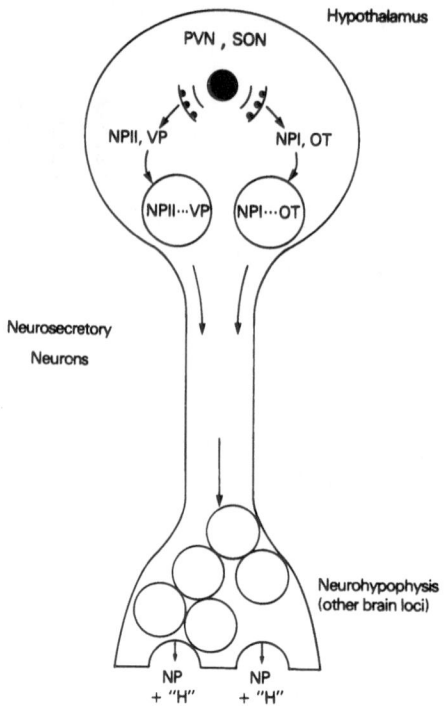

Figure 1. Schematic view of the neuronal origin and fate of the major neurophysins (NPI and NPII) and the associated neuropeptide hormones, oxytocin (OT) and vasopressin (VP). This view portrays hypothalamic synthesis in supraoptic (SON) and paraventricular (PVN) neuronal cells, occurrence of the noncovalently interacting complexes in neurosecretory granules, transport to the neurohypophysis, and synaptic release (Sachs *et al.*, 1969; Livett, 1978; Pickering, 1978). The notation used here of the major neurophysins as I and II is based on the bovine system. The species I and II also have been denoted as neurophysins VLDL and MSEL, respectively, based on systematic amino acid sequence differences between these closely related proteins at residue positions 2, 3, 6, and 7 (Chauvet *et al.*, 1975). [Figure taken from Hough *et al.*, 1980.]

due to scrambling of the disulfides to nonnative combinations (Chaiken *et al.*, 1975). Further, this process is speeded up by the inclusion of microsomal disulfide interchange enzyme. With regard to disulfide bond stability under interchange conditions, the neurophysins behave in much the same manner as do insulin and chymotrypsin and unlike the respective precursors proinsulin and chymotrypsinogen (Givol *et al.*, 1965; Steiner and Clark, 1968). A straightforward (though not obligatory) interpretation of this behavior is that neurophysins represent processed derivatives of originally biosynthesized forms and, further, that their incorporation into noncovalent complexes with hormones derives in some way from these postsynthetic processing events.

The prediction from folding studies of nuerophysin biosynthetic precursors has fit well with several observations obtained by pulse-chase and other biosynthesis analyses. The pivotal experiments of Sachs and his colleagues (Sachs *et al.*, 1969) revealed a kinetic lag phase in the biosynthetic incorporation of [^{35}S]cysteine into neurophysin. This lag phase suggested conversion of an already-formed precursor molecule to mature neurophysin. More recent characterization of pulse-chase events in the rat (Brownstein *et al.*, 1980) has allowed detection and separation of [^{35}S]cysteine-labeled products by size from the hypothalamus as well as from tissue sections anatomically located along the route of transport of neurophysin, within neurosecretory granules, through the infundibular stalk to the posterior pituitary. These latter studies have defined the conversion of two approximately 20,000-dalton proteins that are immunologically cross-reactive with the neurophysins to two approximately 10,000-dalton forms. Based on comparative studies with Brattleboro and normal rats, the 20,000-dalton proteins have been distinguished as "putative precursors" of neurophysins I and II. Recently, ^{35}S containing these 20,000 dalton "putative precursors" have been shown, by high-performance liquid-chromatography peptide mapping, to contain neurophysin-specific peptides (Abercrombie, *et al.*, 1982). Overall, the pulse-chase labeling experiments strongly portray a precursor–protein relationship for the neurophysins.

Several of the persistent questions regarding neurophysin biosynthesis have been addressed productively by cell-free synthesis. It has been shown generally that translation of mRNA isolated from specific tissues can be accomplished by addition of this mRNA to extracts, from unrelated cells, that contain protein-synthesizing machinery (Marcus *et al.*, 1974; Pelham and Jackson, 1976). This approach has allowed tissue-specific proteins to be made, optimally without subsequent processing events. Thus, original translation products can be identified. On the premise that neurophysins are made as parts of biosynthetic precursors in the hypothalamus (Sachs *et al.*, 1969; Livett, 1978; Pickering, 1978), hypothalamic mRNA has been isolated and translated in several established cell-free systems (Guidice and Chaiken, 1979a,b; Lin *et al.*, 1979; Schmale *et al.*, 1979; Richter *et al.*, 1980). Characterization of synthesized protein products has yielded clear-cut identification of neurophysin amino acid sequences in high-molecular-weight species coded by the tissue-specific mRNA and therein chemical evidence that neurophysins are made as parts of precursor molecules. Characterization of neurophysin-specific mRNA and controlled use of exogenously added processing enzymes have begun to provide tools by which the *in vitro* synthesis route should help define the chemical events leading to neurophysin biosynthetic precursors and intermediates and ultimately to intact neurophysin–neuropeptide complexes.

2. Cell-Free Synthesis and Immunological Recognition of Neurophysin-Related Translation Products

The prototypic approach devised for applying *in vitro* synthesis to the isolation of bovine neurophysin biosynthetic products (Guidice and Chaiken, 1979a,b; Hough *et al.*, 1980) is shown schematically in Fig. 2. The

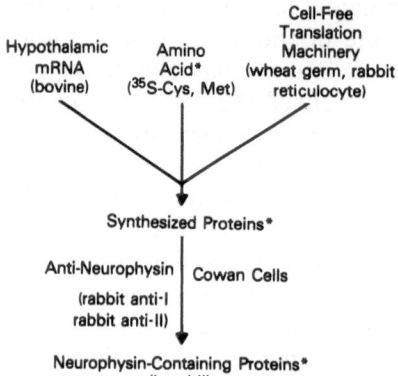

Figure 2. Tactics of *in vitro* synthesis as used for isolation of bovine hypothalamic neurophysin biosynthetic precursors. [Figure taken from Hough *et al.*, 1980.]

basic tactic has been to use such cell-free systems as wheat germ extracts and rabbit reticulocyte lysates to express all proteins coded by total hypothalamic mRNA [used as poly(A)$^+$ RNA, namely the RNA bound to oligo(deoxythymidine)-cellulose] and then to isolate any products that are neurophysin-related using specific antibodies elicited by the major, authentic bovine neurophysins (I or II). The amount of poly(A)$^+$ RNA obtained from bovine hypothalami has generally been low in comparison to amounts from other (nonbrain) tissues (Giudice and Chaiken, 1979a). Nonetheless, this RNA has been active in coding for specific proteins in cell-free systems.

With total poly(A)$^+$ RNA used as the hypothalamic message, the antibody has assumed a central role as the discriminator of neurophysin-specific translation products. In studies in our own laboratory, we have employed purified antibodies elicited in rabbits against neurophysins conjugated to multi-(poly-DL-alanyl)-poly-L-lysine (Fischer *et al.*, 1977) (see Fig. 3). Since the conjugated neurophysins are active, antibodies obtained are directed toward protein of nativelike conformation. Positive antisera, found in essentially all rabbits immunized with the protein conjugates, were fractionated by immunoaffinity chromatography on neurophysin-agarose columns, using matrices with immobilized neurophysin of the same type (I or II) as that used as immunogen. The tightly bound antibody, which eluted with 6 M guanidine hydrochloride (pH 4) but not with 2 M guanidine hydrochloride (pH 7), was used as purified anti-neurophysin I or II. These antibodies are nonprecipitating with the neurophysins. Thus, if one is to isolate biosynthetic products by immunoprecipitation (see Fig. 2), it is necessary to add a precipitating vehicle, such as *Staphylococcus aureus* Cowan strain cells.

As shown by data such as those in Fig. 4, radioimmunoassay of specific antibodies obtained as above indicates that, while each class (anti-I or anti-II) has a preference for the neurophysin used initially as immunogen, there is significant cross-recognition of the other neurophysin (Fischer *et al.,* 1977); Hough *et al.,* 1980). The differences in affinities of a particular antibody for the authentic neurophysins are distinct. While corresponding differences for precursors cannot be assumed, efforts have been made to take advantage of any such affinity differences in order to achieve discrimination in the immunoprecipitation of specific neurophysin I- and II-related proteins from translation mixtures. This has been approached by using a nonsaturating amount of antibody in immunoprecipitation of cell-free synthesis products, as established by titrating these products with the particular antibody (Giudice and Chaiken, 1979a).

The separation and identification of neurophysin-specific proteins from the *in vitro*-synthesized proteins coded by total hypothalamic mRNA has been substantially realized using fractionated antibodies. Labeled proteins immunoprecipitated by anti-neurophysins I and II can be separated by SDS-polyacrylamide gel electrophoresis and visualized by fluorography or gel slicing and counting. As shown in Fig. 5 for wheat germ products, anti-I and -II lead to identification of two independent translated proteins, with

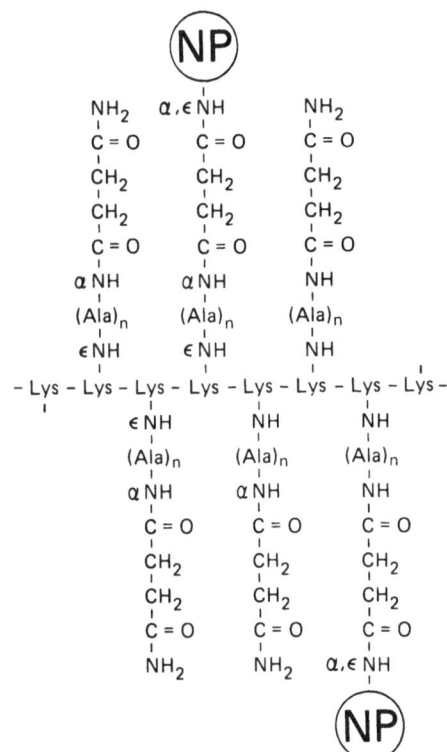

Figure 3. Schematic diagram of neurophysin conjugates used as immunogens to obtain neurophysin antibodies (Fischer *et al.,* 1977). To prepare these soluble conjugates, multi-(poly-DL-alanyl)-poly-L-lysine was activated by succinylation, and conversion of the resulting free carboxyl groups to azido derivatives. These carboxyl-activated forms finally were reacted with the free amino groups of neurophysins. The conjugated neurophysins remain capable of binding the active site ligand methionyl-tyrosyl-phenylalanine amide.

apparent molecular weights of 17,000–19,000 and 23,000–25,000, respectively. Given the apparent nature of the molecular weight values determined for these species, it now seems most convenient to define the smaller and larger neurophysin-related proteins as translation product I and translation product II, respectively. That antibody recognition of the proteins in Fig. 5 is specific has been shown by competition with authentic neurophysins (but not with other proteins such as ovalbumin), lack of immunoprecipitation by anti-neurophysins from translations in which either no exogenous mRNA or mRNA from a nonhypothalamic source has been added, and lack of immunoprecipitation by nonimmune sera or nonneurophysin antibodies. Thus, differential immunoprecipitation has provided a reasonable way to separate neurophysin I- and II-specific translation products for further characterization. Nonetheless, simultaneous immunoprecipitation of both species of translation by a single antibody type can be expected and indeed has been observed, at high enough antibody excess. In most of our own experiments, discrimination of the two neurophysin-related translation

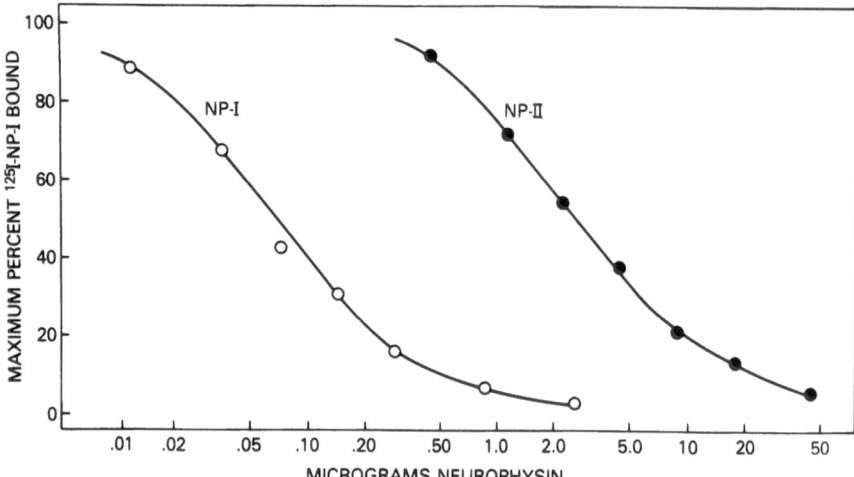

Figure 4. Comparison of immunoreactivities of authentic neurophysins I and II toward anti-neurophysin I by radioimmunoassay. For the basic assay, [^{125}I]neurophysin I, prepared by the method of Bolton and Hunter (1973) and purified by affinity chromatography on methionyl-tyrosyl-phenylalanyl-ω-aminohexylamino-agarose (Chaiken, 1979), was immunoprecipitated via *S. aureus* Cowan strain cells using purified rabbit anti-neurophysin I (Fischer *et al.*, 1977). Nonradioactive bovine neurophysins I (o) and II (●) were used as competitors, in assays containing about 0.4 μg [^{125}I]neurophysin I (4.5 × 10^5 cpm) and 3.2 μg antibody. The data show that neurophysin II cross-reacts with neurophysin I, but that the latter is preferred by the antibody tested by a factor of about 40. Comparisons with other anti-neurophysin I and II populations reveal similar preferences of antibody for the neurophysin used as immunogen. [Figure taken from Hough *et al.*, 1980.]

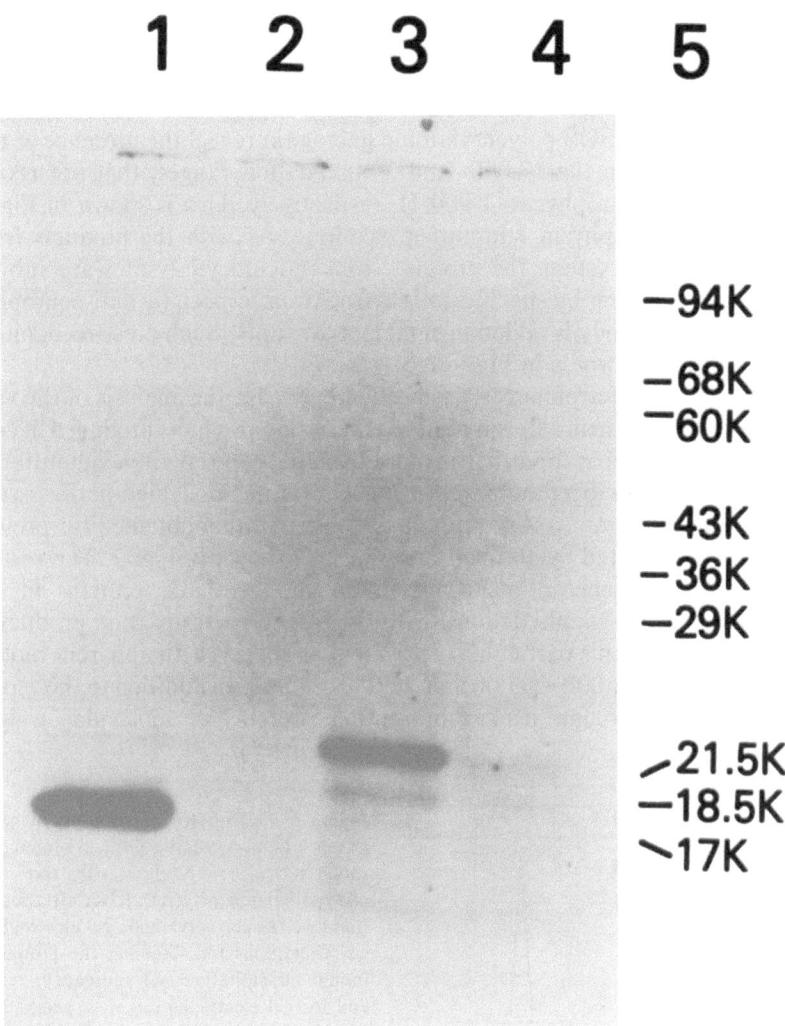

Figure 5. Fluorograph of SDS-polyacrylamide gel electrophoresis identification of neurophysin I- and II-immunoreactive proteins produced by wheat germ *in vitro* translations of bovine hypothalamic poly(A)$^+$ RNA with added [^{35}S]cysteine. Labeled proteins were immunoprecipitated with either anti-neurophysin I (lane 1) or anti-neurophysin II (lane 3). Lane 2 is a sample immunoprecipitated with anti-I in the presence of excess bovine neurophysin I; lane 4, a sample immunoprecipitated with anti-II in the presence of excess bovine neurophysin II. Lane 5 is the scale of molecular weights as defined by standard proteins. [Figure adapted from Giudice and Chaiken, 1979b.]

products has turned out to be more easily achieved using anti-neurophysin II than I.

Examination of translation products from reticulocyte lysate synthesis has shown that essentially the same neurophysin-related proteins are obtained in this cell-free system as from wheat germ extracts. Electrophoretic analyses on SDS-polyacrylamide gels again reveal the presence of two major proteins, in the 17,000- and 25,000-dalton ranges, that are recognized by anti-neurophysins I and II, respectively. This is shown in Fig. 6 for an anti-neurophysin I immunoprecipitate. As with the products from the wheat germ system, the proteins from reticulocyte lysates are subject to cross-recognition by specific anti-neurophysin, especially anti-neurophysin I, if the antibody is added in sufficient amounts. Such cross-recognition is evident in the profile in Fig. 6.

While gel electrophoresis has been effective for the analysis of the very small amounts of neurophysin-related translation products produced in cell-free systems, a separation that is more conducive to repetitive quantitative measurement has been achieved using gel permeation high-performance liquid chromatography. An example of the separation obtained for protein immunoprecipitated by anti-neurophysin II is shown in Fig. 7. The results confirm the presence of neurophysin-specific translated proteins in the region of 17,000–25,000 daltons, with the respective translation products I and II eluting in this particular experiment as species with apparent molecular weights of 18,000–20,000 and 22,000–24,000. In addition to these proteins, a breakthrough peak representing proteins of molecular weight

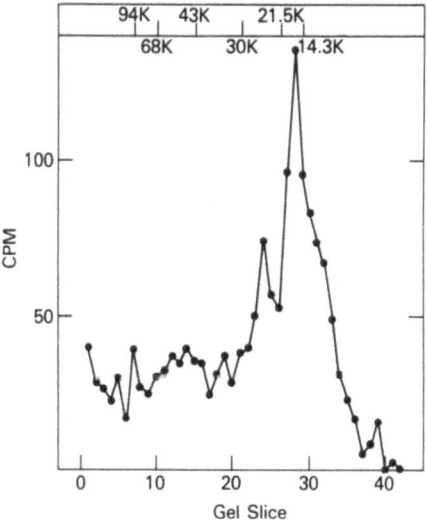

Figure 6. Quantitative distribution of neurophysin I-immunoreactive proteins produced by rabbit reticulocyte lysate *in vitro* translation of hypothalamic poly(A)$^+$ RNA after separation by SDS-polyacrylamide gel electrophoresis. Conditions for obtaining the protein by immunoprecipitation and subsequent release and for gel electrophoresis were generally as described before for the wheat germ system (Giudice and Chaiken, 1979a,b). The gel lane containing translated protein was sliced into approximately 2-mm sections that were transferred to scintillation vials, dissolved in 30% hydrogen peroxide at 60°C overnight, then supplemented with ascorbic acid, and measured for ^{35}S content by scintillation counting in Aquasol. The molecular weight scale at the top was derived from known protein markers that were electrophoresed in parallel. The top of the gel is at the left.

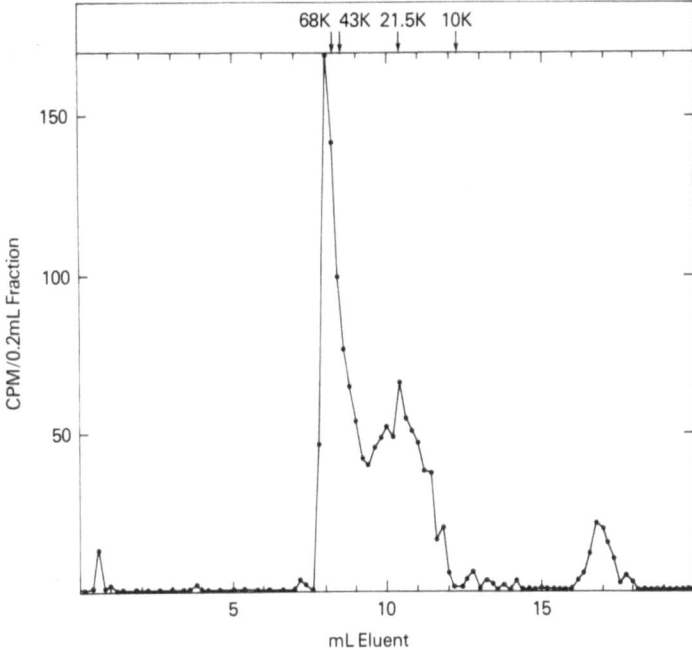

Figure 7. Separation, by gel permeation high-performance liquid chromatography, of neurophysin I-immunoreactive proteins produced by rabbit reticulocyte lysate *in vitro* translation of hypothalamic poly(A)$^+$ RNA. Translation products, obtained in a manner analogous to that used for the sample of Fig. 6, were chromatographed on a TSK 2000SW column (Varian, 0.75 × 50 cm) in 3 M guanidine hydrochloride dissolved in phosphate-buffered saline (final pH 5.5). Elution flow rate was 1 ml/min, at ambient temperature. Fractions were measured for ^{35}S content by scintillation counting in Aquasol. The molecular weight scale at the top was derived from known protein markers eluted in separate experiments and monitored by UV absorbance.

greater than 40,000 can be observed. While the meaning of this larger material has not been deduced unequivocally as yet, comparison with gel electrophoresis patterns suggests that it consists of a set of relatively minor components of varying molecular weight and not of a single protein species. In spite of this ambiguity, the gel permeation high-performance liquid chromatography procedure provides a simple, quick, and quantitative way to visualize translation products I and II. Recovery of the translation products varies depending on the amounts applied but falls generally in the 15–40% range for the picogram amounts of labeled proteins applied here.

Using either gel electrophoresis or gel permeation high-performance liquid chromatography, immunoprecipitated translation products have now been identified from bovine (Giudice and Chaiken, 1979a,b; Schmale *et al.*, 1979; Hough *et al.*, 1980; Richter *et al.*, 1980), rat (Lin *et al.*, 1979), and mouse (Lin *et al.*, 1979) hypothalamic mRNA. For the rat and mouse mRNA cases, a single molecular weight species of 17,500 has been reported to be produced and recognized by antisera to rat neurophysins (Lin *et al.*, 1979). In contrast, observation of the two differently sized translation products from bovine mRNA now has been made not only in our own (Giudice and Chaiken, 1979a,b; Hough *et al.*, 1980) but also in a second laboratory (Schmale *et al.*, 1979; Richter *et al.*, 1980). In the latter studies, molecular weights of 16,500 and 21,000 respectively were reported for what are defined here as translation products I and II. Overall, the major products of *in vitro* synthesis obtained so far from all sources are in the size range of 17,000–25,000 daltons. This correspondence of molecular weights would argue that intact neurophysin-related biosynthetic products coded by mature mRNA are about 1.5 to 2.5 times the size of the 10,000-dalton mature neurophysins.

3. Chemical Verification That Neurophysin-Related Translation Products Contain Neurophysin Amino Acid Sequences

While antibody interaction with translation products is consistent with their identification as neurophysin-containing, such recognition does not provide a rigorous proof for the assertion. In view of this problem, efforts have been made to establish chemically the presence of neurophysin sequence in high-molecular-weight proteins made in cell-free systems. To date, this effort has been made only for the bovine mRNA products and has rested heavily on peptide mapping of proteins translated in the presence of [^{35}S]cysteine. As indicated earlier, the latter amino acid is quite abundant in the neurophysins, to the extent of 14 half-cystine residues per neurophysin chain of 93 residues for neurophysin I or 95 residues for II. In addition, [^{35}S]cysteine is available commercially with high isotopic enrichment, an important feature for any amino acid to be used to detect the very small amounts of protein made in the *in vitro* synthesis systems. Based on the amino acid sequences of the authentic bovine neurophysins, the four cysteic acid-containing authentic tryptic peptides expected from performic acid-oxidized neurophysins I and II contain 89 and 80%, respectively, of the total sequence. Thus, by using [^{35}S]cysteine in mapping, a significant portion of the neurophysin-specific sequence can be identified as [^{35}S]cysteic acid-containing fragments of trypsin digests of performic acid-oxidized translation products.

Initial peptide mapping studies were carried out on paper for the translation products immunoprecipitated from wheat germ extracts by anti-neurophysin II (Giudice and Chaiken, 1979a). A representative map for this case is shown in Fig. 8. This co-map, obtained by treating immunoprecip-

A **B**

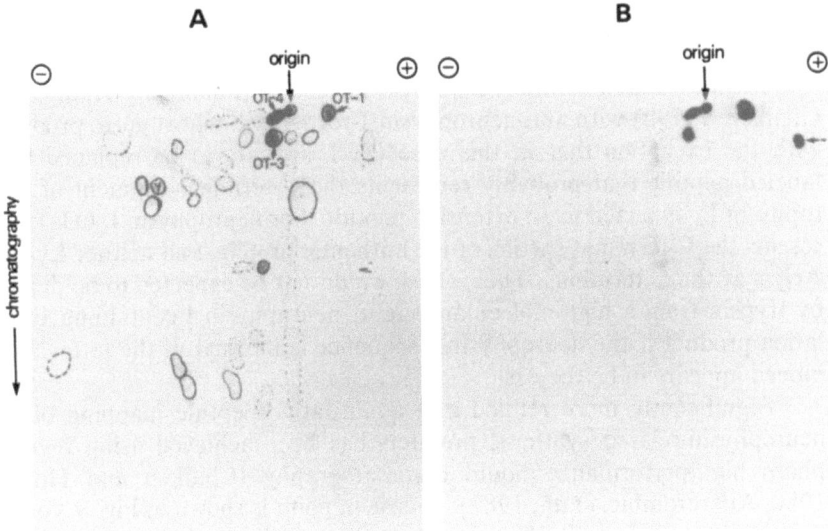

Figure 8. Peptide map comparison on paper of cysteine-containing sequences of authentic neurophysin II and those of the neurophysin translation product immunoprecipitated by anti-neurophysin II from wheat germ translation mixtures. [^{35}S]cysteine-labeled translation product and authentic neurophysin II were mixed, oxidized with performic acid, digested with trypsin, and fractionated by two-dimensional paper chromatography (16 hr, in upper phase of butanol : acetic acid : H$_2$O, 4 : 1 : 5 v/v)/paper electrophoresis (1500 volts, 1 hr, in 0.6 M pyridine acetate, pH 3.6). (A) Ninhydrin-stained map; (B) autoradiograph of the same map. Neurophysin-derived, cysteic acid-containing peptides OT-1, -3, and -4 were identified on the basis of separate experiments on these authentic neurophysin II peptides. The fourth expected cysteic acid-containing peptide, OT-2, was not detected strongly in the map. The components denoted "Y" in panel A initially stained yellow with collidine/ninhydrin reagent before turning blue with time. The dashed arrow in panel B indicates a major nonneurophysin [^{35}S]cysteic acid-containing component. The latter has been identified as free cysteic acid, which was carried nonspecifically in the immunoprecipitate from the translation mixtures. [Figure taken from Giudice and Chaiken, 1979a.]

itated protein to which a large amount of authentic neurophysin II was added, shows the correlation of three cysteic acid-containing neurophysin peptides (OT's 1, 3, and 4, using the nomenclature of Wuu and Crumm, 1976) in the ninhydrin map with major ^{35}S-containing peptides in the autoradiograph. All of the cysteic acid-containing tryptic peptides are internal in the neurophysin II sequence and would be expected to be released even if neurophysin represents an internal sequence, with C- and N-terminal extensions, in the high-molecular-weight translation product. Thus, the complete (contour-for-contour) overlap of the three detected OT peptides in the authentic neurophysin ninhydrin map with major autoradiographic spots from labeled synthesized protein confirms that the anti-neurophysin

II-recognized translation product indeed has sequences identifiable as those of neurophysin II.

Results similar to the above also have been obtained (Giudice and Chaiken, 1979b) with anti-neurophysin I-recognized wheat germ products, with the exception that in this case OT-1 appears to be replaced by a labeled peptide that probably represents the C-terminal segment of neurophysin I connected to an extension peptide. For neurophysin I, OT-1 represents the C-terminal section of the authentic protein, and neither Lys nor Arg is at the C terminus. Thus, OT-1 would not be expected to be derived by trypsin from a high-molecular-weight, neurophysin I-containing translation product if the neurophysin I sequence is internal in the latter. This indeed appears to be the case.

Significantly more refined and quantitative peptide mapping of the neurophysin-related synthesis products has been achieved using reverse-phase high-performance liquid chromatography (Chaiken and Hough, 1980, Abercrombie, *et al.*, 1982). A case in point is shown in Fig. 9, for the tryptic peptides from performic acid-oxidized mixtures of authentic neurophysin II and anti-neurophysin II-immunoprecipitated protein from rabbit reticulocyte lysate translation. Here, one can observe co-migration of all authentic neurophysin cysteic acid-containing peptides (as assigned by amino acid analysis) with ^{35}S-containing peptides from the translation product. This correlation includes OT-2, which is cysteic acid-containing but which was not observed as a clean spot on paper maps, probably due

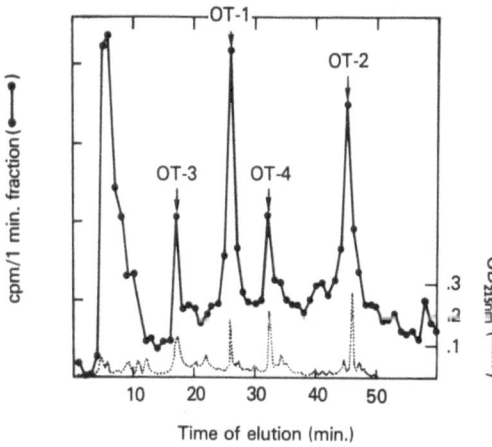

Figure 9. Peptide maps of anti-neurophysin II-recognized *in vitro* translation product, from rabbit reticulocyte lysate in the presence of [^{35}S]cysteine, by reverse-phase high-performance liquid chromatography. Column: Zorbax CN (γ-cyanopropylsilyl, from Dupont), 0.46 × 25 cm; gradient: 100% TEAP (0.25 N phosphoric acid adjusted to pH 3 with triethylamine and containing 0.02% sodium azide) at time zero to 60% TEAP–40% acetonitrile at 60 min; other conditions: ambient temperature, 0.8 ml/min elution flow rate. As in Fig. 8, the sample was an immunoprecipitate to which a large amount of authentic bovine neurophysin II was added, followed by performic acid oxidation and trypsin digestion. Fractions were measured for the presence both of peptides from the large (nanomole) amount of neurophysin carrier (· · · ·) and of ^{35}S-containing peptides from the small (subfemtomole) amount of translation product II (●—●). Ordinate: 50 cpm/division. [Taken from Chaiken and Hough, 1980.]

to chromatographic smearing. Of note, the relative sizes of the ^{35}S-containing peaks assignable as OT's 1, 2, 3, and 4 correlate well with the contents of ^{35}S expected from the cysteic acid contents of the authentic peptides, namely 5, 4, 2, and 3 residues per peptide molecule, respectively. Taken with similar results for reticulocyte lysate products immunoprecipitated by anti-neurophysin I, the occurrence of the neurophysin I and II sequences in translation products I and II is substantiated.

4. Identity of Translation Products as Biosynthetic Precursors

At present, defining translation products I and II as biosynthetic precursors of the neurophysins seems reasonable but rests on circumstantial evidence only. Inasmuch as the neurophysins have been shown by pulse-chase studies to be synthesized in the hypothalamus (Sachs *et al.*, 1969), it can be asserted that any biosynthetic precursors of these proteins should be coded by a species of hypothalamic mRNA. And, since the translation products identified (I and II) represent all of the major neurophysin-containing proteins detectable as coded by total hypothalamic poly(A)$^+$ RNA, it can be concluded that these translation products represent the unprocessed precursors. Finally, since the neurophysins are neurosecretory proteins (they are stored and transported after synthesis in neurosecretory granules), the translation products can be defined as preproneurophysins.

These deductions notwithstanding, the specific processing of translation products to active, authentic neurophysins has not been established as yet. Partial processing in the presence of dog pancreatic microsomes has been observed (Schmale *et al.*, 1979). In these studies, the large neurophysin translation product (II) was found to be converted to a slightly larger species that contained carbohydrate, while the smaller product (I) was converted to a smaller form not containing carbohydrate. The results are consistent with the microsome-mediated removal of prepieces from both species but glycosylation only in the neurophysin II case. Of note, the "putative" proneurophysins reported as 20,000-dalton proteins from pulse-chase studies have been shown to be glycosylated for the II-related but not for the I-related case (Brownstein *et al.*, 1980). Further, the II-related form could be converted, upon gentle trypsin hydrolysis, to a form that has hormone binding activity (Brownstein *et al.*, 1980). It is tempting to correlate these latter findings with those from *in vitro* synthesis. By the view thus produced, translation products I and II, representing *in vitro*-synthesized preproneurophysins of 17,000–19,000 and 21,000–25,000 daltons, respectively, would correspond to the *in vivo* products, proneurophysin forms, which differ by prepiece removal and, in the II-related case, glycosylation. Subsequent to microsomal processing, limited proteolysis of the "pro" forms eventually would produce active, mature neurophysins. As yet, the detailed chemical events ocurring during processing are not defined, and isolation of the biologically relevant processing enzymes is not yet achieved.

5.   Nucleic Acid-Precursor Interrelationships

While the biochemical characterization of neurophysin biosynthesis has centered logically on the identification of protein precursor molecules, helpful correlative information has come from observations made with the nucleic acid species leading to specific translation. The studies of Richter and colleagues (Schmale et al., 1979; Richter et al., 1980) have shown that the neurophysin I- and II-specific in vitro translation products coded by bovine poly(A)$^+$ RNA also are synthesized from the bovine hypothalamic mRNA isolated from polysomes. This finding helps substantiate the conclusion that translation products I and II described above [and defined as 16,500- and 21,000-dalton species, respectively, in the polysomal RNA translation study (Richter et al., 1980)] are the intact translation products of mature neurophysin-specific mRNA. This latter interpretation in turn would necessitate that if larger translated proteins or other precursorlike molecules are observed, they would have to be attributed to forms other than these preproneurophysins. Specific translation products larger than the 17,000- to 25,000-dalton forms have not been identified chemically as yet. Nonetheless, very large (up to 80,000 dalton) forms of neurophysin-related proteins have been found in tissue extracts, as discussed by Béguin et al., 1981. Chemical information obtained for these big molecules should help explain their relationship to in vitro translation products.

A useful perspective of the relationship between precursors and neurophysin-specific mRNA has been derived, at least partially so far, from fractionation of the total hypothalamic poly A$^+$ RNA by electrophoresis on agarose gels in the presence of methyl mercuric hydroxide (Bailey and Davidson, 1976). Through this separation procedure, specific size classes of poly(A)$^+$ RNAs have been isolated that code for the neurophysin biosynthetic products. As shown in Fig. 10, there are in fact at least two major electrophoretically migrating forms of RNA that code for each of the major neurophysins. The smaller forms have apparent molecular weights of 125,000–250,000 (0.4–0.7 kilobase). This size range is close to that expected for an mRNA that would have a coding region for a 17,000- to 25,000-dalton protein in addition to expected poly(A) and other possible nontranslated sequences. In contrast, the larger RNA forms separated electrophoretically (Fig. 10) are in the range of 1–3 $\times$ 10^6 daltons (3–9 kilo-

Figure 10. Fractionation of neurophysin-specific mRNA from total hypothalamic poly(A)$^+$ RNA by electrophoresis on 1.5% agarose gels containing methyl mercuric hydroxide. A 5-μg amount of total poly(A)$^+$ RNA was electrophoresed, the finished gel was sliced into 5-mm zones, and the slices were extracted from mRNA with 0.5 M ammonium acetate (Fuchs and Green, 1979). The mRNA from each fraction was assayed for the ability to code for neurophysin translation products I and II in 30-μl rabbit reticulocyte lysate translations in the presence of [^{35}S]cysteine. Translated protein was immunoprecipitated from the resultant translation mixtures by anti-neurophysin I (upper panel) and subsequently (after the anti-I immunoprecipitate was removed) by anti-neurophysin II (lower panel). Samples of reduced, alkylated proteins released from immunoprecipitates were precipitated in 10% trichloracetic acid, filtered on glass-fiber filters, and an aliquot counted (●—●). Another aliquot was subjected to gel exclusion high-performance liquid chromatography, as described in the legend of Fig. 7, and the amount of labeled protein measured that elutes in the apparent molecular weight range 17,000–25,000 (▲—▲).

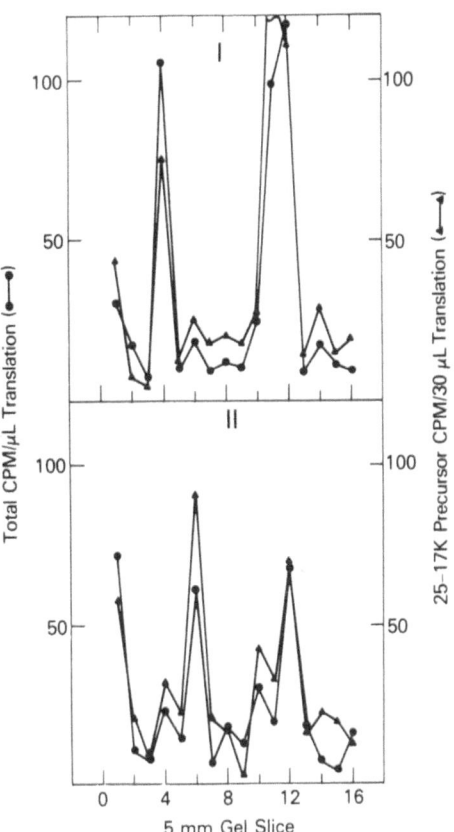

bases). This latter size range is considerably greater than would be expected for an mRNA coding for the 17,000- to 25,000-dalton translation products. Nonetheless, these forms apparently also can code for anti-neurophysin proteins that elute in the size range of 17,000–25,000 daltons upon gel permeation chromatography.

The assignment of the 125,000- to 250,000-dalton poly(A)$^+$ RNA forms as the mature mRNAs for preproneurophysins is reasonable based largely on the correspondence of the sizes of these forms with the sizes of the preproneurophysins for which they code. A possible but unproven view for the larger poly(A)$^+$ RNA forms is that these could represent unprocessed or incompletely processed neurophysin-specific mRNAs. While the possibility cannot be ruled out that the apparent high-molecular-weight RNAs represent persistent aggregates of the smaller mRNAs, the existence of authentic large forms is not unreasonable, since the poly(A)$^+$

RNA used as starting material is from intact tissue and would comprise both cytoplasmic and nuclear RNA. Some species of such big mRNAs could still have an intact coding region for prepropeurophysins. However, other variants of the unprocessed forms might also code for large neurophysin-containing proteins, which would elute in the breakthrough region upon gel permeation chromatography. So far, no unequivocal evidence on this point is available. Nonetheless, such large precursor-related forms, if made, could be related to the large neurophysin-related proteins identified in tissue extracts (Béguin *et al.*, 1981).

6. Relationship between Neurophysin and Neuropeptide Hormone Biosynthesis

One of the more forceful stimuli to the study of neurophysin biosynthesis has been the view that this process is related to the biosynthesis of the neuropeptide hormones, oxytocin and vasopressin. This biosynthetic relatedness has been recognized from several observations on the occurrence of the peptides and proteins in the hypothalamo–neurohypophyseal system. These include co-incorporation of [^{35}S]cysteine into neurophysin and vasopressin after injection of labeled amino acid in the hypothalamus (Sachs *et al.*, 1969) and coordinate loss of vasopressin and one major neurophysin (but not oxytocin and the second neurophysin) in Brattleboro rats (Burford *et al.*, 1971). In addition, preferential segregation is found (Dean *et al.*, 1968) of neurophysin I with oxytocin and neurophysin II with vasopressin in neurosecretory granules in spite of a lack of any obvious driving force due to differential hormone–neurophysin binding affinities [each of the neurophysins can interact noncovalently with either of the hormones and with similar affinities (Breslow, 1979)]. All of these findings are consistent with biosynthesis of neurophysins and hormones in a closely linked fashion. One obvious possibility (Sachs *et al.*, 1969) is that a particular hormone–neurophysin set is made as part of a common precursor, which is processed and packaged in neurosecretory granules at the sites of synthesis (namely the Golgi of neurosecretory neurons). This possibility is supported by the observation that the kinetic lag phase observed for neurophysin in pulse-chase studies also is observed for vasopressin (Sachs *et al.*, 1969).

In spite of the *in vivo* observations made, rigorous chemical evidence to establish the presence of hormone in neurophysin biosynthetic precursors has been elusive. Pulse-chase products, *in vitro* translation products, and large proteins in tissue extracts all have been found to interact with at least certain preparations of hormone antibodies (Fischer *et al.*, 1977; Schmale *et al.*, 1979; Brownstein *et al.*, 1980; Richter *et al.*, 1980; Béguin *et al.*,

1981). Further, limited proteolytic digestion of the pulse-chase neurophysin-related proteins has been found to liberate peptides that not only are immunoreactive with anti-vasopressin sera but also bind to neurophysin-agarose columns (Brownstein *et al.,* 1980). Nonetheless, a detailed inspection of peptide maps of neurophysin translation products as in Figs. 8 and 9 has yet to reveal the presence of any obvious hormone-related peptides. In such tryptic co-maps, hormone-containing peptides should be visible due to the expected content of the ^{35}S from at least two cysteic acid residues. However, as indicated most quantitatively by data such as those in Fig. 9, the only major ^{35}S-containing peptides other than those in the early breakthrough fractions are the peptides from the neurophysin sequence. Any peptides with at least two ^{35}S atoms would be expected to be at least as prominent as the peak for OT-2, a neurophysin peptide with two S atoms. Further, since performic acid-oxidized, trypsin-digested oxytocin and arginine vasopressin products elute at positions corresponding to the retention time range 35–42 min in the profile of Fig. 10, it is unlikely that trypsin-derived peptides containing the hormone sequences would elute in the early breakthrough fractions.

The above results notwithstanding, Schmale and Richter (1981) have reported a peptide mapping experiment which for the first time has allowed detection, at least qualitatively, of hormone-derived peptides in both neurophysin II translation product and processed forms of this translation product. This study employed 20 hour, 30°C tryptic digestion (instead of the 2–3 hour digestion at 37° used above) of performic acid oxidized proteins and mapping on cellulose thin layer plates. This procedure allowed detection of the peptide corresponding to the 1–8 sequence of arginine vasopressin in the glycosylated pro-neurophysin II derived from translation product II but of a modified form of this peptide from intact translation product II. The result gives strong chemical support to the view that pre-pro-neurophysins indeed contain hormone sequences.

More strikingly, Land *et al.* (1982) have used the bovine hypothalamic mRNA which produces translation product II (the mRNA coding for neurophysin II and arginine vasopressin and originally detected through *in vitro* translation analyses) as the basis for producing a neurophysin-specific complementary DNA. Direct nucleotide sequencing of this complementary DNA after selection and proliferation using recombinant DNA methods has revealed the presence of nucleotide sequences that would code for both arginine vasopressin (with a code for Gly at the residue position corresponding to hormone Gly amide 9) and neurophysin. Interestingly, the sequence deduced for translation product also shows that, while the 1–8 sequence would be produced by tryptic cleavage of the partially processed pro-neurophysin, it would be released with an amino-terminal extension (derived from the pre-piece section) from intact translation product II. Overall, the

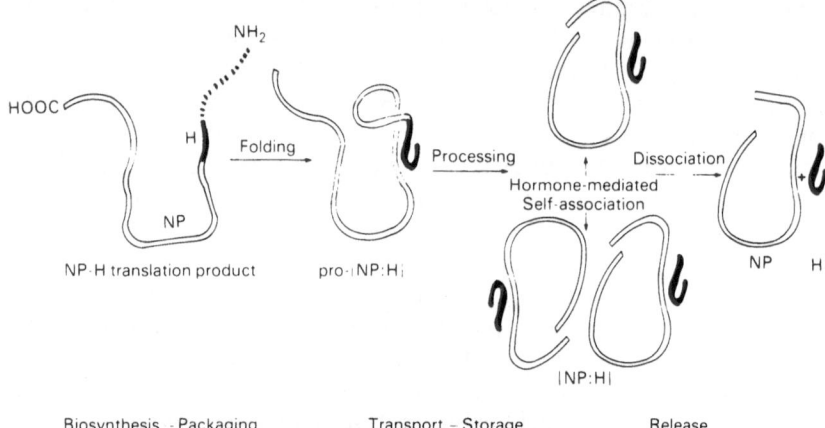

Figure 11. Schematic diagram of the common precursor model for the biosynthetic origin of neurophysin–neuropeptide hormone complexes, placed in the context of noncovalent interactions of mature protein and peptide components. The sequential order of segments in the translation product at the left is taken from Land *et al.* (1982). These segments are denoted, from the amino terminus (NH_2) as follows: prepiece or signal peptide (Blobel and Dobberstein, 1975) is broken line; hormone is filled line; neurophysin is open line; and non-hormone and -neurophysin sequence is stipled line. For the *in vivo* biosynthesis of neurophysin and hormone, prepiece removal likely occurs during the initial stages of polypeptide synthesis, before the polypeptide sequence is completed, folds, and attains correct disulfide bonds. Occurrence of the expected chemical events of folding, processing, self-association, and dissociation is correlated with the overall intra-neuronal stages (see also Figure 1) of hypothalamic biosynthesis, subsequent packaging, transport, and storage in neurosecretory granules, and ultimate release from the neurohypophysis.

results provide compelling evidence that neurophysin and associated neuropeptide hormone originate through a single polypeptide precursor.

Based on the common precursor structure of Land *et al.* (1982) and the assumption that the structure of the neurophysin I-oxytocin precursor is similar, the way in which noncovalent complexes of mature (processed) neurophysins and associated neuropeptide hormones are derived biosynthetically is shown schematically in Fig. 11. This model denotes the coexistence of hormone and neurophysin, in precursor forms, in the sequence (amino to carboxyl terminal) defined by the complementary DNA structure of Land *et al.* (1982). The model also portrays the way in which common precursor, once formed and folded, could be expected to lead to noncovalently interacting hormone and protein through post-translational processing and subsequent interdependent hormone binding and self-association (Nicolas *et al.*, 1980; Angal and Chaiken, 1982). Figure 11 connotes the view that ultimately it should be possible to use emerging information on neurophysin-hormone precursors to more fully understand how the functional properties of the mature proteins and peptides are produced and used in neurosecretory granules.

7. Future Directions—The Chemical Basis of Neurophysin-Neuropeptide Folding and Function

In vitro translation studies have succeeded in establishing that the neurophysins are synthesized as parts of higher-molecular-weights proteins. These have been defined in the bovine system as translation products I and II for the major neurophysins I and II, respectively. The in vitro studies also have allowed identification of mRNA species that code for these proteins. Finally, mRNA obtained through in vitro detection has led to definitive chemical evidence for the common precursor of model of neurophysin-neuropeptide hormone biosynthesis. In spite of the accomplishments so far, several aspects of neurophysin biosynthesis require further elucidation. Centrally, processing events leading from neurophysin-specific DNA to mature mRNA and ultimately to the mature neurophysin proteins themselves are yet to be understood chemically. The interrelationship of such events for the neurophysins with those for the neuropeptide hormones remain to be clarified conclusively. Defining chemically how the neurophysin-related proteins obtained by in vitro synthesis, pulse-chase labeling, and extraction relate to one another and to these processing events would be useful. Ultimately, given the above kinds of information, a fuller understanding of the chemical basis and regulatory mechanisms whereby neurophysins (and associated hormones) fold up and achieve biological function would seem a reasonable goal.

REFERENCES

Abercrombie, D. M., Hough, C. J., Seeman, J. R., Brownstein, M. J., Gainer, H., Russell, J. T., and Chaiken, I. M., 1982, Use of reverse phase high performance liquid chromatography in structural studies of neurophysins, photolabelled derivatives, and biosynthetic precursors, Proc. First Internat. Symp. on HPLC of Proteins and Peptides, in press.

Angal, S., and Chaiken, I. M., 1982, Interdependence of neurophysin self-association and neuropeptide hormone binding as expressed by quantitative affinity chromatography, Biochemistry 21:1574–1580.

Bailey, J. M., and Davidson, N., 1976, Methylmercury as a reversible denaturing agent for agarose gel electrophoresis, Anal. Biochem. 70:75–85.

Béguin, P., Nicolas, P., Bussetta, H., Fahy, C., and Cohen P., 1981, Characterization of the 80,000 molecular weight form of neurophysin isolated from bovine neurohypophysis, J. Biol. Chem. 256:9289–9294.

Blobel, G., and Dobberstein, B., 1975, Transfer of proteins across membranes. I. Presence of proteolytically processed and unprocessed nascent immunoglobulin light chains on membrane-bound ribosomes of murine myeloma, *J. Cell Biol.* **67**:835–851.

Bolton, A. E., and Hunter, W. M., 1973, The labelling of proteins to high specific radioactivities by conjugation to a ^{125}I-containing acylating agent, *Biochem. J.* **133**:529–539.

Breslow, E., 1979, Chemistry and biology of the neurophysins, *Annu. Rev. Biochem.* **48**:251–274.

Brownstein, M. J., Russell, J. T., and Gainer, H., 1980, Synthesis, transport, and release of posterior pituitary hormones, *Science* **207**:373–378.

Burford, G. D., Jones, C. W., and Pickering, B. J., 1971, Tentative identification of a vasopressin-neurophysin and an arginine-neurophysin in the rat, *Biochem. J.* **124**:809–813.

Chaiken, I. M., 1979, Preparative and analytical affinity chromatography of neurophysins on methionyl-tyrosyl-phenylalanyl-aminohexyl-agarose, *Anal. Biochem.* **97**:302–308.

Chaiken, I. M., and Hough, C. M., 1980, Mapping and isolation of large peptide fragments from bovine neurophysins and biosynthetic neurophysin-containing species by high performance liquid chromatography, *Anal. Biochem.* **107**:11–16.

Chaiken, I. M., Randolph, R. E., and Taylor, H. C., 1975, Conformational effects associated with the interaction of polypeptide ligands with neurophysins, *Ann. N.Y. Acad. Sci.* **248**:442–450.

Chauvet, M. T., Chauvet, J., and Acher, R., 1975, Phylogeny of neurophysins: Partial amino acid sequences of a sheep neurophysin, *FEBS Lett.* **52**:212–215.

Dean, C. R., Hope, D. B., and Kazic, T., 1968, Evidence for the storage of oxytocin with neurophysin-I and of vasopressin with neurophysin-II in separate neurosecretory granules, *Proc. Br. Pharmacol. Soc.* **34**:1928–1938.

Fischer, E. A., Curd, J. G., and Chaiken, I. M., 1977, Preparation of biologically active conjugates of bovine neurophysins and other polypeptides with multi-(poly-D,L-alanyl)-poly-L-lysine and their use to elicit antibodies, *Immunochemistry* **14**:595–602.

Fuchs, E., and Green, K., 1979, Multiple keratins of cultured human epidermal cells are translated from different mRNA molecules, *Cell* **17**:573–582.

Giudice, L. C., and Chaiken, I. M., 1979a, Immunological and chemical identification of a neurophysin-containing protein coded by messenger RNA from bovine hypothalamus, *Proc. Natl. Acad. Sci. USA* **70**:3800–3804.

Giudice, L. C., and Chaiken, I. M., 1979b, Cell-free biosynthesis of different high molecular weight forms of bovine neurophysins I and II coded by hypothalamic mRNA, *J. Biol. Chem.* **254**:11767–11770.

Givol, D., DeLorenzo, F., Goldberger, R. F., and Anfinsen, C. B., 1965, Disulfide structure and the three-dimensional structure of proteins, *Proc. Natl. Acad. Sci. USA* **53**:676–684.

Hough, C. J., Hargrave, P. A., and Chaiken, I. M., 1980, On the biosynthetic origin of neurophysin–neurohypophyseal peptide hormone complexes, in: *Biosynthesis, Modification and Processing of Cellular and Viral Polyproteins* (G. Koch and D. Richter, eds.), pp. 29–42, Academic Press, New York.

Land, H., Schulz, G., Schmale, H., and Richter, D., 1982, Nucleotide sequence of cloned cDNA encoding the bovine arginine vasopressin-neurophysin II precursor, *Nature* **295**:299–303.

Lin, C., Joseph-Bravo, P., Sherman, T., Chen, L., and McKelvy, J. F., 1979, Cell-free synthesis of putative neurophysin precursors from rat and mouse hypothalamic poly(A)-RNA, *Biochem. Biophys. Res. Commun.* **89**:943–950.

Livett, B. G., 1978, Immunohistochemical localization of nervous system-specific proteins and peptides, *Int. Rev. Cytol. Suppl.* **7**:53–237.

Marcus, A., Efron, D., and Weeks, D. P., 1974, The wheat germ cell-free system, *Methods Enzymol.* **30**:749–754.

Nicholas, P., Batelier, G., Rholam, M., and Cohen, P., 1980, bovine neurophysin dimerization and neurohypophyseal hormone binding, *Biochemistry* **19**:3565–3573.

Pelham, H. R. B., and Jackson, R. J., 1976, An efficient mRNA-dependent translation system from reticulocyte lysates, *Eur. J. Biochem.* **67**:247–256.

Pickering, B. T., 1978, The neurosecretory neurone: A model system for the study of secretion, *Essays Biochem.* **14**:45–81.

Richter, D., Schmale, H., Ivell, R., and Schmidt, C., 1980, Hypothalamic mRNA-directed synthesis of neuropeptides: Immunological identification of precursors to neurophysin II/arginine vasopressin and to neurophysin I/oxytocin, in: *Biosynthesis, Modification and Processing of Cellular and Viral Polyproteins* (G. Koch and D. Richter, eds.), pp. 43–66, Academic Press, New York.

Sachs, H., Fawsett, P., Takabatake, Y., and Portanova, R., 1969, Biosynthesis and release of vasopressin and neurophysin, *Recent Prog. Horm. Res.* **25**:447–491.

Schmale, H., and Richter, D., 1981, Tryptic release of authentic arginine vasopressin$_{1-8}$ from a composite arginine vasopressin/neurophysin II precursor, *Neuropeptides* **2**:47–52.

Schmale, H., Leipold, B., and Richter, D., 1979, Cell-free translation of bovine hypothalamic mRNA: Synthesis and processing of the preproneurophysins I and II, *FEBS Lett.* **108**:311–316.

Steiner, D. F., and Clark, J. L., 1968, The spontaneous reoxidation of reduced beef and rat proinsulins, *Proc. Natl. Acad. Sci. USA* **60**:622–629.

Wuu, T.-C., and Crumm, S. E., 1976, Amino acid sequence of bovine neurophysin-II: A reinvestigation, *Biochem. Biophys. Res. Commun.* **68**:634–639.

17

Steroid Antagonism of Melanosome Movements Induced by Neuropeptides

Philippa M. Edwards and Guy G. Rousseau

1. Effects of ACTH and Related Peptides and of Glucocorticoids on the Brain

It is now well established that corticotropin (ACTH) is involved in the acquisition and maintenance of conditioned behavioral responses in laboratory animals, possibly by facilitating a selective arousal state (see review by Bohus, 1978). Glucocorticoid hormones also modify behavior and learning performance (Bohus, 1978), alter the threshold to and integration of sensory stimuli in several species (Henkin, 1970; Sakellaris, 1972), and cause psychological changes in man (Woodbury, 1958). Furthermore, glucocorticoids modify the excitability of nerve cells in several brain regions (Chambers *et al.*, 1963; Pfaff *et al.*, 1971; Phillips and Dafny, 1971; Conforti and Feldman, 1975).

Thus, it might be argued that the effects of ACTH on the central nervous system are mediated through its physiologic stimulation of the adrenal cortex. However, there is considerable evidence (see review by de Wied, 1977) that ACTH acts independently of glucocorticoid release. The behavioral abnormalities produced by hypophysectomy are reversed by administration of ACTH but not of dexamethasone. Furthermore, the effects of ACTH are not abolished by adrenalectomy and are mimicked by related peptides such as MSH and other fragments that are not steroidogenic

Philippa M. Edwards and Guy G. Rousseau • International Institute of Cellular and Molecular Pathology, B-1200 Brussels, Belgium.

(Schwyzer and Eberle, 1977; Bohus, 1978; La Hoste *et al.*, 1980). Finally, the behavioral effects of glucocorticoids are, in general, opposite to those of ACTH. This, in turn, might suggest that glucocorticoid actions result from their feedback inhibition of ACTH release. This is not the case. Glucocorticoids modify behavior in the absence of an effect on ACTH release, such as following hypophysectomy or intracerebral administration (de Wied, 1977). Therefore, ACTH-related neuropeptides and glucocorticoids appear to have independent and opposite effects on the brain.

In most tissues, regulation by glucocorticoid hormones is mediated by an alteration in the activity of specific enzymes. This results from the binding of the steroid to an intracellular receptor protein and the subsequent interaction of the hormone–receptor complex with chromatin to modify gene expression (Rousseau, 1975). This mode of action gives rise to the lag period and requirement for RNA and protein synthesis that are typical of such glucocorticoid-mediated effects. In brain, several enzymes, both glial and neuronal, are under glucocorticoid control (see review by Leung and Munck, 1975) and glucocorticoid receptors resembling those in other tissues have been identified (McEwen *et al.*, 1976). However, alteration of enzyme activities by induction or repression appears unlikely to account for all the effects of glucocorticoids on the function of the central nervous system. Certain effects are too rapid to involve a modulation in protein synthesis. Inhibition of the release of corticotropin-releasing factor and ACTH by corticosterone through the fast feedback mechanism begins within 5 min (Jones and Hillhouse, 1976). An almost instantaneous modification of the electrical activity of certain brain neurons has been demonstrated following direct application of glucocorticoids by microiontophoresis (Steiner *et al.*, 1969; Ben Barak *et al.*, 1977). Moreover, structure–activity studies on the influence of steroids on the extinction of avoidance behavior (van Wimersma Greidanus, 1970) suggest that this effect might be mediated by receptors that are neither glucocorticoid nor mineralocorticoid in type. As the central nervous system is not easily amenable to investigation, we sought a simpler model in which to study the antagonism between steroids and neuropeptides.

2. The Amphibian Melanophore as a Model

The amphibian melanophore is embryologically derived from the neural crest (Rawles, 1948) and, as with the neuron, differentiation includes the development of an extensive dendritic field. It contains pigment granules (melanosomes) that, under the influence of various stimuli, are either dispersed throughout the dendritic field of the cell (skin darkening) or aggregated in the cell body (skin lightening) by a process analogous to rapid axonal flow. It appears that melanosome dispersion involves the microfilament system (Malawista, 1971a), whereas aggregation requires intact microtubules (Malawista, 1971b). Melanosome dispersion is brought about by MSH, which plays a physiological role in the adaptation of the skin color of the animal to its environment. Related peptides containing the amino acid residues 4–10 of ACTH/MSH (α-MSH corresponds to residues 1–13 of ACTH) have comparable effects (see review by Schwyzer and Eberle, 1977). Among the substances that cause aggregation of melanosomes are many that also modulate nerve cell function: serotonin, catecholamines, and acetylcholine (Lerner, 1959). Furthermore, although the results are somewhat contradictory, it has been suggested that endorphins and other fragments of β-lipotropin (Novales and Novales, 1979; Carter *et al.*, 1979) can modulate the response of the melanophore to other agents. Pertaining to our interest in glucocorticoids, cortisol has been reported to reverse the effects of MSH (Wright and Lerner, 1960; Malawista, 1965). This is reminiscent of the antagonism between glucocorticoids and ACTH in the extinction of avoidance behavior (de Wied, 1977) and in changes in the firing rate of hypothalamic and midbrain neurons (Steiner *et al.*, 1969). Thus, the interplay of the ACTH/MSH and corticosteroid hormone systems seems to be involved in the response of the organism to its environment. In cold-blooded animals, adaptation involves a change in skin color; in higher animals, a modified behavior pattern is induced.

We have monitored melanosome movements in frog skin melanophores by measuring the skin reflectance essentially as described by Shizume *et al.* (1954). Frogs *(Rana ridibunda)* of both sexes were maintained at 4°C in shallow drinking water under constant illumination and were sacrificed by pithing. Two pieces of dorsal skin from the trunk and one from each thigh were obtained from each animal. Four frogs were normally used for each experiment and the treatments randomly assigned to give a Latin square. The skins were washed in Ringer's solution (NaCl, 115 mM; CaCl$_2$, 1 mM; KHCO$_3$, 2.5 mM; pH 7.8), then mounted in Plexiglas holders with the dermal surface exposed to Ringer's solution and the epidermal surface open to the air. The Ringer's was aerated throughout the experiment. Measurements of the reflectance of the epidermal surface were made by placing the holders on the light unit of a Photovolt 670 reflection meter. The reflection meter was arbitrarily standardized by setting the reflectance of a gray melamine tile to 90%. Stock solutions (10 μg/ml) of synthetic α-MSH were prepared in Ringer's solution containing bovine serum albumin (1 mg/ml), HCl (0.05 N), and bacitracin (10^{-5} M) and stored at -20°C. Steroids (5 $\times$ 10^{-3} M) were dissolved in ethanol and diluted such that the final ethanol concentration (less than 0.2%) did not influence skin reflectance.

3. Effects of Corticosteroids on Frog Skin Melanophores

3.1. Cortisol Reverses MSH-Induced Melanosome Dispersion

The frog skin pieces were washed in three changes of Ringer's solution over a period of 2 hr during which time the reflectance increased to a stable baseline value. Addition of MSH caused a fall in reflectance that reached a plateau at about 60 min. The system was highly sensitive to MSH as shown by the dose–response curve (Fig. 1). A concentration of 3×10^{-11} M was routinely used to study the effects of steroids. Addition of cortisol to skins previously darkened with MSH resulted in a rapid reversal of the darkening effect. A typical result is shown in Fig. 2. It can be seen that the action of cortisol is very rapid, being detectable after 4 to 8 min and maximal at about 16 min. Although, in our hands, MSH gave very reproducible results, the sensitivity to cortisol was highly variable between frogs. This variability is discussed below (Section 3.2). Reversal by cortisol of the MSH effect was rarely 100% and usually around 30% but sometimes there was no response. Using an internal standard (10^{-5} M cortisol) to eliminate inter-frog variations, a dose–response curve for cortisol was obtained (Fig. 3). It can be seen that responses could be obtained with as little as 10^{-8} M cortisol and a concentration of 4×10^{-6} M was required for the maximum effect. If the skin was then washed in Ringer's solution, the reflectance

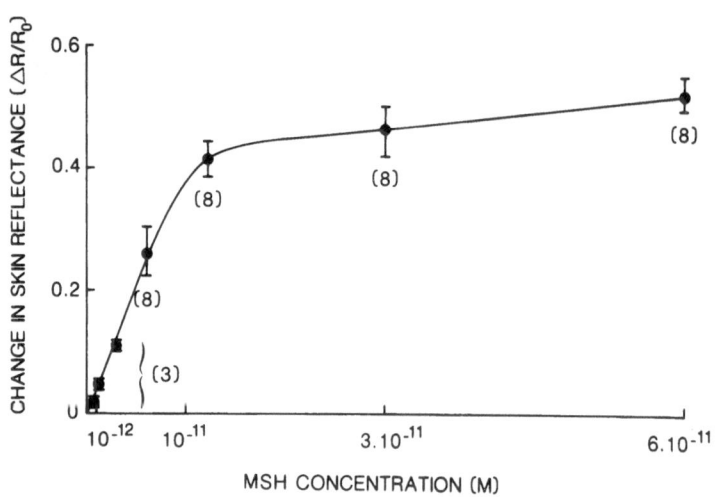

Figure 1. Dose–response curve for MSH in frog skin system. Data (mean ± S.E.M., number of determinations in parentheses) were obtained as described in Section 2. Each skin piece was exposed to a single concentration of MSH. R_0 is the baseline reflectance before addition of MSH. ΔR is the change in reflectance observed 60 min after addition of MSH.

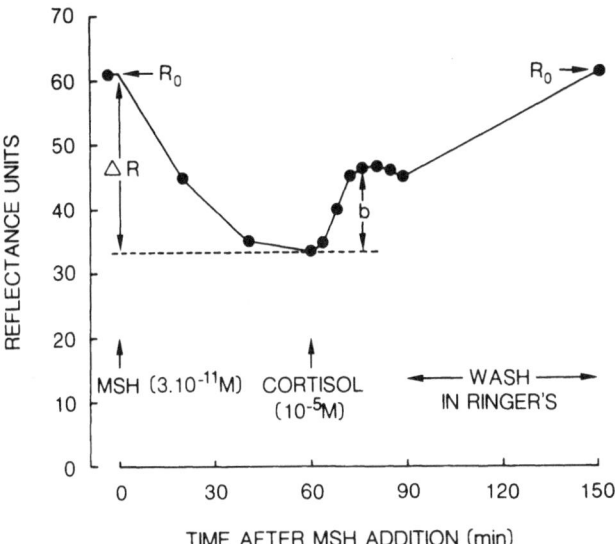

Figure 2. Reversal by cortisol of MSH effect on frog skin. The figure illustrates a typical result obtained from one piece of skin. R_0 is the baseline reflectance, ΔR is the change in reflectance caused by MSH, and b is the change in reflectance caused by cortisol.

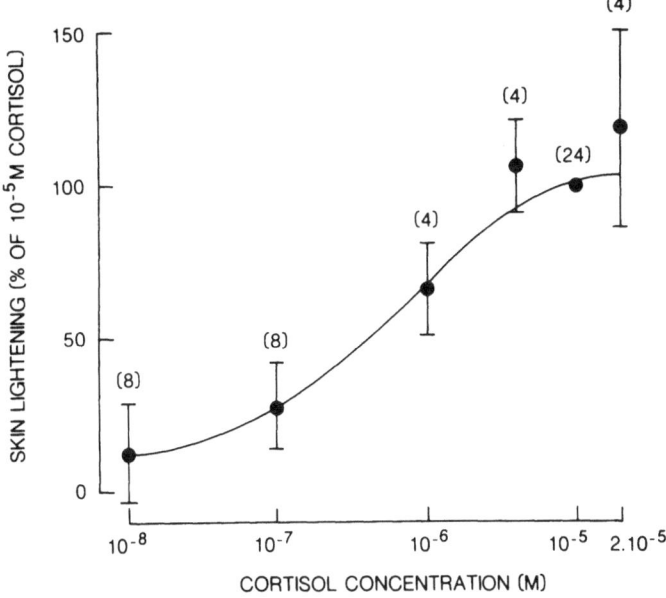

Figure 3. Dose–response curve for cortisol on frog skin. The figure shows the extent of reversal by various cortisol concentrations of the skin darkening produced by MSH (3×10^{-11} M). The protocol was as described in Fig. 2. The reversal (b/ΔR, see Fig. 2) was calculated relative to the value obtained with matched skin pieces exposed to 10^{-5} M cortisol. The data are means $\pm$ S.E.M.; the number of frogs is shown in parentheses.

returned to baseline values and a second response to MSH and cortisol could be obtained (Fig. 4). The second response to cortisol was usually smaller than the first (see Section 3.2) although the response to MSH was unchanged. The effect of cortisol could be directly reversed by washing the skins in Ringer's solution containing the initial concentration of MSH (3 $\times$ 10 $^{-11}$ M) but no cortisol as shown in Fig. 5.

Observation of the whole skin by light microscopy before and after treatment with cortisol (Fig. 6) showed that the most prominent change was a diminution of the apparent dendritic field of the dermal melanophores. The dendrites cannot be seen at this level in the absence of pigment. Therefore, it is likely that cortisol stimulates a centripetal movement of the melanosomes, as has been shown to be the case for other lightening agents (Lerner and Takahashi, 1956).

3.2   Factors That Affect the Cortisol Response

As was mentioned above, sequential trials with cortisol on the same piece of skin gave diminishing responses. This loss of sensitivity was not a tachyphylactic phenomenon; there was no significant difference in response between pieces repeatedly exposed to MSH, cortisol, and washings, and pieces that were maintained *in vitro* untouched for the equivalent length of time. It was thought that the decrease in sensitivity to cortisol might be due to the progressive loss of some salt, nutrient, or other factor (such as another hormone) that is essential for the expression of the cortisol activity.

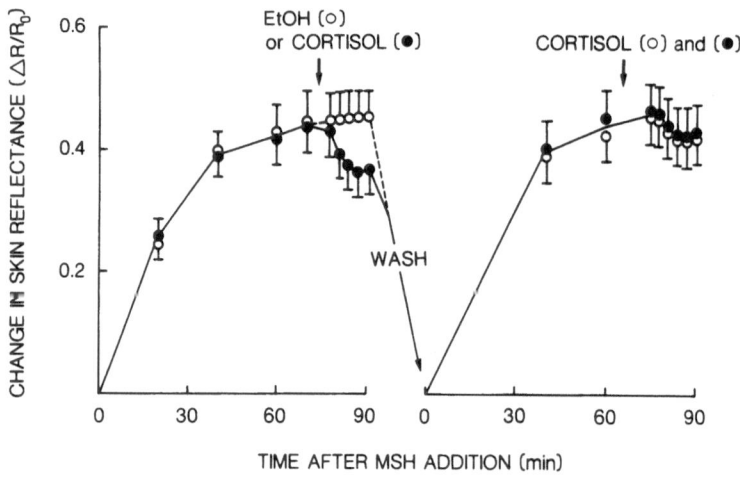

Figure 4. Stability of MSH and cortisol responses. Data are means $\pm$ S.E.M. from four frogs according to the protocol described in Fig. 2. EtOH refers to ethanol, the vehicle for cortisol.

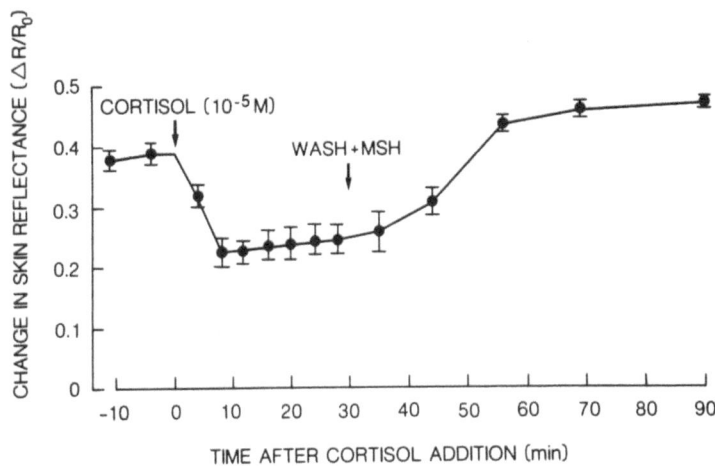

Figure 5. Time course and reversibility of the cortisol effect. The data obtained as in Fig. 2 are means ± S.E.M. from four frogs. MSH concentration was 3×10^{-11} M.

Furthermore, the variability observed between frogs could be related to variations in the concentration of the same permissive factors in the skins from different individuals.

Neither glucose, which is required for maximal expression of mineralocorticoid activity in amphibian skin (Crabbé, 1977), nor calcium, which is involved in the stimulation by steroids of meiosis in *Xenopus* oocytes (Baulieu *et al.*, 1978) and in glucocorticoid-induced lymphocytolysis (Kaiser and Edelman, 1977), had any effect on the melanophore response to cortisol (see Table I). Indeed, even replacement of the Ringer's solution with amphibian culture medium, which contains all the salts and nutrients required for cell growth, had no effect on the initial response to cortisol or on the decrease in sensitivity with time. Insulin is required for induction of meiosis in *Xenopus* oocytes by certain steroids (El-Etr *et al.*, 1979) and is synergistic with glucocorticoids in the stimulation of tyrosine aminotransferase activity (Gelehrter, 1979). Thyroid hormone is also required for certain glucocorticoid effects (Samuels *et al.*, 1977). Neither insulin nor thyroid hormones had any effect on initial or sequential melanophore responses to cortisol.

We also investigated the possibility that the frog-to-frog variations were due to different concentrations of inhibitory factors. The influence of adrenocortical hormones, of hormones or other factors (e.g., catecholamines, neuropeptides, melatonin) deriving from the brain and pituitary, and of sex steroids was studied. This was carried out by physiological adrenalectomy (suppression of adrenal cortical secretion by maintaining the

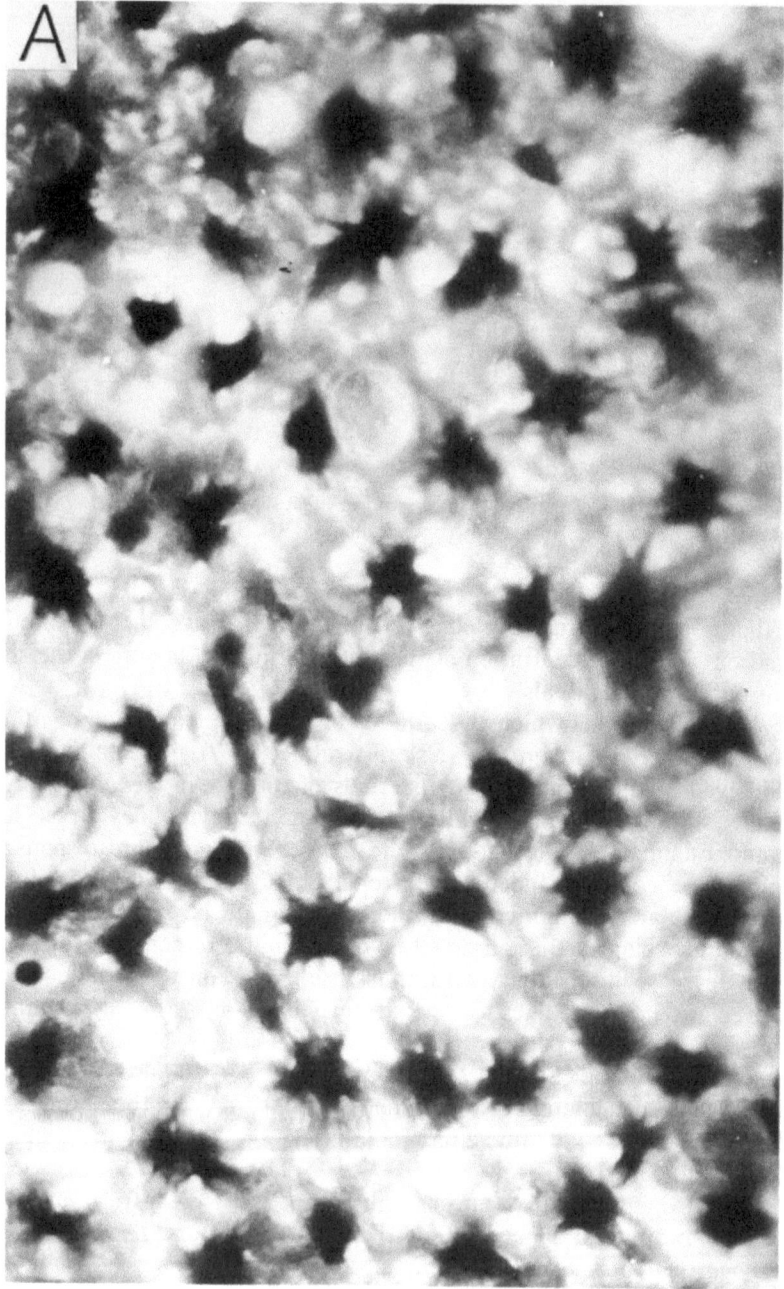

Figure 6. Light micrographs of unfixed frog skin before and after cortisol treatment. Panel A shows a piece of frog skin treated with MSH (3×10^{-11} M) for 60 min. Cortisol (10^{-5} M) was

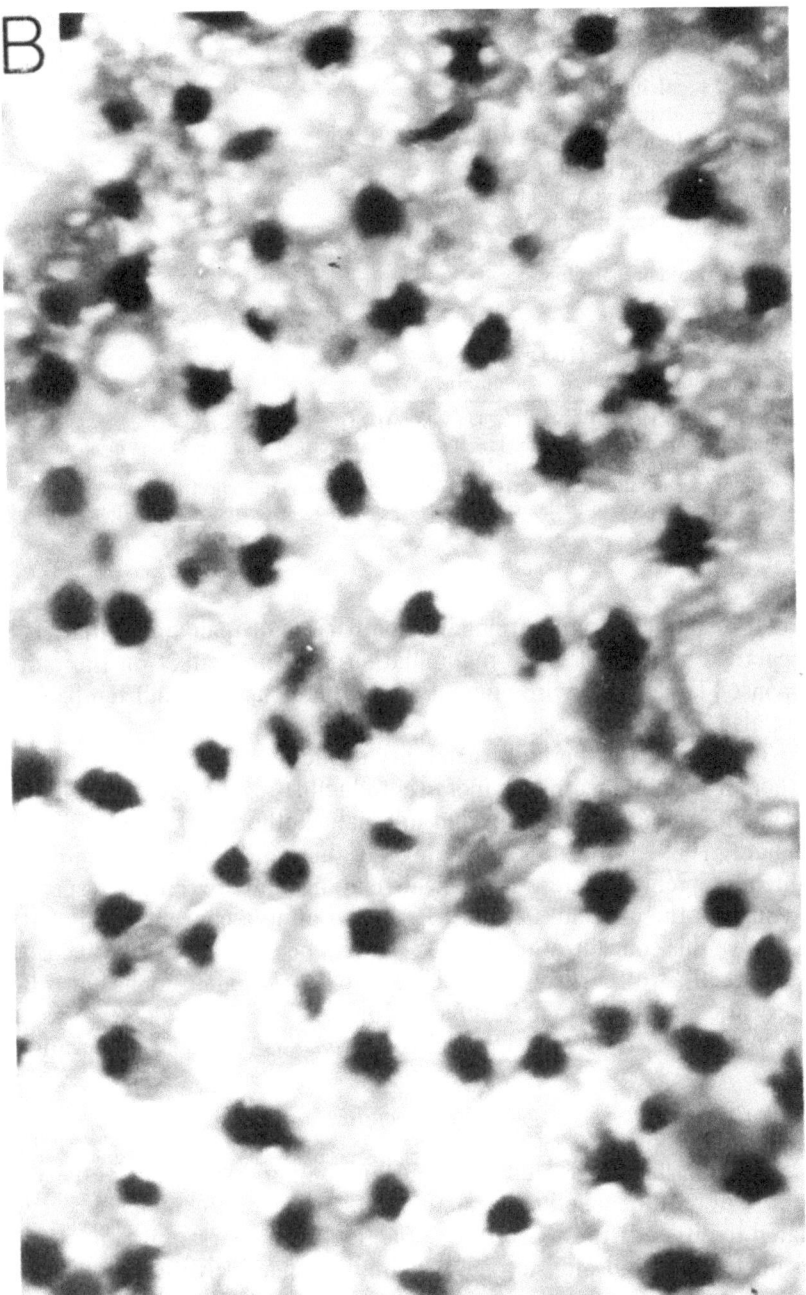

then added and panel B shows the same piece of skin 15 min later. The microscope was focused as well as possible on the dark irregular cells, which are the dermal melanophores.

Table I.   Factors That Do Not Alter the Melanophore Response to Cortisol

In vitro
   Ringer[a] + calcium (10 mM)
   Ringer − calcium
   Ringer + glucose (10 mM)[b]
   Ringer + insulin $(10^{-10}-10^{-6}M)$[b]
   Ringer + triiodothyronine $(10^{-8}-10^{-6}M)$[b]
   Amphibian culture medium[b,c]
In vivo
   Physiological adrenalectomy[d]
   Decapitation[e]
   Sex of frog
   Injection of thyroxine (10 μg/g frog)
   Temperature at which frogs maintained

[a] NaCl 115 mM, $KHCO_3$ 2.5 mM, $CaCl_2$ 1 mM, pH 7.8.
[b] Treatment also had no effect on the loss of sensitivity to cortisol with time of skin incubation.
[c] As supplied by Grand Island Biological Company.
[d] By maintenance in saline for up to 14 days (depresses secretion of corticosteroids).
[e] For up to 7 days prior to testing (removes endogenous hormones from brain, pituitary and pineal glands).

animal in saline), decapitation, and comparison of male and female frogs, respectively. None of these manipulations had any effect on the cortisol response (Table I). Environmental factors, such as light and temperature, were also ineffective.

The reason for the high variability in the response remains unknown. Although we obtained much more reproducible results with other lightening agents such as melatonin and adrenalin, other workers (Hadley and Bagnara, 1969) have reported highly variable sensitivities to a number of lightening agents. There was no clear relationship between the response to cortisol and the concentration of MSH used up to a concentration of 10^{-10} M. Above this concentration, the MSH effect became insensitive to cortisol.

The effect of cortisol is not limited to *R. ridibunda;* it has also been observed using skins from *R. pipiens* (Malawista, 1965) and *R. esculenta, R. temporaria, Xenopus laevis,* and *Bufo marinus* (personal observations).

3.3.   Evidence for a New Type of Receptor

Cortisol has both glucocorticoid and mineralocorticoid activity. In order to determine whether the activity of cortisol on the melanophore was related to either of these two types of action, a comparison was made with dexamethasone and aldosterone, which are respectively more potent than cortisol in their glucocorticoid and mineralocorticoid activities. As is shown in Fig. 7, both of these steroids were less active than cortisol in their ability to reverse MSH-induced darkening of frog skin. The study was therefore extended to the series of steroids shown in Table II. Although the variability was too great to obtain an exact value for the relative potency of the active steroids, a rough estimate was obtained by comparing the activity of each steroid at 10^{-5} M with that of the same concentration of cortisol in another piece of skin from the same frog.

The response was very specific; cortisol and cortisone (the latter has no glucocorticoid activity unless it is hydroxylated to cortisol) were the most potent steroids tested. The sex steroids, estradiol and testosterone, were inactive. We also tested alphaxalone, a potent anesthetic steroid that acts very rapidly, presumably through an influence on nerve cell membranes (Richards and Hesketh, 1975). It was inactive on frog skin. There was no correlation between the activity of steroids in the melanophore system and their relative affinity for the glucocorticoid receptor (Table II). It is unlikely that the differences observed between binding to glucocorticoid receptors in rat hepatoma cells and activity in the melanophore system can

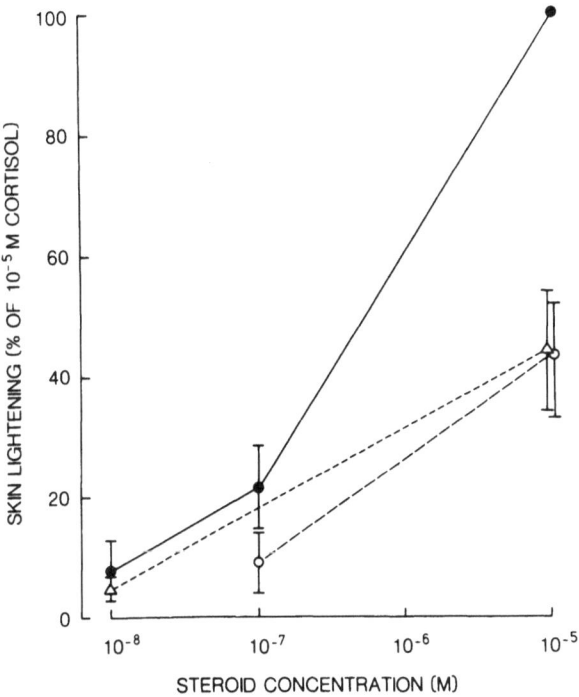

Figure 7. Dose–response relationships for three steroids on frog skin. The experiments were designed as described in Fig. 2 using cortisol (●), dexamethasone (○), or aldosterone (△). Data are means ± S.E.M. from eight frogs and are expressed as a percentage of the value obtained using 10^{-5} M cortisol (16 frogs).

be explained by species or tissue variations since such variations have been shown to be minimal (Rousseau and Baxter, 1979). Table II also indicates that the melanophore activity does not correlate with binding to the mineralocorticoid receptor in frog skin.

Receptor binding is not the only criterion on which predictions of activity can be based, as partial agonists and antagonists may also bind with high affinity. However, a similar lack of correlation is seen with glucocorticoid activity [induction of enzymes in rat hepatoma cells (Rousseau, 1975)]. Here, the relative potency is dexamethasone > corticosterone >

Table II. Comparison of Relative Melanophore Activity of Steroids with Relative Affinity for Gluco- and Mineralocorticoid Receptors and with Relative Lipid Solubility[a]

Steroid[b]	n	Melanophore activity	Receptor binding		Lipid solubility
			Glucocorticoid	Mineralocorticoid	
Cortisone	8	145 ± 51	1	—	11
Cortisol	—	100	53	24	9
DOC	9	64 ± 16	62	45	100
Prednisolone	11	63 ± 9	45	—	7
Epicortisol	12	52 ± 9	0	—	9
6α-Methylprednisolone	12	50 ± 19	360	—	9
Corticosterone	8	49 ± 11	80	—	24
17α-Hydroxyprogesterone	12	48 ± 12	3	—	10
Aldosterone	8	44 ± 10	14	100	15
Dexamethasone	8	43 ± 9	100	0	7
5β-Dihydro-DOC	9	34 ± 14	1	—	33
Progesterone	9	16 ± 4	11	29	100
Tetrahydrocorticosterone	10	6 ± 3	0	—	30
DMB	8	2 ± 2	3	—	49
Testosterone	9	0	1	2	69
Estradiol	9	0	1	0	87
Alphaxalone	6	0	0	—	33

[a] The reversal of MSH-induced darkening of frog skin by the steroids listed was tested at 10^{-5} and 10^{-7} M steroid in the number of frogs indicated (n). The melanophore activity given here is the mean ± S.E.M. of the reversal by each steroid at 10^{-5} M expressed as a percentage of the reversal caused by 10^{-5} M cortisol in the same frog. The results at 10^{-7} M (not shown here) were comparable but less reliable. The lipid solubility was calculated from the formula of each steroid by the method described by Pickup and Beckett (1977) and expressed as a percentage of the highest value obtained. The binding to the glucocorticoid receptor is the K_A (equilibrium association constant) of each steroid determined in rat hepatoma cell cytosol (Rousseau and Schmit, 1977; personal observations) and expressed as a percentage of the value obtained for dexamethasone. Binding to the mineralocorticoid receptor in ventral frog skin (Moguilewsky and Raynaud, 1976) is expressed as a percentage of the highest value obtained.

[b] Steroid nomenclature: cortisone, 17,21-dihydroxy-4-pregnene-3,11,20-trione; cortisol, 11β,17,21-trihydroxy-4-pregnene-3,20-dione; DOC, 21-hydroxy-4-pregnene-3,20-dione; prednisolone, 11β,17,21-trihydroxy-1,4-pregnadiene-3,20-dione; epicortisol, 11α,17,21-trihydroxy-4-pregnene-3,20-dione; corticosterone, 11β,21-dihydroxy-4-pregnene-3,20-dione; dexamethasone, 9α-fluoro-16α-methyl-11β,17,21-trihydroxy-1,4-pregnadiene-3,20-dione; aldosterone, 11β,21-dihydroxy-18-al-4-pregnene-3,20-dione; 5β-dihydro-DOC, 21-hydroxy-5β-pregnane-3,20-dione; progesterone, 4-pregnene-3,20-dione; tetrahydrocorticosterone, 3α,11β,21-trihydroxy-5β-pregnane-20-one; DMB, 9α-fluro-16α-methyl-11β-hydroxy-1,4-pregnadiene-3-one-17β-benzylcarboxamide; testosterone, 17β-hydroxy-4-androsten-3-one; estradiol, 3,17β-dihydroxy-1,3-5(10)-estratriene; alphaxalone, 3α-hydroxy-5α-pregnane-11,20-dione.

cortisol, which is the reverse of what was seen in the melanophore system. Likewise, for mineralocorticoid activity [stimulation of sodium transport across frog skin (Yorio and Bentley, 1978)], the order is aldosterone > corticosterone > cortisol, again the reverse of the order for melanophore activity. Steroids have been shown to bind specifically to amphibian oocyte melanosomes (Coffman *et al.*, 1979) but once again, the relative binding affinity of various steroids does not correlate with their relative activity in our system. Finally, the cortisol effect does not result from a nonspecific lipophilic interaction as there is not correlation with relative lipid solubility (Table II). For example, the highly polar steroid cortisol and the nonpolar deoxycorticosterone were both among the most active steroids.

It seems clear, therefore, that a new mechanism is implicated in the action of cortisol. Moreover, the specificity for certain C_{21}-hydroxypregnane steroids may indicate that a receptor molecule of a type as yet not described is involved in the response to these steroids. One action of steroid hormones that appears to involve the cell surface and not the classical receptors is the stimulation of meiosis in *Xenopus* oocytes (Baulieu *et al.*, 1978). However, this system shows little steroid specificity and steroids that do not produce skin lightening such as testosterone and alphaxalone are active. The steroid structure–activity relationships in the melanophore system do not correlate with those described for steroid effects on the central nervous system. The fast feedback mechanism in the rat hypothalamus is more sensitive to dexamethasone than cortisol and deoxycorticosterone (Jones and Hillhouse, 1976). In studies on the behavioral effect of steroids, corticosterone, dexamethasone, progesterone, and pregnenolone were equally active (van Wimersma Greidanus, 1970).

3.4. Mechanism of Action of Cortisol

The speed of the response to cortisol suggested that it was unlikely that the mechanism of action involved an interaction at the genome followed by new RNA and protein synthesis. We therefore expected that neither actinomycin D nor cycloheximide (inhibitors of RNA and protein synthesis, respectively) would alter the response to cortisol. As shown in Table III, neither treatment with actinomycin D nor short (30 min) exposure to cycloheximide had any effect on the ability of cortisol to reverse darkening produced by MSH. This indicates that the effect of cortisol does not require concomitant RNA or protein synthesis. However, prolonged (80 min) exposure to cycloheximide diminished the cortisol response. Since cycloheximide blocks protein synthesis in intact cells within 1 min (Lin *et al.*, 1966), this suggests that the expression of the cortisol activity in melanophores may depend on adequate concentrations of a protein or proteins with relatively short half lives.

Table III. Effect of Actinomycin D and of Cycloheximide on Cortisol-Induced
Skin Lightening[a]

Treatment	Time (min)	No. frogs	% reversal of MSH Effect		t test
			Control	Treated	
Actinomycin	90	6	10.3 ± 2.1	12.5 ± 5.0	NS
Cycloheximide	30	8	20.1 ± 6.4	35.3 ± 10.4	NS
	80	19	23.3 ± 5.0	13.4 ± 3.0	$0.02 < p < 0.05$

[a] Pieces of skin were incubated with or without cycloheximide (10^{-4} M) or actinomycin D (4×10^{-6} M, 5 µg/ml) for the lengths of time shown and darkened with MSH (3×10^{-11} M) for 1 hr prior to addition of cortisol (10^{-5} M). The values given are the means $\pm$ S.E.M. of the percentage reversal of the MSH effect. Statistical analysis was carried out by Student's paired t test on control and treated pieces from the same frog. Neither cycloheximide nor actinomycin D had any effect on baseline skin reflectance or on the response to MSH.

MSH is believed to act by an initial stimulation of the adenylate cyclase system in the melanophores and the subsequent rise in cyclic AMP causes a dispersion of the melanosomes (Abe *et al.,* 1969). In contrast to what has been reported by Lerner (1959) for progesterone, none of the steroids that we have tested had any effect on skin color in the absence of MSH. However, it is unlikely that cortisol acts by direct competition with MSH for its receptor, since we have shown that this steroid will also reverse the darkening effect of isoproterenol (Edwards *et al.,* 1981). The latter acts via a β-adrenergic receptor that is distinct from the MSH receptor. Likewise, cortisol cannot be acting by interfering with the ability of MSH and isoproterenol to stimulate adenylate cyclase since it is also active in reversing the darkening effect of caffeine, which increases cellular cyclic AMP by inhibiting phosphodiesterase activity. However, this antagonism of cortisol toward caffeine does not rule out an inhibitory effect of the steroid on the cyclase itself. In fact, α-adrenergic agonists that presumably inhibit the cyclase do inhibit the darkening effect of caffeine (Lerner, 1959; Edwards and Rousseau, personal observations). We do not at present know whether cortisol changes cellular cyclic AMP concentrations. Nor do we know whether it acts via receptors for other lightening agents. This could occur either by direct agonist activity at such receptors or indirectly by altering local concentrations of endogenous lightening agents such as acetylcholine, noradrenaline, or serotonin. Evidence for the latter type of effect has been described in other systems (Graefe and Trendelenburg, 1974; Leeuwin *et al.,* 1978).

4. Summary

Cortisol rapidly reverses the darkening effect of MSH by causing a centripetal movement of melanosomes in the dermal melanophores. This effect is not dependent on new RNA and protein synthesis. It is specific to certain C_{21}-hydroxypregnane steroids and the structure required for activity does not correlate with lipid solubility or relative activity in any other known steroid receptor-mediated system. A novel type of action is proposed that may involve a specific receptor and that may be relevant to the known antagonistic effects of corticosteroids and neuropeptides in the brain.

ACKNOWLEDGMENTS. We thank the Queen Elizabeth Medical Foundation (Belgium) for support. G. G. Rousseau is Maître de Recherches of the FNRS (Belgium) and recipient of FRSM Grant 3.4514.75. The assistance of Ch. Rolin Jacquemyns is gratefully acknowledged. This research was supported in part by the Roussel-Uclaf Company (Romainville, France) and an FDS grant from Louvain University. We thank Organon (Oss, Holland) for synthetic α MSH, Schering (Berlin, FRG) for epicortisol, Upjohn (Puurs, Belgium) for 6α-methylprednisolone, Merck, Sharp and Dohme (Rahway, N.J.) for dexamethasone, Roussel-Uclaf (Romainville, France) for 5β-dihydro-DOC, Glaxo (Greenford, U.K.) for alphaxalone, and Dr. P. Formstecher (Lille, France) for DMB. We also thank Patricia Lahy for secretarial help.

REFERENCES

Abe, K., Robison, A., Liddle, G. W., Butcher, R. W., Nicholson, W. E., and Baird, C. E., 1969, Role of cyclic AMP in mediating the effects of MSH, norepinephrine and melatonin on frog skin color, *Endocrinology* **85:**674.

Baulieu, E.-E., Godeau, F., Schorderet, M., and Schorderet-Slatkine, S., 1978, Steroid-induced meiotic division in *Xenopus laevis* oocytes: Surface and calcium, *Nature (London)* **275:**593.

Ben Barak, Y., Gutnick, M. J., and Feldman, S., 1977, Iontophoretically applied corticosteroids do not affect the firing of hippocampal neurons, *Neuroendocrinology* **23:**248.

Bohus, S., 1978, Behavioural effects of pituitary–adrenal system hormones, in: *The Endocrine Function of the Human Adrenal Cortex* (V. H. T. James, M. Serio, G. Giusti, and L. Martini, eds.), pp. 105–114, Academic Press, New York.

Carter, R. J., Shuster, S., and Morley, J. S., 1979, Melanotropin potentiating factor is the C-terminal tetrapeptide of human β-lipotropin, *Nature (London)* **279:**74.

Chambers, W. F., Freedman, S. L., and Sawyer, C. H., 1963, The effect of adrenal steroids on evoked reticular responses, *Exp. Neurol.* **8:**458.

Coffman, G. K., Keem, K., and Smith, L. D., 1979, The progesterone receptor-like properties of *Xenopus laevis* oocyte melanosomes are probably due to eumelanin, *J. Exp. Zool.* **207:**375.

Conforti, N., and Feldman, S., 1975, Effect of cortisol on the excitability of limbic structures of the brain in freely moving rats, *J. Neurol. Sci.* **26:**29.

Crabbé, J., 1977, The mechanism of action of aldosterone, in: *Modern Pharmacology-Toxicology*, Vol. 8: *Receptors and Mechanisms of Action of Steroid Hormones* (J. R. Pasqualini, ed.), Part II, pp. 513–568, Dekker, New York.

de Wied, D., 1977, Pituitary adrenal system hormones and behaviour, *Acta (Copenhagen) Endocrinol.* **85**(Suppl. 214):9.

Edwards, P. M., Rolin Jacquemyns, C., and Rousseau, G. G., 1981, Melanosome aggregation by corticosteroids: Evidence for a novel type of steroid action, *J. Steroid Biochem,* **15**:17.

El-Etr, M., Shorderet-Slatkine, S., and Baulieu, E. E., 1979, Meiotic maturation in *Xenopus laevis* oocytes initiated by insulin, *Science* **205**:1397.

Gelehrter, T. D., 1979, Synergistic and antagonistic effects of glucocorticoids on insulin action, in: *Glucocorticoid Hormone Action* (J. D. Baxter and G. G. Rousseau, eds.), pp. 583–592, Springer-Verlag, Berlin.

Graefe, K. H., and Trendelenburg, U., 1974, The effect of hydrocortisone on the sensitivity of the isolated nictitating membrane to catecholamines, *Naunyn-Schmiedeberg's Arch. Pharmacol.* **286**:1.

Hadley, M. E., and Bagnara, J. T., 1969, Integrated nature of chromatophore responses in the *in vitro* frog skin bioassay, *Endocrinology* **84**:69.

Henkin, R. I., 1970, The effects of corticosteroids and ACTH on sensory systems, *Prog. Brain Res.* **32**:270.

Jones, M. J., and Hillhouse, E. W., 1976, Structure–activity relationships and the mode of action of corticotropin-releasing factor (corticoliberin), *J. Steroid Biochem.* **7**:1189.

Kaiser, N., and Edelman, I., 1977, Calcium dependence of glucocorticoid-induced lymphocytolysis, *Proc. Natl. Acad. Sci. USA* **74**:638.

La Hoste, G. J., Olson, G. A., Kastin, A. J., and Olson, R. D., 1980, Behavioral effects of MSH, *Neurosci. Behav. Rev.* **4**:9.

Leeuwin, R. S., Veldsema-Currie, R. D., and Wolters, E. C. M. J., 1978, The effects of cholinesterase inhibitors and corticosteroids on rat–muscle preparations treated with hemicholinium 3, *Eur. J. Pharmacol.* **50**:393.

Lerner, A. B., 1959, Mechanism of hormone action, *Nature (London)* **184**:674.

Lerner, A. B., and Takahashi, Y., 1956, Hormonal control of melanin pigmentation, *Recent Prog. Horm. Res.* **12**:303.

Leung, K., and Munck, A., 1975, Peripheral actions of glucocorticoids, *Annu. Rev. Physiol.* **37**:245.

Lin, S.-Y., Mosteller, R. D., and Hardesty, B., 1966, The mechanism of sodium fluoride and cycloheximide inhibition of hemoglobin biosynthesis in the cell-free reticulocyte system, *J. Mol. Biol.* **21**:51.

McEwen, B. S., de Kloet, R., and Wallach, G., 1976, Interactions *in vivo* and *in vitro* of corticoids and progesterone with cell nuclei and soluble macromolecules from rat brain regions and pituitary, *Brain Res.* **105**:129.

Malawista, S. E., 1965, On the action of colchicine: The melanocyte model, *J. Exp. Med.* **122**:361.

Malawista, S. E., 1971a, Cytochalasin B reversibly inhibits melanin granule movement in melanocytes, *Nature (London)* **234**:354.

Malawista, S. E., 1971b, The melanocyte model: Colchicine-like effects of other antimitotic agents, *J. Cell Biol.* **49**:848.

Moguilewsky, M., and Raynaud, J. P., 1976, Un modèle d'étude des antiminéralocorticoides: La peau de grenouille, *J. Pharmacol. (Paris)* **7**:211.

Novales, R. R., and Novales, B. J., 1979, Endorphins supersensitise frog skin melanophores to isoproterenol but subsensitise them to α-melanocyte-stimulating hormone, *Gen. Comp. Endocrinol.* **39**:481.

Pfaff, D. W., Silva, M. T. A., and Weiss, J. M., 1971, Telemetered recording of hormone effects on hippocampal neurons, *Science* **172**:394.

Phillips, M. I., and Dafny, N., 1971, Effect of cortisol on unit activity in freely moving rats, *Brain Res.* **25**:651.

Pickup, M. E., and Beckett, A. H., 1977, Steroid adsorption: Use of the substituent constant π', *J. Pharm. Pharmacol.* **29**:715.

Rawles, M. E., 1948, Origin of melanophores and their role in the development of colour patterns in vertebrates, *Physiol. Rev.* **28**:383.

Richards, C. D., and Hesketh, T. R., 1975, Implications for theories of anaesthesia of antagonism between anaesthetic and non-anaesthetic steroids, *Nature (London)* **256**:179.

Rousseau, G. G., 1975, Interaction of steroids with hepatoma cells: Molecular mechanisms of glucocorticoid hormone action, *J. Steroid Biochem.* **6**:75.

Rousseau, G. G., and Baxter, J. D., 1979, Glucocorticoid receptors, in: *Glucocorticoid Hormone Action* (J. D. Baxter and G. G. Rousseau, eds.), pp. 49–77, Springer-Verlag, Berlin.

Rousseau, G. G., and Schmit, J. P., 1977, Structure–activity relationships for glucocorticoids. I. Determination of receptor binding and biological activity, *J. Steroid Biochem.* **8**:911.

Sakellaris, P. C., 1972, Olfactory thresholds in normal and adrenalectomised rats, *Physiol. Behav.* **9**:495.

Samuels, H. H., Horwitz, Z. D., Stanley, F., Casanova, J., and Shapiro, L. E., 1977, Thyroid hormone controls glucocorticoid action in cultured GH_1 cells, *Nature (London)* **268**:254.

Schwyzer, R., and Eberle, A., 1977, On the molecular mechanism of α-MSH receptor interactions, *Front. Horm. Res.* **4**:18.

Shizume, K., Lerner, A. B., and Fitzpatrick, T. B., 1954, *In vitro* bioassay for the melanocyte stimulating hormone, *Endocrinology* **54**:553.

Steiner, F. A., Ruf, K., and Akert, K., 1969, Steroid-sensitive neurones in rat brain: Anatomical localization and responses to neurohumours and ACTH, *Brain Res.* **12**:74.

van Wimersma Greidanus, T. J. B., 1970, Effects of steroids on extinction of an avoidance response in rats: A structure–activity relationship study, *Prog. Brain Res.* **32**:185.

Woodbury, D. M., 1958, Relation between the adrenal cortex and the central nervous system, *Pharm. Rev.* **10**:275.

Wright, M. R., and Lerner, A. B., 1960, On the movement of pigment granules in frog melanocytes, *Endocrinology* **66**:599.

Yorio, T., and Bentley, P. J., 1978, Stimulation of the short-circuit current (sodium transport) across the skin of the frog by corticosteroids: Structure–activity relationships, *J. Endocrinol.* **79**:283.

18

Is VIP a Neuroregulator or a Hormone?

G. Rosselin, W. Rotsztejn, M. Laburthe, and P. M. Dubois

1. Introduction

The vasoactive intestinal peptide (VIP) was first isolated by Said and Mutt from hog duodenum (1972). VIP is a 28-residue peptide structurally and biologically related to secretin, glucagon, and to the peptide having N-terminal histidine and C-terminal isoleucine amide [PHI isolated by Tatemoto and Mutt (1980)] (Table I). It has a wide spectrum of biological action (Said and Mutt, 1970) including stimulation of excretion and endocrine secretion, glycogenolysis and lipolysis, relaxation of smooth muscle and stimulation of neuronal activity. It acts at different levels: on the gastrointestinal, the respiratory, the cardiovascular, the urogenital, the central and peripheral nervous system, on endocrine functions and metabolism (see review by Rosselin *et al.*, 1980).

As first described in liver and fat tissues (Bataille *et al.*, 1974; Desbuquois *et al.*, 1973; Desbuquois, 1974), VIP possesses receptors located at the plasma membranes of the target cells. Later, several tissues, mainly exocrine pancreas (Christophe *et al.*, 1976), gut epithelia (Amiranoff *et al.*, 1978; Broyart *et al.*, 1981; Dupont *et al.*, 1978, 1980b, 1981; Laburthe *et al.*, 1977; Prieto *et al.*, 1979), pituitary (Bataille *et al.*, 1979a,b; Rotsztejn *et al.*, 1980a), and brain (Robberecht, 1978; Taylor and Pert, 1979), were shown to exhibit VIP receptors. The importance of the characterization of

G. Rosselin, W. Rotsztejn, and M. Laburthe • Unité INSERM U. 55, Hôpital Saint-Antoine, 75571 Cedex 12, Paris, France. *P. M. Dubois* • Laboratoire d'Histologie-Embryologie, Faculté de Médecine Lyon-Suc, Oullins, France.

Table I.   Sequence of VIP and Its Natural Analogues[a]

VIP	HSDAVFT DNYTRLR KQMAVKKYLN S I L N⩲
PHI[b]	HADGVFT . . . . L I⩲
Secretin	HSDGTFT S E LSRLRDS ARLQRLLQGLV⩲
Glucagon	HSQGTFT SDYSKYLDS R RAQDFVQWL M N T

[a] The one-letter symbols are recommended by the IUPAC–IUB Commission on Biochemical Nomenclature (*Eur. J. Biochem.* **5**:151, 1961). The symbol ⩲ indicates that the C-terminal amino acid is in amide form.

[b] PHI, peptide having N-terminal histidine and C-terminal isoleucine amide (Tatemoto and Mutt, 1980).

those receptors in assessing the role of VIP as a neuroregulatory peptide will be developed below.

The difference in the definition of a peptide as an endocrine or a neuroregulatory substance depends on: (1) the source of the peptide: either endocrine or neuronal cells; (2) the nature of the transport to target cells: hormones are transported in the bloodstream and neurotransmitters are secreted by nerve endings directly at the vicinity of the target cells (Fig. 1).

VIP was initially considered to be a gastrointestinal hormone (Grossman, 1974): The initial immunocytochemical studies indicated that, like many gut hormones, VIP was present in mammalian and avian endocrine cells (Polak *et al.*, 1974) that are dispersed throughout the epithelial cells of the gastrointestinal tract. In one situation, i.e., the pathological state of the watery diarrhea syndrome, high level of VIP was found in blood due to the release of immunoreactive (Bloom *et al.*, 1973; Said and Faloona, 1975) and bioreactive (Laburthe *et al.*, 1980) VIP by secreting-tumors. Thereafter, strong evidence accumulated indicating that VIP was not a hormone but rather a neuroregulatory substance.

2.   Source of VIP

The origin of VIP is compatible with its function as a neuroregulatory substance. The presence of a VIP immunoreactivity was found in the brain of dog (Giachetti *et al.*, 1977; Said and Rosenberg, 1976) and various other species including rat, human, pig, mouse (Besson *et al.*, 1979a; Emson *et al.*, 1978; Fahrenkrug and Schaffalitzky de Muckadell, 1978; Fuxe *et al.*, 1977; Larsson *et al.*, 1976). VIP was localized by immunocytochemical techniques or by subcellular fractionation followed by radioimmunoassay in brain nerve endings, but also in neuronal cell bodies and fibers (Besson *et al.*, 1979a; Fuxe *et al.*, 1977; Larsson *et al.*, 1976). In the periphery, several studies have shown that the localization in neuronal structure accounts for the wide distribution of VIP in the body. Neuronal localization of VIP was first demonstrated in gut (Larsson *et al.,* 1976) (Fig. 2) and subsequently in the urogenital tract (Larsson *et al.,* 1977a,b), the tracheo-bronchial tract, and the nasal mucosa (Uddman *et al.,* 1978). For example, in rat less than 1% of the total VIP content of the gut was found in the epithelium that contains the other gut hormones; the majority of the peptide was confined to the deeper strata of the gut wall (Besson *et al.,* 1978).

Discrepancies of the results obtained by immuno-chemical studies on the localization of VIP in the gut probably occurred because VIP, like other peptide hormones, belongs to a family of peptides having different molecular forms. Consequently, the evidence of VIP immunoreactivity in a tissue depends upon the specificity of the antibody. The VIP present in mucosa underlying the epithelial cells of the gut is a true VIP whereas endocrine cells of gut or the brain might contain other immunoreactive VIP-like peptides (Dimaline and Dockray, 1979; Maletti *et al.,* 1980). For example, antibodies to the C-terminal part of VIP can cross-react with fragments such as VIP-(10–28), and even with GIP. In contrast, PHI, a native agonist of VIP, as discovered by receptor studies (Bataille *et al.,* 1978, 1980), has a very low affinity for such antibody, but could react to antibodies directed to the initial part of VIP (see Table I). Secretin can cross-react with antibodies directed to the C-terminal part or to the initial part of VIP. Several forms of immunoreactive VIP have been shown after passage through dextran gel or DEAE-cellulose (Dimaline and Dockray, 1979; Maletti *et al.,* 1980). The existence of a large biosynthetic precursor of VIP has not as

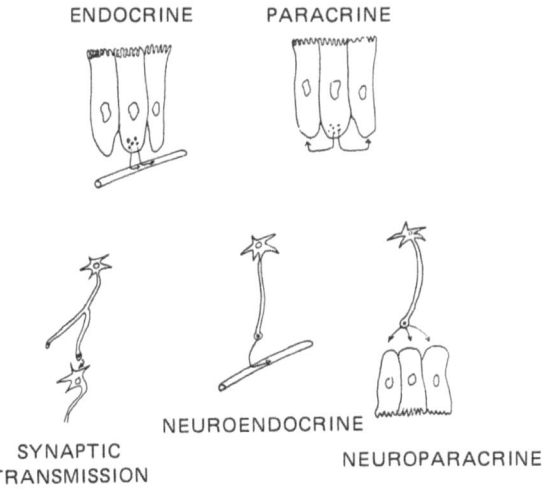

Figure 1. Scheme illustrating the difference between hormonal and neuroregulatory peptides: the regulatory peptide originated from endocrine cells (above) or neuronal cells (below). The nature of the transport to target cells is indicated together with the corresponding names.

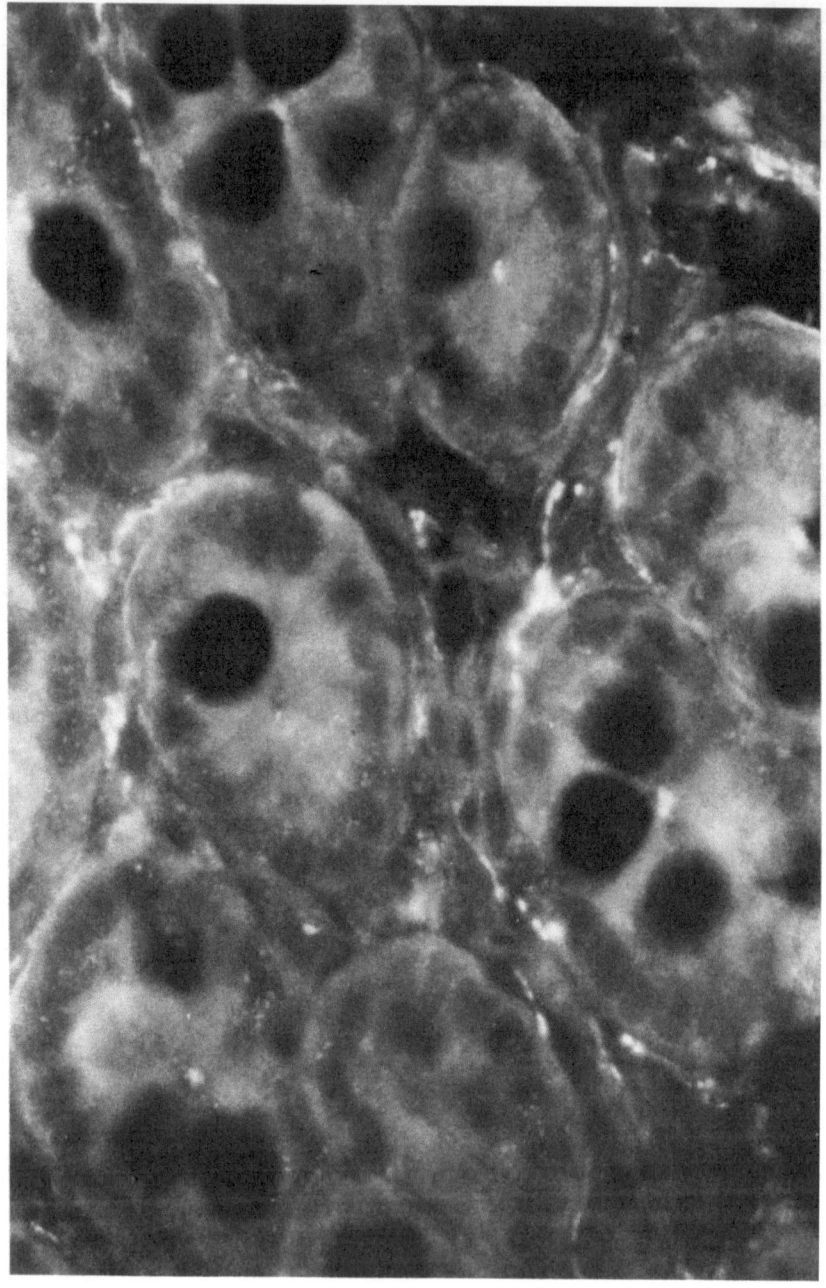

Figure 2. Transversal cut of rat villosity in intestinal part of the gut. Specific immunofluorescence of VIP. [P. M. Dubois.]

yet been demonstrated. All these substances that are VIP-like immuno-reactive substances are candidates for giving a VIP-like reaction in cells.

The conditions of VIP release are also compatible with those of a neuroregulatory substance. Various depolarizing agents acting on calcium or sodium channels such as potassium, veratridine, and batrachotoxin stimulated VIP release from brain synaptosomal fractions or slices (Giachetti *et al.*, 1977; Emson *et al.*, 1978; Besson *et al.*, 1981). VIP can be released from the vagus nerve after electrical stimulation (Fahrenkrug *et al.*, 1978a,b; Bitar *et al.*, 1980) or after stimulation of ileum by an electrical field (Hubel *et al.*, 1978). VIP, as other neurotransmitters, has been shown to be inactivated in the synaptic regions at the proximity of the receptor by enzymes which are partially specific (Keltz *et al.*, 1980). In addition, arguments related to the presence of specific receptor of VIP will be developed below in the text.

3. VIP as a Regulatory Peptide

The physiological role of VIP is still unknown. Administration to intact animals results in a large variety of biological effects and it is difficult to say what is directly due to the action of the peptide and what is a secondary effect. The general distribution by the bloodstream of pharmacological doses of the peptide gives no indication of the local effect of physiological doses released after specific and discrete stimulation of the nerves.

In vitro experiments on systems described in Sections 4.1 and 4.2 give supplementary arguments concerning the question of the physiological relevance of a regulatory peptide:

1. The peptide is implicated in a sequence of biochemical events including its release after a physiological stimulus, its specific binding to a receptor, and its effect on the target cells.

2. The kinetic and stoichiometric characteristics of each of the reactions implicated in this cascade must obey precise quantitative rules. For example, release must occur for concentrations of substances that are compatible with their physiological levels; receptors of the peptide must have affinity high enough to be compatible with physiological concentrations of the peptide at the vicinity of the cell. However, it remains more difficult to prove a physiological effect for a neuropeptide than for a classical hormone. For peptide hormones, the quantitative requirement implicated in the sequence stimulus–release is well known since the work of Yalow and Berson (1960) who gave the first system for assay of hormone in blood: most of the stimuli that specifically release hormones in blood are known and precise dose–response relationships have been established. Final biological effects observed can be correlated with a rise in the hormonal concentration

in blood. Receptors can be measured in target cells that are located at a distance from the peptide. Much less is known concerning the physiological effect of the neuroregulatory peptides (Guillemin, 1977) because they act close to the site of their release. The immediate variations in the concentration of a neuropeptide released during different states of neuronal excitation are difficult to appreciate since the neuropeptide is secreted, acts, and is destroyed at the same place in a short period of time. Moreover, the measurement of the specific receptors in the target cells can be blunted by the high concentrations of endogenous VIP (Laburthe and Dupont, 1981). Therefore: (1) special preparations are necessary to study the release; (2) the use of a target tissue that is not contaminated by the peptide is a prerequisite for the characterization of the receptor, as will be described below in two examples.

4.   Biological Effects and Receptors of VIP

4.1.   Evidence for a VIPergic Control in the Function of Gut Epithelial Cells

4.1.1.   Studies of the Biological Effect

Experiments carried out *in vivo* or in different preparations of organ indicated that VIP was biologically active in gut: intravenous injection of VIP evoked a net hydroelectrolytic secretion of water and sodium in the small intestine (Barbezat and Grossman, 1971) due to a reduction of the lumen-to-plasma movements and to a rise in the plasma-to-lumen movements (Coupar, 1976; Mailman, 1978). This effect was demonstrated not to be due to the vasodilation elicited by VIP (Eklund *et al.*, 1979): intraarterial perfusion of small quantities of VIP was able to alter the hydroelectrolytic transport in the small intestine without modifying the intestinal blood flow. The action of VIP was also observed in an *in vitro* model where a vascular component of VIP action was absent (Schwartz *et al.*, 1974).

However, these systems were not able to demonstrate the physiological action of VIP nor its mechanism for the following reasons. (1) Biological effect was obtained with considerable concentrations of VIP. Even in the most careful studies, the effective concentration of VIP when perfused through mesenteric arterial flow was about 4×10^{-9} M (Krejs *et al.*, 1978), i.e., at concentrations still higher than those found in the case of the VIP-secreting tumor (Said and Faloona, 1975). In normal men, VIP levels in peripheral blood, when detected, did not exceed 2–100 pg/ml, i.e., 0.6 to 30×10^{-12} M (Said, 1980; Fahrenkrug and Schaffalitzky de Muckadell,

1977; Mitchell and Bloom, 1977). (2) Comparison of these doses is an indirect argument suggesting that VIP is acting after release by nerve endings and not by a hormonal effect. Indeed, stimulation of the peripheral ends of the vagus nerve in the cat produces together with a biological effect a venous VIP release from a control value of 48 to a peak of 76×10^{-12} M in the stomach, 90 to 224×10^{-12} M in small intestine, 104 to 360×10^{-12} M in colon (Fahrenkrug *et al.*, 1978b). After stimulation of the vagus nerve in calves, the rise in the concentration of VIP was observed to be higher in the intestinal lymph than in blood (Edwards *et al.*, 1978), but not exceeding a difference of 40 fmoles/kg per min or 30×10^{-12} M. In the study by Eklund *et al.*, (1978) the arterial VIP concentration necessary for producing the same gastrointestinal effect as in the experiment of Fahrenkrug and colleagues reveals that the nervous release of VIP into the bloodstream was only 1/10,000 of the arterial VIP concentration needed to induce quantitatively similar physiological effect upon intraarterial VIP administration (Eklund *et al.*, 1979). As pointed out by Fahrenkrug and colleagues, these large differences in concentrations are likely to be due to the slow effect of diffusion of VIP between the synapse and the blood. The concentrations of VIP at the nerve endings as compared to that of VIP in blood may be partly inferred from those data: it appears to be higher than for noradrenaline. Since only about 10% of the release of noradrenaline reaches the blood as a transmitter overflow, the concentration of VIP at nerve endings is at least 10 times higher than its concentration in blood after vagal stimulation, i.e., higher than 0.25 to 3.6×10^{-9} M. (3) The effect of VIP was found to be similar to that of cyclic AMP or of agents (e.g., methylxanthine, cholera toxin, prostaglandins E) that alter the mucosal cyclic AMP system (Field, 1974). However, when tested, the cyclic AMP increases were small, not localized in a given cell population, and the effect of VIP was only observed at concentrations much higher than those quoted above. For example, in rat colon (Waldman *et al.*, 1977) the concentrations that were effective on intestinal secretion were lower by three orders of magnitude than those effective on cyclic AMP.

4.1.2. Evidence and Characterization of VIP Receptor and Effect in Epithelial Intestinal Cells

The isolation of pure intestinal cells and the demonstration of epithelial VIP receptors were a clue for understanding VIP physiology. Inasmuch as intestinal epithelial cells from rat were described as the first model exhibiting specific binding sites of VIP together with adenylate cyclase highly sensitive to VIP, the characteristics of the stimulation and of the coupling between VIP binding sites and adenylate cyclase have been extensively analyzed (see Fig. 3) (Laburthe *et al.*, 1977, 1979a; Amiranoff *et al.*, 1978,

1980a; Prieto *et al.*, 1979). The properties of the system are as follows: VIP was active at doses as low as 0.1 nM. The maximal stimulation (30 nM VIP) represented an 11-fold increase above basal cyclic AMP production. The quantitative relationship between binding to specific receptors and adenylate cyclase indicated a nonlinear coupling; i.e., when about 8% of the total binding sites were occupied, 50% of maximal cyclic AMP production was elicited. *In vivo,* the quantity of VIP available for the epithelial cells of a villosity can be roughly calculated: The number of intestinal epithelial cells, as measured in jejuno-ileum after isolation, amounted to about 2×10^8 (Prieto *et al.*, 1979), or 3×10^8 if the cryptic cells that remain nonisolated are taken into account. This feature is in the range of the 1.75×10^8 cells per small intestine found by others (Harrer *et al.*, 1964; Yousef and Kuksis, 1972). The amount of VIP in the wall of jejuno-ileum is about 10 μg (Besson *et al.*, 1978) with half in the smooth muscle and half in the subepithelial mucosa. Assuming that the basal and maximal values of VIP release are 1 and 10% of the storage of VIP in the nerves, respectively, the amount of VIP released at the vicinity of the 3×10^8 epithelial cells is in the range of 0.05–0.5 μg, i.e., 30,000 to 300,000 molecules of VIP per cell. If we assume 2800 epithelial cells per villosity (Meslin *et al.*, 1974) with a volume of VIP distribution in the conjunctivo-vascular axis of the villosity of 7×10^6 μ^3, the concentration of VIP released may be estimated to be between 1.2×10^{13} and 1.2×10^{14} molecules of VIP per milliliter. Given the K_d of the reaction (Laburthe *et al.*, 1979a,b,c), basal release of VIP would represent occupancy of about 20% of the high-affinity sites and maximal release of VIP would result in nearly total occupancy of the 140,000 high-affinity receptor sites per cell resulting in more than 50% of the maximum stimulation of cAMP.

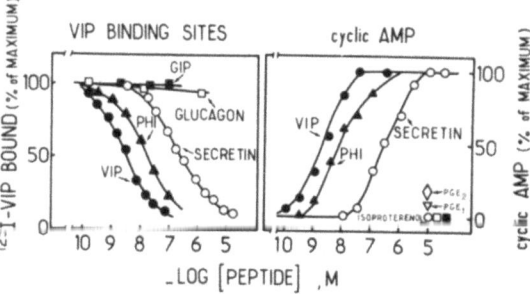

Figure 3. Evidence and stoichiometric characteristics of the effect of VIP in isolated intestinal cells of rat. (Left) Binding sites of VIP or receptor; (right) effect of different substances on cyclic AMP content in those cells. Conditions of the experiment were as previously detailed (Laburthe *et al.*, 1979a; Prieto *et al.*, 1979; Bataille *et al.*, 1980). PHI, peptide having N-terminal histidine and C-terminal isoleucine amide; GIP, gastroinhibitory peptide; PGE, prostaglandins.

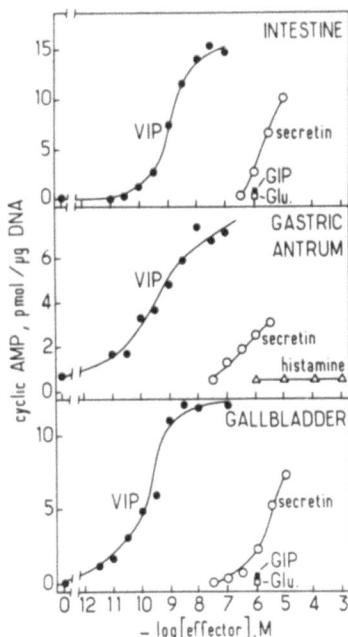

Figure 4. Effect of VIP on cyclic AMP content in differential epithelial cells of man. For experimental details see Dupont *et al.* (1980a,b, 1981).

The specificity of VIP–receptor interaction was demonstrated by testing peptides of the same family as VIP, namely glucagon, secretin, and PHI (Bataille *et al.*, 1980). In rat intestinal epithelium, binding of [^{125}I]-VIP was inhibited by VIP, PHI, and secretin in a parallel way, but PHI and secretin were respectively 10 and 100 times less potent than native VIP. The stimulation of adenylate cyclase system by VIP, PHI, and secretin were parallel, but PHI and secretin were respectively 10 and 100 times less potent than VIP. Other active substances that have been demonstrated to stimulate cyclic AMP in gut epithelium (e.g., prostaglandins E, isoproterenol) were also much less potent than VIP (see Fig. 3).

The monophasic effect of VIP on cyclic AMP production, its high efficiency and potency in regulating the adenylate cyclase of epithelial cells from rat jejuno-ileum, and its specificity of action were demonstrated throughout the intestinal tract (Laburthe *et al.*, 1979b). Other gut epithelia such as those from human colon (Dupont *et al.*, 1980b), stomach (Dupont *et al.*, 1980a), gallbladder (Dupont *et al.*, 1981) (see Fig. 4), hamster (Gaginella *et al.*, 1978) or guinea pig (Binder *et al.*, 1980) small intestine, exhibit a comparable VIP-sensitive cyclic AMP production system. The expression of VIP binding sites and of a VIP-sensitive cyclic AMP production system is not confined to the differentiated epithelial cells of the small-intestinal villi, since VIP receptors are also present in the undifferentiated epithelial cells in the rat small intestine (Laburthe *et al.*, 1979c).

4.1.3.   Mechanism of VIP Action (Fig. 5)

Further indication of the step involved in the VIP action on gut epithelium was given by studies indicating that the effect of VIP on cyclic AMP production in cells was due to the activation of adenylate cyclase. This has been demonstrated in intestinal cell membranes (Amiranoff *et al.*, 1978). It has also been proved that VIP receptor and adenylate cyclase are distinct entities, as demonstrated by transfer experiment (Laburthe *et al.*, 1979c). The nature of the coupling between the receptor and the adenylate cyclase has been elucidated following the study of Rodbell *et al.* (1971) on the glucagon-sensitive adenylate cyclase of liver. The requirement of guanyl nucleotides (GTP) for the activation by VIP of adenylate cyclase was first demonstrated using membranes obtained from isolated intestinal epithelial cells (Amiranoff *et al.*, 1980a). The effect of GTP in this system consists mainly in a potentiation of the adenylate cyclase response to VIP (Amiranoff *et al.*, 1980a). This effect is likely to be due to a GTP regulatory protein that is coupled with a catalytic unit of adenylate cyclase. It is associated with a change in the configuration of the VIP–receptor complex resulting in a decrease of the binding of VIP (Amiranoff *et al.*, 1980b).

INTESTINAL EPITHELIAL CELLS

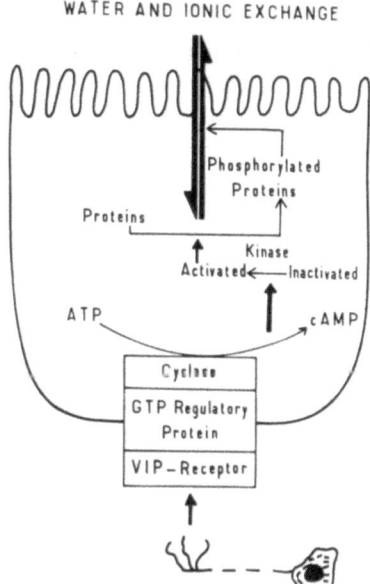

WATER AND IONIC EXCHANGE

Figure 5. Scheme of the action of VIP on intestinal epithelial cells.

The effect of VIP can also be modulated beyond the adenylate cyclase system in the cell; for example, the VIP-induced stimulation of cyclic AMP is more important and more sustained in gallbladder (Dupont *et al.*, 1981) than in cells from colon (Dupont *et al.*, 1980a). This fact is well correlated to a difference in the phosphodiesterase activity, which is much lower in gallbladder than in colonic cells, and suggests that the cyclic AMP-dependent phosphodiesterase is one of the modulating steps of VIP biological effect. It has also been demonstrated that VIP is effective in stimulating cyclic AMP-dependent protein kinases in gut epithelium (Laburthe *et al.*, 1979d). On the other hand, specific proteins phosphorylated by cyclic AMP, possibly related to transport function have been detected in cell membranes (Shlatz *et al.*, 1978).

4.1.4. Hypothesis on VIPergic Control

From these different data, the binding of VIP to specific receptors in intestinal epithelial cells initiates a sequence of events, finally resulting in biological effects that seem to occur in those cells dependent on cyclic AMP, such as transport of water, electrolytes (Strewler and Orloff, 1977), and amino acids (Kinzie *et al.*, 1973), synthesis of glycoproteins (Forstner *et al.*, 1973). However, it remains to be demonstrated which type of stimulation is involved in the VIP release by the nerve endings of the mucosa below the intestinal epithelial cells (Figs. 2 and 5). The only data are related to the chemical and electrical depolarization of vagal nerves (Fahrenkrug and Schaffalitzky de Muckadell, 1978; Fahrenkrug *et al.*, 1978a; Hubel *et al.*, 1978; Schaffalitzky de Muckadell *et al.*, 1977). The release of VIP can be elicited by high-threshold electrical stimulation of vagus fibers (Fahrenkrug and Schaffalitzky de Muckadell, 1978). The mechanism by which VIP is released from peripheral neurons seems to be non-cholinergic and nonadrenergic since atropine and adrenergic blocking agents do not influence the response (Fahrenkrug *et al.*, 1978). According to Bitar *et al.* (1980), VIP is released from intrinsic neurons of the gut under preganglionic cholinergic control. Much less is known about the conditions of stimulation of VIPergic fibers that are intrinsic to the gut wall and have been found very abundant in the conjunctivo-vascular axis (Larsson *et al.*, 1976). These fibers constitute the most precise network in the lamina propria (Schultzberg *et al.*, 1980). Many VIP nerves of the intestinal wall extend into the most superficial part of the mucosa and form a network around the epithelial crypts, extending into the villi, coming close to and even in contact with the intestinal epithelial cells (Dubois, unpublished observation) (Fig. 2). It is not known at present whether these plexuses are under preganglionic control, cholinergic or not, or if they are stimulated by modifications of different components in the interstitial medium

(Lundberg *et al.*, 1979). In any case, neuronal control of VIP release is well appropriate to the involvement of the different parts of the gut by successive stimulations. It also fits well with the fact that in gut, the function of the cells must obey some kind of segmental regulation. The presence of VIP receptors in different types of epithelial cells (crypts, villus cells) (Laburthe *et al.*, 1979b) indicates that the activity of cells of different types could be coordinated by VIP during the physiological work in the intestine. Indeed, by its action, VIP may coordinate both the secretion of intestinal cells and the modification of smooth muscle tonus (Kachelhoffer *et al.*, 1976), facilitating the propulsion of the alimentary bolus, the vasodilation, and resulting in an easier absorption of nutriments.

4.2. Evidence for a VIPergic Control of the Pituitary Release of Prolactin (Table II)

4.2.1. Biological Effect of VIP

VIP has been detected in hypothalamic nerve endings (Besson *et al.*, 1979a; Emson *et al.*, 1978; Giachetti *et al.*, 1977) and in hypophyseal portal blood (Said and Porter, 1979; Shimatsu *et al.*, 1981). This suggests a possible involvement of VIP in the regulation of pituitary secretion. However, as summarized in Table II, findings from different laboratories in some cases are conflicting: for example, only high doses of VIP cause prolactin release *in vivo* in ovariectomized rats (Vijayan *et al.*, 1979), whereas it is found more effective *in vivo* in male rats (Kato *et al.*, 1978). Preliminary experiments using different *in vitro* preparations result also in contradictory results: with no effect (Kato *et al.*, 1978) or small effect (Ruberg *et al.*, 1978) on prolactin release, only visible at high concentrations of the peptide.

The reasons why these *in vitro* systems did not demonstrate the possibility of a physiological action of VIP are likely to be the following: (1) inactivation of the VIP added by endogenous peptidase activity, which may be favored by the long duration of the incubation procedure, sometimes exceeding 4 hr (Kato *et al.*, 1978; Vijayan *et al.*, 1979). It was previously described that bacitracin prevents the degradation of VIP and glucagon by liver membranes (Desbuquois *et al.*, 1973; Bataille *et al.*, 1974). Thus, when bacitracin was added to a medium containing hemipituitaries together with albumin, VIP induced a better release of prolactin (Shaar *et al.*, 1979). (2) Furthermore, in heterogeneous preparations of cells a direct effect on the pituitary cannot be restricted to one particular cell type. (3) The quantity of endogenous VIP present in pituitary preparations could also be responsible for the apparent absence of VIP effect although to a lesser extent than in gut due to the low content of VIP in pituitary (Besson

Table II. Effect of VIP on Pituitary Secretion

Reference	Range of doses — In vivo (μG/100 g body wt)	Range of doses — In vitro (M)	VIP route of administration	Species	Final effect studied — Nature	Final effect studied — Magnitude S max/basal	Final effect studied — Sensitivity
Kato et al. (1978)	1–10		intravenous	male rat	PRL	10	1
	2–5		intraventricular		PRL	40	0.2
		10^{-7}	dispersed cells		PRL	0	NS[a]
		10^{-7c}	dispersed cells		PRL	5	NS
Ruberg et al. (1978)		10^{-7}	hemipituitary	male rat	PRL	1.2–1.8	NS
Vijayan et al. (1979)	0.004–1		intraventricular	ovariectomized rat	LH	1.7	0.004
					PRL	2	0.04
					GH	2	0.04
	0.04–1		intravenous		PRL		1
		3.1×10^{-9}–3×10^{-6}	hemipituitary		PRL, GH, LH	0	
Bataille et al. (1979b)		10^{-11}–10^{-9}	plasma membranes	human	binding		10^{-10} M
		10^{-8}			cyclase	5	1.2×10^{-10} M
Gourdji et al. (1979)			GH$_3$/B$_6$	cultures	PRL	3	
Robberecht et al. (1979)			plasma membranes	rat	cyclase		3×10^{-9} M
Shaar et al. (1979)		10^{-8}–10^{-5}	hemipituitary	rat	PRL	4	10^{-8} M
					cAMP	2.4	
Rotsztejn et al. (1980b)			purified cells	male rat	PRL[b]	5	2×10^{-9} M
					binding		2×10^{-9} M

[a] NS, not shown.
[b] No effect on LH or GH—FSH or TSH. LH, FSH, GH, ACTH = 0.
[c] + dopamine (VIP suppresses dopamine inhibitory effect on prolactin release).

et al., 1979a; Samson *et al.*, 1979). Consequently, the demonstration of the possibility of VIP action on prolactin at physiological doses required homogeneous cellular preparations devoid of damaging effect and still containing functional receptors for the peptide.

4.2.2. Evidence and Characterization of VIP Receptors in Pituitary

The conditions required to evidence the receptors of VIP were fulfilled in two cases: Using a clone of rat prolactin-secreting cells in culture, VIP was shown to increase prolactin release together with cyclic AMP accumulation. The concentrations that were found to be efficient ranged between 10^{-10} and 10^{-8} M with an ED_{50} of 3.8×10^{-10} M (Bataille *et al.*, 1979a; Gourdji *et al.*, 1979). This remarkable potency of VIP on prolactin release was not due to the fact that it was studied in transformed cells, since VIP was also shown to be active in normal prolactin-secreting cells (Rotsztejn *et al.*, 1980b): Different cell populations from normal male rat pituitary were separated by fractionation (Hymer *et al.*, 1973). VIP was effective at concentrations ranging from 10^{-10} to 10^{-7} M (Rotsztejn *et al.*, 1980b) only in the fractions containing purified lactotrophs.

From these two studies, the steps of VIP effect on prolactin-secreting cells are likely the following. Prolactin-secreting cells contain specific binding sites for VIP (Bataille *et al.*, 1979b; Rotsztejn *et al.*, 1980a). VIP activates the adenylate cyclase of prolactin-secreting cells (Bataille *et al.*, 1979a; Gourdji *et al.*, 1979; Rotsztejn *et al.*, 1980b, 1981b) and induces prolactin secretion both in normal (Rotsztejn *et al.*, 1980b) and in transformed prolactin cells (Gourdji *et al.*, 1979). The stoichiometry of these reactions is compatible with physiological action as attested by the quantity of VIP released in portal pituitary blood, the affinity of the receptor, the K_m of the adenylate cyclase and of prolactin secretion (Table II).

4.2.3. VIP in Humans

The effect of VIP on prolactin secretion has been observed not only in laboratory animals but in humans as well. Indeed, in humans, specific receptors are also present in plasma membranes isolated from prolactin adenoma. At doses of VIP as low as 10^{-10} M, 50% of the labeled VIP is displaced (Bataille *et al.*, 1979b). Those receptors are coupled to an adenylate cyclase activity. The half-maximal inhibition of the binding and of the stimulation of the cyclase occurred at 10^{-10} and 3.5×10^{-10} M, respectively (Bataille *et al.*, 1979b). These results were supported by those found in three different human prolactinomas (Nicosia *et al.*, 1980). However, in this study, the half-maximal stimulation of adenylate cyclase occurred only at 4.5×10^{-7} M VIP (Nicosia *et al.*, 1980).

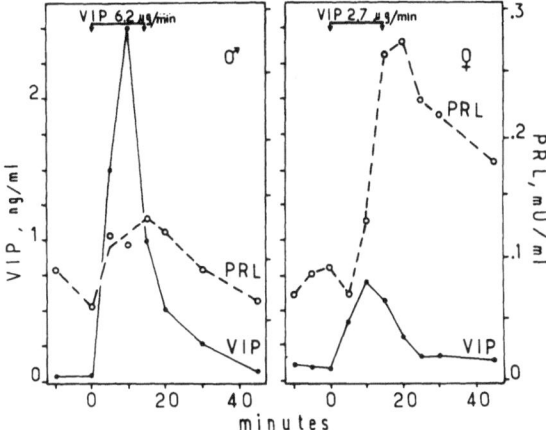

Figure 6. Effect of VIP on prolactin levels in man. Perfusions of VIP were given at the concentrations and for the times indicated above. [From Bataille *et al.*, 1981.]

Investigations carried out on human subjects indicate also that VIP is likely to be a releasing factor for prolactin *in vivo* in humans (Bataille *et al.*, 1981). After perfusion of VIP at doses in the microgram range, a stimulation of prolactin release is observed in blood. Figure 6 shows the kinetics of prolactin release as compared to the increase of VIP during the perfusion. The rise of prolactin above the basal level is significant after 5 min.

4.2.4. Regulation of Prolactin Release Induced by VIP

The presence of immunoreactive VIP in the anterior pituitary (Besson *et al.*, 1979a) has recently been substantiated by cytoimmunological techniques (Morel *et al.*, 1981).As controlled by ultrastructural studies, the VIP immunoreactivity is only found in prolactin cells (Morel *et al.*, 1981).

As the concentration of VIP in the pituitary depends on the rate of VIP release in the portal pituitary blood and on its presence on specific receptors in the pituitary gland, it has to be determined which neural structure is responsible for this release of VIP. Systematic study of the VIP content of hypothalamus indicates that it decreases from the anterior to the posterior part (Besson *et al.*, 1979a), the maximum occurring in the suprachiasmatic nucleus (Loren *et al.*, 1979; Samson *et al.*, 1979; Rotsztejn *et al.*, 1981b). Selective transections of neuronal fibers reaching the hypothalamus have shown the following: complete or posterior deafferentations of the mediobasal hypothalamus result in a 40% decrease of the VIP content in the posterior part of the mediobasal hypothalamus (Besson *et al.*,

1979). Transection of the stria terminalis between the amygdala and the suprachiasmatic nucleus results in a 60% decrease of VIP in the latter (Palkovits *et al.*, 1981; Roberts *et al.*, 1980). From these experiments, it appears that (1) VIP is present in nerve endings originating from neuronal cell bodies located outside the hypothalamus as well as from neuronal cells present in the mediobasal hypothalamus; (2) the amygdaloid complex may partly control hypothalamic VIP content via its direct connection with the suprachiasmatic nucleus and perhaps also with other parts of the hypothalamus. It remains to understand the significance of high VIP levels in the part of the hypothalamus that is not in direct anatomical connection with the portal system. The suprachiasmatic nucleus may control VIP release in the portal system either directly in the anterior part of the hypothalamus, for example through tanycytes (Scott and Paull, 1979), or by its connection with the posterior part of the mediobasal hypothalamus. There may also be species differences, since in humans the highest VIP content is found in median eminence, the opposite of what is found in the rat (Samson *et al.*, 1978). However, other arguments propose that, in the rat, the anterior hypothalamus is of importance in control in VIP release: a direct regulation of VIP content in the anterior hypothalamus has recently been found (Maletti *et al.*, 1981); VIP concentrations in the anterior hypothalamus are decreased by more than 50% in rats with hyperprolactinemia induced by estradiol and by pituitary graft beneath the kidney capsule. This is well due to the hyperprolactinemia and not to a change in the prolactin-secreting cells, since in pregnant rats, which exhibit large development of prolactin-secreting cells but without hyperprolactinemia, there is no change in the VIP content of the anterior hypothalamic area.

4.2.5. Interaction of VIP with Other Agents Regulating Prolactin Secretion (Table III)

Much data suggest that thyrotropin-releasing factor (TRH) is able to stimulate prolactin release both *in vivo* and *in vitro* (Tashjian *et al.*, 1971; Vale *et al.*, 1977). The effects of VIP and TRH as tested on GH_3 cells or on rat hemipituitaries are additive (Gourdji *et al.*, 1979; Enjalbert *et al.*, 1980), suggesting that receptors of VIP and TRH are different. Furthermore, VIP is much more effective in stimulating cyclic AMP than TRH, in agreement with other works suggesting that TRH acts on prolactin release through a non-cyclic AMP calcium-dependent system (Gershengorn *et al.*, 1980).

The interaction of VIP, dopamine, and opiates is at present less clear. A precursor of dopamine, L-Dopa, completely inhibits the VIP stimulation of prolactin release *in vivo* (Kato *et al.*, 1978). *In vitro*, the inhibitory effect of dopamine on prolactin can be suppressed by VIP (Kato *et al.*, 1978;

Table III.   Inhibitors and Stimulators of
Prolactin Secretion

Inhibitors	Stimulators
Dopamine	Serotonin
Acetylcholine	Histamine
Diketopiperazine	TRH
GABA	Substance P
	Neurotensin
	Opioid peptides
	VIP

Enjalbert *et al.*, 1980). The effect of naloxone, an opiate receptor antagonist, is different *in vivo* and *in vitro*. *In vivo,* naloxone treatment significantly blunted the plasma prolactin response to VIP (Kato *et al.*, 1978),
whereas *in vitro,* naxolone has no effect on VIP-induced secretion of prolactin (Enjalbert *et al.*, 1980).

In vivo, prolactin release is tonically inhibited by the release of tuberoinfundibular dopamine in the pituitary portal system (McLeod, 1976).
Enkephalins and endogenous opioid peptides are potent stimulators of prolactin secretion in the rat (Meites *et al.*, 1979); this effect is related, in part,
to a decrease in tuberoinfundibular dopamine release (Ferland *et al.*,
1977). Naloxone *in vivo* suppressed the effect of opiates by increasing the
release of dopamine with a final decrease of prolactin secretion. Therefore,
the dependence of prolactin production on opiates occurs at the hypothalamic level, via tuberoinfundibular dopamine, in agreement with the observations of Kato *et al.* (1978). However, opioid receptors that are present in
prolactin-secreting cells have no effect on the VIP-induced prolactin secretion, since they act only on the dopaminergic system receptors (Enjalbert
et al., 1979) and not on the specific VIP receptors. Thus, prolactin secretion
is dependent on a balance between VIP and TRH stimulation on one hand
and dopamine inhibition on the other hand, those three factors acting
directly on the cell in different ways. The importance of their relationship
in physiology is not well established since there have been no studies regarding the dose–effect of VIP at physiological concentrations of dopamine.
Finally, no data have been reported as yet on physiological factors that are
able to modulate VIP release from the hypothalamus.

4.2.6.   Interaction of VIP and Steroids

4.2.6.1.   On the Pituitary.   The effect of estradiol on prolactin secretion
is well known (see review by Brunet *et al.*, 1980). However, previous studies
both in humans and animals have shown that glucocorticoids can also modify the release of prolactin (Harms *et al.*, 1975; Sowers *et al.*, 1977). In

this respect, it was recently demonstrated that on enriched prolactin cells dexamethasone, which has no effect by itself, inhibits prolactin secretion induced by VIP. This effect is observed with a concentration of dexamethasone of 10^{-9} M (Rotsztejn et al., 1980c).

4.2.6.2. On the Brain. The effect of steroid hormones was also observed on the VIP-containing structures of the central nervous system. Whereas castration of male rats has no effect on VIP concentrations of various brain and peripheral structures, adrenalectomy was shown to produce a decrease of VIP concentrations in the hippocampus, together with an increase of VIP concentrations in the adenohypophysis, two structures known to contain glucocorticoid receptors (Rotsztejn et al., 1980d). VIP concentrations return to normal when corticosterone or dexamethasone is given to the adrenalectomized rats (Rotsztejn et al., 1980d). This observation together with the fact that VIP can depolarize neurons in the area of the hippocampus (Dodd et al., 1979) where most of the corticosterone receptors are located (Rotsztejn et al., 1980d) suggests the existence of a possible relationship between VIP and corticosteroids in the neural activity of the hippocampus.

4.2.7. Effect of VIP on Other Pituitary Secretions

The effect of VIP on several pituitary secretions observed in vivo (Vijayan et al., 1979) was not found on in vitro preparations. In fact, the different cell populations from normal rat pituitaries producing ACTH, GH, LH, and FSH were isolated and not stimulated by VIP (Rotsztejn et al., 1980b); the discrepancy between the results obtained in vivo and in vitro can be explained by an indirect action of VIP. Indeed, it was shown that VIP could in part modulate GH secretion by acting on hypothalamic somatostatin release. Addition of VIP to the incubation medium of rat mediobasal hypothalamic slices results in a dose-dependent inhibition of SRIF effective at doses ranging from 3×10^{-11} to 10^{-9} M (Epelbaum et al., 1979). This inhibitory effect of VIP is restricted to the mediobasal hypothalamus. Secretin has a similar effect but with 600-fold higher concentrations. However, the stimulatory effect of VIP on GH observed in vivo (Vijayan et al., 1979) could also be explained by a modulatory effect of VIP which counteracts the inhibitory action of SRIF on GH directly on the pituitary (Tapia-Arancibia et al., 1980). In contrast, VIP does not act on gonadotropic hormone secretion through a hypothalamic or pituitary site of action, since VIP does not induce any LH-RH release (Besson et al., 1979b; Drouva et al., 1981) and has no effect on the pituitary (Rotsztejn et al., 1980b). The mechanism by which VIP might increase plasma LH in vivo (Vijayan et al., 1979) remains unknown.

Finally, VIP effect seems to be restricted to the adenohypophysis since

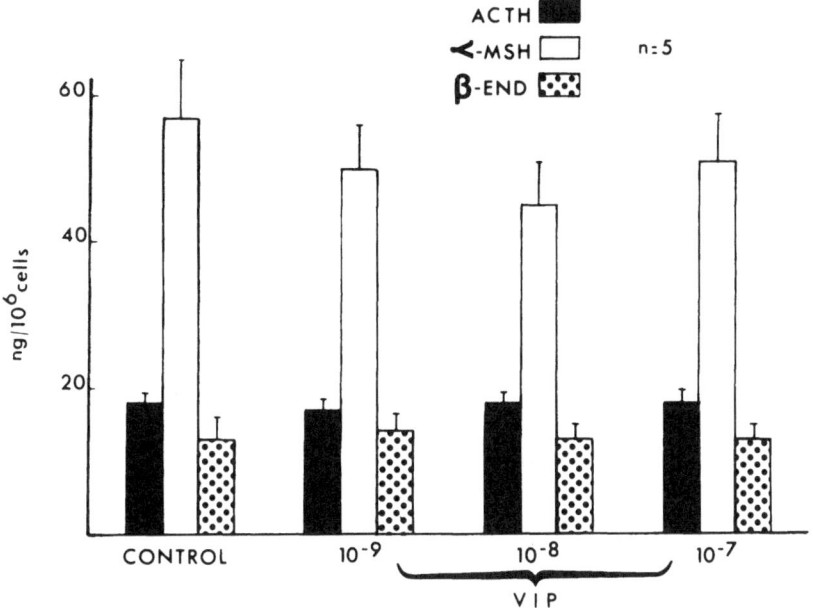

Figure 7. Effect of VIP on dispersed cells from rat intermediate lobe. Experiments were performed using the method of Briaud *et al.* (1978).

it has no effect on the release of ACTH, β-endorphin, and α-MSH from the intermediate lobe (Rotsztejn *et al.*, 1981a) (Fig. 7).

5. Conclusion: VIP as Neuroregulator

Arguments indicating that VIP functions as a neuroregulatory substance in both central nervous system and periphery have been given with two examples, in which the kinetic and stoichiometric characteristics of each reaction involved by VIP effect are compatible with physiology.

The place of VIP among the other regulatory peptides found in gut and brain is indicated in Table IV. It appears that VIP, glucagon, and secretin—structurally related molecules—differ in their mechanism of regulation. Glucagon and secretin act as hormones since they are released by endocrine glands, and VIP acts as a neuroregulator. The current view on the families of peptides is that each peptide has been differentiated through the duplication of an ancestral gene. Thereafter, daughter genes followed independent evolution. The occurrence of either a neuronal or an endocrine cell release of peptides belonging to the same family indicates a link between both types of cells. With regard to this possible evolution, transitional state between endocrine and neuronal cells can be observed taking as example the somatostatin-secreting cells (Larsson *et al.*, 1979): somatostatin endocrine cells can develop protoplasmic prolongation that reaches target cells in their vicinity. This protoplasmic prolongation could be a primitive expression of the axonal structure.

Development in the target cells of the expression of specific receptors is also a clue for the final expression of the regulatory control. For example, as indicated by receptor studies, the regulation of the function of gastric antral glands varies according to species: there is a considerable difference in the effect of VIP and secretin on the production of cyclic AMP in gastric antral glands of rat and man (Fig. 8). In man the effect of VIP is prevalent, secretin acting only through VIP receptors. It suggests that in man, the function of antral gastrin cells, mainly mucous and pepsin cells, is regulated by a neuromodulator. In contrast, in rat, the effect of secretin is prevalent, indicating a different type of regulation, i.e., a regulation of a hormonal type. The appearance of a dominant VIP receptor in human antral glands indicates the increasing role of a predominant neuroregulatory control of the gut functions in this species.

In any case, the conditions by which a neuroregulatory control appears instead of a hormonal control are still largely unknown. In gut, the receptor of VIP is present in different types of cells, the functions of which have to be coordinated during the digestion, in a segmentary way. To what extent the previous presence of receptors in a cell type is responsible for the establishment of specific cell-to-cell interaction remains to be elucidated. Indeed, several arguments deriving from the study of CNS neurons in culture suggest that the presence of postsynaptic receptors could occur before the development of the nervous connection (Monroy 1980). Moreover, co-culture experiments using muscle and nerve cells indicate that when smooth muscle cells of gut are present, the development of nerve fibers becomes

Table IV. Origin of Different Regulatory Peptides Found in the Gastroenteropancreatic System

Endocrine	Intestinal nerves and brain	Endocrine, intestinal nerves and brain
Insulin[a]	Enkephalin	Cholecystokinin
Glucagon[a]	VIP	Gastrin
Enteroglucagon		Bombesin
GIP		Pancreatic polypeptide
Secretin[a]		Neurotensin
Motilin[a]		Substance P
		Somatostatin

[a] Also found in brain.

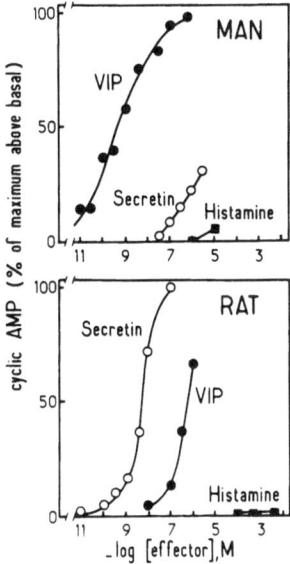

Figure 8. Species specificity of VIP and secretin effect on gastric antral glands.

predominant in the direction of the muscle cells, finally resulting in the creation of synaptic junctions (Burnstock *et al.,* 1981). The conditions that permit the same neuronal cells to be connected with different types of other neurons are also unknown. The presence of a specific receptor could represent a mechanism through which selective cell-to-cell connections might occur and therefore favor the development of neuron-to-neuron-specific junction.

ACKNOWLEDGMENTS. This work was supported by the Institut National de la Santé et de la Recherche Médicale (CRL 80 70 21) and the Fondation pour la Recherche Médicale Française.

DISCUSSION

KANAZIR: 1. If we accept that all effects of steroids are mediated via specific receptors, how do you explain your results that dexamethasone does not act on basal release of prolactin but suppresses the effect of VIP on prolactin release?

2. What is the time interval of dexamethasone action?

3. Does the VIP bind to the same sites as cholera toxin?

ROSSELIN: 1. The presence of specific receptors (in the case of steroids could be cytosolic, nuclear, or even membrane receptors) does not mean that the substance which acts on it directly stimulates or inhibits hormone release. It is well known that many substances can act as modulators. Such modulatory action has been defined as "any substance which has no effect by itself on the release but is able to change the capacity of the cell to respond to a given stimulus" (Rotsztejn, *Trends Neurosci.* **3**:67, 1980). Many examples can be given to illustrate this action. For instance, VIP has no effect by itself on the release of GH [though receptors for VIP have been described in the pituitary (see text)], but is able to counteract the inhibitory effect of somatostatin on GH secretion (Tapia-Arancibia *et al., Eur. J. Pharmacol.* **63**:235, 1980). An example is given by sexual steroids: estradiol, for instance, doesn't modify basal level of LH *in vitro* but interacts with LH release induced by LH-RH (Labrie *et al., Recent Prog. Horm. Res.* **34**:25, 1978).

2. Preliminary experiments show that the effect of dexamethasone is rapid and doesn't need a long-term incubation period.

3. Binding site of VIP is on plasma membrane. Cholera toxin is acting on the catalytic part of the adenylate cyclase, probably in inhibiting the GTPase. Recent work of Swedish laboratories indicated that cholera toxin could also act by increasing VIP release.

MAKAVA: Most of the neurons in the suprachiasmatic nucleus project to places other than the stalk–median eminence region of the hypothalamus.

ROSSELIN: One of the major problems for the putative involvement of VIP as a hypothalamic PRF concerns its hypothalamic localization. High amount of VIP was demonstrated in hypophyseal portal blood as compared to systemic blood by Said and Porter (*Life Sci.* **24**:227, 1979). A recent work of Shimatsu *et al.* (*Endocrinology* **108**:395, 1981) using the same preparation but devoid of pituitary gland, strongly suggests that portal VIP originates only from the hypothalamus.

It is true that VIP concentrations in the median eminence are low, especially in the rat, and the highest concentration is observed in the suprachiasmatic nucleus (NSC). Though the anatomical pathways linking the NSC to the mediobasal hypothalamus are not well known, we postulated some hypotheses (see above in the text) for the presence of high levels of VIP in the portal blood. Either: (1) VIP turnover in the median eminence is very rapid which could explain the low VIP immunoreactivity found in this structure; (2) VIP released from nerve terminals of the NSC reaches directly the portal blood or goes first to the CSF; (3) VIP pathways from the NSC to the posterior part of the mediobasal hypothalamus can also be considered (Besson *et al., Brain Res.* **165**:79, 1979).

REFERENCES

Amiranoff, B., Laburthe, M., Dupont, C., and Rosselin, G., 1978, Characterization of a vasoactive intestinal peptide-sensitive adenylate cyclase in rat intestinal epithelial cell membranes, *Biochim. Biophys. Acta* **544**:474.

Amiranoff, B., Laburthe, M., and Rosselin, G., 1980a, Potentiation by guanine nucleotides of the VIP-induced adenylate cyclase stimulation in intestinal epithelial cell membranes, *Life Sci.* **26**:1905.

Amiranoff, B., Laburthe, M., and Rosselin, G., 1980b, Characterization of specific binding sites for vasoactive intestinal peptide in rat intestinal epithelial cell membranes, *Biochim. Biophys. Acta* **627**:215.

Barbezat, G. O., and Grossman, M. I., 1971, Intestinal secretion: Stimulation by peptides, *Science* **174**:422.

Bataille, D., Freychet, P., and Rosselin, G., 1974, Interactions of glucagon, gut glucagon, vasoactive intestinal polypeptide and secretin with liver and fat cell plasma membranes: Binding to specific sites and stimulation of adenylate cyclase, *Endocrinology* **95**:713.

Bataille, D., Laburthe, M., Dupont, C., Tatemoto, K., Vauclin, N., and Rosselin, G., 1978,

VIP-like effects of a newly isolated intestinal peptide (PIHIA), 2nd International Symposium on Gastrointestinal Hormones, Oslo, 1978.

Bataille, D., Gourdji, D., Maletti, M., Vauclin, N., Grouselle, D., Tixier-Vidal, A., and Rosselin, G., 1979a, Vasoactive intestinal peptide (VIP): Concomitant stimulation of prolactin release and cyclic AMP production in a rat anterior pituitary cell line (GH3/B6): Comparison with thyroliberin (TRH), in *Hormone Receptors in Digestion and Nutrition* (G. Rosselin, P. Fromageot, and S. Bonfils, eds.), Vol. 1, p. 465, Elsevier/North-Holland, Amsterdam.

Bataille, D., Peillon, F., Besson, J., and Rosselin, G., 1979b, Vasoactive intestinal peptide (VIP): Récepteurs Spécifiques et activation de l'adénylate cyclase dans une tumeur hypophysaire humaine à prolactine, *C.R. Acad. Sci.* **288:**1315.

Bataille, D., Gespach, C., Laburthe, M., Amiranoff, B., Tatemoto, K., Vauclin, N., Mutt, V., and Rosselin, G., 1980, Porcine peptide having N-terminal histidine and C-terminal isoleucine amide (PHI): Vasoactive intestinal peptide (VIP) and secretin-like effects in different tissues from the rat, *FEBS Lett.* **114:**240.

Bataille, D., Talbot, J. N., Milhaud, G., Mutt, V., and Rosselin, G., 1981, Effet du peptide intestinal vasoactif (VIP) sur la sécrétion de prolactine chez l'homme, *C.R. Acad. Sci.* **292:**511.

Besson, J., Laburthe, M., Bataille, D., Dupont, C., and Rosselin, G., 1978, Vasoactive intestinal peptide (VIP): Tissue distribution in the rat as measured by radioimmunoassay and radioreceptorassay, *Acta Endocrinol. (Cophehagen)* **87:**799.

Besson, J., Rotsztejn, W., Laburthe, M., Epelbaum, J., Beaudet, A., Kordon, C., and Rosselin, G., 1979a, Vasoactive intestinal peptide (VIP): Brain distribution, subcellular localization and effect of deafferentation of the hypothalamus in male rats, *Brain Res.* **165:**79.

Besson, J., Rotsztejn, W., and Ruberg, M., 1979b, Involvement of VIP in neuroendocrine functions in the rat, in: *Hormone Receptors in Digestion and Nutrition* (G. Rosselin, P. Fromageot, and S. Bonfils, eds.) Vol. 1, p. 79, Elsevier/North-Holland, Amsterdam.

Besson, J., Rotsztejn, W., Lhiaubet, A. M., Poussin, B., and Rosselin, G., 1981, Potassium-veratridine and batrachotoxin-induced release of vasoactive intestinal peptide (VIP) from brain cortical and amygdala slices, International Symposium on Brain–Gut Axis, Florence, 1981.

Binder, H. J., Lemp, G. F., and Gardner, J. D., 1980, Receptors for vasoactive intestinal peptide and secretin on small intestinal epithelial cells, *Am. J. Physiol.* **238:**G190.

Bitar, K. N., Said, S. I., Weir, G. C., Saffouri, B., and Makhlouf, G. M., 1980, Neural release of vasoactive intestinal peptide from the gut, *Gastroenterology* **79:**1288.

Bloom, S. R., Polak, J. M., and Pearse, A. G. E., 1973, Vasoactive intestinal peptide and watery diarrhea syndrome, *Lancet* **2:**14.

Briaud, B., Koch, B., Lutz-Bucher, B., Milahe, C., 1978, *In vitro* regulation of ACTH release from neurointermediate lobe of rat hypophysis. I. Effect of crude hypothalamic extract. *Neuroendocrinology* **25:**47.

Broyart, J. P., Dupont, C., Laburthe, M., and Rosselin, G., 1981, Characterization of vasoactive intestinal peptide (VIP) receptors in human colonic epithelial cells, *J. Clin. Endocrinol. Metab.* **52:**715.

Brunet, N., Gourdji, D., and Tixier-Vidal, A., 1980, Effect of 17β-estradiol on thyroliberin responsiveness in GH3/B6 rat prolactin cells, *Mol. Cell. Endocrinol.* **18:**123.

Burnstock, G., 1981, Physiology of gastrointestinal nerves, in: Gut Hormones, S. R. Blomm and J. M. Polak eds., p. 482, Churchill Livingstone, London.

Christophe, J. P., Conlon, T. P., and Gardner, J. D., 1976, Interaction of porcine vasoactive intestinal peptide with dispersed pancreatic acinar cells from the guinea pig: Binding of radioiodinated peptide, *J. Biol. Chem.* **251:**4629.

Coupar, I. M., 1976, Stimulation of sodium and water secretion without inhibition of glucose absorption in the rat jejunum by vasoactive intestinal peptide (VIP), *Clin. Exp. Pharmacol. Physiol.* 3:615.

Desbuquois, B., 1974, The interaction of vasoactive intestinal polypeptide and secretin with liver cell membranes, *Eur. J. Biochem.* 46:439.

Desbuquois, B., Laudat, M. H., and Laudat, P., 1973, Vasoactive intestinal polypeptide and glucagon: Stimulation of adenylate cyclase activity via distinct receptors in liver and fat cell membranes, *Biochem. Biophys. Res. Commun.* 53:1187.

Dimaline, R., and Dockray, G. J., 1979, Molecular variants of vasoactive intestinal polypeptide in dog, rat and hog, *Life Sci.* 25:1893.

Dodd, J., Kelly, S., and Said, S. I., 1979, Excitation of CA1 neurones of the rat hippocampus by the octacosapeptide, VIP, *Br. J. Pharmacol.* 66:125P.

Drouva, S. V., Epelbaum, J., Tapia-Arancibia, L., Laplante, E., and Kordon, C., 1981, Opiate receptors modulate LHRH and SRIF release from mediobasal hypothalamic neurons, *Neuroendocrinology* 32:163.

Dupont, C., Amiranoff, B., Laburthe, M., and Rosselin, G., 1978, Récepteurs du peptide intestinal vasoactif (VIP) dans les membranes d'adénocarcinome colique humain: Liaison spécifique et stimulation de l'adénylate cyclase, *C.R. Acad. Sci.* 286:209.

Dupont, C., Gespach, C., Chenut, B., and Rosselin, G., 1980a, Regulation by vasoactive intestinal peptide of cyclic AMP accumulation in gastric epithelial glands: A characteristic of human stomach, *FEBS Lett.* 113:25.

Dupont, C., Laburthe, M., Broyart, J. P., Bataille, D., and Rosselin, G., 1980b, Cyclic AMP production in isolated colonic epithelial crypts: A highly sensitive model for the evaluation of vasoactive intestinal peptide action in human intestine, *Eur. J. Clin. Invest.* 10:67.

Dupont, C., Broyart, J. P., Broer, Y., Chenut, B., Laburthe, M., and Rosselin, G., 1981, Importance of the vasoactive intestinal peptide receptor in the stimulation of cyclic adenosine 3':5'-monophosphate in gallbladder epithelial cells of man: Comparison with the guinea pig, *J. Clin. Invest.* 67:742.

Edwards, A. V., Bircham, P. M. M., Mitchell, S. J., and Bloom, S. R., 1978, Changes in the concentration of vasoactive intestinal peptide in intestinal lymph in response to vagal stimulation in the calf, *Experientia* 34:1186.

Eklund, S., Jodal, M., Lundgren, O., and Sjoqvist, A., 1979, Effect of vasoactive intestinal polypeptide on blood flow, motility and fluid transport in the gastrointestinal tract of the cat, *Acta Physiol. Scand.* 105:461.

Emson, P. C., Fahrenkrug, J., Schaffalitzky de Muckadell, O. B., Jessel, T. M., and Iversen, L. L., 1978, Vasoactive intestinal peptide (VIP): Vesicular localization and potassium evoked release from rat hypothalamus, *Brain Res.* 143:174.

Enjalbert, A., Ruberg, M., Arancibia, S., Priam, M., and Kordon, C., 1979, Endogenous opiates block dopamine inhibition of prolactin secretion *in vitro, Nature (London)* 280:595.

Enjalbert, A., Arancibia, S., Ruberg, M., Priam, M., Bluet-Pajot, M. T., Rotsztejn, W., and Kordon, C., 1980, Stimulation of *in vitro* prolactin release by vasoactive intestinal peptide, *Neuroendocrinology* 31:200.

Epelbaum, J., Tapia-Arancibia, L., Besson, J., Rotsztejn, W., and Kordon, C., 1979, Vasoactive intestinal peptide inhibits release of somatostatin from hypothalamus *in vitro, Eur. J. Pharmacol.* 58:493.

Fahrenkrug, J., and Schaffalitzky de Muckadell, O. B., 1977, Radioimmunoassay of vasoactive intestinal polypeptide (VIP) in plasma, *J. Lab. Clin. Med.* 89:1379.

Fahrenkrug, J., and Schaffalitzky de Muckadell, O. B., 1978, Distribution of vasoactive intestinal polypeptide (VIP) in the porcine central nervous system, *J. Neurochem.* 31:1445.

Fahrenkrug, J., Galbo, H., Holst, J. J., and Schaffalitzky de Muckadell, O. B., 1978a, Influence of the autonomic nervous system on the release of VIP from the porcine gastrointestinal tract, *J. Physiol. (London)* **280:**405.

Fahrenkrug, J., Haglund, U., Jodal, M., Lundgren, O., Olbe, L., and Schaffalitzky de Muckadell, O. B., 1978b, Nervous release of vasoactive intestinal polypeptide in the gastrointestinal tract of cats: Possible physiological implications, *J. Physiol. (London)* **284:**291.

Ferland, L., Fuxe, K., Eneroth, P., Gustafsson, J. A., and Skett, P., 1977, Effects of methionine-enkephalin on prolactin and catecholamine levels and turnover in the median eminence, *Eur. J. Pharmacol.* **43:**89.

Field, M., 1974, Intestinal secretion, *Gastroenterology* **66:**1063.

Forstner, G., Shih, M., and Lukie, B., 1973, Cyclic AMP and intestinal glycoprotein synthesis: The effect of beta adrenergic agents, theophyllin and dibutyryl cyclic AMP, *Can. J. Physiol.* **51:**122.

Fuxe, K., Hökfelt, T., Said, S. I., and Mutt, V., 1977, Vasoactive intestinal polypeptide and the nervous system: Immunohistochemical evidence for localization in central and peripheral neurons, particularly intracortical neurons of the cerebral cortex, *Neurosci. Lett.* **5:**241.

Gaginella, T. S., Phillips, S. F., Dozois, R. R., and Go, V. L., 1978, Stimulation of adenylate cyclase in homogenates of isolated intestinal epithelial cells from hamsters, *Gastroenterology* **74:**11.

Gershengorn, M. C., Rebecchi, M. J., Geras, E., and Arevalo, C. D., 1980, TRH action in mouse thyrotropic tumor cells in culture: Evidence against a role for adenosine $3':5'$-monophosphate as a mediator of TRH-stimulated thyrotropin release, *Endocrinology* **107:**665.

Giachetti, A., Said, S. I., Reynolds, R. C., and Koniges, F. C., 1977, Vasoactive intestinal polypeptide in brain: Localization in and release from isolated nerve terminals, *Proc. Natl. Acad. Sci. USA* **74:**3424.

Gourdji, D., Bataille, D., Vauclin, N., Grouselle, D., Rosselin, G., and Tixier-Vidal, A., 1979, Vasoactive intestinal peptide (VIP) stimulates prolactin (PRL) release and cAMP production in a rat pituitary cell line (GH3/B6): Additive effects of VIP and TRH on PRL release, *FEBS Lett.* **104:**165.

Grossman, M. I., 1974, Candidate hormones of the gut, *Gastroenterology* **67:**730.

Guillemin, R., 1977, The expanding significance of hypothalamic peptides, or, is endocrinology a branch of neuroendocrinology?, *Recent Prog. Horm. Res.* **33:**1.

Harms, P. J., Langlier, P., and McCann, S. M., 1975, Modification of stress-induced prolactin release by dexamethasone or adrenalectomy, *Endocrinology* **96:**475.

Harrer, D. S., Stern, B. K., and Reilly, R. W., 1964, Removal and dissociation of epithelial cells from the rodent gastrointestinal tract, *Nature (London)* **203:**319.

Hubel, K. A., Gaginella, T. S., and O'Dorisio, T. M., 1978, Release of vasoactive intestinal peptide by electrical field stimulation of rabbit ileum, *Gastroenterology* **74:**1127.

Hymer, W. C., Evans, W. H., Kraicer, J., Mastro, A., Davis, J., and Griswold, E., 1973, Enrichment of cell types from the rat adenohypophysis by sedimentation at unit gravity, *Endocrinology* **92:**275.

Kachelhoffer, J., Mendel, C., Dauchel, J., Hohmatter, D., and Grenier, J. F., 1976, The effects of VIP on intestinal motility: Study on ex vivo perfused isolated canine jejunal loops, *Am. J. Dig. Dis.* **21:**957.

Kato, Y., Iwasaki, Y., Iwasaki, J., Abe, H., Yanaihara, N., and Imura, H., 1978, Prolactin release by vasoactive intestinal polypeptide in rats, *Endocrinology* **103:**554.

Keltz, T. N., Straus, E., and Yalow, R. S., 1980, Degradation of vasoactive intestinal polypeptide by tissue homogenates, *Biochem. Biophys. Res. Commun.* **92:**669.

Kinzie, J. L., Ferrendelli, J. A., and Alpers, D. H., 1973, Adenosine cyclic $3':5'$-monophosphate mediated transport of neutral and dibasic acids in jejunal mucosa, *J. Biol. Chem.* **248**:7018.

Krejs, G. J., Barkley, R. M., Read, N. W., and Fordtran, J. S., 1978, Intestinal secretion induced by vasoactive intestinal peptide: A comparison with cholera toxin in the canine jejunum *in vivo, J. Clin. Invest.* **61**:1337.

Laburthe, M., and Dupont, C., 1981, VIPergic control of intestinal epithelium in health and disease, in *Monograph on VIP* (S. I. Said, ed.), p. 407, Raven Press, New York.

Laburthe, M. Besson, J., Hui Bon Hoa, D., and Rosselin, G., 1977, Récepteurs du peptide intestinal vasoactiv (VIP) dans les entérocytes: Liaison spécifique et stimulation de l'AMP cyclique, *C.R. Acad. Sci.* **284**:2139.

Laburthe, M., Prieto, J. C., Amiranoff, B., Dupont, C., Hui Bon Hoa, D., and Rosselin, G., 1979a, Interaction of vasoactive intestinal peptide with isolated intestinal epithelial cells from rat. 2. Characterization and structural requirement of the stimulatory effect of vasoactive intestinal peptide on production of adenosine $3':5'$-monophosphate, *Eur. J. Biochem.* **96**:239.

Laburthe, M., Prieto, J. C., Amiranoff, B., Dupont, C., Broyart, J. C., Hui Bon Hoa, D., Broer, Y., and Rosselin, G., 1979b, VIP receptors in intestinal epithelial cells: Distribution throughout the intestinal tract, in: *Hormone Receptors in Digestion and Nutrition* (G. Rosselin, P. Fromageot and S. Bonfils, eds.) Vol. 1, p. 241, Elsevier/North-Holland, Amsterdam.

Laburthe, M., Rosselin, G., Rousset, M., Zweibaum, A., Korner, M., Selinger, Z., and Schramm, M., 1979c, Transfer of the hormone receptor for vasointestinal peptide to an adenylate cyclase system in another cell, *FEBS Lett.* **96**:237.

Laburthe, M., Mangeat, P., Marchis-Mouren, G., and Rosselin, G., 1979d, Activation of cyclic AMP-dependent protein kinase by vasoactive intestinal peptide (VIP) in isolated intestinal epithelial cells from rat, *Life Sci.* **25**:1931.

Laburthe, M. C., Dupont, C. M., Besson, J. D., Rousset, M., and Rosselin, G. E., 1980, A new bioassay of VIP: Results in watery diarrhea syndrome, *Gut* **21**:619.

Larsson, L. I., Fahrenkrug, J., Schaffalitzky de Muckadell, O. B., Sundler, F., Hakanson, R., and Rehfeld, J. F., 1976, Localization of vasoactive intestinal polypeptide (VIP) to central and peripheral neurons, *Proc. Natl. Acad. Sci. USA* **73**:3197.

Larsson, L. I., Fahrenkrug, J., and Schaffalitzky de Muckadell, O. B., 1977a, Occurrence of nerves containing vasoactive intestinal polypeptide immunoreactivity in the male genital tract, *Life Sci.* **21**:503.

Larsson, L. I., Fahrenkrug, J., and Schaffalitzky de Muckadell, O. B., 1977b, Vasoactive intestinal polypeptide occurs in nerves of the female genito-urinary tract, *Science* **197**:1374.

Larsson, L. I., Goltermann, N., De Magistris, L., Rehfeld, J. F., and Schwartz, T. W., 1979, Somatostatin cell process as pathways for paracrine secretion, *Science* **205**:1393.

Loren, I., Emson, P. C., Fahrenkrug, J., Bjoklund, A., Alumets, J., Hakanson, R., and Sundler, F., 1979, Distribution of vasoactive intestinal polypeptide in the rat and mouse brain, *Neuroscience* **4**:1953.

Lundberg, J. M., Hökfelt, T., Schutzberg, M., Uvnas-Wallenstein, K., Kohler, C., and Said, S. I., 1979, Occurrence of VIP-like immunoreactivity in certain cholinergic neurons of the cat: Evidence from combined immunohistochemistry and acetylcholinesterase staining, *Neuroscience* **4**:1539.

McLeod, R. M., 1976, Regulation of prolactin secretion, in: *Frontiers in Neuroendocrinology* (L. Martini and W. F. Ganong, eds.), Vol. 4, p. 169, Raven Press, New York.

Mailman, D., 1978, Effects of vasoactive intestinal polypeptide on intestinal absorption and blood flow, *J. Physiol. (London)* **279**:121.

Maletti, M., Besson, J., Bataille, D., Laburthe, M., and Rosselin, G., 1980, Ontogeny and immunoreactive forms of vasoactive intestinal peptide (VIP) in rat brain, *Acta Endocrinol. (Copenhagen)* **93**:479.

Maletti, M., Rotsztejn, W., and Rosselin, G., 1981, Differential effect of hyperprolactinemia and pregnancy on vasoactive intestinal peptide (VIP) concentration in various brain structures of the female rat, International Symposium on Brain–Gut Axis, Florence, 1981.

Meites, J., Bruni, J. F., Van Vugt, D. A., and Smith, A. F., 1979, Relation of endogenous opioid peptides and morphine to neuroendocrine functions, *Life Sci.* **24**:1325.

Meslin, J. C., Sacquet, E., and Guenet, J. L. 1974, Action d'une flore microbienne qui ne déconjugue pas les sels biliaires sur la morphologie et le renouvellement cellulaire de la muqueuse de l'intestin grêle du rat. *Ann. Biol. Anim. Biochim. Biophys.* **14**:709.

Mitchell, S. J., and Bloom, S. R., 1977, Measurement of fasting and postprandial VIP in man, *Gut* **19**:1043.

Monroy, A., 1980, Cell–cell recognition, *Medicine* **32**:103.

Morel, G., Besson, J., Rosselin, G., and Dubois, P. M., 1982, Ultrastructural evidence for endogenous immunoreactive vasoactive intestinal peptide (VIP)-like in the pituitary gland, *Neuroendocrinology,* **34**:85.

Nicosia, S., Spada, A., Borghi, C., Cortelazzi, L., and Giannattasio, G., 1980, Effect of vasoactive intestinal polypeptide (VIP) in human prolactin (PRL) secreting pituitary adenomas: Stimulation of PRL release and activation of adenylate cyclase, *FEBS Lett.* **112**:159.

Palkovits, M., Besson, J., and Rotsztejn, W., 1981, Distribution of vasoactive intestinal polypeptide in intact, stria terminals transected and cerebral cortex isolated rats, *Brain Res.* **213**:455.

Polak, J. M., Pearse, A. G. E., Garaud, J. C., and Bloom, S. R., 1974, Cellular localization of a vasoactive intestinal peptide in the mammalian and avian gastrointestinal tract, *Gut* **15**:720.

Prieto, J. C., Laburthe, M., and Rosselin, G., 1979, Interaction of vasoactive intestinal peptide with isolated intestinal epithelial cells from rat. 1. Characterization, quantitative aspects and structural requirements of binding sites, *Eur. J. Biochem.* **96**:229.

Robberecht, P., De Neef, P., Lammens, M., Deschodt-Lanckman, M., Christophe, J. P., Specific binding of vasoactive intestinal peptide to brain membranes from the guinea pig. *Eur. J. Biochem.* **90**:147.

Robberecht, P., Deschodt-Lanckman, M., Camus, J. C., De Neef, P., Lambert, M., and Christophe, J. P., 1979, VIP activation of rat anterior pituitary adenylate cyclase, *FEBS Lett.* **103**:229.

Roberts, G. W., Woodhams, P. L., Crow, T. J., and Polak, J. M., 1980, Loss of immunoreactive VIP in the bed nucleus following lesions of the stria terminalis, *Brain Res.* **195**:471.

Rodbell, M., Birnbaumer, L., Pohl, S., and Krans, M. J., 1971, The glucagon-sensitive adenylate cyclase system in plasma membranes of rat liver. V. An obligatory role of guanyl nucleotides in glucagon action, *J. Biol. Chem.* **246**:1877.

Rosselin, G., Laburthe, M., Bataille, D., Prieto, J. C., Dupont, C., Amiranoff, B., Broyart, J. P., and Besson, J., 1980, Receptors and effectors of vasoactive intestinal peptide in cell function and differentiation, in: *Hormones and Cell Regulation* (J. Dumont and J. Nunez, eds.), Vol. 4, p. 311, Elsevier/North-Holland, Amsterdam.

Rotsztejn, W. H., Benoit, L., Besson, J., Béraud, G., Bluet-Pajot, M. T., Kordon, C., Rosselin, G., and Duval, J., 1980a, Effect of vasoactive intestinal peptide (VIP) on the release of adenohypophyseal hormones from purified cells obtained by unit gravity sedimentation, *Regul. Pept. Suppl.* **1**:S94.

Rotsztejn, W. H., Benoit, L., Besson, J., Béraud, G., Bluet-Pajot, M. T., Kordon, C., Rosselin, G., and Duval, J., 1980b, Effect of vasoactive intestinal peptide (VIP) on the release of adenohypophyseal hormones from purified cells obtained by unit gravity sedimentation: Inhibition by dexamethasone of VIP-induced prolactin release, *Neuroendocrinology* **31**:282.

Rotsztejn, W., Benoit, L., Besson, J., and Duval, J., 1980c, Régulation par la dexaméthosane de la libération de prolactine induite par le peptide vasoactif intestinal (VIP), *C.R. Acad. Sci.* **290**:791.

Rotsztejn, W. H., Besson, J., Briaud, B., Gagnant, L., Rosselin, G., and Kordon, C., 1980d, Effect of steroids on vasoactive intestinal peptide in discrete brain regions and peripheral tissues, *Neuroendocrinology* **31**:287.

Rotsztejn, W. H., Besson, J., Briaud, B., Benoit, L., Béraud, G., Giraud, P., and Duval, J., 1981a, Implication des glucocorticoides dans la régulation des effets centraux et hypophysaires du peptide vasoactif intestinal (VIP), *J. Physiol. (Paris)*, in press.

Rotsztejn, W. H., Dussaillant, M., Nobou, F. and Rosselin, G. 1981b, Rapid glucocorticoid inhibition of vasoactive intestinal peptide-induced cyclic AMP accumulation and prolactin release in rat pituitary cells in culture. *Proc. Natl. Acad. Sci. USA* **78**:7584.

Ruberg, M., Rotsztejn, W., Arancibia, S., Besson, J., and Enjalbert, A., 1978, Stimulation of prolactin release by vasoactive intestinal peptide (VIP), *Eur. J. Pharmacol.* **51**:319.

Said, S. I., 1980, Vasoactive intestinal peptide (VIP): Isolation, distribution, biological actions, structure–function relationships and possible functions, in: *Gastrointestinal Hormones* (G. B. Jergy Glass, ed.), Vol. 1, p. 245, Raven Press, New York.

Said, S. I., and Faloona, G. R., 1975, Elevated plasma and tissue levels of vasoactive intestinal peptide in the watery diarrhea syndrome, *N. Engl. J. Med.* **293**:155.

Said, S. I., and Mutt, V., 1970, Polypeptide with broad biological activity: Isolation from small intestine, *Science* **169**:1217.

Said, S. I., and Mutt, V., 1972, Isolation from porcine intestinal wall of a vasoactive octacosapeptide related to secretin and to glucagon, *Eur. J. Biochem.* **28**:199.

Said, S. I., and Porter, J. C., 1979, Vasoactive intestinal polypeptide: Release into hypophyseal portal blood, *Life Sci.* **24**:227.

Said, S. I., and Rosenberg, R. N., 1976, Vasoactive intestinal polypeptide: Abundant immunoreactivity in neural cell lines and normal nervous tissue, *Science* **192**:907.

Samson, W. K., Said, S. I., Graham, J. W., and McCann, S. M., 1978, VIP concentrations in median eminence of hypothalamus, *Lancet* **2**:901.

Samson, W. K., Said, S. I., and McCann, S. M., 1979, Radioimmunologic localization of vasoactive intestinal polypeptide in hypothalamic and extrahypothalamic sites in the rat brain, *Neurosci. Lett.* **12**:265.

Schaffalitzky de Muckadell, O. B., Fahrenkrug, J., and Holst, J. J., 1977, Release of vasoactive intestinal polypeptide (VIP) by electric stimulation of the vagal nerves, *Gastroenterology* **72**:373.

Schultzberg, M., Hökfelt, T., Nilsson, G., Terenius, L., Rehfeld, J. F., Brown, M., Elde, R., Goldstein, M., and Said, S. I., 1980, Distribution of peptide- and catecholamine-containing neurons in the gastrointestinal tract of rat and guinea pig: Immunohistochemical studies with antisera to substance P, vasoactive intestinal polypeptide, enkephalin, somatostatin, gastrin/cholecystokinin, neurotensin and dopamine beta-hydroxylase, *Neuroscience* **5**:689.

Schwartz, C. J., Kimberg, D. V., Sheerin, H. E., Field, M., and Said, S. I., 1974, Vasoactive intestinal peptide stimulation of adenylate cyclase and active electrolyte secretion in intestinal mucosa, *J. Clin. Invest.* **54**:536.

Scott, D. E., and Paull, W. K., 1979, The tanycyte of the rat median eminence, *Cell Tissue Res.* **200**:329.

Shaar, C. J., Clemens, J. A., and Dininger, N. B., 1979, Effect of vasoactive intestinal polypeptide on prolactin release *in vitro, Life Sci.* **25**:2071.

Shimatsu, A., Kato, Y., Matsusha, N., Katakami, H., Yanaihara, N., and Imura, H., 1981, Immunoreactive VIP in rat hypophyseal portal blood, *Endocrinology* **108**:395.

Shlatz, L. J., Kimberg, D. V., and Cattieu, K. A., 1978, Cyclic nucleotide-dependent phosphorylation of rat intestinal microvillus and basal-lateral membrane proteins by an endogenous protein kinase, *Gastroenterology* **75**:838.

Sowers, J. R., Carlson, H. E., Brautbar, N., and Hershman, J. M., 1977, Effect of dexamethasone on prolactin and TSH responses to TRH and metoclopramide in man, *J. Clin. Endocrinol. Metab.* **44**:237.

Strewler, G. J., and Orloff, J., 1977, Role of cyclic nucleotides in the transport of water and electrolytes, in: *Advances in Cyclic Nucleotide Research* (M. Greengard and G. A. Robison, eds.), Vol. 8, p. 311, Raven Press, New York.

Tapia-Arancibia, L., Arancibia, S., Bluet-Pajot, M. T., Enjalbert, A., Epelbaum, J., Priam, M., and Kordon, C., 1980, Effect of VIP on somatostatin inhibition of pituitary growth hormone secretion *in vitro, Eur. J. Pharmacol.* **63**:235.

Tashjian, A. H., Osborne, R., Moina, D., and Knaian, A., 1977, Hydrocortisone increases the number of receptors for thyrotropin releasing hormone on pituitary cells in culture, *Biochem. Biophys. Res. Commun.* **79**:333.

Tatemoto, K., and Mutt, V., 1980, Isolation of two novel candidate hormones using a chemical method for finding naturally occurring polypeptides, *Nature (London)* **285**:417.

Taylor, D. P., and Pert, C. B., 1979, Vasoactive intestinal polypeptide: Specific binding to rat brain membranes, *Proc. Natl. Acad. Sci. USA* **76**:660.

Uddman, R., Alumets, J., Densert, O., Hakanson, R., and Sundler, F., 1978, Occurrence and distribution of VIP nerves in the nasal mucosa and tracheobronchial wall, *Acta Oto-Laryngol.* **86**:443.

Vale, W., Rivier, C., and Brown, M., 1977, Regulatory peptides of the hypothalamus, *Annu. Rev. Physiol.* **39**:473.

Vijayan, E., Samson, W. K., Said, S. I., and McCann, S. M., 1979, Vasoactive intestinal peptide: Evidence for a hypothalamic site of action to release growth hormone, luteinizing hormone and prolactin in conscious ovariectomized rats, *Endocrinology* **104**:53.

Waldman, D. B., Gardner, J. D., Zfass, A. M., and Makhlouf, G. M., 1977, Effects of vasoactive intestinal peptide, secretin and related peptides on rat colonic transport and adenylate cyclase activity, *Gastroenterology* **73**:518.

Yalow, R. S., and Berson, S. A., 1960, Immunoassay of endogenous plasma insulin in man, *J. Clin. Invest.* **30**:1157.

Yousef, I. M., and Kuksis, A., 1972, Release of chylomicrons by isolated cells of rat intestinal mucosa, *Lipids* **7**:380.

19

Gonadotropin-Releasing Hormone

Release into Hypophyseal Portal Blood and Mechanism of Action

George Fink, Mukund Aiyer, Sharon Chiappa, Simon Henderson, Murray Jamieson, Victor Levy-Perez, Anthony Pickering, Dipak Sarkar, Nancy Sherwood, Alison Speight, and Alan Watts

1. Introduction

Geoffrey Harris' criteria for a pituitary hormone-releasing factor (Harris, 1972) have been satisfied for gonadotropin-releasing hormone (GnRH). That is, this decapeptide is present in large amounts in the hypothalamus, stimulates the release of luteinizing hormone (LH) and follicle-stimulating hormone (FSH), and its output into hypophyseal portal vessel blood under experimental and physiological conditions correlates well with the output of LH from the anterior pituitary gland. GnRH has other actions in addition to stimulating the release of gonadotropins. The peptide is necessary for the structural and functional integrity of the pituitary gonadotrophs, stimulates the synthesis of both LH and FSH, and has the apparently unique quality of being able to increase the responsiveness of the pituitary

George Fink, Mukund Aiyer, Sharon Chiappa, Simon Henderson, Murray Jamieson, Victor Levy-Perez, Anthony Pickering, Dipak Sarkar, Nancy Sherwood, Alison Speight, and Alan Watts • MRC Brain Metabolism Unit, University Department of Pharmacology, Edinburgh EH8 9JZ, Scotland.

to itself (the priming effect of GnRH). In addition, GnRH has dramatic and apparently direct effects on the ovary and testis, and these effects may be related to the presence in the organs of a substance that is biologically, but not immunologically, similar to GnRH (Ying and Guillemin, 1980; Sharpe and Fraser, 1980; Sharpe *et al.,* 1981).

Knowledge of the mechanisms of GnRH release and action is, of course, essential for our understanding of the control of gonadotropin release and the relationship of this to infertility and fertility control. But, in addition, because the gonadotropins and GnRH are released in the form of clear-cut signals that are modulated by steroids and involve interactions between central peptidergic and monoaminergic neurons, the hypothalamic–gonadotropin system provides a convenient "window" on central neurotransmission (Fink and Geffen, 1978). The hypothalamic–gonadotropin system does not act in a simple linear fashion; rather, the anterior pituitary acts as an amplifier for the neural signal mediated by GnRH. The degree of amplification of the signal is modified dramatically by steroid hormones and the priming effect of GnRH. There have been several recent general reviews on gonadotropin control (e.g. Brown-Grant, 1977; Fink, 1979a,b; Knobil, 1974; Knobil and Plant, 1978; Sawyer, 1975; Yen, 1978), and, therefore, we shall focus attention mainly on studies carried out on GnRH release and action in the rat.

2. Gonadotropin Release into Hypophyseal Portal Blood Evoked by Electrical Stimulation

Fink and Jamieson (1976) found that blood collected from the cut pituitary stalk of the rat contained immunoreactive GnRH as measured with the aid of a highly specific anti-GnRH serum (generously provided by Dr. Terry Nett and Dr. Gordon Niswender), and this was confirmed by Eskay *et al.* (1977). Neither group found significant amounts of GnRH in peripheral plasma. Preliminary to studies on hypothalamic stimulation and GnRH release, we (Jamieson and Fink, 1976) determined the optimal parameters for stimulating LH release. Using bipolar, platinum, glass-insulated electrodes implanted into the medial preoptic area, we found that for an accurately balanced biphasic square-wave pulse the optimum parameters were 500 μA (pulse height), 1 msec (pulse width), and 60 Hz (frequency). An increase in any of the parameters beyond these values resulted in diminished LH release.

Electrical stimulation of the medial preoptic area according to the above parameters resulted in a significant increase in the output of GnRH into portal blood collected from both male and female rats (Fink and

Jamieson, 1976). The effects of stimulation were area-specific in that the increment in GnRH release was greater when the stimulus was applied to the median eminence than to the preoptic area, and preoptic stimulation was more effective than stimulation of the suprachiasmatic nucleus (Chiappa *et al.,* 1977). Stimulation of the anterior hypothalamus, at a coordinate only 0.8 mm caudal to the preoptic area, resulted in an increment in GnRH release that was only 18% of that evoked by preoptic stimulation. Stimulation of either the amygdala or the dorsal or ventral hippocampus had no effect on GnRH output (Chiappa *et al.,* 1977), although stimulation of the ventral hippocampus before and during preoptic stimulation in the same animals resulted in a diminished increment in the GnRH response to preoptic stimulation. This showed that although the ventral hippocampus does not have a direct effect on GnRH release, the "responsiveness" of preoptic neurons to electrical stimulation may be influenced, at least experimentally, by ventral hippocampal projections to the preoptic area.

A survey of the responsiveness to stimulation of the preoptic–GnRH system throughout the estrous cycle of the rat showed that the response increased between diestrus and proestrus to reach a peak at 1300 hr of proestrus after which the response fell to a relatively low level at 1800 hr of proestrus. This profile of response raised the possibility that the responsiveness of preoptic neurons was altered by steroid hormones, and indeed Sherwood *et al.* (1976) showed that estradiol increased while progesterone decreased the GnRH response to preoptic stimulation. Testosterone propionate also increased the GnRH response to preoptic stimulation but 5α-dihydrotestosterone had no effect, suggesting that the effect of testosterone may depend upon its conversion to estradiol. How steroids exert their effects on the response of preoptic neurons is not clear, but the results obtained by Sherwood *et al.* (1976) are consistent with the fact that estrogen increases while progesterone decreases the firing of single neurons.

We (Chiappa *et al.,* 1977) also carried out studies on the effect of placing a horizontal knife cut at the level of the anterior commissure with a diameter of 4.6 mm. This cut, which severed all the dorsal afferents of the hypothalamus and caused widespread terminal degeneration throughout the brain, had no long-lasting effect on the occurrence of regular estrous cycles, and had no effect on the GnRH response to preoptic stimulation when tested after the resumption of regular estrous cycles (10–47 days after the placement of the lesion). However, the potentiating effect of estrogen on the GnRH response to preoptic stimulation was less, though not abolished, in lesioned compared with intact animals. These results show that the responsiveness of the preoptic–GnRH system does not depend upon the dorsal (including major limbic) afferents of the hypothalamus, but that these afferents may be important for estrogen to exert its full effect.

3. Spontaneous Release of GnRH into Hypophyseal Portal Vessel Blood

3.1. Models and Anesthetics

In their studies of changes in GnRH output during the estrous cycle of the rat, neither Fink and Jamieson (1976) using urethane nor Eskay *et al.* (1977) using sodium pentobarbitone were able to demonstrate any significant increase in GnRH release at the time of the preovulatory surge of GnRH, and so we (Sarker *et al.*, 1976; Sherwood *et al.*, 1980) surveyed several anesthetics. We confirmed that no change in GnRH release occurred under either sodium pentobarbitone, urethane, ketamine hydrochloride, or chloralose anesthesia. However, using the steroid anesthetic Althesin (Glaxo) (alphaxalone plus alphadolone acetate), a clear-cut surge of GnRH was detected in portal blood at the time of the spontaneous surge of LH. In preliminary studies we (Sarker *et al.*, 1976) showed that Althesin did not block the spontaneous proestrous surge of LH, although it reduced the surge by about 50%. This was confirmed in subsequent experiments (Sarker and Fink, 1979a,b; Sherwood *et al.* 1980). We also showed that Althesin did not stimulate LH release (Sarker *et al.* 1976; Sarkar and Fink, 1980). We have always administered Althesin by intraperitoneal injection, and this may be one explanation for the fact that our results differed from those of Rabii *et al.* (1978), who found that intravenous infusion of Althesin did block the LH surge. Recently, we (Fink *et al.*, 1980) demonstrated a marked sex difference in the anesthetic effect of Althesin, in that the dosage required to induce anesthesia in the male was four times that in the female.

3.2 The Spontaneous Preovulatory Surge of GnRH

In all the experiments described in this and subsequent sections, hypophyseal portal blood was collected from rats anesthetized with Althesin. The animals were all of the Wistar strain and maintained under controlled lighting (lights on 0500–1900 hr) and temperature (22 °C) with free access to food and tap water. Figure 1 shows that the concentrations of GnRH in portal plasma remained low throughout most of the estrous cycle, but that a surge of GnRH occurred at about the time of the spontaneous surge of LH at about 1800 hr of proestrus. A second, smaller peak of GnRH occurred at about 0200 hr of estrus. This smaller peak is not related to a change in LH output, but may be related to the prolonged surge of FSH. The second peak occurs about the time of maximal sexual activity in the rat and also at about the time of ovulation. The values in Fig. 1 show a

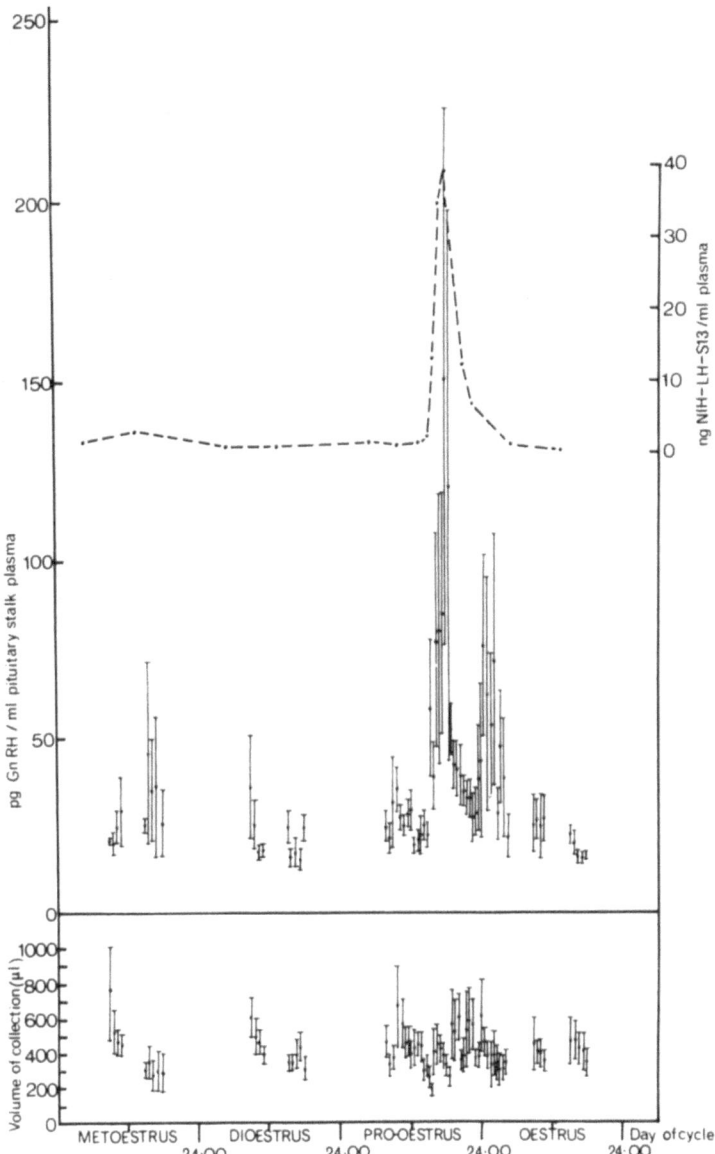

Figure 1. Upper panel: Mean (± S.E.M.) concentrations of GnRH in hypophyseal portal (stalk) plasma collected from rats anesthetized with Althesin at different times of the estrous cycle. Dashed line shows mean concentrations of LH in jugular venous plasma (data from Aiyer *et al.*, 1974). Lower panel: Mean (± S.E.M.) volumes of collections of portal blood. [From Sarkar *et al.*, 1976, with permission of the Editor of *Nature (London)* and Macmillan Journals Ltd.]

relatively large scatter and the reason for this is illustrated in Fig. 2, which shows that the concentrations of GnRH in portal plasma varied considerably within and between animals. At the time of the expected spontaneous surge of LH (1500–1900 hr), the concentrations reached levels above 40 pg GnRH/ml in 15 animals and above 100 pg/ml in 12 animals. Concentrations remained below 40 pg/ml in 14 animals. Figure 3 shows that we (Sarkar and Fink, 1979b) were also able to detect the first (pubertal) surge

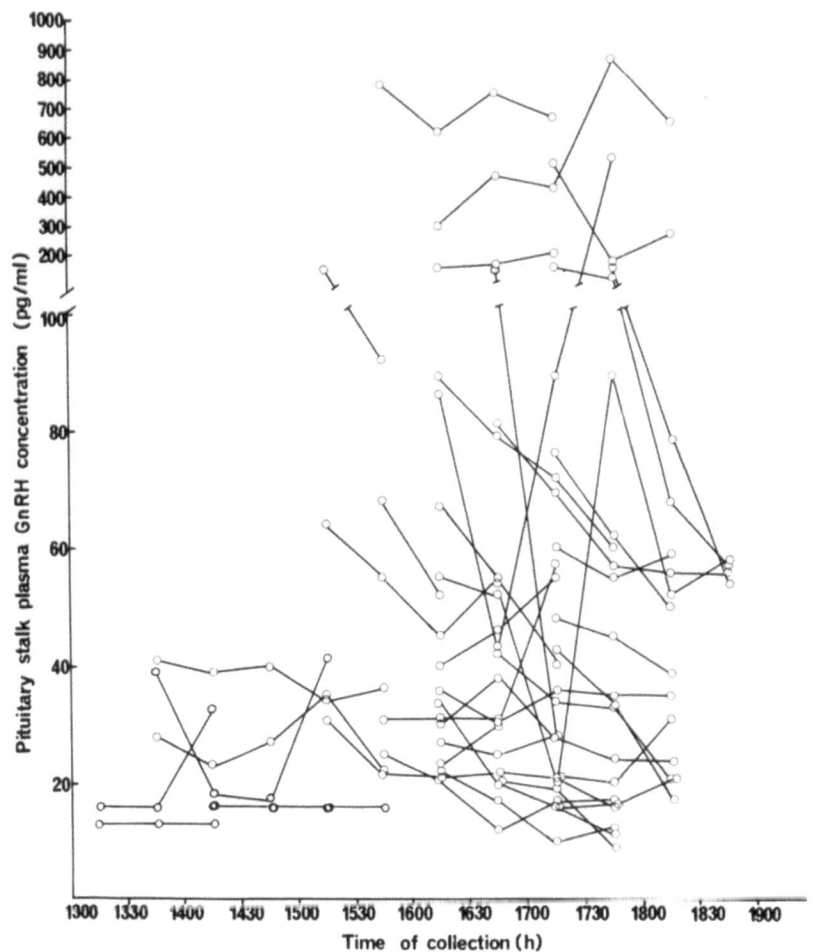

Figure 2. Concentrations of GnRH in hypophyseal portal (stalk) plasma in individual rats anesthetized with Althesin on the day of proestrus. [From Sherwood *et al.*, 1980, with permission of the Editor of *Endocrinology* and the Williams & Wilkins Co.]

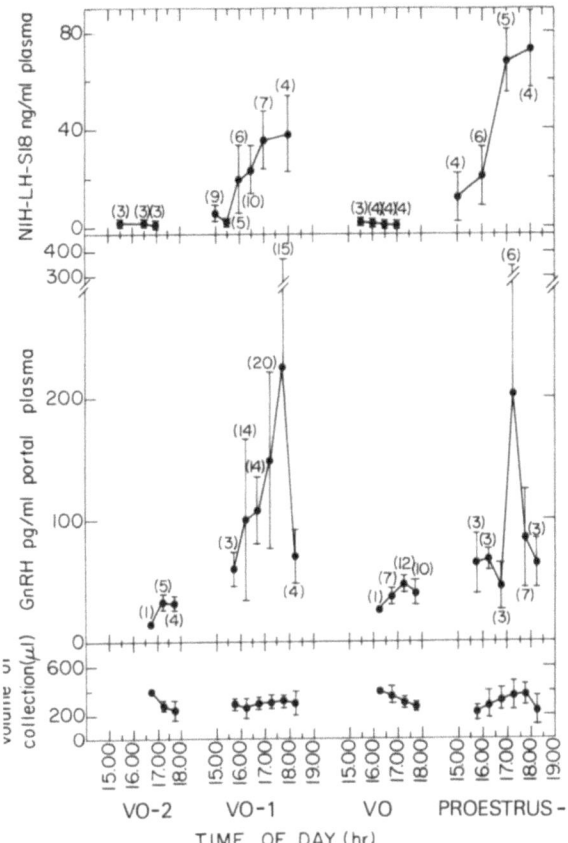

Figure 3. Mean ± S.E.M. (*n* in parentheses) concentrations of LH in peripheral plasma and GnRH in hypophyseal portal plasma, and volumes of portal blood collected from animals anesthetized with Althesin either on the day of vaginal opening (VO), or one (VO − 1) or two days (VO − 2) before vaginal opening. Concentrations and volumes obtained on the second proestrus are also shown. [From Sarkar and Fink, 1979b, with permission of *The Journal of Endocrinology*.]

of GnRH, which occurred at about the time of the first spontaneous LH surge.

3.2.1. Role of Steroid Hormones

Experiments carried out by acute manipulation of steroid hormones or the administration of antiestrogen showed that the GnRH surge depends upon the preovulatory rise in the plasma concentration of estradiol-17β

(Sarkar and Fink, 1979a,b). Progesterone, perhaps unexpectedly, had either no effect or tended to inhibit the stimulatory effect of estrogen on the GnRH surge. The inhibitory effect of progesterone was not found in animals that had been adrenalectomized. The way in which estrogen triggers the GnRH surge and the reason for the apparent inhibitory action of progesterone are not clear, but these observations are perhaps consistent with the effect that these steroids have on the firing of single neurons (mentioned in Section 2).

3.2.2. Role of Central Monaminergic Systems

The possible role of central monaminergic systems in the generation of the LH surge has been the subject of many studies since the early investigations at Duke University, North Carolina (Sawyer, 1975). The evidence shows that a noradrenergic system stimulates LH release. However, the evidence is conflicting with respect to the role of central dopaminergic neurons. We have studied the modulation of the release of GnRH by central monoaminergic neurons (Sarkar and Fink, 1981). The test system was the GnRH surge that occurs at 1700 hr of Day 32 in female rats injected on Day 30 with PMSG. The release of GnRH was correlated with the release of LH measured by radioimmunoassay of external jugular venous plasma.

The height of the PMSG-induced surges of GnRH and LH was reduced in animals that had been injected with either α-methyl-p-tyrosine (inhibitor of tyrosine hydroxylase) or diethyldithiocarbamate (DDC) or fusaric acid (inhibitors of dopamine-β-hydroxylase). The action of all three inhibitors was reversed by administering dihydroxyphenylserine, and the inhibitory action of DDC was potentiated by L-DOPA. The height of the PMSG-induced LH surge was reduced by phenoxybenzamine, but was not affected by either phentolamine, yohimbine, DL- or D-propranolol, or clonidine. Phenoxybenzamine also reduced significantly the height of the PMSG-induced GnRH surge. The height of the GnRH and LH surge was increased by pimozide and domperidone and reduced by haloperidol. The action of all three of these dopamine receptor blockers could be reversed by apomorphine. Domperidone administered before the critical period on the afternoon of proestrus in adult rats also increased significantly the height of the spontaneous LH surge. The results obtained with domperidone are illustrated (Figs. 4 and 5) because they were the most dramatic. Domperidone is thought to act primarily at D_2 (not coupled to adenylate cyclase) receptors. In a separate study (Sarkar *et al.*, 1981) it was shown that the administration of desipramine followed by 6-hydroxydopamine (a regime that lesions dopaminergic but not noradrenergic neurons) caused a marked (about four fold) potentiation of the GnRH surge in PMSG-treated immature rats.

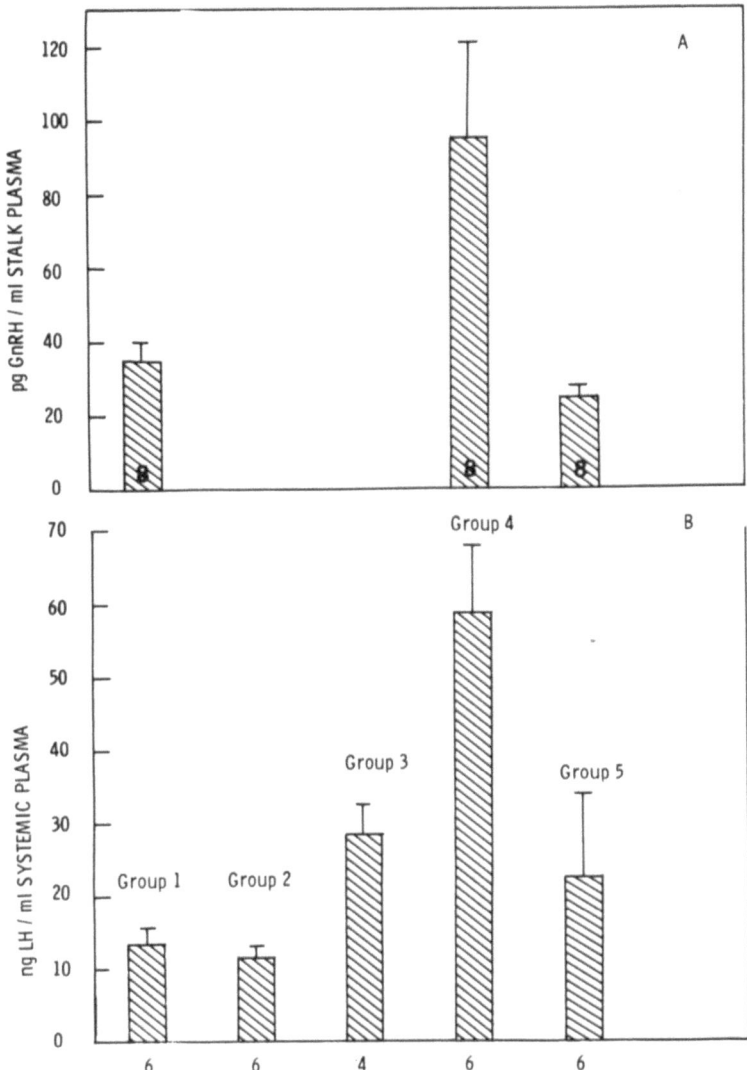

Figure 4. Effect of dopamine antagonist domperidone with or without apomorphine on the PMSG-induced surge of GnRH into portal (stalk) plasma (A) and LH (B). The control group of animals (Group 1) was injected with 0.9% saline/ethanol vehicle (20/1, v/v) at 1400 hr of Day 32. Group 2, 3, and 4 animals were injected with domperidone s.c.: 0.5 mg/kg at 0930 hr, 0.1 mg/kg at 1400 hr, and 0.5 mg/kg at 1400 hr of Day 32, respectively. Group 5 was injected with domperidone (0.5 mg/kg, s.c.) at 1400 hr followed by apomorphine (0.1 mg/kg, s.c.) at 1600 hr of Day 32. The values are the mean + S.E.M., and the number of animals is shown at the base of each bar. The mean (± S.E.M.) volumes of pituitary stalk blood, which ranged from 246 ± 29 (Group 1) to 311 ± 31 (Group 2), did not differ significantly. [From Sarkar and Fink, 1981, with permission of the Editor of *Endocrinology* and the Williams & Wilkins Co.]

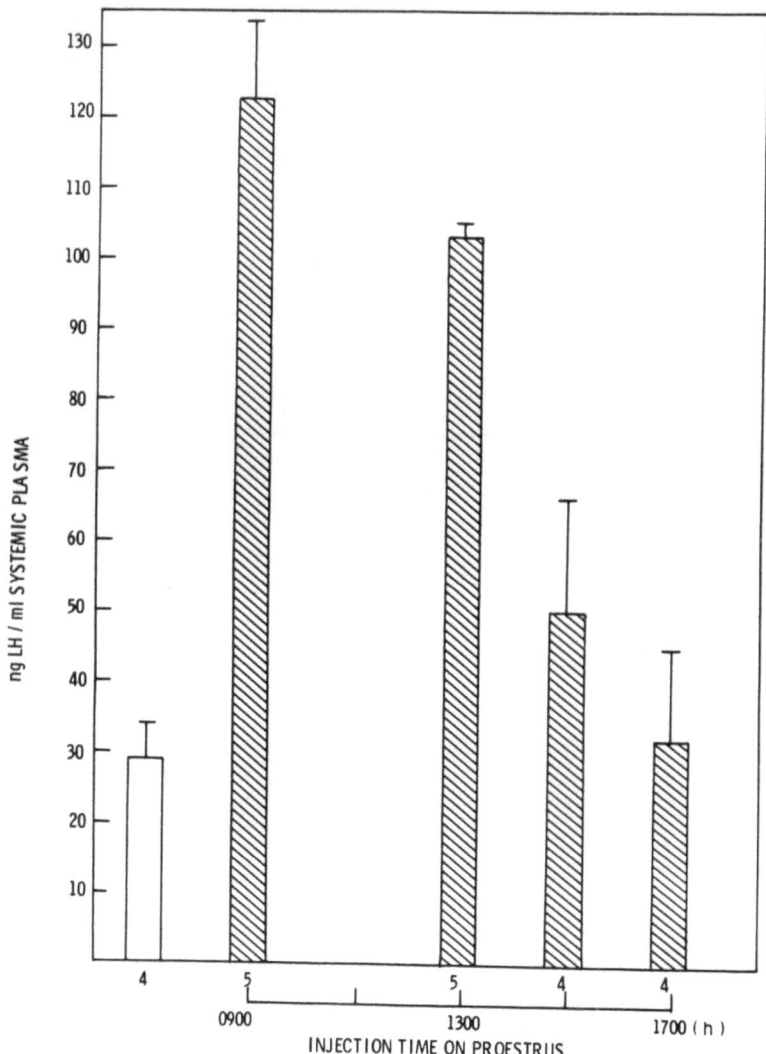

Figure 5. Effect of domperidone on the proestrous surge of LH. Each hatched bar represents the mean ± S.E.M. concentration of LH in jugular venous plasma taken at 1800 hr of proestrus from animals that had been injected with domperidone (0.5 mg/kg, s.c.) at the times shown at the base. The open bar represents the mean + S.E.M. concentration of LH in jugular venous plasma taken at 1800 hr of proestrus from untreated control animals. The number of animals is shown at the base of each bar. [From Sarkar and Fink, 1981, with permission of the Editor of *Endocrinology* and the Williams & Wilkins Co.]

These results show that the GnRH surge depends upon the functional integrity of central noradrenergic neurons, which facilitate the GnRH release through α adrenoreceptors. The GnRH surge can be inhibited by dopamine acting on receptors that are blocked by pimozide and domperidone, and facilitated by dopamine acting on receptors that are blocked by haloperidol.

3.3. Release of GnRH in Long-Term Ovariectomized Rats

Figure 6 shows that after ovariectomy at estrus the mean plasma LH concentrations rise after Day 4 and correlate well with the mean concentrations of GnRH in portal plasma. These results were confirmed in a separate study by Sherwood and Fink (1980). The concentrations in individual animals are shown in Fig. 7. Immediately after ovariectomy there is marked spiking of GnRH release. At Days 2 and 4 after ovariectomy the concentrations of GnRH remain low with few spikes, but from Day 8 onwards the height of the spikes increases markedly. This spiking or episodic release of GnRH corresponds with the episodic release of LH shown first by Gay and Midgley (1969) and Gay and Sheth (1972). The administration of estradiol i.v. reduces both the episodic release and the mean plasma concentration of LH, and Fig. 8 shows that estradiol i.v. has a similar effect on the portal plasma concentrations of GnRH.

The administration of relatively large amounts of estrogen to long-term ovariectomized animals results in a diurnal release of LH with peaks occurring at about 1700 hr of the afternoon (Caligaris *et al.,* 1971). We (Sarkar and Fink, 1980) found that the diurnal release of LH corresponded with a diurnal release of GnRH, and that the afternoon surge of GnRH (and LH) could be blocked by sodium pentobarbitone. In animals so blocked, the surges of GnRH and LH were significantly increased on the afternoon of the next day. However, the mean peaks of GnRH were remarkably small (About 40–80 pg/ml portal plasma) in relation to the peaks of LH. The reason for this discrepancy is discussed in Section 4.1.2, as is the fact that we (Sarkar and Fink, 1980) were unable to detect a surge of GnRH in animals treated with estrogen followed by progesterone, a regime that results in a massive surge of LH (Caligaris *et al.,* 1968).

3.4. Generation of the GnRH Surge in Primates

In the human female and in female rhesus monkeys, elevated plasma concentrations of estradiol trigger a surge of LH (Knobil, 1974; Knobil and Plant, 1978; Yen *et al.,* 1975). Neill *et al.* (1976) demonstrated the occurrence of a spontaneous surge of GnRH around the expected time of the spontaneous surge of LH. The administration of estradiol benzoate to rhe-

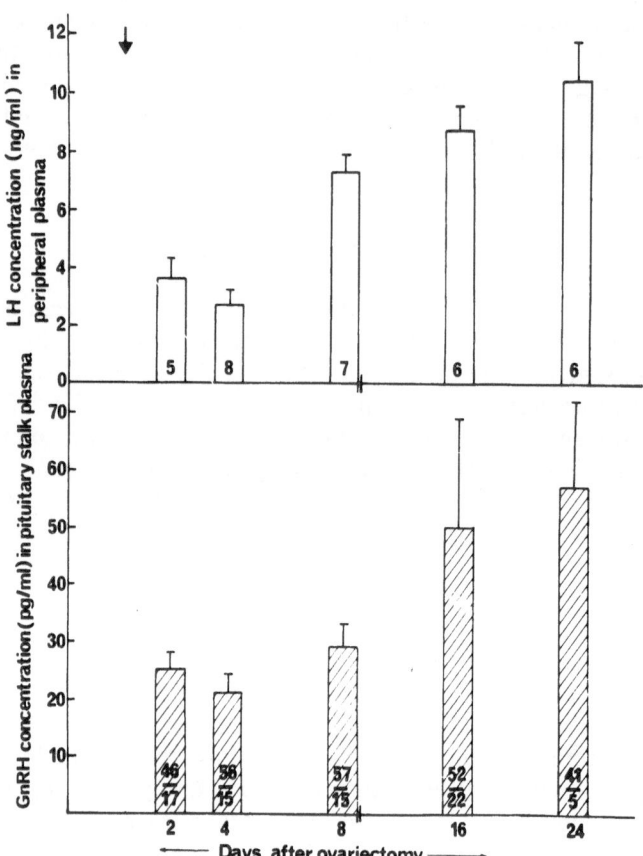

Figure 6. Mean (+ S.E.M.) concentrations of LH in jugular venous plasma (ng NIH-LH-S18/ ml) and GnRH in portal (stalk) plasma after ovariectomy carried out at 1000–1100 hr of estrus (shown by arrow). The pituitary stalk blood was collected between 1100 and 1500 hr on the days shown and the mean concentrations were calculated from the values in Fig. 7. Blood for LH determination (0.25 ml) was withdrawn from the external jugular vein from the same animals before exposing the pituitary stalk. [From Sarkar and Fink, 1980, with permission of the Editor of *The Journal of Endocrinology.*]

sus monkeys in the follicular phase of the cycle induced a surge of GnRH in portal blood that correlated with a surge of LH in peripheral plasma (Neill *et al.,* 1977). The release of GnRH into portal blood of ovariecto- mized rhesus monkeys was episodic, but in contrast to the rat, episodic release could not be blocked by the administration of estradiol (Carmel *et al.,* 1976).

The most important difference between the rhesus monkey and the rat

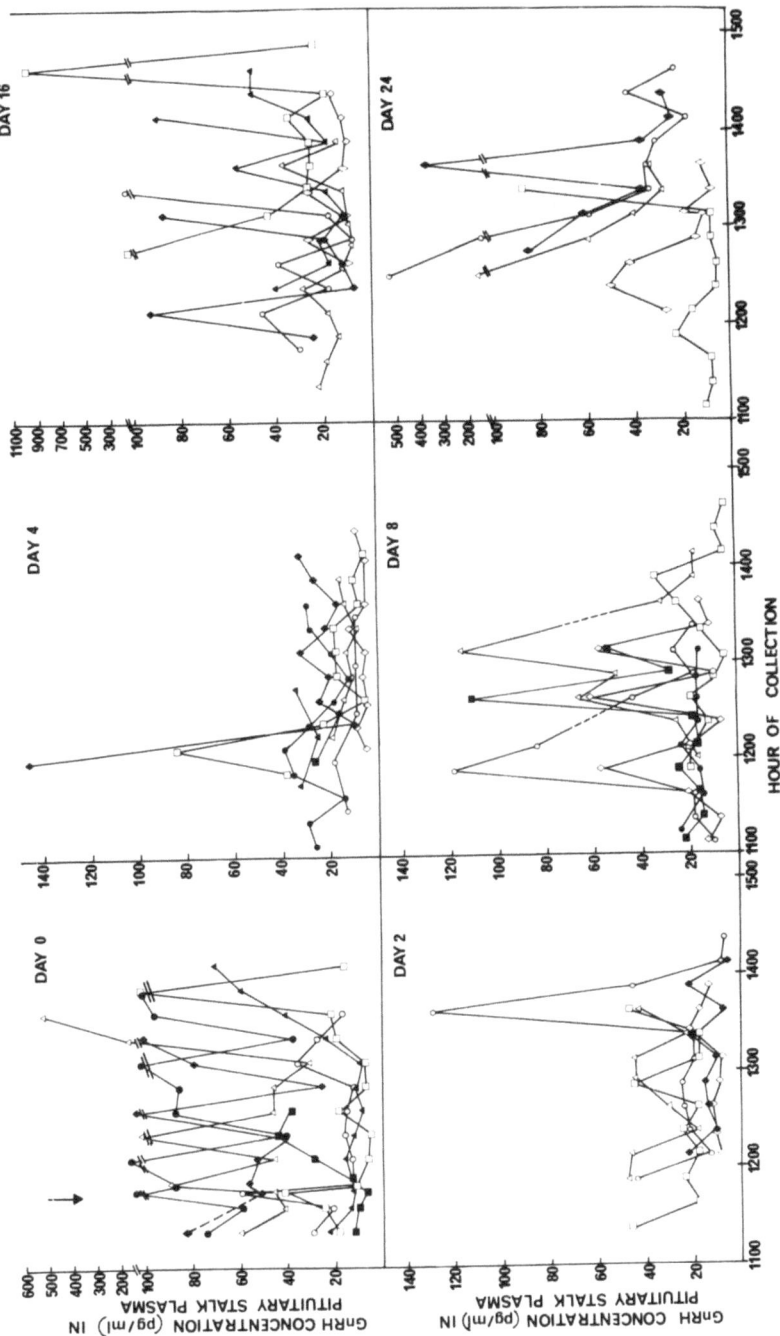

Figure 7. Concentrations of GnRH in portal (stalk) plasma from individual animals at various times before and after ovariectomy carried out between 1000 and 1200 hr of estrus (day 0; shown by arrow). Most samples were collected during periods of 15 min and a few of 30 min. [From Sarkar and Fink, 1980, with permission of the Editor of *The Journal of Endocrinology.*]

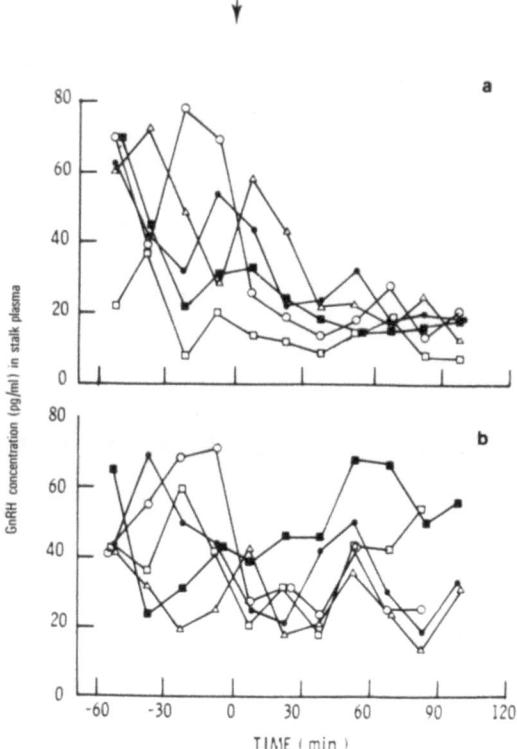

Figure 8. Concentration of GnRH in portal (stalk) plasma in individual rats ovariectomized 3–4 weeks previously and anesthetized with Althesin. Stalk blood was collected during consecutive periods of 15 min before and after the i.v. injection (shown by arrow) of either (a) 1 µg estradiol-17β or (b) 0.5 ml ethanol/saline (vehicle). [From Sarkar and Fink, 1980, with permission of the Editor of *The Journal of Endocrinology.*]

is that in the former, spontaneous- and estrogen-induced surges of LH occur after total deafferentation of the medial basal hypothalamus (Krey *et al.,* 1975), a procedure which blocks ovulatory LH surges in the rat. Estradiol was found to stimulate a surge of LH in ovariectomized monkeys in which a lesion of the arcuate nucleus had been made and in which pulses of GnRH were infused intravenously (Nakai *et al.,* 1978). These studies showed that in contrast to the rat where the stimulatory effect of estradiol appears to be exerted at the preoptic level (Goodman, 1978), the site of action in the rhesus monkey may be at the medial basal hypothalamic–pituitary unit. Indeed, studies on ovariectomized rhesus monkeys in which the pituitary stalk had been sectioned (Ferin *et al.,* 1979) or in which the arcuate nuclei had been lesioned (Wildt *et al.,* 1981) showed that at least for a short period after the section (Ferin *et al.,* 1979) or the cessation of GnRH infusion (Wildt *et al.,* 1981) estrogen can induce a surge of LH, probably by a direct action on the anterior pituitary gland. In apparent contradiction to these findings, Norman *et al.* (1976) found that, in the rhesus monkey, lesions of the ventral preoptic area abolished spontaneous- and estrogen-induced surges of LH.

4. Changes in Pituitary Responsiveness to GnRH

Before and during the spontaneous, preovulatory LH surge there is a dramatic increase in the responsiveness of the anterior pituitary to GnRH. Although this has been reported in several mammals, it has been investigated in greatest detail in the rat (Fink, 1979a,b) and the human (Yen *et al.,* 1975), where, allowing for differences in cycle length, the magnitude (20- to 50-fold) and timing of the increase are remarkably similar.

The increase in pituitary responsiveness is initiated by and depends upon the preovulatory rise in the plasma concentration of estradiol-17β. The increase is then reinforced by the increased secretion of progesterone (at least in the rat) that occurs almost immediately after the beginning of the LH surge, and by the priming effect of GnRH (Fink and Pickering, 1980).

4.1. Relative Importance of the Increase in Pituitary Responsiveness to GnRH

4.1.1. The Spontaneous Surge

As assessed by radioimmunoassay the amount of GnRH released into portal blood during the spontaneous surge is very small. Sherwood *et al.* (1980) calculated that during the spontaneous surge in the rat only about 130 pg GnRH was released (between 1500 and 2200 hr of proestrus). This value is similar to the values found during the spontaneous (Neill *et al.,* 1977) surges of GnRH in the rhesus monkey, and the surge of GnRH induced by cupric acetate in the rabbit (Tsou *et al.,* 1977). These amounts of GnRH and the corresponding portal plasma concentrations are far too small to induce an LH surge other than when the responsiveness of the pituitary has been increased. This is illustrated in Fig. 9, where it may be seen that the infusion of GnRH at a rate sufficient to raise plasma concentrations to those seen about the peak of the spontaneous surge of GnRH in portal blood increased plasma LH concentrations to spontaneous surge levels in proestrous but not diestrous rats. This point was also established by studies on the release of LH in proestrous rats by prolonged electrical stim-

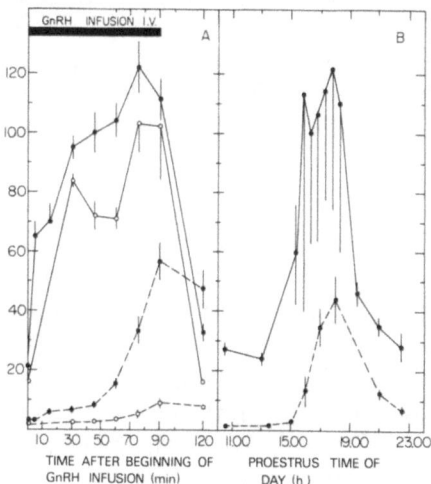

Figure 9. (A) The mean ($\pm$S.E.M.) peripheral plasma concentrations of LH (NIH-LH-S18) (dashed lines) and GnRH (solid lines) in either proestrous ($\bullet$; $n = 11$) or diestrous (o; $n = 5$) rats that were infused i.v. with synthetic GnRH at a rate of 0.167 ng/100 g[1] per min for a period of 90 min after anesthesia with sodium pentobarbitone at 1330 hr. The data were obtained by Fink et al. (1976). (B) The mean ($\pm$S.E.M.) concentrations of GnRH in hypophyseal portal plasma (solid line; $n = 9$–52) and LH (NIH-LH-S13) in peripheral plasma (dashed line; $n = 6$–9) during the day of proestrus. The concentrations of GnRH were obtained by Sarkar et al. (1976) and Sherwood et al. (1980) and the LH concentrations were obtained by Aiyer et al. (1974).

ulation of the preoptic area (Sherwood et al., 1980). Stimulation with biphasic square-wave pulses with a height of 200 μA, duration 1 msec, and frequency of 60 Hz resulted in a surge of LH comparable in height to that of the spontaneous surge. However, the total amount of GnRH released into portal blood collected from similarly treated animals was even less than that released during the spontaneous surge of GnRH.

As assessed by measuring the amount of GnRH released by electrical stimulation of the preoptic area or median eminence, the total amount of GnRH released during the spontaneous surge in the rat is similar to the totally releasable pool (Sherwood et al., 1980). This pool, which is an operational term and may depend upon factors other than the nature or amount of stored GnRH (e.g., the state of Ca^{2+} permeability channels), constitutes only about 2% of the total hypothalamic content of GnRH.

4.1.2. Steroid-Evoked LH Release in Long-Term Ovariectomized Rats

In Section 3.3. it was pointed out that the GnRH peaks were surprisingly low in long-term ovariectomized rats treated with high doses of estrogen, and that in ovariectomized rats treated with estrogen followed by progesterone no surge of GnRH was detected in spite of the massive surge of LH that occurs in similarly treated animals (Sarkar and Fink, 1980). This may be explained by the massive increase in pituitary responsiveness that occurs in these animals. In long-term ovariectomized animals treated with estrogen the LH-response to GnRH is 1.5–2.3 times greater than at 17.00 hr of proestrus (the time of maximal pituitary responsiveness) (Henderson *et al.*, 1977b). After treatment with estrogen followed by progesterone the LH-response is 2.2 times that at 17.00 hr of proestrus (Sarkar and Fink, 1980). Conceivably, in the latter model, the timing and height of the massive LH surge is determined by the increase in pituitary responsiveness (2.2 fold) which follows the administration of progesterone to estrogen-primed, long-term ovariectomized rats.

4.1.3. The Site of Action of Estrogen

When it was first discovered that estrogen produced a marked increase in pituitary responsiveness, it was assumed that this was due to an action at the level of the pituitary gland. Thus, it seemed that in the rat estrogen had two sites of action; on the brain to trigger the surge of GnRH and on the pituitary gland to increase pituitary responsiveness to GnRH. However, the following experiments show that the latter also depends upon an action of estrogen at the level of the hypothalamus.

In acutely ovariectomized animals, estrogen first reduced and then increased pituitary responsiveness to GnRH (Cooper *et al.*, 1974; Vilchez-Martinez *et al.*, 1974; Henderson *et al.*, 1977a). This increase, which reached a peak 12 hr after the implantation of a silicone elastomer capsule containing crystals of estradiol-17β, could be blocked by sodium pentobarbitone (Henderson *et al.*, 1977a; Speight *et al.*, 1981). Speight *et al.*, (1981) showed that (1) the pentobarbital block could be overcome by infusing GnRH at a rate which barely produced an increase in the peripheral plasma concentrations of GnRH and LH, and (2) the increased concentrations of GnRH in portal blood which occur immediately after ovariectomy (Fig. 7) were maintained for 12 hr in animals implanted with an estradiol-containing, but not an empty capsule. These results suggested that the increase in pituitary responsiveness produced by estradiol depends, at least to a large extent, on the release of GnRH and, therefore, the priming effect of GnRH. The results of Speight *et al.* (1981) do not exclude a pituitary site of action because estrogen can increase responsiveness in animals in which the pituitary stalk has been severed (Greeley *et al.*, 1975,1978; Fink and Henderson, 1977a), but they do show that estrogen, for its full effect, depends upon the functional integrity of the hypothalamic–pituitary system. The evidence outlined above also explains why Cooper *et al.* (1975) found an increased LH response to GnRH immediately after ovariectomy. The fact that for its full effect on pituitary responsiveness estrogen requires the integrity of the hypothalamic–pituitary system also explains why there is such a dramatic sex difference in pituitary responsiveness to GnRH which cannot be significantly altered by the manipulation of gonadal steroids (Fink and Henderson, 1977b).

In the rhesus monkey estrogen may act predominantly on the anterior pituitary gland (Wildt *et al.,* 1981) from which it may actually release LH without necessarily changing responsiveness to GnRH. In the guinea pig, as in the rat, estradiol facilitates pituitary responsiveness by an action on the medial basal hypothalamus as well as on the anterior pituitary gland (Terasawa *et al.,* 1980).

5.   The Neural Mechanism Responsible for Diurnal and Episodic Release of LH

Apart from the spontaneous preovulatory surge, LH is released in an episodic manner and also in the form of a diurnal rhythm. The importance of episodic release of the gonadotropins may vary from species to species. In the rat, this type of release can only be found consistently after ovariectomy, whereas in the human (Yen *et al.,* 1972) and sheep (Lincoln, 1979) clear-cut episodic release occurs in intact animals and varies with the time of the menstrual cycle (human) or the season (sheep). The physiological importance of episodic release of LH, which is probably due to episodic release of GnRH, may be related to (1) the sensitizing effect of GnRH (priming) and (2) the fact that the pituitary becomes refractory to prolonged continuous but not episodic exposure to GnRH (e.g., Knobil and Plant, 1978). Clear-cut diurnal gonadotropin release has only been shown experimentally; that is, in long-term ovariectomized rats treated with high doses of estradiol. However, the significance of the latter model is that it revealed the fact that there probably is a daily neural signal for gonadotropin release generated by a neural mechanism in which the suprachiasmatic nucleus is a key component (Everett, 1977; Fink, 1979b). For this signal to be expressed in the form of a GnRH followed by an LH surge, the brain must be exposed to elevated levels of estradiol for at least 24 hr (Fink, 1979a,b).

We (Watts and Fink, 1981) investigated whether the same neural mechanism subserves both the diurnal and episodic release of gonadotropin in long-term ovariectomized rats exposed to constant light, a condition that in intact animals leads to acyclicity, persistent estrus, and conversion from a spontaneous to a reflex ovulatory mode. Figure 10 shows that episodic release of LH remained unchanged in animals exposed to constant light. However, in animals implanted with capsules containing crystalline estradiol-17β, a regime that produces a clear-cut diurnal rhythm of LH release, no diurnal rhythm was detected in the animals exposed to constant light. These results together with those obtained from animals subject to either hypothalamic deafferentation (Blake and Sawyer, 1974) or lesions of the suprachiasmatic nuclei (Arendash and Gallo, 1979) show that the neural mechanisms that subserve the two modes of gonadotropin release are dis-

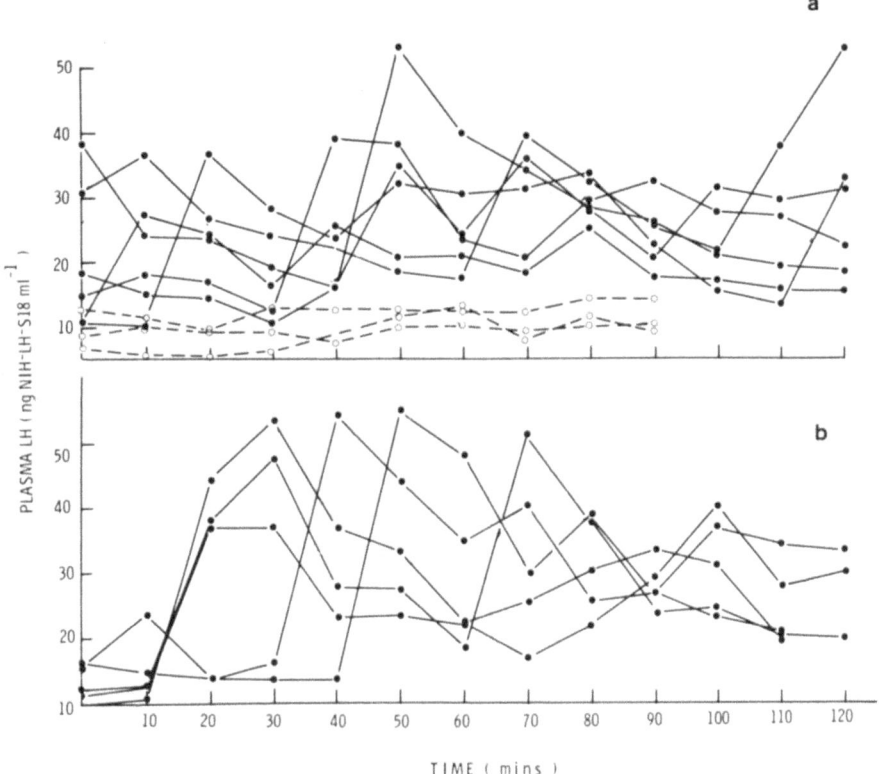

Figure 10. Plasma concentrations of LH in (a) individual rats ovariectomized and then exposed to control lighting (lights on 0500–1900 hr) for 28 days or (b) control lighting for 7 days then constant lighting for 21 days and anesthetized with Althesin (●—●) or sodium pentobarbitone (o- - -o). [From Watts and Fink, 1981, with permission of the Editor of *The Journal of Endocrinology*.]

tinct, and suggest that episodic release of GnRH is not under the control of suprachiasmatic nuclei, which are crucial for most neuroendocrine diurnal rhythms (Rusak and Zucker, 1979).

6. Mechanism of Action of GnRH

6.1. Releasing Action and Priming Effect

The mechanism of action of GnRH and the way this is modified by steroids has been reviewed in detail (e.g., Fink, 1979a,b; Fink and Picker-

George Fink et al.

Table I. Comparison of the Releasing and Priming Actions of GnRH in the Female
Rat

	Releasing	Priming
Ontogenetic development	early as Day 5	only after Day 17
Can be repeated frequently	yes	no (once only in 3-hr period)
Enhanced by estrogen	yes	yes
Dependent on extracellular Ca^{2+}	yes	no
Mimicked by high extracellular K^+	yes	no
Mimicked by Ca^{2+} ionophores	yes	no
Dependent on protein synthesis	no	yes
Dependent on integrity of microfilaments	no	yes
Cyclic AMP acts as second messenger	no	no
Elicited in dispersed cells	yes	not so far[a]

[a] but has been demonstrated in cells obtained from bovine pituitary glands (Padmanabhan et al., 1981)

ing, 1980; Turgeon and Waring, 1981; Waring and Turgeon, 1980). However, in order to provide a basis for understanding the data considered above on the release of GnRH, and some of the more recent data, we have summarized below and in Table I the differences between the releasing action and priming effect of GnRH.

1. The priming effect of GnRH is the capacity of this decapeptide to increase pituitary responsiveness so that further exposure to GnRH leads to an accentuated release of LH and FSH. This effect has been demonstrated in man as well as in experimental animals.
2. Steroids modulate but do not mediate the priming effect.
3. Studies on the mechanism of the priming effect carried out *in vitro* show that:
   a. Priming by GnRH involves a significant qualitative as well as a quantitative change in pituitary responsiveness.
   b. Within a 3-hr period, the priming effect can be elicited only once.
   c. Although high extracellular potassium concentrations or calcium ionophores stimulate acute gonadotropin release, these nonselective secretagogues cannot prime the pituitary.
   d. Acute gonadotropin release in response to GnRH requires normal extracellular Ca^{2+}, but priming can be elicited in the presence of a medium free of Ca^{2+}.
   e. Adenosine $3':5'$ monophosphate (cyclic AMP) does not appear to act as second messenger for either the releasing (e.g. Conn et al., 1979) or the priming action of GnRH (Pickering and Fink, 1979a). Van Rees and De Koning (1979), however, suggest that cyclic AMP may be involved in the synthesis of the protein necessary for the priming effect (see below).

f. The priming, but not the LH-releasing action of GnRH appears to depend upon protein synthesis. The nature of the new protein has not yet been established, but cytochalasin B can abolish the priming (but not the LH-releasing) action of GnRH, raising the possibility that priming may involve a change in contractile proteins (Pickering and Fink, 1979a). The potentiation by progesterone of the LH response to GnRH in proestrous pituitary glands is also inhibited by cycloheximide (Turgeon and Waring, 1981).

g. Extracts of hypothalamic tissues can prime pituitary glands with respect to LH and FSH, but not adrenocorticotropin, thyrotropin, or prolactin (Pickering and Fink, 1979b).

6.1.1. Ontogeny of Priming Effect

Recently we (Meiden *et al.*, 1981) have shown that priming develops well before puberty (as early as Day 17 after birth), and that estrogen administration to prepubertal animals, while increasing considerably the total responsiveness of the glands *in vitro,* does not increase the relative magnitude of the priming effect.

6.1.2. Studies on Dispersed Cell Systems

For the purpose of carrying out more sophisticated biochemical studies, we (Speight and Fink, 1981a,b; V. Levy-Perez and G. Fink, unpublished) have tried to elicit the priming effect in pituitary cell cultures (gonadotroph enriched) and in dispersed cell systems (using virtually all the described methods for dispersion including nonenzymatic methods) suspended in Biogel columns. However, although we have been able to elicit a biphasic release in response to GnRH (in agreement with Walker and Hopkins, 1978), and have been able to demonstrate differences in responsiveness in cells obtained from animals at different stages of the estrous cycle, with the maximum reponse occurring on proestrus (Speight and Fink, 1981a), we have not so far been able to elicit a clear-cut priming effect (i.e., an enhanced response to further exposure to GnRH). Indeed, the fact that a priming effect cannot be elicited in a dispersed cell system makes this an especially useful bioassay for GnRH in tissue and for GnRH analogues. The explanation for the fact that we have not been able to prime dispersed cells with GnRH is not clear, but perhaps the integrity of normal contacts between pituitary cells is crucial for the effect.

Other workers (e.g., Dr. Jean Martin, Washington University, St. Louis, Missouri, personal communication) have also failed to elicit priming from dispersed cells obtained from rat pituitary glands. However, Padmanabhan *et al.,* (1981) were able to demonstrate priming in enzymatically dispersed cells from bovine pituitary glands.

6.1.3. GnRH Receptors and Pituitary Responsiveness and GnRH Priming

Clayton *et al.* (1980) and Clayton and Catt (1981) have shown that steroids have potent effects on the density of pituitary GnRH receptors. Thus orchidectomy and ovariectomy produced an increase in receptor density which correlated with the increase in plasma LH concentration. The increase could be inhibited by testosterone, 5α dihydrotestosterone, diethystilbestrol, and estradiol in the male, and by estradiol in the female. During the normal estrous cycle receptor density increased between metestrus and diestrus and remained high until just before the spontaneous LH surge when the density fell to low levels. The pattern of GnRH receptor density during the cycle could be related to plasma concentrations of estradiol-17β, but the fall in density at the time of the LH surge and the fact that it took 40 hr before a significant increase in GnRH receptor density occurred after administration of a potent GnRH agonist (Clayton *et al.,* 1980) suggests that the priming effect of GnRH may not necessarily involve an increase in pituitary-binding sites for GnRH. Clayton *et al.,* (1980) speculate that the decrease in GnRH receptor density at the time of the LH surge may be due to increased receptor processing consequent upon increased occupancy produced by the GnRH surge. A definitive conclusion on the relationship between GnRH receptors and priming must await further work.

6.2 Effect of GnRH on the Synthesis of Gonadotropins

6.2.1 Acute Effects

Most of the investigations of the possible acute stimulation of gonadotropin synthesis by GnRH have been based on studies in which the total amount of gonadotropin in the system has been determined by measuring the amount of gonadotropin released and that left in the pituitary. Although a definitive answer must await more sophisticated studies, none of the data show that GnRH, or for that matter estrogen, increases pituitary responsiveness by increasing the total amount of gonadotropin (e.g., Chiappa and Fink, 1977a; Pickering and Fink, 1979c; Meidan *et al.,* 1981; Speight and Fink, 1981b). Thus, increased pituitary responsiveness produced by GnRH (and estrogen) is probably due to an apparent increase in the readily releasable pool of LH. The mechanisms involved in producing this increase in the releasable pool have not been established, but the obvious possibilities that should be considered include (1) changes in the receptor-release apparatus of the gonadotrophs that facilitate hormone release, (2) a transfer of hormone from a "storage" pool, and (3) changes in the hormone that may render it more susceptible to release. With respect to (1) the data of Clayton *et al.* (1980) have been discussed in Section 6.1.3.; that is, the increase in pituitary responsiveness on proestrus produced by estradiol may be brought about by increased pituitary receptor content, but there is so far no evidence that GnRH priming depends upon a change in pituitary-receptor density. With respect to (3), it has been shown that GnRH stimulates glycosylation of LH (Liu and Jackson, 1978; Azhar *et al.*, 1978), and that, depending upon its concentration, GnRH can preferentially stimulate either synthesis (low concentrations) or release (high concentrations) of LH (Liu and Jackson, 1979).

6.2.2. Long-Term Effects: The Mutant Hypogonadal Mouse

That the synthesis of gonadotropins was dependent upon the integrity of the hypothalamic–pituitary system was suggested by the classical studies on pituitary grafts carried out by Harris and Jacobsohn (1952) and Nikitovitch-Winer and Everett (1958, 1959), and confirmed by studying the effects of anti-GnRH sera on pituitary gonadotropin content (Koch, 1977; Fraser and Baker, 1978) and the ontogenesis of the hypothalamic–pituitary system (Chiappa and Fink, 1977b).

Recently, with Dr. Harry Charlton, we have initiated studies on a mutant hypogonadal mouse discovered by Dr. Bruce Cattanach (MRC Radiobiology Unit, Harwell). The inheritance of the hypogonadal trait is autosomal recessive. The hypothalamus of the mutant contains no or very little immunoreactive GnRH (Cattanach *et al.*, 1977). Studies with bioassay and HPLC (G. Fink, A. Speight, H. M. Charlton and A. Harmar, unpublished) have shown that the small amount of immunoreactive GnRH present in some mutant hypothalami is not authentic GnRH. As might have been expected, electrical stimulation of the hypothalamus of the mutant did not elicit an increase in plasma LH concentration; however, perhaps surprisingly, there was a reasonable LH response (about 60% of that in the wild mouse) to exogenously administered GnRH and we were able to elicit in the mutant a remarkably good priming effect of GnRH (Iddon *et al.*, 1980; G. Fink, J. Sheward, and H. M. Charlton, unpublished data). The response to GnRH suggested that (1) GnRH receptors are present in the mutant pituitary and therefore are not dependent upon GnRH for their development, (2) some of the small amount of LH in the mutant pituitary (about 5–10% of that in the wild-type pituitary) is releasable, and (3) the release apparatus in the mutant pituitary is intact.

Our early studies on the synthesis of gonadotropins in the mutant revealed perhaps the most surprising results. We (H. M. Charlton, C. A.

Iddon, and G. Fink, unpublished data) administered synthetic GnRH to the mutant either by implanting GnRH containing silicone elastomer capsules or by regular daily injection. The pituitary content of LH increased by 2- to 3-fold, but the FSH content increased by 17-fold. This together with measurements of FSH in plasma demonstrated clearly that at least in the mouse GnRH is certainly a very potent FSH-releasing and synthesis-stimulating factor.

7. Summary

1. Studies carried out with electrical stimulation of the brain showed that (a) stimulation of the median eminence, medial preoptic area, and suprachiasmatic nucleus, but not the amygdala or hippocampus, caused a significant increase in the output of GnRH into hypophyseal portal vessel blood, and (b) the responsiveness of the preoptic–GnRH system can be altered significantly by estradiol, testosterone, and progesterone, but appears to be only moderately influenced by hypothalamic afferents, especially from the limbic system.

2. Using a steroid anesthetic (Althesin) it is possible to show that a spontaneous surge of GnRH occurs at about the time of the spontaneous, preovulatory surge of LH. This surge is very small (a total of only about 130 pg of GnRH being released) and can only cause an LH surge when acting on a pituitary whose responsiveness has been markedly increased above the levels found throughout most of the estrous cycle. The amount of GnRH released during the spontaneous surge in the rat is remarkably similar to that in the rhesus monkey and the cupric acetate-evoked surge in the rabbit, and, as assessed by electrical stimulation of the preoptic area and median eminence, is roughly equivalent to the readily releasable pool of GnRH. This pool is only about 2% of the total hypothalamic GnRH content in the rat. The spontaneous GnRH surge is blocked by sodium pentobarbitone, urethane, chloralose and ketamine hydrochoride. The surge is triggered by the preovulatory rise of the plasma concentration of estradiol-17β. Progesterone has either no effect or tends to inhibit the stimulatory effect of estradiol. As assessed by studies on PMSG-treated immature rats, the generation of the GnRH surge depends upon the functional integrity of a noradrenergic system and α adrenoreceptors. Central dopaminergic systems, depending upon which type of dopamine receptor is activated, may either facilitate or inhibit the GnRH surge.

3. A good correlation was found between the pattern of GnRH and LH release in long-term ovariectomized rats, although it was not possible to demonstrate a surge of GnRH in animals primed with estrogen and injected with progesterone. The massive surge of LH that occurs in the

latter preparation is probably due to a massive increase in pituitary responsiveness to GnRH.

4. The stimulatory effect of estrogen on pituitary responsiveness to GnRH is due not only to direct action on the pituitary, but also on an action to the hypothalamus that by increasing GnRH output or maintaining an already increased GnRH output results in an increase in pituitary responsiveness by way of the priming effect of GnRH.

5. The site and mechanism of action of estradiol in the rat differs from that in the rhesus monkey in which experiments carried out with hypothalamic lesions suggest that providing the pituitary is exposed to regularly occurring pulses of GnRH, estrogen can evoke a surge of LH by a direct action on the anterior pituitary gland.

6. As determined by studies in long-term ovariectomized rats exposed to constant light, the neural mechanism responsible for the episodic release of LH differs from that responsible for the diurnal release of LH that occurs in animals treated with large amounts of estradiol.

7. The mechanism that subserves the releasing action of GnRH differs markedly from the mechanism which subserves the priming effect of GnRH. The priming effect of GnRH involves the synthesis of a new protein that does not appear to be gonadotropin. The new protein may therefore be concerned more with the receptor-release apparatus, a shift of hormone from a storage to a readily releasable pool, and/or changes in hormone that render it more susceptible to release.

8. GnRH is essential for the long-term synthesis of the gonadotropins as shown, for example, by studies on a mutant hypogonadal mouse whose hypothalamus is deficient in GnRH.

ACKNOWLEDGEMENTS. We are most grateful to Norma Brearley and Gill Hay for the careful preparation of the manuscript. The original work reported herein, carried out mainly in the Department of Human Anatomy, Oxford, was supported by the Medical Research Council, and could not have been carried out without the generous supply of radioimmunoassay materials provided by Drs. G. D. Niswender, T. M. Nett, L. E. Reichert, Jr., and A. F. Parlow and the NIAMDD (Bethesda) and the gift of synthetic GnRH from ICI Pharmaceuticals, Cheshire.

REFERENCES

Aiyer, M. S., Fink, G., and Greig, F., 1974, Changes in the sensitivity of the pituitary gland to luteinizing hormone releasing factor during the oestrous cycle of the rat, *J. Endocrinol.* **60**:47.

Arendash, G. W. and Gallo, R. V., 1979, Effects of lesions in the suprachiasmatic nucleus—retrochiasmatic area on the inhibition of pulsatile luteinizing hormone release induced by electrical stimulation of the midbrain dorsal raphe nucleus, *Neuroendocrinology* **28**:349.

Azhar, S., Reel., J. R., Pastushok, C. A. and Menon, K. M. J., 1978, LH biosynthesis and secretion in rat anterior pituitary cell cultures: Stimulation of LH glycosylation and secretion by GnRH and an agonistic analogue and blockade by an antagonistic analogue, *Biochem. Biophys. Res. Commun.* **80**:659.

Blake, C. A. and Sawyer, C. H., 1974, Effects of hypothalamic deafferentation on the pulsatile rhythm in plasma concentrations of luteinizing hormone in ovariectomized rats, *Endocrinology* **94**:730.

Brown-Grant, K., 1977, Physiological aspects of the steroid hormone–gonadotropin interrelationship, in: *International Review of Physiology, Reproductive Physiology II*, vol. 13, (R. O. Greep, ed.), pp. 57–83, University Park Press, Baltimore.

Caligaris, L., Astrada, J. J., and Taleisnik, S., 1968, Stimulating and inhibiting effects of progesterone on the release of luteinizing hormone, *Acta Endocrinol. (Copenhagen)* **59**:177.

Caligaris, L., Astrada, J. J., and Taleisnik, S., 1971, Release of luteinizing hormone induced by estrogen injection into ovariectomized rats, *Endocrinology* **88**:810.

Carmel, P. W., Araki, S., and Ferin, M., 1976, Pituitary stalk portal blood collection in rhesus monkeys: Evidence for pulsatile release of gonadotropin-releasing hormone (GnRH), *Endocrinology* **99**:243.

Cattanach, B. M., Iddon, C. A., Charlton, H. M., Chiappa, S. A., and Fink, G., 1977, Gonadotrophin-releasing hormone deficiency in a mutant mouse with hypogonadism, *Nature (London)* **269**:338.

Chiappa, S. A., and Fink, G., 1977a, Hypothalamic luteinizing hormone releasing factor and corticotrophin releasing activity in relation to pituitary and plasma hormone levels in male and female rats, *J. Endocrinol.* **72**:195.

Chiappa, S. A., and Fink, G., 1977b, Releasing factor and hormonal changes in the hypothalamic–pituitary–gonadotrophin and–adrenocorticotrophin systems before and after birth and puberty in male, female and androgenized female rats, *J. Endocrinol.* **72**:211.

Chiappa, S. A., Fink, G., and Sherwood, N. M., 1977, Immunoreactive luteinizing hormone releasing factor (LRF) in pituitary stalk plasma from female rats: Effects of stimulating diencephalon, hippocampus and amygdala, *J. Physiol. (London)* **267**:625.

Clayton, R. N., and Catt, K. J., 1981, Regulation of pituitary gonadotropin-releasing hormone receptors by gonadal hormones, *Endocrinology* **108**:887.

Clayton, R. N., Solano, A. R., Garcia-Vela, A., Dufau, M. L., and Catt, K. J. 1980, Regulation of pituitary receptors for gonadotropin-releasing hormone during the rat estrous cycle, *Endocrinology* **107**:699.

Conn, P. M., Morrell, D. V., Dufau, M. L., and Catt, K. J., 1979, Gonadotropin-releasing hormone action in cultured pituicytes: Independence of luteinizing hormone release and adenosine 3′,5′-monophosphate production, *Endocrinology* **104**:448.

Cooper, K. J., Fawcett, C. P., and McCann, S. M., 1974, Inhibitory and facilitatory effects of estradiol-17β on pituitary responsiveness to a luteinizing hormone- follicle stimulating hormone releasing factor (LH-RF/FSH-RF) preparation in the ovariectomized rat, *Proc. Soc. Exp. Biol. Med.* **145**:1422.

Cooper, K. J., Fawcett, C. P., and McCann, S. M., 1975, Augmentation of pituitary responsiveness to luteinizing hormone/follicle stimulating hormone-releasing factor (LH-RF) as a result of acute ovariectomy in the four-day cyclic rat, *Endocrinology* **96**:1123.

Eskay, R. L., Mical, R. S., and Porter, J. C., 1977, Relationship between luteinizing hormone releasing hormone concentration in hypophysial portal blood and luteinizing hormone in intact, castrated and electrochemically-stimulated rats, *Endocrinology* **100**:263.

Everett, J. W., 1977, The timing of ovulation, *J. Endocrinol.* **75**:3p.

Ferin, M., Rosenblatt, H., Carmel, P. W., Antunes, J. L., and Vande Wiele, R. L., 1979, Estrogen induced gonadotropin surges in female rhesus monkeys after pituitary stalk section. *Endocrinology* **104**:50.

Fink, G., 1979a, Feedback actions of target hormones on hypothalamus and pituitary with special reference to gonadal steroids, *Annu. Rev. Physiol.* **41**:571.

Fink, G., 1979b, Neuroendocrine control of gonadotrophin secretion, *Br. Med. Bull.* **35**(2):155.

Fink, G., and Geffen, L. B., 1978, The hypothalamo–hypophysial system: model for central peptidergic and monoaminergic transmission, in: *International review of physiology, neurophysiology III,* Vol. 17, (R. Porter, ed.), pp. 1–48, University Park Press, Baltimore.

Fink, G., and Henderson, S. R., 1977a, Site of modulatory action of oestrogen and progesterone on gonadotrophin response to luteinizing hormone releasing factor, *J. Endocrinol.* **73**:165.

Fink, G., and Henderson, S. R., 1977b, Steroids and pituitary responsiveness in female, androgenized female and male rats, *J. Endocrinol.* **73**:157.

Fink, G., and Jamieson, M. G., 1976, Immunoreactive luteinizing hormone releasing factor in rat pituitary stalk blood: Effects of electrical stimulation of the medial preoptic area, *J. Endocrinol.* **68**:71.

Fink, G., and Pickering, A., 1980, Modulation of pituitary responsiveness to gonadotrophin releasing hormone, in: *Synthesis and Release of Adenohypophyseal Hormones,* (M. Jutisz and K. W. McKerns, eds.), pp. 617–638, Plenum Press, New York.

Fink, G., Malnick, S., Sarkar, D. K., and Twine, M., 1980, Sex difference in the anaesthetic dose of a steroid anaesthetic Althesin in the rat: effect of oestradiol-17β, *J. Physiol.* **307**:27p.

Fraser, H. M., and Baker, T. G., 1978, Changes in the ovaries of rats immunization against luteinizing hormone releasing hormone, *J. Endocrinol.* **77**:85.

Gay, V. L., and Midgley, A. R., Jr., 1969, Response of the adult rat to orchidectomy ovariectomy as determined by radioimmunoassay, *Endocrinology* **84**:1359.

Gay, V. L., and Sheth, N. A., 1972, Evidence for a periodic release of LH in castrated male and female rats, *Endocrinology* **90**:158.

Goodman, R. L., 1978, The site of the positive feedback action of estradiol in the rat, *Endocrinology* **102**:151.

Greeley, G. H., Jr., Allen, M. B., and Mahesh, V. B., 1975, Potentiation of luteinizing hormone release by estradiol at the level of the pituitary, *Neuroendocrinology* **18**:233.

Greeley, G. H., Jr., Volcan, I. J., and Mahesh, V. B., 1978, Direct effects of estradiol benzoate and testosterone on the response of the male pituitary to luteinizing hormone releasing hormone (LHRH), *Biol. Reprod.* **18**:256.

Harris, G. W., 1972, Humours and hormones, *J. Endocrinol.* **53**:ii.

Harris, G. W., and Jacobsohn, D., 1952, Functional grafts of the anterior pituitary gland,*Proc. R. Soc. London Ser. B.,* **139**:263.

Henderson, S. R., Baker, C. and Fink, G., 1977a, Oestradiol-17β and pituitary responsiveness to luteinizing hormone releasing factor in the rat: A study using rectangular pulses of oestradiol-17β monitored by non-chromatographic radioimmunoassay, *J. Endocrinol.* **73**:441.

Henderson, S. R., Baker, C., and Fink, G. 1977b, Effect of estradiol-17β exposure on the spontaneous secretion of gonadotrophins in chronically gonadectomized rats, *J. Endocrinol.* **73**:455.

Iddon, C. A., Charlton, H. M., and Fink, G., 1980, Gonadotrophin release in hypogonadal and normal mice after electrical stimulation of the median eminence or injection of luteinizing hormone releasing hormone, *J. Endocrinol.* **85:**105.

Jamieson, M. G., and Fink, G., 1976, Parameters of electrical stimulation of the medial preoptic area for release of gonadotrophins in male rats, *J. Endocrinol.* **68:**57.

Knobil, E., 1974, On the control of gonadotropin secretion in the rhesus monkey, *Recent Prog. Horm. Res.,* **30:**1.

Knobil, E., and Plant, T. M., 1978, Neuroendocrine control of gonadotropin secretion in the female rhesus monkey, in: *Frontiers in Neuroendocrinology,* Vol. 5, (W. F. Ganong and L. Martini, eds.), pp 249–264, Raven Press, New York.

Koch, Y., 1977, in: *Endocrinology* (V. H. T. James, ed.), Vol. 1, pp. 374–378, Effects of antibodies against luteinizing hormone-releasing hormone on reproduction, Excerpta Medica, Amsterdam.

Krey, L. C., Butler, W. R., and Knobil, E., 1975, Surgical disconnection of the medial basal hypothalamus and pituitary function in the rhesus monkey. I. Gonadotropin secretion, *Endocrinology* **96:**1073.

Lincoln, G. A., 1979, Pituitary control of testicular activity, *Br. Med. Bull.* **35**(2):167.

Liu, T.-C., and Jackson, G. L., 1978, Modifications of luteinizing hormone biosynthesis and release by gonadotropin-releasing hormone, cycloheximide, and actinomycin D, *Endocrinology* **103:**1253.

Liu, T.-C., and Jackson, G. L., 1979, Comparison of the biosynthesis and release of luteinizing hormone by rat pituitaries *in vitro* in response to gonadotropin-releasing hormone analogs, *Endocrinology* **104:**962.

Meidan, R., Fink, G., and Koch, Y., 1981, Ontogeny of the sensitizing effect of oestradiol and luteinizing hormone releasing hormone on the anterior pituitary gland of the female rat, *J. Endocrinol.* **91:**347.

Nakai, Y., Plant, T. M., Hess, D. L., Keogh, E. J., and Knobil, E., 1978, On the sites of negative and positive feedback actions of estradiol in the control of gonadotropin secretion in the rhesus monkey, *Endocrinology* **101:**1008.

Neill, J. D., Dailey, R. A., Tsou, R. C., Patton, J., and Tindall, G., 1976, Control of the ovarian cycle in the monkey, in: *Ovulation in the Human* (P. G. Crosignani and D. R. Mishell, eds.), pp. 115–125, Academic Press, New York.

Neill, J. D., Patton, J.M., Dailey, R. A., Tsou, R. C., and Tindall, G. T., 1977, Luteinizing hormone releasing hormone (LHRH) in pituitary stalk blood of rhesus monkeys: Relationship to level of LH release, *Endocrinology* **101:**430.

Nikitovitch-Winer, M., and Everett, J. W., 1958, Functional restitution of pituitary grafts re-transplanted from kidney to median eminence, *Endocrinology* **63:**916.

Nikitovitch-Winer, M., and Everett, J. W., 1959, Histocytologic changes in grafts of rat pituitary on the kidney and upon re-transplantation under the diencephalon, *Endocrinology* **65:**357.

Norman, R. L., Resko, J. A., and Spies, H. G., 1976, The anterior hypothalamus: How it affects gonadotropin secretion in the rhesus monkey, *Endocrinology* **99:**59.

Padmanabhan, V., Kesner, J. S., and Convey, E. M., 1981, A priming effect of luteinizing hormone releasing hormone on bovine pituitary cells *in vitro, J. Anim. Sci.* **52:**1137.

Pickering, A. J. M. C., and Fink, G., 1979a, Priming effect of luteinizing hormone releasing factor *in vitro:* role of protein synthesis, contractile elements, Ca^{2+} and cyclic AMP, *J. Endocrinol.* **81:**223.

Pickering, A. J. M. V., and Fink, G. 1979b, Do hypothalamic regulatory factors other than luteinizing hormone releasing factor exert a priming effect?, *J. Endocrinol.* **81:**235.

Pickering, A. J. M. C., and Fink, G., 1979c, Variation in size of the 'readily releasable pool' of luteinizing hormone during the oestrous cycle of the rat, *J. Endocrinol.* **83:**53.

Rabii, J., Clifton, D. K., and Sawyer, C. H., 1978, Effect of intravenous infusion of a steroid anaesthetic on the preovulatory surge of luteinizing hormone and ovulation in intact rats, *J. Endocrinol.* **76**:361.

Rusak, B., and Zucker, I., 1979, Neural regulation of circadian rhythms, *Physiol. Rev.* **59**:449.

Sarkar, D. K., and Fink, G., 1979a, Effects of gonadal steroids on output of luteinizing hormone releasing factor into pituitary stalk blood in the female rat, *J. Endocrinol.* **80**:303.

Sarkar, D. K., and Fink, G., 1979b, Mechanism of the first spontaneous gonadotrophin surge and that induced by pregnant mare serum, and effects of neonatal androgen in rats, *J. Endocrinol.* **83**:339.

Sarkar, D. K., and Fink, G., 1980, Luteinizing hormone releasing factor in pituitary stalk plasma from long-term ovariectomized rats: Effects of steroids, *J. Endocrinol.* **86**:511.

Sarkar, D. K., and Fink, G., 1981, Gonadotropin releasing hormone surge: possible modulation through postsynaptic α adrenoreceptors and two pharmacologically distinct dopamine receptors, *Endocrinology* **108**:862.

Sarkar, D. K., Chiappa, S. A., Fink, G., and Sherwood, N. M., 1976, Gonadotropin-releasing hormone surge in pro-oestrous rats, *Nature (London)* **264**:461.

Sarkar, D. K., Smith, G. C., and Fink G., 1981, Effect of manipulating central catecholamines on puberty and the surge of luteinizing hormone and gonadotropin releasing hormone induced by pregnant mare serum gonadotropin in female rats, *Brain Res.,* **213**:335.

Sawyer, C. H., 1975, Some recent developments in brain–pituitary–ovarian physiology, *Neuroendocrinology* **17**:97.

Sharpe, R. M., and Fraser, H. M., 1980, HCG stimulation of testicular LHRH-like activity, *Nature (London)* **287**:642.

Sharpe, R. M., Fraser, H. M., Cooper, I., and Rommerts, F. F. G., 1981, Sertoli-Leydig cell communication via on LHRH-like factor, *Nature (London)* **290**:785.

Sherwood, N. M., and Fink, G., 1980, Effect of ovariectomy and adrenalectomy on luteinizing hormone-releasing hormone in pituitary stalk blood from female rats, *Endocrinology* **106**:363.

Sherwood, N. M., Chiappa, S. A., and Fink, G., 1976, Immunoreactive luteinizing hormone releasing factor in pituitary stalk blood from female rats: Sex steroid modulation of response to electrical stimulation of preoptic area or median eminence, *J. Endocrinol.* **70**:501.

Sherwood, N. M., Chiappa, S. A., Sarkar, D. K., and Fink, G., 1980, Gonadotropin releasing hormone (GnRH) in pituitary stalk blood from proestrous rats; effects of anesthetics and relationship between stored and released GnRH and LH, *Endocrinology* **107**:1410.

Speight, A., and Fink, G., 1981a, Changes in responsiveness of dispersed pituitary cells to luteinizing hormone releasing factor at different times of the oestrous cycle of the rat, *J. Endocrinol.* **89**:129.

Speight, A., and Fink, G., 1981b, Comparison of steroid and LHRH effects on the responsiveness of hemipituitary glands and dispersed pituitary cells, *J. Molec. Cell. Endocrinol.,* **24**:267–281.

Speight, A., Popkin, R., Watts, A. G., and Fink, G., 1981, Oestradiol-17β increases pituitary responsiveness by a mechanism that involves the release and the priming effect of luteinizing hormone releasing factor, *J. Endocrinol.* **88**:301.

Terasawa, E., Bridson, W. E., Weishaar, D. J., and Rubens, L. V., 1980, Influence of ovarian steroids on pituitary sensitivity to luteinizing hormone-releasing hormone in the ovariectomized guinea pig, *Endocrinology* **106**:425.

Tsou, R. C., Dailey, R. A., McLanahan, C. S., Parent, A. D., Tindall, G. T., and Neill, J. D., 1977, Luteinizing hormone releasing hormone (LHRH) levels in pituitary stalk plasma during the preovulatory gonadotropin surge of rabbits, *Endocrinology* **101**:534.

Turgeon, J. L., and Waring, D. W., 1981, Acute progesterone and 17β-estradiol modulation of luteinizing hormone secretion by pituitaries of cycling rats superfused *in vitro, Endocrinology* **108**:413.

van Rees, G. P., and De Koning, J., 1979, Pattern of luteinizing hormone release induced by luteinizing hormone releasing hormone, *J. Endocrinol.* **80**:26P.

Vilchez-Martinez, J. A., Arimura, A., Debeljuk, L., and Schally, A. V., 1974, Biphasic effect of estradiol benzoate on the pituitary responsiveness to LH-RH, *Endocrinology* **94**:1300.

Walker, A. M., and Hopkins, C. R., 1978, Dissociation of the porcine anterior pituitary: The kinetics of luteinizing hormone release in response to luteinizing hormone releasing hormone, *Mol. Cell. Endocrinol.* **12**:177.

Waring, D. W., and Turgeon, J. L., 1980, Luteinizing hormone-releasing hormone-induced luteinizing hormone secretion *in vitro:* Cyclic changes in responsiveness and self-priming, *Endocrinology* **106**:1430.

Watts, A. G., and Fink, G., 1981, Constant light blocks diurnal but not pulsatile release of luteinizing hormone in the ovariectomized rat, *J. Endocrinol.* **89**:141.

Wildt, L., Hausler, A., Hutchison, J. S., Marshall, G., and Knobil, E., 1981, Estradiol as a gonadotropin releasing hormone in the rhesus monkey, *Endocrinology* **108**:2011.

Yen, S. S. C., 1978, The human menstrual cycle, in: *Reproductive Endocrinology* (S. Yen and R. Jaffe, eds.) pp 126–151, Saunders, Philadelphia.

Yen, S. S. C., Tsai, C. C., Naftolin, F., Vandenberg, G., and Ajabor, L., 1972, Pulsatile patterns of gonadotropin release in subjects with and without ovarian function, *J. Clin. Endocrinol. Metab.* **34**:671.

Yen, S. S. C., Lasley, B. L., Wang, C. F., Leblanc, H., and Siler, T. M., 1975, The operating characteristics of the hypothalamic pituitary system during the menstrual cycle and observations of biological action of somatostatin, *Recent Prog. Horm. Res.* **31**:321.

Ying, S.-Y., and Guillemin, R., 1980, Gonadocrinins: Peptides in ovarian follicular fluid stimulating the secretion of pituitary gonadotropins, Proceedings of the 62nd Annual Endocrine Society Meeting, abstract No. 158.

20

Biodegradation of Luteinizing Hormone-Releasing Hormone

Károly Nikolics, György Kéri, Balázs Szőke, Anikó Horváth, and István Teplán

1. Introduction

The concept that the proteolytic degradation of peptide hormones could be one of the key mechanisms regulating the homeostasis of the hormone, as proposed by Knights *et al.* (1973), is widely accepted. However, current knowledge about the physiological mechanisms involved is incomplete. The biodegradation of luteinizing hormone-releasing hormone (LH-RH) has been extensively studied, yet the complexity of the information accumulated does not allow a clear conclusion. In this chapter we shall review current information on the biodegradation of LH-RH and discuss the relevance of the data obtained. We also make an attempt to interpret the biochemical mechanisms involved in the degradation of LH-RH that could be of physiological importance.

2. Disappearance of LH-RH in Vivo

Exogenous LH-RH undergoes a very rapid turnover *in vivo*. Plasma disappearance curves can be described by double-exponential function. Half-disappearance times ($t_{1/2}$) between 3 and 7 min for the first compo-

---

Károly Nikolics, György Kéri, Balázs Szőke, Anikó Horváth, and Istvan Teplán • 1st Institute of Biochemistry, Semmelweis University Medical School, Budapest, Hungary.

nent and 30–60 min for the second have been reported in rats, guinea pigs, and mice (Redding and Schally, 1973; Dupont *et al.*, 1974; Sandow, 1978). Similar data were obtained in ewes (3.5 min and 22 min) by Swift and Crighton (1979a). In humans, half-disappearance times of 2.5–8 min and 20–50 min, respectively, have been reported (Redding *et al.*, 1973; Kelch *et al.*, 1975; Pimstone *et al.*, 1977; Huseman and Kelch, 1978).

Interestingly, highly active analogues of LH-RH {[D-Ser(tBu)^6]-LH-RH-(1–9)-ethylamide and [D-Ala^6]-LH-RH-(1–9)-ethylamide} did not show any delayed plasma elimination times as compared to LH-RH (Sandow, 1978; Sandow and König, 1978; Swift and Crighton, 1979a).

Patients with liver disease had $t_{1/2}$ values identical to those of normal subjects, while patients with renal failure had prolonged $t_{1/2}$ values (12–16 min for the first component) (Pimstone *et al.*, 1977). This would indicate that the kidney is an important degrading organ for LH-RH, *in vivo*. Intact LH-RH is found in urinary excretion only in traces, while LH-RH-immunoreactive material (fragments) has been detected in the urine of both untreated and LH-RH-injected individuals (Copeland *et al.*, 1977).

Serum is reported to have a high binding capacity for LH-RH (Tharandt *et al.*, 1977), which may also contribute to the rapid disappearance of exogenous LH-RH. Serum also contains active proteases capable of degrading LH-RH (Marks, 1977).

The metabolic clearance rate (MCR) of exogenous LH-RH was found to be in linear negative correlation with the amount of LH under various endocrine conditions, and thus the MCR could be valid for endogenous LH-RH disappearance as well (Tharandt *et al.*, 1979). However, the fact that LH-RH is found only in traces in peripheral blood could suggest that it is already degraded before reaching the peripheral circulation. Thus, data on *in vivo* metabolism and excretion of LH-RH are of pharmacological and clinical rather than biochemical significance.

3.   Biodegradation of LH-RH in the CNS and Pituitary

According to the models proposed by Knights *et al.* (1973), the feedback signal regulating the activity of the degrading enzyme(s) could occur both at the site of production of the peptide and at the site of its action.

According to immunocytochemical data, LH-RH as a hypophysiotropic hormone is produced in the preoptic area of the rat brain. Axonal connections reach into the median eminence, where LH-RH is released into the portal circulation (Sétáló *et al.*, 1975; Dubois, 1976). After binding to specific receptors in the anterior pituitary, LH-RH and/or its metabolites are excreted through the peripheral circulation. Other brain areas (olfactory tubercle, septum, etc.) also contain LH-RH-producing cell bodies and widespread axonal connections. It is assumed, as for other CNS peptides, that LH-RH may also play a neuromodulatory role in the CNS (Renaud *et al.*, 1975; Moss *et al.*, 1978). Consequently, studies on the biodegradation of LH-RH regulated by feedback signals were carried out with soluble and particulate-bound peptidases of pituitary, hypothalamic, and CNS origin.

3.1. Soluble LH-RH-Degrading Peptidases

Early experiments using tissue homogenates, extracts, or supernatants from whole brain, hypothalamus, and pituitary revealed very high LH-RH-degrading enzyme activities. Very low peptidase activity was found in the particulate fractions as compared to the soluble enzymes, and therefore little attention was given to particulate-bound peptidases degrading LH-RH. Griffiths *et al.* (1974, 1975b), Koch *et al.* (1974), Marks and Stern (1974), and Kuhl and Taubert (1975a) among others demonstrated degradation of LH-RH by hypothalamic or whole brain supernatants. It was shown that the activity of the soluble enzyme(s) was lower after gonadectomy, while sex steroids were able to reverse this effect (Griffiths *et al.*, 1975d; Kuhl *et al.*, 1977; Fridkin *et al.*, 1977; Swift and Crighton, 1979b; however, Loudes *et al.* (1978) were not able to find this regulation. The activity of the LH-RH-degrading enzyme(s) was found to change during the estrous cycle in both pituitary and hypothalamus (Kuhl *et al.*, 1978, 1979a). As a mediator of the short-loop feedback mechanism between pituitary and hypothalamus, LH was found to activate the soluble hypothalamic LH-RH-degrading enzyme(s) (Kuhl and Taubert, 1975b).

Other brain areas were reported to contain levels of degrading enzyme activity approximately identical with that of the hypothalamus (Griffiths *et al.*, 1975b,c, 1978; Marks and Stern, 1974; Kochman *et al.*, 1975; Loudes *et al.*, 1978), suggesting a nonspecific, uniform LH-RH-degrading system in the CNS. Also, arylamidases capable of degrading LH-RH (Kuhl and Taubert, 1975a) were found to show a uniform distribution in different brain areas (Wenn and Kamberi, 1977).

Soluble enzymes from liver and kidney homogenates also degraded LH-RH at a similar rate as hypothalamic and pituitary extracts (Sundberg and Knigge, 1978; Stetler-Stevenson *et al.*, 1980).

The Nature of Soluble LH-RH-Degrading Enzymes. The LH-RH molecule contains pyroglutamic acid at the N- terminus and glycinamide at the C-terminus. Because of the blocked termini, the enzymes degrading LH-RH, with the exception of pyroglutamate aminopeptidase, can be considered to be endopeptidases. Figure 1 summarizes the cleavage sites of soluble LH-RH-degrading peptidases, while some of the major characteristics of the enzymes are listed in Table I. The effect of various enzyme inhibitors on the activity of LH-RH-degrading enzymes is summarized in Table II.

Table I. Properties of Soluble LH-RH-Degrading Peptidases from the Pituitary and the CNS

Enzyme	Origin[a]	Cleavage site	Purification	Molecular weight	pH optimum	K_m (LH-RH)	Reference
Pyroglutamate aminopeptidase	Bovine HT, AP	Glp^1-His^2		28,000			Bauer et al. (1979a,b)
Nonchymotrypsin-like endopeptidase	Bovine AP	His^2-Trp^3, Tyr^5-Gly^6	900×	83,000	8.0	180 µM	Horsthemke and Bauer (1980)
Neutral endopeptidase	Bovine HT	Tyr^5-Gly^6, Gly^6-Leu^7	100×	40,000	7.6	13 µM	Galoyan et al. (1978), Akopyan et al. (1979)
LH-RH-degrading enzyme	Rat HT, AP	$(Ser^4$-$Tyr^5)$, Gly^6-Leu^7	10×		7.6	0.64 µM, HT: 0.87 µM. AP	Fridkin et al. (1977), Koch and Fridkin (1979)
Post-proline cleaving enzyme	Rabbit brain	Pro^9-Gly^{10}	600×	66,000	8.3		Orlowski et al. (1979)
TRF-deamidating enzyme	Bovine AP	Pro^9-Gly^{10}	110×	76,000	7.4–7.6	410 µM (TRH)	Knisatschek and Bauer (1979)
Prolyl endopeptidase	Rat brain	Pro^9-Gly^{10}	7800×	70,000	5.8–6.5		Kato et al. (1980)

[a]Abbreviations used: HT, hypothalamus; AP, anterior pituitary.

Table II. Effect of Inhibitors on the Activity of Soluble LH-RH-Degrading Peptidases[a]

Enzyme	PMSF	DFP	CMB	NEM	IA	BT	EDTA	Ca^{2+}	Zn^{2+}	Co^{2+}	Mn^{2+}	Reference
Nonchymotrypsin-like endopeptidase			+	+	−		−	−				Horsthemke and Bauer, (1980)
Neutral endopeptidase	−	−	+		+		−	−	+	+	−	Galoyan et al. (1978), Akopyan et al. (1979)
LH-RH-degrading enzyme		+	+	+		+	−					Fridkin et al. (1977), Koch and Fridkin (1979)
Post-proline cleaving enzyme	−	+	+	+	−			−	+	+	+	Orlowski et al. (1979)
TRF-deamidating enzyme	−	+	+	+	+	+		−	+	+	−	Knisatschek and Bauer (1979)
Prolyl endopeptidase	−	+	+	+		+		−	+		+	Kato et al. (1980)

[a]Abbreviations used: PMSF, phenylmethanesulfonyl-fluoride; DFP, diisopropyl fluorophosphate; CMB, p-chloromercuribenzoate; NEM, N-ethylmaleimide; IA, iodoacetate; BT, bacitracin.

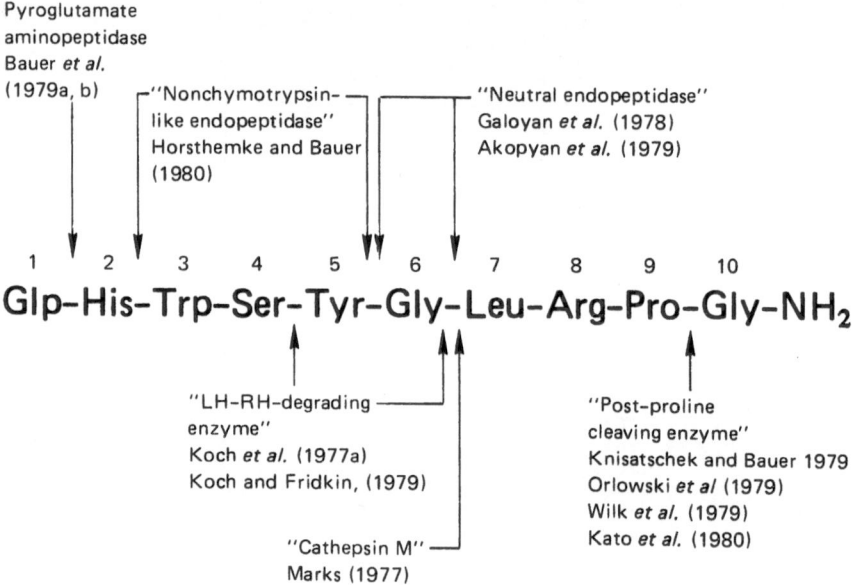

Figure 1. Cleavage sites of LH-RH-degrading soluble enzymes.

Bauer *et al.* (1979a,b) demonstrated that [³H]-LH-RH, during incubation with hypothalamic and pituitary tissue extracts, was cleaved between Glp¹ and His², due to the action of pyroglutamate aminopeptidase.

A "nonchymotrypsin-like endopeptidase" of pituitary origin was purified and characterized by Horsthemke and Bauer (1980). The enzyme cleaves LH-RH between His²-Trp³ and Tyr⁵-Gly⁶.

Akopyan *et al.* (1979) and Galoyan *et al.* (1979) purified a "neutral endopeptidase" from bovine hypothalami. This enzyme is able to cleave the Tyr⁵-Gly⁶ and the Gly⁶-Leu⁷ bonds.

Another LH-RH-degrading enzyme cleaving the decapeptide between Gly⁶-Leu⁷ was isolated and partially purified from both rat hypothalami and pituitaries (Koch *et al.*, 1974, 1977a; Fridkin *et al.*, 1977). The hypothalamic peptidase can also cleave the Ser⁴-Tyr⁵ bond, while the pituitary enzyme cannot (Koch and Fridkin, 1979).

Cathepsin M, which may be identical to the hypothalamic "neutral endopeptidase," also attacks the molecule at Gly⁶-Leu⁷ (Marks, 1977).

Post-proline cleaving enzyme, purified from rabbit; (Orlowski *et al.*, 1979; Yoshimoto *et al.*, 1978) and rat brain (Kato *et al.*, 1980; Tate, 1978), is able to hydrolyze LH-RH between Pro⁹-Gly¹⁰-Ng₂ (Knisatschek and Bauer, 1979; Wilk *et al.*, 1979).

Parallelism between L-cystine-bis-4-nitroanilide (Cys-NA) hydrolysis and degradation of LH-RH was found by Kuhl and Taubert (1975a). The hydrolysis of Cys-NA could be inhibited by LH-RH, TRH, oxytocin, and vasopressin; conversely, the degradation of LH-RH could be inhibited by Cys-NA (Kuhl *et al.,* 1979b). Consequently, the enzyme termed "L-cystine arylamidase" has been extensively studied in correlation with LH-RH degradation; however, its nature is not as yet known. Since arylamidases were found to lack protease or endopeptidase activity, it remains questionable at which site this enzyme would attack LH-RH.

Additional to the data summarized in Table I, McKelvy *et al.* (1976) found a K_m value of 6.3 $\times$ 10^{-6} M for LH-RH degradation by soluble neutral peptidases from guinea pig brain. Loudes *et al.* (1978) reported K_m values between 2.1 and 3.3 $\times$ 10^{-5} M in rat median eminence, cortex, and pituitary, which did not change after gonadectomy. Sundberg and Knigge (1978) measured a K_m of 2.1 $\times$ 10^{-8} M for rat hypothalamic soluble LH-RH-degrading enzyme activity at a pH optimum of 6.9–7.4. Hersh and McKelvy (1979) partially purified two LH-RH-degrading enzymes, one of which seemed to be identical with post-proline cleaving enzyme. A K_m of 12.4 μM was measured for degradation of LH-RH by hypothalamic and pituitary supernatants (Kuhl *et al.,* 1979b).

The similar conditions of preparation of tissue homogenates and extracts and the neutral pH optimum of the soluble LH-RH-degrading peptidases suggest that parallel cleavage at several sites would occur during incubation. Bauer *et al.* (1979a,b) reported concomitant release of at least three fragments originating from specifically radiolabeled LH-RH during incubation with hypothalamic and pituitary extracts.

Highly active analogues of LH-RH substituted by various D-amino acids in position 6 and by ethylamide in position 10 were found to be highly resistant to degradation by soluble pituitary and CNS enzymes (Marks and Stern, 1974; Koch *et al.,* 1977b; Sandow *et al.,* 1979; Griffiths and Hopkinson, 1979). Analogues containing D-amino acid[6] or ethylamide[10] substitutions only, were readily degraded by brain tissue homogenates, further suggesting concomitant cleavage of several bonds (Marks, 1977).

3.2. Subcellular Distribution of LH-RH-Degrading Enzymes in the CNS

The hypothalamus and other brain areas contain very active LH-RH-degrading peptidases in the soluble fractions or supernatants of homogenates (Griffiths *et al.,* 1974, 1975a, 1975b, 1978; Griffiths and Kelly, 1979; Loudes *et al.,* 1978; Sundberg and Knigge, 1978). Joseph-Bravo *et al.* (1979) and Parker *et al.* (1979) investigated the subcellular localization of peptidases degrading LH-RH. Hypothalamic and cortex cytosol contained very active LH-RH-degrading peptidases, while intact synaptosomes and mitochondria showed very little proteolytic activity. Disruption of the synaptosomes revealed very high degrading enzyme activity in the synaptosol, while membrane-bound activity was very low.

Dopamine in high concentrations (10^{-5}–10^{-4} M) stimulated the activity of LH-RH-degrading enzymes of intact synaptosomes of hypothalamic origin (Marcano de Cotte et al., 1980; Edwardson et al., 1980). This mechanism was found to be dependent on the endocrine status of female rats and required the presence of Ca^{2+}. Bacitracin completely inhibited the degradation of LH-RH by synaptosomes. It is still questionable whether dopamine induces the release of peptidases from synaptosomes or membrane-bound peptidases are being activated.

Sundberg and Knigge (1978) and Powers and Johnson (1980) reported the release of peptidases from intact mediobasal hypothalamic tissue into the culture medium, capable of degrading LH-RH. The activity of the enzyme(s) could be inhibited by bacitracin.

3.3 Membrane-Bound LH-RH-Degrading Enzymes in the Pituitary

The pituitary action of LH-RH involves its specific binding to plasma membrane receptors of pituitary gonadotrophs. The specific binding process has been extensively studied by several laboratories (e.g., Spona, 1973; Baumann and Kuhl, 1979; Clayton et al., 1979b; Wagner et al., 1979). Two distinct binding sites were found: one with low affinity and high capacity, and one with high affinity and low capacity.

Several authors observed degradation of radiolabeled LH-RH during binding experiments with purified pituitary plasma membranes (Clayton et al., 1977a,b; Pedroza et al., 1977; Wagner et al., 1979). Further analyzing this phenomenon, Clayton et al. (1979a) clearly demonstrated that the activity of the LH-RH-degrading enzyme was associated with the plasma membranes. The authors assumed that the low-affinity binding of LH-RH would represent binding to the degrading enzyme. Degradation was found to be very rapid and it could be inhibited by bacitracin (Clayton et al., 1979a; Pedroza et al., 1977), trasylol and aprotinin (Wagner et al., 1979). Analogues of LH-RH substituted by D-amino acids in position 6 and ethylamide for Gly^{10}-NH$_2$ were found to be resistant to degradation by pituitary plasma membranes (Clayton et al., 1979b).

Recently, Baumann and Kuhl (1980) also found degradation of LH-RH by isolated pituitary plasma membranes. Interestingly, similar to the soluble enzyme(s), the degradation of LH-RH by plasma membranes also coincided with "cystine-arylamidase" activity. Chromatographic purification of the receptor protein revealed one binding protein with a molecular weight of approximately 80,000. Although not finding degradation of LH-RH, Zolman and Valenta (1980) also isolated and highly purified an LH-RH-binding protein with a molecular weight of 60,000. Degradation of labeled LH-RH by pituitary plasma membranes was further reported by

Wagner (1980). Degrading activity to a smaller extent was also associated with plasma membranes of median eminence, cortex, and liver origin. Studying the inhibitory activity on LH-release of synthetic derivatives of LH-RH, containing three or more D-amino acids, we observed their prolonged inhibitory effect in pituitary cell culture (Nikolics *et al.*, 1981). Since this prolonged action may be a result of resistance to degrading peptidases, we investigated the degradation of LH-RH itself by pituitary cell culture. Degradation was followed by RIA in the case of native LH-RH and HPLC analysis in the case of ^{125}I-labeled LH-RH. In both cases complete and rapid degradation was found. [^{125}I]-LH-RH, as measured by HPLC analysis incubation media, was rapidly degraded, with a $t_{1/2}$ of approximately 15 min (Nikolics *et al.*, 1980).

It seems from these experiments that degradation of LH-RH by pituitary cells would play an important regulatory role in the action of the hormone on pituitary gonadotrophs.

4. Physiological Significance of LH-RH Biodegradation

Due to the dual function of LH-RH as a hypophysiotropic hormone and as a neuromodulator in the CNS, it appears reasonable to treat the proteolytic degradation of LH-RH by the pituitary and the CNS separately.

LH-RH, as a hypophysiotropic hormone, binds to specific receptors of anterior pituitary cells. Radiolabeled LH-RH has been localized on cell surface receptors by autoradiography (Hopkins and Gregory, 1977). Specific binding of endogenous LH-RH was demonstrated to occur only to plasma membranes, while no LH-RH could be detected in pituitary cytosol (Pedroza-Garcia *et al.*, 1980). Exogenous LH-RH also binds to plasma membranes and secretory granules only (Wagner *et al.*, 1979; Zolman and Valenta, 1980; Pedroza-Garcia *et al.*, 1980). These findings strongly support that no internalization of LH-RH occurs in pituitary cells. With intact pituitary cells we were also not able to detect internalization of radiolabeled LH-RH (Kéri and Nikolics, unpublished).

The subcellular localization of the pituitary LH-RH-degrading enzyme activity showed high peptidase activity in the plasma membrane fraction but also very high peptidase activity was contained in the pituitary cystosol (Clayton *et al.*, 1979a).

Earlier work on pituitary LH-RH-degrading peptidases focused interest on the soluble enzymes due to their relatively high activity and their activation by sex steroids. However, the very complex nature of the soluble peptidases and their uniform organ distribution suggest their possible lysosomal origin.

Additionally, the degradation of LH-RH by pituitary plasma membranes (Clayton et al., 1977a, 1979a; Baumann and Kuhl, 1980; Wagner, 1980) and intact pituitary cells (Nikolics et al., 1980b) strongly suggests that the physiologically significant degradation of LH-RH occurs on the cell surface rather than within the cell. The possibility that intracellular peptidases can be secreted remains to be investigated.

Although membrane-bound enzymes degrading LH-RH were found to be uneffected by sex steroids (Clayton et al., 1977a), the feedback regulation at the pituitary level by steroids could occur via direct regulation of specific binding of LH-RH (Spona, 1975; Clayton et al., 1980).

It is conceivable that a structural relationship exists between soluble and membrane-bound LH-RH-degrading peptidases. Highly active analogues were resistant to both soluble (Marks and Stern, 1974; Koch et al., 1977b; Sandow et al., 1979; Griffiths and Hopkinson, 1979) and membrane-bound peptidases (Clayton et al., 1979b). Bacitracin and trasylol were found to inhibit LH-RH degradation by both soluble (Koch et al., 1974; Kochman et al., 1975; Knisatschek and Bauer, 1979) and membrane-bound enzymes (Pedroza et al., 1977; Clayton et al., 1979a). The parallelism between LH-RH degradation and Cys-NA hydrolysis by both soluble (Kuhl and Taubert, 1975a; Kuhl et al., 1979b) and membrane-bound peptidases (Baumann and Kuhl, 1980) further suggests a structural similarity between the two families of enzymes. Thus, studies with soluble LH-RH-degrading peptidases may provide useful information for studies with membrane-bound peptidases. The regulatory function of the effect of LH-RH on the pituitary gonadotrophs by the membrane-bound degrading system requires further clarification. The phenomenon of degradation by target cell plasma membranes has already been observed with glucagon and insulin (Desbuquois and Cuatrecasas, 1972; Freychet et al., 1972; Pohl et al., 1972).

The nature of membrane-bound enzymes is unknown yet, and further studies are needed to reveal the mechanism of action of membrane-bound LH-RH-degrading enzymes and their regulation by various factors.

In the CNS, LH-RH is assumed to function as a neuromodulator (Renaud et al., 1975; Moss et al., 1978). LH-RH has been localized by immunocytochemical methods within axon terminals (Ajika, 1979). The LH-RH-degrading enzyme activity in the hypothalamus and cortex was found to be highest in the synaptosol, while LH-RH was highest in the synaptosomal membrane (Joseph-Bravo et al., 1979; Parker et al., 1979). These data would suggest that the degradation of LH-RH in the CNS occurs in the synaptosomal membrane.

However, it has recently been reported that LH-RH specifically binds to hypothalamic subsynaptosomal fractions. Hypothalamic cytosol, in contrast to pituitary cytosol, also showed specific binding of LH-RH (Pedroza et al., 1980).

Intact hypothalamic tissue was found to release LH-RH-degrading enzymes into the incubation medium (Sundberg and Knigge, 1978; Powers and Johnson, 1980). Dopamine in low concentrations (10^{-9}–10^{-7} M) stimulated the release of LH-RH from synaptosomes, while in high concentrations 10^{-5}–10^{-4} M) activated the degradation of LH-RH (Marcano de Cotte *et al.*, 1980; Edwardson *et al.*, 1980), suggesting the possibility of a complex regulatory system.

In conclusion, the data are not sufficient to propose a unique degrading mechanism of LH-RH of physiological relevance. It seems, in view of recent data, that in the CNS, soluble enzymes in the cytosol or synaptosol may be of physiological importance in the regulation of overall tissue levels of LH-RH. Further evidence is needed for the localization of the degrading enzyme(s) involved in the regulation of LH-RH homeostasis in the CNS.

DISCUSSION

AAKVAAG: Your very active LH-RH inhibitory peptides seem to be carrying out a chemical castration. Have you tested these in male animals by measuring testosterone in serum?

NIKOLICS: Indeed, there are data in the literature showing a lowering of testosterone in male animals. Inhibitory analogues of LH-RH have been expected to be effective male contraceptives as well. The idea of using the inhibitors as a way of chemical castration is very challenging; however, the amount of material presently required for chronic administration is too high (considering the price of the peptides) to introduce it as a general practice.

PECK: I am fascinated by your tetramer and its potential use as an immunogen. However, the antigenic site is likely to be at the carboxyl end in that case. Can an analogue tetramer be constructed through carboxyl ends of LH-RH?

NIKOLICS: Practically any kind of tetramer could be constructed from LH-RH derivatives. LH-RH itself, having no free carboxyl or amino groups, cannot be coupled to carriers like BSA by carbodimide. Either Glu[1]- or Gly[10]-LH-RH is being used for this purpose. Glu[1]-LH-RH can be readily coupled with EDTA-tetra-activated esters. Gly[10]-LH-RH through a corresponding handle could also be coupled with EDTA. The real advantage of the tetramers lies in their well-characterized structure. That could be another way of raising less diverse populations of antibodies.

McKERNS: What precautions do you take in the storing of your nonlabeled precursor peptides to assure stability?

NIKOLICS: Peptides as lyophilized powders can usually be stored for long periods of time at 0 or $-20°$C. The same applies for the precursor peptides we have been using for tritiation. Storing peptides as solids is certainly more advantageous than storing them in solution.

KRAICER: You implied that the degrading enzyme in the cell culture studies is plasma membrane bound. What evidence do you have for this? Why would it not be released and found in the medium?

NIKOLICS: We have not tested whether the incubation medium after standing for 4 hr with cells degrades LH-RH or not. We have planned this experiment already, and I hope that very soon we can report on this. However, as reported by Clayton et al., [Endocrinology 104:1484 (1979)], the LH-RH-degrading activity was highest in the plasma membrane-containing fractions of pituitary cells and no activity was found in the secretory granules or in the endoplasmic reticulum. Although pituitary cytosol is also reported to be very active in degrading LH-RH [Kochman et al., FEBS Lett. 50:190 (1975); Koch et al., Biochem. Biophys. Res. Commun. 74:488 (1977); Loudes et al., Biochem. Biophys. Res. Commun. 83:921 (1978); Clayton et al., Endocrinology 104:1484 (1979)], we do not find incorporation of [^{125}I]-LH-RH or [^{3}H]-LH-RH in pituitary cells and culture. Thus, the cytosol LH-RH-degrading activity may not be specific, but due to the high nonspecific neutral proteinase activity.

KRAICER: I have a related physiological question. What would be the role of such a membrane-bound enzyme? What would be its relation to the LH-RH receptor? What role might it play in regulating the action of LH-RH on gonadotrophs? The membrane receptor binding of [^{125}I]-LH-RH has been reported to be influenced by sex steroids [Drouin et al., Endocrinology 99:1477 (1976); Spona, FEBS Lett. 35:59 (1973) and Endocrinol. Exp. 9:167 (1975). Kuhl et al., Acta Endocrinol. (Copenhagen) 87:476 (1978), reported the stimulation of LH-RH-degrading "pituitary L-cystine arylamidase" by sex steroids. (However, this is again an enzyme of cytosolic origin.) The regulation of the levels of LH-RH is probably a complex system functioning at more than one level. It is well conceivable that the degradation of LH-RH at the pituitary level is of physiological significance in controlling the responsiveness of the gonadotroph cells.

HERBERT: What part of the pituitary have you used for your experiments in the degradation of LH-RH?

NIKOLICS: Primary cell cultures of rat anterior pituitary were used. Anterior pituitaries were dispersed after removal by trypsinization at 37°C, 20 min under CO_2 atmosphere; washed by medium 199, and plated into multiwells in a concentration of 6×10^5/ml. Cells were grown in medium 199 containing 10% fetal calf serum. On the day of incubation (day 4 of culture), cells were washed with serum-free medium 199, and incubated with the same medium containing radiolabeled LH-RH.

HERBERT: How stable are the tritiated peptides on storage and what are the most favorable conditions for storage?

NIKOLICS: Degradation (autoradiolysis) of the peptides during long periods of time shows great variations. The processes involved in radiolysis are radical reactions, and the theory of understanding these is very unclear. However, as a general rule, storage between −40 and −80°C is worse than 0 or −190°C (liquid nitrogen), probably due to the prolonged lifetime of the radicals formed. Usually, the addition of scavengers such as ethanol, benzyl, alcohol, etc., improves the stability of the peptides; e.g., [^{3}H]bradykinin, in a dilute 50% methanolic solution, is stable for 2 years at 0°C; [^{3}H]-LH-RH, in a dilute aqueous solution containing ethanol or methanol, has been found to be stable for 6 months. Radiolysis is also dependent on the specific activity of the peptides. Solutions with a radioactive concentration below 1 m Ci/ml are to be used for storage. Another method for storage is the use of a "carrier," e.g., [^{3}H]substance P (27 Ci/mol) was nearly completely stable when stored absorbed onto a paper strip for 6 months.

KANAZIR: Does the degradation of hormones also play a role in the mechanism of down-control of the numbers of receptors by hormones?

NIKOLICS: I don't think that the degradation of LH-RH is associated with a decrease in the number of receptors. In fact, Clayton et al. [Endocrinology 104:1484 (1979)] reported that the quantitation of specific (high affinity) binding of [^{125}I]-LH-RH to purify pituitary plasma membranes could only be done with a correction of LH-RH degradation. This did not imply a decrease in the number of receptors.

KANAZIR: Could we speculate further that some other regulatory mechanisms are involved in the controlled cleavage?

NIKOLICS: LH-RH-degrading enzymes show a very wide distribution. Whether or not these enzymes are identical with cathepsin M (neutral proteinase) and/or with prolyl endopeptidase or other enzymes is not clear yet. LH-RH is being released from secretory granules of the median eminence into the portal blood, and this transports it into the anterior pituitary where it is bound to plasma membrane receptors. Degradation of LH-RH could already start in the secretory granules, or at the site of release, in the portal blood system and in the anterior pituitary. So far we do not have data for the degradation by the earlier sites, but we find degradation in the pituitary cell culture. This degradation is certainly influenced by several factors, e.g., levels of sex steroids, etc.

MCKERNS: I would like to make a comment relative to the inhibitory effects of potent LH-RH analogues. The administration of a high level of LH-RH or its more potent agonists leads to an eventual loss of response of the corpus luteum to LH or to the response of the developing follicles to FSH. This has been thought to be due possibly to depletion of precursor metabolites, a loss of LH receptors (down-regulation), or to some intrinsic inhibitory action of LH-RH on the corpus luteum or follicular function.

It is interesting that the N-terminal sequence of LH-RH:Leu-Arg-Pro-Gly-NH$_2$ has similarities to my E$_2$ inhibitory peptide and both peptides have sequences similar to hCG beta 134–138, namely, -Leu-Pro-Gly-Pro-, and to similar sequences in LII beta.

REFERENCES

Ajika, K., 1979, Simultaneous localization of LH-RH and catecholamines in rat hypothalamus, *J. Anat.* **128**:331.

Akopyan, T. N., Arutunyan, A. A., Oganisyan, A. I., Lajtha, A., and Galoyan, A. A., 1979, Breakdown of hypothalamic peptides by hypothalamic neutral endopeptidase, *J. Neurochem.* **32**:629.

Bauer, K., 1980, Degradation of neuropeptides, in: *Brain and Pituitary Peptides* (W. Wuttke, A. Weindl, K. H. Voigt, and R.-R. Dries, eds.) pp. 213–222, Karger, Basel.

Bauer, K., Horsthemke, B., Knisatschek, H., Nowak, P., and Kleinkauf, H., 1979a, Degradation of luliberin by brain and pituitary tissue enzymes, *Hoppe-Seyler's Z. Physiol. Chem.* **360**:229.

Bauer, K., Knisatschek, H., Horsthemke, B., and Nowak, P., 1979b, Enzymatic degradation of neuropeptides, *Acta Endocrinol. (Copenhagen)* **91**:422.

Baumann, R., and Kuhl, H., 1979, Interaction of ^{125}I-LH-RH and other oligopeptides with plasma membranes of rat anterior pituitaries, *Acta Endocrinol. (Copenhagen)* **92**:228.

Baumann, R., and Kuhl, H., 1980, LH-RH receptors on isolated pituitary plasma membranes: An LH-RH degrading enzyme?, *Acta Endocrinol. (Copenhagen)* **94**:80.

Clayton, R. N., Shakespear, R. A., and Marshall, J. C., 1977a, Effect of testosterone and oestradiol on LH-RH degradation by purified plasma membranes, *Acta Endocrinol. (Copenhagen)* **85**:59.

Clayton, R. N., Shakespear, R. A., and Marshall, J. C., 1977b, LH-RH degrading activity assoicated with a purified pituitary plasma membrane fraction, *J. Endocrinol.* **73**:P34.

Clayton, R. N., Shakespear, R. A., Duncan, J. A., and Marshall, J. C., 1979a, LH-RH inactivation by purified pituitary plasma membranes: Effects on receptor-binding studies, *Endocrinology* **104**:1484.

Clayton, R. N., Shakespear, R. A., Duncan, J. A., and Marshall, J. C., 1979b, Radioiodinated nondegradable Gn-RH analogs: New probes for the investigation of pituitary Gn-RH receptors, *Endocrinology* **105**:1369.

Clayton, R. N., Solano, A. R., Garcia-Vela, A., Dufau, M. L., and Catt, K. J., 1980, Regulation of pituitary receptors for Gn-RH during the rat estrous cycle, *Endocrinology* **107**:699.

Copeland, K. C., Aubert, M. L., Paunier, L., and Sizonenko, P. C., 1977, Measurement of urinary LH-RH *Acta Endocrinol. (Copenhagen)* **85**:60.

Desbuquois, B., and Cuatrecasas, P., 1972, Independence of glucagon receptors and glucagon inactivation in liver cell membranes, *Nature (New Biol.)* **237**:202.

Dubois, M.-P., 1976, Immunocytological evidence of LH-RF in hypothalamus and median eminence: A review, *Ann. Biol. Anim. Biochim. Biophys.* **16**:177.

Dupont, A., Labrie, F., Pelletier, G., Puviani, R., Coy, D. H., Coy, E. J., and Schally, A. V., 1974, Organ distribution of radioactivity and disappearance of radioactivity from plasma after administration of ^{3}H-LH-RH to mice and rats, *Neuroendocrinology* **16**:65.

Edwardson, J. A., Marcano de Cotte, D., and De Menezes, C. E. L., 1980, A complex hypothalamic mechanism for the degradation of gonadotrophin-releasing hormone, *J. Physiol. (London)* **301**:P89.

Freychet, P., Kahn, R., Roth, J., and Neville, D. M., 1972, Insulin interactions with liver plasma membranes, *J. Biol. Chem.* **247**:3953.

Fridkin, M., Hazum, E., Baram, T., Lindner, H. R., and Koch, Y., 1977, Hypothalamic and pituitary LRF-degrading enzymes: Characterization, purification and physiological role, in: *Peptides 1977* (M. Goodman and J. Meienhofer, eds.), pp. 193–196, Wiley, New York.

Galoyan, A. A., Akopyan, T. N., Karapetyan, R. O., Arutunyan, A. A., and Oganissyan, A. I., 1978, Enzymatic mechanisms of the formation of biologically active peptides in the hypothalamus, in *Endorphins –78* (L. Gráf, M. Palkovits, and A. Z. Rónai, eds.), pp. 37–59, Akadémiai Kiadó, Budapest.

Griffiths, E. C., and Hopkinson, C. R. N., 1979, Inactivation of two hyperactive LH-RH analogues by rat hypothalamic peptidases, *Hormone Res.* **10**:233.

Griffiths, E. C., and Kelly, J. A., 1979, Mechanisms of inactivation of hypothalamic regulatory hormones, *Mol. Cell. Endocrinol.* **14**:3.

Griffiths, E. C., Hooper, K. C., Jeffcoate, S. L., and Holland, D. T., 1974, The presence of peptidases in the rat hypothalamus inactivating LH-RH, *Acta Endocrinol. (Copenhagen)* **77**:435.

Griffiths, E. C., Hooper, K. C., and Hopkinson, C. R. N., 1975a, Further studies on enzymic inactivation of LH-RH by peptidases in the rat hypothalamus, *Acta Endocrinol. (Copenhagen)* **79**:7.

Griffiths, E. C., Hooper, K. C., Hutson, D., Jeffcoate, S. L., and Holland, D. T., 1975b, The presence of peptidases in the rabbit hypothalamus capable of inactivating LH-RH, *Fertil. Steril.* **26**:802.

Griffiths, E. C., Hooper, K. C., Jeffcoate, S. L., and Holland, D. T., 1975c, Peptidases in different areas of the rat brain inactivating LH-RH, *Brain Res.* **85**:161.

Griffiths, E. C., Hooper, K. C., Jeffcoate, S. L., and Holland, D. T., 1975d, The effects of gonadectomy and gonadal steroids on the activity of hypothalamic peptidases inactivating LH-RH *Brain Res.* **88**:384.

Griffiths, E. C., Forbes, R., Jeffcoate, S. L., and Holland, D. T., 1978, Local degradation of LH-RH in the rat CNS, *Neurosci. Lett.* **7**:49.

Hersh, L. B., and McKelvy, J. F., 1979, Enzymes involved in the degradation of TRH and LH-RH in bovine brain, *Brain Res.* **168**:553.

Hopkins, C. R., and Gregory, H., 1977, Topographical localization of the receptors for LH-RH on the surface of dissociated pituitary cells, *J. Cell Biol.* **75**:528.

Horsthemke, B., and Bauer, K., 1980, Characterization of a nonchymotrypsin-like endopeptidase from anterior pituitary that hydrolyzes LH-RH at the Tyr-Gly and His-Trp bonds, *Biochemistry* **19**:2867.

Huseman, C. A., and Kelch, R. P., 1978, Gonadotropin responses and metabolism of synthetic Gn-RH during constant infusion of Gn-RH in men and boys with delayed adolescence, *J. Clin. Endocrinol. Metab.* **47**:1325.

Joseph-Bravo, P., Loudes, C., Charli, J. L., and Kordon, C., 1979, Subcellular distribution of brain peptidases degrading LH-RH and TRH, *Brain Res.* **166**:321.

Kato, T., Nakano, T., Kojima, K., Nagatsu, T., and Sakakibara, S., 1980, Changes in prolyl endopeptidase during maturation of rat brain and hydrolysis of substance P by the purified enzyme, *J. Neurochem.* **35**:527.

Kelch, R. P., Clemens, L. E., Markovs, M., Westhoff, M. H., and Hawkins, D. W., 1975, Metabolism and effects of synthetic Gn-RH in children and adults, *J. Clin. Endocrinol. Metab.* **40**:53.

Knights, E. B., Baylin, S. B., and Foster, G. V., 1973, Control of polypeptide hormones by enzymatic degradation, *Lancet* **2**:719.

Knisatschek, H., and Bauer, K., 1979, Characterization of "thyroliberin-deamidating enzyme" as a post-proline-cleaving enzyme, *J. Biol. Chem.* **254**:10936.

Koch, Y., and Fridkin, M., 1979, Biodegradation of luteinizing hormone-releasing hormone, *Acta Endocrinol. (Copenhagen)* **91**:468.

Koch, Y., Baram, T., Chobsieng, P., and Fridkin, M., 1974, Enzymic degradation of LH-RH by hypothalamic tissue, *Biochem. Biophys. Res. Commun.* **61**:95.

Koch, Y., Baram, T., Hazum, E., and Fridkin, M., 1977a, Purification of the Gn-RH-degrading enzyme by affinity chromatography, *Endocr. Res. Commun.* **4**:247.

Koch, Y., Baram, T., Hazum, E., and Fridkin, M., 1977b, Resistance to enzymic degradation of LH-RH analogues possessing increased biological activity, *Biochem. Biophys. Res. Commun.* **74**:488.

Kochman, K., Kerdelhué, B., Zor, U., and Jutisz, M., 1975, Studies of enzymatic degradation of LH-RH by different tissues, *FEBS Lett.* **50**:190.

Kuhl, H., and Taubert, H.-D., 1975a, Inactivation of LH-RH by rat hypothalamic L-cystine arylamidase, *Acta Endocrinol. (Copenhagen)* **78**:634.

Kuhl, H., and Taubert, H.-D., 1975b, Short-loop feedback mechanism of LH: LH stimulates hypothalamic L-cystine arylamidase to inactivate LH-RH in the rat hypothalamus, *Acta Endocrinol. (Copenhagen)* **78**:649.

Kuhl, H., Rosniatowski, C., and Taubert, H.-D., 1977, The regulatory function of a pituitary LH-RH-degrading enzyme system in the feedback control of gonadotrophins, *Acta Endocrinol. (Copenhagen)* **86**:60.

Kuhl, H., Rosniatowski, C., and Taubert, H.-D., 1978, The activity of the LH-RH-degrading enzyme in the anterior pituitary during the rat oestrus cycle and its alteration by injections of sex hormones, *Acta Endocrinol. (Copenhagen)* **87**:476.

Kuhl, H., Rosniatowski, C., and Taubert, H.-D., 1979a, Effect of sex hormones on LH-RH-degrading hypothalamic enzyme system during estrus cycle in rats, *Endocrinol. Exp.* **13**:29.

Kuhl, H., Sandow, J., Krauss, B., and Taubert, H.-D., 1979b, Enzyme kinetic studies and inhibition by oligopeptides of LH-RH degradation in rat hypothalamus and pituitary, *Neuroendocrinology* **28**:339.

Loudes, C., Joseph-Bravo, P., Leblanc, P., and Kordon, C., 1978, Specific activity of LH-

RH degrading enzymes in various tissues of normal and castrated male rats, *Biochem. Biophys. Res. Commun.* **83**:921.

McKelvy, J. F., Leblanc, P., Loudes, C., Perrie, S., Grimm-Jorgensen, Y., and Kordon, C., 1976, The use of bacitracin as an inhibitor of the degradation of TRH and LH-RH, *Biochem. Biophys. Res. Commun.* **73**:507.

Marcano de Cotte, D., De Menezes, C. L. E., Bennett, G. W., and Edwardson, J. A., 1980, Dopamine stimulates the degradation of Gn-RH by rat synaptosomes, *Nature (London)* **283**:487.

Marks, N., 1977, Conversion and inactivation of neuropeptides, in: *Peptides in Neurobiology* (H. Gainer, ed.), pp. 221–258, Plenum Press, New York.

Marks, N., and Stern, F., 1974, Enzymatic mechanisms for the inactivation of LH-RH, *Biochem. Biophys. Res. Commun.* **61**:1458.

Moss, R. L., Dudley, C. A., and Vale, W., 1978, Hypothalamic peptides: Putative modulators of neural activity, in: *Brain–Endocrine Interaction*, Vol. III: *Neural Hormones and Reproduction* (P. Scott, G. P. Korlowski, A. Weindl, eds.), pp. 313–326, Karger, Basel.

Nikolics, K., Bieglmayer, C., Spona, J., Sepródi, J., and Teplán, I., 1981, Inhibition of LH-release by LH-RH analogues in rat pituitary cell culture, *Peptides,* **2**:65.

Nikolics, K., Szóke, B., and Teplán, I., 1980, Enzymic degradation of luteinizing hormone-releasing hormone in the pituitary and hypothalamus, *Proc. Int. Union Physiol. Sci.* **14**:612.

Orlowski, M., Wilk, E., Pearce, S., and Wilk, S., 1979, Purification and properties of a prolyl endopeptidase from rabbit brain, *J. Neurochem.* **33**:461.

Parker, C. R., Jr., Foreman, M. M., and Porter, J. C., 1979, Subcellular localization of LH-RH degrading activity in the hypothalamus, *Brain Res.* **174**:221.

Pedroza, E., Vilchez-Martinez, J. A., Fishback, J., Arimura, A., and Schally, A. V., 1977, Binding capacity of luteinizing hormone-releasing hormone and its analogues for pituitary receptor sites, *Biochem. Biophys. Res. Commun.* **79**:234.

Pedroza, E., Piyachaturawata, P., Vilchez-Martinez, J. A., Coy, D. H., and Schally, A. V., 1980, LH-RH receptors in the hypothalamus: Effect of superactive and inhibitory LH-RH analogues, *Endocrinology,* in press.

Pedroza-Garcia, E., Vilchez-Martinez, J. A., Piyachaturawata, P., Coy, D. H., and Schally, A. V., 1980, Binding of LH-RH to various anterior pituitary fractions: Interaction with a superactive and an inhibitory analog, *Endocrinology,* in press.

Pimstone, B., Epstein, S., Hamilton, S. M., LeRoith, D., and Hendricks, S., 1977, Metabolic clearance and plasma half disappearance time of exogenous Gn-RH in normal subjects and in patients with liver disease and chronic renal failure, *J. Clin. Endocrinol. Metab.* **44**:356.

Pohl, S. L., Krans, H. M. J., Birnbaumer, L., and Rodbell, M., 1972, Inactivation of glucagon by plasma membranes of rat liver, *J. Biol. Chem.* **247**:2295.

Powers, C. A., and Johnson, D. C., 1980, Hypothalamic peptidase regulation of extracellular levels of Gn-RH *in vitro, Fed. Proc.* **39**:594.

Redding, T. W., and Schally, A. V., 1973, The distribution half-life and excretion of tritiated luteinizing hormone-releasing hormone (LH-RH) in rats, *Life Sci.* **12**:23.

Redding, T. W., Kastin, A. J., Gonzalez-Barcena, D., Coy, D. H., Coy, E. J., Schalch, D. S., and Schally, A. V., 1973, The half-life, metabolism and excretion of tritiated LH-RH in man, *J. Clin. Endocrinol. Metab.* **37**:626.

Renaud, L. P., Martin, J.-B., and Brazeau, P., 1975, Depressant action of TRH, LH-RH and somatostatin on activity of central neurones, *Nature (London)* **255**:233.

Sandow, J., 1978, Studies on the biological activity and metabolism of LH-RH and its analogues, *Horm. Res.* **9**:68.

Sandow, J., and König, W., 1978, Chemistry of the hypothalamic hormones, in: *The Endocrine Hypothalamus* (S. L. Jeffcoate and J. S. M. Hutchinson eds.), Academic Press, New York.

Sandow, J., Kuhl, H., and Krauss, B., 1979, Studies on enzyme stability of LH-RH analogues, *J. Endocrinol.* **81**:P157.

Sétáló, G., Vigh, S., Schally, A. V., Arimura, A., and Flerkó, B., 1975, Immunohistological study of the origin of LH-RH containing nerve fibers in the rat hypothalamus, *Brain Res.* **103**:597.

Spona, J., 1973, LH-RH-stimulated gonadotrophin release mediated by two distinct pituitary receptors, *FEBS Lett.* **35**:59.

Spona, J., 1975, Sex steroids influence LH-RH-receptor interaction, *Endocrinol. Exp.* **9**:167.

Stetler-Stevenson, M. A., Yang, D. C., McCartney, L., Peterson, D., Lipkowski, A. W., and Flouret, G., 1980, Synthesis and characterization of the metabolic products of LH-RH breakdown by renal tissue, in: *Peptides 1980* (K. Brunfelt, ed.), pp. 725–730, Scriptor Publisher ApS, Copenhagen.

Sunberg, D. K., and Knigge, K. M., 1978, LH-RH production and degradation by rat medial basal hypothalami *in vitro, Brain Res.* **139**:89.

Swift, A. D., and Crighton, D. B., 1979a, Relative activity, plasma elimination and tissue degradation of synthetic LH-RH and certain of its analogues, *J. Endocrinol.* **80**:141.

Swift, A. D., and Crighton, D. B., 1979b, The effects of certain steroid hormones on the activity of ovine hypothalamic LH-RH-degrading enzymes, *FEBS Lett.* **100**:110.

Tate, S. S., 1978, Purification and properties of brain thyrotropin-releasing factor deamidase, *Fed. Proc.* **37**:1780.

Tharandt, L., Schulte, H., Benker, G., Hackenberg, K., and Reinwein, D., 1977, Some characteristics of the binding of ^{125}I-LH-RH in human sera, *Acta Endocrinol. (Copenhagen)* **85**:60.

Tharandt, L., Rosanowski, C., Grapow, A., Koch, H. W., Benker, G., Hackenberg, K., and Reinwein, D., 1979, The metabolic serum clearance of LH-RH as an indicator of endogenous LH-RH level, *Acta Endocrinol. (Copenhagen)* **91**:228.

Wagner, T. O. F., 1980, The effect of iodination of Gn-RH on its degradation and affinity to the pituitary receptor, *Acta Endocrinol. (Copenhagen)* **94**:81.

Wagner, T. O. F., Adams, T. E., and Nett, T. M., 1979, Gn-RH interaction with anterior pituitary. I. Determination of the affinity and number of receptors for Gn-RH in ovine anterior pituitary, *Biol. Reprod.* **20**:140.

Ween, R. V., and Kamberi, I. A., 1977, The distribution of arylamidases in the hypothalamus of rat and the effect of exogenous testosterone, *Horm. Metab. Res.* **9**:285.

Wilk, S., Benuck, M., Orlowski, M., and Marks, N., 1979, Degradation of LH-RH by brain prolyl endopeptidase with release of des-Gly-NH$_2$-LH-RH and Gly-NH$_2$, *Neurosci. Lett.* **14**:275.

Yoshimoto, T., Fischl, M., Orlowski, R. C., and Walter, R., 1978, Post-proline cleaving enzyme and post-proline dipeptidyl aminopeptidase, *J. Biol. Chem.* **253**:3708.

Zolman, J. C., and Valenta, L. J., 1980, Gn-RH receptor binding in bovine anterior pituitary, *Biochim, Biophys. Acta* **627**:172.

21

Ultrastructural Characteristics of Proteinergic Synthesis in the Pineal Gland

Josif R. Milin

1. Introduction

Histophysiology of the pineal gland, the unique sensoneuroendocrine organ of the central nervous system (Miline *et al.,* 1968), is under the resultant monoamine-hormonal inputs which determine its neuroendocrine activity by synthesis rate regulation of metoxy indoles and specific polypeptides. Although the presence of melatonin, biologically the most potent metoxy indole, has been established in the pineal gland of all examined species, this pineal effector principle cannot be qualified as a proper messenger of glandular activity, in view of the fact that the synthesis of this metoxy indole occurs in some other organs too: retina (Cardinali and Rosner, 1971), Harderian gland (Bubenik *et al.,* 1976), and intestine (Bubenik *et al.,* 1977). Moreover, every attempt of selecting one of the specific polypeptides for the major candidate of pineal hormonal principle would be very complex. Namely, the multiple classes of pineal biologically active polypeptides isolated to date (Milcu *et al.,* 1963; Orts and Benson, 1973; Ebels *et al.,* 1975; Blask *et al.,* 1976; Benson, 1977; Reiter and Vaughn, 1977; Thiéblot *et al.,* 1979) demonstrate nonequalized effects, either of inhibitory or of stimulatory nature, particularly on the histophysiology of the organs of reproductive system. The presence of vasopressin- and oxytocin-containing fibers was also established (Buijs and Pévet, 1980). By isolating the biologically

---

Josif R. Milin • Institute of Histology and Embryology, Medical Faculty, Novi Sad, Yugoslavia.

active biopterin (van der Have-Kirchberg *et al.,* 1977), the events connected with exploration of pineal effector compound(s) were only more
complicated. Not going into details while discussing the monopinealocyte
or polypinealocyte composition of glandular parenchyma, the fact that
metoxy indoles and specific polypeptides are produced within the same
mother cell in the main species of laboratory animals is generally accepted.
In close connection to the above-stated fact, the question as to the rate of
quantitative relations of the synthesis, or secretion, of indoles and specific
polypeptides remains unknown. In laboratory animals kept in controlled
microclimate conditions, a diurnal rhythm of melatonin synthesis/secretion
was established. The numerical presence of dense-cored vesicles, which
were considered to be indole amine–protein carrier depots (Collin, 1979;
Juillard, 1979), is also governed by a diurnal rhythm. On the other hand,
in wild, polar, or subterranean animals which live in environmental conditions that play a role of natural stimuli of the pineal gland activity, the
evident engagement of GER in polypeptidergic material synthesis has been
demonstrated (Miline *et al.,* 1974; Miline, 1979; Pévet, 1979). A similar
phenomenon was also seen in laboratory animals exposed to stressogenic
living conditions (Miline *et al.,* 1968; Milin, 1975). In the study of ultrastructural characteristics of the pineal gland of rats in the first half of lactation or gravidity period, it has been emphasized that an intensified
engagement of GER in the synthesis of polypeptidergic material could be
a reliable evidence of pineal gland-stimulated neuroendocrine activity:
caused by temporal oscillatory exteroreceptive stimuli (lactating animals)
or probably by peculiar endocrine inputs (gravid animals) (Milin, 1973a,b;
Milin *et al.,* 1981).

   This paper is devoted to the further perception of GER-engaged polypeptidergic material synthesis in pinealocytes in the case of physiologically
stimulated pineal gland activity, with respect to the reproductive physiology. The results obtained support a hypothesis that in rats in the scope of
long-term stimulation of pineal gland activity, whether physiological or
experimental-stressogenic, the specific polypeptides, or only one of them,
may represent the main messenger(s) of its neuorendocrine proper influences.

2. Materials and Methods

2.1. Experimental Animals

   Cyclic animals. twenty-one-week-old Wistar rats were used. The ani-
mals were sacrificed by decapitation in estrus, 0800–1000 hr.
   Lactating animals. Twenty-one-week-old lactating animals were used.
On the first postpartum day, the litter was reduced to five sucklings. The

animals were sacrificed on the fifth postpartum day, in the same time and in the same manner as cyclic animals.

Litter-deprived lactating animals. Twenty-one-week-old lactating animals were used. On the first postpartum day, the litter was reduced to five sucklings as well. The animals were sacrificed on the third postpartum day. A litter was removed 24 hr before sacrificing the lactating animals.

Gravid animals. Nineteen-week-old gravid animals were used. The animals were sacrificed on the 10th and 11th day of gestation.

2.2. Electron Microscopy

The fixation consisted in immersions of 4% glutaraldehyde/0.2 M cacodylate buffer at pH 7.4; the postfixation was in 2% osmic acid/0.2 M cacodylate buffer at pH 7.4. The solutions contained 3% saccharose. The tissue samples were embedded in Durcopan-Fluka. Sections were contrasted with uranyl acetate and lead citrate. For cytoenzymological investigations the pineal gland was incubated before immersion in glutaraldehyde in appropriate incubative media for 2–4 hr: alkaline phosphatases, *secundum* Mizutani–Barrnett (Mizutani and Barrnett, 1965), and acid phosphatases, *secundum* Weissenfels (Pearse, 1968). The examination was performed with JEM 100 B and JEM 100 C electron microscopes.

3. Results

3.1. Lactating Animals

The basic ultrastructural features of light pinealocytes—the well—developed network of GER tubules which populated the metabolically active parts of cytoplasmic gel, and the occurrence of numerous secretory granules, larger than in cyclic animals—directly pointed to the existence of an engaged polypeptidergic material synthesis in the studied period of lactation (Milin, 1973a). The separated branches of GER tubules of short and uneven course, encircled by sparse ribosome necklaces, came into a very close relation with building elements of Golgi zone, solitarily distributed mitochondria, or larger lipid droplets. Within the cystically expanded terminal spurs of these tubules deprived of ribosomes, the accumulation of fine-grained content was noticed (Fig. 1). Further increase in the mass of this content, as well as the consecutive swarming of terminal spurs, led to the formation of visicles or cisternae, i.e., structures which we termed as prosecretory granules (Fig. 2). These structures reached the neighboring parts of the cytoplasmic gel, invaded with Golgi vesicles. There was an inclination towards their mutual confluence and uniting of their matrix into a probably more complex molecular pool of newly formed secretory gran-

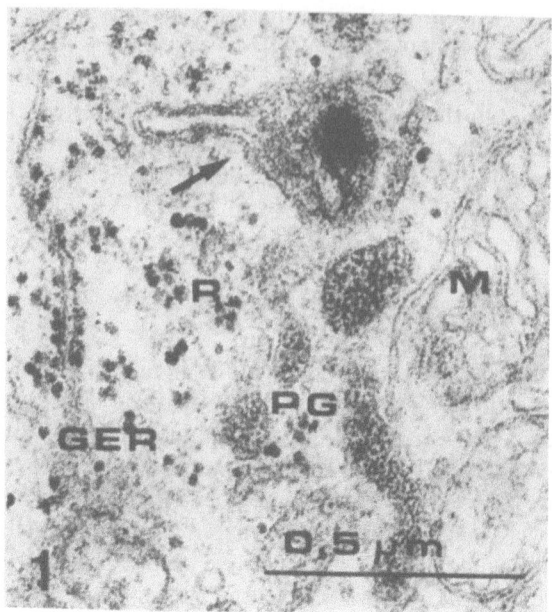

Figure 1. Lactating animal, light pinealocyte. Cystically dilated terminal spur of GER tubule (arrow), deprived of ribosomal necklace, showing the accumulation of fine-grained content. Prosecretory granules (PG) are in close vicinity of the terminals of separated GER tubules (GER). Clusters of solitarily scattered ribosomes (R) are noticeable. M, mitochondria.

ules. These structures were characterized by picturesque polymorphism and uneven distribution within the metabolically active parts of cytoplasmic gel. The fine-grained matrix of secretory granules was interwoven with some solitarily present electron-dense grains which showed a positive reaction for acid or alkaline phosphatases (Figs. 7, 8, and 9).

While following the further fate of secretory granules, the inclination towards their progressive confluence into "container bodies" was observed. These bodies, being structures of different appearance, possessed some solitarily distributed dense-cored vesicles and electron-dense grains (Fig. 3). "Container bodies" were present within all topographical parts of cytoplasmic gel. Their occurrence in the interior of gulf-shaped pericellular brim, which showed a tendency towards opening itself into intercellular pathways, pointed out that the polypeptidergic material originating from GER predominantly resulted in the massive extrusion of "container bodies" (Fig. 4). Since we did not find any evidence for exocytosis of prosecretory granules, the "container bodies" could be considered to be transitional morphodynamic forms, i.e., secondary polypeptidergic structures which were formed in the course of polypeptidergic material secretion. This process was followed by the release of lipid droplets as well (Fig. 5). Both events

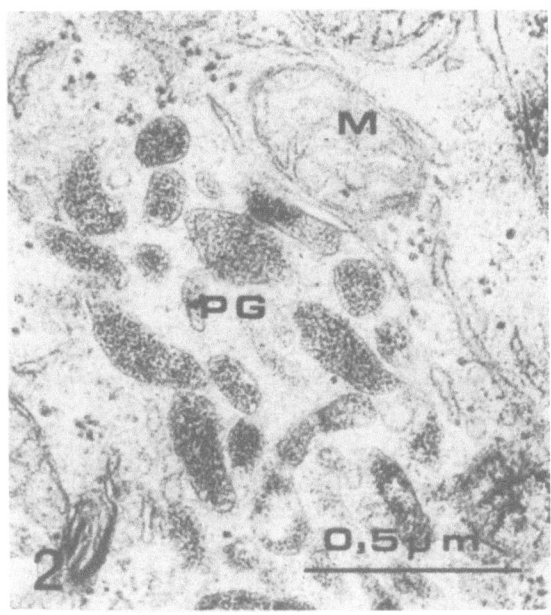

Figure 2. Lactating animal, light pinealocyte. Survey to the typical form of prosecretory granules (PG). They were predominantly found within GER area. Prosecretory granules, the structures of picturesque polymorphism, are composed of monolayer membrane and fine-grained matrix. M, mitochondria.

impelled us to the conclusion that two independent morphodynamic processes of neurosecretory material discharge might be involved in light pinealocytes in the course of their stimulated activity.

Cytoplasmic gel of dark pinealocytes was dominated by a well-ramified network of GER tubules with dilated lumen filled with rough-grained material. Golgi zone was poorly developed. Terminal spurs of GER tubules also inclined towards the cystic dilatation and consecutive separation of vesicles or cisternae. The progressive accumulation in the mass of their matrix led to the formation of prosecretory granules. Contrary to the light pinealocytes, these prosecretory granules, together with electron-dense grains, possessed a large number of electron-dense patches. Although the occurrence of melting of prosecretory granules with Golgi vesicles was not the common finding, the absence of acid phosphatases testified that these structures should be considered as secretory granules. Special attention was attracted by the finding of the process of "lipidization" of secretory granules. This phenomenon encompasses the progressive spreading of lipidlike mass, or lipid nucleus, through the whole interior of their matrix (Fig. 6). The results obtained were indicative for electron-dense patches that probably represented the lipid nodal buds which would bloom in lipidlike lumps.

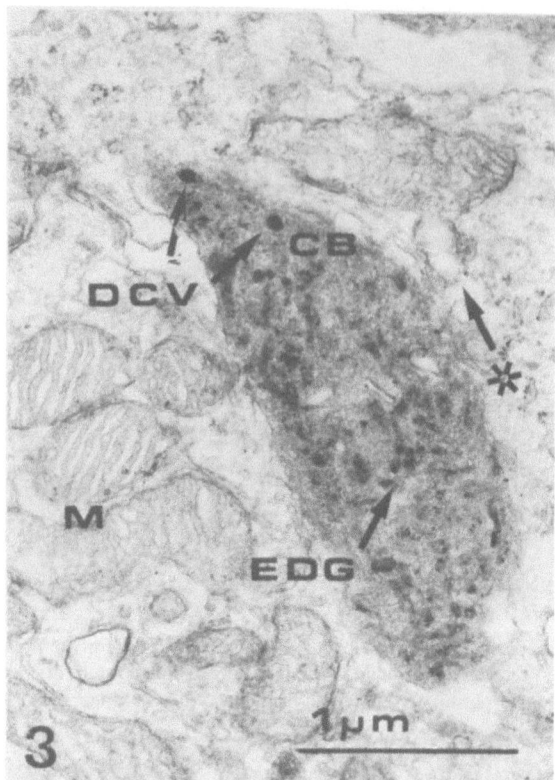

Figure 3. Lactating animal, light pinealocyte. CB, "container body." Fine-grained matrix of this secondary polypeptidergic structure is interwoven with solitarily distributed dense-cored vesicles (DCV) and a great deal of electron-dense grains (EDG). The vicinity of the "container body" is populated by mitochondria (M). There are variations in mitochondrial size or granularity of their intercrystal matrix. The enlarged intercrystal matrix indicates that these granules are in "energized" stage of their activity. *, mitochondrial herniation.

After the final morphological homogenesis of dispersed lipidlike mass, lipid droplets were made with their proper features. The presence of the positive precipitate of acid and alkaline phosphatases around the brim of arising lipid droplets testified that both hydrolases were involved in the process of "lipidization" (Figs. 7–9). The occurrence of secretory granules or extrusion of lipid droplets was not noticed in this cell population.

3.2. Litter-Deprived Lactating Animals

Holocrine disintegration of cell bodies followed by massive lipid droplet extrusion, on one hand, and the surplus made secretory granules auto-

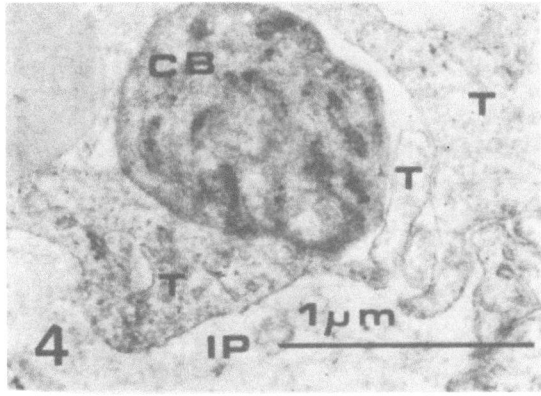

Figure 4. Lactating animal, light pinealocyte. The presence of a "container body" inside the gulf-shaped pericellular depression formed by tongues (T) of cytoplasmic protrusions indicates that secretion of polypeptidergic material results in "container body" extrusion into intercellular pathways (IP). During the process of "container body" discharge, there is an inclination of electron-dense grains towards melting into fine-grained matrix.

phagic destruction of lipofuscin-like bodies, on the other hand, were the main morphodynamic characteristics of light pinealocytes in the secession of lactation caused by litter removal (Milin, 1973a). These results showed that secretory material, originating from GER tubules and synthesized in considerable amount during nondisturbed lactating period, was immediately structurally destroyed when suckling stimuli were brutally omitted. The removal of overproduced secretory granules was predominantly based on their autophagic disintegration. The prosecretory granules were

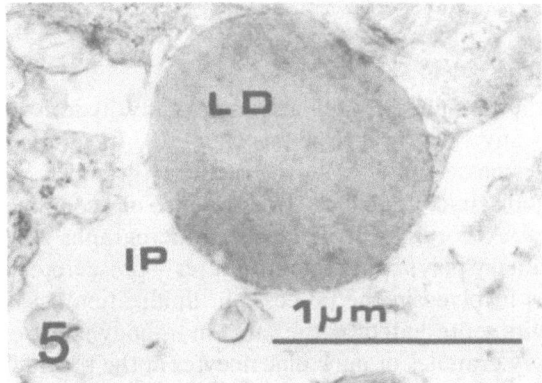

Figure 5. Lactating animal, light pinealocyte. The process of lipid droplet (LD) release into intercellular pathway (IP).

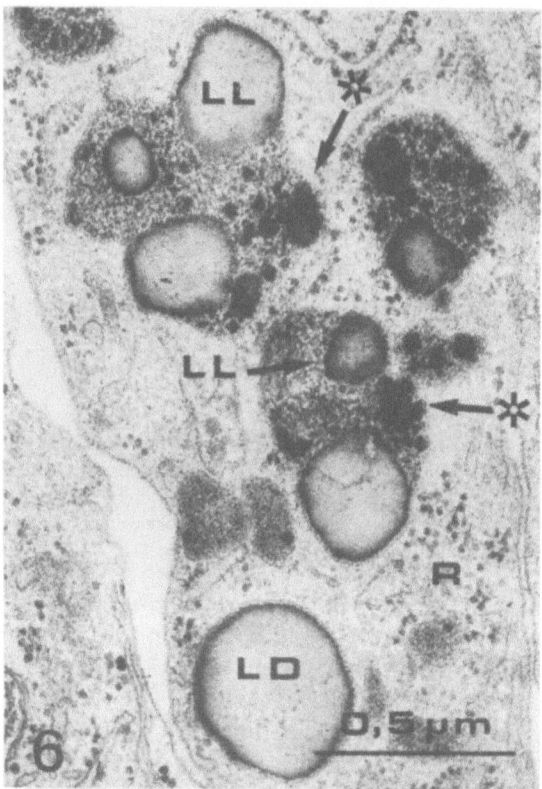

Figure 6. Lactating animal, dark pinealocyte. The process of "lipidization" of secretory granules. This event encompasses the progressive growth of electron-dense patches (asterisk), which might represent the lipid nodal buds which would bloom in lipidlike lumps (LL). The further progressive growth and spreading of lipidlike lumps, i.e., lipid nuclei, lead to the genesis of lipid droplet (LD) in its proper appearance. R, ribosomes.

engulfed by the autophagic vacuoles, and totally hydrolyzed without the appearance of any structural residuals (Fig. 10). In some cases, the autophagic vacuoles possessed, besides the secretory granules, the functionally exhausted cytoplasmic organelles. The presence of "container bodies" was not established. According to the electron micrographs studied, a certain quantity of primary polypeptidergic structures—prosecretory and secretory granules—was involved in the process of "lipidization" as well (Fig. 11). This process was represented by the same morphodynamic events as noted for the secretory granules of dark pinealocytes in the previous group of lactating animals. Their relatively fine-grained matrix was also interwoven with electron-dense grains or lipid nuclei. Lipofuscin-like bodies of smaller size and different shape were also the common constituents of light pinealocytes. The occurrence of the remnants of GER tubules and some other

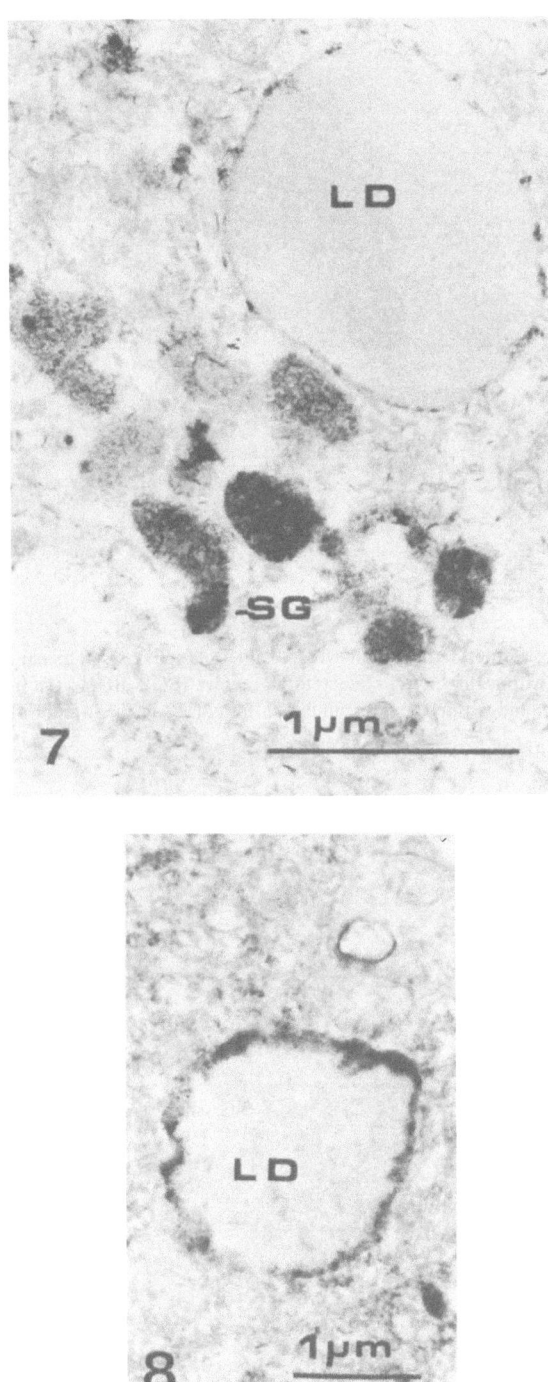

Figures 7 and 8. Lactating animal, light pinealocyte. Acid phosphatases. Fine-grained enzyme-positive precipitate interweaves the matrix of secretory granules (SG) and encircles the brim of arising lipid droplet (LD).

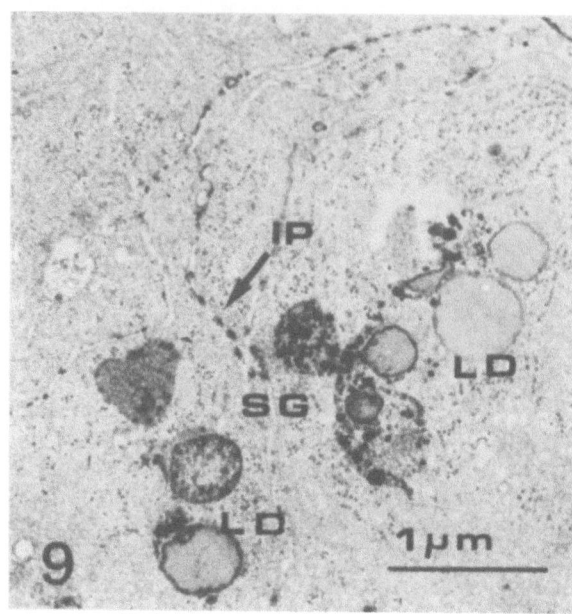

Figure 9. Lactating animal, light pinealocyte. Alkaline phosphatases. The enzyme-positive precipitate is present within the matrix of secretory granules (SG), around the brim of appearing lipid droplets (LD), and within the intercellular pathway (IP) in the form of rosary.

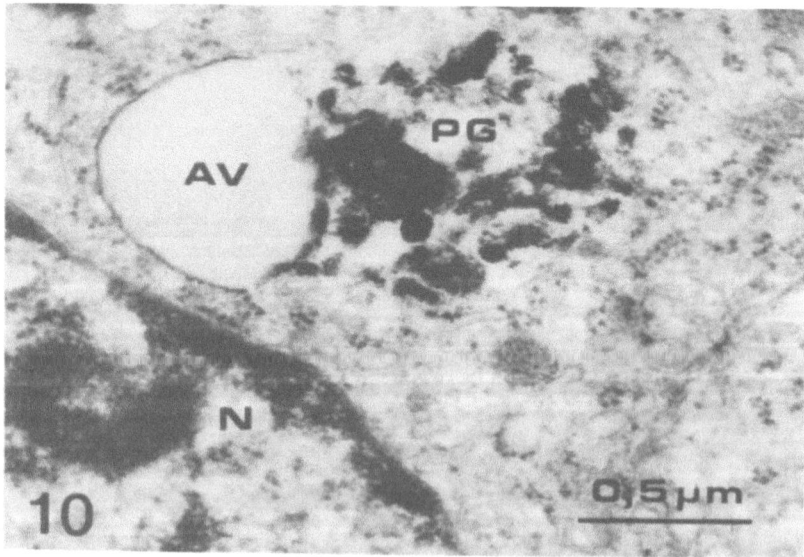

Figure 10. Litter-deprived lactating animal, light pinealocyte. Autophagic disintegration of primary polypeptidergic structures. The prosecretory granules (PG) of condensed matrix are engulfed by the autophagic vacuole (AV). Note the absence of structural residuals. N, nucleus.

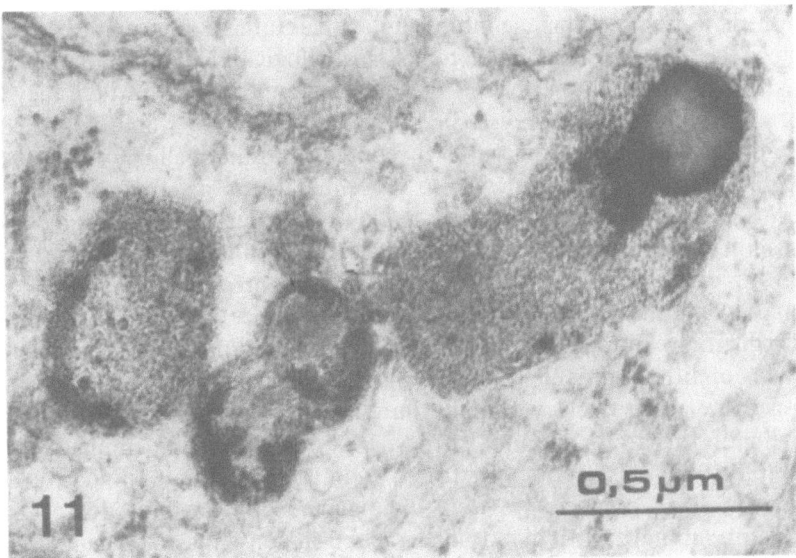

Figure 11. Litter-deprived lactating animal, light pinealocyte. The process of "lipidization" of primary polypeptidergic structures.

tubular structures within the cytoplasmic area in which these bodies were noticed, substantiated that lipofuscin-like bodies were predominantly formed within the previously metabolically active parts of cytoplasmic gel (Fig. 12). The concurrent appearance of an altered picture of lipid nuclei, in the form of cogged brim, within the secretory granules involved in the process of "lipidization," also pointed to the presumption that the slowing down of the process of "lipidization" of secretory granule matrix led to the genesis of lipofuscin bodies.

3.3. Gravid Animals

Judging by the well-developed and ramified GER tubules and by the occurrence of a considerable number of gathered secretory granules, light pinealocytes were also involved in an intensive polypeptide synthesis in the gestation period studied (Milin *et al.*, 1981). Genesis of primary polypeptidergic structures developed in a similar manner as in the respective cells of lactating animals. In the close vicinity of these structures, a moderate amount of enlarged vesicles was noticed. The presence of well-developed Golgi zone within the neighboring parts of cytoplasmic gel permitted us to assume with certainty that these vesicles were of Golgi descent. According to the results obtained, the largest number of secretory granules inclined towards mutual confluence into gigantic structures of appropriate appearance. These structures seemed to be morphologically identical, in a certain manner, with "container bodies" in lactating animals, from which they differed by the absence of dense-cored vesicles. Since they were found next to the cellular brim, pouring out their content into intercellular pathways, it could be supposed that they also presented the proper morphodynamic forms of fused secretory granules made in the course of discharge of neurosecretory material (Fig. 13). There was also the finding of secretory granule "lipidization" (Fig. 14). This process represented the same morphodynamic events that were characteristic for the process of secretory granule "lipidization" in dark pinealocytes of lactating animals, or in light pinealocytes of litter-deprived lactating animals. The immediate pouring in of prosecretory or secretory granules into lipid droplets was also the common finding in this group of animals (Figs. 15, 16). The finding emphasized the complexity of polypeptidergic structures in their relation to lipid droplets.

The dark pinealocytes were distinguished by the presence of large secretory granules with homogeneous electron-dense matrix. The finding of GER tubules only in small amounts indicated that this cell population was not fully involved in neuroendocrine material synthesis. Furthermore, a diminished appearance of secretory granule "lipidization" also supported this conclusion. Lipid droplets of relatively equal size were more frequent than in the previous group of animals.

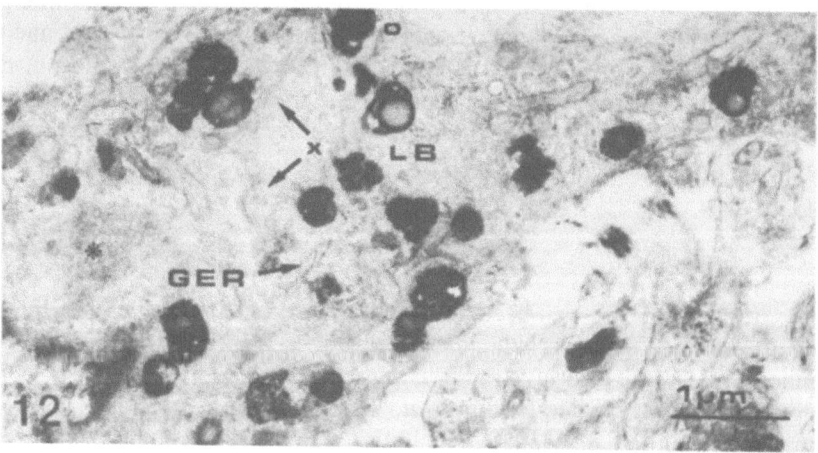

Figure 12. Litter-deprived lactating animal, light pinealocyte. The occurrence of lipofuscin-like bodies (LB) within the previously metabolically active parts of cytoplasmic gel. This area is also populated by the remnants of GER tubules (GER) and some other tubular structures of unknown origin (x). *, condensed bits of cytoplasmic gel.

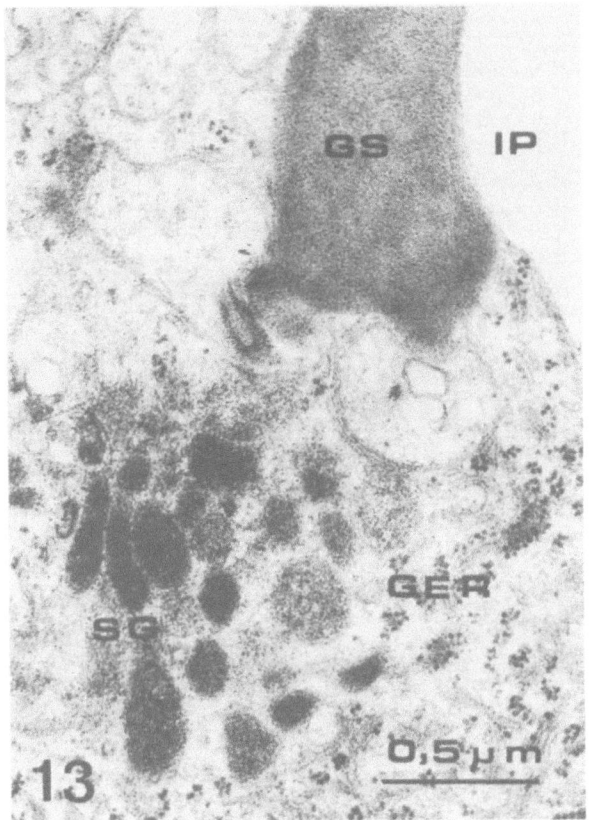

Figure 13. Gravid animal, light pinealocyte. An amassment of primary polypeptidergic struc-tures within the periceliular zone. The mutual confluence of these structures, particularly of secretory granules (SG), leads to the formation of gigantic structure (GS). This newly formed structure, characterized by spread, fine-grained matrix, is formed next to the pericellular brim. The position of gigantic structure is strictly suggestive for the conclusion that these secondary polypeptidergic structures pour their content into intercellular pathways (IP). GER, GER tubules.

4. Discussion

By summing up the established morphodynamic characteristics—the abundantly ramified network of GER tubules, the presence of more numer-ous secretory granules than in cyclic animals—one can conclude that in conditions of physiologically stimulated neuroendocrine activity of the pineal gland (lactating and gravid animals), the enhanced polypeptidergic material synthesis occurs in light pinealocytes. Both the ultrastructural and

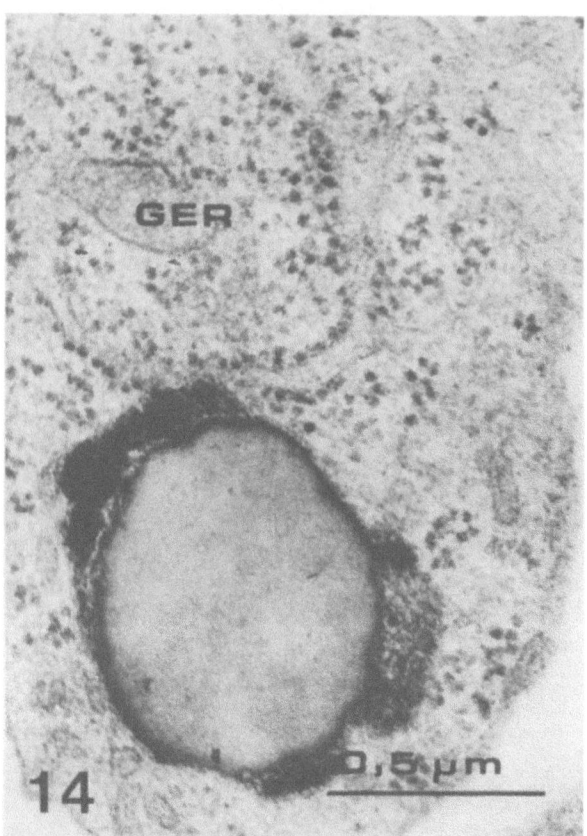

Figure 14. Gravid animal, light pinealocyte. The final stage in the process of "lipidization" of secretory granule. GER, GER tubules.

fluorescence studies carried out by Pévet *et al.* (1976) provided good evidence that the fine-grained material present within the cisternae of GER origin, i.e., prosecretory granules, was of polypeptidergic nature. Contemporary finding of the fusion of prosecretory granules and vesicles of Golgi descent pointed to the whole complexity of the morphodynamics of neuroendocrine material synthesis. As the current study comprised only ultrastructural investigations of pineal gland activity, we can only speculatively elucidate all the events that occurred in the interior of secretory granules as the result of the union of prosecretory granules and Golgi vesicles. According to the knowledge that Golgi zone is involved in metoxy indole synthesis (Sheridan and Sladek, 1975), together with the observation that "container bodies" were quartered by dense-cored vesicles (structures which probably represented the metoxy indole–protein complex) (Collin, 1979; Juillard, 1979), we may suppose that the union of prosecretory granules and Golgi vesicles resulted in the formation of more complex pool of matrix of secretory granules. Owing to the view that the synthesis of serotonin and other metoxy indoles takes place in cytosol rather than cell organelles (Ariëns Kappers, 1978), we may explain the diminished cytoplasmic number of dense-cored vesicles, in comparison to the cyclic animals, also as the consequence of constant melatonin secretion (or of some related metoxy indole) provoked by suckling stimuli in lactating animals, or by the proper endocrine equilibrium in gravid animals. This assumption is fully corroborated by other investigation that have provided reliable evidence that stress can result in considerable genesis of polypeptidergic material (Miline *et al.*, 1968; Milin, 1975) and in increase production of melatonin (Lynch *et al.*, 1979; Vaughan *et al.*, 1978). From this aspect, it would be very interesting to investigate whether melatonin, in the studied periods of reproductive physiology, demonstrates a diurnal rhythm of its synthesis and secretion. When dealing with the concept of formation of a more complex molecular pool of matrix of secretory granules, one should not disregard the probability that neuroepiphysin–SP carrier complex (Legros *et al.*, 1975; Reinharz *et al.*, 1975) is made after the fusion of prosecretory granules and Golgi vesicles. Moreover, on the basis of the finding of some hydrolytic enzymes within the secretory granules (Milin, 1973b), we can also consider the assertion that Golgi zone provides the enzymes needed in the process of further synthesis of neurosecretory material. Since biologically active polypeptides of low and high molecular weight have been isolated from the pineal gland (Orts and Benson, 1973; Ebels *et al.*, 1975), we cannot neglect the possibility that the cleaving of the basic polypeptidergic chain, arising from GER tubules, takes place within the secretory granules or "container bodies," with the help of Golgi enzyme system. Regradless, the established production of polypeptidergic material by GER may represent valid morphodynamic evidence of an enhanced SP synthesis in light pinealocytes during the lactational or gestational period studied. Structures strikingly similar to prosecretory or secretory granules, termed grumose inclusions (Wolfe, 1965) or single-membrane-limited inclusions (Arstila, 1967) (which are potentially involved in secretory processes), were also described.

The existence of GER-engaged polypeptidergic synthesis of pinealocytes of stressed laboratory rats (Miline *et al.*, 1968; Miline, 1980) or in polar and subterranean animals (whicn live in environmental conditions that represent the natural stimuli for pineal neuroendocrine activity (Miline *et al.*, 1974; Pévet, 1974, 1979) also supports the working hypothesis that in the case of long-term stimulated pineal gland histophysiology, enhanced synthesis of polypeptidergic material takes place. Since during

lactation and gestation the rate of synthesis and secretion of many hormones, especially gonadotropins, was changed, one could take into account
that the prevalence of endocrine influences led towards the increased polypeptidergic synthesis. Some results obtained by other authors directly support this assertion (Pévet and Smith, 1975; Karasek *et al.,* 1976; Karasek
and Marek, 1978). As melatonin, in prolonged application, exerted different effects from pineal extracts on the histophysiology of some endocrine
glands (Šćepović *et al.,* 1972; Reiter, 1978), one may speculate that pineal
polypeptides may be considered as synchronizers of endocrine equilibrium
in relation to the environmental living conditions, mediating certain exteroreceptive factors in long-term organism adaptation, particularly of its
reproductive physiology. This assumption in no case excludes the participation of melatonin in pineal gland activity. On the contrary, the view that
this metoxyindole affects the synthesis of polypeptides, within its own
mother cell (Quay, 1974) by increasing amino acid incorporation (Orsi *et*

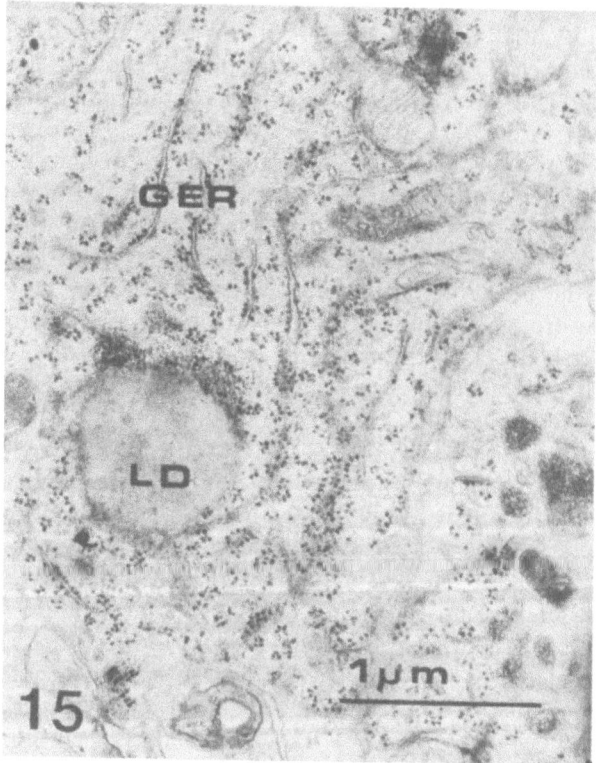

Figures 15 and 16. Gravid animal, light pinealocyte. The process of pouring the fine-grained
content from GER tubules (GER) or prosecretory granules (PG) into lipid droplets (LD).

al., 1973), illustrates the nondivisible role of metoxy indoles and polypeptides in neuroendocrine determinism of the pineal gland.

Generally speaking, the further morphodynamic fate of primary polypeptidergic structures—prosecretory and secretory granules—was not identical in lactating and gravid animals. Namely, in lactating animals the extracellular discharge of polypeptidergic material in the form of "container bodies" was not followed by the occurrence of "lipidization" of matrix of prosecretory granules or "container bodies," contrary to the case in gravid animals. Moreover, the finding of the phenomenon of "lipidization" of secretory granules in light pinealocytes, besides their autophagic disintegration, after removal of suckling stimuli, smooths the way towards a very interesting discussion. If we accept the view that the pineal gland, in the first days of lactation, is in an immediate dependence on the temporal oscillatory suckling stimuli which, at the precise moment, lead pinealocytes into the crescendo of their cytophysiology, or that the histophysiology of

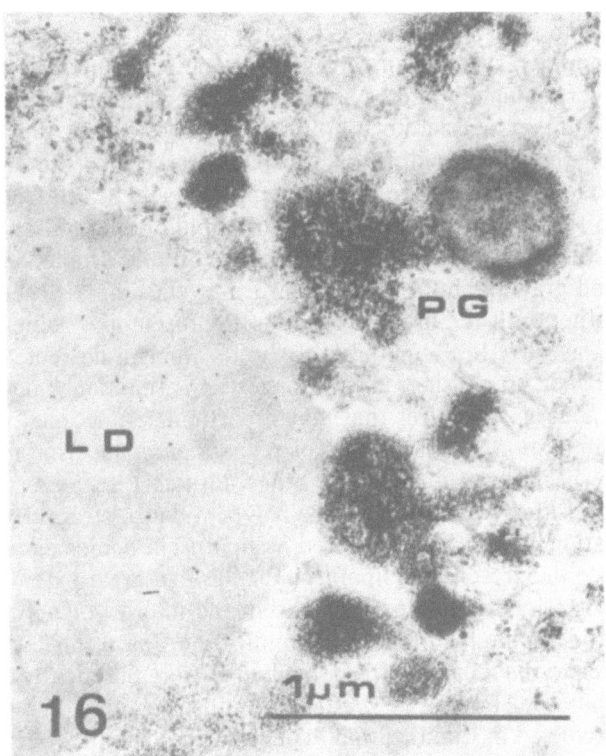

the pineal gland in gravidity is mainly dependent on the gestational endo-crine equilibrium, it can be considered that "container bodies" and lipid droplets represent the particular organoleptic aspects of intracellular accu-mulation of neuroendocrine material. Since the presence of "container bod-ies" was more common in lactating animals, on one hand, and since these structures were absent in litter-deprived lactating animals, on the other hand (together with the appearance of autophagic destruction of overpro-duced primary polypeptidergic structures), we were impelled to the conclu-sion that Golgi zone furnished enzymes for their role in the final step of polypeptidergic material synthesis rather than the catabolic processes. Thus, "container bodies" cannot be considered as secondary lysosomes, but as transitional, secondary polypeptidergic structures arisen by mutual melt-ing of of secretory granules in the scope of discharge of polypeptidergic material. Furthermore, when the elaborative-secretory rate of neurosecre-tory material was in disharmony, the process of "lipidization" of a certain mass of overproduced polypeptidergic structures began. In this way, the lipidlike content, in the form of lipid droplets, is most probably made. In considering this view, one should bear in mind the fact that in the case of turnover of previously stimulated cytophysiology of pinealocytes (litter-deprived animals), two extreme events were noticed simultaneously: auto-phagy of overproduced secretory granules and "lipidization" of the matrix of smaller number of secretory granules (Milin, 1973a). Appropriate to this, with the help of "lipidization," or pouring the prosecretory granules' content into lipid droplets, specific polypeptides could be retained for longer time within the mother cell in the form of "lipid droplets," not losing any-thing from their hormonal value. The change of lipid droplets' presence in lactating and gravid animals, with regard to cyclic animals (Milin, 1973a, 1975), together with the usual absence of lipofuscin-like bodies, is in har-mony with this view. If a hypothesis of "lipidization" is correct, "lipid drop-lets" should not be considered any longer as a functional postmetabolic byproducts, but as structures that play a particular functional role in syn-thesis, accumulation, and secretion of pineal neuroendocrine material, as has already been stressed by some other authors (Arstila *et al.,* 1971). From a very critical point of view, the polypeptidergic structures disturbed by "lipidization" could be considered as lipofuscin bodies as well, all the more so since they resembled lipofuscin bodies to a great extent. However, the presence of hydrolytic enzymes within the interior of polypeptidergic structures undergoing the process of "lipidization," or at the level of arising lipid droplets, to the fact that lipofuscin bodies were practically deprived of acid phosphatases (Deams *et al.,* 1972), suggested that polypeptidergic structures involved in "lipidization" could not be considered as lipofuscin bodies, but as transient structures which could bloom into structural forms known as lipid droplets. Furthermore, closely related to this part of the discussion, since lipofuscin-like bodies were commonly found in the litter-deprived lactating animals, they could be regarded as structures created in a case of a disturbed, or, even more, of a thwarted advanced process of prosecretory or secretory granule "lipidization" in the moment of the disappearance of stimulus inputs. Summarizing, we should like to point out that the phenomenon of "lipidization" of polypeptidergic structures is a proper morphodynamic or morphobiochemical phenomenon, which occurs during the genesis of the pinealocyte neurosecretory structures. On the whole, this assumption, together with the results of our current investigations, and the occurrence of lipofuscin-like bodies, should be understood, as morphological evidence of an interrupted process of "lipidization" of the primary polypeptidergic structures.

Regarding the results obtained in the course of this study, it was observed that the extrusion of polypeptidergic material from the mother cell occurred in two different and independent morphodynamic ways: a "prompt discharge" of just-synthesized material ("container bodies") and a "late discharge" of previously synthesized material ("lipid droplets"). This view completes the working hypothesis, and at the same time points out that secretory granules and lipid droplets do not contain essentially different compounds, but represent distinctive morphodynamic features of synthesized and stored neurosecretory material. According to some other authors, polypeptidergic material is released from the mother cell in the form of neuroepiphysin active principle. At that moment, calcium is exchanged for the active principle with the secretion of the latter into the blood vascular system (Lukaszyk and Reiter, 1975). Making use of the findings obtained, together with some other results of ours, we may consider that the "late discharge" of lipid droplet content would occur to meet the increased demands of neurosecretory material, as the answer imposed by the input influences on pineal gland activity at the moment when these demands could not be met by actual synthesis of polypeptidergic material.

Regarding the dark pinealocytes, ultrastructural changes were also seen in both periods of reproductive physiology studied. However, these changes were not as intensive as in light pinealocytes. Moreover, the branching of GER tubule network, the morphodynamics of genesis of polypeptidergic structures, and the phenomenon of secretory granule "lipidization" testified that both populations of pinealocytes possessed the same biocellular mechanisms, regarding the morphodynamic distinctiveness of their elaborative opus. In the scope of current investigations, any attempt to consider the existence of possible similarities or differences in the molecular pool between the polypeptidergic structures of light and dark pinealocytes would be obscure.

464   *Josif R. Milin*

DISCUSSION

C. VILLEE: Do you have any evidence regarding the identity of the protein or polypeptide material that accumulates in these studies you have described?

J. MILIN: Unfortunately, nobody knows exactly how many polypeptidergic compounds having the specific neuroendocrine influences, are really synthesized in the pineal gland. Although the ultrastructural investigations showed an enhanced synthesis of polypeptidergic material within the total pinealocyte populations (in the case of physiologically or experimental stress-stimulated pineal gland), there was no knowledge about biochemical or biological properties of pineal specific polypeptides. Moreover, we do not know whether there was one or several polypeptide compounds with neuroendocrine characteristics synthesized in pinealocytes. The application of imunocytochemical investigations is limited due to difficulties in isolation of pure "peptidergic" compounds. The studies of Buijs and Pévet (1980), which are in progress, may add some answers.

C. VILLEE: My second question concerns the nature of the lipids in these pineal lipid droplets: Are they phospholipids or neutrolipids? What is their function in the cell?

J. MILIN: Histochemical techniques showed an accumulation of triglycerides and phospholipids inside the lipid droplets. The excellent work done by Quay (1974) showed the light pinealocytes to contain ethanol-soluble lipid droplets with high amounts of neutrolipids, whereas dark pinealocytes possess choline-containing phospholipids. However, no biologically active lipids or lipid compounds have been isolated. The pineal lipids may be involved in the general metabolism of the pinealocytes or lipid droplets could be considered as structural forms.

KANAZIR: Is the giant structure containing the vesicles discharged as a whole? What is the size of the vesicles? Does the synthesis of polypeptides take place in the vesicles?

J. MILIN: The electro micrographs indicated they were discharged as a whole into intracellular pathways. It is generally accepted that polypeptide synthesis takes place in GER tubules.

REFERENCES

Ariëns Kappers, J., 1978, Localization of indoleamine and protein synthesis in the mammalian pineal gland, *J. Neural Transm. Suppl.* **13**:13.
Arstila, A., 1967, Electron microscopic studies on the structure and histochemistry of the pineal gland of the rat, *Neuroendocrinology* **2**(Suppl.):1.
Arstila, A., Kalimo, H. O., and Hyyppä, M., 1971, Secretory organelles of rat pineal gland: Electron microscopic and histochemical studies *in vivo* and *in vitro,* in: *The Pineal Gland* (G. E. W. Wolstenholme and J. Knight, eds.), pp. 147–163, Churchill/Livingstone, Edinburgh.
Benson, B., 1977, Current status of pineal peptides, *Neuroendocrinology* **24**:241.
Blask, D. E., Vaughan, M. K., Reiter, R. J., Johnson, L. Y., and Vaughan, G. M., 1976, Prolactin-releasing and release-inhibiting factor activities in the bovine, rat and human pineal gland: *In vivo* and *in vitro* studies, *Endocrinology* **99**:152.
Bubenik, G. A., Brown, G. M., and Grota, L. J., 1976, Immunohistochemical localization of melatonin in the rat Harderian gland, *J. Histochem. Cytochem.* **24**:1173.

Bubenik, G. A., Brown, G. M., and Grota, L. J., 1977, Immunohistochemical localization of melatonin in the rat digestive system, *Experientia* **33**:663.

Buijs, R. M., and Pévet, P., 1980, Vasopressin and oxytocin-containing fibres in the pineal gland and subcommissural organ of the rat, *Cell Tissue Res.* **205**:11.

Cardinali, D. P., and Rosner, J. M., 1971, Retinal localization of the hydroxyindole-O-methyl transferase (HIOMT) in the rat, *Endocrinology* **89**:301.

Collin, J. P., 1979, Recent advances in pineal cytochemistry. Evidence of the production of indoleamines and proteinaceous substances by rudimentary photoreceptor and pineal-ocytes of amniota, in: *The pineal gland of Vertebrates Including Man* (J. Ariëns Kappers and P. Pévet, eds.), pp. 271–296, Elsevier/North-Holland, Amsterdam.

Daems, W. T., Wisse, E., and Brederoo, P., 1972, Electron microscopy of the vacuolar apparatus, in: *Lysosomes: A Laboratory Handbook* (J. T. Dingle, ed.), pp. 150–189, North-Holland, Amsterdam.

Ebels, E. I., Citharel, A., and Moszkowska, A., 1975, Separation of pineal extract by gel filtration. III. Sheep pineal factors acting either on the hypothalamus or on the anterior hypophysis of mice and rats in "in vitro" experiments, *J. Neural Transm.* **36**:281.

Juillard, M.-T., 1979, The proteinaceous content and possible physiological significance of dense-cored vesicles in hamster and mouse pinealocytes, *Ann. Biol. Anim. Biochim. Biophys.* **19**:413.

Karasek, M., and Marek, K., 1978, Influence of gonadotropic hormones on the ultrastructure of rat pinealocytes, *Cell Tissue Res.* **188**:133.

Karasek, M., Pawlikowski, M., Ariëns Kappers, J., and Stepien, H., 1976, Influence of castration followed by administration of LH-RH on the ultrastructure of rat pinealo-cytes, *Cell Tissue Res.* **167**:325.

Legros, L. L., Louis, F., Grotchel-Stewart, U., and Franchimont, P., 1975, Presence of immunoreactive neurophysin-like material in human target organs and pineal gland: Physiological meaning, *Ann. N.Y. Acad. Sci.* **248**:157.

Lukaszyk, A., and Reiter, R. J., 1975, Histophysiological evidence for the secretion of poly-peptides by the pineal gland, *Am. J. Anat.* **143**:451.

Lynch, H. J., Ho, M., and Wurtman, R. J., 1979, The adrenal medulla may mediate the increase in pineal melatonin synthesis induced by stress but not caused by exposure to darkness, *J. Neural Transm.* **40**:87.

Milcu, S. M., Pavel, S., and Neacsu, C., 1963, Biological and chromatographic character-ization of a polypeptide with pressor and oxytocic activities isolated from bovine pineal gland, *Endocrinology* **72**:563.

Milin, J., 1973a, The pineal gland in lactation, Master's thesis, Novi Sad.

Milin, J., 1973b, Electron microscopic study on localization of alcaline and acid phospha-tases in the pineal gland in rat, XV Kongres Udruženja antoma Jugoslavije, Rijeka.

Milin, J., 1975, Correlation between the pineal gland and prolactin cells of the adenohy-pophysis in the first half of the period of lactation under auditive stress, Doctoral dis-sertation, Novi Sad.

Milin, J., Marić, D., and Krsmanović, L., 1981, Cytophysiological characteristics of the pineal gland in the first period of gravidity, *Folia Anat. Iugos.,* in press.

Miline, R., 1979, Different populations of pinealocytes in the pineal gland of the mole-rat (*Spalax leucodon,* Nordmann), in: *The Pineal Gland of Vertebrates Including Man* (J. Ariëns Kappers and P. Pévet, eds.), pp. 207–212, Elsevier/North-Holland, Amster-dam.

Miline, R., 1980, The role of the pineal gland in stress, *J. Neural. Transm.* **47**:191.

Miline, R., Krstić, R., and Devečerski, V., 1968, Sur le comportement de la glande pinéale dans des conditions de stress, *Acta Anat.* **71**:352.

Miline, R., Devečerski, V., Milin, J., and Marjanov, M., 1974, Histophysiologie de la glande pinéale chez *Spalax leucodon, Bull. Assoc. Anat.* **58**:692.

Mizutani, A., and Barrnett, R. J., 1965, Fine structural demonstration of phosphatase at pH 9, *Nature (London)* **206**:1001.

Orsi, L., Denari, J. H., Nagle, C. A., Cardinali, D. P., and Rosner, J. M., 1973, Effects of melatonin on the synthesis of proteins by rat hypothalamus, hypophysis and pineal organ, *J. Endocrinol.* **58**:131.

Orts, R. J., and Benson, B., 1973, Inhibitory effects on serum and pituitary LH by a melatonin-free extract of bovine pineal galnds, *Life Sci.* **12**:513.

Pearse, A. G., 1968, *Histochemistry,* Vol. I, p. 729, Churchill/Livingstone, Edinburgh.

Pévet, P., 1974, The pineal gland of the mole (*Talpa europea* L.). The fine structure of pinealocytes, *Cell Tissue Res.* **153**:277.

Pévet, P., 1979, Secretory processes in the mammalian pinealocyte under natural and experimental conditions, in: *The Pineal Gland of Vertebrates Including Man* (J. Ariëns Kappers and P. Pévet, eds.), pp. 149–194, Elsevier/North-Holland, Amsterdam.

Pévet, P., and Smith, A. R., 1975, The pineal gland of the mole (*Talpa europea* L.). II. Ultrastructural variations in the pinealocytes during different parts of the sexual cycle, *J. Neural Transm.* **36**:227.

Pévet, P., Juillard, M.-T., Smith, A. R., and Ariëns Kappers, J., 1976, The pineal gland of the mole (*Talpa europea* L.). A fluorescence histochemical study, *Cell Tissue Res.* **165**:297.

Quay, W. B., 1974, *Pineal Chemistry in Cellular and Physiological Mechanisms,* pp. 137–201, Thomas, Springfield, Ill.

Reinharz, A. C., Caernichow, P., and Vallotton, M. B., 1975, Neurophysins I and II from the bovine posterior pituitary lobe and neurophysin-like proteins from bovine pineal gland, *Ann. N.Y. Acad. Sci.* **248**:172.

Reiter, R. J., 1978, The pineal, vol. III, pp. 122–144, Churchill/Livingstone, Edinburgh.

Reiter, R. J., and Vaughan, M. K., 1977, Pineal antigonadotropic substances: Polypeptides and indoles, *Life Sci.* **21**:159.

Šćepović, M., Hrnjičević, M., and Deronja, N., 1972, The significance of melatonin in the activity regulation of the thyroid gland, *Folia Med. Fac. Med. Univ. Saraeviensis* **8**:105.

Sheridan, M. N., and Sladek, J. R., Jr., 1975, Histofluorescence and ultrastructural analysis of hamster and monkey pineal, *Cell Tissue Res.* **164**:145.

Thiéblot, L., Grizard, G., Dastugue, B., Gachon, A. M., and Thiéblot, P., 1979, Purification du facteur antigonadotrope de la glande pinéale, *Ann. Endocrinol. (Paris)* **40**:519.

Van der Have-Kirchberg, M. L. L., de Morée, A., van Laar, J. F., Gerwig, G. J., Versluis, C., Ebels, I., Hus-Citharel, A., L'Héritier, A., Roaeau, S., Zurburg, W., and Moszkowska, A., 1977, Separation of pineal extracts by gel filtration. VI. Isolation and identification from sheep pineal of biopterin; comparison of the isolated compound with some synthetic pteridines and the biological activity "in vitro" and "in vivo" bioassays, *J. Neural Transm.* **40**:205.

Vaughan, G. M., Allen, J. P., Tullis, W., Sackman, J. W., and Vaughan, M. K., 1978, Stress-induced increase of pineal N-acetyltransferase activity in intact rats, *Neurosci. Lett.* **9**:83.

Wolfe, D. E., 1965, The epiphyseal cell: An electron-microscopic study of its intercellular relationships and intracellular morphology in the pineal body of the albino rat, in: *Structure and Function of the Ephiphysis Cerebri* (J. Ariëns Kappers and J. P. Schadé, eds.), pp. 332–389, Elsevier, Amsterdam.

22

Peptidergic Activity of the Pineal Gland in Stress

Radivoy Miline

1. Introduction

According to some recent results, the presence of some peptides was established within the pineal gland: AVT-like substance (Milcu *et al.*, 1963; Ebels *et al.*, 1965; Pavel and Petrescu, 1966; Quay, 1974; Reiter, 1974; Benson *et al.*, 1976a,b; Benson, 1977; Reiter and Vaughan, 1977; Ebels and Benson, 1978; Pavel, 1978; Pavel *et al.*, 1978; Dogterom *et al.*, 1979), an α-MSH-like compound (Rudman *et al.*, 1972; Pévet and Swaab, 1979), and hypoglycemic factor of peptidergic nature called "pinealine" (Nanu *et al.*, 1978).

Adaptation, stress, and behavior represent the central problems in biology and medicine. On the basis of long-term investigations in our laboratories, we obtained some results about the active part of the pineal gland in the phenomena of adaptation and resistance to aggression (Miline, 1957a; Miline *et al.*, 1966, 1968; Miline, 1980).

2. Aim of Investigation

The aim of the present study was to investigate the histophysiological characteristics and behavioral changes in polypeptidergic activity of the pineal gland under the influence of a psychoneurogenic stressor: total iso-

Radivoy Miline ● Institute of Histology and Embryology, Medical Faculty, Novi Sad, Yugoslavia.

lation. With the aim to interpret the results obtained, mention will also be made of correlative reactions in the hypothalamus, nervous part of hypophysis, adrenal glands, and reproductive organs.

3.   Materials and Methods

The experiments were carried out during the summer season on white male laboratory Wistar rats of the same age. The animals were divided into two groups. The experimental group was comprised of rats exposed to total isolation for 2 weeks: each rat was placed in a separate cage in a room totally isolated from its social group so that it could not see, hear, or smell the other rats. The rats from the control group were placed five to a cage, cage to cage, in a stable. The animals were sacrificed by decapitation.

For light microscopy, organs were fixed in Bouin's and Bouin-D'Holland's fluids. Staining methods were: light green-erythrosin-iron-hematoxylin, chrome-alum-hematoxylin, Alcian blue-PAS-orange G (Herlant and Pasteels, 1967; Hurduc *et al.,* 1968). Usual histochemical and enzymological methods were used (Miline, 1980).

For electron microscopy, tissues were fixed by immersion in 6% glutaraldehyde in 0.2 M cacodylate buffered at pH 2.4. Postfixation was with 1% solution of osmic acid. This was followed by acetate dehydration and inclusion in Araldite. The sections were contrasted with uranyl acetate and lead citrate. They were examined using an ISKRA LEM 4C electron microscope.

4.   Results

4.1.   Pineal Gland

In experimental rats, some changes in the structure of the pineal gland were noticed in both the peripheral and the central zone. Both zones of the distal segment of the glandular body contained hypertrophied light pinealocytes, more clearly demarcated than those in the control group of animals. In many pinealocytes the nuclei were hypertrophied (Fig. 1), and eccentrically placed, with hypertrophied and eccentrically placed nucleoli. Nuclei with incisions were prevalent. Within the cytoplasm, on the bases of applied methods of staining, the small grains were either diffusely distributed or clustered in the pericapillary cellular pole, or larger isolated grains were seen either near the nuclei or within their incisions. In both zones, the presence of cells of a dark type, either solitarily impacted between hyperplastic light cells or grouped in irregularly shaped clusters, was emphasized. Dark cells were most numerous in the boundary region of the distal and proximal glandular body. Some cells were clearly demarcated, others accumulated in the form of syncytia. They were very variable in shape, many of them smaller in size than those in the control group of animals, either with small-grained cytoplasm or with entirely very chromophilic and homogeneously stained cytoplasm. The nuclei were of different size and shape. In many cells they were of smaller size and hyperchromatic. Cells with signs of nuclear pyknosis and disintegration were numerous.

In the experimental rats, the more numerous presence of lipid droplets, areally distributed, in contrast to their diffuse distribution in both glandular zones in control animals, was established. The presence of nucleic acids was increased in areally distributed, hypertrophied, and hyperplastic pinealocytes. Activity of nonspecific esterases and succinate dehydrogenases was also increased in these groups of cells.

The ultrastructural characteristics of reaction of light pinealocytes reflect some special changes in their cytodynamics. There were many pinealocytes with deep and numerous incisions in the nuclei (indented nuclei) with prominent electron-dense nucleoli with regular or irregular nucleolonema. The structure of certain nucleoli revealed a wide area of nucleolonema, totally encircled by chromatin. Some nucleoli, both in light and dark pinealocytes, contained chromatin only on one of the nuclear poles, with spreadings in the pars-amorpha region. Some nucleoli were seen to contain nucleolonemae of decreased density, but of increased density in the amorphous part. Within them were found spherically shaped parts similar to inclusions with fine-grained content and ringlike microvesicles (Fig. 2). Nucleolonemae in these nucleoli were usually curved peripherally, present only on a narrow area. Light pinealocytes with hypertrophied granulated endoplasmic reticulum, usually shaped like a mass of parallelly directed membranes of rectilinear or winding course, were numerous, found side by side with a large number of polysomes, diffusely distributed within the cytoplasm. The mitochondria were more numerous, present both in the close vicinity of the nucleus and in other parts of the cytoplasm, often adhering to some lipid droplets, a phenomenon that is partially demonstrated in light pinealocytes. There are great differences in the shape and size of mitochondria in numerous light, as well as dark pinealocytes. In the perikaryon, as within the cell processes in light cells, mitochondria of giant diameters and of oval or spindlelike shapes were present. Prevalent in some cells were thickly compacted crests of variable osmiophilia, directed either longitudinally or transversely, or both, within the same mitochondrion. In some mitochondria the finding of osmiophilic ballets was more frequent, together with the finding of concentrically distributed lamellae of uneven thickness and density. Direct continuities, i.e., the fusion of exterior mitochondrial membranes with those of endoplasmic reticulum, were noticed. In many light cells the Golgi zone was extended in reference to both saccules and vacuoles or microvacuoles. There were a large number of light microvesicles, together with microvesicles of osmiophilic content encircled with an osmiophilic membrane of uneven size, and dense core vesicles that drifted apart from the zone during their maturation and distributed within the other part of the cytoplasm. The presence of secretory, or so-called "grumose" granules or inclusions, described in rats by Wolfe (1965), was more numerous. In some light cells, these grains were of uneven shape and variable size, usually present in clusters (Fig. 3). There were some grains of hyperosmiophilic content, encircled by stretched membrane, together with grains of fine osmiophilic granulations. In some cells the secretory granules were spherical, resembling the grains of neurosecretion (neurosecretlike granules) adhered to the plasma membrane. In light cells, the presence of lipid droplets distributed separately or in grapelike clusters was

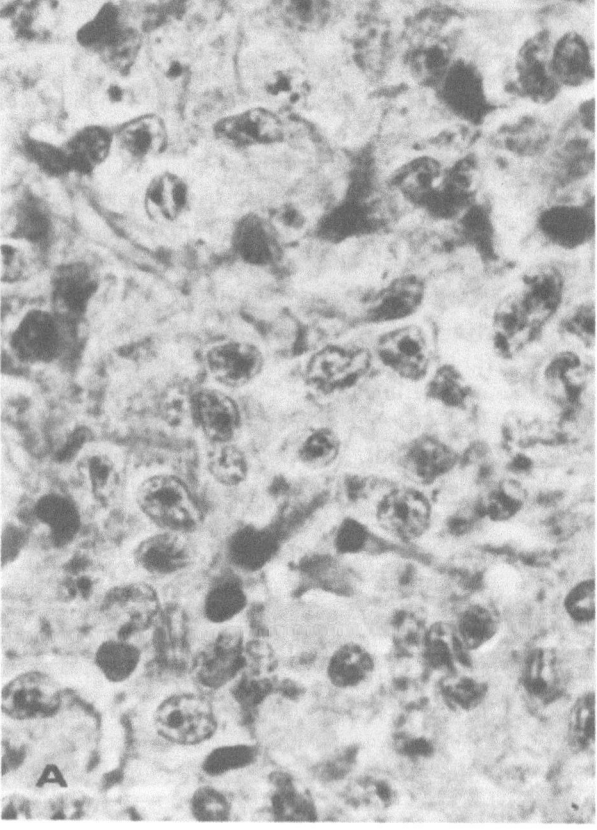

Figure 1. Gl. pinealis, central zone. (A) Control rat. (B) Rat under the influence of isolation: hypertrophy of the light cells, presence of hypertrophic nuclei. × 787.

more numerous. The finding of lipid droplets encircled by an abundance of granulated endoplasmic reticulum and freely distributed ribosomes was more frequent. There were some lipid droplets with osmiophilic clusters, peripherally distributed (Fig. 4), stringed in necklacelike formations, and either solitarily present, or clustered in the center of a droplet. They were encircled by concentrically distributed membranes of granulated endoplasmic reticulum, or by vacuoles and dilated cisternae. Lysosomes were present unevenly from cell to cell. In some cells, especially in those of dark type, nuclei of decreased volume during the atrophy, in which the mitochondria were partially or totally deprived of crests, with the signs of membranous ruptures, or transformed into granular mass, together with numerous secondary lysosomes, phagosomes, and lipopigment bodies, present both in the perikaryon and in the cellular processes, were noticed (Fig. 5).

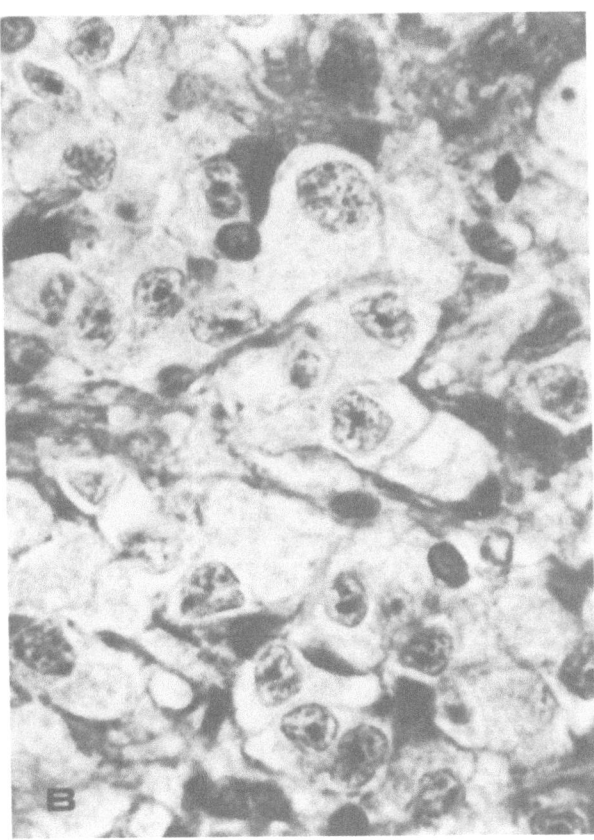

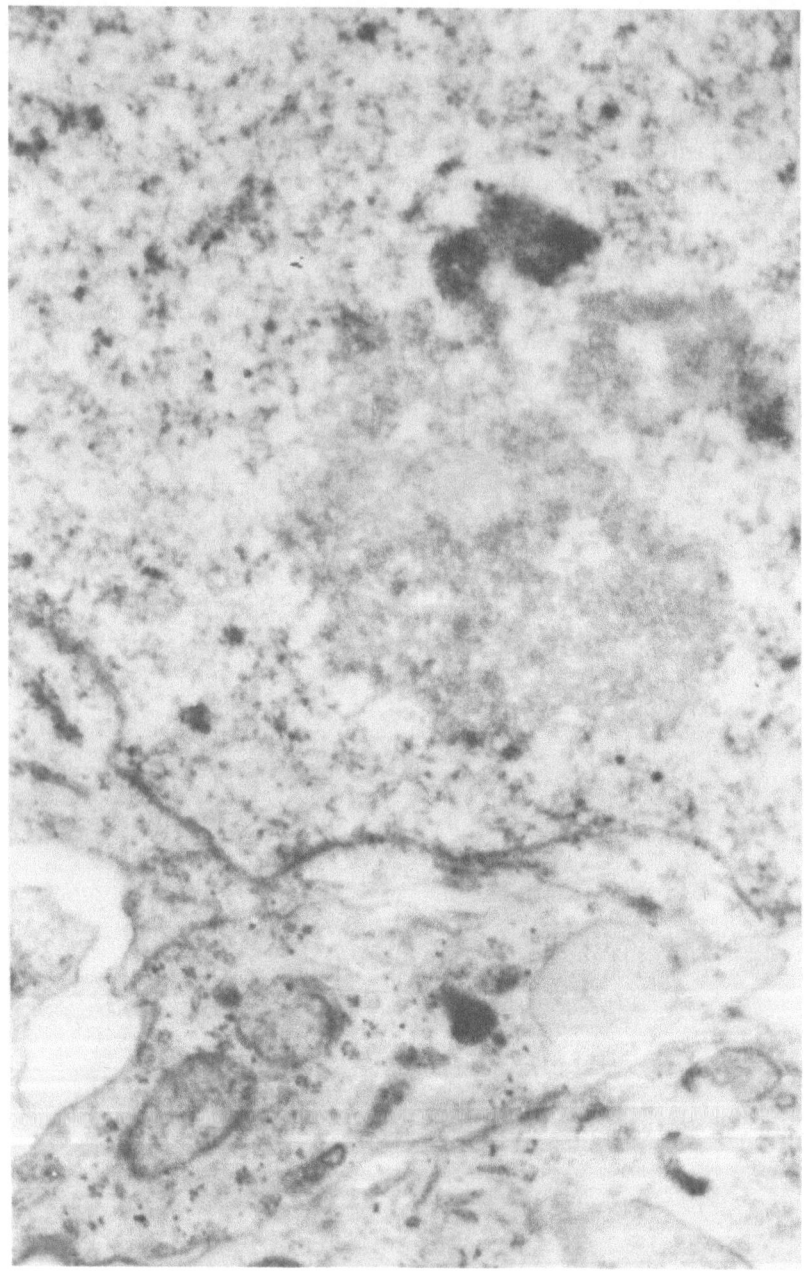

Figure 2. Part of the nucleus of dark cells: dissociation of the nucleolus. × 35,000.

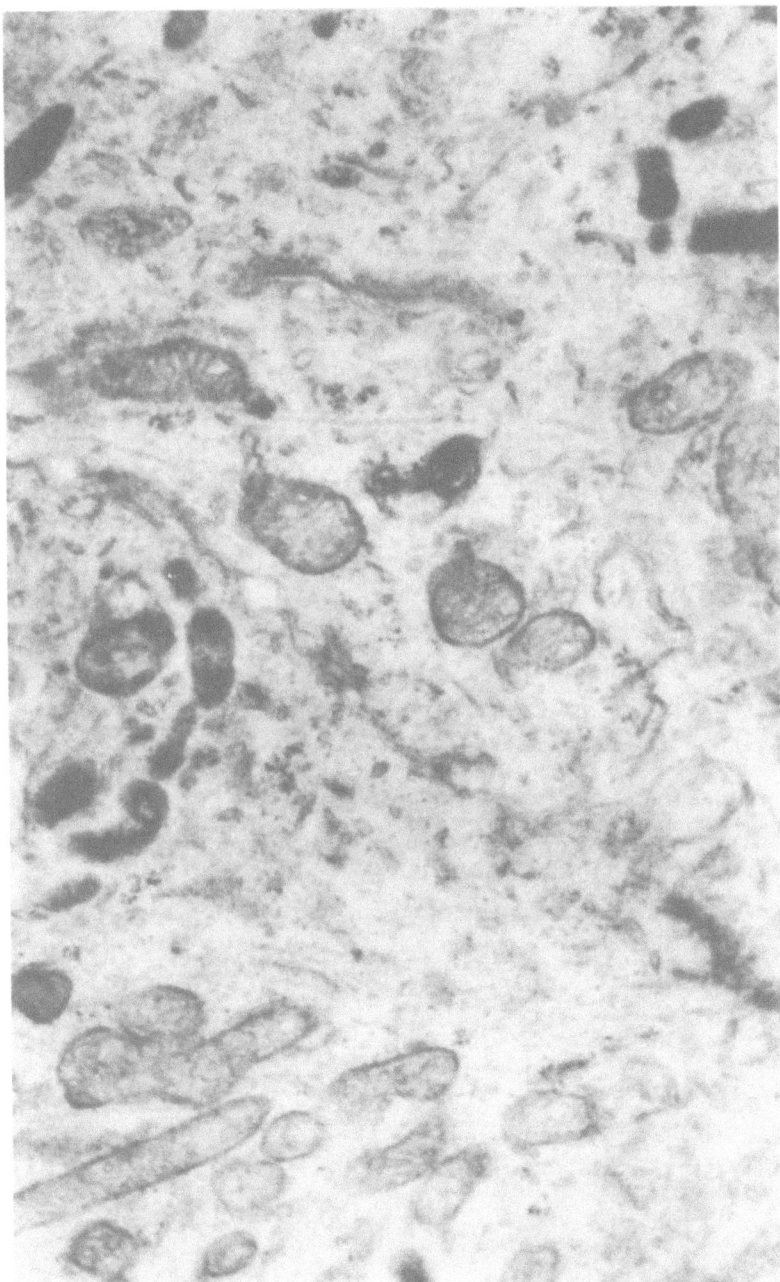

Figure 3. Light pinealocyte: grumose inclusions. × 35,000.

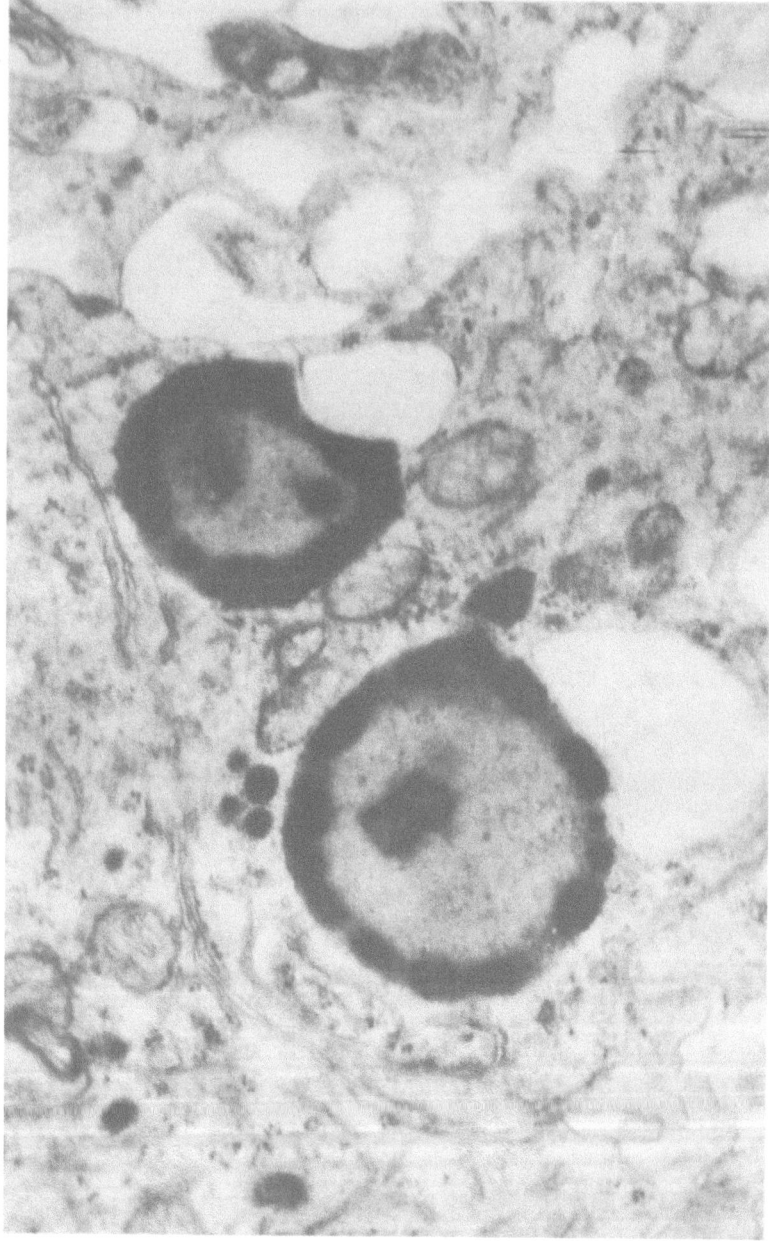

Figure 4. Light pinealocyte: lipid droplets with hyperosmiophilic clusters, vacuoles in the cyto-plasm. × 35,000.

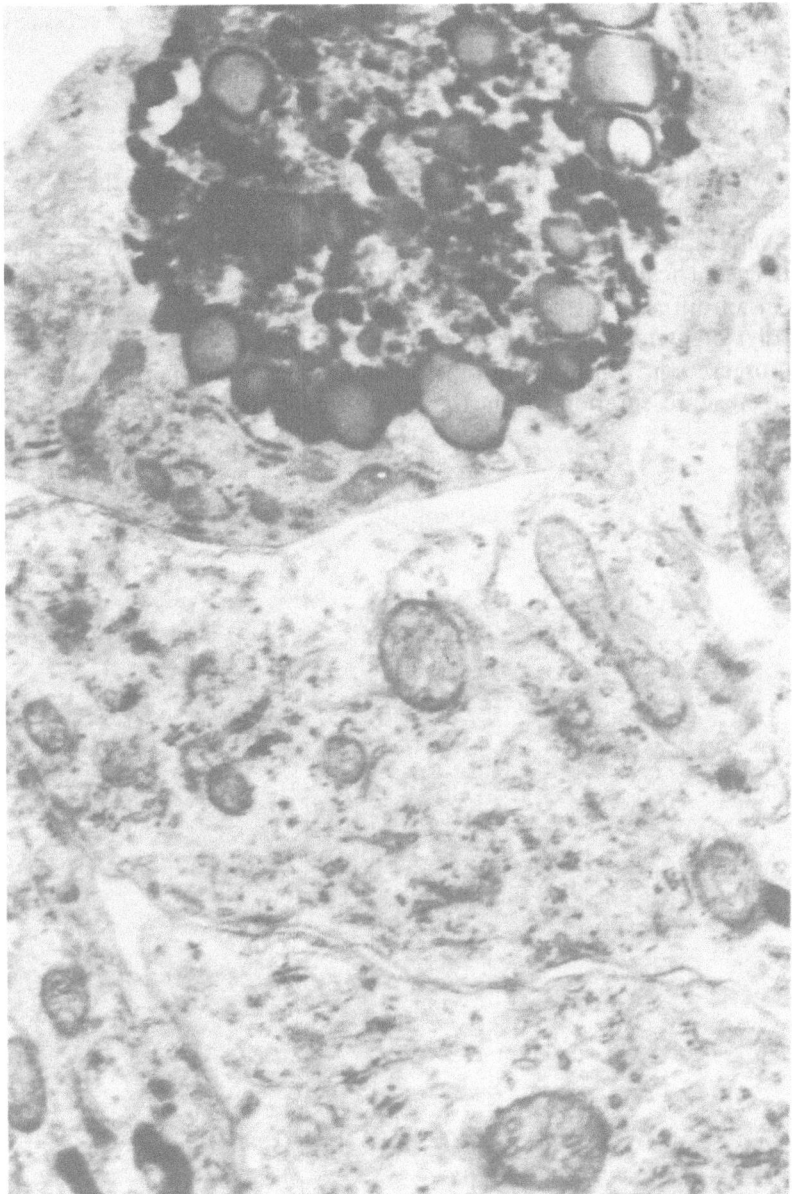

Figure 5. Ending of a dark pinealocyte process with lipopigment bodies; parts of the light pineal-ocytes with numerous polysomes. × 35,000.

The process of vacuolization and disintegration of cellular processes was one of their characteristics.

4.2. Nucleus Supraopticus

Within the supraoptic nuclei of experimental animals the following changes took place: presence of neuroglandular cells of considerably larger diameters than those in the control group of animals, with hypertrophied and eccentrically placed nucleus and nucleolus (Fig. 6), and with fine grains of diffusely distributed hypochromatic neurosecretion. Some cells of smaller diameter and with centrally placed nucleus with an abundance of neurosecretion with hyperchromatic features were also noticed. The presence of neurosecretion within the interstitium was very variable. Satellite

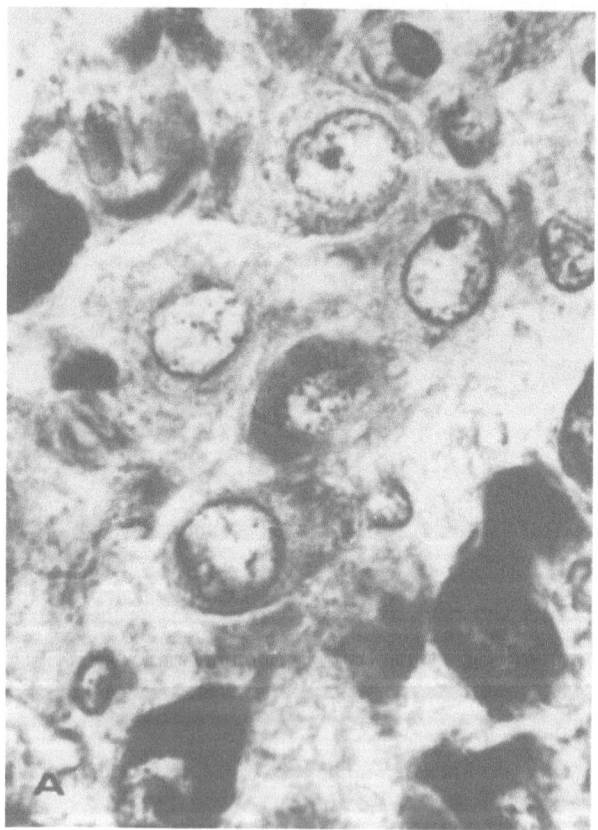

Figure 6. Nucleus supraopticus. (A) Control rat. (B) Rat under the influence of isolation: hypertrophy of the neuroglandular cells, hypertrophy of the nucleus and the nucleolus. × 1250.

neuroglial cells adhering to hypertrophied neuroglandular cells were also with hypertrophied nucleus.

4.3. Pars Neuralis Hypophyseos

Within the pars neuralis hypophyseos, hyperplasia of pituicytes in the whole area of this part of the hypophysis was manifested (Fig. 7). In the cytoplasm of pituicytes, some of which were hypertrophied and with hypertrophied nucleus, the neurosecret was present, a reflection of the phenomenon of phagocytosis. The neurosecretory fibers were of very uneven chromophilia. Herring's bodies were of smaller diameters, and more sparsely represented. The pericapillary accumulation of neurosecretion of hypochromatic features was established.

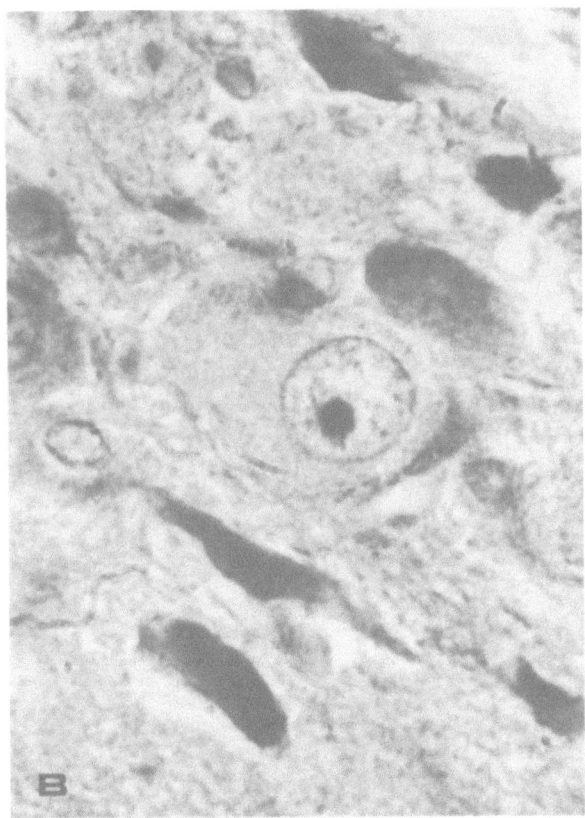

4.4. Adrenal Cortex

Within the structure of the adrenal glands, besides the changes of progressive nature within the first two zones, the phenomena of hypertrophy and hyperplasia in the area of the reticular zone were particularly stressed. The extended parts of this zone deeply penetrated into the medulla.

4.5. Testicles

On the basis of analyzing the serial sections of testicles, some changes in the seminal epithelium, as well as in the formation of interstitial gland, were established. Together with the presence of seminal tubules of normal structure, there were some seminal tubules of smaller diameters, seminal tubules with rarefied seminal epithelium, tubules with thickened basal

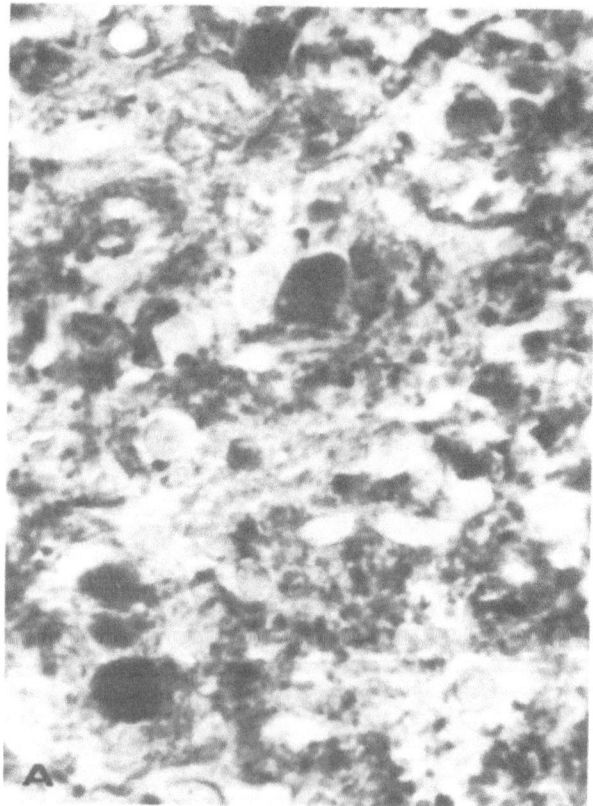

Figure 7. Pars neuralis hypophyseos. (A) Control rat. (B) Rat under the influence of isolation: hyperplasia of pituicytes. × 787.

membrane, tubules totally deprived of spermatoorgenic epithelium, reduced in their structure solely to Sertoli cells (Fig. 8). The interstitial glandular cells were hyperplastic, with hyperbasophilic cytoplasm, especially in the perinuclear part; their fields, between the seminal tubules, were spatially enlarged, the capillary network more numerous. This phenomenon was not displayed diffusely, but areally.

5. Discussion

The results obtained represent the morphological response of the pineal gland to chronic isolation stress. The established cytophysiological characteristics of glandular cells were of two kinds. Hypertrophy and hyperplasia of light pinealocytes with very distorted nucleus of hypertro-

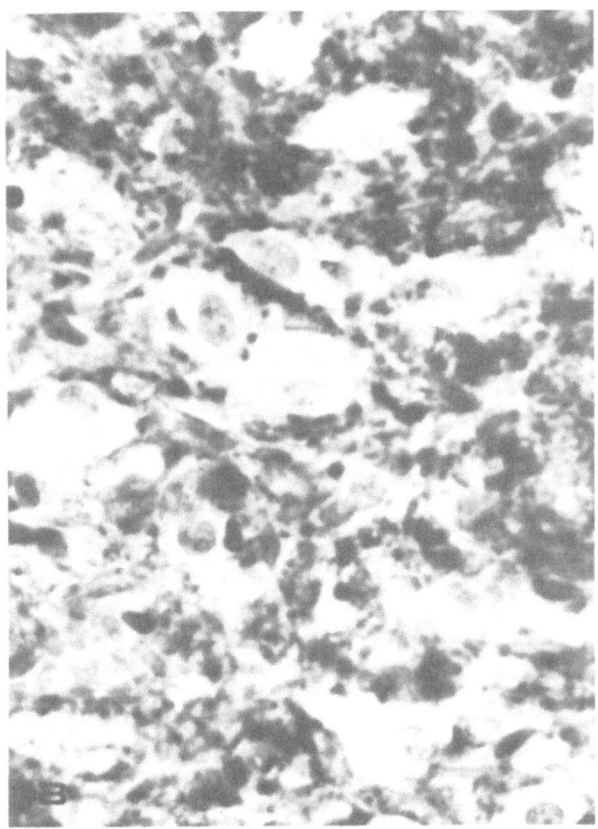

phied nucleoli, with abundance of granulated endoplasmic reticulum and polysomes, with numerous secretory, i.e., grumose, granules, and lipoprotein droplets spoke in favor of their very stimulated activity. Judging by these data, the isolation stress in light cells will be manifested through stimulated DNA–RNA protein and lipoprotein metabolism processes, a reflection of nucleocytoplasmic transfer. The changes described in the dynamics of dark pinealocytes, i.e., atrophy of the nucleus, signs of nuclear dissociation and segregation, phenomena of disintegration both in the perikaryon and in cellular processes, represented the reflection of decreased activity and involution of those cells.

The antigonadotropic, antiadrenocorticotropic, antithyreotropic, and anti-MSH features of the pineal gland are known. On the bases of the described results, the question is posed as to which features of the pineal

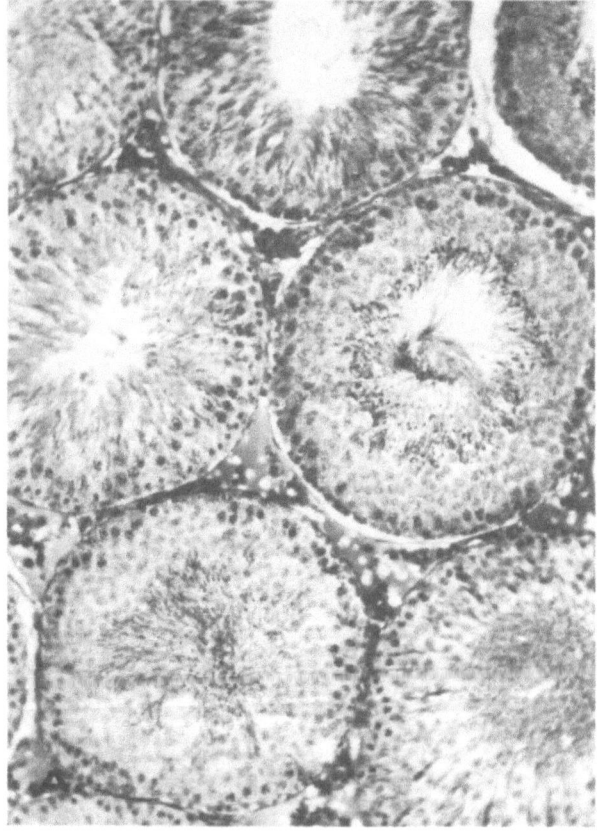

Figure 8. Testicle. (A) Control rat. (B) Rat under the influence of isolation: degeneration of the seminal epithelium, hyperplasia of Leydig cells. × 125.

gland are more or less exposed to the isolation stress. The changes described in the formation of the adrenal reticular zone (secretion of androgenous substances), as well as those in the structure of the testicles, indicate that in isolation stress the anti-LH, the antiandrogenic function of the pineal gland would be decreased, analogously to the results obtained in our laboratories under the influence of other stressors of particularly neurogenic features, i.e., under the influence of emotion and noise (Miline, 1960a). The changes described in the cytodynamics of the supraoptic nucleus, one of the hypophyseoregulator centers, very sensitive to the changes in pineal gland secretion, as well as to the changes described in the nervous part of the hypophysis, speak in favor of a role of the hypothalamus in the organization of behavior under the influence of isolation. Alterations in the hypothalamus releasing factors of peptidergic neurons, especially in the increased

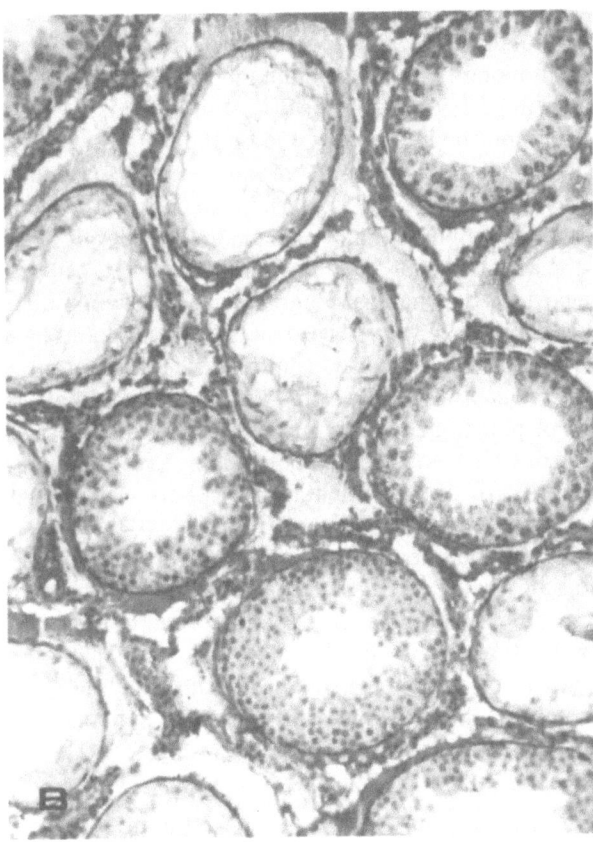

vasopressin secretion, and the presence within the supraoptic nucleus of the cells, of chemioreceptors for testosterone, for androgenic hormones (Dessi-Fulgheri and Di-Prisco, 1979; Orsini and Mei, 1979), suggest that social deprivation produces an increase of testosterone metabolism within the hypothalamus.

The described cytophysiological characteristics of the pineal gland provide evidence for the existence of a more numerous population of pinealocytes. The results obtained stand in confrontation to biochemical findings of a larger number of peptides in the pineal gland. They suggest the hypothesis that the pineal gland could be the center of secretion of a distinctive peptidergic system, somewhat like the peptidergic system in the hypophysis and diencephalon (Guillemin, 1978; Nemeroff and Prange, 1978; Schally, 1978). Having in mind that the hypothalamus represents the strategic relay area for certain activities of the pineal gland (Miline, 1959, 1960a; Kappers *et al.,* 1974; Reiter, 1974), and that the manifestation of pineal peptidergic activity is particularly demonstrated under conditions of stress, the question about the role of its peptides in the defense of organism can be asked: Does the pineal gland, being the biological antagonist of the hypophysis (Miline, 1980), resist the aggression by means of certain peptides of hormonal character, acting both on the hypophysis and on the central nervous system? Is the role of the pineal gland, being the organ of sensoneuroendocrine nature (Miline *et al.,* 1969; Miline, 1973), perhaps to coordinate the mutual relations between the hypophyseal and encephalic peptidergic systems? These are some of the questions yet to be answered.

With regard to the increased aggressiveness of animals deprived of pineal gland (Laborit, 1973), water extracts of pineal gland have a sedative effect and we established decreased combativeness in Bantam cocks and in the fish *Beta splendens* (Miline *et al.,* 1959). The animals treated with water extracts of pineal gland were more resistant to aggression and stress (Miline, 1960b,c, 1973). In human medicine, the favorable effects of pineal extracts in the therapy of some mental diseases are well known (Altschule *et al.,* 1954; Altschule, 1957, 1975).

The described and mentioned results suggest the important role of pineal peptides on the central level: control of behavior. Influencing normally the correlative function of the hypophysis and of the brain, the pineal gland could be particularly involved in modulation of behavior in stress conditions. The described morphophysiological aspects of the gland under the influence of total social deprivation, total isolation, support this hypothesis.

DISCUSSION

KRAICER: During the 2 weeks of isolation, could you describe the general state of the animals, for example, body weight, food and fluid intake, state of general activity and behavior.

MILINE: The animals demonstrated anxiety, increased general activity, and interrupted sleep. We conducted similar experiments on rats placed one by one in cages which were covered by appropriately placed mirrors, so that the rats were under the impression that they were part of a large social group and not isolated. In these animals, neither the changes in the pineal gland, nor those in the structure of the testicles described in the paper, were manifested. The behavior was characterized by tranquility, decreased general activity in relation to the former ones, and by deeper sleep.

C. VILLEE: Are these hypertrophied Leydig cells secreting an increased amount of testosterone? Have you measured circulating levels of testosterone?

MILINE: The evaluation of the activity of Leydig cells was measured by histochemical method, histoenzymology and electron microscopy.

C. VILLEE: I might comment that in hypophysectomized animals injections of LH will maintain the seminiferous tubules by an indirect mechanism: it stimulates the Leydig cells to produce testosterone and testosterone will maintain the seminiferous tubules. Your results, with regression of the seminiferous tubules despite a high production of testosterone by the Leydig cells, are further evidence of an antigonadotroph, actually anti-FSH, produced by the pineal.

NIKOLICS: Recently there have been reports in the literature on the isolation of pineal peptides. Have you had access to such preparations for testing them? Have you tried to isolate specific peptides from the pineal gland?

MILINE: Our experiments have been limited to the total water extracts of the pineal gland, and we have not isolated or tested for specific peptides.

GLEDIĆ: You showed a strong reduction of germ epithelium as a result of isolation stress. Some of the tubules contained only Sertoli cells. What is the number of these tubules in the testes and how long does it take to renew the germ epithelium? Is it a total or partial renewal?

MILINE: On the basis of sections of both testicles, involutive changes in the structure of the seminoepithelium were not manifested diffusely. The changes were of a reversible nature: In a month or two the atrophied seminoepithelium was restored.

KANAZIR: How do you explain the effects of isolation stress on the gastric mucosa?

MILINE: The presence of numerous hemorrhagic erosions in the glandular part of the stomach represents a characteristic of Selye's general adaptation syndrome. The glucocorticoids are involved. The appearance of numerous ulcerations in the stomach was also established in stress under the influence of emotion and influence of noise, being distinctly neurogenic stress. There has long been the concept that gastric and duodenal ulcers are the diseases of stress.

C. VILLEE: You suggest that pineal secretes an antigonadotropic principle, but the testes showed hypertrophied Leydig cells and regressed seminiferous tubules. Is the pineal secreting only an anti-FSH but no anti-LH?

MILINE: The release of FSH and LH is not only under the influence of GnRH in the hypothalamus but it is also under the influence of antigonadotropin factors in the pineal gland. There is an anti-FSH and an anti-LH factor. Inhibin originating from the Sertoli cells is also implicated. On the basis of the results obtained in the isolation stress within the pineal gland, the decreased secretion of anti-LH factor, the insufficiency of the antiandrogenic function of the pineal gland is manifested.

REFERENCES

Altschule, M. D., 1957, Some effects of aqueous extracts of acetone-dried beef-pineal substance in chronic schizophrenia, *N. Engl. J. Med.* **257**:919–922.

Altschule, M. D. (ed.) 1975, Effect of a pineal extract on carbohydrate metabolism in schizophrenic patients, in: *Frontiers of Pineal Physiology,* pp.197–203, MIT Press, Cambridge, Mass.

Altschule, M. D., Siegel, E. P., Goncz, R. M., and Murnate, J. P., 1954, Effect of pineal extracts on blood glutathione level in psychotic patients, *Arch. Neurol. Psychiatry* **71**:615–618.

Benson, B., 1977, Current status of pineal peptides, *Neuroendocrinology* **24**:241–258.

Benson, B., Matthews, M. J., and Hruby, V. J., 1976a, Characterization and effects of bovine pineal antigonadotropic peptide, *Am. Zool.* **16**:17–24.

Benson, B., Matthews, M. J., Hadley, M. E., Powers, S., and Hruby, V. H., 1976b, Differential localization of antigonadotropic and vasotic activities in bovine and rat pineal, *Life Sci.* **19**:747–754.

Dessi-Fulgheri, F., and Di-Prisco, L., 1979, Testosterone metabolism in the hypothalamus of socially deprived male rats, *Neurosci. Lett.* Suppl. 3, p. 170.

Dogterom, J., Snijdewint, F. G. M., Pévet, P., and Buijs, R. M., 1979, On the presence of neuropeptides in the mammalian pineal gland and subcommissural organ, in: *The Pineal Gland of Vertebrates Including Man* (J. Ariëns Kappers and P. Pévet, eds.), pp. 465–470, Elsevier/North-Holland, Amsterdam.

Ebels, I., and Benson, B., 1978, A survey of the evidence that unidentified pineal substance affects the reproductive system in mammals, *Prog. Reprod. Biol.* **4**:51–89.

Ebels, I., Moszkowska, A., and Scemama, A., 1965, Etude in vitro des extraits épiphysaires fractionnés: Résultats préliminaires, *C. R. Acad. Sci.* **260**:5126–5129.

Guillemin, R., 1978, Peptides in the brain: The new endocrinology of the neuron, *Science* **202**:390–402.

Herlant, M., and Pasteels, J. L., 1967, Histophysiology of human anterior pituitary, in: *Methods and Achievements in Experimental Pathology* (E. Bajusz and G. Jasmin, eds.), pp. 250–305, Karger, Basel.

Hurduc, M. J., Dogaru, M., Rostan, R., and Heine, G. A., 1968, A polychrome staining in two stages applicable for mammal adenohypophysis, *Rev. Roum. Embryol. Cytol. Ser. Sytol.* **5**:75–80.

Kappers, A. J., Smith, A. R., and De Vries, R. A. C., 1974, The mammalian pineal gland and its control of the hypothalamic activity, in: *Integrative Hypothalamic Activity* (D. F. Swaab and J. P. Schadé, eds.), pp. 149–174, Elsevier/North-Holland, Amsterdam.

Laborit, H., 1973, *Les comportements,* Masson, Paris.

Milcu, S. M., Pavel, S., and Neascu, C., 1963, Biological and chromatographic characterization of a polypeptide with pressor and oxytocic activities isolated from bovine pineal gland, *Endocrinology* **72**:563–566.

Miline, R., 1957a, Contribution à l'étude de la part de la glande pinéale dans le syndrome d adaptation, *Acta Anat.* **31**:581–582.

Miline, R., 1957b, La part de l'épiphyse dans le syndrome d'adaptation, Congrès National des Sciences Médicales, Recueil des travaux (Académie de la République populaire Roumaine, ed.), pp. 421–444, Académie RP Roumaine, Bucharest.

Miline, R., 1959, Effet de l'extrait épiphysaire sur la structure hypothalamique, *Acta Anat.* **38**:167.

Miline, R., 1960a, Sur la réactivité stressogène du complexe hypothalamohypophysaire, *Symp. Biol. Hung.* **1**:105–130.

Miline, R., 1960b, Influence de l'extrait épiphysaire sur la résistance à l'irradiation, *Acta Anat.* **42**:270.

Miline, R., 1960c, Effet protecteur de l'extrait de la glande pinéale sur la glande lymphatique de rats irradiés à la dose letale de rayons X, *Radovi* **15**:55–63.

Miline, R., 1973, Rezultati novijih instaživanja o funkcijama pinealne žlezde, in: *Savremena endokrinologija* (Srpsko lekarsko društvo, ed.), pp. 5–28, Savremena administracija, Beograd.

Miline, R., 1980, The role of the pineal gland in stress, *J. Neural Transm.* **47**:191–220.

Miline, R., Stern, P., Ciglar, M., and Huković, S., 1959, Beitrag zur Erforschung der antiandogenen Funktion der Pinealdrüse, *Naturwissenschaften* **15**:477–478.

Miline, R., Devečerski, V., and Krstić, R., 1966, Les modifications épiphysaires dans le stress et en particulier dans les névroses expérimentales d'effroie, in: *Symposium International sur la neuroendocrinologie* (P. Klotz, ed.), pp. 229–256, L'expansion Scientifique, Paris.

Miline, R., Krstić, R., and Devečerski, V., 1968, Sur le comportement de la glande pinéale dans des conditions de stress, *Acta Anat.* **71**:352–402.

Miline, R., Devečerski, V., and Krstić, R., 1969, Corpus pineale—glande de nature sensoneuroendocrine, *Radovi* **14**:69–84.

Nanu, L., Ionescu, V., and Marcean-Petrescu, R., 1978, The inhibitory effect of crinofizin (pineal peptide extract) on adrenalin-induced hyperglycemia in rabbits, *Rev. Roum. Med.-Endocrinol.* **16**:11–14.

Nemeroff, C. B., and Prange, A. J., 1978, Peptides and psychoneuroendocrinology, *Arch. Gen. Psychiatry* **35**:999–1010.

Orsini, J. C., and Mei, N., 1979, Réponse précoce des neurons hypothalamiques à une injection de testostérone chez le rat mâle, *C. R. Soc. Biol.* **173**:96–102.

Pavel, S., 1978, Arginine vasotocin as pineal hormone, *J. Neural Transm. Suppl.* **13**:135–155.

Pavel, S., and Petrescu, S., 1966, Inhibition of gonadotropin by highly purified pineal peptide and by synthetic arginine vasotocin, *Nature (London)* **212**:1054.

Pavel, S., Goldstein, R., Gheorgiu, C., and Calb, M., 1978, Pineal vasotocin: Release into cat cerbrospinal fluid by melanocyte-stimulating hormone release-inhibiting factor, *Science* **197**:179–180.

Pévet, P., and Swaab, D. F., 1979, Immunocytochemical evidence for the presence of an alfa-MSH-like compound in rat pineal gland, *J. Physiol.* **75**:101–103.

Quay, W. B., 1974, *Pineal Chemistry in Cellular and Physiological Mechanisms*, Thomas, Springfield, Ill.

Reiter, R. J., 1974, Pineal regulation of hypothalamic–pituitary axis: Gonadotropins, in: *Handbook of Physiology–Endocrinology*, Vol. IV (E. Knobil and W. A. Astwood, eds.), pp. 519–550, Waverly Press, Baltimore.

Reiter, R. J., and Vaughan, M. K., 1977, Pineal antigonadotropic substance: Polypeptides and indoles, *Life Sci.* **21**:159–172.

Rudman, D., Del Rio, A. E., Hollins, B. M., House, D. H., Sutin, J., and Mosteller, R. C., 1972, Comparison of lipolytic and melanotropic factors in bovine choroid plexus and in bovine pineal gland, *Endocrinology* **90**:1139–1146.

Schally, A., 1978, Aspects of hypothalamic regulation of the pituitary gland, *Science* **202**:18–28.

Wolfe, D. E., 1965, The epiphyseal cell: An electron microscopic study of its intercellular relationship and intracellular morphology in the pineal body of the albino rat, in: *Structure and Function of the Epiphysis Cerebri* (J. Ariëns Kappers and J. P. Schadé, eds.), pp. 332–376, Elsevier, Amsterdam.

23

Effect of Colchicine on the Stress-Type ACTH Release in Rats

Kalina I. Vaptzarova, Mikhail S. Davidov, Dimitar Strashimirov, and Panteley G. Popov

1. Introduction

The adenohypophysis is an endocrine gland that cumulates the synthesized hormones in membrane-bound organelles—secretory granules. Secretion of cell products is a complex process involving the packaging of cell secretions in membrane to form vesicles, the transport of the vesicles to the cell membrane and their release into the extracellular space.

Recently, many experiments have been reported (for a review see Trifaró, 1977) that tend to show the involvement of microtubules (MTs) in the secretion of cell products, probably participating in the vesicle orientation and movement. In most experiments of this kind the secretion rate of a particular cell product was measured before and after the administration of a MT-disrupting agent, such as colchicine (COL) or vinblastine.

The possible involvement of MTs has been suggested for insulin secretion (Lacy *et al.*, 1968), TSH-stimulated thyroid secretion (Williams and Wolff, 1970), norepinephrine and dopamine-β-hydroxylase release by exocytosis from sympathetic nerve terminals (Poisner and Bernstein, 1971), the secretion of very-low-density lipoproteins (Stein and Stein, 1973), etc.

Kalina I. Vaptzarova and Mikhail S. Davidov • Regeneration Research Laboratory, Bulgarian Academy of Sciences, 1431 Sofia, Bulgaria. *Dimitar Strashimirov* • Department of Physiology, Medical Faculty, Pleven, Bulgaria. *Panteley G. Popov* • State Institute for Control of Drugs, 1040 Sofia, Bulgaria.

Several experiments on the effect of COL on the secretion of pituitary hormones by cultured tumor cells have been reported. It has been shown that COL inhibits the secretion of growth hormone and prolactin from such cells (Gautvik *et al.*, 1973; Sheterline *et al.*, 1975). Kraicer and Milligan (1971) have demonstrated *in vitro* that COL decreases ACTH release induced by high potassium concentration. Some controversy exists about the effect of antimicrotubular agents on the secretion of hormones by pituitary cells. No change (Sundberg *et al.*, 1973; Temple and Wolff, 1973) or even a potentiation (Sundberg *et al.*, 1973) of the *in vitro* release of some pituitary hormones by such agents have been observed.

The effect of COL on corticosteroid secretion has been investigated *in vitro*. Inhibition (Reiterman *et al.*, 1980), stimulation (Ray and Strott, 1978), and no effect (O'Hare, 1976) of corticosteroid release have been reported.

The study of the influence of COL on corticosteroid and ACTH secretion *in vivo* could supplement the *in vitro* data and give more insight on the role of MTs in hormone secretion.

In a series of experiments the effect of COL on glucocorticoid and ACTH secretion was studied by following the changes in plasma corticosterone level and liver tyrosine aminotransferase (TAT) activity. Electron microscopic study of the relevant organs after COL was also performed.

2. Experimental Procedures

All experiments were performed on male Wistar albino rats, weighing 150 to 180 g. Three groups of experimental animals were used: intact, adrenalectomized, and hypophysectomized. Adrenalectomy was performed under light ether anesthesia. Adrenalectomized rats were given 1% NaCl solution to drink *ad libitum*. Transauricular hypophysectomy was performed according to Strashimirov (1967). The completeness of the operation was verified by inspecting the sella region for hypophysis remnants, using a magnifying glass. Hypophysectomized rats were given 5% glucose solution to drink *ad libitum*. The operated animals were used in experiments on the fourth day after surgery.

COL (Fluka) was dissolved in saline and injected intraperitoneally in a dose of 5 mg/kg body wt. A microcrystalline suspension of hydrocortisone (HCS) acetate (Gedeon Richter, Budapest) was injected intraperitoneally in a dose of 100 mg/kg body wt. The drugs were administered during the period of the lowest circadian activity, i.e., between 8 and 9 A.M. In the case when TAT activity was followed, COL or HCS acetate was administered 4 hr before the animals were sacrificed; when given together, COL was injected 3 hr before the HCS application, and the rats were sacrificed 4 hr afterwards.

The animals were killed by decapitation, the liver was chilled on ice and homogenized in a Potter–Elvehjem glass–Teflon homogenizer with 0.125 M potassium phosphate buffer, pH 7.6. A 10% homogenate was prepared and centrifuged at 12,000g for 15 min at 4°C and the supernatant fluid was used for TAT activity assay, according to Diamondstone (1966). The absorbancy at 331 nm was measured, a molar extinction of 19,900 being adopted. The results for TAT activity obtained are expressed as units per gram wet weight, one unit being defined as the amount of enzyme that yielded 1 μmole p-hydroxybenzaldehyde/hr at 37°C.

The plasma corticosterone concentration (in μg/100 ml blood) was determined fluorimetrically according to Silber *et al.*, (1958). The experimental animals for coritcosterone determination were habituated to handling for 2 weeks and they were killed within 30 sec after removal from the cage.

All control animals were injected with a volume of saline equal to that of the drug injected in the experimental rats.

Student's t test was applied for statistical evaluation of the results obtained. Each experimental point represents the mean $\pm$ S.E.M.

Fixation by perfusion through the aorta with 6.5% glutaraldehyde in 0.14 sodium cacodylate buffer (pH 7.2) for 20 min was performed, and material from adenohypophysis, neurohypophysis, hypothalamus, adrenals, and liver was taken for electron microscopic study. After fixation for one more hour in the same solution at 4°C, small pieces of tissue were postfixed in 2% osmic acid dissolved in the buffer. The material was embedded in Durcupan-ACM (Fluka), sectioned with an ultramicrotome, and examined on a JEM 100 B electron microscope.

3. Results and Discussion

3.1. Experiments with Intact Rats

Rat liver TAT is an enzyme inducible by glucocorticoids. The assay of TAT activity in rat liver has the advantage that it reflects the integrated effect of the raised level of corticosterone for a more prolonged period of time. In our previous experiments it was found that stress induces a two- to threefold rise of TAT activity at the fourth hour (Vaptzarova *et al.*, 1973).

In the rats with intact adrenals, COL induced a 2.7-fold increase of TAT activity 4 hr after the application of the alkaloid (Fig. 1). This increase may be due to:

1. Stress reaction as a consequence of the toxic influence of COL on the pituitary–adrenal axis.

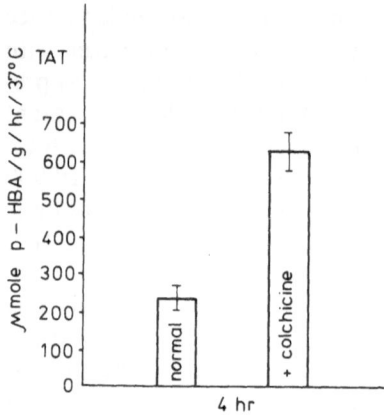

Figure 1. TAT activity in rats with intact adrenals 4 hr after treatment with saline (controls) or COL.

2. Direct stimulation of glucocorticoid secretory cells.
3. Direct influence of COL on the TAT synthesizing system in the hepatocytes (e.g., derepression of the genome; prolonging the longevity of TAT mRNA; increase of the rate of TAT mRNA translation).

In order to elucidate the mechanism of the observed increase of rat liver TAT activity after COL administration we performed a series of experiments.

3.2. Experiments with Adrenalectomized Rats

Adrenalectomy itself decreased TAT activity significantly ($P < 0.01$) and abolished entirely the rise of TAT activity caused by COL in intact animals (Fig. 2). Therefore, the increase of enzyme activity after COL administration observed in the intact animals could be the result of an elevation of the concentration of corticosterone, the main glucocorticoid of the rat.

3.3. Plasma Corticosterone after COL Treatment

The plasma corticosterone concentration was determined 1, $2\frac{1}{2}$, and 4 hr after COL administration. As shown in Fig. 3, COL has elicited a marked elevation of plasma corticosterone with a maximum at $2\frac{1}{2}$ hr. Ray and Strott (1978) have demonstrated that COL stimulates steroid synthesis in cultured adrenocortical cells. It is possible therefore that the raised level of plasma corticosterone is due to a direct effect of COL on the adrenocortical cells. In order to check this possibility we performed the next experiments.

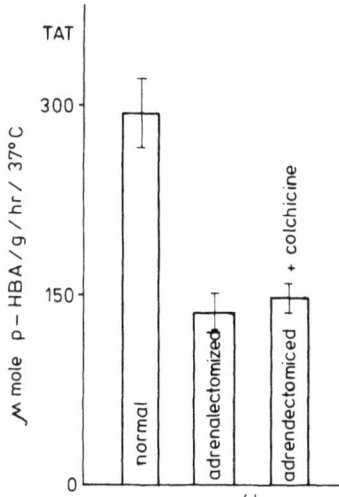

Figure 2. TAT activity in adrenalectomized rats 4 hr after treatment with saline ("adrenalectomized") or COL.

3.4. Experiments with Hypophysectomized Rats

Hypophysectomy alone did not influence significantly TAT activity (Fig. 4). In hypophysectomized rats COL not only did not increase liver TAT but even lowered it. Therefore, the increase of enzyme activity observed in rats with intact adrenals could not be explained by a direct influence of COL on the adrenocortical cells. Most probably, it was mediated through the pituitary.

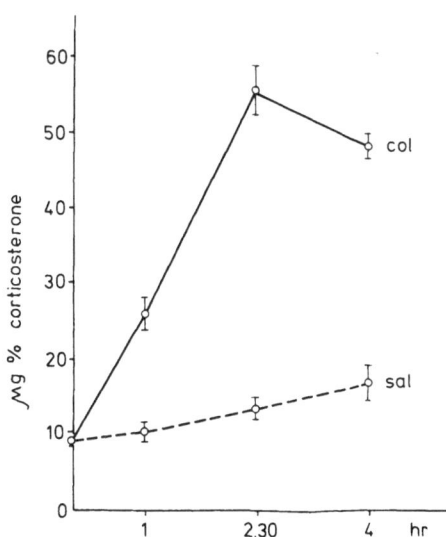

Figure 3. Plasma corticosterone level 1, 2½, and 4 hr after treatment with COL or saline.

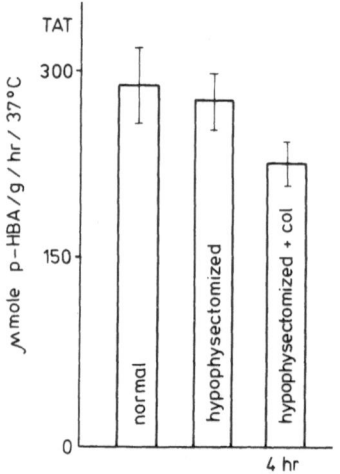

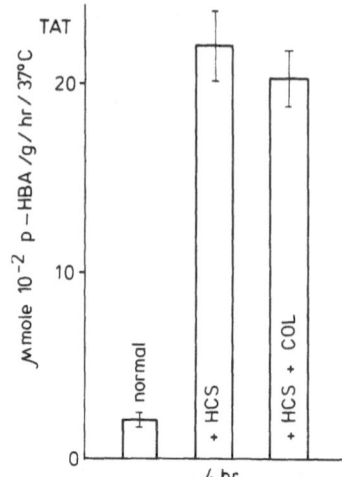

Figure 4. TAT activity in hypophysectomized rats 4 hr after treatment with saline or COL.

Figure 5. TAT activity in intact rats 4 hr after treatment with HCS or COL + HCS. COL in this case is injected 3 hr before the hormone.

Butcher and Perdue (1973) have reported that cytochalasin B, a microfilament-disrupting agent, inhibited the induction of rat liver TAT by HCS. It was intriguing, therefore, to examine this effect for COL.

3.5. The Effect of COL on HCS Induction of Liver TAT

COL was administered 3 hr before the injection of HCS. The rats were sacrificed 4 hr after the application of the hormone. In the rats treated only with HCS, a near 10-fold increase of TAT activity was found. COL treatment did not influence significantly ($p > 0.05$) the HCS induction of TAT (Fig. 5). Therefore, in contrast to cytochalasin B, COL did not interfere with the TAT synthesizing mechanism of the liver cells.

Our results from experiments with adrenalectomized and hypophysectomized rats show that the increased liver TAT activity after COL treatment is due to an activation of the pituitary–adrenal axis of stress type.

The increase of the plasma corticosterone level and TAT induction by COL suggests that COL does not inhibit the stress-type ACTH and glucocorticoid secretion *in vivo* and, hence, MTs do not take part in the secretory process. This conclusion, however, should not be preemptive. The participation of MTs in the secretory process of the cells is still questionable (Trifaró, 1977). Secretion is evidently a complex process and depends on a variety of cell components. The effect of COL on MTs itself depends on many factors such as permeability of cell and/or organ barriers for the alkaloid, differential sensitivity of MTs from various cell types to COL, etc.

Valuable information in this respect could be obtained from electron microscopy.

3.6. Results from the Electrom Microscopic Investigations

Several hypothalamic nuclei (n. arcuatus, n. supraopticus, and n. suprachiasmaticus), neurohypophysis, adenohypophysis, adrenal cortex, and liver have been investigated ultrastructurally.

3.6.1. Hypothalamic Nuclei and Neurohypophysis

In the cells of the hypothalamic nuclei as well as in the axons in the neurohypophysis from COL-treated animals, MTs are not affected probably because of the poor permeability for the alkaloid of the blood–brain barrier.

3.6.2. Adenohypophysis

Moriarti and Moriarti (1973) have described the electron microscopic appearance of ACTH-producing cells with an immunohistochemical technique. Similar cells were found in our preparations. These cells have processes surrounding the neighboring cells and contain little ergastoplasmic reticulum (ER). The secretory granules are of regular shape with a narrow halo around the membrane, their size ranging from 170 to 250 nm. These secretory granules are often observed in a row under the cell membrane (Fig. 6).

After COL administration the cells of adenohypophysis (pars distalis) and those of the intermediate lobe show typical alterations. In most of the cells the MTs disappear. Strong dilation of the rough ER can be seen during the first hours after COL application. The dilated cisternae of the rough ER are filled with finely granulated material of moderate electron density. Secretory granules in the ACTH-like cells appear increased in number and show a great variety in size and in the electron density of their cores. Part of the Golgi complexes appear to be disintegrated. Multilamellar (myelin-like) and autophagic bodies are observed, as well as a great number of electron-transparent vesicles. At the first hour after COL treatment MTs can be seen only in some cells of pars intermedia and pars distalis of the hypophysis (Fig. 7). Two and a half hours after the injection of the alkaloid all MTs disappear. Exceptionally strong dilation of the ER can be observed at the first hour after COL administration in the cells of pars ventro-centralis and pars distalis. At the fourth hour ER is narrower, autophagic vacuoles are fewer, but secretory granules and light vesicles appear increased in number. At the same time a tendency to normalization can be observed, but this process is far from completion (Fig. 8).

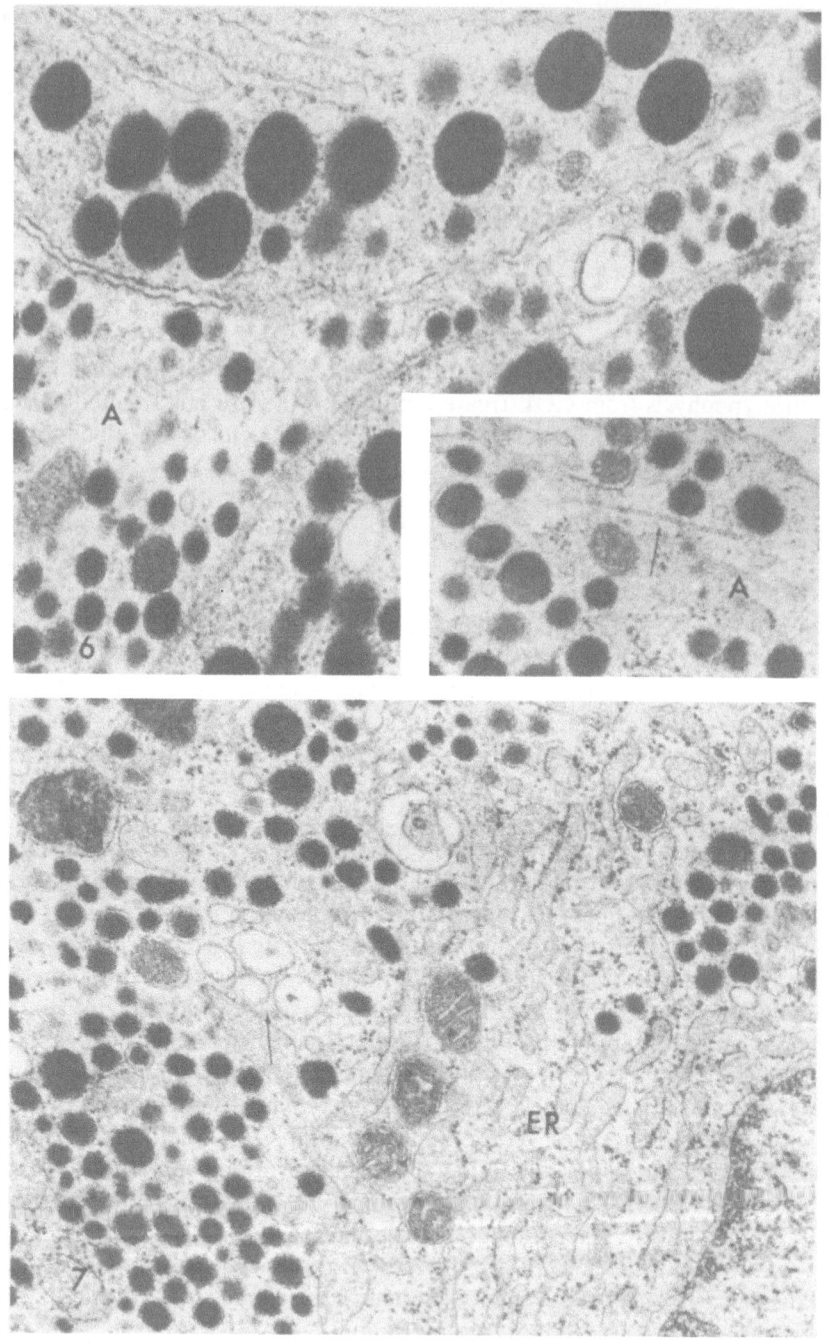

Figures 6 to 8. Pars distalis of rat hypophysis.

Like adrenocorticotrophs, other cell species of the adenohypophysis show similar or more strongly pronounced alterations.

3.6.3. Adrenals

There is a striking difference between the adrenals of control (Figs. 9 and 10) and COL-treated (Fig. 11) rats. Total disappearance of MTs and disorganization of the Golgi complex are observed in the COL-treated animals. Two and a half hours after the alkaloid administration the cytoplasm is filled with closely packed mitochondria. ER is arranged densely around the mitochondria. Many lipid droplets are accumulated in the cytoplasm

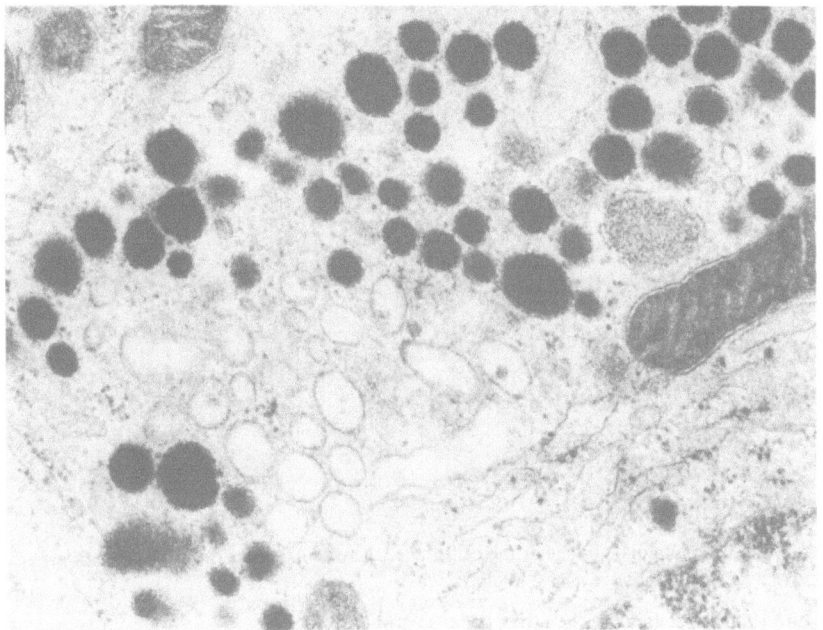

Figure 8. Four hours after the injection of COL. A tendency to narrowing of the ER, accumulation of many translucent vesicles and secretory granules of varied size and electron density can be observed. × 36,000.

←——

Figure 6. Control animal. ACTH-like cell with a process and peripherally situated secretory granules. The arrow points to a MT. × 38,000. A, ACTH-like cells.

Figure 7. One hour after the injection of COL. Dilation of the ER, accumulation of translucent vesicles (arrow) and of secretory granules. × 34,000.

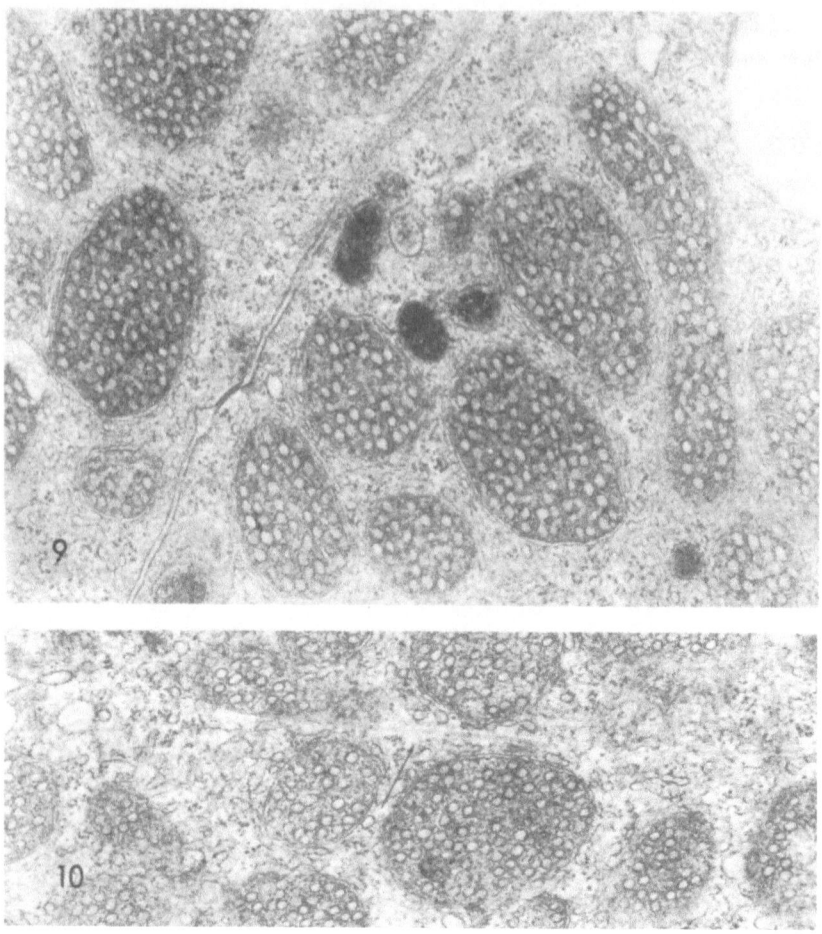

Figures 9 to 11. Rat adrenal cortex.

Figure 9. Control animal. In one of the cells under the plasmalemma, accumulation of granules of varied density can be seen. × 34,000.

Figure 10. Control animal. The arrow points to a MT. × 25,000.

forming aggregates. Four hours after the COL treatment a clearing of the cytoplasm and a dispersion of the mitochondria are observed. Many autophagic bodies and translucent vesicles of varied size can be seen in the cytoplasm (Fig. 11).

3.6.4. Liver

The normal electron microscopic structure of the liver is shown in Figs. 12 and 13. In the liver the alterations after COL are greatly marked. The

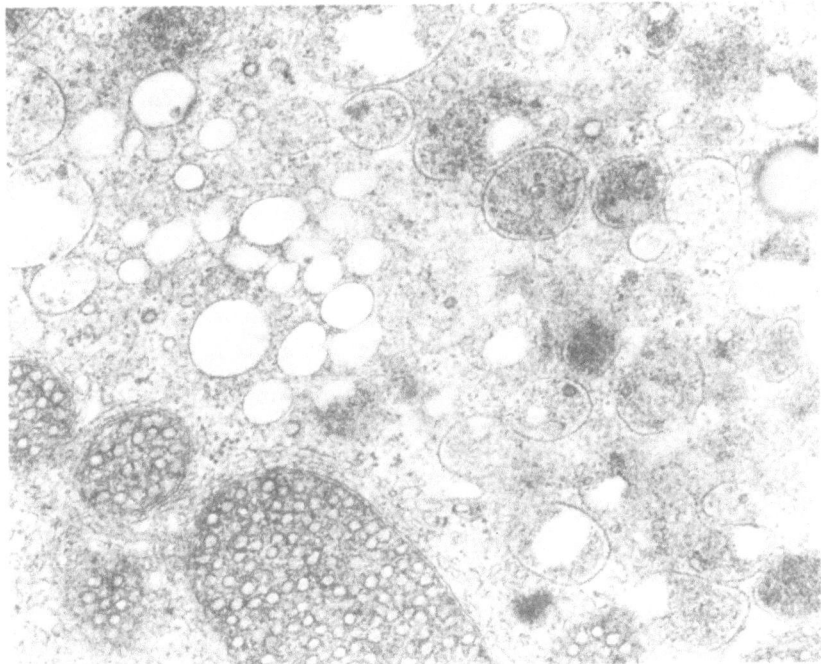

Figure 11. Four hours after the injection of COL. In the cytoplasm are accumulated many trans-lucent vesicles and vesicles with varied electron-dense content. × 36,000.

MTs disappear and the Golgi complexes are disorganized. Storing vesicles containing electron-translucent droplets appear in the cytoplasm. The pre-dominant part of the deposited material in these vesicles resembles very-low-density lipoproteins. The number of autophagic vacuoles and glycogen rosettes is greatly increased (Figs. 14 and 15).

3.6.5. General Discussion

In all investigated organs there are more or less strongly manifested signs of a blockade of the secretory activity and exocytosis of substances deposited in vesicles. In all cases (except hypothalamus and neurohypophy-seal axons) a disappearence of the MTs is seen.

The increased, stress-type secretion of glucocorticoids observed after COL treatment led us to assume that ACTH secretion is not lowered by the alkaloid. COL caused in the same time an extensive destruction of MTs in the investigated organs. This leads us to infer that the postulated partic-ipation of MTs in the process of ACTH secretion should be considered uncertain.

Our results suggest that ACTH, considered as being secreted by exo-

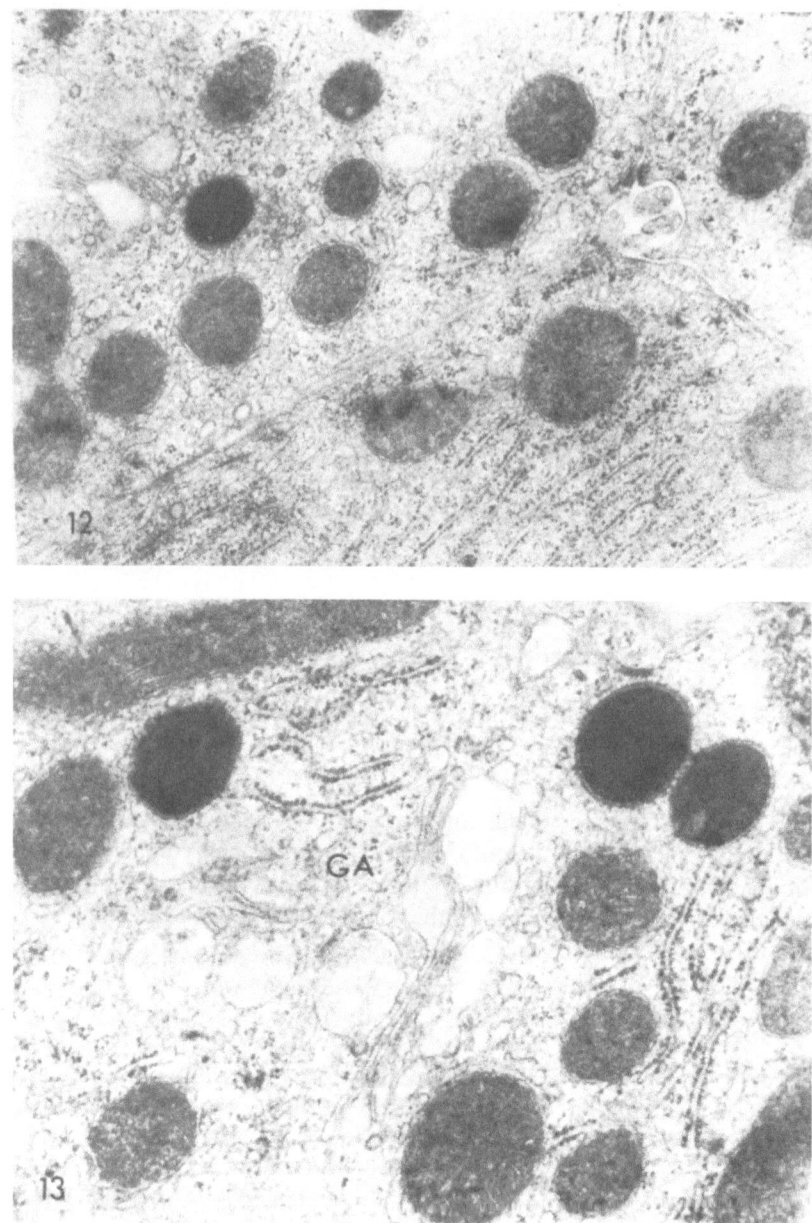

Figures 12 to 15. Rat liver.

Figures 12 and 13. Control animals. GA, Golgi area. Notice the relatively uniform distribution of cell organelles in the cytoplasm. Figure 12, × 30,000; Figure 13, × 35,000.

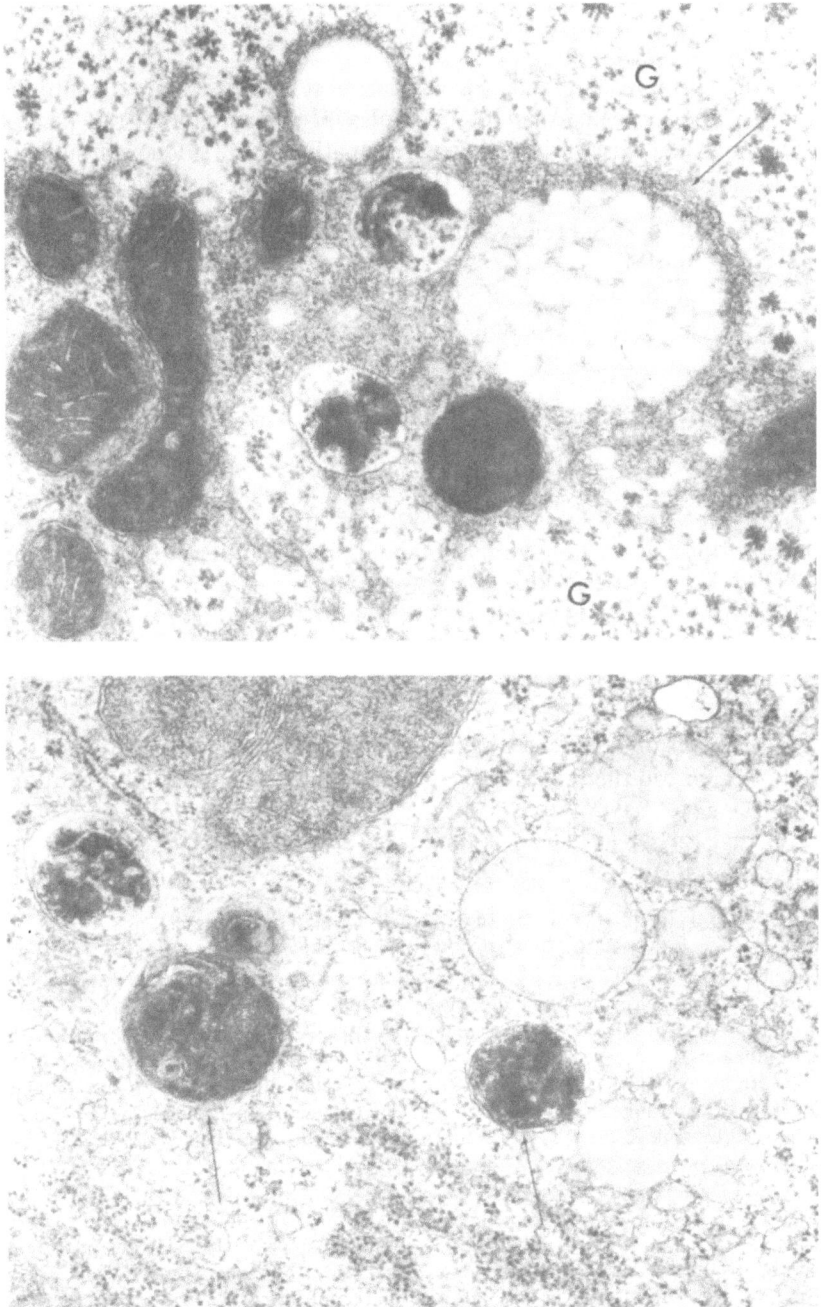

Figure 14. Two and a half hours after the injection of COL. Large glycogen areas (G), many autophagic vacuoles, and a body containing very-low-density lipoprotein particles (arrow) can be seen. × 40,000.

Figure 15. Four hours after the injection of COL. Cytoplasm is returning to its normal appearance. Glycogen is almost not observed. The number of autophagic bodies (arrows) and of the structures containing very-low-density lipoprotein particles are increased. × 40,000.

cytosis, is synthesized in the ER and a part of it is released directly into the cytoplasm, thus evading the Golgi complex. It is possible that this part of the hormone is released without the participation of MTs. One can suppose the existance of two pools of ACTH in the secretory cells:

1. A pool for rapid release without the participation of the Golgi complex and MTs.
2. A pool for slower release of secretory granules containing the hormone, with participation of the Golgi apparatus and MTs.

4. Conclusion

The effects of COL on the plasma corticosterone level and TAT activity in rat showed that COL *in vivo* caused a stress-type plasma corticosterone elevation and TAT induction and hence it did not lower the stress-type ACTH secretion.

Electron microscopic study showed that under the same experimental conditions occurred the complete disappearance of MTs in liver, adrenals, and, at least in the later terms, in the pituitary cells, adrenocorticotrophs included.

The results obtained once again demonstrate that interpretation of the effects of COL on the process of secretion is difficult and requires caution.

REFERENCES

Butcher, F. R., and Perdue, J. F., 1973, Cytochalasin B: Effect on hormone-mediated responses in cultured cells, *J. Cell Biol.* **56**:857.

Diamondstone, T. I., 1966, Assay of tyrosine aminotransferase activity by conversion of *p*-hydroxyphenylpyruvate to *p*-hydroxybenzaldehyde, *Anal. Biochem.* **16**:395.

Gautvik, K. M., Hoyt, R. F., and Tashyan, A. H., Jr., 1973, Effects of colchicine and 2-Br-alpha-ergocryptine-methanesulfonate (CB 154) on the release of prolactin and growth hormone by functional pituitary tumor cells in culture, *J. Cell Physiol.* **82**:401.

Kraicer, J., and Milligan, J. V., 1971, Effect of colchicine on *in vitro* ACTH release induced by high K^+ and hypothalamus–stalk–median eminence extract, *Endocrinology* **89**:408.

Lacy, P. E., Howell, S. L., Young, D. A., and Fink, C. J., 1968, New hypothesis of insulin secretion, *Nature (London)* **219**:1177.

Morlartl, G. C., and Moriarti, C. M., 1973, Immunocytochemical localization of ACTH and MSH in rat and human hypophyses, *Anat. Rec.* **175**:393 (abstract).

O'Hare, M. J., 1976, Monolayer cultures of normal adult rat adrenocortical cells: Steroidogenic responses to nucleotides, bacterial toxins, and antimicrotubular agents, *Experientia* **32**:251.

Poisner, A. M., and Bernstein, J., 1971, A possible role of microtubules in catecholamine release from the adrenal medulla: Effect of colchicine, vinca alkaloids and deuterium oxide, *J. Pharmacol. Exp. Ther.* **177**:102.

Ray, P., and Strott, C. A., 1978, Stimulation of steroid synthesis by normal rat adrenocortical cells in response to antimicrotubular agents, *Endocrinology* **103**:1281.

Reiterman, R. W., Li Chen, Wen, S. C., and Harding, B. W., 1980, The effect of colchicine and microtubule-associated proteins on cAMP-dependent cytosolic protein kinase in bovine adrenal cortical cells, *Fed. Proc.* **30**:721.

Sheterline, P., Schofield, J. G., and Mira, F., 1975, Colchicine binding to bovine anterior pituitary slices and inhibition of growth hormone release, *Biochem. J.* **148**:435.

Silber, R. H., Busch, R. D., and Oslapas, R., 1958, Practical procedure for estimation of corticosterone and hydrocortisone, *Clin. Chem.* **4**:278.

Stein, O., and Stein, J., 1973, Colchicine-induced inhibition of very low density lipoprotein release by rat liver *in vitro, Biochim. Biophys. Acta* **306**:142.

Strashimirov, D., 1967, A method for transauricular hypophysectomy in rats, *Exp. Med. i Morfol.* **6**:109.

Sundberg, D. K., Krulich, L., Fawcett, C. P., Illner, P., and McCann, S. M., 1973, The effect of colchicine on the release of rat anterior pituitary hormones in vitro, *Proc. Soc. Exp. Biol. Med.* **142**:1097.

Temple, R., and Wolff, J., 1973, Stimulation of steroid secretion by antimicrotubular agents, *J. Biol. Chem.* **248**:2691.

Trifaró, J. M., 1977, Common mechanisms of hormone secretion, *Annu. Rev. Pharmacol. Toxicol.* **17**:27.

Vaptzarova, K. I., Popov, P. G., and Galabov, G. P., 1973, Tyrosine-alpha-ketoglutarate aminotransferase activity in rat liver after spinal cord section, *J. Neurochem.* **21**:291.

Williams, J. A., and Wolff, J., 1970, Possible role of microtubules in thyroid secretion, *Proc. Natl. Acad. Sci. USA* **67**:1901.

24

Genesis and Properties of Pituitary ACTH, MSH, Prolactin, and GH Producing Cells

Vladimir R. Pantić

1. Introduction

The cells of both pars intermedia (PI) and anterior (distalis) (PA) pituitary lobe originate from Rathke's pouch as a single cell type and their further differentiation is dependent on signals originating from terminals of developing neurosecretory and adrenergic fibers, inductors present in the intercellular microenvironment and on the neighboring cell plasma membrane surface. Lowry and Scott (1975) pointed out that α-MSH, β-MSH, and corticotropinlike intermediate lobe peptides (CLIP) are derived from the same genes as adrenocorticotropic hormone (ACTH) and β-lipotropin (LPH), suggesting that the cells of both lobes synthesize intact ACTH and β-LPH, while the cells of the PI lobe produce proteolytic enzymes that cleave and modify ACTH and β-LPH to form the smaller peptides characteristic for this lobe.

Two cell types are identified in the adenohypophysis of a primitive organism, as for example the hagfish (*Myxine glutinosa*): one produces a hormone similar to ACTH and MSH, and in the second a hormone similar to prolactin (PRL) and growth hormone (GH) is synthesized (Fernholm, 1972). Common general properties for these two cell groups and specific

Vladimir R. Pantić ● Institute for Biological Research 'Siniša Stanković,' Belgrade, Yugoslavia

characteristics for each cell type have been described in most fish, birds, and mammals (Pantić, 1974). Additional data have recently been accumulated and some of these will be mentioned in this article.

It is well known that stressogenic stimuli increase MSH and ACTH release, and as a result of stressful stimuli ACTH, PRL, and GH are released. Different brain neurons, especially hypothalamic ones, are responsible for the regulation of these hormones' release. Acute stress and other agents provoke the release of ACTH within the first few minutes (Saffran and Schally, 1977). Serum concentration of PRL is increased by stress 5–10 min after the onset of the stress. This effect is stimulated by prolactin-releasing factor (PRF) or inhibited by prolactin-inhibiting factor (PIF) or by both mechanisms. Chronic stress reduces PRL release and causes gonadal atrophy (Taché et al., 1978). The activation of PRL release induced by stress appears to be mediated by neurotransmitters, through a noradrenergic, β-adrenergic, or other routes.

The release and synthesis of adenohypophyseal hormones are regulated by brain peptides, biogenic amines, the feedback effect of peripheral endocrine "target" cell hormones, and autofeedback. The mechanisms of their effects are expressed through the receptors localized on the external surface of the plasma membrane or in the cell cytoplasm after the entry of hormones into corresponding target cells. After binding to the receptor sites, a hormone–receptor complex is formed, a cascade of events occurs, and new proteins are synthesized.

At the present time attention is being paid to the role of releasing (RH) and inhibiting (IH) hormones in the regulation of release and synthesis of ACTH, MSH, GH, PRL, and other pituitary cell hormones, under in vivo and in vitro conditions (Briaud et al., 1979; Jones and Gillham, 1980). Using molecular biologist language, it means that each of these cells has a receptor for a specific RH and for IH.

The release of MSH in intact rats is tonically inhibited by dopaminergic arcuato-hypophyseal neurons (Tilders and Smelik, 1975). O'Donohue et al. (1979) observed a decrease in α-MSH concentration 4 weeks after hypophysectomy. They interpreted it as the brain α-MSH system being distinct from, but related to, that of the pituitary. The existence of a so-called mass action, i.e., autofeedback, as a regulatory mechanism, has often been mentioned (Meurling et al., 1969), and is present in both neurons and pituitary cells. However, mass-action direct feedback control of MSH was not observed by Huntington and Hadley (1974).

There are no data showing that target cell hormones play a role in the feedback mechanism involved in the regulation of MSH and somatotropic hormone (STH) release. On the other hand, the release and synthesis of both hormones are regulated by specific RH and IH.

PRL- and GH-producing cells are good models for the examination of the mechanism of peptide hormone action on gene expression (Evans and

Rosenfeld, 1980). These cells are characterized as cells with good recycling of membranes forming compartmentation characteristic for granular endoplasmic reticulum (GER) and Golgi complex. Their products of synthetic activity are accumulated in specific granules bound by membranes.

PRL and STH are synthesized and released in separate specific cell types; specific intercellular mechanisms regulate the biosynthesis of each hormone, which is stored in secretory granules and released mainly by exocytosis. However, release and synthesis, as very complex processes, depend on extracellular factors with a hormonal role involved in the mechanisms of the regulation of their activities, as: thyrotropin-releasing hormone (TRH), somatostatin (SS), neurotransmitters, opiate molecules, biogenic amines, and even some amino acids. The concentration of ions, pH, and various enzymes are involved as well (Baertschi *et al.*, 1980; Borgeat *et al.*, 1973, 1974).

It is well known that TRH stimulates thyroid-stimulating hormone (TSH), LTH, and GH release and synthesis from corresponding pituitary cells (Bowers *et al.*, 1971; Fleischer *et al.*, 1970). However, recent data have shown that SS has an opposite role to that of TRH, inhibiting the release of PRL and GH (Brazeau *et al.*, 1973; Schally *et al.*, 1973).

On the other hand, TRH stimulates the release of both LTH and (at a low level) GH from transplantable pituitary tumor cells induced by chronic estrogen treatment (Tashjian *et al.*, 1968). It seems that the regulation of PRL synthesis is at least in part at the transcriptional and post-transcriptional level (Morin, 1980). The rates of PRL and GH synthesis induced by TRH, estrogen, or estrogen analogues are precisely described by Evans and Rosenfeld (1980).

TRH and SS are synthesized by brain and spinal cord neurons (Krulich *et al.*, 1977; Elde and Hökfelt, 1979). They have a specific influence on pituitary glandular cells and exert extrahypothalamic effects as neurotransmitters or neuromodulators. As neurotransmitters they are released from nerve terminals, proceed diffusively across a narrow cleft, and act on the postsynaptic membrane receptors. As modulators they are circulating substances interacting with receptors on central nervous system (CNS) neurons (Moss, 1979).

The mechanism of TRH effect is expressed in an increase of the mRNA coding for PRL (Evans *et al.*, 1978), and this peptide regulates the number of its own receptors (Gourdji, 1980).

2. The Aim of This Article

Having in mind recent advances in molecular endocrinology, it seems reasonable that cytology, even improved methods as immunocytochemistry and others, should follow the pathway of differentiation of specific pituitary cells enabling these cells to synthesize prohormones and split them into smaller fragments, such as LPH, ACTH, α- and β-MSH, and opiate molecules. For that reason, both biochemical and cytological results will be dealt with in this article, focusing our attention mainly on pituitary cell properties and reactions to stressful stimuli, brain peptides, and steroids, from the following aspects:

- What is the role of neurosecretory and adrenergic nerve fibers in the differentiation of PI glandular cells?
- Is the PI lobe, as a poorly vascularized lobe, closely related to brain peptides supplied by long and short pituitary portal vessels?
- To what extent is the differentiation of these cells influenced by steroids and at what stage of development are they most sensitive?
- Is the response of PI cells to the deficiency and sufficiency of steroid hormones permanent or temporary?
- Are the development, size, and shape of PI cells species specific?
- What is the difference between the cytology of ACTH/MSH cells differentiated in PI and outside this lobe?
- What common properties are pronounced in the cytology of PRL and GH cells, especially from the point of view of the recycling of plasma and cytoplasmic membranes, nature of specific granules and other organelles?
- What similarities and dissimilarities exist in the cells of various vertebrates, from some fish to the highest mammals?
- Which brain peptides are common or specific for the regulation of PRL and GH cell activities?
- What is the difference between the reaction of PRL and GH cells to gonadal steroids?

3. Origin and Nature of ACTH and MSH

The origin and nature of these two peptides are very complex both from the phylogenetic and ontogenetic point of view. As biologically active peptides they are produced by endocrine and neuroendocrine cells and neurons (Guillemin, 1978; Moriarty and Garner, 1977). Neuropeptides, identical with or immunologically related to ACTH, α-MSH, β-LPH, and endorphins, probably arising from a common precursor molecule similar to that found in pituitary ACTH cells, are synthesized in rat brain neurons as well (Pelletier and Dubé, 1977). These peptides might be involved in basic processes of CNS as neurotransmitters or neuromodulators (Bloch et al., 1979; Jones and Gillham, 1980).

At the present time, extra cisternal peptides may be divided into two groups, each having a parent or aparent precursor molecule, ACTH and β -LPH. In mammals a sequence of 7 amino acids is common, Met-Glu-His-

Phe-Arg-Trp-Gly, and occurs at position 4–10 in α-MSH and ACTH, 47–53 in β- and γ-LPH, and 7–13 in β-MSH (Lowry and Scott, 1977).

β-MSH in most mammals contains 18 amino acid residues and is found at position 41–58 of both β- and γ-LPH. β-LPH is a peptide of some 91 or 95 residues (Chrétien *et al.*, 1976). γ-LPH is composed of its N-terminal sequence of 58 residues (Lowry and Scott, 1977). α-MSH has exactly the same sequence as the N-terminal 13 amino acids of ACTH. These molecules are stored within the small vesicles of a few cell bodies in the arcuate nucleus and numerous brain nerve fibers are widely distributed throughout the brain and they could be considered as neurotransmitters. CLIP is identical to the C-terminal 18–39 sequence of ACTH, and γ-LPH is identical with β-LPH (1–58) (Lowry and Scott, 1977). These authors pointed out that only ACTH has a most clearly defined role in mammals, eliciting corticosteroidogenesis.

Bioactive ACTH has been identified in both lobes, while α-MSH and CLIP appeared to be restricted to PI (Scott and Lowry, 1974). ACTH, β-MSH, and endorphin are stained in all granules of rat PI lobe (Martin *et al.*, 1979). These two groups of peptides, ACTH and β-LPH, have probably been formed by the duplication of a single gene with subsequent progressive substitutions leading to their divergence. The smaller peptides of each group are derived by posttranscriptional specific proteolytic cleavage of ACTH and β-LPH (Lowry and Scott, 1977).

MSH activity is significant and might be detected in PA and PI. ACTH, α-MSH, and β-endorphin have been found in all PI cells of rat and other vertebrate pituitaries (Lowry *et al.*, 1977; Martin, *et al.*, 1979).

Peptides resembling α-MSH and CLIP are dominant hormones in the human fetal pituitary during the second half of pregnancy and are only replaced by ACTH before parturition. In the human fetal PA, corticotropin, β-LPH, α- and β-endorphins are identified (Li *et al.*, 1979).

Considering the role of these peptides, Silman *et al.* (1976) hypothesized that α-MSH and CLIP are the tropic hormones for the fetal zone of the adrenals, while intact ACTH drives the definitive cortex and is the principal hormone in the regulation of adrenal glucocorticoid synthesis. Melatonin has been considered as a potential contributor to the physiologic regulation of adrenal steroidogenesis (Ogle and Kitay, 1977). The N-terminal portion of the ACTH molecule, which is responsible for corticosteroidogenesis, has remained relatively unaltered through evolution, while the C-terminal has been most liable to changes (Lowry and Scott, 1975).

4. Pars Intermedia

Owing to the complexity in the origin and nature of these peptides originating from two precursors, β-LPH and ACTH, or their promolecules, we paid attention to the genesis of PI lobe cells, their specificities in various species and under different experimental conditions. Some of our data will be briefly summarized.

Examining the structure, staining affinity and characteristics of PI cells in the examined fish, birds, and mammals, there is no doubt that the development of this lobe, the differentiation of MSH and ACTH cells, and the number of "ectopic" cells are closely related to the development of the hypothalamo–hypophyseal tract and neurohypophysis, on one side, and to some extent to PA, on the other.

From the evolutionary point of view PI cells are closely dependent on nerve stimuli and microenvironmental factors. In the lizard, an animal that is able to change color rapidly, PI constitutes 70% of its total volume (Saint Gerons, 1961). In wild animals, such as the deer, this lobe is extremely well developed. Undifferentiated, MSH, intermediate, and ACTH cells have been described in the deer (Pantić and Šimić, 1974). Cell hyperplasia and hypertrophy were evident during the hunting and mating season, showing clearly that stressogenic stimuli are involved (Fig. 1). Both the increased cell number and PI size decline during the winter period.

4.1 Vascularization and Innervation

It has been established that PI cells differentiate through Rathke's pocket wall and are closely dependent on the infundibulum; in species with well-developed connective tissue between PA and the infundibular process, no PI, nor even remnants of the original lumen of Rathke's pocket wall, appear in the adult. For the differentiation of MSH, ACTH, endorphin, and LPH cells see Dupouy (1980). The primary bed of capillaries and the long and short portal vessels transmit blood to most mammalian hypophyses. The opposite circulation, to the hypothalamus, has also been established (Palkovits, 1979).

It is an established fact that PI is in some species, as in rats and other animals, virtually avascular (Fig. 2). Between the endocrine cells, stellate cells and the channel system are described (deBold et al., 1980). However, innervation of PI is very rich with many unmyelinated axons mostly at the periphery of the lobules with some synapticlike contact with PI cells (deBold et al., 1980).

4.2. Specific Properties of PI Cells

Some specific properties of ACTH and MSH cell vesicles and granules, as cytoplasmic organelles involved in the synthesis, transport, storage, and release of this hormone, will be mentioned.

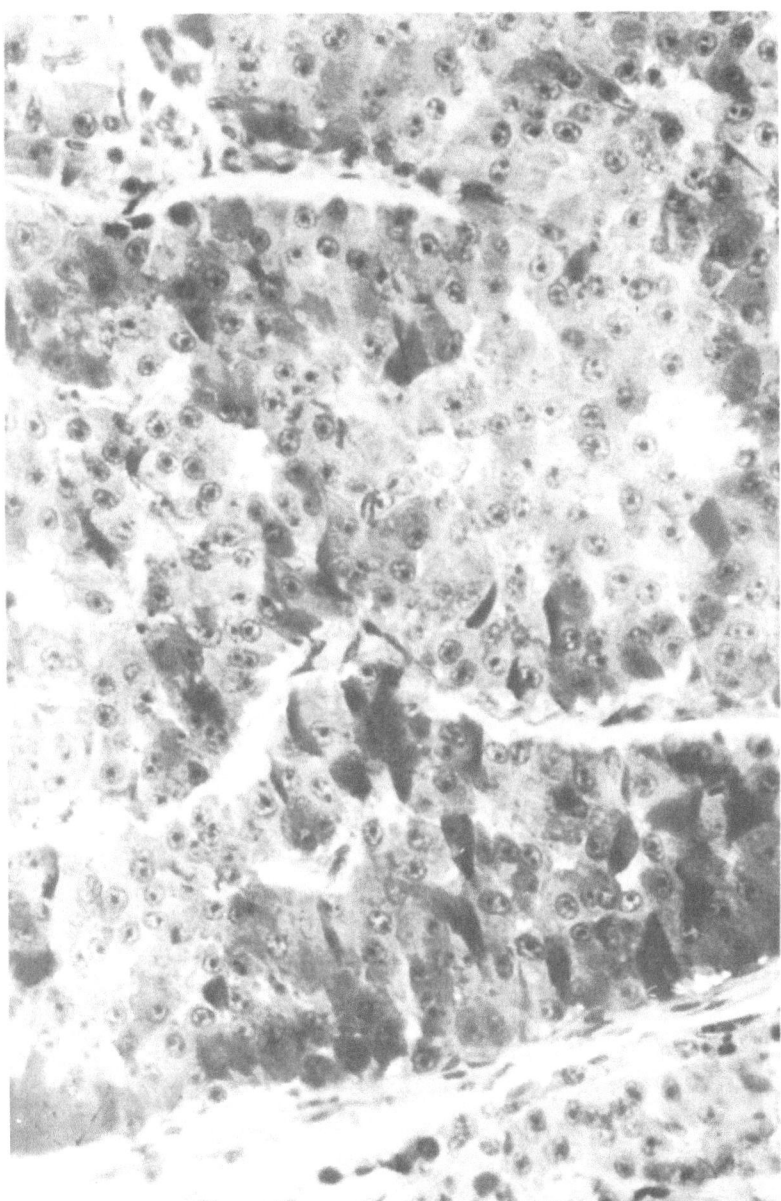

Figure 1. PI cells of deer pituitary showing variability of cells in size and staining affinity.

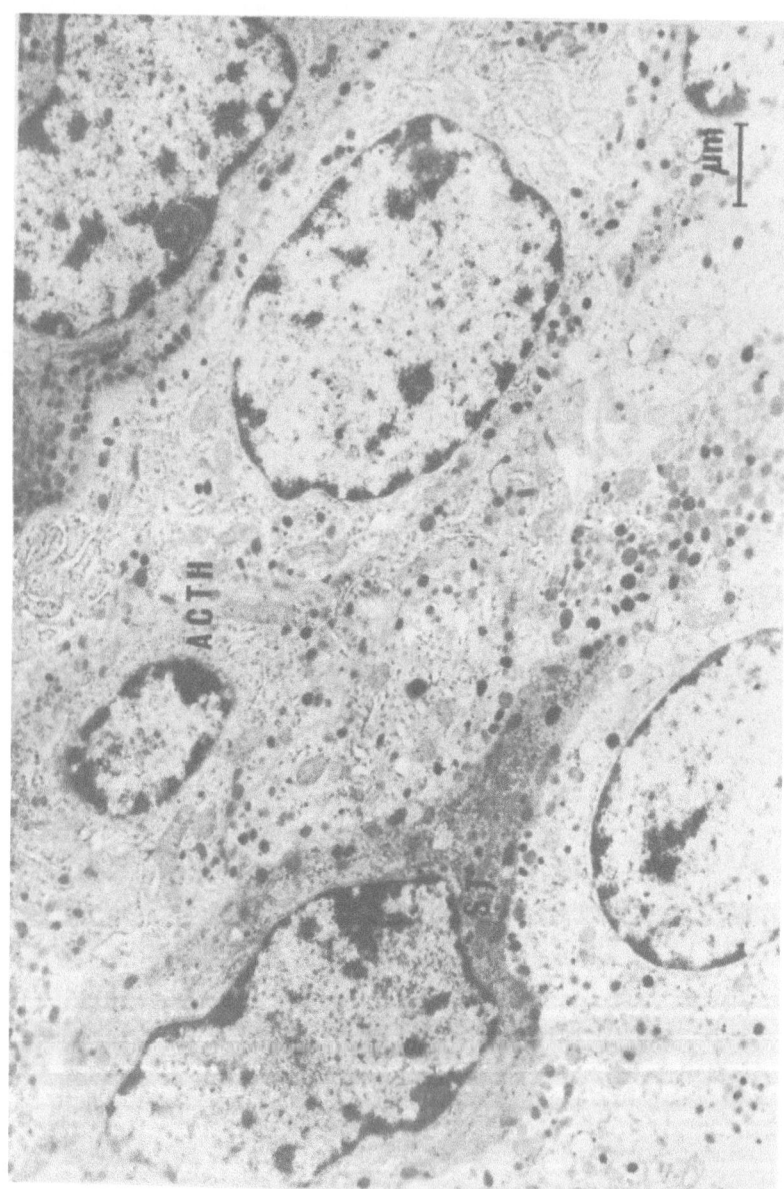

Figure 2. ACTH and stellate (ST) cells in Pl of 6-month-old intact rat.

The PI consists of highly specialized cell types able to synthesize prohormones and split them into smaller peptides as ACTH, α-MSH, CLIP, endorphins, and peptides with N-terminal tryptophan (Pelletier and Dubé, 1977; Martin *et al.*, 1979; Crine *et al.*, 1980). Lowry and Scott (1975) pointed out that the small amount of ACTH in the PI is the precursor of α-MSH and CLIP.

It has been established that ACTH cells, as mainly chromophobic or weakly basophilic cells, contain peripherally localized granules about 200–220 nm in diameter. It is known that these granules contain glycoproteins and hormones. Bowie *et al.* (1978) pointed out that the morphology of granules is the single best criterion for cell identification in the rat adenohypophysis (Fig. 3).

The immunoreactivity obtained by applying antibodies against α-MSH, ACTH, and endorphin has shown that all three hormones are localized in the same granules (Martin *et al.*, 1979). It appears that α-MSH is a typical PI lobe peptide, its occurrences being 85% greater in this lobe than in PA (Usategui-Echeverria *et al.*, 1975).

Actin is present in endocrine cells and is bound to secretory granules (Ooslund *et al.*, 1978). It seems that this contractile protein might play a role in the intracellular transport of secretory granules, i.e., in one of the subcellular mechanisms regulating exocytosis or in the subsequent membrane reinternalization.

4.3. Reaction of PI Cells to Gonadal Steroids

PI cells are sensitive to brain peptides during the early fetal and neonatal period. If a single or three repeated doses of estrogen are administered into neonatal rats, the development of brain neurons incorporating the hormone is retarded. As a result, nerve terminals in the neurohypophysis are undeveloped, the amount of neurosecret is very small, and a small number of Hering's bodies are present (Figs. 4, 5). The pituicytes are altered, the whole neurohypophysis appears undeveloped, and the capillaries at the boundary to PI are less developed (Fig. 6). As a result, the retardation of PI development is expressed up to the onset of puberty and this lobe consists in some cases only of some cell layers (Pantić and Šimić, 1977a). The reaction of both neurohypophysis and PI was much stronger if the animals were treated with three repeated doses of estrogen instead of a single dose, and the reaction was more pronounced if they were treated during the so-called "critical" period rather than after 15 days of age. There is no doubt that the development of both neurohypophysis and PI is dependent on hypothalamic nerve fibers which are more sensitive to estrogen during the neonatal period than at the age of 15 days and later.

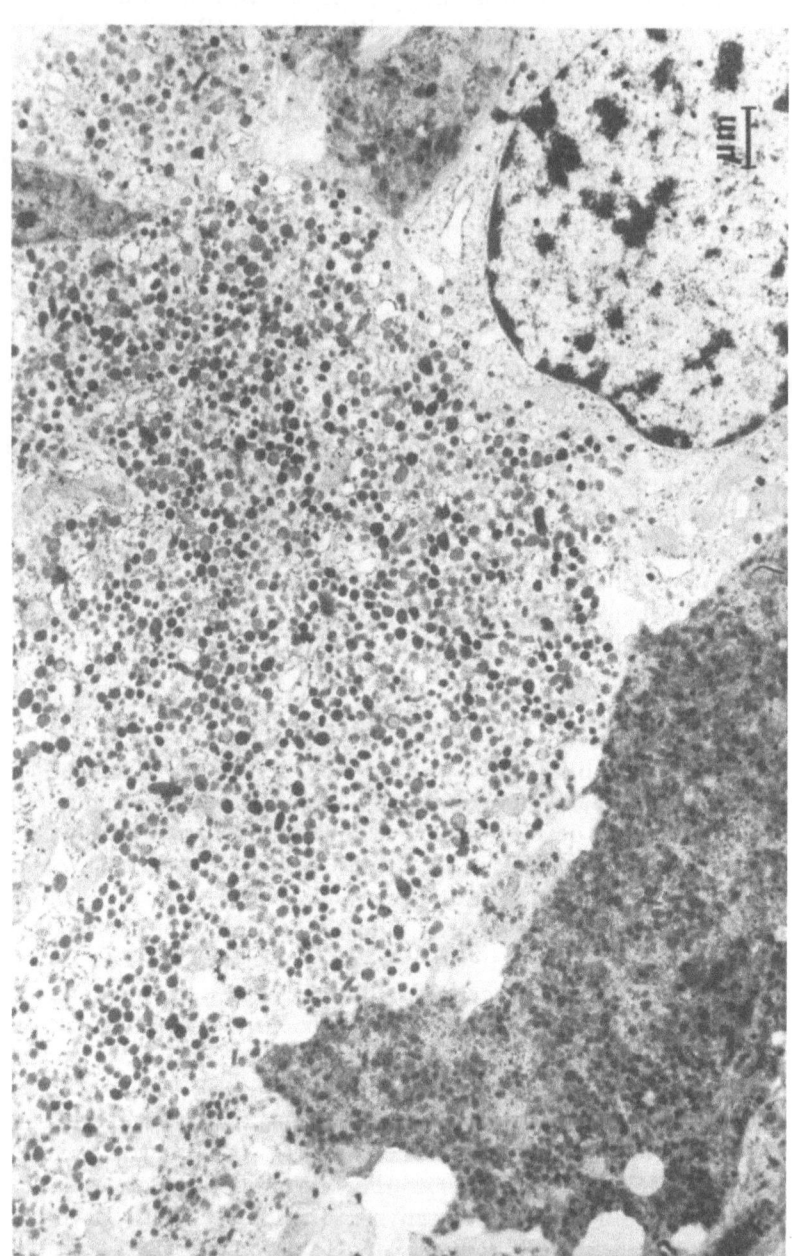

Figure 3. Granules similar in size and density in PI cells of 6-month-old intact rat.

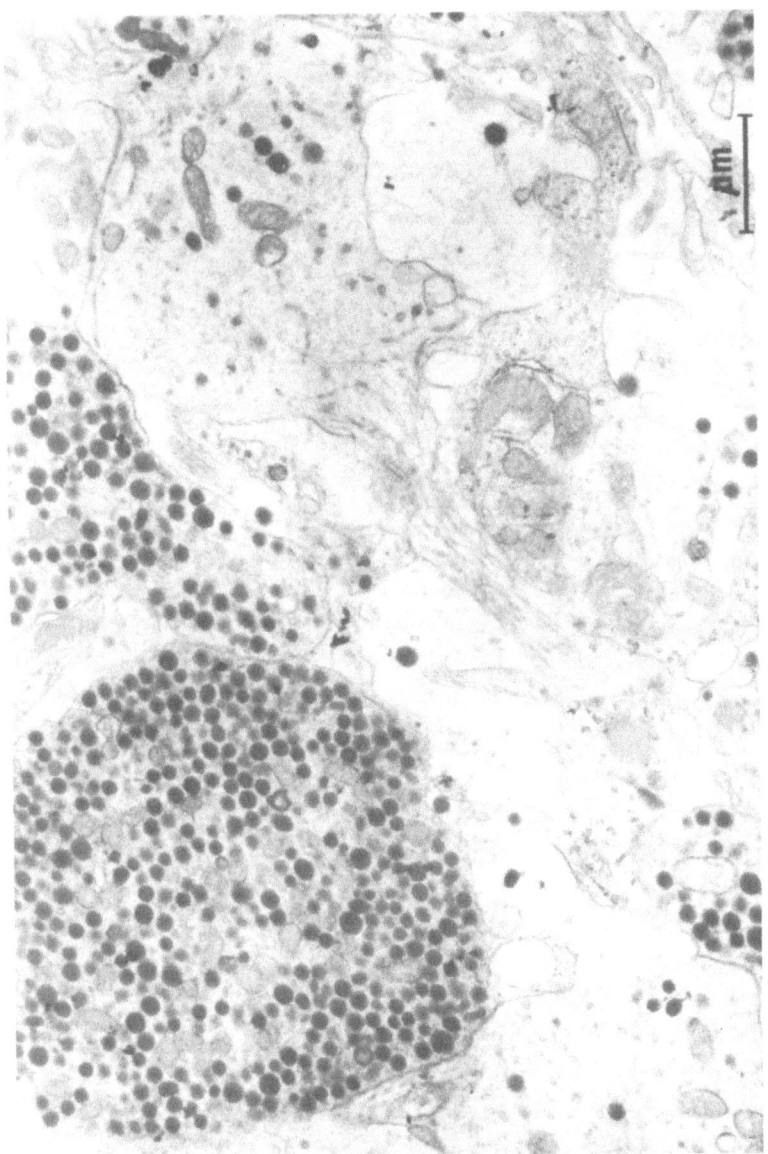

Figure 4. Nerve terminals in the pars nervosa adjacent to PI lobe in 16-day-old intact rat.

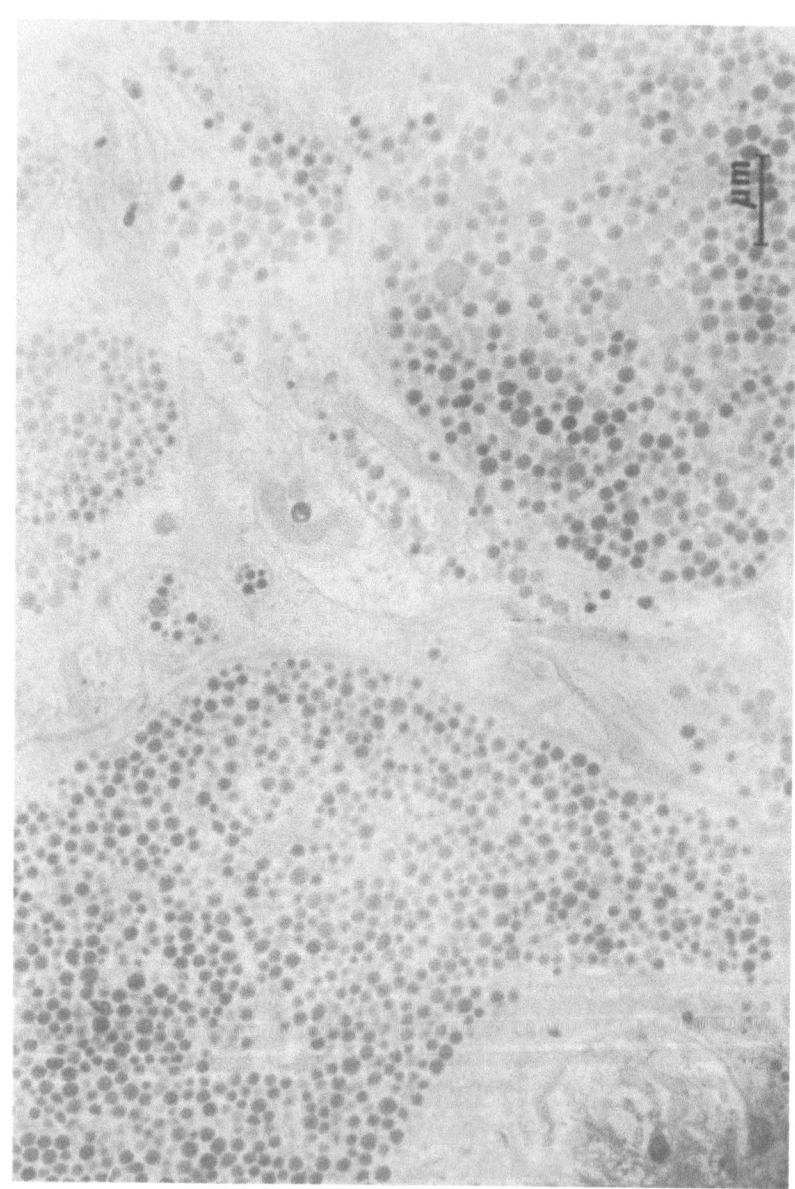

Figure 5. Nerve terminals in the neurohypophysis of 38-day-old intact rat.

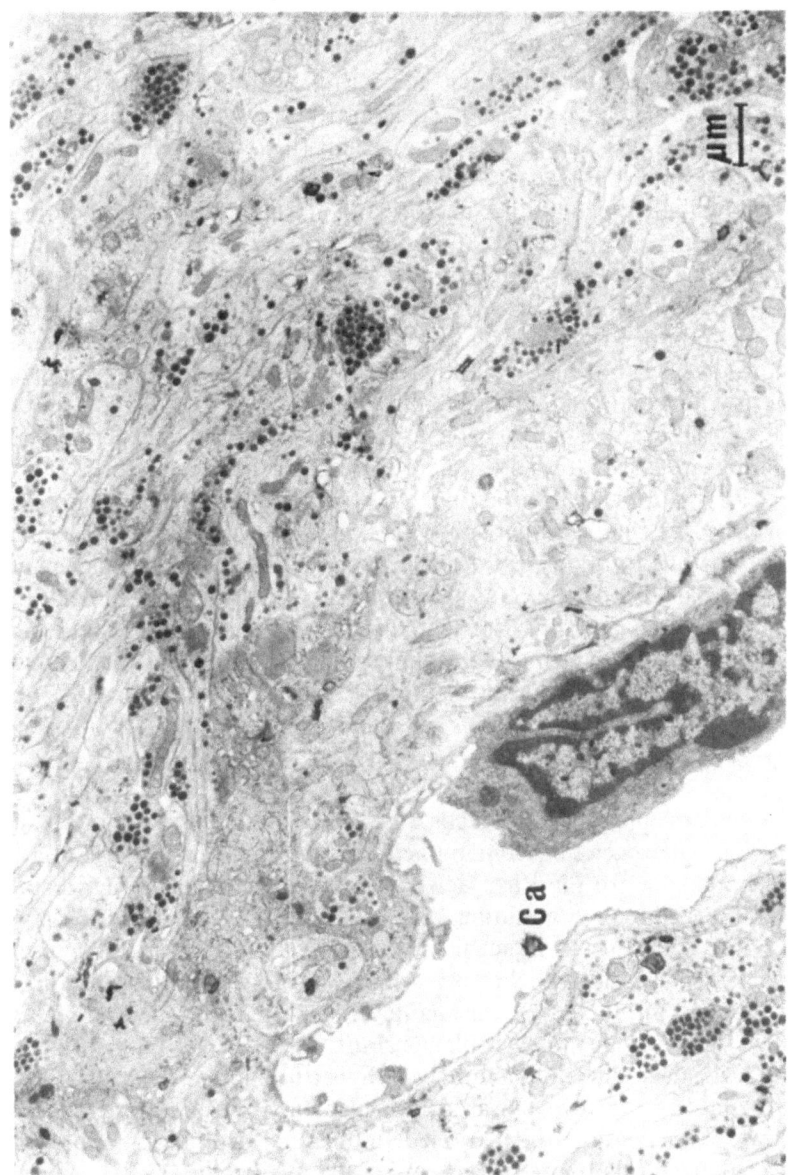

Figure 6. Reduced amount of granules in nerve terminals and capillary (Ca) in that part of the neurohypophysis adjacent to PI in 38-day-old rat neonatally treated with a single dose of 1 mg estrogen.

With development of the animal, the long-term effect of neonatally administered gonadal steroids becomes less pronounced. It seems that brain neurons have recovered, brain peptides are synthesized and transported to their terminals stimulating PI cells, so that at the period of puberty, PI lobe layers almost reach the size of the controls (Pantić and Šimić, unpublished data). However, brain neurons and their terminals in the neurohypophysis and PI cells do not have the same structural properties as in intact non-treated animals. These facts undoubtedly show that the development of PI lobe and the properties of its cells are closely connected with the development of hypothalamic and other neurons and are under the control of nervous stimuli.

Examining the properties of various parts of PI, besides species specificities, we observed differences in cell structure in animals of the same species. In addition, a number of "ectopic" cells varied in various animals. Gonadotropins (GTH), GH, and other specific cells were observed in fish, rat, and pig PI lobe (Pantić and Šimić, 1977a).

The most sensitive cells to gonadal steroids, under our experimental conditions, were cells of the rostral part of PI. For example, the rostral part of piglet PI is more vascularized than other parts of this lobe, and the connection of PA, PI, and neurohypophysis is expressed. The cells in this region are very sensitive to brain peptides and to the deficiency or sufficiency of gonadal steroids during the fetal and neonatal periods of development. In growing piglets neonatally castrated, this part of the lobe is increased, gonadotropic cells are increased in number, hypertrophic and vacuolated, and their properties are the same as in the PA (Pantić and Šimić, 1977b).

4.4. Reaction of ACTH Cells to Corticosteroids

The hypertrophy of ACTH cells was clearly pronounced in the PA of all fetuses and newborn rats from mothers adrenalectomized during the last week of pregnancy. In the PI of newborn rats from mothers adrenalectomized on the 14th day of gestation, ACTH-like cells were noticed (Hristić et al., 1978). ACTH cells in neonatal rats were stimulated in response to maternal adrenalectomy (Nickerson et al., 1978. Hypertrophy of the adrenal gland and an increased plasma level of corticosterone were also observed after adrenalectomy (Milković et al., 1973, 1976). If 3-month-old rats are adrenalectomized, the number of hypertrophic ACTH cells in PA increases and from 15 days onwards adrenalectomy cells are more numerous. However we consider the increased number and density of granules in adrenalectomized rats treated with 1 mg of hydrocortisone (HCS) to be a result of the slowed-down release by exocytosis (Fig. 7). In ACTH cells of adrenalectomized animals, 15 and 30 days after head irradiation with 1200 R, a decreased amount of GER, regressive Golgi complex, and larger gran-

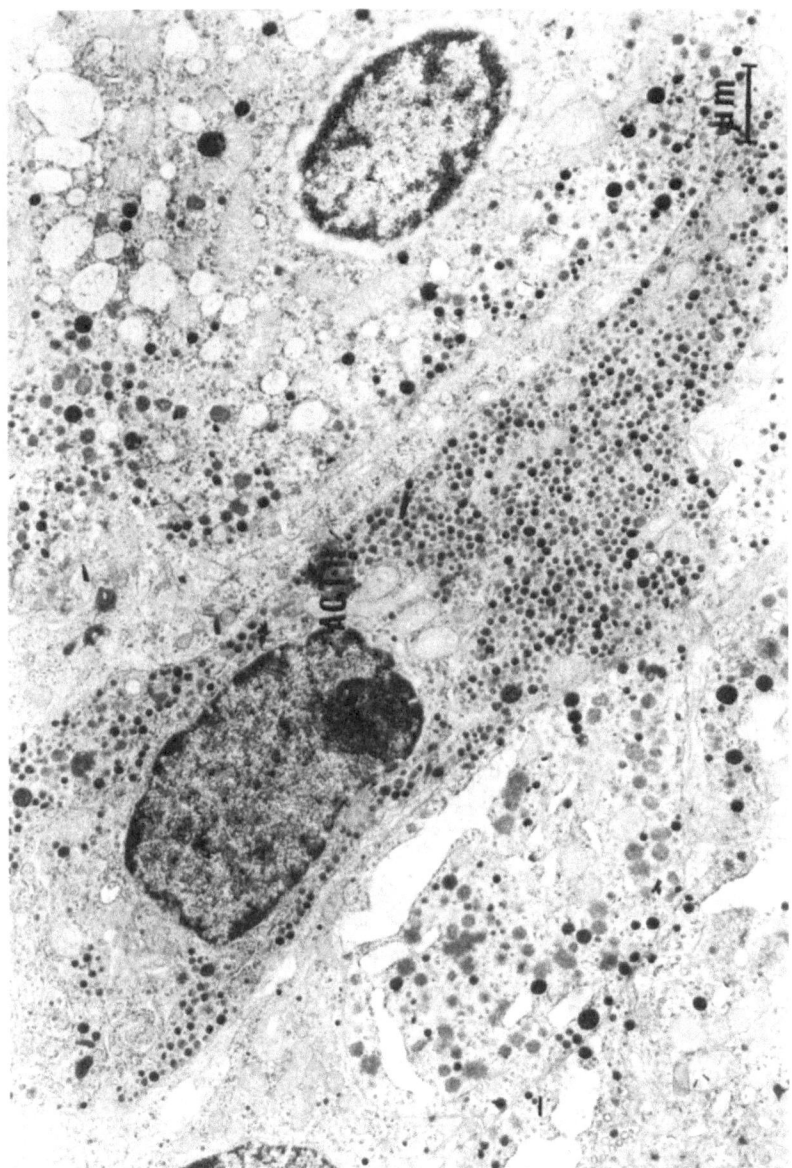

Figure 7. ACTH cell in PA of 3-month-old rat neonatally treated with 1 mg HCS/10 g body wt.

ules were pronounced. This suggests that both protein synthesis and release are decreased. If such animals (adrenalectomized + irradiated) were treated with deoxycorticosterone acetate (DOCA), the smaller number of ACTH cells degenerated (Pantić and Hristić, 1975). Neurosecretory cells of both nucleus supraopticus and nucleus paraventricularis are sensitive to irradiation: they are vacuolated and the quantity of neurosecretory substance is decreased. In both intact and adrenalectomized rats the changes are less pronounced after treatment with adrenocortical steroids (HCS + DOCA) (Hristić and Pantić, 1974).

Much more [³H] thymidine was incorporated in PA cells of adrenalectomized DOCA-treated rats than in the controls, and the smallest number of labeled cells was established in adrenalectomized HCS-treated animals. Similar findings but a smaller number of labeled cells were observed in PI cells. [³H]-HCS was observed in median eminence, PA, and PI lobe, but there were far fewer labeled cells in PI than PA (Hristić and Pantić, 1976).

There is no doubt that the negative feedback effect of glycocorticoids is expressed at the hypothalamic and pituitary level (Figs. 8, 9). After adrenalectomy, glycocorticoid binding sites are increased in brain neurons (Olpe and McEwen, 1976). It is possible that ACTH cells contain nonspecific receptors capable of ACTH release to peptides including vasopressin (VP). Lutz-Bucher et al. (1977) suggested that VP and corticotropin-releasing factor (CRF) have different receptor sites in the anterior pituitary cells and that VP has an additive effect on hypothalamic CRF activity.

In summation, ACTH cells in PA are highly specific in reaction to the deficiency or sufficiency of glycocorticoids and the response is closely dependent on brain peptides known as CRF. Signs of the reaction to corticosteroids are evident but ACTH cells in PI lobe were never observed as adrenalectomy cells.

4.5. Brain Peptides in Regulation of ACTH/MSH Cell Activities

It has been known for a long time that in mammalian nucleus supraopticus and nucleus paraventricularis, two nonapeptides, VP and oxytocin, in association with specific neurophysin, are synthesized. They are transported through neurosecretory cell axons to the neurohypophysis and are stored in nerve terminals from which they are released (Figs. 4, 5).

Peptidergic and adrenergic fibers innervate PI cells of rat and mouse hypophysis (Stoeckel et al., 1973). The direct inhibitory effect of glycocorticoids on the concentration of CRF and on ACTH has been established (Hristić and Pantić, 1978; Mulder and Smelik, 1977; Arimura et al., 1969).

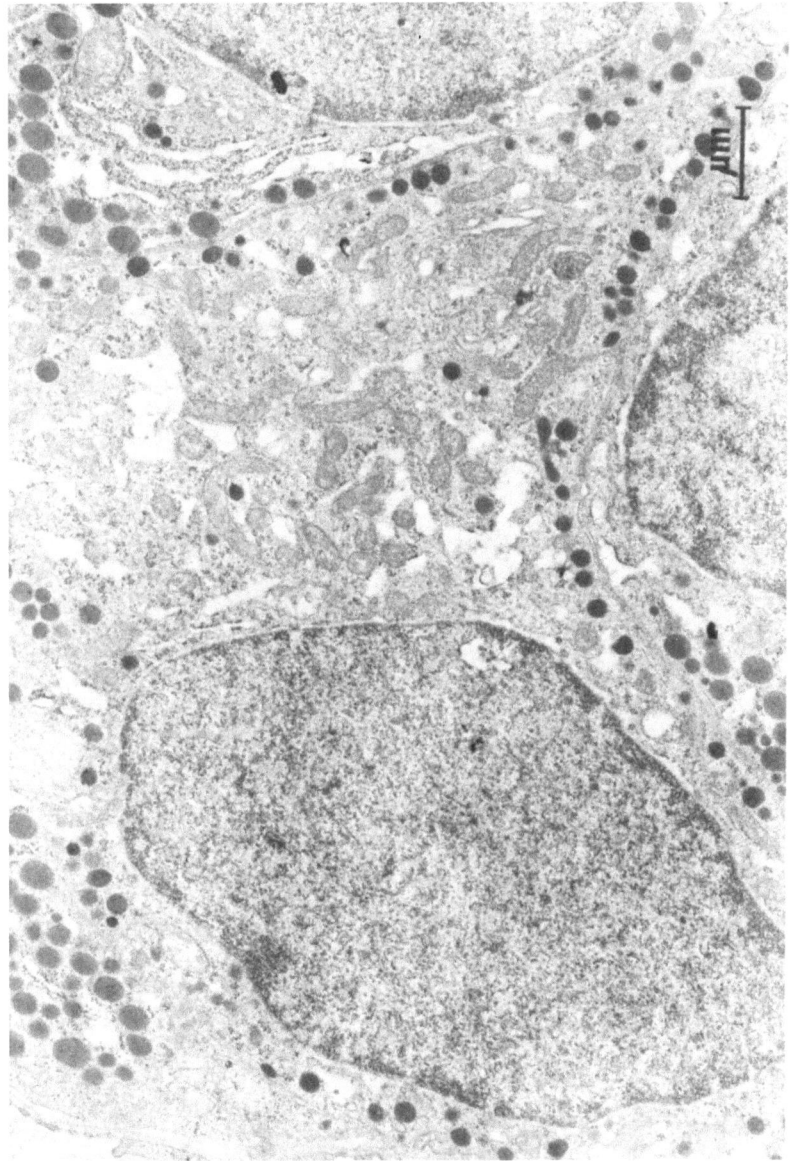

Figure 8. ACTH cell in PA of 3-month-old rat with granules of high density localized in the vicinity of the plasma membrane.

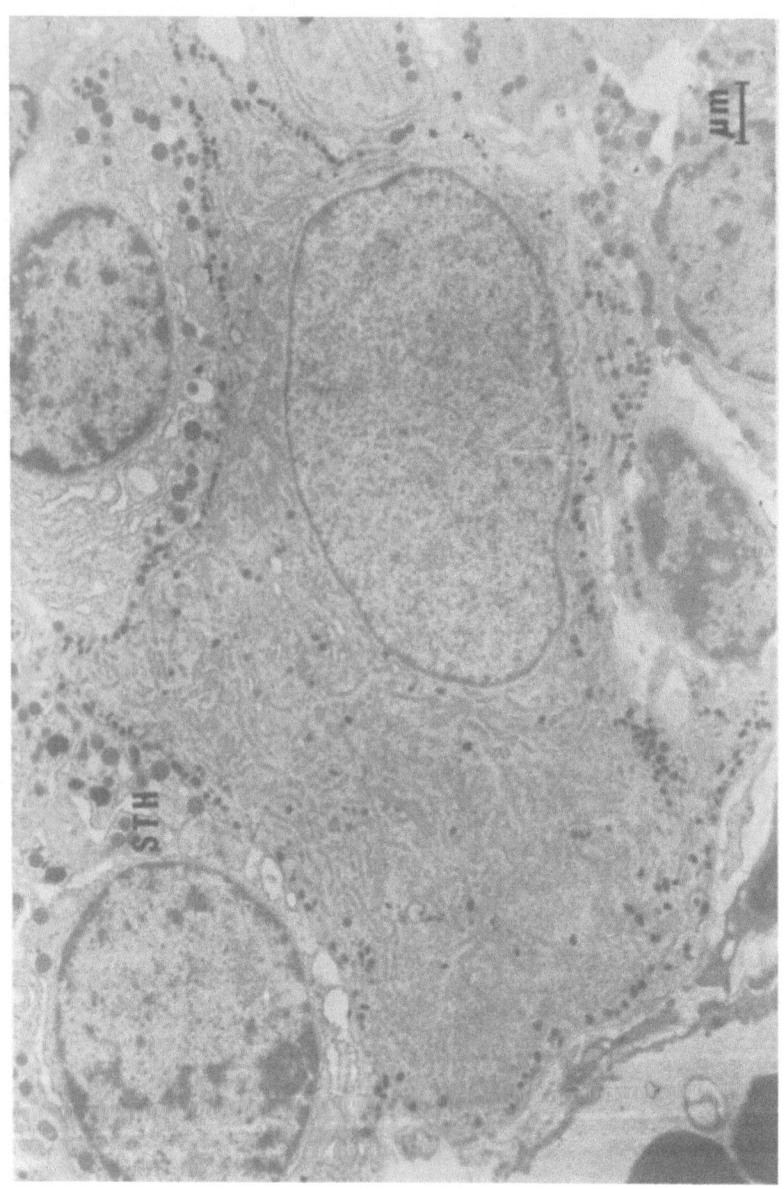

Figure 9. ACTH cell in PA of 3-month-old rat in contact with GH (STH) cells and capillary.

It has been shown that ACTH cells contain nonspecific receptors and are capable of ACTH release in response to many peptides (Dallman and Yates, 1969, Koch *et al.*, 1979, 1980). However, ACTH release is stimulated less by VP than by mediobasal hypothalamic extract (Lutz-Bucher *et al.*, 1977).

The human pituitary produces a small amount of ACTH and relatively larger quantities of α-MSH and CLIP during fetal development, but not after birth, when the PI lobe involutes (Silman *et al.*, 1976).

It must be noted that one neuron may synthesize, store, and release more than one substance involved in transmission and/or modulation at chemical synapses (Hökfelt *et al.*, 1979).

Rat hypothalamic neurons are producers of a common precursor molecule from which neuropeptides identical or immunologically related to ACTH, α-MSH, β-LPH, and endorphins arise. These peptides, similar to those synthesized in ACTH cells, are involved in basic processes of CNS as neurotransmitters or neuromodulators (Bloch *et al.*, 1979)

4.6. What Does the Term MSH Cells Imply?

The intention is not to consider whether the term MSH or LPH cells is more convenient from the point of view of terminology, but to point out the complexity of events occurring during phylogenesis and ontogenesis and evolution of prohormone for LPH, as a large common molecule originating from a common heptapeptide.

Recently, it has been shown that MSH, ACTH, and opioid molecules are fragments of prohormone β-LPH and are stored and split in the same granule (Fig. 3). The release and synthesis of all of these peptides are regulated by melanotropin-releasing factor (MRF) or melanotropin-inhibiting factor (MIF). After these small peptides are bound to the receptor, a hormone–receptor complex is formed, the further sequence of events occurs, and the content of the granule is released by exocytosis. Consequently, a new prohormone and splitting enzymes are synthesized and stored in specific granules. As a result of the release of this content containing ACTH, MSH, LPH, and opioid molecules, corticosteroidogenesis, melanogenesis, lipogenesis, and PRL and GH release are affected, respectively.

Accepting that all these hormones are synthesized and stored in the same granules, that their release and inhibition are regulated by signals originating from MRF and MIF, respectively, and that the above-mentioned hormones are fragments of β-LPH as a larger molecule, the question arises whether we have so precisely developed cytochemical methods in light and electron microscopy that could allow us to characterize these cells, especially as producers of additional hormones and nonhormonal peptides.

5.   Common and Specific Properties of PRL and GH Cells

Common general and specific properties of PRL and GH (STH) cells have been summarized by Pantić (1975). However, some additional observations will be mentioned.

PRL and STH are detected in mouse and rat PA between 16 and 21 days of gestation. Both SS and PRF are detected at the end of the gestation period. However, hyperplasia of PRL cells and the release of PRL increase during fetal development depending on the amount of estrogens in fetal blood plasma, pituitary cell sensitivity to stimuli, and the degree of the maturation of IF-producing brain neurons. Both hormones are usually stored in specific granules. In fish, PRL cell granules are mainly ovoid or round in form, while in mammals irregularity of granules is one of the characteristics of these cells (Figs. 10, 11). Porcine, rat, and other mammalian pituitary GH-producing cells contain round secretory granules, 300–500 nm in diameter (Fig. 12). Separated GH-producing cells were examined *in vitro* and *in vivo* by Hymer *et al.* (1980).

PRL and GH cells are a good model for understanding the role of cytoplasmic membranes in compartmentation forming GER, Golgi complex, and secretory granules (Fig. 10). Both types of cells are characterized as cells with good recycling of membranes during endocytosis and exocytosis. The ER in both cell types is more developed than in other pituitary cells. In GH cells GER could surround the nucleus, but the proliferative rate was not expressed to such an extent as in PRL cells (Figs. 10, 11). Vila-Porcile and Olivier (1980) suggested the role of lysosomal destruction in an internalization of plasma receptors.

PRL and various glycoproteins are stored in membrane-bound granules, and released by exocytosis. This content is released to the extracellular space, or by crinophagy to the lysosomes (Farquhar, 1977; Farquhar *et al.*, 1978; Wilbur and Spicer, 1980).

Dual secretory capacity for GH and PRL was ascribed to cells from mammotropic pituitary tumors (Pantić *et al.*, 1971a,b). Stratmann *et al.* (1974) suggested the transformation of STH into PRL in estrogen-treated rats. These authors observed that these cells may produce either PRL or GH at a certain time, and the other hormone at another time.

6.   Regulation of PRL and GH Release and Synthesis

Many papers, monographs, and reviews have dealt with the mechanisms of central neurotransmitter effects on PRL and GH release. Some of these will be briefly mentioned.

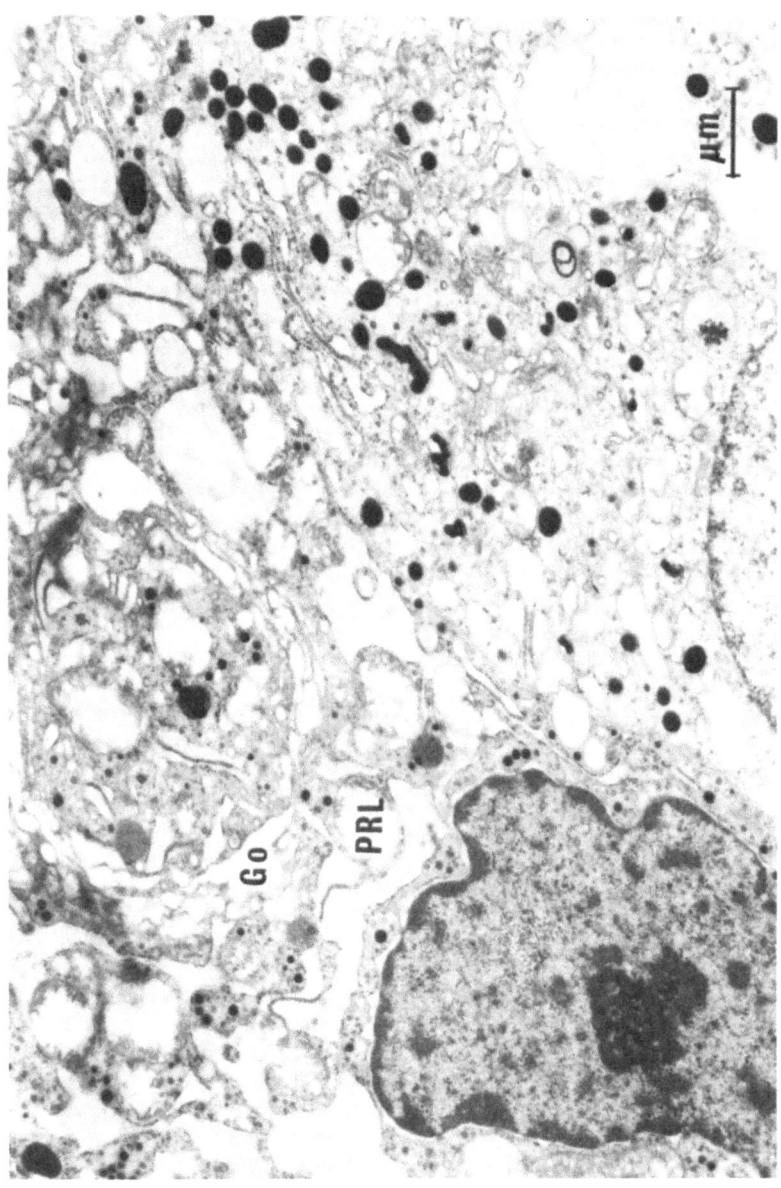

Figure 10. PRL cell with well-developed Golgi complex (Go) in the pituitary of *Torpedo ocellata*.

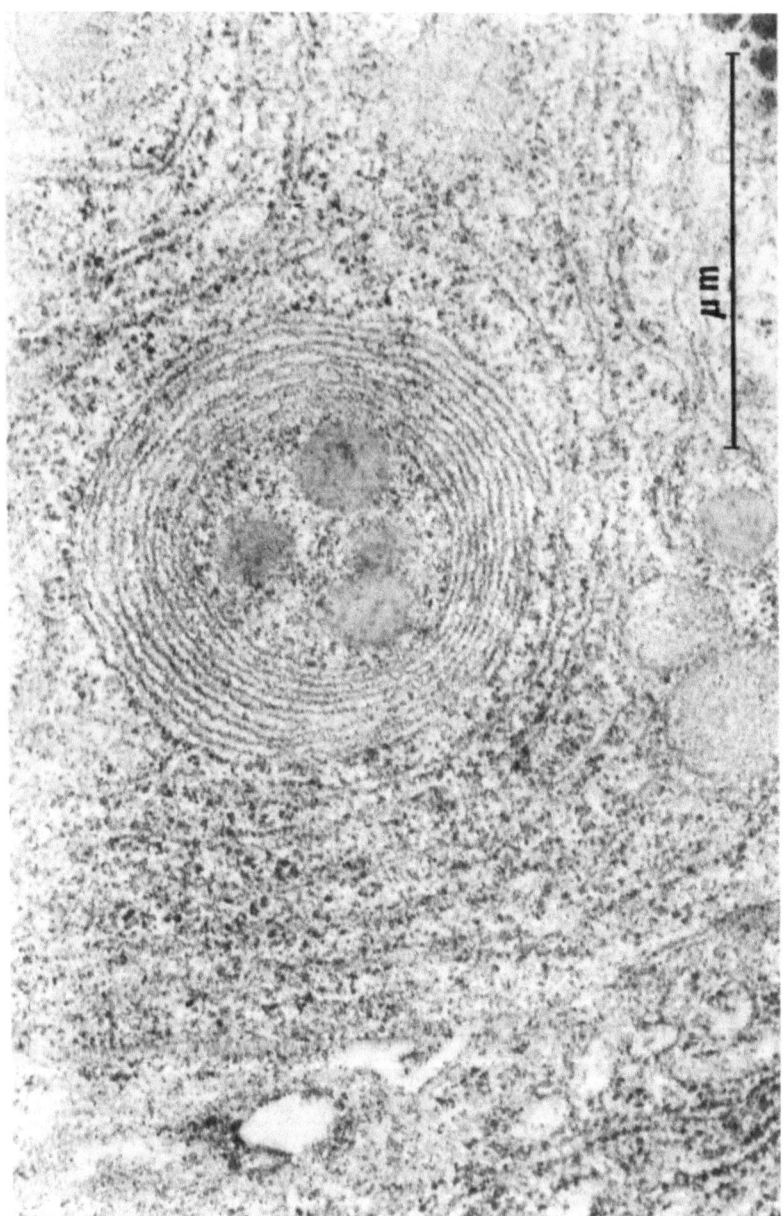

Figure 11. Portion of PRL cell in rat neonatally treated with estrogen. Note concentrically oriented GER.

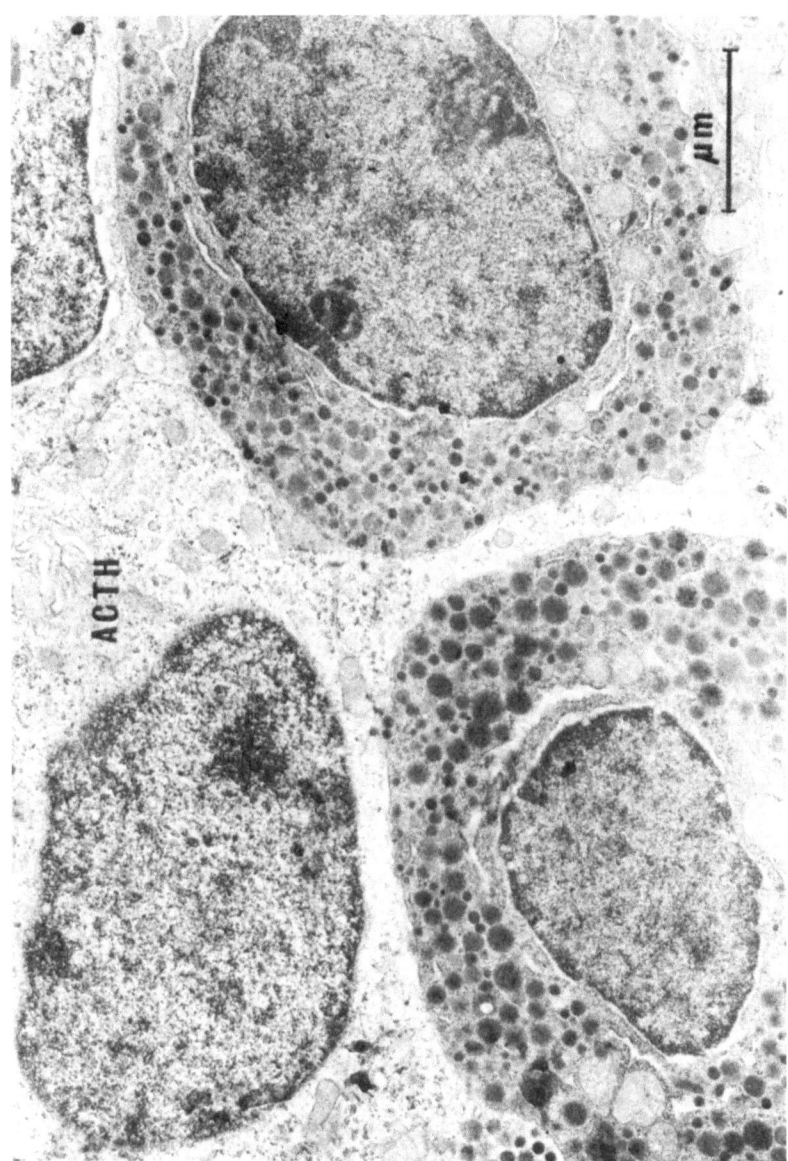

Figure 12. GH (STH) cells surrounding ACTH cell in pituitary of 3-month-old rat.

PRL and GH inhibiting factors and their analogues have their predominant inhibitory effect on PRL and GH release and synthesis, respectively (Sandow and König, 1978).

TRH stimulates TSH, PRL, and GH release from corresponding specific pituitary cells (Bowers et al., 1971). SS has an opposite role to that of TRH, inhibiting the release of GH and TSH (Bélanger et al., 1974). Somatoliberin, as growth hormone-releasing factor (GRF), stimulates the release of GH (Sawano et al., 1974). The stimulatory role of TRH on the release of PRL and (to a low level) GH is observed in GH3B6 cells, as well. This strain of cells is derived from a transplantable pituitary tumor induced by chronic estrogen treatment (Tashjian et al., 1968). It seems that the regulation of de novo PRL synthesis is at least in part at a transcriptional level increasing the mRNA coding for PRL (Evans et al., 1978), and at a posttranscriptional level (Morin, 1980). Induction of PRL synthesis byTRH in GH cells provides a model system for studying the action of polypeptide hormones on gene expression (Evans and Rosenfeld, 1980).

TRH and SS, as brain peptides, distributed in neurons of the brain and spinal cord, have a neurotransmitter or a modulatory role (Moss, 1979). As neurotransmitters they are released from nerve terminals, affecting postsynaptic membrane receptors; as modulators they interact with receptors on neurons.

Various stressful stimuli affect an increase of PRL and GH release (Reichlin, 1974; Krulich, 1979). These stimuli act via the hypothalamo–pituitary axis by the conversion of the stimulus into a nervous impulse affecting receptors. However, only PRL is released in excess after the interception of the hypothalamo–pituitary connection, and secretion of all other PA hormones is reduced. An increased release of PRL induced by stress might be inhibited by naloxone (Van Vugt et al., 1978).

DA and GABA are most often mentioned as PRL-inhibiting factors (Krulich, 1979). These neurotransmitters affect PRL secretion through independent receptors (Enjalbert et al., 1980). DA receptors were identified on pituitary cell membranes by Cronin et al. (1978). As inhibiting factors, nondopaminergic PIF and degradation product of TRH have PIF activity in vitro (Bauer et al., 1978).

PRL release is under the tonic inhibition of DA as a hypothalamic PIF and is delivered by the long portal bloodstream to the receptors of PRL cells and median eminence (Yen, 1979).

Serotonin-containing neurons, possibly projecting to the suprachiasmatic nucleus, are involved in the regulation of PRL and GH release as well (Cocchi et al., 1977).

PRL may have a key role in modifying estrogen and progesterone feedback action on GTH secretion. Negative feedback effect of progesterone on

PRL release was mentioned by Libertun *et al.* (1979). Apart from some differences in sensitivity, both PRL and TSH cells have quantitatively similar mechanisms in reaction to estrogens and thyroid hormones (Labrie *et al.*, 1979).

Opioid molecules, as β-endorphin, Met-enkephalin, and several analogues of Met-enkephalin, may stimulate PRL and GH release (Bruni *et al.*, 1977; Cesselin and Peillon, 1980; Dupont *et al.*, 1977, Krulich, 1979). Their locus of action appears to be through rostral hypothalamic neurons (Spies *et al.*, 1980).

Despite some controversy as to whether the used peptides are neurotransmitters or neuromodulators, under which experimental conditions and in which animals they influence release or inhibition of GH or PRL, in most cases each of these peptides has a dual role. PRL is inhibited; GH is released.

Apart from peptides, biogenic amines are mainly considered as factors playing a role in the regulation of PRL and GH release and synthesis (Scheibel *et al.*, 1980; Stuart *et al.*, 1976). At the present time, some amino acids are mentioned as important factors in the control of hypothalamo–pituitary function. For example, glycine, as the simplest amino acid, significantly increases serum level of GH. The effect is dose dependent (Kasai *et al.*, 1980), and is probably expressed via glutathione (GSH). GSH pervades the internal milieu of the cell, affecting many of the biochemical reactions and functions including neurotransmitter release and transport of proteins within the cell membrane. One of the important roles is the regulation of many of the enzymatic activities (Kosower and Kosower, 1978).

7. The Role of Gonadal Steroids in PRL Cell Genesis and Hormone Release

7.1. Genesis and Structure

PRL cells and other pituitary cell types differentiate very early during fetal development. As differentiation proceeds, characteristics of very active protein-synthesizing cells are clearly developed. GER is proliferating and the Golgi complex is prominent. No clear differences are evident between some properties of PRL and GH granules in fetal pituitaries. Specific granules do not show polymorphism even in the pituitary of newborn animals. As development proceeds, PRL cells in mammals have distinct round and ovoid secretory granules of varying size; most are large, between 450 and 750 nm in diameter. However, some PRL cells display smaller secretory granules (250–500 nm). PRL cells differentiate as a distinct cell type. PRL and GH are synthesized and stored in different cells, and under normal physiological conditions a single cell cannot synthesize and store both hormones simultaneously (Dacheux, 1980).

7.2. PRL Content

PRL content and concentration are low during the postnatal period of life of both male and female rats. They rise in females on day 17 and reach peak values on day 35, decline until the day of vaginal opening, and increase again (with major fluctuations) during the first estrogen cycle.

The summer evaluation in peripheral PRL levels coincides with the increase in plasma testosterone levels that precedes the onset of the breeding season (Ravault, 1976). This stimulation probably involves several effects on Leydig cells, each contributing to the increase in their ability to respond to luteinizing hormone (LH) stimulation.

PRL content in the pituitary homogenate is increased in rats neonatally treated with estrogen (Pantić and Genbačev, 1969, 1970).

In PRL cells, GER is more developed and other signs of stimulation of protein synthesis are clearly evident. A prolonged stimulatory effect on PRL synthesis and release expressed in cell hypertrophy, proliferation of cytomembranes, higher rate of [^{14}C]leucine incorporation, and an increased corresponding band density, was more pronounced in males than in females (Pantić and Genbačev, 1972).

7.3. PRL Cell Specificity and Reaction to Gonadal Steroids

In order to examine the possibility of using the long-term effect of gonadal steroids injected into some fish, chickens and some mammals for the control of reproductive and productive characteristics, different experimental approaches have been employed. In summation, hyperplasia of chromophobes, hypertrophy of PRL cells, and reduced size of GH cells were pronounced in all examined teleosts kept in aquaria and treated repeatedly with estrogen. The effect of this steroid is more clearly expressed in chickens treated soon after hatching with a single dose of estrogen. However, sensitivity was higher in rats and pigs (Vigh et al., 1978). The effect was clearly pronounced if piglets were neonatally treated with a single or repeated doses of estrogen and progesterone.

[^{3}H]estrogen and neurophysin 1 have been simultaneously localized in the nucleus supraopticus and nucleus paraventricularis neurons of the mouse (Sar and Stumpf, 1979). This hormone is concentrated in GH, PRL, GTH, and probably TSH cells as well (Sar and Stumpf, 1979; Stumpf and Sar, 1976).

Without any pretension even to speculate on the key mechanisms involved in the regulation of events occurring after the administration of estrogen, some of the data obtained up to now will be briefly summarized.

During the neonatal period of life, up to the end of the second week, the median eminence is immature in chickens, rats, and some other mammals. The portal blood capillaries are still undeveloped. Consequently, only a few microvesicles are present and the synapticlike contact of axons can be observed. Signs of maturation of the median eminence are clearly observed during the second week of postnatal life. The incorporation of estrogen in such undifferentiated brain neurons, especially hypothalamic ones, undoubtedly modulates their further development. As a result, peptidergic and aminergic fibers are at least retarded in development, appearing even as undeveloped nerve terminals in the pars nervosa with a sloweddown capacity for presynaptic membrane recycling.

The response of pituitary cells of neonatally estrogen-treated rats is characterized by an increase in the proliferative rate of chromophobic and PRL cells, and differentiation of PRL cells. In chromophobes and PRL cells, the cells most sensitive to estrogen, proliferating ER often forms concentrically oriented lamellae in chickens, rats, and pigs (Pantić, 1980). The formation of intracellular membranous whorls in PRL cells of the pituitary of female Mongolian gerbils has also been observed (Nakayama and Nickerson, 1972). There is no doubt that this phenomenon is a result of the direct effect of gonadal steroids (Pantić, 1971).

An increased GER in both male and female rats with a tendency to be concentrically oriented, represents cytoplasmic sites for the biosynthesis of PRL molecules and lysosomal enzymes. This is a result of the high sensitivity of chromophobes and differentiating pituitary PRL cells to estrogen. At present it is not clear whether there is a close correlation between the ability of specific cells for estrogen uptake and the degree of reaction.

7.4. Origin and Fate of Cytomembranes

There is no doubt that the proliferation of cytomembranes occurs by sequential addition of lipid and protein molecules at various sites on preexisting membranes. Having in mind that ER is the source of lipids and proteins, the question arises as to the mechanisms involved in the increased ER proliferation in PRL cells in the pituitaries of estrogen-treated rats. In an attempt to contribute to further discussion related to membrane biogenesis, the following fact might be considered: biosynthesis of cytoplasmic proteins occurs on free polysomes, and macromolecules on polysomes bound to ER. Lipids, as phospholipids and other membrane components, are synthesized on agranular ER. The older and newly synthesized lipid and protein molecules are circulated, transferred, and selectively fused with the preexisting membrane, intermixing membrane components and contributing to the proliferation of cytomembranes. However, depending on the character of cytoplasmic streaming, both GER and agranular ER become concentrically oriented and a whorled lamellar structure is formed (Fig. 11). Finally, lysis of PRL cells occurs.

8. Concluding Remarks

Proliferation of pituitary anlage cells and differentiation into specific cell types occur during the fetal period depending on inductive interaction, autonomously in some period of fetal development and under *in vitro* conditions, and to some extent are dependent on RF or IF.

The highly specific cells of PI, differentiating later than PA cells, as multihormone-producing cells, are able to synthesize prohormones and to cleave them into LPHs, MSHs, ACTH, and endorphins as hormones containing a common heptapeptide. This lobe development is closely connected with the maturation of brain neurons, especially hypothalamic ones, and with the establishment of hypothalamo–pituitary and other target cell feedback mechanisms.

Retardation of brain neuron maturation, as is the case after neonatal administration of gonadal steroids, results in retardation of PI development. This effect is prolonged, dose dependent, and most clearly expressed in repeatedly treated animals. However, with the recovery of brain neurons, both the neurohypophysis and PI are reestablished, almost reaching the size and other characteristics of corresponding controls.

As species-specific cells, PI cells differ in sensitivity to various stimuli. However, the rostral part of this lobe, being more vascularized, is more sensitive to the sufficiency and deficiency of brain peptides and gonadal steroids. In neonatally orchidectomized piglets the cells in this lobe respond to brain peptides, and hyperplasia and hypertrophy of GTH cells are pronounced.

Signs of PI cell reaction to the deficit of corticosteroids are noticeable, but no adrenalectomy cells as a characteristic reaction of ACTH in PA were observed.

Apart from the fact that PRL and GH cells have common stimulators and inhibitors, PRL cells respond specifically to some substances and their reaction to gonadal steroids could be used as a model for the further explanation of pituitary cell multiplication and better understanding of the proliferative rate, role, and fate of cytomembranes.

ACKNOWLEDGEMENT. The authors research presented in this report is supported by a grant from the Serbian Academy of Sciences and Arts, Belgrade.

REFERENCES

Arimura, A., Bowers, C. V., Schally, A. V., Saito, M., and Miller, M. C., 1969, Effects of corticotropin releasing factor, dexamethasone and actinomycin D on the release of ACTH from rat pituitaries *in vivo* and *in vitro, Endocrinology* **85**:300.

Baertschi, A. J., Friendli, M., Munoz, J., Tsacopoulos, M., and Coles, J. A., 1980, Extracellular potassium change in the rat adenohypophysis: An indicator of neurohypophyseal–adenohypophyseal communication, in: *Biosynthesis and Release of Adenohypophyseal Hormones* (M. Jutisz and K. W. McKerns, eds.), pp. 775–782, Plenum Press, New York.

Bauer, K., Gräf, K. J., Faivre-Bauman, A., Beier, S., Tixier-Vidal, A., and Klein-Kauf, H., 1978, Inhibition of prolactin secretion by histidyl-proline-diketo-piperazine, *Nature* (*London*) **274**:174.

Bélanger, A., Labrie, F., Borgeat, P., Savary, M., Götz, J., Drouin, J., Schally, A. V., Coy, D. H., Immer, H., Sestanj, K., Nelson, V., and Gloz, M., 1974, Inhibition of growth hormone and thyrotropin release by growth hormone-release inhibiting hormone, *J. Mol. Cell. Endocrinol.* **1**:329.

Bloch, B., Bugnon, C., Fellmann, D., Lenys, D., and Gouget, A., 1979, Neurons of the rat hypothalamus reactive with antisera against endorphins, ACTH, MSH, and β-LPH, *Cell Tissue Res.* **204**:1.

Borgeat, P., Labrie, F., Poizier, G., Chavancy, G., and Schally, A. V., 1973, Stimulation of adenosine 3', 5'-cyclic monophosphate accumulation in anterior pituitary gland by purified growth hormone-releasing hormone, *Trans. Assoc. Am. Physicians* **86**:284.

Borgeat, P., Labrie, F., Drouin, J., Bélanger, A., Immer, I., Sestanj, K., Nelson, V., Götz, M., Schally, A. V., Coy, D. H., and Coy, E. J., 1974, Inhibition of adenosine 3', 5'-monophosphate accumulation in anterior pituitary gland *in vitro* by growth hormone release-inhibiting hormone, *Biochem. Biophys. Res. Commun.* **56**:1052.

Bowers, C. Y., Friesen, H. G., Hwang, P., Guyda, H. J., and Folkers, K., 1971, Prolactin and thyrotropin release in man by synthetic pyroglutamylhistadyl-prolineamide, *Biochem. Biophys. Res. Commun.* **45**:1033.

Bowie, E. P., Ishikawa, H., Shiino, M., and Rennels, E. G., 1978, An immunocytochemical study of a rat pituitary multipotential clone, *J. Histochem. Cytochem.* **26**:94.

Brazeau, P., Vale, W., Burgus, R., Ling, N., Butcher, M., Rivier, J., and Guillemin, R., 1973, Hypothalamic polypeptide that inhibits the secretion of immunoreactive pituitary growth hormone, *Science* **179**:77.

Briaud, B., Koch, B., Lutz-Bucher, B., and Mialke, C., 1979, *In vitro* regulation of ACTH release from neurointermediate lobe of rat hypophysis, *Neuroendocrinology* **28**:377.

Bruni, J. F., Van Vugt, D., Marschall, S., and Meites, J., 1977, Effects of naloxone, morphine and methionine enkephalin on serum prolactin, luteinizing hormone, follicle stimulating hormone, thyroid stimulating hormone and growth hormone, *Life Sci.* **21**:461.

Cesselin, F., and Peillon, F., 1980, *In vitro* studies on the secretion of human prolactin and growth hormone, in: *Synthesis and Release of Adenohypophyseal Hormones* (M. Jutisz and K. W. McKerns, eds.), pp. 677–722, Plenum Press, New York.

Chrétien, M., Benjannet, S., Dragon, N., Seidah, N. G., and Lis, M., 1976, Isolation of peptides with opiate activity from sheep and human pituitaries: Relationship to beta-lipotropin, *Biochem. Biophys. Res. Commun.* **72**:472.

Cocchi, D., Gil-Ad, I., Panerai, A. E., Locatelli, V., and Müller, E. E., 1977, Effect of 5-hydroxytryptophan on prolactin and growth hormone release in the infant rat: Evidence for different neurotransmitter mediation, *Neuroendocrinology* **24**:1.

Crine, P., Gossard, F., Seidah, N. G., Gianoulakis, C., Lis, M., and Chrétien, M., 1980, Biosynthesis of β-endorphin in the rat pars intermedia, in: *Biosynthesis and Release of Adenohypophyseal Hormones* (M. Jutisz and K. W. McKerns, eds.), pp. 263–283, Plenum Press, New York.

Cronin, M. J., Roberts, J. M., and Weiner, R. I., 1978, Dopamine and dihydroergocryptine binding to the anterior pituitary and other brain areas of the rat and sheep, *Endocrinology* **103**:302.

Dacheux, F., 1980, Ultrastructural localization of LH and FSH in the porcine pituitary, in: *Synthesis and Release of Adenohypophyseal Hormones* (M. Jutisz and K. W. McKerns, eds.) pp. 187–195, Plenum Press, New York.

Dallman, M. F., and Yates, F. E., 1969, Dynamic asymmetrics in the corticosteroid feedback pathways and distribution, binding and metabolism elements of the adrenocortical system, *Ann. N.Y. Acad. Sci.* **156**:696.

de Bold, A. J., de Bold, M. L., and Kraicer, J., 1980, Structural relationships between parenchymal and stromal elements in the pars intermedia of the rat adenohypophysis as demonstrated by extracellular space markers, *Cell Tissue Res.* **207**(3):347.

Dupont, A., Cusan, L., Labrie, F., Coy, D. H., and Hao-Li, C., 1977, Stimulation of prolactin release in the rat by intraventricular injection of β-endorphin and methionine enkephalin, *Biochem. Biophys. Res. Commun.* **75**:76.

Dupouy, J.-P., 1980, Differentiation of MSH-, ACTH-, endorphin-, and LPH-containing cells in the hypophysis during embryonic and fetal development, *Int. Rev. Cytol.* **68**:197.

Elde, R., and Hökfelt, T., 1979, Localization of hypophysiotropic peptides and other biologically active peptides within the brain, *Ann. Rev. Physiol.* **41**:587.

Enjalbert, A., Ruberg, M., Arancibia, S., Priam, M., and Kordon, C., 1980, Pharmacological studies of pituitary receptors involved in the control of prolactin secretion, in: *Synthesis and Release of Adenohypophyseal Hormones* (M. Jutisz and K. W. McKerns, eds.), pp. 525–542, Plenum Press, New York.

Evans, G. A., and Rosenfeld, M. G., 1980, Hormonal regulation of prolactin mRNA, in: *Synthesis and Release of Adenohypophyseal Hormones* (M. Jutisz and K. W. McKerns, eds.), pp. 295–310, Plenum Press, New York.

Evans, G. A., David, D. N., and Rosenfeld, M. G., 1978, Regulation of prolactin and somatotropin mRNAs by thyroliberin, *Proc. Natl. Acad. Sci. USA* **75**:1294.

Farquhar, M. G., 1977, Secretion and crinophagy of prolactin cells, in: *Comparative Endocrinology of Prolactin* (H. D. Dellman, J. A. Johnson, and D. M. Klachko, eds.), pp. 37–91, Plenum Press, New York.

Farquhar, M. G., Reid, J. J., and Daniell, L. W., 1978, Intracellular transport and packaging of prolactin: A quantitative electron microscope autoradiographic study of mammotrophs dissociated from rat pituitaries, *Endocrinology* **102**:296.

Fernholm, B., 1972, Ultrastructure of adenolypophysis of ilyxine glutinosa, *Z. Zellforsch. Mikrosk. Anat.* **132**:451.

Fleischer, N., Burgus, R., Vale, W., Dunn, T., and Guillemin, R., 1970, Preliminary observations on the effect of synthetic thyrotropin releasing factors on plasma thyrotropin levels in man, *J. Clin. Endocrinol. Met.* **31**:109.

Gourdji, D., 1980, Characterization of thyroliberin (TRH) binding sites and coupling with prolactin and growth hormone secretion in rat pituitary cell lines, in: *Synthesis and Release of Adenohypophyseal Hormones* (M. Jutisz and K. W. McKerns, eds.), pp. 463–494, Plenum Press, New York.

Guillemin, R., 1978, Biochemical and physiological correlates of hypothalamic peptides: The new endocrinology of the neuron, in: *Peptides in Neurobiology* (S. Gainer, ed.), Plenum Press, New York.

Hökfelt, T., Johansson, O., Ljungdahl, A., Lundberg, J., Schultzberg, M., Fuxe, K., Goldstein, M., Steinbusch, H., Verhofstand, A., and Elde, R., 1979, Neurotransmitters and neuropeptides: Distribution pattern and cellular localization as revealed by immunocytochemistry, in: *Central Regulation of the Endocrine System* (K. Fuxe, T. Hökfelt, and R. Luft, eds.), Plenum Press, New York.

Hristić, M., and Pantić, V., 1974, Hypothalamic nuclei of rats after head irradiation and adrenocortical hormone treatment, *Int. J. Radiat. Biol.* 26(2):147.

Hristić, M., and Pantić, V., 1976, Autoradiographic examinations of hypothalamus and hypophysis of adrenalectomized rats treated with corticosteroids, *Folia Anat. Yugosl.* 5:13.

Hristić, M., and Pantić, V., 1978, Hypothalamic nuclei of rats treated with corticosteroids, in: *Neurosecretion and Neuroendocrine Activity: Evolution, Structure and Function* (W. Bargmann, A. Oksche, A. Polenov, and B. Scharrer, eds.), pp. 205–208, Springer-Verlag, Berlin.

Hristić, M., Pantić, V., and Nickerson, P. A., 1978, Influence of maternal adrenalectomy on pituitary ACTH-cells of fetal and neonatal rats, *Acta Vet. (Belgrad)* 28(6):231.

Huntington, T., and Hadley, M. E., 1974, Evidence against mass action direct feedback control of melanophore*stimulating hormone (MSH) release, Endocrinology* 96 (2):472.

Hymer, C. W., Page, R., Kelsey, R. C., Augustine, E. C., Wilfinger, W., and Ciolkosz, M., 1980, Separated somatotrophs: Their use *in vitro* and *in vivo,* in: *Synthesis and Release of Adenohypophyseal Hormones* (M. Jutisz and K. W. McKerns, eds.), pp. 125–166, Plenum Press, New York.

Jones, M. T., and Gillham, B., 1980, Corticotropin secretion, in: *Synthesis and Release of Adenohypophyseal Hormones* (M. Jutisz and K. W. McKerns, eds.), pp. 587–616, Plenum Press, New York.

Kasai, K., Suzuki, H., Nakamura, T., Shiina, H., and Shimoda, S.-I., 1980, Glycine stimulates growth hormone release in man, *Acta Endocrinol. (Copenhagen)* 93 (3):283.

Koch, B., Lutz-Bucher, B., Briaud, B., and Mialhe, C., 1979, Relationship between ACTH secretion and corticoid binding to specific receptors in perifused adenohypophysis, *Neuroendocrinology* 28:169.

Koch, B., Lutz-Bucher, B., and Mialhe, C., 1980, Interaction of glucocorticoids, CRF, and vasopressin in the regulation of ACTH secretion, in: *Synthesis and Release of Adenohypophyseal Hormones* (M. Jutisz and K. W. McKerns, eds.), pp. 561–586, Plenum Press, New York.

Kosower, N. S., and Kosower, E. M., 1978, The glutathion status of cells, *Int. Rev. Cytol.* 54:109.

Krulich, L., 1979, Central neurotransmitters and the secretion of prolactin, GH, LH, and TSH, *Annm. Rev. Physiol.* 41:603.

Krulich, L., Quijada, M., Wheaton, J., Illner, P., and McCann, S. M., 1977, Localization of hypophysiotropic neurohomones by assay of section from various brain areas, *Fed. Proc.* 36:1953.

Labrie, F., Borgeat, P., Drouin, J., Beaulieu, M., Lagacé, L., Ferland, L., and Raymond, V., 1979, Mechanism of action of hypothalamic hormones in the adenohypophysis, *Ann. Rev. Physiol.* 41:555.

Li. J. Y., Dubois, M. P., and Dubois, P. M., 1979, Ultrastructural localization of immunoreactive corticotropin, β-lipotropin, α- and β-endorphin in cells of the human fetal anterior pituitary, Cell Tissue Res. 204:37.

Libertun, C., Kaplan, S. E., and de Nicola, A. F., 1979, Progesterone negative feedback on prolactin secretion: Importance of the brain control and of estradiol, *Neuroendocrinology* 28:64.

Lowry, P. J., and Scott, A. P., 1975, The evolution of vertebrate corticotrophin and melanocyte stimulating hormone, *Gen. Comp. Endocrinol.* 26:16.

Lowry, P. J., and Scott, A. P., 1977, Structural relationship and biosynthesis of corticotropin, lipotropin, and melanotropin, in: *Frontiers of Hormone Research* (T. B. van
Wiersma-Greidanus, ed.), Vol. 4, pp. 11–17, Karger, Basel.

Lowry, P. J., Silman, R. E., and Hope, J., 1977, Structure and biosynthesis of peptides
related to corticotropins and β-melanotropins, in: *ACTH and Related Peptides: Structure, Regulation and Action* (D. T. Krieger and W. F. Ganong, eds.), *Ann. N.Y. Acad.
Sci.* **297**:49.

Lutz-Bucher, B., Koch, B., and Mialhe, C., 1977, Comparative *in vitro* activity of vasopressin and hypothalamic median eminence extract, *Neuroendocrinology* **23**:181.

Martin, J. B., Tolis, G., Woods, I., and Guyda, H., 1979, Failure of naloxone to influence
physiological growth hormone and prolactin secretion, *Brain Res.* **168**:210.

Meurling, P., Fremberg, M., and Björklund, A., 1969, Control of MSH release in the intermediate lobe of *Raja radiata* (Elasmobranchii), *Gen. Comp. Endocrinol.* **13**:520.

Milković, S., Milković, K., and Paunović, J., 1973, The initiation of fetal adreno-cortico-
trophic activity in the rat, *Endocrinology* **92**:380.

Milković, K., Joffe, J., and Levine, S., 1976, The effect of maternal and fetal corticosteroids
on the development and function of the pituitary–adrenocortical system, *Endokrinologie* **68**:60.

Moriarty, G. C., and Garner, L. L., 1977, Immunoelectronmicroscopical localization of
ACTH/MSH peptides in rat and human pituitaries, in: *Frontiers of Hormone
Research* (T. B. Van Wiersma Griedanus, ed.), Vol. 4, pp. 26–41, Karger, Basel.

Morin, A., 1980, Effect of TRH on the synthesis and stability of cytoplasmic RNAs of
GH₃B₆ cells, in: *Synthesis and Release of Adenohypophyseal Hormones* (M. Jutisz
and K. W. McKerns, eds.), pp. 285–294, Plenum Press, New York.

Moss, R. L., 1979, Action of hypothalamic hypophysiotropic hormones on the brain, *Ann.
Rev. Physiol.* **41**:617.

Mulder, G. H., and Smelik, P. G., 1977, A superfusion system technique for the study of
the sites of action of glucorcorticoids in the rat hypothalamus–pituitary–adrenal system *in vitro*. I. Pituitary cell superfusion, *Endocrinology* **100**:1143.

Nakayama, I., and Nickerson, P. A., 1972, Intranuclear inclusions in mammotrophs of the
female Mongolian gerbil, *Am. J. Anat.* **135**:93.

Nickerson, P. A., Hristić, M., and Pantić, V., 1978, Influence of maternal adrenalectomy
on the ultrastructure of the adrenal gland in neonatal rats, *Cell Tissue Res.*
189:277.

O'Donohue, T. L., Miller, R. L., Pendleton, R. C., and Jacobowitz, D. M., 1979, A diurnal
rhythm of immunoreactive α-melanocyte-stimulating hormone in discrete regions of
the rat brain, *Neuroendocrinology* **29** (4):281.

Ogle, T. F., and Kitay, J. I., 1977, Effects of melatonin and an aqueous pineal extract on
adrenal secretion of reduced steroid metabolites in female rats, *Neuroendocrinology*
23 (2):113.

Olpe, H. S., and McEwen, B. S., 1976, Glucocorticoid binding to receptor-like protein in
rat brain and pituitary: Ontogenetic and experimentally induced changes, *Brain Res.*
105:121.

Oostlund, R. E., Jr., Leung, J. T., Vaerewyck Hajek, S., Winokur, T., and Melman, M.,
1978, Acute stimulated hormone release from cultured GH₃ pituitary cell, *Endocrinology* **103**:245.

Palkovits, M., 1979, Blood supply of the hypophysiotrophic area, Proceedings of the International Symposium on Neuroendocrine Regulatory Mechanisms, Belgrade (May,
1978, Vol. VI (2), P. 143), Serbian Academy of Sciences and Arts, Scientific Assem-
blies.

Pantić, V., 1971, Ultrastructure of the endocrine cells and protein synthesis, *Glas SANU*
CCLXXXI **24**:273.

Pantić, V. R., 1974, Gonadal steroids and hypothalamo–pituitary–gonadal axis, International Symposium on Sexual Endocrinology of the Perinatal Period, Vol. 32, p. 97, INSERM, Paris.

Pantić, V. R., 1975, The specificity of pituitary cells and regulation of their activities, *Int. Rev. Cytol.* **40**:153.

Pantić, V. R., 1981, Sensitivity of pituitary gonadotropic cells and gonads to hormones, in: *Reproductive Processes and Contraception* (K. W. McKerns, ed.), pp. 47–89, Plenum Press, New York.

Pantić, V., and Genbačev, O., 1969, Ultrastructure of pituitary lactotropic cells of oestrogen treated male rats, *Z. Zellforsch. Mikrosk. Anat.* **95**:280.

Pantić, V., and Genbačev, O., 1970, Ultrastructure of pituitary luteotropic (LTH) and somatotropic (STH) cells of rats neonatally treated with oestrogen, Septieme Congresse Internationalle de la Microscopie Electronique, Grenoble (August, 1970), p. 569.

Pantić, V., and Genbačev, O., 1972, Pituitary of rats neonatally treated with oestrogen. I. Luteotropic and somatotropic cells and hormones content, *Z. Zellforsch. Mikrosk. Anat.* **126**:41.

Pantić, V., and Hristić, M., 1975, Adrenocorticotropic (ACTH) cells of rats after head irradiation, *Int. J. Radiat. Biol.* **28**(1):53.

Pantić, V., and Šimić, M., 1974, Pars intermedia of deer pituitary, *Arh. Biol. Nauka.* **26**(1–2):15.

Pantić, V., and Šimić, M., 1977a, Effect of gonadal steroids on pituitary pars intermedia cells of some Teleostea and rat, *Bull. Acad. Serbe Sci.* **T.LX**(16):23.

Pantić, V., and Šimić, M., 1977b, Sensitivity of pituitary pars intermedia to castration or gonadal steroids, *Bull. Acad. Serbe Sci. Arts* **T.LX**(16):67.

Pantić, V., Ožegović, B., Genbačev, O., and Milković, S., 1971a, Ultrastructure of transplantable pituitary tumor cells producing luteotropic and adrenocorticotropic hormones, *J. Microsc. (Paris)* **12** (2):225.

Pantić, V., Genbačev, O., Milković, S., and Ožegović, B., 1971b, Pituitaries of rats bearing transplantable MtT mammotropic tumor, *J. Microsc. (Paris)* **12** (3):405.

Pelletier, G., and Dubé, D., 1977, Electron microscopic immunohistochemical localization of α-MSH in the rat brain, *Am. J. Anat.* **150**:201.

Ravault, J. P., 1976, Prolactin in the ram: Seasonal variations in the concentration of blood plasma from birth until three years old, *Acta Endocrinol. (Copenhagen)* **83**:720.

Reichlin, S., 1974, Neuroendocrinology, in: *Textbook of Endocrinology* (R. H. Williams, ed.), pp. 774–831, Saunders, Philadelphia.

Saffran, M., and Schally, A. V., 1977, The status of the corticotropin releasing factor (CRF), *Neuroendocrinology* **24** (5–6):359.

Saint Gerons, H., 1961, Particularites anatomiques et histologiques de l'hypophyse chez les Squamata, *Arch. Biol.* **72**:211.

Sandow, J., and König, W., 1978, Chemistry of the hypothalamic hormones, in: *The Endocrine Hypothalamus* (S. L. Jeffcoate and J. S. M. Hutchinson, eds.), pp. 150–211, Academic Press, New York.

Sar, M., and Stumpf, W. E., 1979, Simultaneous localization of steroid and peptide hormone in rat pituitary by combined Thaw-Mount autoradiography and immunohistochemistry: Localization of dihydrosterone in gonadotropes, thyrotropes and pituicytes, *Cell Tissue Res.* **203**:1.

Sawano, S., Matsuo, I., Yamazaki, M., Baba, Y., and Takahashi, K., 1974, Studies on growth hormone-releasing activity in bovine hypothalami, in: *Psychoneuroendocrinology* (N. Natotami, ed.), p. 232, Karger, Basel.

Schally, A. V., Arimura, A., and Kastin, A. J., 1973, Hypothalamic regulatory hormones, *Science* **179**:341.

Scheibel, J., Elsasser, T., and Ondo, J. G., 1980, Stimulation of prolactin secretion by taurine, a neurally depressant amino acid, *Neuroendocrinology* **30**:350.

Scott, A. P., and Lowry, P. J., 1974, Adrenocorticotrophic and melanocyte-stimulating peptides in the human pituitary, *Biochem. J.* **139**:593.

Silman, R. E., Chard, T., Lowry, P. J., Smith, I., and Young, I. M., 1976, Human foetal pituitary peptides and parturation, *Nature (London)* **260**:716.

Spies, H. G., Quadri, S. K., Chappel, S. C., and Norman, R. L., 1980, Dopaminergic and opioid compounds: Effects on prolactin and LH release after electrical stimulation of the hypothalamus in ovariectomized rhesus monkey, *Neuroendocrinology* **30** (4):249.

Stoeckel, M. E., Dellmann, H. D., Porte, A., Klein, M. J., and Stutinsky, F., 1973, Corticotrophic cells on the rostral zone of the pars intermedia and in the adjacent neurohypophysis of the rat and mouse, *Z. Zellforsch. Mikrask. Anat.* **136**:97.

Stratmann, I. E., Ezrin, C., and Sellers, A., 1974, Estrogen induced transformation of somatotrophs into mammotrophs in the rat, *Cell Tissue Res.* **152**:229.

Stuart, M., Lazarus, L., Smythe, G. A., Moore, S., and Sara, V., 1976, Biogenic amine control of growth hormone secretion in the fetal and neonatal rat, *Neuroendocrinology* **22** (4):337.

Stumpf, W. E., and Sar, M., 1976, Steroid hormone target sites in the brain: The differential distribution of estrogen, progesterone, androgen and glucocorticoid, *J. Steroid Biochem.* **7**:1163.

Taché, Y., du Ruisseau, P., Ducharme, J. R., and Collu, R., 1978, Pattern of adenohypophyseal hormone changes in male rats following chronic stress, *Neuroendocrinology* **26**:208.

Tashjian, A. H., Jr., Yasumura, Y., Levine, L., Sato, G. H., and Parker, M. L., 1968, Establishment of clonal strains of rat pituitary tumor cells that secrete growth hormone, *Endocrinology* **82**:342.

Tilders, F. J. H., and Smelik, P. G., 1975, A diurnal rhythm in melanocyte-stimulating hormone content of the rat pituitary gland and its dependence from the pineal gland, *Neuroendocrinology* **17** (4):289.

Usategui-Echeverria, R., Oliver, C., Vaudry, H., Lombardi, G., Mourre, A. M., Rozenberg, I., and Vague, J., 1975, Radioimmunological determination of rat plasma α-MSH and ACTH, *Acta Endocrinol. (Copenhagen)* **85** (Suppl. 199):73.

Van Vugt, D. A., Bruni, J. F., and Meites, J., 1978, Naloxone inhibition of stress induced increase of prolactin secretion, *Life Sci.* **22**:85.

Vigh, S., Sétáló, G., Török, A., Pantić, V., Flerkó, B., and Gledić, D., 1978, Deficiency of FSH and LH cells in rats treated with oestradiol in the early postnatal life, *Bull. Acad. Serbe Sci. Arts* **T.LXI**(17):1.

Vila-Porcile, E., and Olivier, L., 1980, Exocytosis and related membrane events, in: Synthesis and Release of Adenohypophyseal Hormones (M. Jutisz and K. W. McKerns, eds.), pp. 67–104, Plenum Press, New York.

Wilbur, D. L., and Spicer, S. S., 1980, Pituitary secretory activity and endocrinophagy, in: *Synthesis and Release of Adenohypophyseal Hormones* (M. Jutisz and K. W. McKerns, eds.), pp. 167–186, Plenum Press, New York.

Yen, S. S. C., 1979, Studies of the role of dopamine in the control of prolactin and gonadotropin secretion in humans, in: *Central Regulation of the Endocrine System* (K. Fuxe, T. Hökfelt, and R. Luft, eds.), Plenum Press, New York.

25

Interaction between Thyrotropin-Releasing Hormone and Prolactin-Producing Cells

E. Haug, K. M. Gautvik, O. Sand, J.-G. Iversen, and M. Kriz

1. Introduction

In 1968 Tashjian and co-workers reported the establishment of three strains of epithelial cells from a transplantable rat pituitary tumor (MtT/W5) originally induced in a Wistar–Furth rat by radiation (Takemoto *et al.*, 1962). The different cell strains are referred to collectively as GH cells, and the specific strains have appropriate designations. In culture, GH cells spontaneously synthesize and secrete into the culture medium large quantities of prolactin (PRL) and/or growth hormone (GH). The hormones are immunologically indistinguishable from authentic rat hormones, and they have retained their biological activity (Tashjian *et al.*, 1968; Tashjian and Hoyt, 1972; Haug and Gautvik, 1976, 1978). The GH cells are aneuploid. In 1970 the modal chromosome number of the GH_3 cell strain was 69 per cell (Sonnenschein *et al.*, 1970). Seven years later a significant reduction in modal chromosome number to 62–64 was found (Clausen *et al.*, 1977), indicating that during these years of continuous culture either a loss or a

E. Haug • Institute of Physiology, University of Oslo, Oslo 1, Norway, and Hormone and Isotope Laboratory, Aker Hospital, Oslo 5, Norway. K. M. Gautvik, J.-G. Iversen, and M. Kriz • Institute of Physiology, University of Oslo, Oslo 1, Norway. O. Sand • Department of Physiology, Veterinary College of Norway, Oslo 1, Norway.

rearrangement of chromosomes had occurred. The observed reduction in chromosome number is not associated with dramatic changes in cell growth and hormone production (Tashjian and Hoyt, 1972; Clausen et al., 1977).

Thyrotropin-releasing hormone (TRH; L-pyroglutamyl-L-histidyl-L-prolinamide) stimulates the release and synthesis of PRL from the anterior pituitary. The stimulatory effect of TRH on PRL secretion was first shown in the GH cell system (Tashjian et al., 1971), and GH cells have served as model systems for studies of the molecular mechanisms involved in the control of PRL synthesis and secretion (Tashjian and Hoyt, 1972; Gautvik and Tashjian, 1973a,b; Haug et al., 1977; Clausen et al., 1977). Although GH cells are of tumor origin and thus differ from normal pituitary cells in many ways, these cells also have several features in common with normal pituicytes: (1) They respond in the same manner as normal rat pituitary cells to glucocorticoids (Tashjian and Hoyt, 1972; Naess et al., 1980) estrogens (Tashjian and Hoyt, 1972; Haug et al., 1977; Haug, 1979), and TRH (Tashjian et al., 1971; Gautvik et al., 1977; Gourdji, 1980). (2) Calcium ions play an important role in PRL synthesis and secretion (Gautvik and Tashjian, 1973a; Tashjian et al., 1978; Ostlund et al., 1978; Gautvik et al., 1980; Sand et al., 1980). (3) The ergot alkaloid bromocriptine, which is widely used to suppress pituitary PRL secretion (Varga et al., 1972), also causes a rapid reduction in PRL secretion by GH cells (Gautvik et al., 1973). (4) TRH stimulation of PRL synthesis is accompanied by morphological changes determined by light and electronmicroscopic studies (Tixier-Vidal, 1975; Fossum and Gautvik, 1977). (5) It is generally accepted that membrane or cytoplasmic receptors are a prerequisite for hormonal action, and functional receptors for estrogens (Mester et al., 1973; Haug et al., 1978), glucocorticoids (Naess et al., 1980), somatostatin (Schonbrunn and Tashjian, 1978; Aulie et al., 1979), and TRH (Hinkle and Tashjian, 1973, 1974; Gourdji, 1980) have been characterized in GH_3 cells, and the receptors have physicochemical and binding properties identical to those of receptors occurring in other rat tissues.

GH cells are thus obviously capable of carrying out several different functions, and respond normally to a variety of physiological signals. This cell system therefore appears as a suitable model for studying the control of PRL synthesis and secretion. This report describes some of the properties of the TRH receptor and the cyclic nucleotide phosphodiesterases from GH_3 cells, and the effects of TRH on these enzyme activities as well as on the electrical membrane properties of GH_3 cells are described.

2. Binding of [³H]-TRH to GH₃ Cells

The binding of [³H]-TRH to GH_3 cells has been studied by several groups (Hinkle and Tashjian, 1973, 1974; Faivre-Bauman et al., 1975;

Gourdji, 1980), and it is established that specific binding sites for [³H]-TRH exist in these cells. There are, however, differences in the character-istics of these sites, probably reflecting the fact that the above-mentioned studies were performed with cellular homogenate, the particulate fraction, or intact cells at temperatures varying from 0°C to 37°C. Since the bio-logical actions of TRH have most often been studied at 37°C, we have studied binding of [³H]-TRH to living cells at 37°C and pH 7.4.

2.1. Methods

The binding studies were performed either with monolayer cultures or with cell suspensions, made freshly from stock monolayer cultures after mechanical or enzymatic (trypsin–EDTA) detachment of cells. The bind-ing reaction was terminated by separation of cell-bound and free [³H]-TRH within 30 sec. The radioactivity bound to the GH_3 cells was deter-mined after dissolution of the cells in 0.1 N NaOH and a parallel sample was taken for protein determination.

Nonspecific binding was measured in the presence of a 100-fold molar excess of unlabeled TRH and represented 10–15% of total binding. Specific binding was calculated as the difference between total and nonspecific bind-ing. There was a linear relationship between the concentration of cells and the binding of [³H]-TRH, and the binding was similar in monolayer cul-tures and in suspension cultures using cells detached either mechanically or enzymatically. However, pretreatment of the cells with trypsin but not collagenase for 30 min reduced the maximal specific binding by about 30%.

The [³H]-TRH present in the culture medium after incubations up to 240 min at 37°C had unaltered chromatographic behavior and binding properties, demonstrating that the tracer was stable under culture condi-tions.

2.2. Binding of [³H]-TRH at Steady State

Figure 1 shows that the binding between [³H]-TRH and intact GH_3 cells is rapid, and the amount of specifically bound [³H]-TRH increased only slowly after 60 min. When very short incubation periods were included, the binding curves did not show a smooth hyperbolic form. To determine the disociation constant at equilibrium (K_d) of [³H]-TRH bind-ing to GH_3 cells, we incubated monolayer cultures with increasing concen-trations of [³H]-TRH for 2 hr. Figure 2 shows that the specific binding never really reached a plateau, and the Scatchard (1949) analysis of these data gave a downward curvilinear graph (Fig. 3), indicating that [³H]-TRH did not bind to one homogeneous population of binding sites. When the binding data were replotted as shown in Fig. 4, a nonlinear relationship was also observed. Assuming the existence of two different classes of bind-

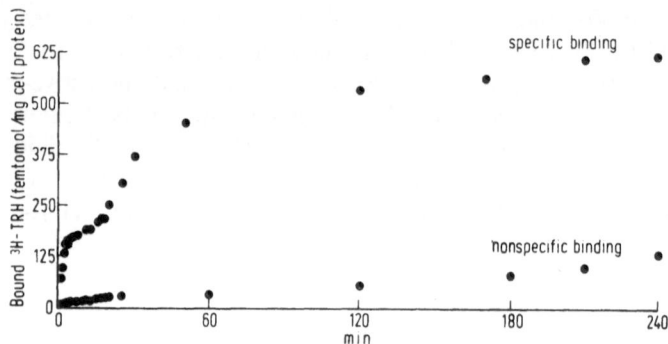

Figure 1. Time course of the binding of [³H]-TRH to monolayer cultures of GH₃ cells at 37°C. The cells were incubated for the indicated times with 10 nM [³H]-TRH alone (total binding) or in the presence of 1 μM TRH (nonspecific binding). Then [³H]-TRH was removed, the cells were washed and cell-bound radioactivity determined. Mean values of duplicate determinations from seven different experiments are shown, and the variation of each point between the individual experiments was less than 24% of the mean value. [From Gautvik and Lystad, 1981.]

ing sites, we adjusted two straight lines to the points giving dissociation constants of 0.9 nM (range 0.35–1.4 nM) and 23 nM (range 18–30 nM). The same complicated pattern of binding has also been described in intact GH₃ cells by Gourdji (1980) and in intact mouse TSH-secreting cells (Grant *et al.*, 1973). The total binding capacity obtained from the slope of

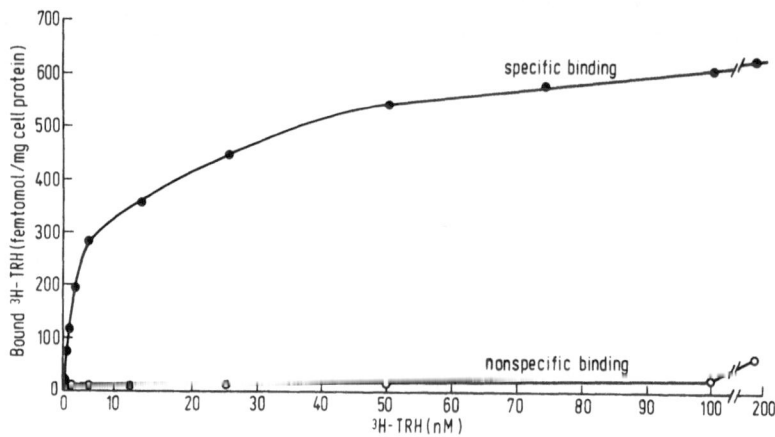

Figure 2. Binding of [³H]-TRH to GH₃ cells in monolayer culture at 37°C for 2 hr. The results are given for six different experiments. Each point represents the mean of 12–16 separate determinations, and the S.D. were less than 10% of the mean values. The specific binding was calculated as described in legend to Fig. 1. [From Gautvik and Lystad, 1981.]

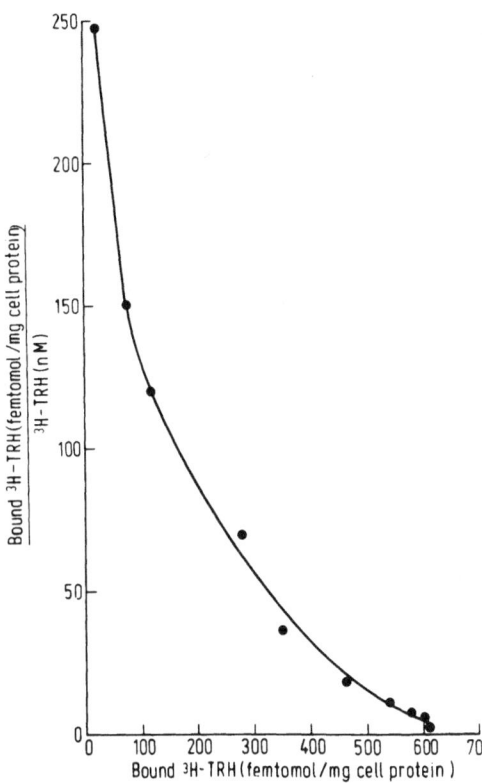

Figure 3. Scatchard analysis of the binding data shown in Fig. 2.

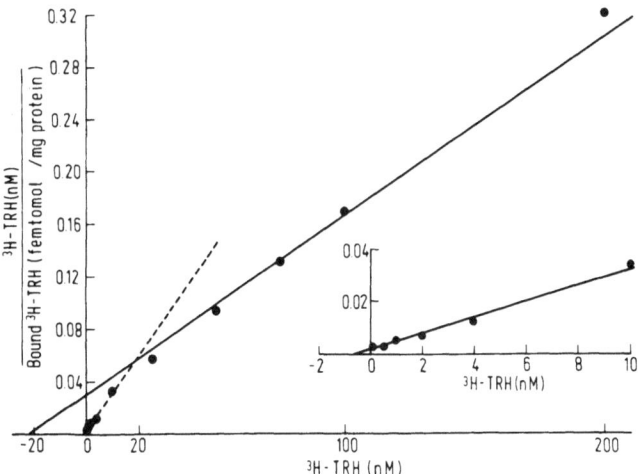

Figure 4. A Hanes–Wolf plot of the binding data shown in Fig. 2. Assuming two different classes of binding sites for TRH, two straight lines were adjusted to the points. The intercepts on the x axis give the $-K_d$ values, and the binding capacity was calculated from the slopes of the curves. [From Gautvik and Lystad, 1981.]

the curves was 750 moles/mg cell protein, corresponding to about 80,000 receptors per cell (Gautvik and Lystad, 1981).

2.3. Binding Kinetics

A downward curvilinear Scatchard plot probably reflects either a heterogeneity of the TRH binding sites or the existence of negative cooperativity. To distinguish between these possibilities the dissociation of [³H]-TRH was studied in experiments in which cell suspensions were preincubated with [³H]-TRH for various time periods. At the end of the prein-

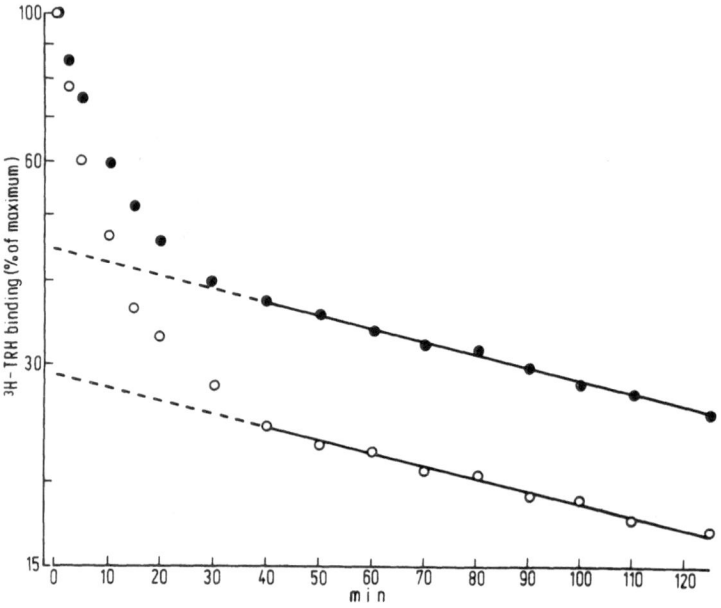

Figure 5. Dissociation of [³H]-TRH from GH₃ cells at 37°C. Suspensions of cells were incubated with [³H]-TRH (10 nM) in the absence and presence of 1 μM TRH. After association for 5 min (o) and 60 min (●), the incubation medium was rapidly diluted with a 100-fold excess medium and a sample (2.5 ml) immediately taken representing the zero point of dissociation. The sample was spotted rapidly on filter paper to separate bound and free [³H]-TRH and washed immediately with ice-cold buffer containing 0.1% albumin. At the different time intervals indicated, duplicate samples were removed and cell-bound radioactivity remaining on the filters quantitated. The nonspecific binding (binding in the presence of 1 μM TRH) at zero-point dissociation was less than 10% and dissociated immediately, so that no corrections of the data had to be made. The dashed lines were drawn by linear regression analysis through a portion of the curves which was used for calculation of the dissociation rate constant (K_{-1}) for the slowly dissociating site. The extensions of these lines were used for correcting the other points in order to obtain the curve describing the release of [³H]-TRH from the rapidly dissociable site. [From Gautvik and Lystad, 1981.]

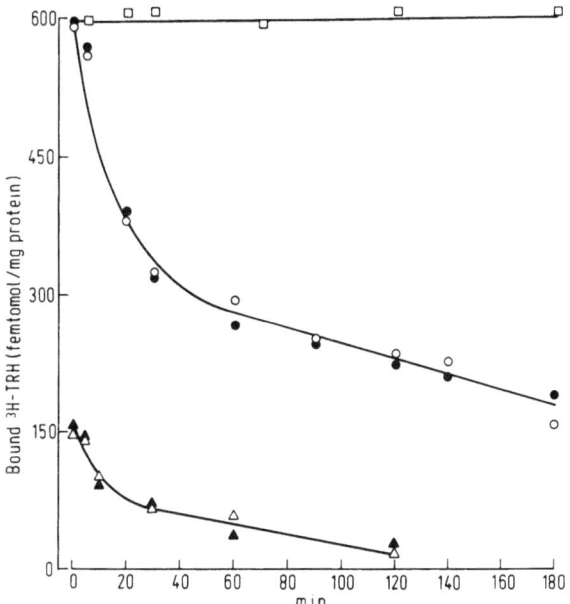

Figure 6. Dissociation of [³H]-TRH from intact GH₃ cells in the absence and presence of excess TRH. Suspensions of GH₃ cells were incubated at 37°C with 10 nM (○, ●, □) and 2 nM (△, ▲) [³H]-TRH for 2 hr. A 100-fold excess of medium without (○, △) or with 1 µM TRH (●, ▲) was added to tubes and aliquots taken immediately to represent the zero point of dissociation. Samples were simultaneously taken from tubes that were incubated for another 3 hr without receiving excess medium (□). The dissociation of [³H]-TRH was followed at the time points indicated. Nonspecific binding has been corrected as described previously. The results represent mean values of duplicates from one typical experiment. [From Gautvik and Lystad, 1981.]

cubation period the medium was diluted rapidly with a 100-fold excess medium, and the amount of specifically bound [³H]-TRH was measured at different time intervals thereafter. Figure 5 shows that the release of specifically bound [³H]-TRH did not follow a monoexponential time course characteristic for one homogeneous type of binding sites. Assuming two different classes of binding sites, the rate constants for dissociation (K_{-1}) were found to be $0.73 \pm 0.41 \times 10^{-4}$ and $11.5 \pm 3.3 \times 10^{-4}$ M sec^{-1} (mean $\pm$ S.D., $n = 7$) for the two affinity sites, respectively. These rate constants correspond to half times of dissociation from the rapidly and slowly dissociating sites of about 5 and 160 min, respectively. When the association time was increased from 5 min to 60 min, the dissociation rate constants were unaltered, but the relative distribution of bound [³H]-TRH between the rapidly and slowly dissociating components was changed.

Figure 6 shows the release of specifically bound [³H]-TRH from GH₃

cells preincubated with 2 or 10 nM [³H]-TRH for 2 hr. At that time duplicate tubes received a 100-fold molar excess medium without or with 1 μM
TRH added, and the amount of specifically bound [³H]-TRH was measured immediately and at different periods of time thereafter. The dissociation in the absence or presence of excess TRH was similar at high and low
receptor occupancy. De Meyts et al. (1973, 1976) found that the negative
cooperativity observed for insulin binding to liver was more pronounced at
lower temperatures. The dissociation experiments were therefore repeated
at 20 °C, and the results were similar to those described in Fig. 6. Our data
describing the interaction between [³H]-TRH and intact GH₃ cells at 37 °C
show that the presence of a heterogeneous population of binding sites cannot be explained by site–site interactions among the TRH receptors.

The association rate constant could not be calculated directly from the
time course of binding (Fig. 1), since the GH₃ cells did possess more than

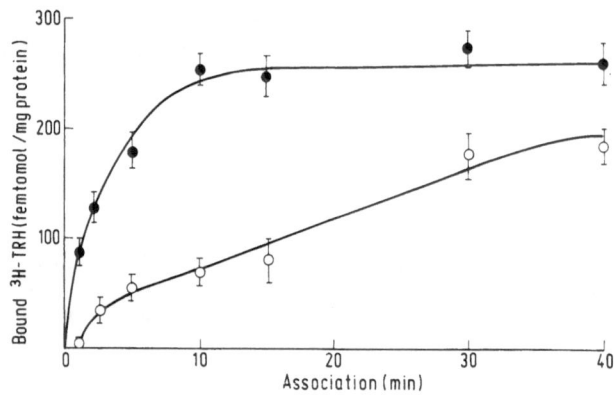

Figure 7. The time course of binding of [³H]-TRH to high- and low-affinity binding sites of GH₃
cells. Suspension of cells were incubated for the indicated times at 37°C with 10 nM [³H]-TRH.
At the end of the incubation the cells received a 100-fold excess of medium and duplicate samples
were immediately taken representing the end point of association or zero point of dissociation.
Duplicate samples were thereafter obtained at 30 sec, 2 min, and 5 min in order to measure the
rapidly dissociable component. A second series of samples were taken after 60, 90, and 120 min,
respectively, for measurements of the slowly dissociable component (see Fig. 5). The mean values
of the amount of [³H]-TRH bound at the different times of dissociation were plotted as described
in the legend to Fig. 5. The amount of [³H]-TRH radioactivity bound to the rapidly dissociable
site was then calculated as the difference between the amount of total label bound at each time
of association and the amount calculated to be bound at the slowly dissociable site at zero point
of dissociation (end of association) (see legend to Fig. 5).

Specific binding of [³H]-TRH to the high (●) and low (○) affinity site as a function of time
of association is shown (mean values ± S.D. of three different experiments, each carried out in
duplicate). [From Gautvik and Lystad, 1981.]

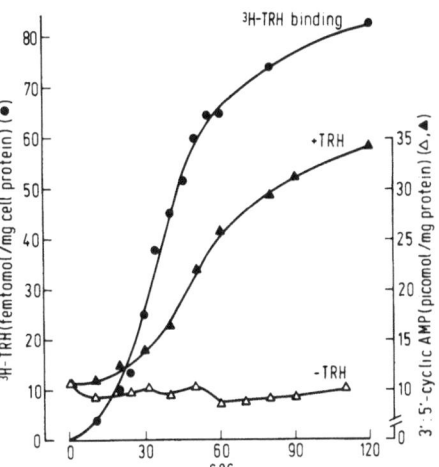

Figure 8. The initial rates of [³H]-TRH binding and cAMP accumulation in suspensions of GH₃ cells. Duplicate tubes were incubated with 10 nM [³H]-TRH at 37°C for the times indicated and the binding reaction stopped by spotting the cell suspensions to the filter paper. Parallel duplicate tubes received 20 nM TRH for measurements of cAMP. Reaction was stopped after the time periods indicated by the addition of 30% TCA. cAMP was measured by radioimmunoassay. [From Gautvik and Lystad, 1981.]

one class of binding sites for TRH. A series of dissociation experiments, like those described in Fig. 5, was therefore carried out using association times from 1 to 40 min. For each time period of association, dissociation experiments were performed after diluting the medium 100-fold. The amount of specifically bound [³H]-TRH was determined at 0.5, 2, and 5 min and after 60, 90, and 120 min, representing the rapidly and the slowly dissociable sites, respectively. The fraction of specifically bound [³H]-TRH representing binding to the different sites was calculated as described in the legend to Fig. 5. Figure 7 shows the time course of the calculated binding to the rapidly and the slowly dissociable sites. The rate constants of association (K_{+1}) for the high- and the low-affinity binding sites were estimated to be about 1.6×10^5 and about 0.4×10^5 M^{-1} sec^{-1}, respectively. The dissociation constants calculated from the two rate constants were 0.5 and 29 nM for the high- and the low-affinity sites, respectively, and consistent with those found in binding experiments performed at or close to steady-state conditions (vide supra).

Figure 8 shows the initial rates of [³H]-TRH binding and cAMP accumulation in GH₃ cell suspensions at 37°C. The binding as well as the accumulation of cAMP started without any significant time lag, and the maximal rates of the two processes occurred within the first minute. The initial stimulation of cAMP formation is therefore mediated via binding to the high-affinity sites. Since maximal cAMP concentrations were obtained with less than 10% receptor occupancy, it is possible that the stimulatory effect of TRH on cAMP formation was mediated solely through binding to the high-affinity sites.

3. Adenosine 3',5'-Cyclic Nucleotide Phosphodiesterase(s) in GH₃ Cells

In the GH₃ cell system TRH stimulates both PRL synthesis and secretion (Hinkle and Tashjian, 1974; Gautvik et al., 1977). Treatment of GH₃ cells with TRH elicits stimulation of the adenylate cyclase activity (Gautvik et al., 1981), increases cAMP formation (Dannies et al., 1976; Gautvik et al., 1977), and activates a cAMP-dependent protein kinase (Gautvik et al., 1977), suggesting involvement of cAMP in the mode of action of TRH. Since the cAMP phosphodiesterase(s) may play an important role in the regulation of intracellular cAMP concentrations, we have studied the effect of TRH on enzyme activity in GH₃ cells.

3.1. Methods

The assay of phosphodiesterase activity was carried out as described by Thompson and Appleman (1971) and Boudreau and Drummond (1975). Briefly, 5'-AMP, the end product of the cAMP phosphodiesterase reaction, was converted to adenosine by addition of 5'-nucleotidase, and adenosine was separated from the substrate by a cation-exchange resin. The recovery of adenosine was $92 \pm 4\%$ (mean $\pm$ S.D., $n = 8$), calculated from the added [³H]cAMP, and the blanks containing no enzyme protein amounted to less than 5% of the radioactivity added. The conversion of radioactively labeled substrate to [³H] adenosine by the snake venom was quantitative. The enzyme preparations used were prepared either from GH₃ cell monolayer cultures for from GH₃ cell tumors induced in rats (Haug, 1979).

3.2. Properties of GH₃ Cell cAMP Phosphodiesterase(s)

GH₃ cell homogenates and 105,000g cytosol fractions from GH₃ tumors hydrolyzed cAMP and gave rise to nonlinear Lineweaver–Burke plots. The results were identical giving two sets of K_m and V_{max} values: K_m 28.6 and 26.3 μM with corresponding V_{max} of 1540 and 1304 pmoles/mg protein per min, and K_m 0.7 and 0.7 μM with V_{max} of 220 and 220 pmoles/ mg protein per min. These kinetic constants were in agreement with those described by Hinkle and Tashjian (1977) in the same cell system, and with results obtained studying cyclic nucleotide phosphodiesterases from normal rat anterior pituitary (Azhar and Menon, 1977). After homogenization, most of the enzyme activities were present in the soluble fraction, and the particulate and soluble form(s) of the enzyme showed similar K_m values (Table I).

Table I. Cyclic AMP Hydrolysis by GH$_3$ Cell and Tumor Homogenates and Subcellular Fractions[a]

	cAMP hydrolysis (pmoles/mg protein per min)	
Fraction	Low K_m (0.5 μM cAMP)	High K_m (50 μM cAMP)
Cell homogenate	85.0	1397
Nuclear	6.6	80
Mitochondrial	13.0	217
Microsomal	17.0	262
Supernatant (105,000g)	51.0	931
Tumor homogenate	96.3	1412
Supernatant (105,000g)	73.0	1110
Sediment	26.0	380

[a] The cells and tumors were homogenized in 0.32 M sucrose.

It is generally accepted that multiple forms of cyclic nucleotide phosphodiesterase activity exist in mammalian tissue (Wells and Hardman, 1977). When GH$_3$ cytosol was subjected to DEAE-cellulose chromatography, three peaks were detected using low substrate concentration. The third of these peaks, eluting at high salt concentration, was free from high-K_m activity. The major part of the high-K_m form of the enzyme coeluted with the cGMP phosphodiesterase activity and with the second peak of the low-K_m cAMP phosphodiesterase activity. This chromatographic distribution was similar to that described for the corresponding enzymes from normal rat anterior pituitary (Azhar and Menon, 1977).

During conditions of low [Ca^{2+}], the high- and low-K_m enzyme activities were reduced by about 50%, while low [Mg^{2+}] had no effect on the enzyme activity. When using an enzyme preparation "free" of activator and Ca^{2+}, the high- and low-K_m activities were inhibited by about 70 and 90%, respectively. When only the activator was added, a stimulation of the cAMP phosphodiesterase activity occurred, and this effect appeared to be partly independent of Ca^{2+}.

3.3 Effect of Theophylline and TRH on cAMP Phosphodiesterase Activity

Enzyme activities characterized *in vitro* may not relate directly to the *in vivo* situation. To examine if regulation of cAMP phosphodiesterase(s) occurred during TRH-stimulated PRL release, intact cells were incubated with TRH and/or theophylline during different experimental conditions.

Table II. The Effect of Theophylline *in Vivo* on PRL Release, cAMP Accumulation, and Phosphodiesterase Activity[a]

Additions	PRL release (dpm/mg protein)	cAMP (pmoles/mg protein)	cAMP phosphodiesterase (pmoles/mg protein per min)	
			Low K_m	High K_m
None	2.3 ± 0.2	18.1 ± 0.1	222 ± 19	1590 ± 56
TRH (3 μM)	3.9 ± 0.3[b] (p < 0.01)	30.2 ± 0.02 ($p < 0.01$)	230 ± 20 (N.S.)	2666 ± 150 (p <0.01)
Theophylline 0.1 nM	2.2 ± 0.1 (N.S.)	20.1 ± 0.2 (N.S.)	190 ± 21 (N.S.)	1501 ± 61 (N.S.)
1.0 mM	3.3 ± 0.02 ($p < 0.01$)	29.5 ± 0.5 ($p < 0.01$)	152 ± 31 ($p < 0.01$)	1200 ± 42 ($p < 0.01$)
TRH (3 μM) + theophylline (0.1 mM)	4.1 ± 0.4 ($p < 0.01$)	34.6 ± 0.5 ($p < 0.01$)		
THR (3 μM) + theophylline (1.0 mM)	4.4 ± 0.3 ($p < 0.01$)	43.0 ± 0.7 ($p < 0.01$)		

[a] GH$_3$ cells were preincubated with 20 μCi/ml [^{3}H] leucine with and without theophylline for 60 min, then incubated for a 30-min period in medium without [^{3}H] leucine and without or with TRH and theophylline added. Extracellular [^{3}H]-PRL was immunoprecipitated and quantitated (Gautvik and Kirz, 1976).

[b] Mean ± S.E.M. of three experiments, $n = 6$.

Table II shows that theophylline enhanced PRL release by the GH$_3$ cells, and that the cells have increased cAMP concentrations and reduced cyclic nucleotide phosphodiesterase activity. It seems reasonable to conclude that the effect of theophylline on PRL release is secondary to the cAMP elevation brought about by reduced hydrolysis of the cyclic nucleotide. The effects of TRH on cAMP accumulation and activation of cAMP-dependent protein kinase were transient, and showed a similar time course (Gautvik et al., 1977). Figure 9 shows that the stimulation of cAMP hydrolysis after TRH treatment was also biphasic and peak activity occurred at 30 min and somewhat later than the rise in cAMP, which was maximal from 2 to 15

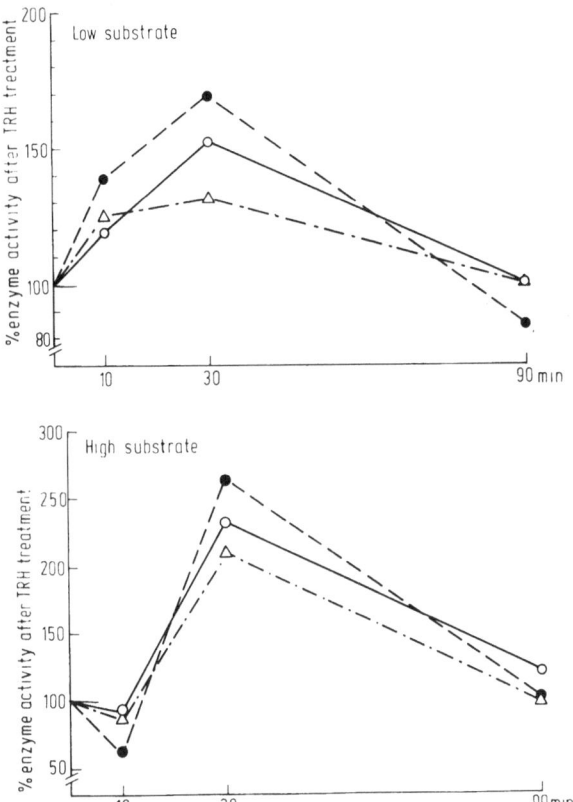

Figure 9. The *in vivo* time course effect of TRH on the low (top) and on the high (bottom) K_m cAMP phosphodiesterase activity. The cells were treated with three different concentrations of the tripeptide at 37°C, and the enzyme activity measured in cell homogenate. Each point represents the mean of triplicate measurements, and three different experiments were performed. The control activities were set equal to 100% and the TRH-treated groups calculated accordingly. ○, 3 nM; ●, 30 nM; △, 300 nM. [From Gautvik *et al.*, 1981.]

min after the addition of TRH (Gautvik *et al.*, 1977). The stimulatory effect of TRH on cAMP hydrolysis was dose dependent, but the highest concentrations of the tripeptide had less effects. Hinkle and Tashjian (1977) previously found that TRH (1 μM) had a small stimulatory effect on the high-K_m phosphodiesterase activity, while the same concentration showed no or a small inhibition of the low-K_m phosphodiesterase activity. Other hormones have also been reported to change the cyclic nucleotide phosphodiesterase activity in intact cells (Means *et al.*, 1978).

The reduced stimulatory effect by high concentrations of TRH was more pronounced for the low-K_m enzyme. A possible explanation is an inhibition of the enzyme(s) occurring at high substrate (cAMP) concentrations. Since cAMP concentrations in the GH cells increased after TRH treatment in the presence of an unchanged or increased activity of the cAMP phosphodiesterase(s), a stimulated formation of the cyclic nucleotide was probable. This notion was substantiated by the finding of a TRH-stimulated adenylate cyclase in broken GH$_3$ cell preparations (Gautvik *et al.*, 1981).

4.   Studies on the Role of Calcium in the Control of PRL Secretion

Calcium ions play a fundamental role in stimulus–secretion coupling in a variety of secretory systems from nerve cells to endocrine and exocrine cells (Douglas, 1968, 1978). It is generally believed that the appropriate stimulation causes influx of Ca^{2+} into the cell, leading to events that initiate exocytosis of the secretory granules. Recent electrophysiological experiments have shown that anterior pituitary cells are electrically excitable, and the generation of action potentials is at least partly dependent on Ca^{2+} (Kidokoro, 1975; Taraskevich and Douglas, 1977; Ozawa and Sand, 1978). Extracellular Ca^{2+} is a critical requirement for the secretion of many anterior pituitary hormones (Geschwind, 1971), and a causal relationship between the action potential-dependent Ca^{2+} uptake and hormone secretion has been suggested (Kidokoro, 1975). We have therefore examined the role of Ca^{2+} in the basal and stimulated release of PRL from GH$_3$ cells and correlated these studies with intracellular recordings from the same cell line.

4.1.   Ionic Requirements for Basal and Stimulated PRL Release

4.1.1.   Methods

PRL release was measured as the amount of hormone that accumulated in the culture medium during 30 min. Previous studies have shown that when radiolabeled amino acids are added to the culture medium, newly synthesized radioactive PRL will appear in the medium 30 min later (Gautvik and Kriz, 1976). Extracellular PRL was measured by radioimmunoassay (Haug and Gautvik, 1976). At the beginning of an experiment the cells were washed to remove extracellular PRL. Then the cells were resuspended in serumless medium and incubated in the presence or absence of hormone. After 30 min incubation PRL release was stopped by transferring the tubes to ice water and centrifuging (3000g) at 0°C for 5 min.

4.1.2. Results

Gautvik and Tashjian (1973a) have previously demonstrated the significance of extracellular Ca^{2+} for basal and stimulated PRL release by GH$_3$ cells. Figure 10 presents the effects on basal and TRH-stimulated PRL release of the Ca^{2+} ionophore A23187 and the Ca^{2+}-chelating substance NaEGTA. The Ca^{2+} concentration in the incubation medium was 0.4 mM. When NaEGTA was used, the cells were preincubated for 1 hr with 4 mM NaEGTA, washed, and suspended in serumless medium containing 4 mM NaEGTA before the addition of TRH and A23187 for 30 min. Removal of extracellular Ca^{2+} decreased basal PRL release to about 60% of controls. A23187 stimulated PRL release significantly in the presence of extracellular Ca^{2+}, but the effect was somewhat less than for TRH or the two combined. On the other hand, NaEGTA treatment inhibited the stimulatory effects of both A23187 and TRH. Replacement of NaEGTA with equimolar concentrations of CaEGTA did not affect basal PRL secretion, nor did it interfere with the stimulatory effects of TRH and A23187, showing that EGTA was not toxic to the GH$_3$ cells (data not shown).

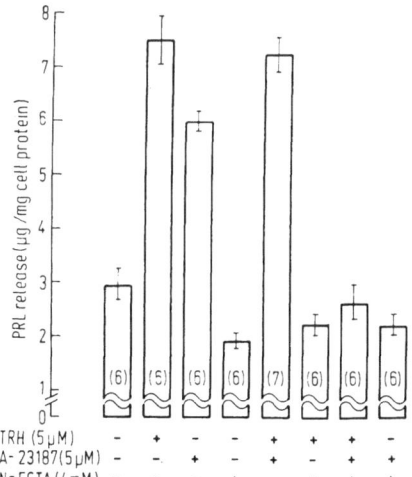

Figure 10. The effects of TRH, NaEGTA, and A23187 on PRL release from GH$_3$ cells. The incubations were carried out in 300 μl buffer (150 mM NaCl, 10 mM Tris–HCl, pH 7.5) to which was added 100 μl of cells suspended in Neumann–Tytell serumless medium. The final concentration of Ca^{2+} was 0.36 mM. When NaEGTA was used, the cells were preincubated with 4 mM NaEGTA (1 hr), centrifuged, and resuspended in Neumann–Tytell serumless medium containing 4 mM NaEGTA before addition of TRH or A23187 for 30 min. Mean values (± S.E.M.) are shown, and number of tubes are given in parentheses. [From Gautvik *et al.*, 1980.]

Figure 11 shows the effects of Ca²⁺ readdition on PRL release. The cell cultures were preincubated for 1 hr with 4 mM NaEGTA to complex free Ca^{2+} and Ca^{2+} loosely attached to the cell surface. After washing, the cells were resuspended in buffer containing $CaCl_2$, $MgCl_2$ or TRH as indicated, and PRL and cAMP levels were measured after incubation periods of 30 and 15 min, respectively. TRH was unable to provoke PRL release in the absence of Ca^{2+}, but it still increased cAMP concentrations. The addition of 4 mM $CaCl_2$ on the other hand stimulated PRL release without having any effect on cAMP production. Mg^{2+} could be substitute for Ca^{2+} and the presence or absence of 4 mM $MgCl_2$ had no effect on either basal and TRH-stimulated PRL release or on basal cAMP formation.

Our results show that TRH-induced PRL release is specifically and critically dependent on the presence of extracellular Ca^{2+}. When Ca^{2+} is complexed to EGTA, or when Ca^{2+} is replaced with Mg^{2+}, the basal release of PRL is partially inhibited and the TRH-stimulated hormone release is completely abolished. Increasing extracellular K^+ concentration to 50 mM caused a rapid release of PRL from the GH_3 cells (Table III). This response requires Ca^{2+} since it was inhibited by the Ca^{2+} antagonist Co^{2+} as well as by EGTA (Gautvik and Tashjian, 1973b; Ostlund *et al.*, 1978; Tashjian *et*

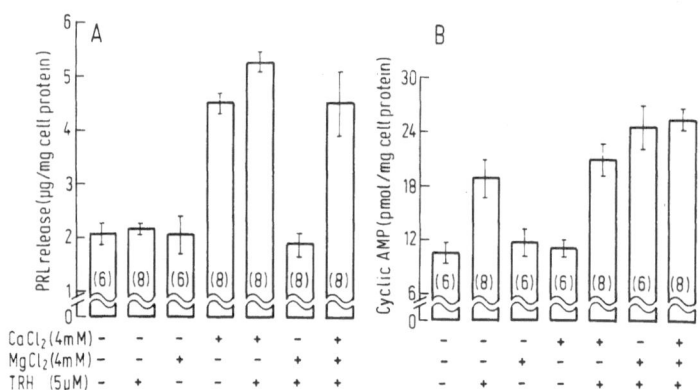

Figure 11. The effects of calcium and magnesium readdition on PRL release (A) and cAMP production (B) by GH₃ cells. The cells were preincubated in buffer (150 mM NaCl, 10 mM Tris–HCl, pH 7.5) containing 4 mM NaEGTA for 1 hr. After washing with buffer without NaEGTA the cells were aliquoted into tubes containing CaCl₂, MgCl₂, and/or TRH in the final concentrations indicated. These procedures were carried out at 0°C, and the experiments were started by transferring the tubes to a water bath at 37°C. After incubation for 30 min the reaction was terminated by rapid centrifugation, and PRL was measured in the culture medium. cAMP was measured in parallel tubes. The reaction was stopped after 15 min by addition of 100 μl 30% TCA, and cAMP was measured in the neutralized supernatant. Mean values ($\pm$ S.E.M.) are shown, and number of tubes are given in parentheses. Control tubes containing cells not pretreated with NaEGTA had basal and TRH-stimulated PRL values (μg/mg protein) of 3.7 $\pm$ 0.3 (8) and 7.3 $\pm$ 0.3 (8), respectively. [From Gautvik *et al.*, 1980.]

Table III. Acute PRL Release[a]

Treatment	Incubation time		
	0–5 min	5–15 min	15–30 min
TRH (1 μM)	272 $\pm$ 10	272 $\pm$ 12	226 $\pm$ 16
KCl (50 mM)	352 $\pm$ 48	298 $\pm$ 18	164 $\pm$ 14
SRIF[b]	92 $\pm$ 8	90 $\pm$ 10	74 $\pm$ 8
TRH + SRIF	280 $\pm$ 14	132 $\pm$ 16	112 $\pm$ 12

[a] Results expressed as percentage of the amount of PRL released from untreated controls during the different periods of incubation (mean $\pm$ S.D., $n = 4$).
[b] SRIF, somatostatin.

al., 1978). Ca^{2+} therefore seems to be a final common pathway of stimulation of PRL release for both TRH and K^+.

It is likely that elimination of external Ca^{2+} has several severe effects on the cells. The 35% reduction of the basal PRL release measured under these circumstances (Fig. 10) therefore does not permit a conclusion about the importance of external Ca^{2+} for basal PRL secretion. However, Tashjian *et al.* (1978) have shown that in Ca^{2+}-containing medium the competitive calcium antagonist Co^{2+} reduced basal PRL release to 60% of controls, indicating some involvement of Ca^{2+} in basal hormone release. The readdition of Ca^{2+} not only restored the TRH responsiveness, but Ca^{2+} alone had a stimulatory effect on PRL secretion (Fig. 11). This could be due to the fact that GH_3 cells spontaneously synthesize and secrete PRL. Termination of a reversible inhibition of the release may thus lead to a rebound effect.

TRH stimulated cAMP formation in the absence of extracellular Ca^{2+} (Fig. 11). The cAMP response was, however, blunted since the increment was less than 50% of that obtained under control conditions. These results also show that it is possible to dissociate the TRH-stimulated PRL release from the TRH-stimulated cAMP formation, since the former process is the most sensitive to reduction in the concentration of extracellular Ca^{2+}.

4.2. Studies on Cellular Ca^{2+} Transport

The activation of GH_3 cells by TRH may be associated with changes in transmembrane fluxes of different cations. We have therefore studied the efflux of $^{45}Ca^{2+}$ and the influx of $^{86}Rb^+$ as representative for K^+ transport.

4.2.1. Methods

$^{45}Ca^{2+}$ efflux studies were performed using suspension cultures ($\sim 2 \times 10^6$ cells) preincubated at 37°C for 1 hr in the presence of 4 μCi $^{45}Ca^{2+}$ (700 Ci/mole) in the following saline (mM): NaCl 140, KCl 5,

MgCl$_2$1.2, glucose 11, buffered by Hepes to pH 7.35. After washing, the cells were resuspended in the same saline supplemented with 1 mM CaCl$_2$, and apportioned into tubes with and without hormone. At the end of the incubation period, aliquots (2 $\times$ 100 μl) were taken from each tube, and the cells rapidly separated from the supernatant by centrifugation (10,000g, 30 sec) through silicone oil. The radioactivity of the supernatant fluid was determined by scintillation counting.

4.2.2. Results

The results presented above (Section 4.1) suggest that Ca^{2+} is critically important for TRH-stimulated PRL release. We were, however, unable to measure a TRH-stimulated uptake of ^{45}Ca^{2+} by the GH$_3$ cells, probably because the high extracellular ^{45}Ca^{2+} concentrations disturb the registrations of small intracellular changes. Studies of ^{45}Ca^{2+} efflux are not hampered by these difficulties, and Fig. 12 shows that TRH stimulated ^{45}Ca^{2+} efflux from preloaded GH$_3$ cells in a time- and dose-dependent manner. The effect of TRH was rapid since the efflux of ^{45}Ca^{2+} was maximal already after 2 min. At this time cAMP and PRL release were also stimulated (see previous section).

Figure 13 shows the effects of the hypothalamic tetradecapeptide somatostatin (SRIF) on ^{45}Ca^{2+} efflux and PRL release. In the ^{45}Ca^{2+} efflux studies the SRIF and TRH treatment period was 5 min and in the PRL

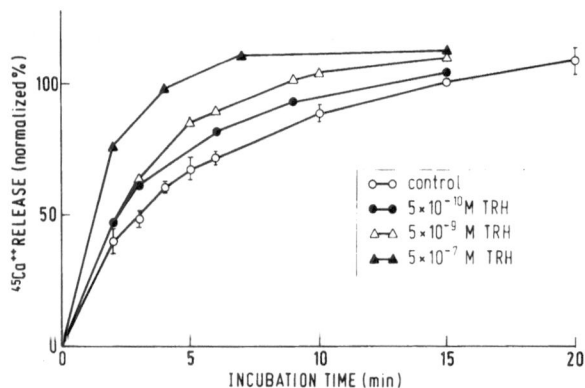

Figure 12. Effect of TRH on ^{45}Ca^{2+} efflux from GH$_3$ cells. GH$_3$ cells were incubated at 37° C in buffer (NaCl 140 mM, KCl 5 mM, MgCl$_2$ 1.2 mM, glucose 11 mM, Hepes 3 mM, pH 7.35) supplemented with ^{45}Ca^{2+} (S.A. 700 Ci/mol). After 1 hr the cells were washed three times and resuspended in buffer and apportioned to replicate tubes. TRH was added to give the concentrations indicated, and ^{45}Ca^{2+} efflux was measured at times ranging from 2 min to 20 min. The values are expressed as percentage of the ^{45}Ca^{2+} efflux measured in untreated controls after 15 min incubation (100%). Mean values ($\pm$ S.E.M.) of quadruplicate determinations are shown.

Figure 13. Effects of TRH and SRIF on $^{45}Ca^{2+}$ efflux and PRL release from GH₃ cells. The efflux of $^{45}Ca^{2+}$ from GH₃ cells was measured after 5 min treatment (see legend to Fig. 12), and PRL release was measured after 30 min treatment of the GH₃ cells with TRH and/or SRIF at the concentrations indicated. All values are expressed as percentage of the values obtained in untreated controls (100%). Mean values ($\pm$ S.E.M.) are shown, and number of tubes are given.

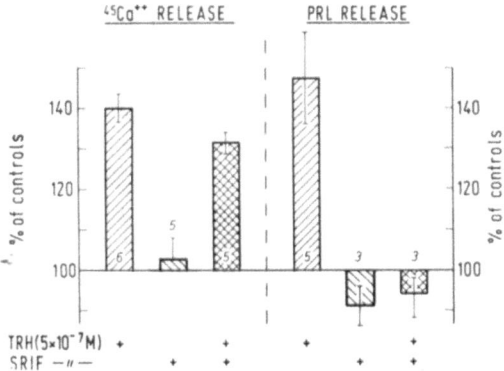

release experiments 15 min. SRIF has previously been shown to inhibit GH synthesis and secretion by different GH cell strains, and functional receptors for SRIF have been demonstrated in these cells (Schonbrunn and Tashjian, 1978; Aulie *et al.*, 1979). Table III shows that SRIF in equimolar concentrations almost completely inhibited TRH-stimulated PRL release after a lag period of 5 min without having any significant effect on TRH-induced $^{45}Ca^{2+}$ efflux (Fig. 13).

The $^{45}Ca^{2+}$ efflux studies indicate that TRH treatment increases the entry of extracellular Ca^{2+} into the cell. However, increased rate of $^{45}Ca^{2+}$ efflux induced by TRH may also reflect an increased release from intracellular Ca^{2+} stores. It is known that cAMP promotes $^{45}Ca^{2+}$ efflux from isolated mitochondria (Borle, 1974; Debnam and Snart, 1976). Since SRIF inhibited TRH-induced PRL release without having any significant effect on the $^{45}Ca^{2+}$ efflux evoked by TRH, SRIF probably inhibits PRL release at a step after Ca^{2+} uptake. SRIF alone has little effect on PRL release (Table I) and no effect on PRL synthesis (Aulie *et al.*, 1979). Recently, Schofield and Bicknell (1978) found that SRIF also inhibited K^+-induced GH release from isolated pituitary cells without modifying K^+-induced $^{45}Ca^{2+}$ efflux. Taken together, these observations suggest that SRIF interferes with the action of Ca^{2+} at a step after Ca^{2+} uptake.

We have also shown that TRH in concentrations known to give maximal PRL release had no effect on the cellular uptake of $^{86}Rb^+$, thus suggesting that the transmembrane transport of K^+ is not influenced by TRH (data not shown).

4.3. Effects of TRH and 4-Aminopyridine on Action Potentials and PRL Release

Several studies have shown that both normal anterior pituitary cells and GH₃ cells are capable of generating action potentials that are at least partially dependent on Ca^{2+} (Kidokoro, 1975; Biales et al., 1977; Ozawa and Sand, 1978; Ozawa and Miyazaki, 1979). Recent studies have demonstrated that TRH increases the frequency of the action potentials of GH_3 cells (Kidokoro, 1975; Dufy et al., 1979; Ozawa and Kimura, 1979; Gautvik et al., 1980; Taraskevich and Douglas, 1980). This increased firing rate of action potentials with a Ca^{2+} component will necessarily lead to an elevated Ca^{2+} influx, but it was still possible that TRH could induce Ca^{2+}-triggered PRL release independently of the action potentials (Gershengorn, 1980). However, this possibility was weakened by Ozawa and Kimura (1979), who showed that specific blockers of voltage-sensitive Ca^{2+} channels (verapamil and cobalt) suppress the action potentials of GH_3 cells and inhibit both TRH- and K^+-stimulated PRL release.

We have investigated the significance of the action potential activity for the TRH-induced PRL release from GH_3 cells by using the drug 4-aminopyridine (4AP) (Sand et al., 1980). 4AP has no effect on the resting membrane permeabilities, but inhibits the late K^+-current in a variety of cells (Llinás et al., 1976; Molgo et al., 1977; Ulbricht and Wagner, 1976). This drug will thus act selectively on the action potentials of excitable cells, and the effects of 4AP were therefore compared with those of TRH on both the electrical properties and the PRL secretion of GH_3 cells.

4.3.1. Methods

The recordings were obtained 3–10 days after subculture. Prior to an experiment the F-10 medium was removed and replaced with the following saline (mM): NaCl 140, KCl 5, $CaCl_2$ 10, $MgCl_2$ 1.3, glucose 10, buffered by Tris–HCl 5 to pH 7.4. Since high Ca^{2+} concentrations are known to facilitate stable penetrations (Kidokoro, 1975), the experiments regarding the effects of TRH and the initial recordings in 4AP solution were performed in saline containing 20 mM $CaCl_2$ (125 mM NaCl). TRH was administered through a micropipete with a tip diameter between 5 and 8 μ m. When a stable recording was achieved from a cell displaying electrical excitability, the tip of the pipete was positioned at a distance of 30–50 μm from the cell and TRH slowly pressure injected into the saline. The experiments involving 4AP were performed by comparing cells in saline with cells in saline containing 5×10^{-4} and 3×10^{-3} M 4AP.

The glass microelectrodes used for intracellular recording were filled from behind by capillary action with 3 M KCl or 5 M K-acetate adjusted to pH 7.2 with acetic acid. The electrode resistance ranged between 150 and 250 MΩ. The input resistance of the preamplifier was more than 10^{12} MΩ and the input offset current less than 10^{-12} A (Sand et al., 1980).

PRL synthesis was measured as the amount of hormone that accumulated in the culture medium during 48 hr. Culture medium concentrations of PRL were measured by radioimmunoassay (Haug and Gautvik, 1976). The intracellular stores of PRL are very small and there is no intracellular degradation of PRL (Haug *et al.*, 1977), and PRL is stable in the medium for as long as 48 hr under culture conditions (Tashjian *et al.*, 1970). Therefore, the extracellular amount of PRL represents more than 97% of the total amount of hormone synthesized by the cells during 48 hr (Haug *et al.*, 1977)

4.3.2   Results

Figure 14 demonstrates how TRH altered the electrical membrane properties of these cells. Record A shows an action potential evoked at the termination of a hyperpolarizing current pulse prior to TRH stimulation. Record B was obtained from the same cell just after positioning of the TRH pipet 40 μm from the cell. The retarded falling phase of the action potential at this stage might be due to leakage of TRH from the pipet. Record C demonstrates spontaneous action potentials in the formerly silent cell 20 sec after ejection of TRH from the pipet. In this particular cell the firing

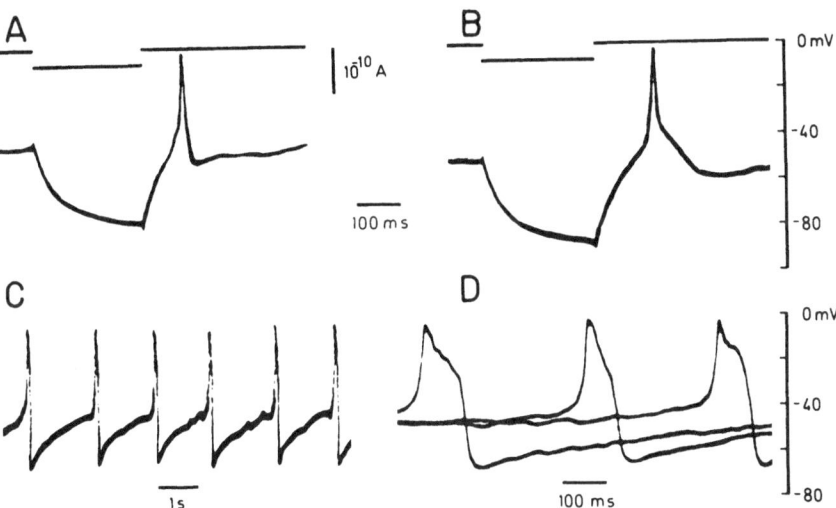

Figure 14. Electrophysiological effects of TRH. (A) Action potential induced in control solution by anodal break stimulation. (B) Corresponding recording immediately after positioning of the TRH pipet close to the cell. Note the retarded falling phase of the action potential. (C) Spontaneous activity recorded 20 sec after ejection of TRH. (D) Three superimposed sweeps showing the prolongation of the spontaneous action potentials relative to the electrically evoked action potentials in control solution. Same sweep speed as in A and B. All recordings are from the same cell. [From Sand *et al.*, 1980.]

rate was about 0.7 Hz. In record D three superimposed sweeps show prolongation of the spontaneous action potentials relative to the electrically
evoked action potentials in control solution (record A). The main conclusion regarding the electrophysiological effects of TRH on the GH_3 cells is
thus that this peptide increases the spontaneous firing rate and prolongs the
action potentials. Both these effects will increase the Ca^{2+} influx.

Figure 15 shows that 4AP also increased the spontaneous firing rate
and prolonged the action potentials of the GH_3 cells, thus mimicking the
effects of TRH. Action potentials induced by current injections were
recorded from 12 cells in 1–3 mM 4AP solutions. Record A shows that all-
or-none action potentials were followed by a pronounced after-hyperpolarization in control solution, but this was abolished in the presence of 4AP,
as seen in record B. Moreover, most of the cells in 4AP solution showed in
addition a clear prolongation of the repolarization phase of the action
potentials. Both these alterations of the action potentials are consistent with
an inhibitory effect of 4AP on the late K^+-current, which is well developed
in GH_3 cells (Ozawa et al., 1979). Eleven cells were successfully penetrated
in 0.5 mM 4AP solution containing 10 mM Ca^{2+} and four displayed spontaneous action potentials with a firing rate of 0.3–1 Hz as recorded in C.
The shape of one of the action potentials is shown at higher sweep speed in
record D. A pronounced shoulder is evident on the repolarizing phase of

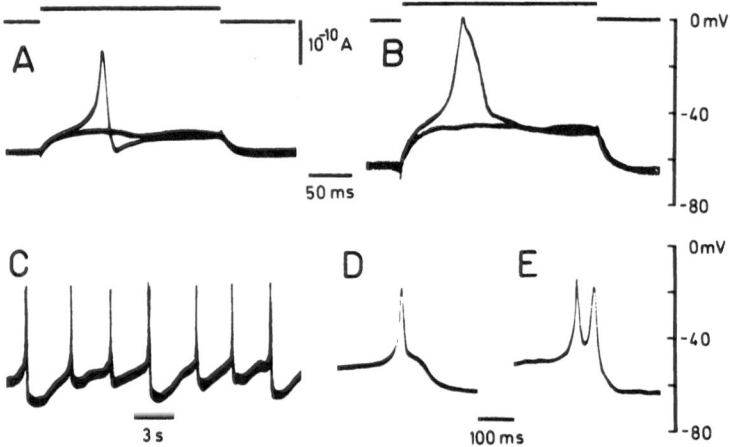

Figure 15. Electrophysiological effects of 4-aminopyridine (4AP). (A) All-or-none action potential induced by depolarizing current in control solution. (B) Corresponding recording from a
different cell in solution containing 1.5×10^{-3} M 4AP. Note the retarded falling phase and
absence of after-hyperpolarization. (C) Spontaneous action potentials in 5×10^{-4} M 4AP solution. Recording from the same cell at higher sweep speed showed single action potentials with a
pronounced shoulder on the repolarizing phase (D) or action potentials with double peaks (E).
[From Sand et al., 1980.]

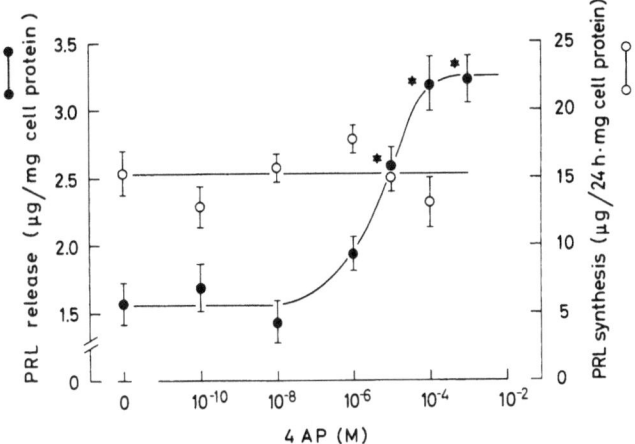

Figure 16. Effects of 4AP on PRL release and synthesis. The treatment periods were 30 min for the release and 2 days for the synthesis experiments. Data are given as mean values ($\pm$ S.E.M.) of quadruplicate determinations. Values significantly different from controls ($p < 0.05$) are marked with asterisks. TRH (10^{-6} M) caused a 1.9-fold increase in PRL release and synthesis (data not shown). [From Sand *et al.*, 1980.]

the action potentials, and a second action potential was occasionally initiated from this retarded falling phase. Recordings were also obtained from nine electrically excitable cells in control solution containing 10 mM Ca^{2+}. None of the cells displayed spontaneous firing, and shoulders and double peaks were never seen in the electrically induced action potentials.

The close similarity between the electrophysiological effects on GH$_3$ cells caused by TRH and 4AP does not necessarily reflect a common mode of action of the two substances. However, if TRH stimulates PRL release due to its effect on the action potentials, then 4AP should have a similar effect on the release. If TRH causes release mainly through different mechanisms, 4AP would not be expected to stimulate hormone release significantly. Figure 16 shows that 4AP stimulated PRL release in a dose-dependent manner, and the effect was maximal at 10^{-4} M. The maximal effect of 4AP on PRL release was a 2.0-fold stimulation. In the same experiment the maximal effect of TRH (10^{-6} M) was a 1.9-fold stimulation, and a significant stimulatory effect on PRL release was found already after 5 min for both TRH and 4AP. 4AP is thus able to duplicate the effect of TRH on PRL release, but in contrast to TRH, which also stimulates PRL synthesis, 4AP had no effect on hormone synthesis (Fig. 16).

These observations support the idea that TRH enhances PRL release from GH$_3$ cells via its facilitating effects on Ca^{2+}-dependent action potentials. Moreover, since PRL secretion was stimulated by 4AP without a concomitant increase in synthesis, the latter process is not necessarily a consequence of hormone release. Therefore, TRH regulation of PRL secretion and synthesis is probably exerted through independent mechanisms.

5. Conclusions

The interaction of [³H]-TRH with intact GH₃ cells does not follow a simple bimolecular reaction pattern. The dissociation of bound [³H]-TRH did not follow a monoexponential time course characteristic for one homogeneous type of binding sites, and Scatchard (1949) analysis of binding data obtained at or close to steady state resulted in a downward curvilinear graph. Assuming two different classes of binding sites, the apparent K_d's were 0.9 and 23 nM, respectively. Since the dissociation of bound [³H]-TRH in the absence or presence of TRH did not differ, the heterogeneity in binding sites could not be explained by the phenomenon of negative cooperativity (DeMeyts et al., 1973, 1976). Internalization of bound [³H]-TRH probably does not explain the heterogeneity, since bound [³H]-TRH dissociated rapidly from GH₃ cells after from 5 to 60 min of association, and there was no degradation of [³H]-TRH in the medium or at the receptor sites during the experiments (Section 2.3).

When the initial rates of [³H]-TRH binding and cAMP production were compared, it was found that binding as well as cAMP formation started without any significant time lag (< 10 sec). The maximal rates of both processes occurred within the first minute after addition of TRH. During the first minute of association there is virtually no contribution to [³H]-TRH binding by the low-affinity sites. It seems possible, therefore, that the initial stimulation of cAMP formation is mediated mainly through binding to the high-affinity sites (Section 2.3).

In many secretory cells extracellular Ca^{2+} is pivotal for the normal release of secretory products to occur. In GH₃ cells both basal PRL secretion and TRH-stimulated PRL release are critically dependent on this cation, since low extracellular $[Ca^{2+}]$ or the presence of specific blockers of Ca^{2+} channels inhibit TRH-stimulated PRL release (Gautvik and Tashjian, 1973a; Ozawa and Kimura, 1979). The molecular mechanism by which extracellular Ca^{2+} acts is unknown. It seems, however, that Ca^{2+} is involved in the control of cyclic nucleotide formation (Rasmussen et al., 1976). The TRH-stimulated efflux of $^{45}Ca^{2+}$, which probably occurs secondary to increased cellular Ca^{2+} influx, may be associated with the effect of the tripeptide on the action potentials of the GH₃ cells. TRH increases the frequency and probably also the duration of the Ca^{2+}-dependent action potentials. The observation that SRIF inhibited TRH-induced PRL release without having effect on TRH-stimulated $^{45}Ca^{2+}$ efflux suggests that cellular uptake of Ca^{2+} is not necessarily followed by hormone release (Sections 4.2 and 4.3).

REGULATION OF PRL SECRETION

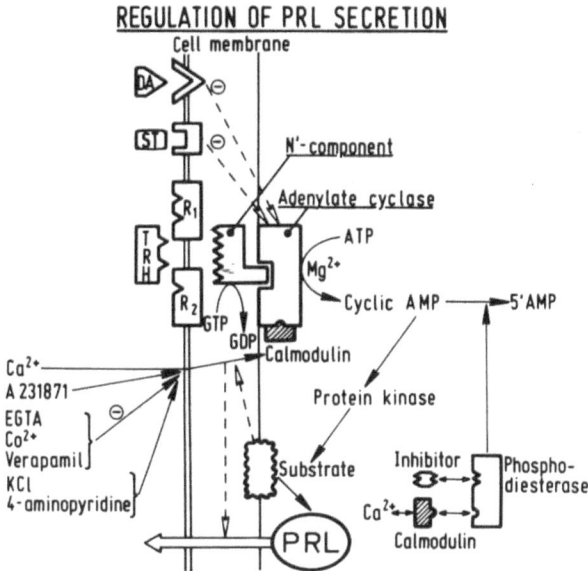

Figure 17. Schematic diagram illustrating the proposed sequence of events that couple TRH binding, dopamine (DA) binding, and somatostatin (ST) binding to PRL secretion. The details are given in the text.

Figure 17 represents our working hypothesis on the regulation of PRL secretion by TRH. The binding of TRH to surface receptors on the lactotropes initiates:

1. Activation of membrane-bound adenylate cyclase. This activation is dependent on the presence of guanyl nucleotides (N'-component) and Mg^{2+} (Gautvik *et al.*, 1981).
2. Depolarization of the cell occurs during which extracellular Ca^{2+} is taken up by the cell (Sand *et al.*, 1980; Gautvik *et al.*, 1980).
3. The cAMP formed activates a protein kinase that together with intracellular Ca^{2+} stimulates PRL release (Gautvik *et al.*, 1976).
4. A transient increase in cyclic nucleotide phosphodiesterase activity occurs due either to removal of an inhibitor or to the action of calmodulin. This transient, TRH-induced rise in phosphodiesterase activity may play a physiological role by terminating the biological effect(s) of intracellular cAMP (Gautvik *et al.*, 1981).

The following questions remain unanswered.

1. Which receptor entity is responsible for activation of adenylate cyclase?
2. What are the characteristics of the isolated TRH receptor, the N'-component, and the adenylate cyclase?

3. What is (are) the endogenous substrate(s) for cAMP-dependent protein kinase?
4. How are cAMP phosphodiesterase, inhibitor, and calmodulin interacting?

ACKNOWLEDGMENTS. This chapter includes previously published and unpublished results of work supported by grants from Norsk Forening til Kreftens Bekjempelse, Norway, from Nordisk Insulin Fond, Gentofte, Denmark, from the Anders Jahre Foundation for the Promotion of Science, Norway, and from Landsforeningen mot Kreft, Norway. PRL for iodination (rPRL-I-1) and standards (rPRL-RP-1)was donated by the National Institute of Arthritis, Metabolic and Digestive Diseases. The authors also thank Mrs. Aasa Stokland and Mrs. Grete Steensland for skilled technical help and Miss Kjersti Gunneng for typing the manuscript.

REFERENCES

Aulie, A., Haug, E., and Gautvik, K. M., 1979, Effects of somatostatin on growth hormone and prolactin secreting rat pituitary tumour cells in culture, *Acta Endocrinol. (Copenhagen) Suppl.* **225**:295.

Azhar, S., and Menon, K. M. J., 1977, Cyclic nucleotide phosphodiesterases from rat anterior pituitary: Characterization of multiple forms and regulation by protein activator and Ca^+, *Eur. J. Biochem.* **73**:73.

Biales, B., Dichter, M. A., and Tischler, A., 1977, Sodium and calcium action potential in pituitary cells, *Nature (London)* **267**:172.

Borle, A. B., 1974, Cyclic AMP stimulation of calcium efflux from kidney, liver, and heart mitochondria, *J. Membr. Biol.* **16**:231.

Boudreau, R. J., and Drummond, G. I., 1975, A modified assay of 3'5'-cyclic-AMP phosphodiesterase, *Anal. Biochem.* **63**:388.

Clausen, O. P., Gautvik, K. M., and Lindmo, T., 1977, Cell cycle distributions, growth characteristics, and variations in prolactin and growth hormone production in cultured rat pituitary cells, *Virchows Arch. B* **23**:195.

Dannies, P. S., Gautvik, K. M., and Tashjian, A. H., Jr., 1976, A possible role of cyclic AMP in mediating the effects of thyrotropin-releasing hormone on prolactin release and on prolactin and growth hormone synthesis in pituitary cells in culture, *Endocrinology* **98**:1147.

Debnam, E. S., and Snart, R. S., 1976, Effect of cyclic AMP on calcium uptake by rat kidney mitochondria, *J. Endocrinol.* **71**:67P.

De Meyts, P., Roth, J., Neville, D. M., Jr., Gavin, J. R., III, and Lesniak, M. A., 1973, Insulin interactions with its receptors: Experimental evidence for negative cooperativity, *Biochem. Biophys. Res. Commun.* **55**:154.

De Meyts, P., Bianco, A. R., and Roth, J., 1976, Site–site interactions among insulin receptors: Characterization of the negative cooperativity, *J. Biol. Chem.* **251**:1877.

Douglas, W. W., 1968, Stimulus–secretion coupling: The concept and clues from chromaffin and other cells, *Br. J. Pharmacol.* **34**:451.

Douglas, W. W., 1978, Stimulus–secretion coupling: Variations on the theme of calcium-activated exocytosis involving cellular and extracellular sources of calcium, *Ciba Found. Symp.* **54**:61.

Dufy, B., Vincent, J.-D., Fleury, H., du Pasquier, P., Gourdji, D., and Tixier-Vidal, A., 1979, Membrane effects of thyrotropin-releasing hormone and estrogen shown by intracellular recording from pituitary cells, *Science* **204**:509.

Dufy, B., Fleury, H., Gourdji, D., Tixier-Vidal, A., du Pasquier, P., and Vincent, J.-D., 1980, Intracellular recordings from prolactin-secreting pituitary cells in culture: Evidence for a direct action of estrogen on the cell membrane, in: *Synthesis and Release of Adenohypophyseal Hormones* (M. Jutisz and K. W. McKerns, eds.), pp. 765–773, Plenum Press, New York.

Faivre-Bauman, A., Gourdji, D., Grouselle, D., and Tixier-Vidal, A., 1975, Binding of thyrotropin releasing hormone and prolactin release by synchronized GH₃ rat pituitary cell line, *Biochem. Biophys. Res. Commun.* **67**:50.

Fossum, S., and Gautvik, K. M., 1977, Sterological and biochemical analysis of prolactin and growth hormone producing rat pituitary cells in culture: Sterology combined with nonparametrical statistics, *Cell Tissue Res.* **184**:169.

Gautvik, K. M., and Kinz, M., 1976, Measurements of prolactin synthesis and secretion by rat pituitary cells in culture, *Endocrinology* **98**:344.

Gautvik, K. M., and Lystad, E., 1981, Demonstration of a heterogenous population of binding sites for thyroliberin in prolactin producing tumour cells and their possible functional significance, *Eur. J. Biochem.*, **116**:235.

Gautvik, K. M., and Tashjian, A. H., Jr., 1973a, Effects of Ca^{++} and Mg^{++} on secretion and synthesis of growth hormone and prolactin by clonal strains of pituitary cells in culture, *Endocrinology* **92**:573.

Gautvik, K. M., and Tashjian, A. H., Jr., 1973b, Effects of cations and colchicine on the release of prolactin and growth hormone by functional pituitary tumor cells in culture, *Endocrinology* **93**:793.

Gautvik, K. M., Hoyt, R. F., Jr., and Tashjian, A. H., Jr. 1973, Effects of colchicine and 2-Br-α-ergocryptine-methane-sulfonate (CV 154) on the release of prolactin and growth hormone by functional pituitary tumor cells in culture, *J. Cell. Physiol.* **82**:401.

Gautvik, K. M., Iversen, J.-G., and Sand, O., 1980, On the role of extracellular Ca^{2+} for prolactin release and adenosine 3′5′-monophosphate formation induced by thyroliberin in cultured rat pituitary cells, *Life Sci.* **26**:995.

Gautvik, K. M., Kriz, M., Jahnsen, T., Haug, E., and Hansson, V., 1982, Relationship between stimulated prolactin release from GH cells and cyclic AMP degradation and formation, *Mol. Cell Endocrinol.*, in press.

Gautvik, K. M., Walaas, E., and Walaas, O., 1977, Effect of thyroliberin on the concentration of adenosine 3′5′-phosphate and on the activity of adenosine 3′5′-phosphate-dependent protein kinase in prolactin-producing cells in culture, *Biochem. J.* **162**:379.

Gershengorn, M. C., 1980, Thyrotropin releasing hormone stimulation of prolactin release, *J. Biol. Chem.* **255**:1801.

Geschwind, I., 1971, Mechanism of release of anterior pituitary hormones: Studies *in vitro*, *Mem. Soc. Endocrinol.* **19**:221.

Gourdji, D., 1980, Characterization of thyroliberin (TRH) binding sites and coupling with prolactin and growth hormone secretion in rat pituitary cell lines, in: *Synthesis and Release of Adenohypophyseal Hormones* (M. Jutisz and K. W. McKerns, eds.), pp. 463–491, Plenum Press, New York.

Grant, G., Vale, W., and Guillemin, R., 1973, Characteristics of the pituitary binding sites for thyrotropin-releasing factor, *Endocrinology* **92**:1629.

Haug, E., 1979, Progesterone suppression of estrogen-stimulated prolactin secretion and estrogen receptor levels in rat pituitary cells, *Endocrinology* **104**:429.

Haug, E., and Gautvik, K. M., 1976, Radioimmunoassay of rat prolactin and its use in measuring prolactin production by cultured pituitary cells, *Acta Endocrinol. (Copenhagen)* **82**:282.

Haug, E., and Gautvik, K. M., 1978, Effects of sex steroids on growth hormone production in cultured rat pituitary cells, *Acta Endocrinol. (Copenhagen)* **87**:40.

Haug, E., Tjernshaugen, H., and Gautvik, K. M., 1977, Variations in prolactin and growth hormone production during cellular growth in clonal strains of rat pituitary cells, *J. Cell. Physiol.* **91**:15.

Haug, E., Naess, O., and Gautvik, K. M., 1978, Receptors for 17β-estradiol in prolactin-secreting rat pituitary cells, *Mol. Cell. Endocrinol.* **12**:81.

Hinkle, P. M., and Tashjian, A. H., Jr., 1973, Receptors for thyrotropin-releasing hormone in prolactin-producing rat pituitary cells in culture, *J. Biol. Chem.* **248**:6180.

Hinkle, P. M., and Tashjian, A. H., Jr., 1974, Interaction of thyrotropin-releasing hormone with pituitary cells in culture, in: *Hormones and Cancer* (K. W. McKerns, ed.), pp. 203–227, Academic Press, New York.

Hinkle, P. M., and Tashjian, A. H., Jr., 1977, Adenylyl cyclase and cyclic nucleotide phosphodiesterases in GH-strains of rat pituitary cells, *Endocrinology* **100**:934.

Kidokoro, V., 1975, Spontaneous calcium action potentials in a clonal pituitary cell line and their relationship to prolactin secretion, *Nature (London)* **258**:741.

Llinás, R., Walton, K., and Bohr, V., 1976, Synaptic transmission in squid giant synapse after potassium conductance blockage with external 3- and 4-amino-pyridine, *Biophys. Chem.* **16**:83.

Means, A. R., Dedman, J. R., Tindall, D. J., and Welsh, M. J., 1978, Hormonal regulation of Sertoli cells, *Int. J. Androl. Suppl.* **2**:403.

Mester, J., Brunelle, R., Jung, I., and Sonnenschein, C., 1973, Estrogen-sensitive cells: Hormone receptors in tumours and cells in culture, *Exp. Cell Res.* **81**:447.

Molgo, J., Lemeignan, M., and Lechat, P., 1977, Effects of 4-aminopyridine at the frog neuromuscular junction, *J. Pharmacol. Exp. Ther.* **203**:653.

Naess, O., Haug, E., and Gautvik, K. M., 1980, Effects of glucocorticoids on prolactin and growth hormone production and characterization of the intracellular hormone receptors in rat pituitary tumour cells, *Acta Endocrinol. (Copenhagen)* **95**:319.

Ostlund, R. E., Jr., Leung, J. T., Hajek, S. V., Winokur, T., and Melman, M., 1978, Acute stimulated hormone release from cultured GH$_3$ pituitary cells, *Endocrinology* **103**:1245.

Ozawa, S., and Kimura, N., 1979, Membrane potential changes caused by thyrotropin-releasing hormone in the clonal GH$_3$ cell and their relationship to secretion of pituitary hormone, *Proc. Natl. Acad. Sci. USA* **76**:6017.

Ozawa, S., and Miyazaki, S., 1979, Electrical excitability of the rat clonal pituitary cell and its relation to hormone secretion, *Jpn. J. Physiol.* **29**:411.

Ozawa, S., and Sand, O., 1978, Electrical activity of rat anterior pituitary cells *in vitro*, *Acta Physiol. Scand.* **102**:330.

Ozawa, S., Miyazaki, S., and Sand, O., 1979, Electrical activity of anterior pituitary cells and its functional implication, in: *Neurobiology of Chemical Transmission* (M. Ohtsuka and Z. W. Hall, eds.), pp. 253–265, Wiley, New York.

Rasmussen, H., Jensen, P., Lake, W., and Goodman, D. B. P., 1976, Calcium ion as second messenger, *Clin. Endocrinol* **5**(Suppl):11s.

Sand, O., Haug, E., and Gautvik, K. M., 1980, Effects of thyroliberin and 4-aminopyridine on action potentials and prolactin release and synthesis in rat pituitary cells in culture, *Acta Physiol. Scand.* **108**:247.

Scatchard, G., 1949, The attractions of proteins for small molecules and ions, *Ann. N.Y. Acad. Sci.* **51**:660.

Schofield, J. G., and Bicknell, R. J., 1978, Effects of somatostatin and verapamil on growth hormone release and ^{45}Ca fluxes, *Mol. Cell. Endocrinol.* **9**:255.

Schonbrunn, A., and Tashjian, A. H., Jr., 1978, Characterization of functional receptors for somatostatin in rat pituitary cells in culture, *J. Biol. Chem.* **253**:6473.

Sonnenschein, C., Richardson, U. I., and Tashjian, A. H., Jr., 1970, Chromosomal analysis, organ-specific function and appearance of six clonal strains of rat pituitary tumor cells, *Exp. Cell Res.* **61**:121.

Takemoto, H., Yokoro, K., Furth, J., and Cohen, A. I., 1962, Adrenotropic activity of mammosomatotrophic tumors in rats and mice. I. Biological aspects, *Cancer Res.* **22**:917.

Taraskevich, P. S., and Douglas, W. W., 1977, Action potentials occur in cells of the normal anterior pituitary gland and are stimulated by the hypophysiotropic peptide thyrotropin-releasing hormone, *Proc. Natl. Acad. Sci. USA* **74**:4064.

Taraskevich, P. S., and Douglas, W. W., 1980, Electrical behaviour in a cell line of anterior pituitary cells (GH cells) and the influence of the hypothalamic peptide, thyrotrophin releasing factor, *Neuroscience* **5**:421.

Tashjian, A. H., Jr., and Hoyt, R. F., Jr., 1972, Transient controls of organ-specific functions in pituitary cells in culture, in: *Molecular and General Developmental Biology* (M. Sussman, ed.), pp. 353–387, Prentice–Hall, Englewood Cliffs, N.J.

Tashjian, A. H., Bancroft, F. C., and Levine, L., 1970, Production of prolactin and growth hormone by clonal strains of rat pituitary tumor cells, *J. Cell. Biol.* **47**:61.

Tashjian, A. H., Jr., Yasumura, Y., Levine, L., Sato, G. H., and Parker, M. L., 1968, Establishment of clonal strains of rat pituitary tumor cells that secrete growth hormone, *Endocrinology* **82**:342.

Tashjian, A. H., Jr., Barowsky, N. J., and Jensen, D. K., 1971, Thyrotropin releasing hormone: Direct evidence for stimulation of prolactin production by rat pituitary cells in culture, *Biochem. Biophys. Res. Commun.* **43**:516.

Tashjian, A. H., Jr., Lomedico, M. E., and Maina, D., 1978, Role of calcium in the thyrotropin-releasing hormone-stimulated release of prolactin from pituitary cells in culture, *Biochem. Biophys. Res. Commun.* **81**:798.

Thompson, W. J., and Appleman, M. M., 1971, Multiple cyclic nucleotide phosphodiesterase activities from rat brain, *Biochemistry* **10**:311.

Tixier-Vidal, A., 1975, Ultrastructure of anterior pituitary cells in culture, in: *The Anterior Pituitary* (A. Tixier-Vidal and M. G. Farquhar, eds.), pp. 181–229, Academic Press, New York.

Ulbricht, W., and Wagner, H. H., 1976, Block of potassium channels of the nodal membrane by 4-aminopyridine and its partial removal on depolarization, *Pfluegers Arch.* **367**:77.

Varga, L., Luterbeck, P. M., Pryor, J. S., Wenner, R., and Erb, H., 1972, Suppression of puerperal lactation with an ergot alkaloid: A double-blind study. *Br. Med. J.* **2**:743.

Wells, J. N., and Hardman, J. G., 1977, Cyclic nucleotide phosphodiesterases, *Adv. Cyclic Nucleotide Res.* **8**:119.

26

Cell-Free Synthesis of Rat Lutropin Subunits

Raymond Counis, Maïthé Corbani, Geneviève Ribot, and Marian Jutisz

1. Introduction

Lutropin (LH) is a member of a family of glycoprotein hormones that includes follitropin (FSH), thyrotropin (TSH), and a placental hormone, human chorionic gonadotropin (hCG). Each one of these hormones is built up of two dissimilar subunits α and β, attached noncovalently in the native hormone. Within a given species, the α subunit is common to all glycoprotein hormones; only the β subunits, which confer to a hormone its biological and immunological specificity, differ (Giudice and Pierce, 1978).

Although considerable information is available concerning the structure of LH and the regulation of its release, little is known about the cellular mechanisms involved in the biosynthesis of LH, and contradictory results have been reported on the effect of gonadoliberin (GnRH) on these mechanisms (Khar and Jutisz, 1980a).

The role of GnRH in the biosynthesis of LH and FSH by rat anterior pituitary cells in culture has been investigated during the last few years in our laboratory (Khar et al., 1978; Khar and Jutisz, 1980a,b). Our results suggest that (1) GnRH stimulates the biosynthesis of the polypeptide chain(s) of gonadotropins, but does not stimulate their glycosylation; (2) GnRH may act at the transcriptional level since actinomycin D inhibited

Raymond Counis, Maïthé Corbani, Geneviève Ribot, and Marian Jutisz • Laboratoire des Hormones Polypeptidiques, CNRS, 91190 Gif-sur-Yvette, France.

the GnRH-induced incorporation of labeled amino acids into LH and FSH. Our data contradict the results of other groups (Liu *et al.*, 1976; Liu and Jackson, 1978; Azhar *et al.*, 1978) who reported that GnRH stimulated the glycosylation of LH.

As a continuation of this work, we recently developed an alternative approach that consists of studying LH biosynthesis in cell-free conditions (Counis *et al.*, 1981a). The ultimate objective of this research is to determine the role of GnRH and other substances at the molecular level in LH biosynthesis.

The aim of this chapter is to discuss, in view of the available data, the problems of the translation of pituitary mRNA in cell-free media and identification of the primary forms of LH subunits synthesized *in vitro*.

2.   Cell-Free Synthesis of LH Subunits

2.1.   Preparation of Pituitary mRNA

Rat anterior pituitary glands were collected in the laboratory and stored in liquid nitrogen. Occasionally, we also used frozen beef pituitaries, purchased from Cofracor (Paris) and stored at $-20°C$. Total RNA was prepared by phenol extraction of a pituitary homogenate (Palmiter, 1974) and poly(A)$^+$ RNA was isolated by two successive affinity chromatographies on oligo(dT)-cellulose (Aviv and Leder, 1972). We currently obtain yields of about 100 μg poly(A)$^+$ RNA/g fresh rat glands.

Because total pituitary poly(A)$^+$ RNA coded for very low levels of LH subunits (see Table II), the possibility of raising RNA from pituitary cells in culture, enriched in gonadotropin cells (Counis *et al.*, 1981b) was also investigated. We developed a one-step micromethod consisting in extraction of RNA from enriched gonadotrophs obtained by sedimentation of dispersed pituicytes in a BSA gradient (Hymer *et al.*, 1973; Denef *et al.*, 1976). A lysate from 10^5–10^6 cells was centrifuged in a small centrifugation tube on a CsCl cushion (Glisin *et al.*, 1974), and the translation step was achieved in the same tube, containing the RNA pellet.

2.2   Translation of Pituitary mRNA

mRNA was translated in the presence of [^{35}S]-Met or [^{3}H]-Pro using the wheat germ extract as the protein synthesizing system (Marcu and Dudock, 1974). We first determined the most favorable conditions of translation of rat and beef mRNA by investigating several parameters of cell-free synthesis such as the quality and amount of wheat germ extract, the amount of mRNA, the time of incubation, and the concentrations of spermidine, K^+, and Mg^{2+}. These conditions were determined by measuring the incorporation of [^{3}H]-Pro into TCA-precipitable proteins, assuming that a direct correlation exists between incorporation of a labeled amino acid into the total protein and into LH subunits. Table I summarizes the optimized conditions of translation of pituitary mRNA.

In order to lower the proportion of wheat germ proteins in the translation medium, a wheat germ S-30 extract was pretreated with micrococcal nuclease under conditions usually employed for depleting rabbit reticulocyte lysate of endogenous mRNA (Pelham and Jackson, 1976) (see legend to Fig. 1). Using nuclease-pretreated extracts, the mRNA-stimulated (60 μg rat pituitary mRNA/ml) increase in ^{35}S radioactivity incorporated into total proteins was 140- to 160-fold over the control media *vs.* 10- to 12-fold for untreated extracts (Counis *et al.*, 1981a). Endogenous wheat germ proteins represented 10 to 20% of total translated proteins in the case of untreated extracts and only less than 1% in the nuclease-pretreated extracts. As shown in Fig. 1, starting from 9 min of treatment with nuclease, the media contained undetectable amounts of endogenous wheat germ proteins (lanes 8 to 10), thus enabling a better identification of translation products raised in response to bovine pituitary mRNA (lanes 1 to 3).

Table I. Optimized Conditions of Translation of Rat and Beef Pituitary mRNA[a]

Hepes (pH 7.4)	30 mM
Dithioerythritol	2 mM
GTP	0.2 mM
ATP	2.0 mM
Creatine phosphate	8.0 mM
Creatine phosphokinase	3.5 U/ml
Spermidine	0.4 mM
Unlabeled amino acids	40 μM
(except Met or Pro), each	
[^{35}S]-Met	1.2 mCi/ml
or	
[^{3}H]-Pro	0.4 mCi/ml
Potassium acetate	106 mM
Magnesium acetate	3.5 mM
mRNA preparation	100 μg/ml
Nuclease-treated wheat germ extract[b]	30% (v/v)

[a] The final volume was 25–150 μl. The reaction was allowed to proceed for 60 min at 30°C and initiated by addition of wheat germ extract.
[b] Conditions for this treatment are given in the legend to Fig. 1. Time period of treatment, 10 min.

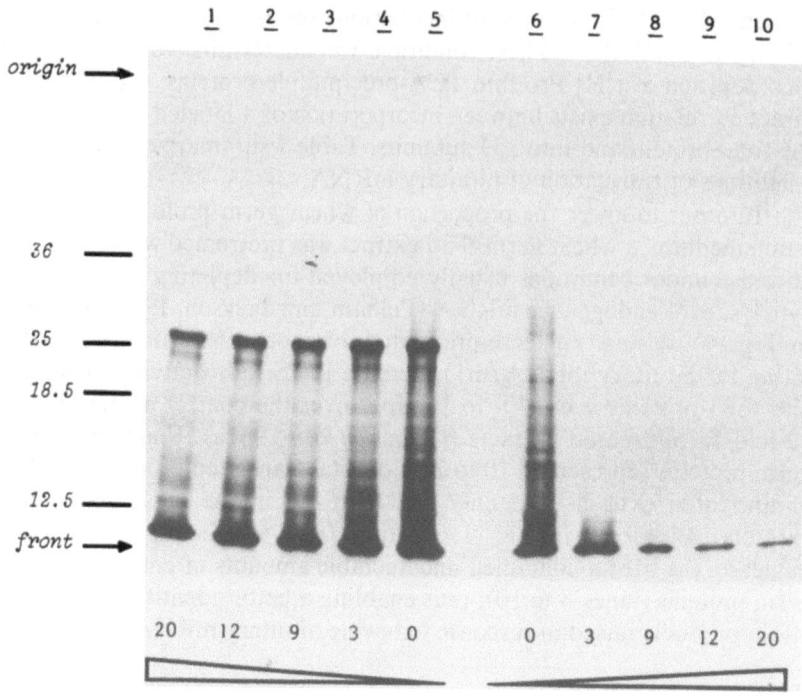

treatment with nuclease (min)

Figure 1. Influence of treatment of a wheat germ extract with nuclease on the electrophoretic profile of translation products encoded by endogenous mRNA (lanes 6 to 10) or by bovine pituitary mRNA (60 µg/ml, lanes 1 to 5). Wheat germ cell-free extract was incubated at 22°C for variable periods of time (0 to 20 min) in the presence of micrococcal nuclease (75 U/ml) + 1 mM CaCl₂. At the end of the incubation period, EGTA was added to 2 mM and subsequent translational incubation proceeded under the conditions summarized in Table I. The SDS-polyacrylamide gels (at 15%) were revealed by fluorography (delay of exposure, 5 days). The amount of radioactivity applied varied in the range of 10,000–300,000 cpm, depending on the biosynthetic activity in 25 µl translational mixture. Protein markers (on the left) were lactate dehydrogenase (36K), chymotrypsinogen A (25K), ferritin (18.5K), cytochrome *c* (12.5K).

2.3.   Isolation and Identification of Translation Products

In order to improve the immune reaction and lower the nonspecific radioactivity, the translation products were reduced and carboxymethylated (Pierce *et al.*, 1976). Translation products were then precipitated with 10% TCA (w/v), solubilized in 1% SDS, 10 mM Tris–HCl (pH 8), and the solution was heated in boiling water for 2 min. Aliquots were then submitted to immunoprecipitation in 25 mM Tris–HCl (pH 8), 10⁻⁵ M bacitracin, 10 mM EDTA, 150 mM NaCl, 1% Triton X-100, 0.5% deoxycho-

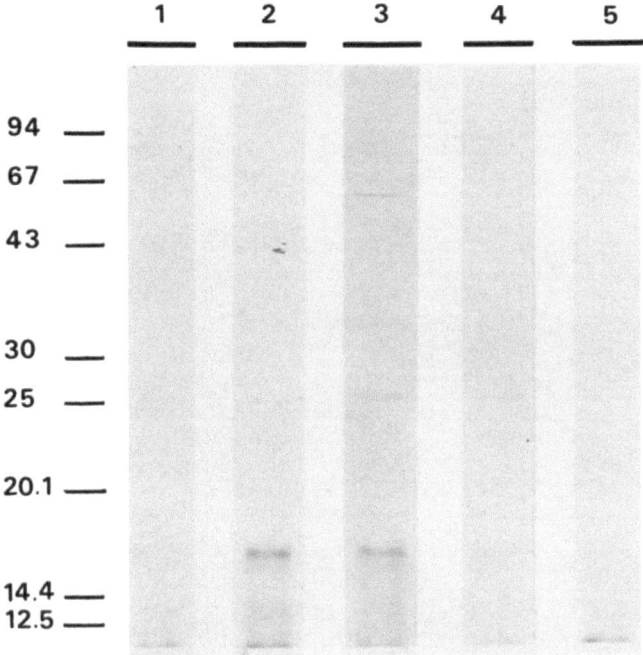

Figure 2. Fluorograph of ^{35}S-labeled polypeptides synthesized in a wheat germ cell-free system in response to rat pituitary mRNA and immunoprecipitated with different antisera to LHα in the conditions given in the text. Lane 1, nonimmune serum; lane 2, anti-RCXM bovine LHα (Dr. Pierce); lane 3, anti-RCXM ovine LHα (laboratory preparation); lane 4, anti-ovine LHα (native, linked to horseradish peroxidase, laboratory preparation); lane 5, anti-ovine LHα (native, Dr. B. Kerdelhué, this laboratory).

Aliquots of cell-free translation products (1.8×10^6 cpm) were subjected to immunoprecipitation with different antisera to LHα, and the resulting isolated, labeled peptides were applied on 15% polyacrylamide gels. Apparent M_r (at the left, in thousands) were calculated from linear plots of the logarithm of M_r of protein markers vs. the distance covered. The following protein markers were used: Phosphorylase b (94K), BSA (67K), ovalbumin (43K), carbonic anhydrase (30K), chymotrypsinogen A (25K), trypsin inhibitor (20.1K), α-lactalbumin (14.4K), and cytochrome c (12.5K).

late, 0.1% SDS. The use of this "cocktail" of detergents is necessary to lower the nonspecific radioactivity adsorbed on the immunoprecipitate.

Immunoprecipitation of labeled LH subunits was investigated using a number of antisera raised either against native or against denatured subunits. Figure 2 shows the electrophoretic pattern of the translation products obtained in response to rat pituitary mRNA after immunoprecipitation with different antisera. Only antisera directed against reduced, S-carboxymethylated (RCXM) subunits of LH gave reliable results. It is probable that precursors of rat LH subunits do not cross-react with antisera to

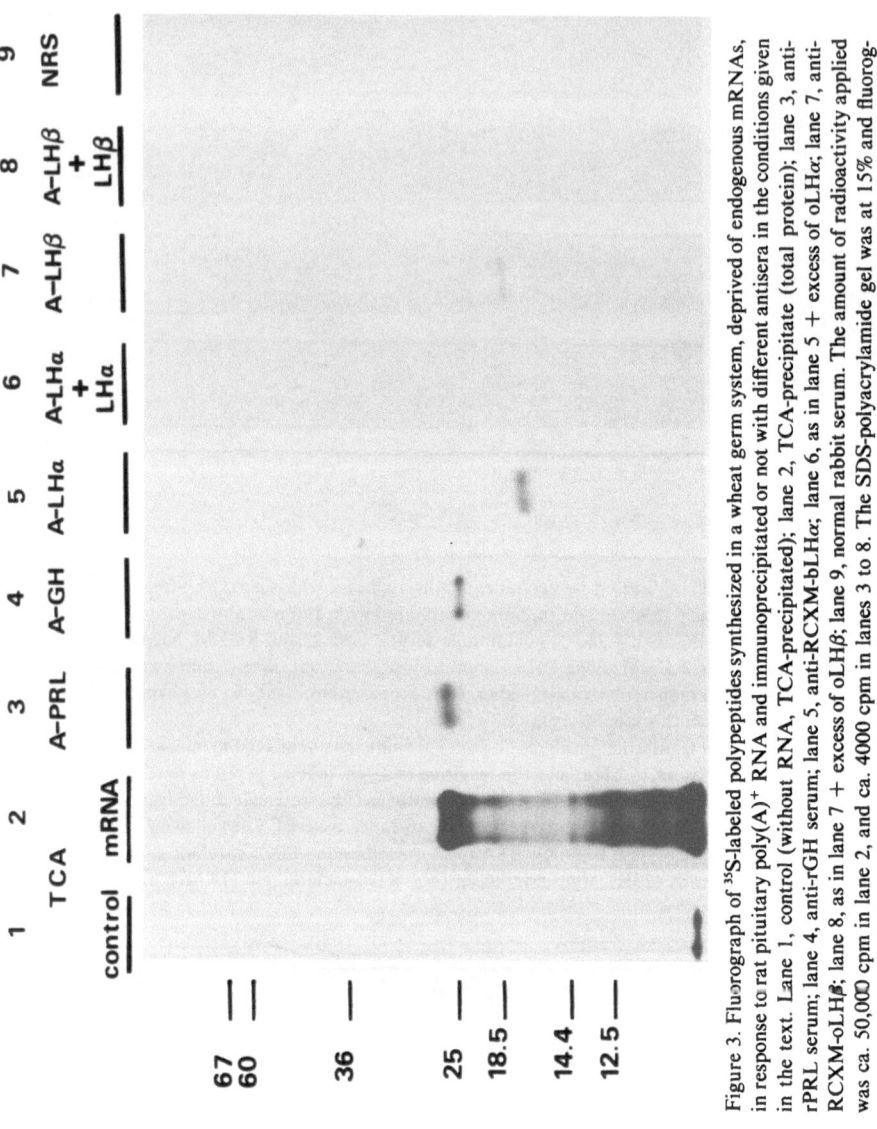

Figure 3. Fluorograph of ³⁵S-labeled polypeptides synthesized in a wheat germ system, deprived of endogenous mRNAs, in response to rat pituitary poly(A)⁺ RNA and immunoprecipitated or not with different antisera in the conditions given in the text. Lane 1, control (without RNA, TCA-precipitated); lane 2, TCA-precipitate (total protein); lane 3, anti-rPRL serum; lane 4, anti-rGH serum; lane 5, anti-RCXM-bLHα; lane 6, as in lane 5 + excess of oLHα; lane 7, anti-RCXM-oLHβ; lane 8, as in lane 7 + excess of oLHβ; lane 9, normal rabbit serum. The amount of radioactivity applied was ca. 50,000 cpm in lane 2, and ca. 4000 cpm in lanes 3 to 8. The SDS-polyacrylamide gel was at 15% and fluorography was revealed in 15 days. Apparent Mᵣ were calculated from linear plots of the logarithm of Mᵣ of protein markers vs. the distance covered (in thousands, at the left of the figure). The following reduced and S-carboxymethylated protein markers were used: BSA (67K), catalase (60K), lactate dehydrogenase (36K), chymotrypsinogen A (25K), ferritin (18.5K), α-lactalbumin (14.4K), cytochrome c (12.5K).

native subunits as their conformation may differ from those of native subunits (Giudice and Weintraub, 1979).

We routinely used antisera to bovine RCXM-LH subunits provided by Dr. J. G. Pierce or antisera to ovine RCXM-LHα and RCXM-LHβ raised in our laboratory. Antigen–antibody complexes were precipitated with Pansorbin (Kessler, 1976).

Total TCA-precipitable polypeptides and immunoprecipitable polypeptides were analyzed by electrophoresis on SDS-polyacrylamide gels and revealed by fluorography.

Figure 3 shows a fluorograph of ^{35}S-labeled polypeptides synthesized in response to rat pituitary mRNA. Lane 2 depicts a profile of total TCA-precipitated proteins and lanes 3 to 8 show products immunoprecipitated with specific antisera. Antisera against rat prolactin (supplied by Dr. A. Tixier-Vidal) and against rat growth hormone (supplied by Dr. Parlow and NIAMDD) each precipitated one main product of M$_r$ 27,000 and 24,000 respectively (lanes 3 and 4). These products corresponded to precursors of PRL and GH, already identified by others in cell-free translation media (Maurer *et al.*, 1976; Evans and Rosenfeld, 1976; Lingappa *et al.*, 1977), thus demonstrating the fidelity of translation of our rat pituitary RNA preparations. As seen in lane 5, antiserum to RCXM-bLHα precipitated mainly one product, with an apparent M$_r$ of 17,000–17,500, which did not precipitate in the presence of an excess of unlabeled oLHα (lane 6). Lane 7 shows that antiserum to RCXM-oLHβ also precipitated one product, with an apparent M$_r$ of 18,500–19,000, which did not precipitate in the presence of an excess of unlabeled oLHβ (lane 8). The fact that antisera against LH subunit did not precipitate corresponding translation products in the presence of an excess of unlabeled subunit confirmed their specificity.

Figure 4 shows a fluorograph of ^{35}S-labeled polypeptides synthesized in response to RNA from enriched gonadotrophs. The profile of total TCA-precipitated polypeptides is shown in lane 5, and products precipitated with specific LHα and LHβ antisera are shown in lanes 2 and 3, respectively. Again, each specific antiserum precipitated a single product in the range of molecular weights mentioned above. The profile of TCA-precipitated polypeptides in response to RNA from enriched gonadotrophs shows a major band in the range of 27,000, which was identified as pre-PRL, and, in addition, another band at 17,000–17,500. This latter band, which corresponds to the product precipitated with anti-LHα, is almost not discernible on a fluorograph showing the translation products of total poly(A)$^+$ RNA prepared directly from rat pituitaries (Fig. 3).

In Table II we have collected some quantitative data concerning an evaluation of the radioactivity incorporated into polypeptides immunoprecipitated with anti-bLHα and anti-oLHβ. These data suggest that: (1) the

amounts of products synthesized in response to both preparations of RNA
and precipitated with anti-bLHα were 5–6 times higher than in the case of
anti-oLHβ; (2) the proportion of mRNAs coding for the subunits of LH in
poly(A)$^+$ RNA preparations must be relatively low compared to total pituitary poly(A)$^+$ RNA; and (3) the use of enriched rat pituitary gonadotrophs for preparing RNA presented an interesting step in the concentrations of mRNAs coding for LH subunits. In that case, the yields of the
synthesis of rLHα and rLHβ increased 6–7 times relative to the proportion
of the subunits obtained in response to the pituitary poly(A)$^+$RNA.

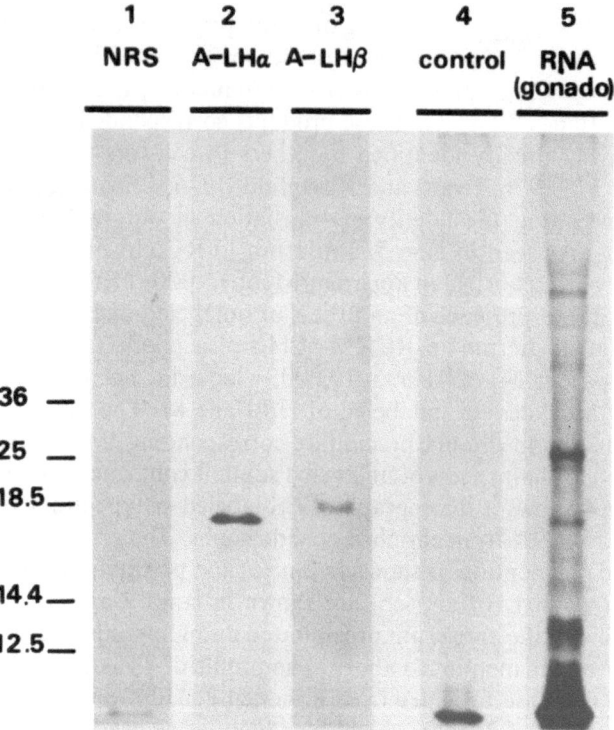

Figure 4. Fluorograph of ^{35}S-labeled polypeptides synthesized in a wheat germ system, deprived
of endogenous mRNAs, in response to enriched gonadotrophs from rat pituitaries. Lane 1, precipitation with normal rabbit serum; lane 2, immunoprecipitation with anti-RCXM-bLHα, lane
3, immunoprecipitation with anti-RCXM-oLHβ; lane 4, control (without RNA, TCA-precipitated); lane 5, TCA-precipitate (total protein). The conditions for RNA preparation and for
immunoprecipitation are given in the text. RNA used for translations was prepared from 10^6 rat
pituitary cells enriched in gonadotrophs, per lane. The amount of radioactivity applied was about
3000 cpm in lanes 2 and 3 and 50,000 cpm in lane 5. The gels (at 15%) were exposed for 15
days. Apparent M_r (at the left, in thousands) were calculated as in the legend to Fig. 3.

Table II. Quantitative Evaluation of the Radioactivity Incorporated into Products Immunoprecipitated with Antisera to LHα and LHβ[a]

RNA	^{35}S incorporated (cpm x 10^3/50 μl medium) into:		
	Total protein[b]	pre-LHα[c]	pre-LHβ[c]
1st experiment			
Control (no RNA)	36	0	0
Poly(A)$^+$ RNA[d] (3.5 μg)	2764	7.6 (0.28%)[e]	1.3 (0.05%)[e]
2nd experiment			
Control (no RNA)[d]	43	0	0
Poly(A)$^+$ RNA (5μg)	3720	11.3 (0.31%)[e]	2.2 (0.06%)[e]
3rd experiment			
Control (no RNA)	26.3	0	0
Cell RNA[f] (10^6 cells)	322.5	5.3 (1.79%)[e]	1.1 (0.37%)[e]

[a] See conditions in the text.
[b] TCA-precipitate.
[c] Products immunoprecipitated with specific antisera to reduced and S-carboxymethylated subunits, bLHα and oLHβ.
[d] Prepared from rat pituitaries.
[e] Percent of radioactivity vs. total protein radioactivity.
[f] Prepared from rat pituitary cells enriched in gonadotrophs.

3. Conclusions and Projections

This work demonstrates that rat pituitary mRNA preparations direct the cell-free synthesis of two products immunoprecipitable with specific antisera to LHα and LHβ. These products showed on SDS-polyacrylamide gel electrophoresis apparent M_r of 17,000–17,500 and 18,500–19,000, respectively. As the M_r of the authentic sugar-free rat LHα and LHβ are approximately 11,000 and 12,500, respectively (Ward *et al.*, 1971), our data suggest that rat LH subunits are synthesized as precursors and that LHα and LHβ are encoded by separate mRNAs. As the difference in M_r between precursor and apopeptide forms of the subunits is 6000 to 6500 in both cases, this may suggest a common extension sequence. This extension sequence would be somewhat larger than that expected from the known sequences of the signal peptides of subunits of the other glycoprotein hormones, which are in the range of 2500–3000 (Hussa, 1980; Giudice *et al.*, 1979).

The M_r of the precursors of rat LHα and LHβ established in our present work are somewhat higher than those of LH pre-subunits from other species reported by others. An apparent M_r of 14,000 has been determined for bovine pre-LHα (Landefeld, 1979; Keller *et al.*, 1980) and for mouse thyrotropic tumor pre-TSHα (Chin *et al.*, 1978; Kourides and Weintraub, 1979). Nevertheless, in a recent work, Giudice *et al.* (1979) compared bovine and mouse pituitary glycoprotein hormone pre-α subunits, synthesized in cell-free conditions, and found in both cases a M$_r$ of 17,000. On the other hand, a value of 16,000–18,000 has been reported for bovine pre-LHβ (Landefeld and Kepa, 1979; Keller *et al.*, 1980).

We do not understand at present the reasons for discrepancies between the M$_r$ of precursors of LH subunits established in our work and those reported by others. We have carefully verified that this is not due to different protein markers for M$_r$ determinations used by us and by others, nor is it due to reduction and carboxymethylation of immunoprecipitated LH subunits carried out in our procedure. Another possibility would be the anomalous mobilities on SDS-polyacrylamide gels exhibited by the precursors of rat LH subunits, as already shown for other proteins (Lambin, 1978). Whatever the actual M$_r$ of these precursors, the relative difference in M$_r$ between authentic α and β apopeptides, which is about 1500, is reflected in the size of their precursors as determined in the present work.

As shown in Table II, poly(A)$^+$ RNA from normal rat pituitary glands coded for very low levels of LH subunits. In order to increase in the translation medium the proportion of LH subunits vs. total translated proteins, we investigated the possibility of preparing RNA from enriched gonadotrophs. Translation of RNA preparations from enriched gonadotrophs resulted in increased yields of synthesized LH subunits. On the other hand, as can be seen from the profiles of TCA-precipitated proteins, in this case the band corresponding to pre-PRL (M$_r$ 27,000) was less abundant than with poly(A)$^+$ RNA from rat pituitaries (Figs. 3 and 4). Furthermore, many additional bands were perceptible in the case of RNA from enriched gonadotrophs as compared to poly(A)$^+$ RNA, in particular the band corresponding to pre-α subunit.

Since we have no tumor cells producing LH, enriched gonadotroph cultures represent in our opinion a useful system for studying the synthesis and processing of gonadotropin subunits. We are currently using this system in our work. Another problem that remains to be solved is the regulation of the synthesis and glycosylation of gonadotropin subunits. We hope that cell cultures enriched in gonadotrophs will also be useful as a tool to investigate this problem.

4. Summary

Rat anterior pituitary mRNAs were prepared from two sources, anterior pituitary glands and dispersed pituitary cells enriched in gonadotrophs.

mRNAs were then translated in a wheat germ cell-free system deprived of endogenous RNAs by nuclease treatment. Using antisera directed against reduced and S-carboxymethylated LHα and LHβ, two different products were isolated by immunoprecipitation from translation media. One of these products, showing on SDS-polyacrylamide gel electrophoresis an apparent M_r of 17,000–17,500, corresponded to a precursor form of rat LHα. The second product, giving an apparent M_r of 18,500–19,000, corresponded to a precursor of rat LHβ. Quantitative evaluation of the two precursors showed that in the same translation medium the amount of pre-LHα was 5–6 times higher than that of pre-LHβ. On the other hand, the use of RNA from enriched gonadotrophs increased the yields of translated LH subunits by about 6 times compared with poly(A)$^+$ RNA from whole anterior pituitaries. These results suggest that rat LHα and LHβ are synthesized by different RNAs with a polypeptide extension of about 6000 as compared with the authentic apopolypeptide moiety of the subunits. The apparent M_r of the precursors of rat LH subunits as determined in our work are somewhat higher than the size of bovine and mouse LH subunit precursors reported by others. We have no explanation at present for this discrepancy.

ACKNOWLEDGEMENTS. We are grateful to Dr. J. G. Pierce for his generous supply of antisera to reduced, S-carboxymethylated bovine LH subunits, to Dr. A. Tixier-Vidal and Mr. D. Grouselle for supplying anti-rat PRL, and to Dr. A. F. Parlow and NIAMDD for anti-rat GH. This work was supported by the CNRS (France) and a contract from the DGRST (France) (No. 79 7 1400).

REFERENCES

Aviv, H., and Leder, P., 1972, Purification of biologically active globin messenger RNA by chromatography on oligothymidylic acid-cellulose, *Proc. Natl. Acad. Sci. USA* **69:**1408–1412.

Azhar, S., Reel, J. R., Pastushok, C. A., and Menon, K. M. J., 1978, LH biosynthesis and secretion in rat anterior pituitary cell cultures: Stimulation of LH glycosylation and secretion by GnRH and an agonistic analogue and blockade by an antagonistic analogue, *Biochem. Biophys. Res. Commun.* **80:**659–666.

Chin, W. W., Habener, J. F., Kieffer, J. D., and Maloof, F., 1978, Cell-free translation of the messenger RNA coding for the α subunit of thyroid-stimulating hormone, *J. Biol. Chem* **253:**7985–7988.

Counis, R., Ribot, G., Corbani, M., Poissonnier, M., and Jutisz, M., 1981a, Cell-free translation of the rat pituitary messenger RNA coding for the precursors of α and β subunits of lutropin, *FEBS Lett.* **123:**151–155.

Counis, R., Corbani, M., Berault, A., Theoleyre, M., Jansen de Almeida Catanho, M. T., and Jutisz, M., 1981b, A microscaled method for the extraction and translation of mRNA from anterior pituitary cells in culture, *C. R. Acad. Sci. Paris* **293:**115–118.

Denef, C., Hautekeete, E., and Rubin, L., 1976, A specific population of gonadotrophs purified from immature female rat pituitary, *Science* **194:**848–850.

Evans, G. L., and Rosenfeld, M. G., 1976, Cell-free synthesis of a prolactin precursor directed by mRNA from cultured rat pituitary cells, *J. Biol. Chem.* **251**:2842–2847.

Giudice, L. C., and Pierce, J. G., 1978, Glycoprotein hormones: Some aspects of studies of secondary and tertiary structure, in: *Structure and Function of the Gonadotropins* (K. W. McKerns, ed.), pp. 81–110, Plenum Press, New York.

Giudice, L. C., and Weintraub, B. D., 1979, Evidence for conformational differences between precursor and processed forms of thyroid-stimulating hormone β subunit, *J. Biol. Chem.* **254**:12679–12683.

Giudice, L. C., Waxdal, M. J., and Weintraub, B. D., 1979, Comparison of bovine and mouse pituitary glycoprotein hormone pre-α subunits synthesized *in vitro*, *Proc. Natl. Acad. Sci. USA* **76**:4798–4802.

Glisin, V. Crkvenjakov, R., and Byus, C., 1974, Ribonucleic acid isolated by cesium chloride centrifugation, *Biochemistry* **13**:2633–2637.

Hussa, R. O., 1980, Biosynthesis of human chorionic gonadotropin, *Endocr. Rev.* **1**:268–294.

Hymer, W. C., Evans, W. H., Kraicer, J., Mastro, A., Davis, J., and Griswold, E., 1973, Enrichment of cell types from the rat adenohypophysis by sedimentation at unit gravity, *Endocrinology* **92**:275–287.

Keller, D., Fetherston, J., and Boime, I., 1980, Isolation of mRNA from bovine pituitary: The cell-free synthesis of the α and β subunits of luteinizing hormone, *Eur. J. Biochem.* **108**:367–372.

Kessler, S. W., 1976, Cell membrane antigen isolation with the staphylococcal protein A-antibody adsorbent, *J. Immunol.* **117**:1482–1490.

Khar, A., and Jutisz, M., 1980a, Role of gonadotropin-releasing hormone in the biosynthesis of LH and FSH by rat anterior pituitary cells in culture, in: *Synthesis and Release of Adenohypophyseal Hormones* (M. Jutisz and K. W. McKerns, ed.), pp. 217–235, Plenum Press, New York.

Khar, A., and Jutisz, M., 1980b, Effect of actinomycin D and cycloheximide on the release and biosynthesis of gonadotropins by pituitary cell cultures, *Mol. Cell. Endocrinol.* **17**:85–93.

Khar, A., Debeljuk, L., and Jutisz, M., 1978, Biosynthesis of gonadotropins by rat pituitary cells in culture and in pituitary homogenates: Effect of gonadotropin releasing hormone, *Mol. Cell. Endocrinol.* **12**:53–65.

Kourides, I. A., and Weintraub, B. D., 1979, mRNA-directed biosynthesis of α subunit of thyrotropin: Translation in cell-free and whole-cell systems, *Proc. Natl. Acad. Sci. USA* **76**:298–302.

Lambin, P., 1978, Reliability of molecular weight determination of proteins by polyacrylamide gradient gel electrophoresis in the presence of sodium dodecyl sulfate, *Anal. Biochem.* **85**:114–125.

Landefeld, T. D., 1979, Identification of *in vitro* synthesized pituitary glycoprotein α subunit, *J. Biol. Chem.* **254**:3685–3688.

Landefeld, T. D., and Kepa, J., 1979, The cell-free synthesis of bovine lutropin β subunit, *Biochem. Biophys. Res. Commun.* **90**:1111–1118.

Lingappa, V. R., Devillers-Thiery, A., and Blobel, G., 1977, Nascent prehormones are intermediates in the biosynthesis of authentic bovine pituitary growth hormone and prolactin, *Proc. Natl. Acad. Sci. USA* **74**:2432–2436.

Liu, T. C., and Jackson, G. L., 1978, Modification of luteinizing hormone biosynthesis and release by GnRH, cycloheximide and actinomycin D, *Endocrinology* **103**:1253–1263.

Liu, T. C., Jackson, G. L., and Gorski, J., 1976, Effect of synthetic GnRH on incorporation of radioactive glucosamine and amino acids into LH and total protein by rat pituitaries *in vitro*, *Endocrinology* **98**:151–163.

Marcu, K., and Dudock, B., 1974, Characterization of a highly efficient protein synthesizing system derived from commercial wheat-germ, *Nucleic Acid Res.* **11**:1385–1397.

Maurer, R. A., Stone, R., and Gorski, J., 1976, Cell-free synthesis of a large translation product of prolactin messenger RNA, *J. Biol. Chem.* **251**:2801–2807.

Palmiter, R. D., 1974, Magnesium precipitation of ribonucleoprotein complexes: Expedient techniques for the isolation of undegraded polysomes and messenger ribonucleic acid, *Biochemistry* **13**:3606–3615.

Pelham, H. R. B., and Jackson, R. J., 1976, Dependent translation system for reticulocyte lysate, *Eur. J. Biochem.* **67**:247–256.

Pierce, J. G., Giudice, L. C., and Reeve, J. R., 1976, Studies on the disulfide bonds of glycoprotein hormones: Course of reduction of bovine luteinizing hormone, bovine thyroid-stimulating hormone, and their subunits, *J. Biol. Chem.* **251**:6388–6391.

Ward, D. N., Reichert, L. E., Jr., Fitak, B. A., Nahm, H. S., Sweeney, C. M., and Neill, J. D., 1971, Isolation and properties of subunits of rat pituitary luteinizing hormone, *Biochemistry* **10**:1796–1802.

27

Cellular Polyproteins from Pituitary and Hypothalamus

Composite Precursors to Oligopeptide Hormones

Dietmar Richter, Hartwig Schmale, Richard Ivell, and Holger Kalthoff

1. Introduction

The recent discoveries that several different oligopeptide hormones may occur within the same protein precursors has added a new complexity to protein biosynthesis and its regulation at the level of transcription, translation, and posttranslation. In the present report we summarize our experience in studying the biosynthesis of cellular polyproteins present in the pituitary and hypothalamus as well as their processing and modifications at the posttranslational level. The intention of this paper is not to review the endocrinological aspects of the oligopeptide hormones involved but to concentrate on the nature and significance of those precursors, their identification, and their implications for future work. So far the bovine polyproteins corticotropin/β-lipotropin (pro-opiocortin), oxytocin/neurophysin I, and arginine vasopressin/neurophysin II have been studied. The experimental concept adopted for investigating cellular polyproteins is the *in vitro* translation of mRNA from hypothalamus or pituitary in cell-free systems

Dietmar Richter, Hartwig Schmale, Richard Ivell, and Holger Kalthoff • Institut für Physiologische Chemie, Abteilung Zellbiochemie, Universität Hamburg, 2 Hamburg 20,

and the characterization of the precursors by immunological and biochemical methods. The *in vitro* strategy has helped to elucidate several neuropeptide precursors, but there are a number of limitations, which are discussed in the light of the failure to find precursors to oligopeptide hormones present in the respective tissues at very low concentrations. A useful tool for studying posttranslational events, such as processing or segregation, has been found to be the *Xenopus laevis* oocyte system which can be microinjected with hypothalamic mRNA.

2. Definition and Significance of Cellular Polyproteins

Polyproteins, first found after virus infections, are unique in that they are composed of several distinct viral entities such as coat proteins, replicases, and probably also proteases (for review see Korant, 1980). The *in vivo* maturation of viral polyproteins is generated by a variety of proteolytic enzymes, some of which are cellular in origin, others virally coded (Rueckert *et al.*, 1980).

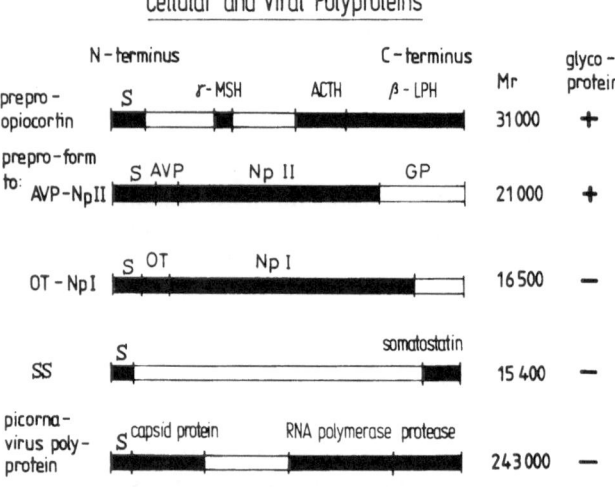

Figure 1. Viral and cellular polyproteins. The precursors to pro-opiocortin (Nakanishi *et al.,* 1979) to arginine vasopressin/neurophysin II (Land *et al.,* 1982), and to fish pancreatic somatostatin (Hobart *et al.,* 1980; Goodman *et al.,* 1980) show roughly the relative positions of the oligopeptides involved in the respective molecule; the positions of the oligopeptides in the other precursors have not yet been firmly established. Within a given precursor the dark bar represents the known, the light bar the unknown part, relative to the overall size of the molecule. S, signal peptide; SS, somatostatin; AVP, arginine vasopressin; Np II, bovine neurophysin II; GP, glycoprotein; OT, oxytocin; Np I, bovine neurophysin I; β-LPH, β-lipotropin; γ-MSH, γ-melanotropin; ACTH, corticotropin.

Only recently have cellular counterparts to viral polyproteins been identified (Fig. 1). These are the precursors to certain oligopeptide hormones and have so far been found mainly in the pituitary and hypothalamus. To date, the best studied example of a cellular polyprotein is pro-opiocortin (Rubinstein *et al.*, 1978), also called pro-opiomelanocortin (Chrétien *et al.*, 1979), isolated from the pituitary and a derived tumor cell line (Roberts and Herbert, 1977a, b; Mains *et al.*, 1977; Nakanishi *et al.*, 1977; Chrétien *et al.*, 1979; Rubinstein *et al.*, 1978). It is composed of 265 amino acids and contains the polypeptides corticotropin (ACTH) and β-lipotropin (β-LPH); β-LPH in turn includes the endorphins and β-melanotropin. Two other precursors of this kind have been found in the hypothalamus, comprising either neurophysin I/oxytocin or neurophysin II/vasopressin (Brownstein and Gainer, 1977; Giudice and Chaiken, 1979; Lauber *et al.*, 1979a,b; Lin *et al.*, 1979; Schmale *et al.*, 1979; Schmale and Richter, 1980; Richter *et al.*, 1980a,b). The bovine precursor to AVP/Np II consists of 166 amino acids and contains, besides the AVP- and neurophysin II-sequences, a glycopeptide of 39 amino acids at the carboxyl end of the molecule (Fig. 1, Land *et al.*, 1982). Whether the recently sequenced precursors to the tetradecapeptide somatostatin from fish pancreas can also be regarded as polyproteins remains to be seen. The sequences apparently do not match any known oligopeptide sequences other than somatostatin (Hobart *et al.*, 1980; Goodman *et al.*, 1980). The hypothalamic releasing hormones may also be potential candidates for such cellular polyproteins; for example, in the case of the decapeptide luteinizing hormone-releasing hormone (LH-RH) there is a report of a high-molecular-weight precursor from sheep (Millar *et al.*, 1977).

The existence of high-molecular-weight forms as precursors to oligopeptide hormones much smaller in size immediately raises the question of the function of the remaining part of the precursor molecule (Fig. 1), which in the case of pro-opiocortin, comprises more than 40% of the overall sequence. It might, for example, have a purely structural role allowing an oligopeptide to traverse a membrane, an operation that otherwise may be prevented by the shortness of the molecule. On the other hand, these parts may contain oligopeptides with biological functions yet to be discovered. With the elucidation of the amino acid sequence of pro-opiocortin it became evident that a fragment with a melanotropin (MSH)-like amino acid sequence was present in the N-terminal part of the molecule (Nakanishi *et al.*, 1979). By analogy with the finding that the biological entities (e.g., ACTH, β-LPH) present in the polyprotein are framed by pairs of basic amino acids, a so-called γ-MSH was predicted (Nakanishi *et al.*, 1979). Recently, Pedersen *et al.* (1980) synthesized the dodecapeptide γ-MSH and found that it enhanced corticosterone and aldosterone production. In one of the precursors to somatostatin from fish pancreas (Goodman *et al.*, 1980) there is also a hexapeptide flanked by pairs of basic residues; its biological relevance has not yet been established. It may well be that, based on these predictions, a new potential tool will emerge for detecting oligopeptides with unknown functions. Immunological studies with antibodies raised against the synthesized oligopeptides combined with receptor experiments should be of help in identifying their biological roles.

3. General Comments on the Synthesis and Identification of Precursors to Oligopeptide Hormones

Since the pituitary and hypothalamus may well contain other as yet undiscovered polyproteins, what sort of strategy should be considered for such precursor hunting expeditions? One approach is to analyze extracts of the respective tissues, or, if available, tumor cell lines, for large-molecular-weight material. Such an *in vivo* strategy, furthermore, allows one to follow precursor synthesis, as well as processing by pulse-chase experiments. A complementary approach is to take advantage of cell-free synthesizing systems and translate, *in vitro*, mRNA extracted from the potential precursor tissue. An added advantage of this approach is that the mRNA could be used for cloning and sequencing experiments. This approach has recently been adopted for the indirect elucidation of the primary structure of the preproform of bovine AVP/Np II by sequencing cloned cDNA complementary to hypothalmic mRNA Land *et al.*, 1982).

3.1. Quality and Quantity of mRNA

The resolving power of an *in vitro* translation system will depend on the quality of the mRNA used to direct synthesis. Since it is also essential to get sufficient quantities of mRNA, we routinely use calf or bovine tissue. It should be stressed that it is extremely important to obtain the material as soon after the death of the animal as possible. Experience with bovine hypothalami from commercial sources indicates that there can be considerable polysome degradation, frequently with mRNA yields less than 5% those of fresh tissue. To test the quality of the mRNA the polysomal profiles from hypothalamus and pituitary are routinely compared; most of the hypothalamic ribosomes are present as tetra- and pentasomes whereas those from the pituitary are mostly decasomes, a difference that may reflect the real *in vivo* length of the mRNA. One can slightly increase the specificity of the mRNA to be used by, for example, isolation of polysomes bound to the membranes of the endoplasmic reticulum, which, as with other peptide hormones, serve as the site of polyprotein synthesis; the membrane-bound polysomes are then extracted by the SDS/phenol/chloroform method (Palmiter, 1974). Further enrichment of the mRNA preparation can be achieved by oligo(dT) affinity chromatography (Richter *et al.*, 1980b) and by size fractionation steps using sucrose density gradient centrifugation (Richter *et al.*, 1980b) or agarose gel electrophoresis (Bailey and Davidson, 1976). The average yield of poly(A)$^+$ RNA was 0.5 μg/g hypothalamic tissue [after oligo(dT) chromatography]. Translation experiments using the tritium-labeled amino acids leucine, phenylalanine, lysine, proline, and tyrosine indicated that about 1% of the radioactivity incorporated was present in material immunoreactive to antibodies against bovine neurophysin II. On the basis of these data, the prepro AVP/Np II precursor mRNA probably represents not more than 1% of the poly(A) + RNA, or about 0.03% of the total RNA of the tissue. In general, it has been found that, instead of extracting isolated polysomes, the direct extraction of, hypothalamic tissues by SDS/phenol/chloroform or by guanidinium thiocyanate (Chirgwin *et al.*, 1979) does not give as large yields nor allows one to handle as much tissue per single extraction. The method used by Chirgwin *et al.* (1979), however, was found to be very useful for extracting rat hypothalmic tissues.

3.2. Synthesis and Analysis of Precursors

mRNA from pituitary or hypothalamus is translated in cell-free systems from wheat germ or reticulocyte lysate using [^{35}S]methionine, [^{35}S]cysteine, or [^{3}H]leucine; the identification of the synthesized polyproteins is initially made by immunological means. The immunoreactive products are precipitated with *Staphylococcus aureus* immunoabsorbent and electrophoresed on SDS/polyacrylamide gels. Biochemical analysis of the immunologically identified products, such as peptide mapping, isoelectric focusing, or high-performance liquid chromatography, will be the next steps in characterizing the respective polyproteins. Final analysis of a polyprotein will come from amino acid sequence data either directly from the precursor or indirectly via the cDNA sequencing technique.

Specific antibodies for identification of the *in vitro* product are rather sensitive tools for a preliminary characterization of a polyprotein; their use depends on the assumption that antibodies directed against an oligopeptide can also recognize the antigenic determinant within a precursor. Antibodies appropriate in radioimmunoassays for quantifying a given oligopeptide may not necessarily be suitable for identifying its precursor, particularly if the antibodies recognize "blocked" C or N termini of the respective oligopeptide. Experiments in this laboratory, for instance, indicated that most of the antibodies against LH-RH, obtained commercially or prepared by ourselves, recognized the C-terminal Gly-NH$_2$ of LH-RH and are, therefore, of little use in identifying a hypothetical LH-RH precursor.

antibodies
against: AVP NpII OT   NpI

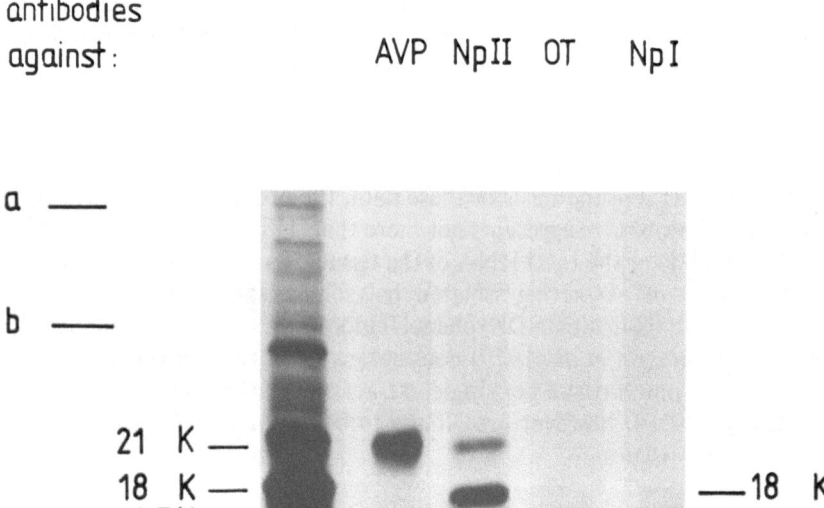

Figure 2. Translation of bovine hypothalamic poly(A)⁺ RNA in a reticulocyte lysate system using [³⁵S]cysteine as labeled amino acid. The synthesized products were immunoprecipitated with the antibodies indicated and analyzed on SDS/polyacrylamide gels. For details see Richter *et al.* (1980b). Lane 1 shows the total translation products before immunoprecipitation; pSS indicates the position where the somatostatin precursor should run (no discrete band can be seen in less exposed fluorograms or when using SDS/polyacrylamide gradient gel systems). 21K, precursor to arginine vasopressin (AVP) and neurophysin II (Np II); 18K, precursor cross-reacting with anti-neurophysin II but not with anti-AVP (Richter *et al.*, 1980b); 16.5K, precursor to oxytocin (OT) and neurophysin I (Np I). In this and subsequent fluorograms lanes are derived from different gels and aligned according to the position of marker proteins; a, ovalbumin, 46,-000; b, carbonic anhydrase, 30,000; c, lysozyme, 14,300.

It is well known that some of the oligopeptide hormones, particularly of the hypothalamus, are present *in vivo* only in minute quantities (Schally *et al.*, 1978). Based on the assumption that the mRNA for a given poly-protein containing one of these oligopeptides would also be present at very low concentrations the identification of such a precursor could be below the limit of detection. Somatostatin, for example, occurs in the hypothalamus at only about 700–2000 pg/mg tissue (Maeda and Frohman, 1980; Spiess and Vale, 1980) whereas vasopressin or oxytocin are present at concentrations an order of magnitude greater (Sachs and Takabatake, 1964; Sachs *et al.*, 1969). These differences in concentration may explain why when a total translation system is analyzed on an SDS/polyacrylamide gel a protein band can be assigned to the arginine vasopressin/neurophysin II but not to the somatostatin precursor (Fig. 2, lane 1); alternatively, the lack of the somatostatin band could be due to different turnover numbers and to a low incorporation of the radioactive amino acid. Oligopeptides occurring at even lower concentrations than somatostatin (e.g., LH-RH) could meet resolution problems and therefore may not be detectable.

To a certain extent the resolution problem can be overcome by scaling up the translation system, though the success of such a procedure depends on a number of factors including the quality of the antibodies used. In our experience, the greater quantities of antibodies needed tend to increase the nonspecific background of the fluorogram after SDS/polyacrylamide gel electrophoresis (this is particularly true in the case of the low-titer antibodies for some oligopeptides).

4. The Pituitary Polyprotein Prepro-opiocortin

Like other secreted polypeptides, the cellular polyprotein pro-opiocortin is ribosomally synthesized with a preceding signal peptide. This preproform is converted into the biologically inactive proform by the rough endoplasmic reticulum, modified by glycosylation, and further processed by a cascade of proteolytic events, to yield the final physiological products (Mains *et al.*, 1977; Loh *et al.*, 1980; Roberts and Herbert, 1977a,b; Nakanishi *et al.*, 1977, Leipold *et al.*, 1980; Herbert *et al.*, 1980). Using poly(A)$^+$ RNA to prepare and clone cDNA, Nakanishi *et al.* (1979) succeeded in elucidating the overall amino acid sequence of prepro-opiocortin.

Our attention was drawn to pro-opiocortin because, as a well-characterized polyprotein, it seemed to be quite suitable to test our various cell-free synthesizing systems as well as the immunological identification techniques. When we began this work in 1977 it had not been established whether the precursor to pro-opiocortin initially was synthesized as a preproform via a signal sequence, similar to other peptide hormones. The precise length of the signal sequence preceding a precursor can be determined by comparing the amino acid sequences of the N termini of the prepro- and proforms. *In vitro* prepro- and proforms of secretory proteins can, in general, be obtained by translating the appropriate mRNA in the absence or presence of microsomal membranes.

Using antibodies to ACTH or β-LPH with cell-free reticulocyte lysate

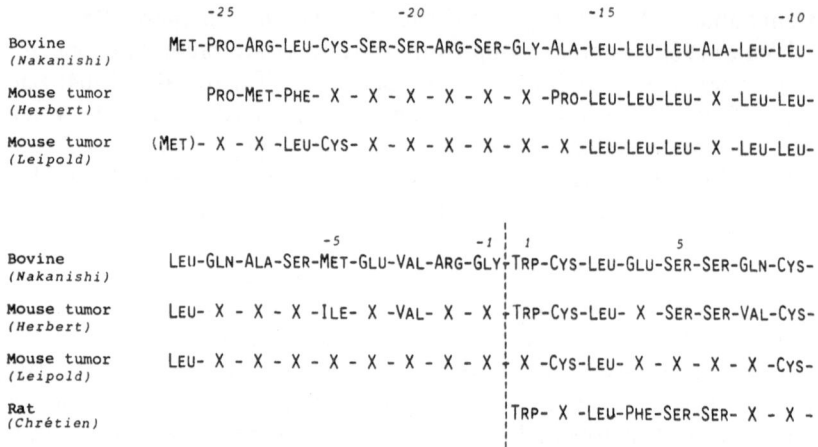

Figure 3. Comparison of the partial amino acid sequence of prepro- and pro-opiocortin. Data are taken from Nakanishi *et al.* (1979), Chrétien *et al.* (1979), and B. Leipold, D. Richter, and K. Beyreuther (unpublished).

systems complemented with mRNA from the AtT-20/D-17v mouse pituitary tumor cell line, the preproform to ACTH and β-LPH could be identified (Leipold *et al.*, 1980). This unglycosylated preproform (M_r 31,000) could be core-glycosylated to a product of M_r 34,000 in the presence of microsomal membrane fractions. The glycosylation step could be inhibited when microsomal membranes derived from tunicamycin-treated ascites tumor cells were used. The antibiotic tunicamycin blocks the formation of activated sugar precursors that serve as donors for the microsomal membrane-dependent core-glycosylation step. Since tunicamycin does not block proteolytic cleavage of the signal sequence, prepro-opiocortin was processed into the unglycosylated proform with an M_r of 28,000.

In order to find out the precise length of the amino acid sequence of the signal peptide preceding the pro-opiocortin, both the prepro-opiocortin and the pro-opiocortin were sequenced at the N terminus with incorporated [³H]leucine, [³⁵S] methionine, or [³⁵S]cysteine labeled amino acids. The immunoreactive products were analyzed on SDS/polyacrylamide gel electrophoresis. The radioactively labeled prepro- and pro-opiocortin were eluted from the gel, concentrated, and used for sequencing in a Beckman automatic sequencer (B. Leipold, D. Richter, and K. Beyreuther, unpublished). By comparing the amino acid sequences from both prepro- and pro-opiocortin, the signal peptide was shown to comprise 26 amino acids (Fig. 3). These data are basically similar to those reported by Nakanishi *et al.* (1979) and Herbert *et al.* (1980), except that Herbert's group found a methionine residue in position ⁻24. The methionine in position ⁻26 is based on experiments using the initiator $tRNA_i^{Met}$ charged with *N*-formyl

[³⁵S]methionine known to be incorporated exclusively in the N-terminal position of synthesized proteins. The prepro- but not the proform to opio-cortin was labeled with *N*-formyl [³⁵S] methionine (Leipold *et al.*, 1980).

5. Hypothalamic Polyproteins

More than a decade ago Sachs and co-workers (Sachs *et al.*, 1969) suggested that arginine vasopressin and one of the neurophysins may be synthesized together as a common precursor. The direct proof by isolation of the arginine vasopressin and neurophysin-containing polyprotein was successfully carried out by Gainer and Brownstein (Gainer *et al.*, 1977). They identified the polyprotein as a glycosylated precursor with an $M_r >$ 20,000. The precursor has been isolated from mouse and rat hypothalami and cross-reacts with antibodies raised against one of the neurophysins (Brownstein and Gainer, 1977; Lauber *et al.*, 1979a,b; Nicolas *et al.*, 1980). Detailed studies by Russell *et al.* (1979) strongly supported the composite precursor model of a hypothalamic polyprotein consisting of arginine vasopressin and one of the neurophysins: By trypsin treatment they could release an arginine vasopressin-like oligopeptide as well as one of the neurophysins from the precursor. They also presented evidence for the existence of a second hypothalamic polyprotein consisting of oxytocin and another neurophysin species. This precursor is not glycosylated and has an apparent M_r of 15,000–17,000 (Russell *et al.*, 1979; Brownstein *et al.*, 1980).

A number of groups have reported the cell-free synthesis of neurophysin precursors (Giudice and Chaiken, 1979: Lin *et al.*, 1979; Schmale *et al.*, 1979); however, identification of vasopressin- or oxytocin-like sequences within the reported polyproteins has been demonstrated only recently using appropriate antibodies against the nonapeptides (Fig. 2; see also Schmale and Richter, 1980; Richter *et al.*, 1980b; Schmale and Richter, 1981a).

In the following our results are summarized showing that a polyprotein consisting either of neurophysin II and arginine vasopressin or of neurophysin I and oxytocin can be synthesized from bovine hypothalamic mRNA in cell-free systems. The precursors have been characterized by specific antibodies and by peptide mapping of the immunoreactive products. Microinjection of hypothalamic mRNA into oocytes from *X. laevis* resulted in the synthesis and segregation of the respective polyprotein and in addition in its posttranslational conversion into intermediates.

5.1. Precursor to Arginine Vasopressin/Neurophysin II

As shown in Fig. 2, translation of bovine hypothalamic mRNA in cell-free systems from reticulocyte lysate or wheat germ (Richter *et al.*, 1980b;

Schmale and Richter, 1980) yields a preproform to arginine vasopressin and neurophysin II with an M_r of 21,000. Another product with an M_r of 18,000 cross-reacted with anti-neurophysin II, but not with anti-arginine vasopressin. Tryptic fingerprints also indicated that the 18k precursor did not contain arginine vasopressin sequences (Schmale and Richter, 1981b). Whether the 18k product is synthesized from a mRNA different from that which gives rise to the 21k form, or whether it is derived from a "false" initiation reaction has yet to be elucidated. When translation was performed in the presence of a microsomal membrane fraction from dog pancreas, the 21K preproform was processed into a proform with an M_r of 19,000 and subsequently core-glycosylated to a product with an M_r of 23,000 (see also Fig. 6, lane 5). Core-glycosylation could be prevented when a microsomal membrane fraction was used that had been prepared from tunicamycin-treated ascites tumor cells (Richter *et al.*, 1980b; Schmale and Richter, 1980; Schmale and Richter, 1981a). In order to establish that the 21K product is the preproform, the translation experiments were performed in the presence of *N*-formylmethionyl-tRNA. Figure 4 shows that in the absence of the microsomal membrane fraction the 21K product was labeled; however, in the presence of the membrane fraction little or no label was found in the 23K, 21K, or 19K positions, clearly suggesting that the 21K product is the preproform and that the 19K and 23K products lack the signal sequence.

The specificity of the immune reaction with antibodies raised against arginine vasopressin is demonstrated in Fig. 5A. Using various oligopep-

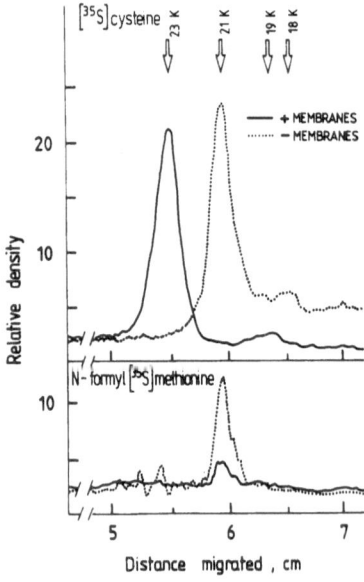

Figure 4. Translation of bovine hypothalamic mRNA in the presence of 1.6 μCi *N*-formyl[^{35}S]methionyl-tRNA/25 μl reticulocyte lysate system (lower panel) or 10 μCi [^{35}S]cysteine/12.5 μl (upper panel). Translation products were immunoprecipitated with antibodies against AVP and analyzed on SDS/PAGE. The gels were dried and fluorographed. The autoradiograms were scanned in a Gilford photometer at 560 nm.

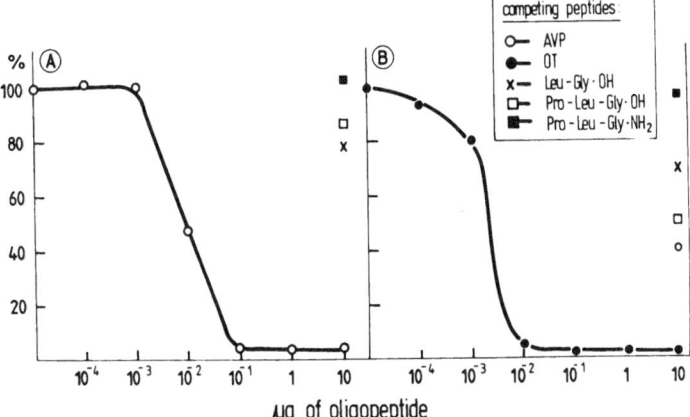

Figure 5. Specificity of the immune reaction. Competition experiments were performed between precursors to arginine vasopressin/neurophysin II or oxytocin/neurophysin I and unlabeled oligopeptides for binding by the respective antibodies. The precursors to arginine vasopressin (AVP)/neurophysin II (Np II) and to oxytocin (OT)/neurophysin I (Np I) were labeled with [^{35}S]cysteine by translating bovine hypothalamic mRNA in the reticulocyte system (Richter *et al.*, 1980b). After translation the synthesized precursors were immunoprecipitated with anti-AVP (A) or anti-OT (B), using *Staphylococcus aureus* immobilized membranes, electrophoresed on SDS/PAGE, and fluorographed (Richter *et al.*, 1980b). The autoradiograms were scanned and recorded in a Gilford photometer at 560 nm; the recorded peaks corresponding to the respective precursors were integrated; background precipitation, obtained with nonimmune IgG's and giving approximately 10% of the value obtained with the specific antibodies, was subtracted. 100% corresponds to the immunoprecipitation with either anti-AVP or anti-OT in the absence of added competing peptide. Where indicated, increasing concentrations of the respective oligopeptide were added together with either anti-OT or anti-AVP to the posttranslation system.

tides as competitors we found that only arginine vasopressin effectively competed.

5.2. Precursor to Oxytocin/Neurophysin I

Similar studies as shown above for the arginine vasopressin neurophysin II precursor were performed for the oxytocin/neurophysin I polyprotein with the respective antibodies (Fig. 2; see also Schmale and Richter, 1980). The *in vitro*-synthesized product has an M_r of 16,500 (Fig. 2, lanes 4, 5); in the presence of microsomal membranes it can be converted into a proform of M_r 15,500 (see also Fig. 6, lane 10). Apparently the proform is not core-glycosylated, which is in line with the oocyte experiments (see below). Competition experiments using oligopeptides related to oxytocin indicated that antioxytocin specifically recognizes amino acid sequences within the 16.5K precursor that are immunologically identical to oxytocin (Fig. 5B). The only slight cross-competition observed was with arginine vasopressin

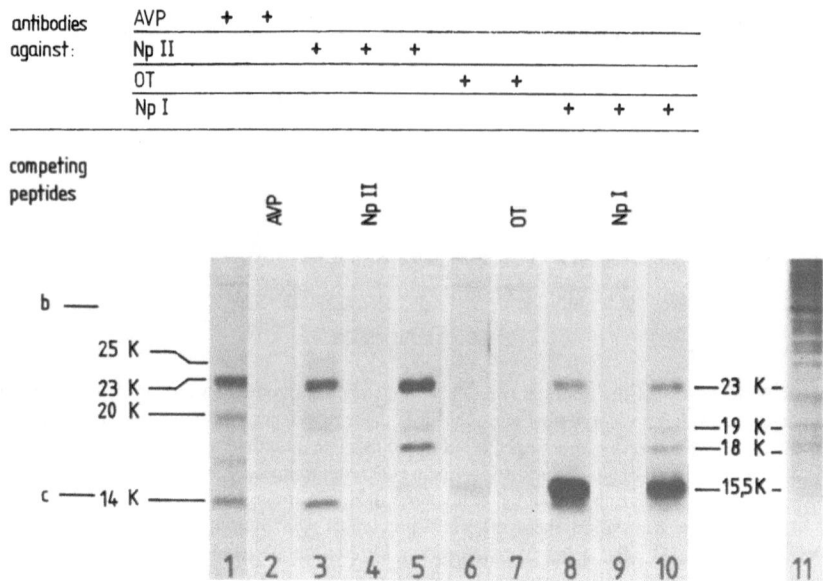

Figure 6. Biosynthesis, modification, and processing of polyproteins after microinjection of bovine hypothalamic mRNA into oocytes from *Xenopus laevis*. *X. laevis* females were induced to ovulate by gonadotropin treatment about 2–3 weeks before the oocytes were used for injection. This was done to ensure that the larger oocytes were growing actively. Dissected oocytes were maintained in modified Barth's solution and oocytes with a diameter of 1.2–1.3 mm were manually defolliculated by means of watchmaker forceps.

The injection was performed with micropipets with a tip diameter of about 40 μm and a calibrated volume of 60 nl. The mRNA solution and batches of five oocytes were placed on an ice-cooled parafilm-coated slide during the injection procedure to avoid evaporation. Injected oocytes were incubated for 9 hr at 18–20°C and then incubation was continued for another 11 hr with fresh medium (15 μl/oocyte) containing [^{35}S]cysteine (312 μCi/ml). Control oocytes were treated in the same way after injection with distilled water. After incubation the medium was removed and the oocytes were washed four times with label-free medium and once with 10 mM Tris–HCl buffer (pH 7.6) containing 100 μg/ml phenylmethyl sulfonyl fluoride. The incubation medium and the washed oocytes were stored in liquid nitrogen until further use. Five oocytes were homogenized, immunoprecipitated with the respective antibodies, and analyzed on SDS/PAGE as shown before (lanes 1–4, 6–9). Lane 11 shows incorporation of [^{35}S]cysteine in total proteins. Note that none of the bands in the total oocyte translation experiments corresponds to the 23 19, 18, 15.5, or 14K precursors. In order to test the specificity of the immunoreaction, 10 μg of the respective competing oligopeptide was added together with the antibodies listed (lanes 2, 4, 7, 9). In lanes 5 and 10 bovine hypothalamic mRNA was translated in the reticulocyte lysate system in the presence of dog pancreas microsomal membranes followed by immunoprecipitation with the indicated antibodies; the 19K precursor represents the unglycosylated proform to arginine vasopressin/neurophysin II.

and the tripeptide Pro-Leu-Gly-OH. Interestingly, Pro-Leu-Gly-NH$_2$ does not compete in the immune reaction. Apparently the antibodies to oxytocin do not recognize the NH$_2$ group at the C terminus.

5.3. Modifications, Processing, and Segregation of Hypothalamic Polyproteins Synthesized in Oocytes

In order to study processing of the hypothalamic polyproteins in more detail, mRNA was injected into oocytes from *X. laevis*. This system potentially allows one to study (1) the glycosylation of the investigated polyprotein under defined *in vivo* conditions and (2) the proteolytic cleavage of the precursor into intermediate and final products. Figure 6 shows that, with antibodies against arginine vasopressin or neurophysin II, the glycosylated 23K precursor is the major product in the oocyte (lanes 1, 3), which could be labeled with [^{3}H] fucose, [^{3}H] glucosamine, and [^{3}H]mannose (Ivell *et al.*, 1981). Although a number of intermediates could be identified using the respective antibodies, a product with an M$_r$ of 14,000 is particularly prominent (Fig. 6, lanes 1, 3). It is immunoreactive with antibodies to AVP and Np II after trypsinization, gives rise to AVP$_{1-8}$ and the Np II tryptides (Schmale and Richter, 1981c). Since neurophysin II runs in SDS/polyacrylamide gels at positions close to an apparent M$_r$ of 12,000 to 13,000, the 14K intermediate represents the direct precursor to both arginine vasopressin and neurophysin II. It also indicates that both peptides must be close neighbors within the polyprotein. This has been confirmed by sequence analysis (Land *et al.*, 1981).

The potency of the oocyte system as a tool for studying posttranslational events such as proteolytic cleavage and/or secretion of the polyproteins is further illustrated in Fig. 7. Apparently the hypothalamic precursors, synthesized in the oocyte, are also secreted into the incubation medium. Interestingly, the major secreted product found in the medium is the 25K precursor, which most likely contains the completed carbohydrate side chains. Whether the 14K intermediate, present as a minor band, is derived from proteolysis of the 25K precursor after secretion or whether it is secreted as an intermediate into the medium remains to be elucidated.

When similar microinjection experiments were performed using antibodies to oxytocin or neurophysin I, only the 15.5K proform could be identified in the oocytes as well as in the medium (Fig. 6, lanes 6–10; Fig. 7, lanes 6–8). Intermediates or final products cross-reacting with anti-oxytocin and/or anti-neurophysin I have not as yet been detected.

6. The Search for Precursors to Somatostatin and LH-RH

In order to study the resolution of the *in vitro* assay system, attention has recently been turned to the precursors of the tetradecapeptide hormone somatostatin. As indicated above, somatostatin is present in the hypothalamus at a concentration two orders of magnitude lower than arginine vasopressin or the neurophysins. Using specific immunoprecipitation followed by SDS/polyacrylamide gel electrophoresis we have been able to identify among the *in vitro* translation products of bovine hypothalamic mRNA a somatostatin precursor with an M_r of 15,400 (Fig. 8, lanes 3, 4). On cotranslational addition of dog pancreas microsomal membranes the pre-proform could be processed into a 14,400 product (lane 7). Apparently there is no concomitant glycosylation of the precursor, implying an extremely short signal sequence (Ivell and Richter, 1981).

Considering that LH-RH *in vivo* is present in the hypothalamus at a concentration one order of magnitude lower than somatostatin (Kalra and Kalra, 1980; Millar *et al.*, 1977; Ramirez *et al.*, 1980), one immediately realizes the problem of identifying a hypothetical LH-RH precursor. We have been trying to isolate precursors to LH-RH by *in vitro* translation of bovine hypothalamic mRNA assuming that LH-RH is synthesized in the

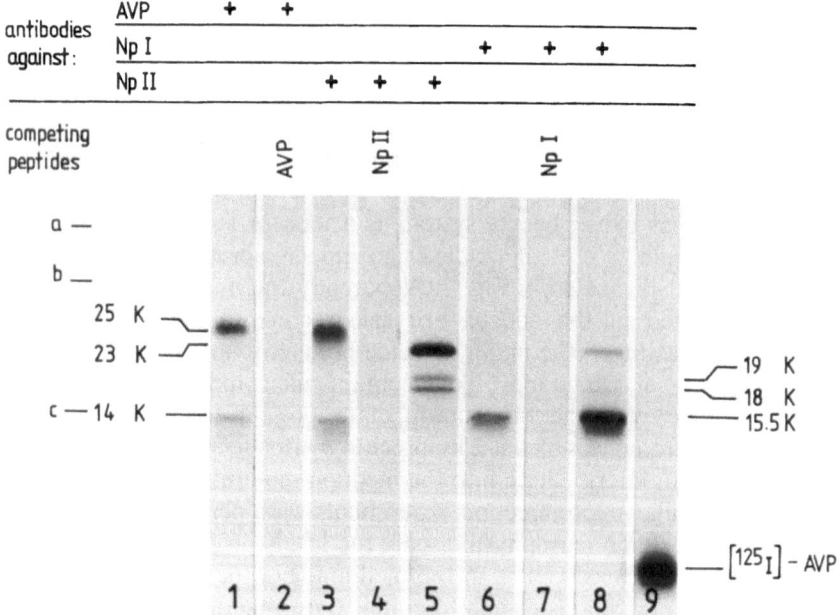

Figure 7. Secretion of hypothalamic polyproteins into the medium after microinjection of hypothalamic mRNA into oocytes. For details see legend to Fig. 6. 0.1 ml of oocyte medium was threefold concentrated and used for immunoprecipitation. Where indicated, 10 µg of the respective competing peptide was added. Lanes 5 and 8 represent cell-free translation in the reticulocyte system in the presence of microsomal membranes followed by immunoprecipitation with the indicated antibodies.

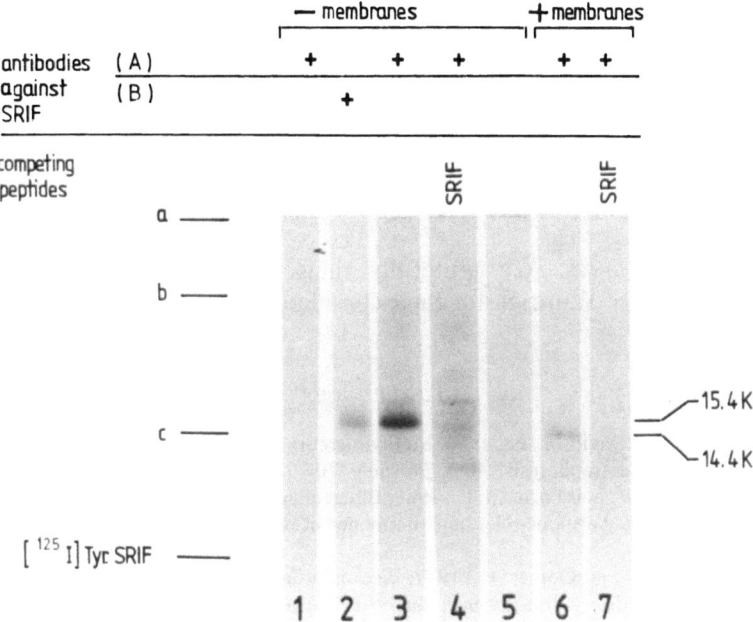

Figure 8. Biosynthesis and processing of the precursor to somatostatin in a cell-free reticulocyte lysate system. Bovine hypothalamic mRNA was translated in a rabbit reticulocyte lysate system in the absence (lanes 1–5) or presence (lanes 6 and 7) of microsomal membranes as previously described (Richter *et al.*, 1980b). Each 25-μl reaction mixture included 20 μCi [³⁵S]cysteine (specific activity 500–800 Ci/mmole). For the immune reaction, in place of the described buffer system, a buffer was used that contained 10 mM NaH₂PO₄, pH 7.6, 1 mM EDTA, 1% (w/v) Nonidet P-40, 1% (w/v) sodium deoxycholate, 0.3% SDS, 2% (v/v) Antagosan (Behring), 2 mM dithiothreitol, and 5 mM unlabeled cysteine. When indicated, 10 μg somatostatin (SRIF) (Bioproducts, Brussels) was used as competing peptide, together with about 50 μg rabbit anti-SRIF IgG from either Bioproducts (A) or Ferring, Kiel (B), or preimmune IgG (lane 5). The specific immunoreactive products were isolated using *Staphylococcus aureus* immunoabsorbent and analyzed by SDS/polyacrylamide gel electrophoresis using a gradient running gel of 10–20% acrylamide, followed by fluorography. Lane 1 is a control, where hypothalamic mRNA was omitted from the translation.

hypothalamus (Carillo, 1980). All experiments so far have proved negative. There are a number of possible reasons for these results. First, the antibodies we have raised may not be capable of recognizing LH-RH within a larger precursor molecule; specificity tests with our anti-LH-RH suggest that the antigenic site on the LH-RH molecule is close to the amide-blocked C terminus. Second, LH-RH is isolated in very small quantities from the hypothalamus; if the corresponding mRNA is also present in only small amounts, any synthesized molecules may be below the level of resolution of the assay system. We are currently engaged in exploring these possibilities and especially in improving the resolution of our assay system. Although all peptide hormones studied to date are synthesized in a ribosome-dependent way, there is no a priori proof that this is the case for LH-RH (Johanson *et al.*, 1973). There is one report (Millar *et al.*, 1977) of a large-molecular-weight LH-RH-like polypeptide from sheep hypothalami, which would imply synthesis on ribosomes, but this work has been questioned by others (Barnea *et al.*, 1980).

ACKNOWLEDGMENTS. We gratefully thank C. Schmidt, I. Kandt, S. Schweim, and S. Wintrich for their considerable help.

REFERENCES

Bailey, J. M., and Davidson, N., 1976, Methylmercury as a reversible denaturing agent for agarose gel electrophoresis, *Anal. Biochem.* **70**:75.

Barnea, A., Cho, G., and Porter, I. C., 1980, Effect of aging on the subneuronal distribution of luteinizing hormone-releasing hormone in the hypothalamus, *Endocrinology* **106**:1980.

Brownstein, M. J., and Gainer, H., 1977, Neurophysin biosynthesis in normal rats and in rats with hereditary diabetes insipidus, *Proc. Natl. Acad. Sci. USA* **74**:4046.

Brownstein, M. J., Russell, J. T., and Gainer, H., 1980, Synthesis, transport, and release of posterior pituitary hormones, *Science* **207**:373.

Carillo, A. I., 1980, The medial basal hypothalamus and luteinizing hormone release in the rat: Where are the LH-RH neurons responsible for tonic gonadotropin secretion?, *Life Sci.* **27**:179.

Chirgwin, J. M., Przybyla, A. E., MacDonald, R. I., and Rutter, W. J., 1979, Isolation of biologically active ribonucleic acid from sources enriched in ribonuclease, *Biochemistry* **18**:5294.

Chrétien, M., Benjannet, S., Gossard, F., Gianoulakis, C., Crine, R., Lis, M., and Seidah, N. G., 1979, From β-lipotropin to β-endorphin and 'pro-opio-melanocortin,' *Can. J. Biochem.* **57**:1111.

Gainer, H. Sarne, Y., and Brownstein, M. J., 1977, Biosynthesis and axonal transport of rat neurohypophysial proteins and peptides, *J. Cell Biol.* **74**:366.

Goodman, R. H., Jacobs, J. W., Chin, W. W., Lund, K., Dee, P. C., and Habener, J. F., 1980, Nucleotide sequence of a cloned structural gene coding for a precursor of pancreatic somatostatin, *Proc. Natl. Acad. Sci. USA* **77**:5869.

Giudice, L. C., and Chaiken, I. M., 1979, Immunological and chemical identification of a neurophysin-containing protein coded by messenger RNA from bovine hypothalamus, *Proc. Natl. Acad. Sci. USA* **76**:3800.

Herbert, E., Budarf, M., Phillips, M., Rosa, P., Policastro, P., and Oates, E., 1980, Presence of a pre-sequence (signal sequence) in the common precursor to ACTH and endorphin and the role of glycosylation in processing of the precursor and secretion of ACTH and endorphin, in: *Precursor Processing in the Biosynthesis of Proteins* (M. Zimmerman, R. A. Mumford, and D. F. Steiner, eds.), pp. 79–93, Academic Press, New York.

Hobart, P., Crawford, R., Shen, L. P., Pictet, R., and Rutter, W. J., 1980, Cloning and sequence analysis of cDNAs encoding two distinct somatostatin precursors found in the endocrine pancreas of anglerfish, *Nature (London)* **288**:137.

Ivell, R., and Richter, D., 1981, *In vitro* mRNA-directed synthesis and processing of an immunologically identified precursor to tetradecapeptide somostatin from bovine hypothalmus, *Biochemistry* **21**:1204.

Ivell, R., Schmale, H., and Richter, D., 1981, Glycosylation of the arginine vasopressin/ neurophysin II common precursor, *Biochem. Biophys. Res. Commun.* **102**:1230.

Johansson, K. N., Currie, B. L., Folkers, K., and Bowers, C. Y., 1973, Biosynthesis of the luteinizing hormone releasing hormone in mitochondrial preparations and by a possible pantetheine-template mechanism, *Biochem. Biophys. Res. Commun.* **53**:502.

Kalra, P. S., and Kalra, S. P., 1980, Modulation of hypothalamic luteinizing hormone-releasing hormone levels by intracranial and subcutaneous implants of gonadal steroids in castrated rats: Effects of androgen and estrogen antagonists, *Endocrinology* **106**:390.

Korant, B. D., 1980, Proteolytic events in viral replication, in: *Proteolysis, Demineralization and Other Degradative Processes in Human Biology and Disease*, Academic Press, New York, in press.

Land, H., Schütz, G., Schmale, H., and Richter, D., Cloning of DNA complementary to bovine hypothalmic mRNA, *Nature* **295**:299.

Lauber, M., Camier, M., and Cohen, P., 1979a, Higher molecular weight forms of immunoreactive somatostatin in mouse hypothalamic extracts: Evidence of processing *in vitro*, *Proc. Natl. Acad. Sci. USA* **76**:6004.

Lauber, M., Camier, M., and Cohen, P., 1979b, Immunological and biochemical characterization of distinct high molecular weight forms of neurophysin and somatostatin in mouse hypothalamus extracts, *FEBS Lett.* **97**:343.

Leipold, B., Schmale, H., and Richter, D., 1980, Processing and core-glycosylation of pre-pro-opiocortin synthesized *in vitro* by RNA from a mouse pituitary tumor cell line, *Biochem. Biophys. Res. Commun.* **94**:1083.

Lin, C., Joseph-Bravo, P., Sherman, T., Chan, L., and McKelvy, J. F., 1979, Cell-free synthesis of putative neurophysin precursors from rat and mouse hypothalamic poly(A) − RNA, *Biochem. Biophys. Res. Commun.* **89**:943.

Loh, Y. P., Jenks, B. G., and Broadwell, R. D., 1980, The role of the carbohydrate in the stabilization, processing and packaging of pro-ACTH/αMSH-endorphin in frog and mouse pituitary neurointermediate lobes, in: *Biosynthesis, Modification, and Processing of Cellular and Viral Polyproteins* (G. Koch and D. Richter, eds.), pp. 151–162, Academic Press, New York.

Maeda, K., and Frohman, L. A., 1980, Release of somatostatin and thyrotropin-releasing hormone from rat hypothalamic fragments *in vitro*, *Endocrinology* **106**:1837.

Mains, R. E., Eipper, B. A., and Ling, N., 1977, Common precursor to corticotropins and endorphins, *Proc. Natl. Acad. Sci. USA* **74**:3014.

Millar, R. P., Aehnelt, C., and Rossier, G., 1977, Higher molecular weight immunoreactive species of luteinizing hormone releasing hormone: Possible precursors of the hormone, *Biochem. Biophys. Res. Commun.* **74**:720.

Nakanishi, A., Inoue, A., Taii, S., and Numa, S., 1977, Cell-free translation product containing corticotropin and β-endorphin encoded by messenger RNA from anterior lobe and intermediate lobe of bovine pituitary, *FEBS Lett.* **84**:105.

Nakanishi, A., Inoue, A., Kita, T., Nakamura, M., Chang, A. C. Y., Cohen, S. N., and Numa, S., 1979, Nucleotide sequence of cloned cDNA for bovine corticotropin-β-lipotropin precursor, *Nature (London)* **278**:423.

Nicolas, P., Camier, M., Lauber, M., Masse, M. O., Möhring, J., and Cohen, P., 1980, Immunological identification of high molecular weight forms common to bovine neurophysin and vasopressin, *Proc. Natl. Acad. Sci. USA* **77**:2587.

Palmiter, R. D., 1974, Magnesium precipitation of ribonucleoprotein complexes: Expedient techniques for the isolation of undegraded polysomes and messenger ribonucleic acid, *Biochemistry* **13**:3606.

Pedersen, R. C., Brownie, A. C., and Ling, N., 1980, Pro-adrenocorticotropin/endorphin-derived peptides: Coordinate action on adrenal steroidogenesis, *Science* **208:**1044.

Ramirez, V. D., Dluzen, D., and Lin, D., 1980, Progesterone administration *in vivo* stimulates release of luteinizing hormone-releasing hormone *in vitro*, *Science* **208:**1037.

Richter, D., Ivell, R., and Schmale, H., 1980a, Neuropolypeptides illustrate a new perspective in mammalian protein synthesis—the composite common precursor, in: *Biosynthesis, Modification, and Processing of Cellular and Viral Polyproteins* (G. Koch and D. Richter, eds.), pp. 5–13, Academic Press, New York.

Richter, D., Schmale, H., Ivell, R., and Schmidt, C., 1980b, Hypothalamic mRNA-directed synthesis of neuropolypeptides: Immunological identification of precursors to neurophysin II/arginine vasopressin and to neurophysin I/oxytocin, in: *Biosynthesis, Modification, and Processing of Cellular and Viral Polyproteins* (G. Koch and D. Richter, eds.), pp. 43–66, Academic Press, New York.

Roberts, J. L., and Herbert, E., 1977a, Characterization of a common precursor to corticotropin and β-lipotropin: Cell-free synthesis of the precursor and identification of corticotropin peptides in the molecule, *Proc. Natl. Acad. Sci. USA* **74:**4826.

Roberts, J. L., and Herbert, E., 1977b, Characterization of a common precursor to corticotropin and β-lipotropin: Identification of β-lipotropin peptides and their arrangement relative to corticotropin in the precursor synthesized in a cell-free system, *Proc. Natl. Acad. Sci. USA* **74:**5300.

Rubinstein, M., Stein, S., and Udenfriend, S., 1978, Characterization of pro-opiocortin, a precursor to opioid peptides and corticotropin, *Proc. Natl. Acad. Sci. USA* **75:**669.

Rueckert, R., Palmenberg, A. C., and Pallansch, M. A., 1980, Evidence for a self-cleaving precursor of virus-coded protease, RNA-replicase and VPg, in: *Biosynthesis, Modification, and Processing of Cellular and Viral Polyproteins* (G. Koch and D. Richter, eds.), pp. 263–276, Academic Press, New York.

Russell, J. T., Brownstein, M. J., and Gainer, H., 1979, Trypsin liberates an arginine vasopressin-like peptide and neurophysin from a M_r 20,000 putative common precursor, *Proc. Natl. Acad. Sci. USA* **76:**6086.

Sachs, H., and Takabatake, Y., 1964, Evidence for a precursor in vasopressin biosynthesis, *Endocrinology* **75:**943.

Sachs, H., Fawcett, P., Takabatake, Y., and Portanova, R., 1969, Biosynthesis and release of vasopressin and neurophysin, *Recent Prog. Horm. Res.* **25:**447.

Schally, A. V., Coy, D. H., and Meyers, C. A., 1978, Hypothalamic regulatory hormones, *Annu. Rev. Biochem.* **47:**89.

Schmale, H., and Richter, D., 1980, *In vitro* biosynthesis and processing of composite common precursors containing amino acid sequences identified immunologically as neurophysin I/oxytocin and as neurophysin II/arginine vasopressin, *FEBS Lett.* **121:**358.

Schmale, H., and Richter, D., 1981a, Immunological identification of a common precursor to arginine vasopressin and neurophysin II synthesized by *in vitro* translation of bovine hypothalamic mRNA, *Proc. Natl. Acad. Sci. USA,* **78:**766.

Schmale, H., and Richter, D., 1981b, Bovine hypothalmic polyA+ RNA-directed synthesis of a common precursor to the nonapeptide arginine vasopressin and neurophysin II. immunological identification and tryptic peptide mapping, *Hoppe Seyler's Z. Physiol. Chemie* **362:**1551.

Schmale, H., and Richter, D., 1981c, Tryptic release of authentic arginine vasopressin$_{1-8}$ from a composite arginine vasopressin/neurophysin II precursor, *Neuropeptides* **2:**47.

Schmale, H., Leipold, B., and Richter, D., 1979, Cell-free translation of bovine hypothalamic mRNA: Synthesis and processing of the prepro-neurophysin I and II, *FEBS Lett.* **108:**311.

Spiess, J., and Vale, W., 1980, Multiple forms of somatostatin-like activity in rat hypothalamus, *Biochemistry* **19:**2861.

28

Tritium Labeling of Brain Peptides

István Teplán, Imre Mező, Károly Nikolics, János Seprődi, György Kéri, Michael Bienert, and Erhard Klauschenz

1. Introduction

During the past decade, radiolabeling of biologically active peptides has become a widely applied and essential technique for a variety of biochemical and physiological investigations.

The simple introduction of iodine into peptides and proteins, the high specific radioactivity and simple detection of iodine-125 have made radioiodination the most popular and routinely applied labeling technique. Also, the revolutionizing introduction of radioimmunoassay has strongly contributed to the wide application of radioiodination in the case of peptides and proteins. Radioiodination, however, is an external labeling method, i.e., it is based on the introduction of a radionuclide not normally present in the peptide structure. External labeling often significantly alters the biological properties of peptides, which may be a serious limitation in more sophisticated investigations. Such studies require labeled peptides with identical chemical and biological properties as the natural compounds. This can only be achieved by internal labeling. In the case of peptides, tritium, carbon-14 (and 11), and sulfur-35 are the most obvious choice as radionuclides for internal labeling. Of these, tritium is the most advantageous because of its

István Teplán, Imre Mező, Károly Nikolics, János Seprődi, and György Kéri ● 1st Institute of Biochemistry, Semmelweis University Medical School, Budapest, Hungary. *Michael Bienert and Erhard Klauschenz* ● Institute of Drug Research, GDR Academy of Sciences, Berlin, GDR.

high specific radioactivity and the variety of methods for its introduction into the peptide structure.

In the present paper we shall review various methods applied for the incorporation of tritium into peptides.

Tritium is a soft beta-emitting radionuclide with a half-life of 12.3 years. One milliatom tritium represents a radioactivity of 29 Ci (1080 GBq). Thus, multiple tritium incorporation into peptides may result in labeled preparations with over 100 Ci/mmole specific radioactivity suitable for studies in physiological concentrations of the peptides.

2. Methods for Tritium Labeling of Peptides

Tritium can be introduced into peptides by various chemical reactions. For practical reasons, they will be reviewed in four major types, as schematically represented in Fig. 1.

2.1. Direct Hydrogen–Tritium Exchange Methods

In earlier attempts to label peptides, direct exchange methods such as the Wilzbach technique were applied (e.g., Redding and Schally, 1972). The method is based on the activation of the C–H bond by several energy sources, such as the disintegration energy of tritium itself. Hembree *et al.* (1975) utilized a more sophisticated version of the direct exchange method

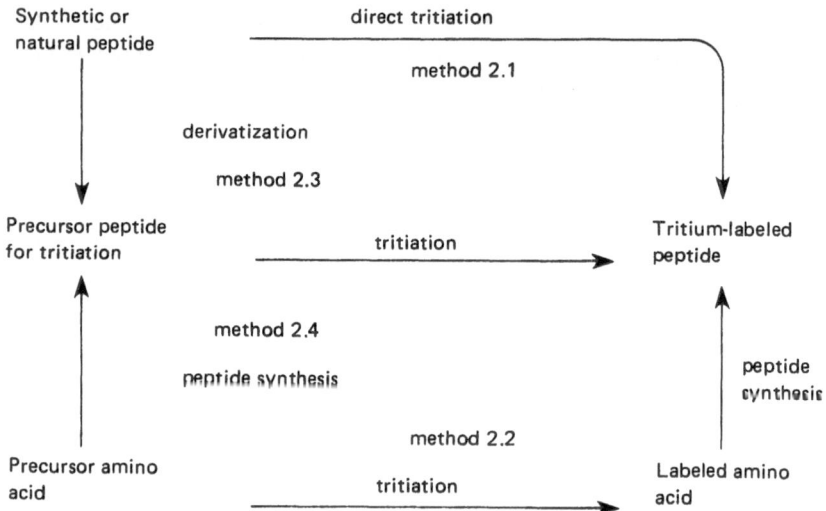

Figure 1. Schematic summary of the methods for tritium labeling of peptides.

for peptides by microwave discharge activation of tritium gas. ACTH labeled by this technique had full biological activity and 1 to 25 Ci/mmole specific radioactivity, while LH-RH labeled by the same method had 3.5 Ci/mmole specific activity (Ehrenkaufer et al., 1977).

The direct hydrogen–tritium exchange methods are relatively easily carried out; however, due to impurities formed during the reaction, purification of the products requires great effort. Further disadvantages, such as random distribution of the label, make the application of this technique very limited.

One promising approach of direct exchange labeling, only recently recognized, is exposure of the peptide to tritium in the presence of catalysts (Levine-Pinto et al., 1980; Klauschenz and Bienert, 1980). Under such conditions, histidine is preferentially labeled to a high level of tritium incorporation. Thus, TRH was obtained with a specific radioactivity of 30 Ci/mmole (Levine-Pinto et al., 1980), while LH-RH had 7–9 Ci/mmole specific activity (Klauschenz and Bienert, 1980).

2.2. Synthesis of Peptides from Labeled Amino Acids

With the availability of tritiated amino acids having high specific radioactivity, it is an obvious choice to prepare the peptides to be labeled from these. In a great number of examples peptides have been labeled by this method. Oxytocin and vasopressin (Sjöholm and Carlsson, 1967) were among the first peptides labeled this way. Angiotensin II was prepared from tritiated tyrosine using conventional methods of peptide synthesis (Mező and Teplán, 1971). ACTH-(1–10) and α-MSH were also synthesized from tritium-labeled tyrosine (Medzihradszky et al., 1972, 1974). For the preparation of [³H]-LH-RH, tritiated pyroglutamic acid was coupled to the C-terminal nonapeptide of LH-RH by the solid-phase method (Coy et al., 1973).

Enkephalins have also been prepared using this methodology (Sosn et al., 1977). A great advantage of synthesizing peptides from labeled amino acids is the availability of characterized intermediates and synthetic technology. However, work with very high doses of radioactivity can be a limiting factor to the application of this method. A further disadvantage, especially in the case of tritium-labeled compounds, is the possible autoradiolytic decomposition of the synthetic intermediates during storage.

2.3. Tritium Labeling of Peptides via Postsynthetically Modified Derivatives

Peptides to be labeled can be converted by a number of chemical modifications into derivatives that can be reduced by tritium to yield specifically labeled compounds. These reactions have been extensively reviewed earlier (Morgat and Fromageot, 1973, Morgat et al., 1977). Thus, only brief mention is made of this widely applied method that still carries great potential for the future.

The chemical modifications started with the iodination of peptides containing tyrosine or histidine residues, such as angiotensin II (Morgat et al., 1970), TRH (Pradelles et al., 1972a), LH-RH (Marche et al., 1972), and [Lys[8]]-vasopressin (Pradelles et al., 1972b). Iodinated derivatives were catalytically tritiated to give peptides with high specific radioactivities. Tyrosine carries the label in positions 3 and 5, while histidine in positions 2 and 5 of the aromatic rings after tritiation of completely iodinated peptide derivatives. Under mild conditions, monoiodination can also be achieved. This way, histidine can be preferentially labeled in position 5 of the imidazole ring (Levine-Pinto et al., 1980).

In the case of peptides containing more than one tyrosine or an additional histidine, nonspecific labeling will be achieved. For this reason a combination of iodination of fragment 11–24 and peptide synthesis (coupling 1–10 and iodo-11–24) was applied for the labeling of ACTH-(1–24) on the Tyr[23] residue by Brundish and Wade (1973) to avoid simultaneous labeling at Tyr[2] and His[6].

The basis for the extension of the method to label the tryptophan residue was the specific reaction of the indole ring with o-nitrophenyl sulfenyl (NPS) chloride. The NPS-tryptophan residue is readily tritiated to give specifically labeled peptides as shown in the case of pentagastrin (Marche et al., 1975), somatostatin, and LH-RH (Morgat et al., 1977).

Axo derivatives of histidine and tyrosine are also suitable for the introduction of tritium into peptides (Morgat et al., 1977). In the case of histidine-containing peptides, the imidazole ring is preferentially substituted in position 2 (Levine-Pinto et al., 1980).

In summary, the method using postsynthetically modified precursor compounds for peptide labeling is widely applicable to a very broad range of peptides. The speed and simplicity of this method of tritium labeling have contributed to the numerous examples realized so far. A great advantage of the method is the availability of a chemically characterized precursor compound, which can be converted into the specifically labeled peptide in one single reaction with carrier-free tritium gas.

2.4. Tritium Labeling of Peptides via Synthetic Precursors

Another approach for the preparation of precursor compounds for tritiation is peptide synthesis starting from suitable amino acid derivatives. These precursor peptides are basically similar to the peptides prepared by derivatization described in Section 2.3. However, peptide synthesis from suitable amino acid derivatives yields uniquely characterized precursor peptides that can be reduced by tritium gas in the presence of catalysts. Halogenated aromatic amino acids have been widely used for the synthesis of peptides to be tritiated.

Dibromotyrosine (Dbt) was applied for the synthesis of tritiation precursors for α-MSH (Nikolics *et al.*, 1976), angiotensin II (Teplán *et al.*, 1977; Mező *et al.*, 1978), and [Lys⁸]-vasopressin (Kéri and Teplán, 1978a).

Diiodotyrosine (Dit) served as precursor amino acid for the tritium labeling of ACTH-(1–24) in different positions (Brundish *et al.*, 1976). [Dit²³]-ACTH (human) was synthesized by the solid-phase method followed by catalytic reduction with tritium (Ramachandran and Behrens, 1977). [Dit²]- and [Dit²³]-ACTH were reported to be suitable for the preparation of ACTH with high specific radioactivity (Brundish and Wade, 1980). Dit was also used for the preparation of precursors for the tritiation of α-MSH and its derivatives (Eberle *et al.*, 1977; Eberle and Hübscher, 1979).

In addition to tyrosine labeling, halogenated derivatives of phenylalanine have also been used for the synthesis of precursor peptides for tritiation. This development was especially useful in those instances where no tyrosine or histidine was present in the peptide structure, thus other labeling methods failed. 4-Chlorophenylalanine (Cpa) was resolved on preparative scale (Kovács *et al.*, 1975) and several derivatives for peptide synthesis were prepared from it. Angiotensin II was labeled via its Cpa⁸-derivative (Teplán *et al.*, 1977). [Cpa⁵]-Bradykinin (Seprődi *et al.*, 1980) served as precursor for bradykinin labeling. [Lys⁸]-vasopressin was also labeled by the same approach in a different position (Kéri and Teplán, 1978a).

4-Iodophenylalanine (Ipa) was also used for the preparation of precursor peptides. ACTH-(1–24) was synthesized containing Ipa in position 7 and then tritiated to give a preparation with high specific radioactivity (Brundish *et al.*, 1976). Also, the entire ACTH molecule containing the Ipa⁷ substitution was synthesized and labeled with tritium (Brundish and Wade, 1980). Somatostatin was also labeled via its Ipa⁶-derivative to yield the highly-labeled tetradecapeptide (Allen *et al.*, 1978). Among halogenated amino acids, 2,5-diiodo-histidine was incorporated into ACTH-(1–24) and the resulting precursor peptide successfully tritiated (Allen *et al.*, 1979). Although not tritiated, 5-bromotryptophan was incorporated into ACTH-(1–24) to increase the biological activity of the molecule (Allen *et al.*, 1980).

Besides halogenated aromatic amino acid derivatives, unsaturated amino acids form another major class of precursors. Dehydronorleucine was used to label pentagastrin (Pande *et al.*, 1969). Dehydronorvaline was applied for ACTH-(1–24) labeling (Schwyzer and Karlaganis, 1973).

Felix *et al.* (1977) applied 3,4-dehydroproline for the synthesis of peptides to be labeled. The saturation of double bonds is usually significantly slower than the reduction of halogenated aromatic amino acids. The longer reaction times required for complete tritiation of unsaturated amino acid residues may lead to side reactions, especially nonspecific labeling. Direct exchange may take place during long exposure to tritium, which was first observed in the case of ACTH-(1–24) labeling (Schwyzer and Karlaganis, 1973).

In summary, the tritium labeling method using precursor compounds prepared by chemical peptide synthesis has found wide application in the labeling of several neuropeptides. With the development of methods of peptide synthesis, this labeling technique has become less tiresome. The precursor compounds prepared synthetically are well characterized and provide specific labeling at very high specific radioactivity.

Further progress can still be expected from this method. New precursor amino acids can be developed, such as dehydrolysine, halogenated tryptophan, etc., to broaden the applicability of this technique.

Finally, amino acid derivatives used as precursors for tritium labeling have found additional applications in studies of structure–activity relationship of several biologically active peptides. Smith *et al.* (1977) prepared 3,4-dehydroproline-containing derivatives of neurohypophyseal peptides and reported interesting endocrine properties. Cpa[2]-containing analogues of inhibitory LH-RH derivatives proved to be very effective antagonists of LH-RH action (Coy *et al.*, 1979; Seprődi *et al.*, 1980). 3,4-Dehydroproline[1] substitution was also found to enhance the inhibitory activity of LH-RH derivatives (Rivier *et al.*, 1980).

3.   Tritium Labeling of Brain Peptides

Using the methods described above, several brain peptides have been labeled with tritium for biochemical and endocrinological studies. Besides labeling work done by the authors, we also review literature data. These examples demonstrate the present possibilities provided by the individual labeling methods.

3.1.   α-MSH and ACTH

For the specific labeling of α-MSH (Ac-Ser-Tyr-Ser-Met-Glu-His-Phe-Arg-Trp-Gly-Lys-Pro-Val-NH$_2$), we first applied the conventional route of incorporating [^{3}H]tyrosine into the peptide chain by classical synthetic methods (Medzihradszky *et al.*, 1974). During the synthesis, numerous operations were carried out with radioactive intermediates; thus, the specific radioactivity was diluted to 2 Ci/mmole to avoid considerable auto-radiolytic decomposition of the peptides. For the preparation of α-MSH with higher specific radioactivity, we prepared its 3′,5′-Dbt[2] derivative by a similar synthetic route as shown in Fig. 2.

The N-terminal tetrapeptide was prepared by condensation of fragments 1–2 and 3–4. Dibromo-α-MSH was then synthesized via coupling fragments 1–4 and 5–10 followed by coupling fragments 1–10 to 11–13. The precursor peptide was tritiated in the presence of Pd/Al$_2$O$_3$ catalyst yielding [^{3}H]α-MSH with a specific radioactivity of 22 Ci/mmole. In earlier experiments using Pd/CaCO$_3$ catalyst, considerably lower specific radioactivities (3–4 Ci/mmole) were obtained, probably due to catalyst poisoning by methionine sulfur. Under the conditions of this synthesis, no oxidation of Met[4] to its sulfoxide was observed. The labeled peptide had full melanotropic activity and amino acid analysis after hydrolysis confirmed the presence of methioine. [^{3}H]α-MSH was used in experiments on the brain distribution and *in vivo* disappearance and excretion of the hormone (Kastin *et al.*, 1976; Redding *et al.*, 1978).

Using 3′,5′-Dit as a precursor amino acid, Eberle and co-workers (Eberle *et al.*, 1977; Eberle and Hübscher, 1979) synthesized α-MSH and several of its derivatives for specific tritium labeling. Using a combination of Pd/C and Rh/CaCO$_3$ catalysts for the tritiation, a specific radioactivity of 34 Ci/mmole was reached for α-MSH.

Due to the structural identity between α-MSH and ACTH for the 13

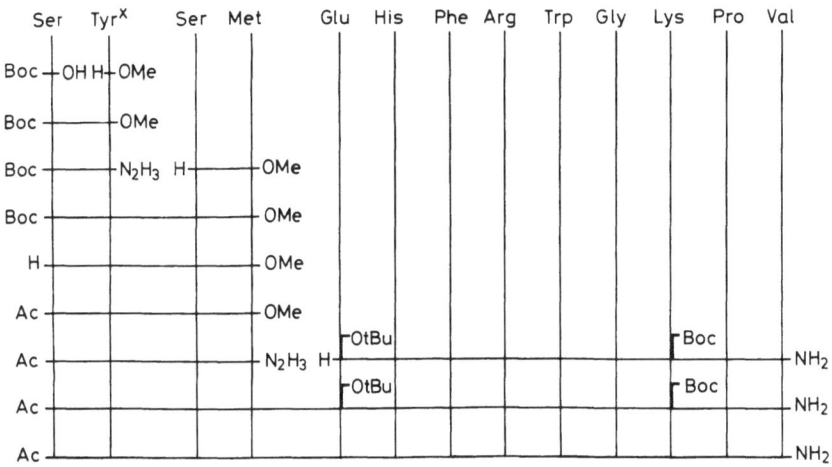

Figure 2. Synthesis of α-MSH labeled with tritium in the Tyr[2] residue. For the preparation of the [Dbt[2]]-α-MSH precursor peptide, 3,5-Dbt was incorporated in place of Tyr[2].

amino acids of α-MSH, the tritium labeling of ACTH requires basically similar methods. Several fragments of ACTH have been prepared incorporating various precursor amino acids for specific tritium labeling. Dehydronorvaline was used as a precursor for position 4 labeling in an ACTH-(1–24) analogue (Schwyzer and Karlaganis, 1973). After extremely long (26 hr) tritiation in the presence of Pd/C, a specific radioactivity of 7.4 Ci/mmole was reached. Various halogenated aromatic amino acids have been used as precursors for ACTH labeling. In a combination of postsynthetic iodination and further synthesis, Brundish and Wade (1973) tritiated [Dit[23]]-ACTH-(1–24) in the presence of Pd/C and Rh/CaCO$_3$ catalysts to yield ACTH-(1–24) with 46 Ci/mmole specific radioactivity. We have also incorporated Dbt^2 into ACTH-(1–10) and ACTH-(1–20)-NH$_2$ partial sequences (Nikolics, 1974). The syntheses are shown in Fig. 3. Similar to the preparation of [Dbt^2]-α-MSH, fragments 1–4 and 5–10 were condensed to yield the decapeptide. The [Dbt^2]-ACTH-(1–20)-NH$_2$ was coupled from fragments 1–4, 5–14, and 15–20-NH$_2$. Both peptides were tritiated in the presence of Pd/CaCO$_3$ to yield specific radioactivities between 2 and 5 Ci/mmole. Using peptide synthesis starting from Dit, Brundish *et al.* (1976) prepared [Dit^2]-ACTH-(1–24) and the Dit^2 derivative of highly active ACTH-(1–18) sequence. Under the same tritiation conditions as given for the Dit[23] derivative, specific radioactivities of 29 and 17 Ci/mmole were reached in the case of the tetracosapeptide and the octadecapeptide, respectively. Brundish and Wade (1976) also prepared the Ipa^7 and the 4-bromophenylalanine^7 analogues of ACTH-(1–24). While the bromo derivative, in the presence of Pd/C and Rh/CaCO$_3$ catalysts, could not be completely reduced, the iodo compound yielded a product with 27 Ci/mmole specific radioactivity.

The entire human ACTH molecule was first specifically tritiated by Ramachandran and Behrens (1977) via the synthetic [Dit[23]]-ACTH derivative. A very high (46 Ci/mmole) tritium incorporation could be attained under the reaction conditions described: 0.1 M acetic acid–hexamethyl-

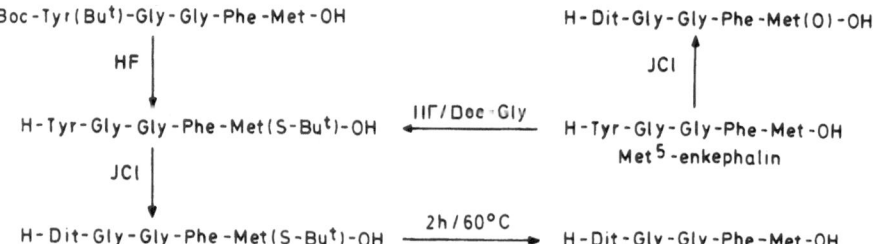

Figure 3. Iodination of [Met^5]-enkephalin and protection of methionine against oxidation by sulfonium salt formation.

phosphoramide–dimethylformamide solvent mixture and PdO catalyst with CaCO$_3$ added.

Also, the human ACTH sequence was labeled in three different positions via iodo derivatives by Brundish and Wade (1980). Synthetic [Dit^2]-, [Dit^{23}]-, and [Phe(I)^7]-ACTH derivatives were tritiated in the same way as the ACTH partial sequences, and resulted in preparations with 49, 25, and 31 Ci/mmole specific radioactivity, respectively. As for α-MSH, methionine sulfur did not interfere with the catalytic reduction in the case of ACTH, either. However, in nearly all cases, a partial tritium incorporation could be observed into His^6. The significance and potential of this side reaction will be discussed later. ACTH was also labeled by microwave discharge activation of tritium gas (Hembree *et al.*, 1975). Specific radioactivities between 1 and 25 Ci/mmole were obtained in different batches. Interestingly, nonrandom labeling of the individual amino acids was observed with preferential tritium incorporation into leucine (30–50% of the total radioactivity).

3.2. Vasopressin

Attempts on specific labeling of vasopressin started with the incorporation of tritiated tyrosine into the nonapeptide through a number of synthetic steps. The synthetic peptide had a specific radioactivity of 1.9 Ci/mmole. The postsynthetic iodination of [Lys^8]-vasopressin served as a basis for the tritium labeling in Tyr^2 as described by Pradelles *et al.* (1972b). Despite the presence of the disulfide bridge, relatively high (10 Ci/mmole) specific radioactivity could be reached after tritiation in the presence of Pd/Al$_2$O$_3$. A similar method for the tritium labeling of neurohypophyseal hormones was applied by Flouret *et al.* (1977).

For the specific tritium labeling of [Lys^8]-vasopressin, Cys-Tyr-Phe-Gln-Asn-Cys-Pro-Lys-Gly-NH$_2$, we prepared the Dbt^2 and the Cpa^3 derivatives of the nonapeptide. The two peptides were prepared by solid-phase peptide synthesis. After cleavage of the peptide from the resin and deprotection, the precursor peptides were chromatographically purified. Tritiation in the presence of Pd/Al$_2$O$_3$ catalyst from the dibromo derivative gave peptides with 8–12 Ci/mmole specific radioactivity. The Cpa derivative yielded peptides with 4–5 Ci/mmole specific activity (Kéri and Teplán, 1978a). The CNS-active derivative of [Lys^8]-vasopressin, des-glycinamide-[Lys^8]-vasopressin, was also labeled similarly, with Dbt^2- and Cpa^3-substituted octapeptides as precursors (Kéri and Teplán, 1978b).

3.3. LH-RH

Several tritium-labeled analogues of LH-RH, Pyr-His-Trp-Ser-Tyr-Gly-Leu-Arg-Pro-Gly-NH$_2$, containing the tritium label in different positions of the peptide chain have been synthesized in recent years. [3,4-^{3}H-Pyr^1]-LH-RH with a specific radioactivity of 18 Ci/mmole was synthesized by the solid-phase procedure using [3,4-^{3}H]-Pyr in the final coupling step of the synthesis (Coy *et al.*, 1973). Iodinated analogues of LH-RH were also used as precursors for catalytic tritiation. Iodination of the LH-RH molecule with iodine monochloride resulted in an iodine substitution of the aromatic moiety of Tyr^5 accompanied by a partial iodination of the histidine imidazole side chain in position 2. Thus, the dehalogenation of this analogue with tritium gas, performed in glacial acetic acid in the presence of Pd/Al$_2$O$_3$ (Marche *et al.*, 1972), led to a partial histidine tritiation besides labeling of the tyrosine residue. To circumvent this problem, iodination of the LH-RH partial sequence 4–10, Ser-Tyr-Gly-Leu-Arg-Pro-Gly-NH$_2$, and subsequent coupling of the iodinated product with the N-terminal tripeptide, Pyr-His-Trp, as well as peptide synthesis using iodine derivatives of tyrosine and histidine, respectively, were performed (Terada *et al.*, 1980). These procedures yield LH-RH analogues containing iodine only in the desired position and avoid a possible oxidative degradation of the tryptophan residue during the iodination procedure. The tritium labeling of these iodinated analogues of LH-RH has not yet been published.

Our aim was to study the tritiation of the LH-RH molecule when proline in position 9 is replaced by 3,4-dehydroproline. The precursor peptide, [Dhp^9]-LH-RH, was synthesized by solid-phase synthesis of the intermediate H-Ser(Bzl)-Try-Gly-Leu-Arg(NO$_2$)-Dhp-Gly-resin and condensation with the N-terminal fragment Pyr-His-Trp-OH. Homogeneous [Dhp^9]-LH-RH was obtained after cleavage and deprotection with hydrogen fluoride followed by final purification using ion-exchange and partition chromatography (Klauschenz and Bienert, 1980; and Klauschenz *et al.*, 1981). A series of tritiation experiments were carried out to find optimal labeling conditions. The results of Felix *et al.* (1977) prompted us to use PdO as catalyst and a reaction time of about 2 hr. Satisfactory tritium incorporation was achieved but further experiments suggested that the use of Pd/Al$_2$O, is more convenient. Radioactivities of 50–60 Ci/mmole were reached even using very short reaction times (10 min). The peptide was successfully used in proteolytic degradation experiments in pituitary cell culture (see Nikolics *et al.*, Chapter 20).

The extremely high tritium incorporation suggested the possibility that other amino acid residues besides proline could have also been labeled. Therefore, the labeled peptides were analyzed by amino acid analysis after acid hydrolysis, showing the incorporation of about 40% of the total radioactivity into histidine and 60% into proline with PdO as catalyst and a reaction time of 2 hr. This corresponds to a specific radioactivity of 12Ci/mmole due to tritium incorporation into the histidine residue. The histidine labeling decreased down to 10% by using Pd/Al$_2$O$_3$ and short reaction times, but even under these conditions 5–6 Ci/mmole is located in the histidine residue. These results are in good correlation with the tritium distribution measured after chymotrypsin treatment of the labeled peptides, yielding the fragments [³H]-Pyr-His-Trp, [³H]-Pyr-His-Trp-Ser-Tyr, and Gly-Leu-Arg-[³H]-Pro-Gly-NH $_2$. These fragments were separated by paper electrophoresis and the distribution of radioactivity was measured (see also Horsthemke and Bauer, 1980, with regard to the cleavage of [^{H}-Pry¹]-LH-RH by chymotrypsin).

The direct catalytic exchange between hydrogen and tritium of the histidine imidazole ring as a side reaction during the catalytic dehalotritiation of ACTH and α-MSH precursors (Table I) observed earlier was found to be negligible or low. The high tritium incorporation into histidine during the tritiation of [Dhp⁹]-LH-RH prompted us to label LH-RH directly. According to the conditions of Felix *et al.* (1977) for the tritiation of [Dhp³]-TRH without simultaneous tritium incorporation into histidine, we exposed a methanolic solution of LH-RH to tritium gas in the presence of

Table I.   Tritium Incorporation into Histidine as a Side Effect during the Catalytic Tritiation of ACTH and α-MSH Sequences

Precursor peptide tritiated[a,b]	Specific radioactivity (Ci/mmole)	Percentage tritium found in histidine[c]	References
[Phe²,Agl⁴]-ACTH-(1–24)	7.4	5[d]	Schwyzer and Karlaganis (1973)
[Dit²³]-ACTH-(1–24)	46	0.5[d]	Brundish and Wade (1973)
[Dit²]-ACTH-(1–24)	29	2.5[e]	
[D-Ser¹,Dit²,Lys¹⁷,¹⁸]-ACTH-(1–24) amide	17	2.5[e]	Brundish *et al.* (1976)
[pIPhe⁷]-ACTH-(1–24)	27	3.3[e]	Brundish and Wade (1976)
[Dit²³]-ACTH-(1–39)	25	1.3[e]	Brundish and Wade (1977)
[Dit²]-ACTH-(1–39)	49	4.7[e]	Brundish and Wade (1980)
[pJPhe⁷]-ACTH-(1–39)	31	8.2[e]	
[Dit²]-α-MSH	34	5[e]	Eberle and Hübscher (1979)

[a] Uncoded amino acids, see references.
[b] For conditions of tritiation see references.
[c] The histidine residue of the ACTH and α-MSH sequences is located in position 6.
[d] Determined by amino acid analysis after acid hydrolysis.
[e] Determined by amino acid analysis after enzymatic hydrolysis.

PdO for 3 hr. After purification by paper electrophoresis, LH-RH with 7–9 Ci/mmole was obtained due to direct tritium incorporation into the histidine residue (Klauschenz and Bienert, 1980). Recently, a direct tritiation of histidine in TRH (Levine-Pinto et al., 1980) has shown by means of ^{3}H NMR that 84% of the tritium was incorporated into the C-2 and 16% into the C-5 position of the imidazole ring.

The preparation of tritium-labeled [Pro^9]-LH-RH via solid-phase synthesis of [D,L-Dhp^9]-LH-RH followed by tritiation of the diastereomers and isolation of the labeled L-isomer by HPLC has been reported (Perrin et al., 1980).

LH-RH was also labeled using microwave discharge activation of tritium gas (Ehrenkaufer et al., 1977). The tritium incorporation was found to be 3.5 Ci/mmole when nanomole amounts of the peptide were placed on a cellulose Millipore filter during the exposure to activated tritium gas. Three separate chromatographic steps were necessary to obtain homogeneous [^{3}H]-LH-RH.

Inhibitory analogues of LH-RH containing Dhp and/or halogenated aromatic amino acids have also been prepared (Coy et al., 1979; Seprődi et al., 1980; Rivier et al., 1980). These peptides showed interesting biological properties and they could also serve as precursors for tritium labeling.

3.4. Substance P

To prepare precursor peptides for the tritiation of substance P(SP), Arg-Pro-Lys-Pro-Gln-Gln-Phe-Phe-Gly-Leu-Met-NH$_2$, it is advantageous to replace the sulfur-containing methioine residue in position 11 by the isosteric norleucine since this substitution does not influence the biological activity of SP (Van Rietschoten et al., 1975). Problems such as the formation of sulfoxides during the synthesis and storage of the peptides as well as the inactivation of the catalyst during the tritiation of the precursor can thus be avoided. Starting from Dhp^{2,4} and Cpa^{7,8} analogues of [Nle^{11}]-SP, we have introduced tritium into the positions Pro^2, Pro^4, and Phe^7, Phe^8, respectively, by catalytic tritiation. The synthesis of [Cpa^{7,8},Nle^{11}]-SP was achieved by condensation of the fragments Boc-Arg(NO$_2$)-Pro-Lys-(Boc)-Pro-OH and H-Gln-Gln-Cpa-Cpa-Gly-Leu-Nle-NH$_2$ according to the synthetic procedure for SP, as published earlier (Bienert et al., 1979a). In contrast, stepwise chain elongation was used to synthesize [Dhp^{2,4},Nle^{11}]-SP. Dhp was introduced as Boc-Dhp-OPfp or Boc-Dhp-ONB (Bienert et al., unpublished). For final purification of the precursor peptides, partition and ion-exchange chromatography were used. The tritiation of [Cpa^{7,8}, Nle^{11}]-SP was performed in aqueous solution in the presence of Pd/Al$_2$O$_3$ catalyst yielding labeled [Phe^{7,8},Nle^{11}]-SP with a specific radioactivity of 20–27 Ci/mmole. Specific radioactivities were based on peptide content obtained by radioimmunoassay, amino acid analysis, and fluorescence measurement after reaction with fluorescamine. Purification of the labeled peptides before use in biological experiments was achieved by paper electrophoresis of aliquots. Recently, similar results in labeling of this precursor have been obtained by Morgat *et al.* (1980). Additionally, a modified partial sequence of [Nle^{11}]-SP, Ac-Lys-Cpa-Phe-Gly-Leu-Nle-NH$_2$, was labeled by catalytic dehalogenation yielding the corresponding tritium-containing analogue with 23 Ci/mmole specific radioactivity (Bienert *et al.*, 1978).

In addition to the specific tritium labeling of the *para* positions of Phe^7 and Phe^8, a nonspecific labeling in other positions of the aromatic rings by direct hydrogen–tritium exchange may take place during catalytic dehalotritiation. The aromatic rings of SP are able to incorporate about 15 Ci/mmole tritium, when exposing the peptide in aqueous solution to tritium gas in the presence of Pd/Al$_2$O$_3$ (Morgat *et al.*, 1980).

Interestingly, methionine-containing SP precursors have also been labeled successfully (Hanley *et al.*, 1980; Morgat *et al.*, 1980). [Boc-Arg^1,Lys(Boc)^3,Phe (I)^8]-SP was dehalogenated in dimethylformamide by using 5% Pd/charcoal and 5% Rh/CaCO$_3$. The same catalysts were found to be suitable for the dehalogenation of other methionine-containing peptides, for example [Dit^{23}]-α-corticotropin-(1–24)-tetracosapeptide (Brundish and Wade, 1973). However, [Cpa^{7,8}]-SP has been tritiated using Pd/Al$_2$O$_3$ catalyst, demonstrating that no special sulfur-resistant catalyst is necessary for labeling SP (Morgat *et al.*, 1980). Nevertheless, the replacement of methionine by norleucine can only be advantageous, because of the high tendency of the methionine sulfur to undergo oxidation. Thus, formation of S-oxides was observed during storage and purification of tritium-labeled SP (Morgat *et al.*, 1980; Hanley *et al.*, 1980). The Dhp-containing precursor, [Dhp^{2,4},Nle^{11}]-SP, allows the incorporation of four tritium atoms into the N-terminal part of the molecule (theoretical specific radioactivity 116 Ci/mmole). In a first labeling experiment, tritium-labeled [Pro^{2,4},Nle^{11}]-SP was obtained with a specific radioactivity of 96 Ci/mmole (Morgat and Bienert, unpublished).

Tritium-labeled [Phe^{7,8},Nle^{11}]-SP and tritium-labeled [Phe^8]-SP have been extensively used to study the enzymatic degradation (Berger *et al.*, 1979) as well as receptor binding of SP (Hanley *et al.*, 1980).

3.5. Enkephalins and Endorphins

During iodine monochloride treatment of [Met^5]-enkephalin, H-Tyr-Gly-Gly-Phe-Met-OH, besides the iodination of the Tyr residue, a complete transformation of the methionine thioether group into sulfoxide takes place (see Fig. 3). A method for the reversible protection of the thioether group to prevent the S-oxidation was developed (Bienert *et al.*, 1979b,

1981) on the basis of S-*tert*-butylsulfonium intermediates, formerly known as by-products in the synthesis of methionine-containing peptides (Noble *et al.*, 1976; Bienert *et al.*, 1979a). The methionine peptide is quantitatively converted into the S-*tert*-butylsulfonium derivative by treatment of the corresponding Boc- and/or Bu^t-protected peptide with liquid hydrogen fluoride (Fig. 3). Alternatively, the sulfonium intermediate was obtained from [Met^5]-enkephalin by the addition of Boc-Gly and liquid hydrogen fluoride, or Boc-Gly and trifluoromethanesulfonic acid, respectively. Iodination of the sulfonium derivative was performed by iodine monochloride and subsequently the thioether group was regenerated by heating the aqueous solution (60°C for 2 hr), yielding the sulfoxide-free Dit peptide, which was purified by chromatography. Figure 3 shows the reaction scheme of the iodination of [Met^5]-enkephalin {the same procedure was also useful in the synthesis of the Dit analogue of [Tyr^8]-SP (Bienert *et al.*, 1981)}.

[Dit^1,Met^5]-enkephalin was tritiated in aqueous solution in the presence of Pd/Al$_2$O$_3$ catalyst for 10 min yielding labeled [Met^5]-enkephalin, which after purification by HPLC exhibited a specific radioactivity of 38 Ci/mmole showing that the presence of the thioether group did not diminish the effectiveness of the catalyst (Bienert *et al.*, 1981).

The preparation of tritium-labeled [Leu^5]-enkephalin via iodination of the tyrosine residue can be performed without difficulties in contrast to [Met^5]-enkephalin. Also, the use of Dit or Dbt in the synthesis of the enkephalin results in suitable precursors for triitiation. The introduction of tritium into the Phe^4 position via synthesis of the Cpa^4 analogue has also been performed (Day and Freer, 1978). It was found that additional replacement of leucine by norleucine results in a more convenient ligand for receptor binding studies because of its higher biological potency (Day and Freer, 1978).

A tritium labeling method in the methionine side chain has been devel-

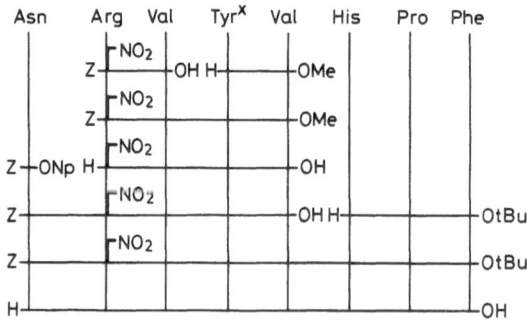

Figure 4. Synthesis of angiotensin II labeled with tritium in the Tyr^4 residue. For the preparation of the Dbt^4 and the Cpa^8 precursor peptides, 3,5-Dbt or 4-Cpa was incorporated into the peptide.

oped consisting of the alkylation of the sulfur atom by use of tritium-labeled methyliodide to form an S-[^{3}H]methylsulfonium intermediate. This has to be dealkylated, resulting in the corresponding methionine peptide bearing a tritium label in the methyl group of the thioether moiety (Morgat and Fromageot, 1979). The application of this procedure to label [Met^5]-enkephalin gave a product with 5 Ci/mmole specific radioactivity (Morgat *et al.*, 1977).

β-Endorphin has been labeled with 50 Ci/mmole specific radioactivity by tritiating the Dit^{27} analogue obtained by solid-phase peptide synthesis. The dehalotritiation was performed in the solvent system hexamethylphosphoric acid amide–dimethylacetamide in the presence of Pd catalyst (Houghten and Li, 1978). The thioether protection method shown in Fig. 3 could also be applied during iodination of endorphins.

3.6. Angiotensin and Bradykinin

Although not directly termed as neuropeptides, angiotensin and bradykinin have been shown to possess CNS effects. Angiotensin II, Asn-Arg-Val-Tyr-Val-His-Pro-Phe, was first synthesized from tritiated tyrosine (Mezõ and Teplň, 1971) to give specifically labeled compound. The direct iodination of the octapeptide followed by catalytic tritiation as applied by Morgat *et al.* (1970) resulted in nearly complete tritium incorporation (specific radioactivity 56 Ci/mmole). For the specific labeling of angiotensin II in two different positions, we synthesized the Dbt^4 and Cpa^8 derivatives of the octapeptide (Teplán *et al.*, 1977) (Fig. 4). The synthesis of the two precursor peptides was based on the condensation of the two tetrapeptides 1–4 and 5–8. The catalytic tritiation of the dibromo derivative in the presence of Pd/Al$_2$O$_3$ resulted in labeled [Tyr^4]-angiotensin II with 30 Ci/mmole, while the chloroderivative gave labeled [Phe^8]-angiotensin II with 23 Ci/mmole specific radioactivity.

For the labeling of bradykinin, Arg-Pro-Pro-Gly-Phe-Ser-Pro-Phe-Arg, we synthesized the Cpa^5 derivative of the nonapeptide as shown in Fig. 5. The molecule was condensed from the three tripeptides 1–3, 4–6 and 7–9 (Seprõdi *et al.*, 1980). Catalytic tritiation of the precursor compound succeeded in very high tritium incorporation to give a specific radioactivity of 19 Ci/mmole.

4. Conclusion

Great progress has been made in the development of methods for tritium labeling of peptides. The techniques reviewed here form a solid basis for the labeling of brain peptides to be used in biochemical, neurochemical, and endocrinological experiments.

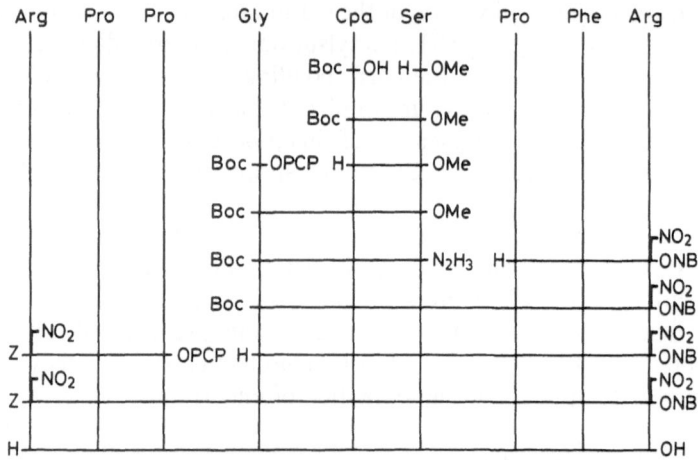

Figure 5. Synthesis of bradykinin labeled with tritium in the Phe⁵ residue via [Cpa⁵]-bradykinin precursor.

The methods (2.3 and 2.4) based on the preparation of precursors for catalytic tritiation provide wide variability in respect to introduction of tritium and position of the label. These methods also carry the potential to reach very high specific radioactivities. Multiple substitution of halogenated or unsaturated amino acids in peptides to be labeled may result in specific activities above 100 Ci/mmole. Using the methods available at present, basically any peptide can be specifically labeled with tritium at a high level of incorporation.

Further progress can be expected in various directions. New derivatization procedures may enrich the spectrum of precursors. New amino acid derivatives capable of capturing tritium would contribute to labeling different amino acid residues in peptides via synthetic precursors. The potential application of direct tritiation in the presence of various catalysts for peptide labeling has yet to be thoroughly investigated. This possibility carries great promise as a simple and rapid technique. Problems of storing tritiated peptides with very high specific activity still limit the applicability of these. More simple tritiation techniques would circumvent the problem of storage, enabling tritiation of precursors at the required time.

REFERENCES

Allen, M. C., Brundish, D. E., Martin, J. R., and Wade, R., 1978, Synthesis of (6-(4-³H) phenylalanine)-somatostatin of high specific radioactivity, *J. Chem. Soc. Chem. Commun.* **1978**:901.

Allen, M. C., Brundish, D. E., and Wade, R., 1979, Tritiated peptides. Part 7. Synthesis of (2,5-^{3}H$_2$-His^6)-α-corticotrophin-(1 –24)-tetracosapeptide, *J. Chem. Soc. Perkin Trans. 1* **1979**:2057.

Allen, M. C., Brundish, D. E., and Wade, R., 1980, Synthesis of 5- and 7-bromotryptophan and of (5-bromotryptophan^9)-α-corticotrophin-(1 –24)-tetracosapeptide, a highly potent corticotrophin analogue, *J. Chem. Soc. Perkin Trans. 1* **1980**:1928.

Berger, H., Fechner, K., Albrecht, E., and Niedrich, H., 1979, Substance P: *In vitro* inactivation by rat brain fractions and human plasma, *Biochem. Pharmacol.* **28**:3173.

Bienert, M., Niedrich, H., Teplán, I., and Medzihradszky, K., 1978, Synthesis of 6-(*N*-acetyl-lysyl)-11-norleucine-substance P (6–11)-hexapeptide labeled with tritium on the phenylalanine residue in position 7, *J. Labelled Compd. Radiopharm.* **14**:577.

Bienert, M., Köller, G., Wohlfeil, R., Mehlis, B., Bergmann, J., Niedrich, H., and Kraft, R., 1979a, Synthesis of substance P and of acylated partial sequences, *J. Prakt. Chem.* **321**:721.

Bienert, M., Klauschenz, E., Ehrlich, A., Katzwinkel, S., Niedrich, H., Tóth, G., and Teplán, I., 1979b, Tritium labelling in two phenylalanine residue of norleucine^{11}-substance P, *J. Labelled Compd. Radiopharm.* **16**:673.

Bienert, M., Klauschenz, E., Nikolics, K., and Niedrich, H., 1981, Protection of methionine in peptides during iodination by sulfonium salt formation, *Int. J. Pept. Protein Res.*, in press.

Brundish, D. E., and Wade, R., 1973, Synthesis of 3,5-^{3}H$_2$-Tyr^{23}-α-corticotrophin-(1 –24)-tetracosapeptide, *J. Chem. Soc. Perkin Trans. Trans. 1* **1973**:2875.

Brundish, D. E., and Wade, R., 1976, Tritiated peptides, Part 3. Synthesis of (4-^{3}H-Phe^7)-α-corticotropin-(1 –24)-tetracosapeptide, *J. Chem. Soc. Perkin Trans. 1* **1976**:2186.

Brundish, D. E., and Wade, R., 1977, The synthesis of tritium-labelled human corticotropin of high specific radioactivity, *Biochem. J.* **165**:169.

Brundish, D. E., and Wade, R., 1980, Tritiated peptides. Part 8. Synthesis of (3,5-^{3}H$_2$-Tyr^2)-, (4-^{3}H-Phe^7)-, and (3,5-^{3}H$_2$-Tyr^{23})-human corticotrophin, *J. Chem. Soc. Perkin Trans. 1* **1980**:462.

Brundish, D. E., Martin, J. R., and Wade, R., 1976, Tritiated peptides. Part 2. Synthesis of two (3,5-^{3}H$_2$-Tyr^2)-analogues of corticotropin, *J. Chem. Soc. Perkin Trans. 1* **1976**:2182.

Coy, D. H., Coy, J. E., Redding, T. W., and Schally, A. V., 1973, Synthesis of LH-RH containing tritium-labelled pyroglutamic acid, *Biochem. Biophys. Res. Commun.* **50**:866.

Coy, D. H., Mező, I., Pedroza, E., Nekola, M. V., Vilchez, J. A., Piyachaturawat, P., Schally, A. V., Seprődi, J. A., and Teplán, I., 1979, LH-RH antagonists with potent antiovulatory activity, in: *Peptides: Structure and Function,* Proceedings of the 6th American Peptide Symposium (E. Gross, ed.), pp. 775–781, Pierce Chemical Company, Rockford, Illinois.

Day, A. R., and Freer, R. J., 1978, Synthesis of an intrinsically radiolabelled enkephalin analog: (*p*-tritio-phenylalanyl)^4-norleucyl^5-enkephalin, *J. Labelled Compd. Radiopharm.* **14**:381.

Eberle, A., and Hübscher, W., 1979, α-Melanotropin labelled at its tyrosine^2 residue: Synthesis and biological activities of 3′-iodotyrosine^2-, 3′-^{125}Iodotyrosine^2-, 3′,5′-diiodotyrosine^2-, and (3′,5′-^{3}H$_2$)-tyrosine^2-α-melanotropin, and of related peptides, *Helv. Chim. Acta* **62**:2460.

Eberle, A., Hübscher, W., und Schwyzer, R., 1977, Synthese von radioaktiv markierten Bromacetyl- und Diazoacetyl-α-melanotropin-Derivaten zum Studium von Kovalenten Hormon-Makromolekül-Komplexen, *Helv. Chim. Acta* **60**:2895.

Ehrenkaufer, R. L. E., Wolf, A. P., and Hembree, W. C., 1977, A novel surface for high specific activity tritium labelling using microwave discharge activation of tritium gas, *J. Labelled Compd. Radiopharm.* **14**:271.

Felix, A. M., Wang, C. T., Liebman, A. A., Delaney, C. M., Mowles, T., Burghardt, B. A., Charnecki, A. M., and Meienhofer, J., 1977, Synthesis, biological activity and tritiation of L-3,4-dehydroproline-containing peptides, *Int. J. Pept. Protein Res.* **10**:299.

Flouret, G., Terada, S., Yang, F., Nakagawa, S. H., Nakahara, T., and Hechter, O., 1977, Iodinated neurohypophyseal hormones as potential ligands for receptor binding and intermediates in synthesis of tritiated hormones. *Biochemistry* **16**:2119.

Hanley, M. R., Sandberg, B. E. B., Lee, C. M., Iversen, L. L., Brundish, D. E., and Wade, R., 1980, Specific binding of ³H-substance P to rat brain membranes, *Nature (London)* **286**:810.

Hembree, W. C., Wolf, A. P., Ehrenkaufer, R. L. E., and Lieberman, S., 1975, Tritium labelling of amino acids in polypeptide hormones, in: *Methods in Enzymology,* Vol. 37 (B. W. O'Malley and J. G. Hardman, eds.), Academic Press, New York.

Horsthemke, B., and Bauer, K., 1980, Characterization of a nonchymotrypsin-like endopeptidase from anterior pituitary that hydrolyzes luteinizing hormone-releasing hormone at the tyrosyl-glycine and histidyl-tryptophan bonds, *Biochemistry* **19**:2867.

Houghten, R. A., and Li, C. H., 1978, Preparation and properties of tritiated human β-endorphin with high specific radioactivity, *Int. J. Pept. Protein Res.* **12**:325.

Kastin, A. J., Nissen, C., Nikolics, K., Medzihradszky, K., Coy, D. H., Teplán, I., and Schally, A. V., 1976. Distribution of ³H α-MSH in rat brain, *Brain Res. Bull.* **1**:19.

Kéri, G., and Teplán, I., 1978a, Specific tritium labelling of lysine-vasopressin in two different aminoacid residues, *J. Labelled Compd. Radiopharm.* **15**:153.

Kéri, G., and Teplán, I., 1978b, Synthesis and specific tritium labelling of the behaviourally active octapeptide des-9-glycinamide-lysine-vasopressin in two different aminoacid residues, *J. Labelled Compd. Radiopharm.* **15**:163.

Klauschenz, E., and Bienert, M., 1980, Tritium labelled analogues of substance P and gonadotropin-releasing hormone, Abstracts, First Conference of Organic and Bioorganic Chemistry of Young Scientists, Bechyne, Czechoslovakia, (P. Kocovsky and P. Drasar, eds.), pp. 150–151.

Klauschenz, E., Berger, H., Bienert, M., Egler, H., Pleiss, U., Nikolics, K., and Niedrich, H., 1981, Tritium labelling of gonadotropin releasing hormone via its dehydroproline analog and by direct tritium exchange, *Peptides,* **2**:684.

Kovács, J., Teplán, I., and Mező, I., 1975, Labelled amino acids and their derivatives. V. Preparation of L-phenylalanine-4-³H through the resolution of p-chloro-D,L-phenylalanine, *Acta Chim. Budapest* **84**:109.

Levine-Pinto, H., Pradelles, P., Morgat, J. L., and Fromageot, P., 1980, Specific tritiation onto C-2 and C-5 positions of histidine-containing peptide: Application to thyroliberin, *J. Labelled Compd. Radiopharm.* **17**:231.

Marche, P., Morgat, J. L., Fromageot, P., Kerdelhué, B., and Jutisz, M., 1972, ³H-Labelling of a synthetic decapeptide having LH and FSH releasing activity (LH-RH/FSH-RH), *FEBS Lett.* **26**:83.

Marche, P., Girma, J. P., Morgat, J. L., Fromageot, P., Ghelis, C., Dubrasquet, M., and Bonfils, S., 1975, Specific tritiation of indole derivatives by catalytic desulfenylation, *Eur. J. Biochem.* **50**:375.

Medzihradszky, K., Nikolics, K., and Seprődi, J., 1972, Synthesis of the N-terminal decapeptide of ACTH labelled with tritium on the tyrosine residue, *Ann. Univ. R. Eötvös* **13**:25.

Medzihradszky, K., Nikolics, K., Teplán, I., Seprődi, J., and Fittkau, S., 1974, Synthesis of α-melanotropin labelled with tritium on the tyrosine residue, *J. Labelled Compd.* **10**:23.

Mező, I., and Teplán, I., 1971, Labelling of Val₅-angiotensin II by tritiated, optically active tyrosine, *Radioisotopy* **12**:551.

Mező, I., Teplán, I., Seprődi, J. A., Morgat, J. L., Fromageot, P., Tóth, G., and Sirokmán, F., 1978, Synthesis of (Tyr-³H)⁴-angiotensin II and (Phe-³H)⁸-angiotensin II via halogen derivatives of angiotensin II, *J. Labelled Compd. Radiopharm.* **14**:557.

Morgat, J. L., and Fromageot, P., 1973, Preparation of tritium-labelled peptidic hormones of high specific radioactivity: Radiopharmaceuticals and labelled compounds, IAEA-SM-171/60, Vienna, **II**:109.

Morgat, J. L., Lam Thanh Hung, and Fromageot, P., 1970, Preparation of highly labelled (³H) angiotensin II, *Biochim. Biophys. Acta* **207**:374.

Morgat, J. L., Girma, J. P., and Fromageot, P., 1977, Tritium labelling of peptidic hormones, *First International Symposium on Hormonal Receptors in Digestive Tract Physiology*, INSERM Symp. No. 3:43, Elsevier, North-Holland, Amsterdam.

Morgat, J. L., Fromageot, P., Michelot, R., and Glowinski, J., 1980, ³H-Labelling of substance P by tritium exchange and dehalogenation, *FEBS Lett.* **111**:19.

Nikolics, K., 1974, Specific tritium labelling of α-MSH and ACTA fragments, Ph. D. dissertation, R. Eötvös University, Budapest.

Nikolics, K., Teplán, I., and Medzihradszky, K., 1976, Synthesis of α-melanotropin labelled with tritium on the tyrosine residue via the dibromo intermediate, *J. Labelled Compd. Radiopharm.* **12**:163.

Noble, R. L., Yamashiro, D., and Li, C. H., 1976, Synthesis of a nonadecapeptide corresponding to residue 37–55 of ovine prolactin: Detection and isolation of the sulfonium form of methionine-containing peptides, *J. Am. Chem. Soc.* **98**:2324.

Pande, C. S., Rudick, J., Ornstein, L., Schwartz, I. L., and Walter, R., 1969, Specific tritium labelling of a potent gastrin analogue, *Mol. Pharmacol.* **5**:227.

Perrin, M. H., Rivier, J. E., and Vale, W. W., 1980, Radioligand assay for gonadotropin-releasing hormone: Relative potencies of agonists and antagonists, *Endocrinology* **106**:1289.

Pradelles, P., Morgat, J. L., Fromageot, P., Oliver, C., Jacquet, P., Gourdji, D., and Tixier-Vidal, A., 1972a, Preparation of highly labelled ³H-thyrotropin releasing hormone by catalytic hydrogenolysis, *FEBS Lett.* **22**:19.

Pradelles, P., Morgat, J. L., Fromageot, P., Camier, M., Bonne, D., Cohen, P., Bockaert, J., and Jard, S., 1972b, Tritium labelling of 8-lysine-vasopressin and its purification by affinity chromatography on Sepharose-bound neurophysins, *FEBS Lett.* **26**:189.

Ramachandran, J., and Behrens, C., 1977, Preparation and characterization of specifically tritiated adrenocorticotropin, *Biochim. Biophys. Acta* **496**:321.

Redding, T. W., and Schally, A. V., 1972, On the half-life of TRH in rats, *Neuroendocrinology* **9**:250.

Redding, T. W., Kastin, A. J., Nikolics, K., Schally, A. V., and Coy, D. H., 1978, Disappearance and excretion of labelled α-MSH in man, *Pharmacol. Biochem. Behav.* **9**:207.

Rivier, J., Porter, J., Galyean, R., Rivier, C., and Vale, W., 1980, Application of the Hansch approach to design LRF and somatostatin analogs using Topliss manual method, in Abstracts, 16th European Peptide Symposium, Helsigør, Denmark, p. 128, Scriptor Kopenhagen.

Schwyzer, R., and Karlaganis, G., 1973, Hormon-Receptor Beziehungen: Synthese und Eigenschaften von (Phenylalanin²)-4,5-dehydro-4,5-ditritio-(norvalin⁴)-adrenocorticotropin-(1–24)-tetracosipeptide, *Liebigs Ann. Chem.* **1973**:1298.

Seprődi, J., Teplán, I., Magócsi, M., and Medzihradszky, K., 1980, Synthesis of bradykinin labelled with tritium, *Acta Chim. Acad. Sci. Hung.* **104**:227.

Sjohölm, I., and Carlsson, L., 1967, Synthesis of [Tyr-³H]-oxytocin, *J. Labelled Compd.* **3**:1.

Smith, C. W., Botos, C. R., and Walter, R., 1977, Conformation–activity approach to the design of vasopressin analogs with high and specific antidiuretic activity, in: *Peptides* (M. Goodman and J. Meienhofer, eds.), p. 161, Wiley, New York.

Sosa, R. P., McKnight, A. T., Hughes, J., and Kosterlitz, H. W., 1977, Incorporation of labelled amino acids into the enkephalins, *FEBS Lett.* **84**:195.

Teplán, I., Mező, I., Seprődi, J., and Kovács, J., 1977, Synthesis of (3,5-dibromo-tyrosine)⁴-angiotensin II and (4-chlorophenylalanine)⁸-angiotensin II for the labelling of angiotensin II, *Acta Chim. Budapest* **95**:285.

Terada, S., Nakagawa, S. H., Yang, D. C., Lipkowski, A., and Fluoret, G., 1980, Iodination of luteinizing hormone-releasing hormone, *Biochemistry* **19**:2572.

Van Rietschoten, J., Tregear, G., Leeman, S., Powell, D., Niall, H., and Potts, J. T., Jr., 1975, Simultaneous synthesis of two peptide analogs on different insoluble supports, in: *Peptides* 1974 (Y.Wolman, ed.), pp. 113–116, Wiley, New York.

Appendix: Members of the International Foundation for Biochemical Endocrinology

Blue Hill Falls, Maine 04615

Kenneth W. McKerns, M.Sc., Ph.D., President

Asbjørn Aakvaag, Ph.D. Oslo Kommune, Aker Sykehus, Oslo 5, Norway

Hélio Aguinaga, M.D., Avenida Vieira Souto 620, apt°. 201, CEP 22420, Rio de Janeiro, Brazil

Etienne-Emile Baulieu, M.D., Ph.D., Professeur, Départment de Chimie Biologique, Faculté de Medecine de Bicêtre 78, Avenue du General Leclerc, Université de Paris Sud 94 Bicetre France

J. D. Baxter, M.D., Professor, University of California, Endocrine Research Division, 671 HSE, San Francisco, California 94143

A. Oriol Bosch, M.D., Professor, Catedra De Endocrinologia Experimental, Facultad De Medicina, Universidad Complutense, Madrid 3, Spain

George S. Boyd, Ph.D. Professor and Head, Department of Biochemistry, University of Edinburgh Medical School, Teviot Place, Edinburgh EH8 9 AG, Scotland

D. W. Bullock, Ph.D., Department of Cell Biology, St. Luke's Episcopal Hospital, 25th Floor, Texas Medical Center, Houston, Texas 77030

Paul Burgess, L.Th., Dr. P.H., 8605 Bradmoor Drive, Bethesda, Maryland 20014

Robert Chatterton, Ph.D., Professor, Department of Obstetrics and Gynecology, 333 East Superior Street, Northwestern University Medical School, Chicago, Illinois 60611

A. Kent Christensen, Ph.D., Professor and Chairman, Department of

Anatomy, Medical Sciences II Building, University of Michigan, Medical School, Ann Arbor, Michigan 48109

James E. Clayton, L.L.D., Clayton, Duncan, Johnson, Quincy, Ireland, Felder and Gadd, Commerce Building, P.O. Box 1090, Gainesville, Florida 32602

Jean Crabbé, M.D., Professeur, Department de Physiologie Endocrinologie–Metabolisme, Faculte de Medecine, Universite Catholique de Louvain, B 12001 Brussels, Belgium

Derek Denton, M.B.B.S., Director, Howard Florey Institute of Experimental Physiology & Medicine, University of Melbourne, Parkville, Victoria 3052, Australia

Donald Exley, Ph.D. Professor, Biochemistry Department, Queen Elizabeth College, University of London, London, England

Béla Flerko, M.D., Professor, Department of Anatomy, University Medical School, 1643 Pécs Hungary

Jean Garnier, Ph.D., Laboratoire De Biochimie Physique, Universite De Paris—Sud, Batiment 433—91405, Orsay Cedex, France

James K. Grant, Ph.D., Reader, Department of Steroid Biochemistry, Royal Infirmary, University of Glasgow, Glasgow G4, OSF, Scotland

Carlos Gual, M.D., Professor of Reproductive Biology, Instituto Nacional de la Nutricion, Departamento de Biologia de la Reproduccion, Centro de Investigacion Clinica de la Organizacion Mundial de la Salud, Viaducto Tlalpan y San Fernando, Mexico 22, D.F., Mexico

Edward Herbert, Ph.D., Professor, Chemistry Department, University of Oregon, Eugene, Oregon 97403

W. C. Hymer, Ph.D., Professor, The Pennsylvania State University, Department of Biochemistry & Biophysics, 401 Althouse Labs, University Park, Pennsylvania 16802

Marian Jutisz, Sc.D., Director, Polypeptide Hormone Laboratory, Centre National de la Recherche Scientifique, 91190 Gif-Sur-Yvette, France

Dušan T. Kanazir, M.D., Vice President, Serbian Academy of Sciences and Arts, 35, Knez-Mihailova, 11000 Beograd, Yugoslavia

Claude Kordon, Ph.D., Professor, Unité de Neurobiologie dl'Inserm, U. 109-2ter rue d'Celésia, 75014, Paris, France

J. Kraicer, M.D., Ph.D., Professor, Department Physiology, Abramsky Hall, Queen's University, Kingston K7L 3Ng, Ontario, Canada

Choh Hao Li, Ph.D., Director, Hormone Research Laboratory, University of California, San Francisco, California 94143

Fernand Labrie, Ph.D., Directeur, Group du Consel de Recherches Medicales, en Endocrinologie Moleculaire, le Centre Hospitalier de L'Universite Laval, 2704 Boul. Laurier, Quebec ClV 4G2, Canada

Shutsung Liao, Ph.D., Professor, The Ben May Laboratory for Cancer Research, The University of Chicago, 950 East 59th Street, Chicago, Illinois 60637

Hans R. Lindner, Ph.D., Professor, Department of Hormone Research, The Weismann Institute of Science, Rehovot, Israel

W.I.P. Mainwaring, M.D., Head, Androgen Physiology Department, Imperial Cancer Research Fund Laboratories, P.O. Box No. 123, Lincoln's Inn Fields, London, WC2A 3PX, England

Samuel M. McCann, M.D., Professor, Department of Physiology, The University of Texas, Health Science Center, 5323 Harry Hines Blvd., Dallas, Texas 75235

Magdelein Mousseron-Canet, Ph.D., Centre National de la Recherche Scientifique, Equipe de Recherche Photobioorganique, Ecole Nationale Superierore de Chimie, 8, rue de l'Ecole Normale, 84075 Montpellier Cedex, France

Allan Munck, Ph.D., Professor of Physiology, Dartmouth Medical School, Hanover, New Hampshire 03755

Milton S. Nakamura, M.D., Director, Family Planning Center, Praca Oswaldo Cruz, 138–Conj. 61, São Paulo, Brazil

Károly Nikolics, M.D., Professor, First Institute of Biochemistry, Semmelweis University Medical School, Budapest, Hungary

Vladimir Pantić, M.D., Professor, Institute for Biological Research, Division of Cytology and Embryology, 29, Novembar 142, Beograd, Yugoslavia

Roland Pattillo, M.D., Professor, Cancer Research & Reproductive Endocrinology Laboratories, Department of Gynecology and Obstetrics, The Medical College of Wisconsin, 8700 West Wisconsin Avenue, Milwaukee, Wisconsin 53226

M. R. Sairam, Ph.D., Director, Reproduction Research Laboratory, Institute de Recherches Cliniques de Montreal, 110 Oest, Avenue Des Pins, Montreal H2W 1R7, Canada

Brian P. Setchell, Ph.D., Waite Agricultural Research Institute, University of Adelaide, Glen Osmond, South Australia 5064, Australia

Harold Spies, Ph.D., Senior Scientist, Department of Reproductive Physiology, Oregon Regional Primate Center, 505 N.W. 185th Avenue, Beaverton, Oregon 97005

Samuel Spicer, M.D., Professor, Department of Pathology, Medical University of South Carolina, 171 Ashley Avenue, Charleston, South Carolina 29403

Clara M. Szego, Ph.D., Professor, Department of Biology, Molecular Biological Institute, University of California, Los Angeles, California 90024

Paul Talalay, M.D., Professor, The Johns Hopkins University School of Medicine, Department of Pharmacology and Experimental Therapeutics, 725 N. Wolfe Street, Baltimore, Maryland 21205

Bun-ichi Tamaoki, Ph.D., Professor, National Institute of Radiological Sciences, 9-1, 4-chrome, Anagawa, Chiba-shi, Japan

A. Tixier-Vidal, Ph.D., Professor, Neuroendocrinologie Cellulaire Collége de France, 11 Place Marcelin Berthelot, 75231 Paris, Cedex 5, France

H. J. van der Molen, Ph.D. Professor, Afdelin Biochemie, Erasmum Universiteit Rotterdam, Postbus 1738, Rotterdam, Holland

Andrew V. Schally, Ph.D., Director, Peptide Hormone Laboratory, The Veterans Administration Medical Center, 1601 Perdido Street, New Orleans, Louisiana 60146

Anna Steinberger, Ph.D., Professor, Department of Reproductive, Medicine and Biology, 6431 Fannin, Texas Medical Center, Houston, Texas 77025

Emil Steinberger, M.D., Professor and Head Department of Reproductive Medicine and Biology, 6431 Fannin, Texas Medical Center, Houston, Texas 77025

John M. Stewart, Ph.D., Professor, Department of Biochemistry, University of Colorado, Health Sciences Center, Denver, CO 80262

Claude A. Villee, Jr., Ph.D., Andelot Professor of Biological Chemistry, Laboratory of Human Reproduction and Reproductive Biology, Harvard Medical School, 45 Shattuck Street, Boston, Massachusetts 02115

Dorothy B. Villee, M.D., Associate in Endocrinology, Assistant Professor of Pediatrics, Harvard Medical School, The Children's Hospital Medical Center, 300 Longwood Avenue, Boston, Massachusetts 02115

Donald L. Wilbur, Ph.D., Associate Professor, Department of Anatomy, Medical University of South Carolina, 171 Ashley Avenue, Charleston, South Carolina 29403

Index

A II, *see* Angiotensin II
Acetylation, of synthetic enkephalin
    analogues, 10–11
Acetylcholine
    $17\beta E_2 S$ effect on, 258
    CRF stimulation of, 105
    lipid structure sensitivity in, 246
ACTH (adrenocorticotropin)
    α helix in, 50
    CD spectroscopy of, 52–53
    CNS and, 349
    in conditioned behavioral response, 349
    conformational analysis of, 49–50
    conformational parameters of, 51
    conformation of in solution, 50–54
    defined, 49
    ether effect on, 166
    glucocorticoid negative feedback on, 265
    from human pituitary, 521
    in inhibition of pituitary ACTH release,
      174
    inhibiting hormones and, 504
    MSH and, 503–530
    origin and nature of, 506–507
    in pars intermedia, 511
    pituitary, *see* Pituitary ACTH
    proline residues in, 53
    releasing hormones and, 504
    statistical sequence analysis of, 51
    TAT activity and, 488–489
    tritium labeling of, 605–607
ACTH conformation
    carbon-13 NMR spectroscopy and, 54
    conformation-activity relationship in, 54–
      56

ACTH level, negative feedback signal and,
    172
ACTH/MSH, 351
ACTH/MSH cell activities, brain peptides in
    regulation of, 518–521
ACTH-(1-19)-NH_2 peptides, residue
    replacement in, 55
ACTH release, afferent pathways or stressful
    stimuli in, 169–171, *see also* ACTH
    secretion
ACTH-releasing potency
    in neural lobe stimulation, 162
    of stalk-median eminence and
      neurohypophyseal extracts, 160–161
ACTH-secreting cells, 158, *see also* ACTH
    release
    hypertrophy of, 516–518
    reaction to corticosteroids, 516–518
ACTH secretion
    COL destruction of microtubules in, 492–
      500
    feedback regulation of, 171–175
    regulation of, 157–175
    stress-type, 487–500
    vasopressin in, 159
ACTH treatment, pituitary–adrenal system
    changes in, 174
Actinomycin D
    in cortisol-induced frog-skin lightening,
      362
    mRNA synthesis and, 189–190
"Active Center," amino acids as, 71–72
Adenohypophysis
    defined, 487
    of hagfish, 503

623